W9-BCL-151

Einar Hille: Classical Analysis and Functional Analysis

Selected Papers

Mathematicians of Our Time

Gian-Carlo Rota, series editor

1. Charles Loewner: Theory of Continuous Groups
   notes by Harley Flanders and Murray H. Protter

2. Oscar Zariski: Collected Papers
   Volume I
   Foundations of Algebraic Geometry and Resolution of Singularities
   edited by H. Hironaka and D. Mumford

3. Collected Papers of Hans Rademacher
   Volume I
   edited by Emil Grosswald

4. Collected Papers of Hans Rademacher
   Volume II
   edited by Emil Grosswald

5. Paul Erdös: The Art of Counting
   edited by Joel Spencer

6. Oscar Zariski: Collected Papers
   Volume II
   Holomorphic Functions and Linear Systems
   edited by M. Artin and D. Mumford

7. George Pólya: Collected Papers
   Volume I
   Singularities of Analytic Functions
   edited by R. P. Boas

8. George Pólya: Collected Papers
   Volume II
   Location of Zeros
   edited by R. P. Boas

9. Stanislaw Ulam: Selected Works
   Volume I
   Sets, Numbers, and Universes
   edited by W. A. Beyer, J. Mycielski, and G.-C. Rota

10. Norbert Wiener: Collected Works
    Volume I
    Mathematical Philosophy and Foundations; Potential Theory; Brownian Movement, Wiener Integrals, Ergodic and Chaos Theories, Turbulence and Statistical Mechanics
    edited by P. Masani

11. Einar Hille: Classical Analysis and Functional Analysis
    Selected Papers
    edited by Robert R. Kallman

Einar Hille: Classical Analysis and Functional Analysis

Selected Papers

Edited by
Robert R. Kallman

The MIT Press
Cambridge, Massachusetts, and London, England

This book was printed on Finch Title 93
and bound in Columbia Millbank Vellum 4657
by The Colonial Press Inc.
in the United States of America

Library of Congress Cataloging in Publication Data

Hille, Einar, 1894-
Classical analysis and functional analysis

(Mathematicians of our time, v. 11)
Bibliography of Einar Hille: p.
1. Mathematical analysis–Collected works.
2. Functional analysis–Collected works. 3. Hille, Einar, 1894- –Bibliography. I. Title. II. Series.
QA300.H532 515'.08 74-18465
ISBN 0-262-08-080-X

Einar Hille

Contents

(Bracketed numbers are from the Bibliography)

Chapter 4. Integral Equations
111

Chapter 5. Summability
117

Chapter 6. Fourier Transforms and Analytic Function Theory
143

Chapter 7. Laplace Integrals
191

Chapter 8. Factorisatio Numerorum and Möbius Inversion
205

Chapter 9. Hermitian Series and Differential Operators of Infinite Order Which Are Entire Functions of Finite Operators
239

Editor's Preface

Einar Hille is one of the very few mathematicians to make significant contributions to both classical analysis and functional analysis. This volume consists of Hille's own selection of a number of papers that highlight his career in mathematical research. One need only glance through the commentaries at the end of this volume to note that most of Hille's work on functional equations, ordinary differential equations, analytic function theory, ergodic theory, semi-groups, partial differential equations, etc., is germane to present-day research. Thus, this volume should serve as a source of guidance and inspiration for today's budding young analyst.

Both Professor Hille and I owe a debt of gratitude to the many mathematicians who enhanced the value of this Selecta by their commentaries. It has been a special pleasure and honor to serve as the editor of this volume, being, via I. E. Segal, a mathematical grandson of Einar Hille.

Robert R. Kallman
University of Florida

In Retrospect *

I wish to thank you for the invitation to address the Colloquium once more. Whoever had the idea of this invitation seems to have had in mind a prepublication reading of Volume I of my Memoirs. There will be no memoirs and I am no good at dramatizing. A few days ago I read Harald Bohr's charming *Looking back,* a lecture he gave to the students on his sixtieth birthday. It is full of his whimsical wit, quiet humor, and warm personality. I cannot imitate him, but his article is recommended reading, Volume I of his Collected Works. What you will get from me here is a monologue where I figure as the red thread. I have tried to string some beads and pearls on this thread, mostly baubles I am afraid. You are supposed to look at the semiprecious stones, not at the thread!

1. *Stockholms Högskola.* I graduated from high school 51 years and one day ago, that is, on May 17, 1911. In September of that year I registered as a student at what I shall call the University of Stockholm. There were no entrance requirements beyond the high school diploma, though some departments such as physics and chemistry gave entrance examinations which had to be passed for admission. The fees had recently been doubled and amounted to about $25 a year. They were raised by 50 percent before I got my final degree.

Actually the title of University was given to my Alma Mater only a few years ago. Up to that time it had been known as Stockholms Högskola, where "högskola" is used in the same sense as the German "Hochschule," that is, an institute of higher learning, an Institute for Advanced Study. It was founded in 1881 as a private institute, privately endowed. The school was originally a faculty of mathematics and the natural sciences. To this had been added chairs in history of literature, history of art, and political economy, and, five years before my entrance, also a law school. The latter was on the budget of the city of Stockholm. The law school radically changed the character of the institute. Originally it had been essentially a research institute, though with very little in the way of laboratories, and it gave no degrees. If a student wanted a degree, he had to pass exams at one of the state universities, either Uppsala or Lund, where the upstart Stockholm was quite unpopular and its students were apt to receive rough treatment by the examiners. As a result many of the better and wealthier students never bothered to take any degrees. This was changed when the law school appeared on the scene, since its students had to have degrees to get anywhere; so, after much resistance

*This paper was presented at the Yale Mathematical Colloquium on May 16, 1962.

on all sides, the institute finally got the right to give degrees on all levels. The price paid for this was rather moderate: the chancellor of the universities now exercised supervision over appointments and courses of study, and appointments to professorships had to be confirmed by the government.

Originally the institute had been housed in two apartments, with large rooms to be sure but still barely a dozen rooms. It got its own building in 1909. In spite of the very inadequate facilities, though, excellent work was done in this small institute during the first thirty years of its existence.

When I came to the university I was not sure of what field I wanted to specialize in, mathematics or organic chemistry. I had a substantial background in both. Since chemistry is a laboratory science, the study handbook suggested starting with it, and so I did. For the first two years I had fairly little time for anything beyond laboratory work, which I soon found I did not like and had little aptitude for. The professor of general and organic chemistry was a Bavarian, Hans von Euler-Chelpin, Nobel prize 1929, born 1873, and still very active scientifically though close to 90. He was at that time already specializing in biochemistry. Hence my first scientific paper was *Über die primäre Umwandlung der Hexosen bei der alkoholischen Gärung,* a joint note with Euler that appeared in Zeitschrift für Gärungsphysiologie, vol. 3, 1913. By that time I had acquired an aversion for biochemistry and laboratory work. Nevertheless, I have kept my interest in organic and metallo-complex chemistry over the years and nowadays when biochemistry has become respectable, highly important, and full of interesting results, I am also interested in biochemistry. To return: I finished chemistry in May 1913 with the equivalent of honors. I then turned to mathematics.

Let me go back once more to the beginning of the university. A chair in pure mathematics had been included in the first deal in 1881, and the first incumbent was Gösta Mittag-Leffler (1846-1927) who retired in 1911. He was a Swede by birth who had studied with Weierstrass and had occupied a chair in mathematics at the University of Helsingfors in the then grand duchy of Finland. This he gave up to accept the new appointment and to start the work in mathematics in Stockholm. To this he devoted his energy and ingenuity, his wife's fortune, and his many connections at home and abroad. In his heyday he was a marvelous lecturer, and he attracted a number of excellent students, some better mathematicians than he was but none as colorful as he. There was no library at the university and no room to build one, so, when he built his home in the fashionable suburb Djursholm, he made it into a library which in its days was probably the best mathematical research library in the world. He founded the journal Acta Mathematica where he could offer quick publication to people like Georg Cantor and Henri Poincaré. Here his good connections with both French and German mathematicians were invaluable. M. L. played a very definite role in making Cantor's revolutionary ideas known and accepted in wide circles, and he played perhaps some role also in the scientific life of Poincaré. I want also to mention that M. L. started the Scandinavian mathematical congresses in 1909 and he established the close relations between mathematics and the insurance companies which is a characteristic feature of Sweden and Scandinavia in general. He was himself an actuary, as were Fredholm, M. Riesz, and Cramér.

A professorship in higher mathematical analysis was created for S. Kovalevski (1850-1891) in 1884, and after her untimely death it was held by E. Phragmén (1863-1937) until 1905 when he resigned to devote himself to life insurance in various forms. He was not lost completely to mathematics; in fact his most famous paper, jointly with Ernst Lindelöf, dates from 1908. Moreover, he created the actuarial society of Sweden, and for almost a quarter of a century this society and not the almost defunct mathematical colloquium at the university was the center of mathematical activity in Stockholm. What Phragmén had started, Cramér carried on. Phragmén was succeeded by another of M. L.'s pupils, Ivar Bendixson (1861-1935), who served as rector (president) of the university from 1913 until 1927. M. L. retired, with at least one foot, in 1911 and was succeeded by his pupil Helge von Koch (1870-1924). There was also a professorship in mechanics and mathematical physics whose first incumbent, in 1893, had been V. F. K. Bjerknes, who was followed by Ivar Fredholm (1866-1927) in 1906. Bjerknes's father had been professor of mathematics at the university of Oslo, and his son is the founder of modern meteorology and is a professor of this subject at UCLA.

Thus, when I entered the university, the chairs were held by Bendixson, Fredholm, and von Koch, all pupils of M. L. There were a number of docents, although the only one who will figure in the following is Marcel Riesz (1886-), who was appointed in 1911. Mathematics was quite informal. There were no entrance requirements. At the beginning of the year there was a meeting where the instructors stated what they were going to give; you signed your name on a list for the course you wanted, and you attended lectures and problem sessions if you wanted. During my first year I attended Bendixson's course and became fascinated with the beauty of mathematics. Bendixson was an excellent lecturer; he had a sense of perspective and could pull you with him. To him I owe that I ultimately became a mathematician. He was a wealthy man, public-spirited, an alderman, and prominent in the liberal party. After he became president of the university, he could devote little time to his teaching. He had given up research some years earlier. (He told me many years later that he did not want to play chess with himself. Perhaps he had not been able to find results of the same calibre as his early work on set theory and the geometry of characteristics of differential equations.) To return to my own experience with him: his topic in 1911-12 was elliptic functions, for which I was by no means prepared. But I hung on as long as I could steal time from the lab, and I returned again in the spring after he had reached the theta functions. There was much that I did not understand, but I got a good introduction to Cauchy function theory and managed to take fairly good notes which made sense in later years.

During my second year I became brave and listened to von Koch's lectures on differential equations. He was much more formal than Bendixson, and his lectures were probably more polished. By this time I understood much more and could even take part in a couple of problem sessions. But you will notice from this that what I went to was what we would call second-year graduate courses, for which I lacked the proper foundation. The elements I woefully neglected. I had no time and much overrated my ability to handle things like three dimensional geometry and the theory of equations. I spent the summer of 1913 reviewing mathematics and learning some mechanics. I passed exams in both in the fall, though not in very good form. This led to the first academic degree, candidate of philosophy. Since I wanted to meet the academic requirements for teaching in a government school, I took physics during 1913-14 as well as two half-year courses in psychology and history of education. This led to an M. A. degree in the spring of 1914.

During 1913-14 I made my first acquaintance with Marcel Riesz, though I had seen him shortly after his arrival in Sweden in 1911. He is, as you know, a brother of Frigyes Riesz, younger by almost seven years. He was a docent at Stockholm until he was appointed to a professorship in Lund in 1926. In the fall of 1914 I started to concentrate on math. I listened to von Koch, Fredholm, Bendixson, and Riesz. Von Koch suggested that I should give a seminar talk on Volterra's equation following Lalesco's book, and thanks to Riesz's help with literature and higher points of view, this became a success, and the first dividend was an assistantship in spring 1915. My first assignment was to give the course in algebraic analysis: set theory, real numbers, series, continued fractions, etc. I was scared stiff when Bendixson came in to listen to my first lecture. This appointment also led to a command visit to Mittag-Leffler. He took a lively and on the whole kind interest in the students, but again I was scared stiff. The interview went fairly well until he asked me what method I used to introduce the real number system. When I said the Dedekind cut, he almost hit the ceiling. I was given a reprint of the way he and Weierstrass had introduced the real numbers. This was never really published and in the meantime I have lost it. I never understood it anyway. But he was kind and wanted to get me started on research by reading a paper by G. N. Watson on asymptotic series. I got nowhere with it. This was a pity because F. Nevanlinna made a beautiful generalization of Watson's results seven years later. Von Koch lectured on conformal mapping that year (1914-15) and I asked him for a topic for an essay for the licentiate. He suggested a study of Koebe's method based on successive extraction of square roots. I worked on it during the summer and got some beautiful curves but no results of any value. Again I am afraid that I spoiled a good topic: Bieberbach carried out such an investigation the following year.

Finally Riesz suggested a problem that was within my reach. Suppose that $f(z)$ is a holomorphic function in the disk $|z| < R$ and suppose to start with that $f'(z) \neq 0$ in the disk. Then $w = f(z)$ maps the disk $|z| \leqslant r < R$ conformally on a region in the w-plane, and the length of the boundary is given by

$$r \int_0^{2\pi} |f'(r e^{i\theta})| \, d\theta = L(r).$$

Riesz asked: What is this integral in the case where

$$f(z) = z - z^2/2, \quad f'(z) = 1 - z?$$

I had no difficulty in showing that

$$L(r) = 2\pi r \, F(-\tfrac{1}{2}, -\tfrac{1}{2}, 1; r^2), \quad r < 1,$$

and a similar expression for $r > 1$. This suggested that $L(r)$ should be a piecewise analytic function of r, if $f'(z)$ does vanish in the original disk. This turned out to be the case, and I could study also the relation between functions for adjacent annuli. On this investigation I got the licentiate degree in May 1916.

The year 1916-17 was an important one for my development. I did my military service, one year, as a typist in a division staff. One learned to type after a fashion very fast, and this has put me in good stead. More important was the time for contemplation and research. I had plenty of free time, since I had a room in town and spent only about half an hour a month in the barracks. The result was three investigations. A continuation of the arclength study was not worth publishing, but one investigation developed into my dissertation a year later, and another on analytic continuation of functions defined by Dirichlet series was published in extended form in 1926. During this year I also attended my first mathematical congress, the fourth Scandinavian one whose proceedings until a few years ago was one of the rarities among mathematical publications. Cramér and I formed the Secretariat and had comparatively little to do. Nobody could attend from Finland owing to the war, but Norway and Denmark were well represented. I met F. Riesz who lectured on the extension of the Fredholm theory of compact operators. Hardy could not come, though his paper, the first on partition functions, is published in the Comptes Rendus. I had met him a year earlier when he came to visit M. Riesz.

After my return to Stockholm in the summer of 1917, I started to work on a dissertation. I was not satisfied with the results obtained on Legendre functions and got interested in difference equations, more precisely so-called q-difference equations

$$\sum_{j=0}^{n} f_j(z)\, w(q^{n-j}z) = 0,$$

where the f's are analytic functions. I developed a considerable theory for such equations using various patterns. By this time I thought that it would be best to take a look at the literature, and I found that most of the things that I had were already in papers by G. D. Birkhoff and F. D. Carmichael. So this investigation was filed and I turned back to Legendre's equation. Legendre polynomials are orthogonal over the interval $[-1, +1]$ and all their zeros belong to the same interval. Since $P_n(x)$ is a hypergeometric function, it is natural to raise similar questions for functions like

$$F(\alpha + 1, -\alpha, 1, \tfrac{1}{2}(1-x)),$$

which reduces to $P_n(x)$ if $\alpha = n$. This was carried out in great detail, and I got my degree on May 30, 1918.

The dissertation contained an idea which had been used for a rather particular purpose only, but which was capable of much extension, as I showed during 1921-1924. This, however, takes us to the second chapter.

2. *Harvard.* Two days after getting my degree I started to earn a living in Swedish government service, first in the Old Age Pension Board, later as an insurance examiner. It was dull. I was allowed time off to do some teaching at the Högskola, and I started on an investigation that was published later;

but it was difficult to see any scientific future in what I was doing. One Sunday morning in the fall of 1919 I read an article about the Swedish-American Foundation and that they were giving out fellowships for work in the United States. As luck would have it, on my morning walk I encountered the then secretary of the foundation, a friend of mine since chemistry days. We started talking, I expressed an interest, and he suggested that I should apply for a fellowship. This I did and ultimately got a fellowship to work on difference equations with G. D. Birkhoff at Harvard in 1920-21. Birkhoff made me welcome, sent reprints, and suggested a problem on differential-difference equations analogous to those satisfied by the hypergeometric series. This I tried during the summer before leaving, but it seemed to be difficult to single out the significant solutions among the large number of possibilities.

I came to Harvard towards the end of August 1920 and was very well received by Birkhoff. Harvard gave me a tuition scholarship ($200 in those days!) and Osgood insisted that I should take courses for credit on account of the scholarship. He was always a formalist, but he gave a good course in algebraic functions and integrals and I learned a lot from him. Most people were very friendly: Birkhoff, Kellogg, Bouton, Graustein, even Coolidge. Harvard was a revelation to me. Of course, it was a shock to find prominent analysts who knew nothing about summability, Fourier series, Dirichlet series, and the various other things that I had got from Riesz. On the other hand, it was very useful to sit at the feet of a geometer. Bouton gave a course in geometric transformations, inversions, Cremona transformations, etc. Coolidge lectured in his inimitable fashion on line geometry. All these topics were strange to me, and, even if I did not learn so much, it widened the horizon. Birkhoff took up differential equations the first term and difference equations the second. Here the most important gain was the matrix point of view.

Birkhoff gave the Chicago Colloquium on Dynamical Systems in September 1920, and as a result he did not have too much time. I could not afford to go to Chicago even if I had really wanted to do so. As a result I got a couple of weeks to crystallize my own ideas on the distribution of the zeros of solutions of second order differential equations with analytic coefficients. A method that I had used in the dissertation could be developed into a tool to study so-called zero free regions. I showed this to Birkhoff, who said that this was worth pursuing and that I had better spend my time on this problem rather than on difference equations. So ended my second attempt to do something with difference equations.

I worked out a general method for using a certain identity, the Green's transform, for the distribution problem, as well as a method of asymptotic integration based on Liouville's transformation. This was applied to various special equations, including those of Legendre and Mathieu and the parabolic cylinder (= Weber-Hermite) equation, in about a dozen papers published between 1921 and 1927. The basic identity is, in the simplest case,

$$[\overline{w(z)}\, w'(z)]_{z_1}^{z_2} - \int_{z_1}^{z_2} |w'(s)|^2\, \overline{ds} + \int_{z_1}^{z_2} P(s)\, |w(s)|^2\, ds = 0$$

if

$$w''(z) + P(z)\, w(z) = 0.$$

Here it is essential that the same path is used in both integrals. In general the integrals are not independent of the path. This shows that, if we integrate between two zeros of $w(z)\, w'(z)$, there must be an identity

$$\int_{z_1}^{z_2} |w'(s)|^2\, \overline{ds} = \int_{z_1}^{z_2} P(s) |w(s)|^2\, ds.$$

This relation puts severe restrictions on the relative position of the two zeros under consideration. Thus, for instance, suppose that z_1 lies in a region where the argument of $P(z)$ lies between 0 and $\pi/2$; then if z_2 lies in the same region, $\arg(z_2 - z_1)$ either lies between 0 and $-\pi/4$ or between $3\pi/4$ and π. If we replace $P(z)$ by $kP(z)$, where k is a large positive constant, and pick a solution which vanishes at z_1, then its other zeros are pulled in toward z_1 as k increases. These zeros describe paths which at z_1 are tangent to the curve

$$\mathfrak{I} \int_{z_1}^{z} [P(s)]^{1/2}\, ds = \text{Const.}$$

passing through z_1. These curves play an important role also in the discussion based on the Liouville transformation.

I wrote a long memoir on these questions in 1924 which was never published. There is much differential geometry in the large left to do for these questions which I think would still be interesting.

Among the graduate students then at Harvard I mention R. E. Langer, Ray Adams, Hazebrouk Van Vleck, and B. O. Koopman. I returned to Harvard in 1921-22, now as a Benjamin Pierce instructor. The appointment had originally gone to Jessie Douglas who, however, resigned owing to ill health. Now I made the acquaintance of Joe Walsh, who was a regular instructor after his return from Paris. Among the undergraduates were D. V. Widder and Marshall Stone. My teaching assignments were somewhat mixed. Math E, Trigonometry, has long since been dropped. It was a bore except for the fact that I had to learn spherical trigonometry. The most interesting course was Math 10b, Heat and Sound, or words to that effect. Here I learned a lot. I also gave Math 18, Differential Equations, as replacement for Bouton, and assisted Huntington in Math 5.

From Harvard I went to Princeton as an instructor, and was promoted to assistant professor the following year. This I probably owed to T. H. Gronwall, whose acquaintance I had made at my first A.M.S. meeting at the end of 1920. Gronwall had been at Princeton. J. W. Alexander actually wrote his dissertation under Gronwall, though he later changed to topology under the inspiration of Veblen. Princeton was somewhat of a disappointment. There were in power the old undergraduate teachers Gillespie, McInnes, and Thompson. I think that during my first term there I had two divisions of trigonometry with endless homework. The second term there was a graduate course in differential equations and I cheered up somewhat. I could talk to Alexander and his students, especially Gleason, Hotelling, and Raynor. Wedderburn was rather friendly in a shy way. I got to know MacDuffee. Eisenhart and Veblen were very active, but somehow I could not get interested in combinatorial topology and tensors; however, thanks to Eisenhart I learned some projective geometry. Lefschetz came in 1924 as a visiting lecturer, but he was then given a permanent appointment and so stayed on. He was an interesting, very colorful personality, completely unpredictable and, in the long run, somewhat trying. Wedderburn wanted to retire from the editorship of the Annals, and in 1929 Lefschetz and I were put in charge. We raised the Annals beyond recognition, but not without friction.

In 1924 I attended the Toronto International Congress. This was the second congress held without the former Central Powers, and the United States, Italy, and the former neutrals were becoming restive. The next congress, held in Bologna in 1928, restored the international character of these meetings.

3. *N. R. C.* Toward 1925 I had a feeling that my research was becoming stale and that some new ideas and contacts were needed. Veblen was sympathetic and I got an N.R.C. fellowship for 1926-27. I was supposed to spend the winter in Copenhagen with Nörlund, making still another attempt with difference equations, and the summer semester in Göttingen. It did not quite

come out that way. I spent three months at the Mittag-Leffler Institute working on a bibliography on analytic representation that was ultimately published. This was not a sheer waste of time, though it may have appeared that way to the N.R.C. I read a large portion of the 1000-odd papers included in the bibliography, and this is still valuable to me. When I came back to Copenhagen, I worked on various problems connected with functions defined by difference equations, but the planned contribution to the theory of partial difference equations never materialized. Thus ended my third attempt, and after this I have not tried to contribute to the theory of difference equations, though it is still a field that interests me.

Göttingen in 1927 was still one of the most lively mathematical centers in the world and I made a number of interesting contacts. Felix Klein had died, and Hilbert was in poor health so that I saw him only once. Courant, Landau, Herglotz, and Felix Bernstein were active. This was before Courant had created the new institute. I met a number of his students: Friedrichs, Neugebauer, Hans Levy, Will Feller. P. S. Alexandroff lectured on topology, and through him I met a number of visiting topologists: Heins Hopf, Knaster, and Kuratowski. Emmy Noether, van der Waerden, and John von Neumann were other interesting acquaintances. Among temporary visitors were the Russians Stepanof (almost periodic functions) and Otto Schmidt, later a leader of Soviet arctic research. Another acquaintance was Aurel Wintner. I had considerable contact with a young German student, Späth, who told me about his work on best approximation of irrational numbers along the lines of Perron. He unfortunately committed suicide the following year while his committee for the examen rigorosum was waiting. This put an end to what I think would have been a very interesting career.

4. *Princeton, Tamarkin.* I came back to Princeton in the fall of 1927. The place was gradually changing. Thompson had retired, MacInnes had died, Gillespie was kicked upstairs to become master of the Graduate College ("Mother Superior" was the phrase used by the inmates). Alexandroff and Hopf spent the year in Princeton and gave good courses. The following year brought Dieudonné, Hardy, and Weyl. Hardy had moved from Trinity, Cambridge, to Oxford's New College (new in 1379!) and the Savilian chair of geometry. Much to his annoyance, he found that he had to lecture, at least occasionally, on geometry. For the year 1928-29 he and Veblen had arranged a private exchange: Veblen would spend the year at Oxford and Hardy would come to Princeton for one term and go to Cal. Tech. for the other. Weyl was intended for the Jones chair of mathematical physics, but he returned to Zürich after a year. He was called to Göttingen as Hilbert's successor, but

after Hitler's arrival in power he finally returned to Princeton, to the new Institute for Advanced Study where Veblen was creating the School of Mathematics.

For me the second period at Princeton and the first few years at Yale are marked by the collaboration with J. D. Tamarkin. We had first met at the summer meeting at Cornell in 1925. He had recently come over and was at that time at Dartmouth. He moved to Brown two years later. There was a time when every, almost every, mathematician in this country knew and liked Tamarkin. But it is 17 years since he passed away and the number who knew him and cherished him is getting steadily smaller. So I should perhaps tell a few things about him. He was born in 1888 and grew up in St. Petersburg, which he always referred to as Petrograd. His mathematical ability showed itself very early and he got a gold medal for an investigation carried out in high school and later published. At the University he became attached to Stekhloff and worked mainly on boundary value problems for linear *n*th order equations. One of his early papers (1913) was a critique of Birkhoff's solution of this problem which forced Birkhoff into doing a more thorough job. He got his degree (a lower form of the doctorate, but probably above the candidate degree as now awarded in the Soviet Union) in 1917. (A condensed version of the dissertation appeared ten years later in Math. Z.) J. D. had a very thorough knowledge of what was then modern analysis as well as of mathematical physics. He was a very skilled teacher who in a few years put Brown on the mathematical map, and he was also an excellent collaborator. I learned ever so much from him of classical analysis, and together we studied and worked on functional analysis as developed by Banach in the 1920s and early 1930s.

Tamarkin's parents had been very wealthy and his apartment in Petrograd had contained two grand pianos, a collection of other musical instruments, and a musical library with several thousands of items, besides an excellent mathematical library. Politically he was a menshevik, equally unpopular with the secret police during the time of the tsars and the communists. In the long run he found the situation not to be endured. Silverman from Dartmouth visited the Soviet Union in 1924 and found him willing to leave if a position could be found for him in the United States and if he could make his escape. Silverman arranged a job for him at Dartmouth, and J. D. got a postcard from him stating that the courses for 1925-26 had now been settled and that one of the new attractions would be an eminent foreign visitor. This was the signal. The next problem was to get out. Where there is a demand there is usually a supply, and there was a fairly efficient underground railroad from the Soviet Union to Latvia, at that time an independent country. But the passage was fairly expensive, something like $800, and the only way of getting money was from the government, which certainly was not prepared to give subvention for this purpose. So the first thing was to earn some extra money in a legitimate manner. This could be done by writing a book, so J. D. induced his friend Smirnoff to collaborate on a textbook on mathematics for engineers and physicists, and his share in the royalties sufficed to pay for the expenses. Actually the book must have been started quite early for the preface of the first edition is dated in May 1923. Tamarkin's name remained on the cover for the 2nd edition of 1926. This is now Smirnoff's *Kurs Vysshey Matematiki.* At any rate, J. D. now had the funds to finance the escape. He and Besicovitch crossed the Latvian border, crawling under a railroad bridge; Besicovitch carried him part of the way on his back. They were arrested on the Latvian side, a little too early for comfort, but were finally sent off to Riga where Nansen passports were waiting. J. D. went to the American Consulate where the officer in charge found it hard to believe that this dirty unshaved tramp could be professor at the Leningrad University or any other respectable place. He therefore proceeded to question J. D. on analytic geometry, but ran out of questions and gave up when Tamarkin threatened to take over.

Our collaboration started in December 1927, when we began working on the frequency of the characteristic values of an integral equation under various assumptions on the kernel. This was followed by papers on summability of Fourier series, Fourier transforms, relations between Hausdorff methods of summability, Laplace transforms, etc. All in all we had 26 joint papers during the ten years 1927-1937 that we collaborated. Several of my own papers during this period were also an outgrowth of joint work.

I would also like to mention two other investigations from this time in which Tamarkin did not figure. The first was the problem of the width of the vertical strip in the complex plane in which an ordinary Dirichlet series could be uniformly convergent without being also absolutely convergent. In 1913 Harald Bohr had shown that the width could be at most 1/2, but he could not show even that there were series for which the width was positive, much less equal to 1/2. The only additional result he had was that, in the case in which the summation extends over the primes, the width is zero. Soon afterwards Otto Toeplitz was able to construct an example in which the width is 1/4. Bohr's results came from a discussion of analytic functions in infinitely many unknowns, combined with beautiful applications of Kronecker's approximation theorem and a discussion of the addition of convex sets.

Toeplitz got his result from the theory of bounded quadratic forms in infinitely many unknowns. This was the state of the problem in 1930. The A.M.S. summer meeting that year was held in Providence; I was there and found in the library a recent copy of the Oxford Journal with an article by Littlewood on multilinear forms in infinitely many unknowns. This seemed to me to provide exactly the needed approach to Bohr's problem. I tried to interest Tamarkin in it, but for once he failed to respond. Back in Princeton, I found that on the basis of Littlewood's results I could prove that, if in an ordinary Dirichlet series the summation extends only over those integers which are the product of n primes, then the width of the strip is at most $(n-1)/(2n)$. This was in agreement with Bohr for $n = 1$, and with Toeplitz for $n = 2$; but effective examples were lacking. These, however, were provided by H. F. Bohnenblust, who had come over from Switzerland with Hermann Weyl and was my research assistant. On this joint work Bohny got his degree in 1931.

The second investigation was an outgrowth of my attempts to do something with the closure of translations in L_p-spaces for $1 < p < 2$. During this work I acquired much information concerning some special transforms, depending upon a parameter, which are used to approximate continuous or integrable functions, such as the Gauss-Weierstrass, the Poisson integral for the half-plane, the Picard transform, the Dirichlet integral, etc. I got interested in their properties qua functions of the parameter and wrote a paper about them which appeared in the TAMS in January 1936. Two of the transforms had the semi-group property, and this marked the beginning of the analytical theory of semi-groups.

I have mentioned three of my collaborators. All told I have had seventeen.* The list is as follows:

1. H. v. Euler
2. G. Rasch
3. J. D. Tamarkin
4. H. F. Bohnenblust
5. A. C. Offord
6. O. Szász
7. G. Szegö
8. J. Shohat
9. J. L. Walsh
10. H. L. Garabedian

*Since 1962 two more additions to this list have been made:
18. S. L. Salas
19. G. L. Curme

11. H. S. Wall
12. M. Zorn
13. W. B. Caton
14. N. Dunford
15. G. Klein
16. R. S. Phillips
17. Paul Erdös

I had hoped that this list would include S. Kakutani to whom I owe so much. I have learned something from all of them and this is more than just mechanical collaboration. I owe them a debt of gratitude.

I am not at the end of my story, but this is a wise point to stop. To an old man the past is ever so much more vivid than the present. Perhaps he glamorizes it, but he can look at it in a more dispassionate mood. The present is full of troubled waters, of unresolved conflicts, one sees only part of the truth, darkly as through a pane. The historian knows that the present cannot be judged objectively by a person who is up to his neck in the struggle, so he leaves it to the future. This I will also do.

This is in a way a farewell. It remains for me to thank you all for the voyage we have had together, and I am not talking about today. I have been an academic teacher since 1915. There must have been thousands of students who have had to listen to me. I hope I gave them something worthwhile, but I am not so sure. I am aware of my shortcomings, or at least of some of them. I have never been able to make people work unless they wanted to. Perhaps I gave them too little encouragement to try their own wings. Be that as it may, my first thanks go to the students here present. I have enjoyed my teaching, and I thank you for keeping me alert and giving me a chance to think aloud. I hope you have profited by it and I wish you the best of luck.

My colleagues I wish to thank for friendship, collaboration, and patience with my foibles. With apologies for my shortcomings I thank you.

Finally thank you for this opportunity to think aloud and forgive my talking too much about myself!

Einar Hille

Bibliography of Einar Hille

(Entries preceded by an asterisk are reprinted in this volume.)

[0] (with Hans von Euler) *Über die primäre Umwandlumg der Hexosen bei der alkoholischen Gärung,* Z. Gärungsphysiologie vol. 3(1913) pp. 235-240.

[1] *Über die Variation der Bogenlänge bei konformer Abbildung von Kreisbereichen,* Ark. Mat. Astronom. Fys. vol. 11, no. 27(1917) 11 pp.

[2] *Some problems concerning spherical harmonics,* dissertation, Stockholm 1918, Ark. Mat. Astronom. Fys. vol. 13, no. 17 (1918) 76 pp.

[3] *Oscillation theorems in the complex domain,* Trans. Amer. Math. Soc. vol. 23 (1922) pp. 350-385.

[4] *An integral equality and its applications,* Proc. Nat. Acad. Sci. U. S. A. vol. 7 (1921) pp. 303-305.

*[5] *On the zeros of Sturm-Liouville functions,* Ark. Mat. Astronom. Fys. vol. 16, no. 17 (1922) 20 pp.

[6] *Convex distribution of the zeros of Sturm-Liouville functions,* Bull. Amer. Math. Soc. vol. 28 (1922) pp. 261-265. *A correction,* ibid., p. 462.

[7] *On the zeros of Legendre functions,* Ark. Mat. Astronom. Fys. vol. 17, no. 22 (1923) 16 pp.

*[8] *A Pythagorean functional equation,* Ann. of Math. (2) vol. 24 (1922) pp. 175-180.

[9] *Note on Dirichlet's series with complex exponents,* Ann. of Math. (2) vol. 25 (1924) pp. 261-278.

[10] *On the zeros of Mathieu functions,* Proc. London Math. Soc. (2) vol. 23 (1924) pp. 185-237.

[11] *On the zeros of the functions of the parabolic cylinder,* Ark. Mat. Astronom. Fys. vol. 18, no. 26 (1924) 56 pp.

*[12] *An existence theorem,* Trans. Amer. Math. Soc. vol. 26 (1924) pp. 241-248.

[13] *A note on regular singular points,* Ark. Mat. Astronom. Fys. vol. 19A, no. 2 (1925) 21 pp.

*[14] *A class of functional equations,* Ann. of Math. (2) vol. 29 (1928) pp. 215-222.

*[15] *On the zeros of the functions defined by linear differential equations of the second order,* Proc. Internat. Math. Cong., Toronto, 1924, vol. 1, pp. 511-519.

[16] *A general type of singular point,* Proc. Nat. Acad. Sci. U.S.A. vol. 10 (1924) pp. 488-493.

*[17] *Some remarks on Dirichlet series,* Proc. London Math. Soc. (2) vol. 25 (1926) pp. 177-184.

[18] *Miscellaneous questions in the theory of differential equations. I. On the Method of Frobenius,* Ann. of Math. (2) vol. 27 (1926) pp. 195-198.

[19]-[21] *On Laguerre's series. First-Third Note,* Proc. Nat. Acad. Sci. U. S. A. vol. 12 (1926) pp. 261-265, 265-269, 348-352.

[22] *A class of reciprocal functions,* Ann. of Math. (2) vol. 27 (1926) pp. 427-464.

*[23] *Zero point problems for linear differential equations of the second order,* Mat. Tidsskrift, B (1927) pp. 25-44.

[24] *On the logarithmic derivatives of the gamma function,* Danske Vid. Selsk. Mat.-Fys. Medd. vol. 8, no. 1 (1927) 58 pp.

[25] (with G. Rasch) *Über die Nullstellen der unvollständigen Gammafunktion* $P(z,\rho)$. *II. Geometrisches über die Nullstellen,* Math. Z. vol. 29 (1928) pp. 319-334.

[26] *Note on the behavior of certain power series on the circle of convergence with application to a problem of Carleman,* Proc. Nat. Acad. Sci. U. S. A. vol. 14 (1928) pp. 217-220.

[27] *Note on the preceding paper by Mr. Peek,* Ann. of Math. (2) vol. 30 (1929) pp. 270-271.

[28] *Note on some hypergeometric series of higher order,* J. London Math. Soc. vol. 4 (1929) pp. 50-54.

*[29] (with J. D. Tamarkin) *On the characteristic values of linear integral equations,* Proc. Nat. Acad. Sci. U. S. A. vol. 14 (1928) pp. 911-914.

[30] (with J. D. Tamarkin) *On the summability of Fourier series,* Proc. Nat. Acad. Sci. U. S. A. vol. 14 (1928) pp. 915-918.

[31] (with J. D. Tamarkin) *On the summability of Fourier series. Second Note,* Proc. Nat. Acad. Sci. U. S. A. vol. 15 (1929) pp. 41-42.

[32] (with J. D. Tamarkin) *Remarks on a known example of a monotone continuous function,* Amer. Math. Monthly vol. 36 (1929) pp. 255-264.

[33] *Note on a power series considered by Hardy and Littlewood,* J. London Math. Soc. vol. 4 (1929) pp. 176-182.

[34] (with J. D. Tamarkin) *Sur une relation entre des résultants de MM. Minetti et Valiron,* C. R. Acad. Sci. Paris vol. 188 (1929) p. 1142.

[35] *On functions of bounded deviation,* Proc. London Math. Soc. (2) vol. 31 (1930) pp. 165-173.

[36] *Bemerkung zu einer Arbeit des Herrn Müntz,* Math. Z. vol. 32 (1930) pp. 422-425.

[37] (with J. D. Tamarkin) *On the theory of linear integral equations. I,* Ann. of Math. (2) vol. 31 (1930) pp. 479-528.

[38] (with J. D. Tamarkin) *On the summability of Fourier series. Third Note,* Proc. Nat. Acad. Sci. U. S. A. vol. 16 (1930) pp. 594-598.

[39] (with J. D. Tamarkin) *On the characteristic values of linear integral equations,* Acta Math. vol. 57 (1931) pp. 1-76.

[40] (with H. F. Bohnenblust) *Sur la convergence absolue des séries de Dirichlet,* C. R. Acad. Sci. Paris vol. 192 (1931) pp. 30-32.

*[41] (with H. F. Bohnenblust) *On the absolute convergence of Dirichlet series,* Ann. of Math. (2) vol. 32 (1931) pp. 600-622.

[42] (with J. D. Tamarkin) *On the summability of Fourier series. Fourth Note,* Proc. Nat. Acad. U. S. A. vol. 17 (1931) pp. 376-380.

[43] (with J. D. Tamarkin) *On the summability of Fourier series. I,* Trans. Amer. Math. Soc. vol. 34 (1932) pp. 757-783.

[44] *Summation of Fourier series,* Bull. Amer. Math. Soc. vol. 38 (1932) pp. 505-528.

[45] (with H. F. Bohnenblust) *Remarks on a problem of Toeplitz,* Ann. of Math. (2) vol. 33 (1932) pp. 785-786.

[46] (with J. D. Tamarkin) *The summation of Fourier series by Hausdorff means,* Verh. Internat. Math. Kong., Zürich, 1932, vol. 2, pp. 131-132.

[47] (with J. D. Tamarkin) *On the summability of Fourier series,* Verh. Internat. Math. Kong., Zürich, 1932, vol. 2, pp. 133-134.

*[48] (with J. D. Tamarkin) *On the summability of Fourier series. II,* Ann. of Math. (2) vol. 34 (1933) pp. 329-348, 602-605.

[49] (with J. D. Tamarkin) *On the summability of Fourier series. III,* Math. Ann. vol. 108 (1933) pp. 525-577.

*[50] (with J. D. Tamarkin) *On a theorem of Paley and Wiener,* Ann. of Math. (2) vol. 34 (1933) pp. 606-614.

[51] (with J. D. Tamarkin) *Questions of relative inclusion in the domain of Hausdorff means,* Proc. Nat. Acad. Sci. U. S. A. vol. 19 (1933) pp. 573-577.

[52] *Über die Nullstellen der Hermiteschen Polynome,* Jber. Deutsch. Math.-Verein. vol. 44 (1934) pp. 162-165.

[53] *On the complex zeros of the associated Legendre functions,* J. London Math. Soc. vol. 8 (1933) pp. 216-217.

*[54] (with J. D. Tamarkin) *On the theory of Fourier transforms,* Bull. Amer. Math. Soc. vol. 39 (1933) pp. 768-774.

*[55] (with J. D. Tamarkin) *A remark on Fourier transforms and functions analytic in a half-plane,* Compositio Math. vol. 1 (1934) pp. 98-102.

[56] (with J. D. Tamarkin) *On moment functions,* Proc. Nat. Acad. Sci. U. S. A. vol. 19 (1933) pp. 902-908.

[57] (with J. D. Tamarkin) *On the theory of Laplace integrals,* Proc. Nat. Acad. Sci. U. S. A. vol. 19 (1933) pp. 908-912.

[58] (with J. D. Tamarkin) *On the theory of Laplace integrals. II,* Proc. Nat. Acad. Sci. U. S. A. vol. 20 (1934) pp. 140-144.

[59] (with J. D. Tamarkin) *On the theory of linear integral equations. II,* Ann. of Math. (2) vol. 35 (1934) pp. 445-455.

[60] (with J. D. Tamarkin) *On the summability of Fourier series. Fifth Note,* Proc. Nat. Acad. Sci. U. S. A. vol. 20 (1934) pp. 369-372.

[61] (with A. C. Offord and J. D. Tamarkin) *Some observations on the theory of Fourier transforms,* Bull. Amer. Math. Soc. vol. 41 (1935) pp. 427-436.

*[62] *On Laplace integrals,* Ått. Skand. Matematikerkong., Stockholm, 1934, pp. 216-227.

*[63] *Notes on linear transformations. I,* Trans. Amer. Math. Soc. vol. 39 (1936) pp. 131-153.

*[64] (with J. D. Tamarkin) *On the absolute integrability of Fourier transforms,* Fund. Math. vol. 25 (1935) pp. 329-352.

[65] (with Otto Szász) *On the completeness of Lambert functions,* Bull. Amer. Math. Soc. vol. 42 (1936) pp. 411-418.

[66] (with Otto Szász) *On the completeness of Lambert functions, II,* Ann. of Math. (2) vol. 37 (1936) pp. 801-815.

*[67] *A problem in "Factorisatio Numerorum,"* Acta Arith. vol. 2 (1936) pp. 134-144.

*[68] *The inversion problem of Möbius,* Duke Math. J. vol. 3 (1937) pp. 549-568.

[69] (with G. Szegö and J. D. Tamarkin) *On some generalizations of a theorem of A. Markoff,* Duke Math. J. vol. 3 (1937) pp. 729-739.

[70] *Bilinear formulas in the theory of the transformation of Laplace,* Compositio Math. vol. 6 (1938) pp. 93-102.

[71] *On the absolute convergence of polynomial series,* Amer. Math. Monthly vol. 45 (1938) pp. 220-226.

*[72] *On semi-groups of transformations in Hilbert space,* Proc. Nat. Acad. Sci. U. S. A. vol. 24 (1938) pp. 159-161.

*[73] *Notes on linear transformations. II. Analyticity of semi-groups,* Ann. of Math. (2) vol. 40 (1939) pp. 1-47.

[74] *Analytical semi-groups in the theory of linear transformations,* Ått. Skand. Matematikerkong., Helsingfors, 1938, pp. 135-145.

[75] *Remarks concerning group spaces and vector spaces,* Compositio Math. vol. 6 (1939) pp. 375-381.

[76] *Contributions to the theory of Hermitian series,* Duke Math. J. vol. 5 (1939) pp. 875-936.

*[77] *Sur les séries associées à une série d'Hermite,* C. R. Acad. Sci. Paris vol. 209 (1939) pp. 714-716.

*[78] *Contributions to the theory of Hermitian series. II. The representation problem,* Trans. Amer. Math. Soc. vol. 47 (1940) pp. 80-94.

*[79] *A class of differential operators of infinite order. I,* Duke Math. J. vol. 7 (1940) pp. 458-495.

[80] (with H. L. Garabedian and H. S. Wall) *Formulations of the Hausdorff inclusion problem,* Duke Math. J. vol. 8 (1941) pp. 193-213.

[81] *On the oscillation of differential transforms. II. Characteristic series of boundary value problems,* Trans. Amer. Math. Soc. vol. 52 (1942) pp. 463-497.

[82] *Gelfond's solution of Hilbert's seventh problem,* Amer. Math. Monthly vol. 49 (1942) pp. 654-661.

*[83] *Representation of one-parameter semi-groups of linear transformations,* Proc. Nat. Acad. Sci. U. S. A. vol. 28 (1942) pp. 175-178.

*[84] *On the oscillation of differential transforms and the characteristic series of boundary-value problems,* Univ. Calif. Publ. Math. (N. S.) vol. 2 (1944) pp. 161-168.

*[85] *On the analytical theory of semi-groups,* Proc. Nat. Acad. Sci. U. S. A. vol. 28 (1942) pp. 421-424.

[86] (with G. Szegö) *On the complex zeros of the Bessel functions,* Bull. Amer. Math. Soc. vol. 49 (1943) pp. 605-610.

[87] (with M. Zorn) *Open additive semi-groups of complex numbers,* Ann. of Math. (2) vol. 44 (1943) pp. 554-561.

*[88] *On the theory of characters of groups and semi-groups in normed vector rings,* Proc. Nat. Acad. Sci. U. S. A. vol. 30 (1944) pp. 58-60.

*[89] *Remarks on ergodic theorems,* Trans. Amer. Math. Soc. vol. 57 (1945) pp. 246-269.

[90] (with W. B. Caton) *Laguerre polynomials and Laplace integrals,* Duke Math. J. vol. 12 (1945) pp. 217-242.

*[91] (with N. Dunford) *The differentiability and uniqueness of continuous solutions of addition formulas,* Bull. Amer. Math. Soc. vol. 53 (1947) pp. 799-805.

*[92] *Non-oscillation theorems,* Trans. Amer. Math. Soc. vol. 64 (1948) pp. 234-252.

[93] *Sur les semi-groupes analytiques,* C. R. Acad. Sci. Paris vol. 225 (1947) pp. 445-447.

[94] *Remarks on a paper by Zeev Nehari,* Bull. Amer. Math. Soc. vol. 55 (1949) pp. 552-553.

[95] *Les semi-groupes linéaires,* C. R. Acad. Sci. Paris vol. 228 (1949) pp. 35-37.

*[96] *Lie theory of semi-groups of linear transformations,* Bull. Amer. Math. Soc. vol. 56 (1950) pp. 89-114.

[97] *On the differentiability of semi-group operators,* Acta Sci. Math. (Szeged) vol. 12B (1950) pp. 19-24.

*[98] *On the integration problem for Fokker-Planck's equation in the theory of stochastic processes,* C. R. Onzième Cong. Math. Scand., Trondheim, 1949, pp. 183-194.

[99] *Les probabilités continues en chaîne,* C. R. Acad. Sci. Paris vol. 230 (1950) pp. 34-35.

*[100] *"Explosive" solutions of Fokker-Planck's equation,* Proc. Internat. Cong. Math., Cambridge, Massachusetts, 1950, vol. 1, p. 453.

[101] *On the generation of semi-groups and the theory of conjugate functions,* Proc. Roy. Physiographical Soc. Lund vol. 21, no. 14 (1952) 13 pp.

*[102] *Behavior of solutions of linear second order differential equations,* Ark. Mat. vol. 2 (1951) pp. 25-41.

[103] *A note on Cauchy's problem,* Ann. Polon. Math. vol. 25 (1952) pp. 56-68.

[104] *Such stuff as dreams are made on—in mathematics,* Amer. Sci. vol. 41 (1953) pp. 106-112.

[105] *Mathematics and mathematicians from Abel to Zermelo,* Math. Mag. vol. 26 (1953) pp. 127-146.

*[106] *Une généralisation du problème de Cauchy,* Ann. Inst. Fourier (Grenoble) vol. 4 (Année 1952) pp. 31-48.

[107] *Sur le problème abstrait de Cauchy,* C. R. Acad. Sci. Paris vol. 236 (1953) pp. 1466-1467.

[108] *Le problème abstrait de Cauchy,* Rend. Sem. Mat. Torino vol. 12 (1953) pp. 95-103.

[109] *The abstract Cauchy problem and Cauchy's problem for parabolic differential equations,* J. Analyse Math. vol. 3 (1954) pp. 81-196.

[110] *An abstract formulation of Cauchy's problem,* C. R. Douzième Cong. Math. Scand., Lund, 1953 (1954) pp. 79-89.

[111] *Some extremal properties of Laplace transforms,* Math. Scand. vol. 1 (1953) pp. 227-236.

[112] (with George Klein) *Riemann's localization theorem for Fourier series,* Duke Math. J. vol. 21 (1954) pp. 587-591.

[113] *On the integration of Kolmogoroff's differential equations,* Proc. Nat. Acad. Sci. U. S. A. vol. 40 (1954) pp. 20-25.

[114] *Quelques remarques sur les équations de Kolmogoroff,* Bull. Soc. Math. Phys. R. P. Serbie vol. 5 (1953) pp. 3-14.

*[115] *Some aspects of Cauchy's problem,* Proc. Internat. Cong. Math., Amsterdam, 1954, vol. 3, pp. 109-116.

[116] *Sur un théorème de perturbation,* Rend. Sem. Mat. Torino vol. 13 (1954) pp. 169-184.

[117] *Perturbation methods in the study of Kolmogoroff's equations,* Proc. Internat. Cong. Math., Amsterdam, 1954, vol. 3, pp. 365-376.

[118] *On a class of orthonormal functions,* Rend. Sem. Mat. Univ. Padova vol. 25 (1956) pp. 214-249.

[119] *Quelques remarques sur l'équation de la chaleur,* Rend. Mat. e Appl. (5) vol. 15 (1956) pp. 102-118.

[120] *Über eine Klasse Differentialoperatoren vierter Ordnung,* S. B. Berlin. Math. Ges. (Jahrgang 1954/1955 u. 1955/1956) pp. 39-44.

[121] *Problème de Cauchy: existence et unicité des solutions,* Bull. Math. Soc. Sci. Math. Phys. R. P. Roumanie (1) vol. 49, no. 2 (1957) pp. 31-33.

[121a] *Problema lui Cauchy: Existenta si unicitatea solutiilor,* Roumanian translation of [121], ibid., pp. 143-145.

*[122] *Remarques sur les systèmes des équations différentielles linéaires à une infinité d'inconnues,* J. Math. Pures Appl. (9) vol. 37 (1958) pp. 375-383.

[123] *Green's transforms and singular boundary value problems,* Abstracts of Short Communications, Internat. Cong. Math., Edinburgh, 1958, pp. 80-81.

*[124] *On roots and logarithms of elements of a complex Banach algebra,* Math. Ann. vol. 136 (1958) pp. 46-57.

*[125] *On the inverse function theorem in Banach algebras,* Calcutta Mathematical Society Golden Jubilee Commemoration Volume, 1958-1959, pp. 65-69.

[126] *An application of Prüfer's method to singular boundary value problems,* Math. Zeit. vol. 72 (1959) pp. 95-106.

*[127] *Sur les fonctions analytiques définies par des séries d'Hermite,* J. Math. Pures Appl. (9) vol. 40 (1961) pp. 335-342.

[128] *What is a semi-group?* MAA Studies in Mathematics, vol. 3, Studies in Real and Complex Analysis, edited by I. I. Hirschman, Jr., Prentice-Hall, Englewood Cliffs, N. J., 1965, pp. 55-66.

[129] *Differential Equations, Ordinary,* Encyclopaedia Britannica, vol. 7, p. 407.

[130] *Fekete polynomials and conformal mapping,* abstract, Second Hungar. Math. Cong., Budapest, 1960.

[131] *Functions, Analytic,* Encyclopaedia Britannica, vol. 9, pp. 1005-1007.

* [132] *Linear differential equations in Banach algebras,* Proc. Internat. Symp. Linear Spaces, Jerusalem, July 1960, Jerusalem Academic Press, 1961, pp. 263-273.

[132a] *Linear differential equations in Banach algebras,* abstract of [132], Studia Math. (Seria Specjalna) Zeszyt 1 (1963) pp. 53-54 (Proc. Conf. Functional Analysis, Warsaw, September 1960).

*[133] *Pathology of infinite systems of linear first order differential equations with constant coefficients,* Ann. Mat. Pura Appl. (4) vol. 55 (1961) pp. 133-148.

[134] *Remarks on transfinite diameters, general topology and its relations to modern analysis and algebra,* Proc. Symposium, Prague, September 1961, Czechoslovak Academy of Sciences, Prague, 1962, pp. 211-220. (Mimeographed corrections, 1962, reproduced in [137], pp. 222-223.)

[135] *Remarks on differential equations in Banach algebras,* Studies in Mathematical Analysis and Related Topics, Essays in Honor of George Pólya, Stanford University Press, Stanford, 1962, pp. 140-145.

*[136] *Green's transforms and singular boundary value problems,* J. Math. Pures Appl. (9) vol. 42 (1963) pp. 331-349.

[137] *A note on transfinite diameters,* J. Analyse Math. vol. 14 (1965) pp. 209-224. (Theorems 4 and 8 are false. See [139].)

[138] *Topics in Classical Analysis,* Lectures on Modern Mathematics, vol. 3, edited by T. L. Saaty, John Wiley, New York, 1965, pp. 1-57.

*[139] *Some geometric extremal problems,* J. Austral. Math. Soc. vol. 6 (1966) pp. 122-128.

*[140] *Some aspects of differential equations in B-algebras,* Functional Analysis, Proc. Conf., University of California, Irvine, 1966, edited by B. R. Gelbaum, Thompson Book Co., Washington, D. C., 1967, pp. 185-194.

*[141] *Some properties of the Jordan operator,* Aequationes Math. vol. 2 (1968) pp. 105-110.

[142] (with G. L. Curme) *Classical analytic representations,* Quart. Appl. Math. vol. 27 (1969) pp. 185-192.

*[143] *On the Thomas-Fermi equation,* Proc. Nat. Acad. Sci. U. S. A. vol. 62 (1969) pp. 7-10.

[144] *Some aspects of the Thomas-Fermi equation,* J. Analyse Math. vol. 23 (1970) pp. 147-170.

[145] *Aspects of Emden's equation,* J. Fac. Sci. Univ. Tokyo, Sect. I, vol. 17 (1970) pp. 11-30.

[146] *Remark on the Landau-Kallman-Rota inequality,* Aequationes Math. vol. 4 (1970) pp. 239-240.

[147] *On the Landau-Kallman-Rota inequality,* J. Approximation Theory vol. 6 (1972) pp. 117-122.

[148] *Pseudo-poles in the theory of Emden's equation,* Proc. Nat. Acad. Sci. U. S. A. vol. 69 (1972) pp. 1271-1272.

[Bibl. 1] *Essai d'une bibliographie de la représentation analytique d'une fonction monogene,* Acta Math. vol. 52 (1929) pp. 1-80.

[Bibl. 2] (with J. Shohat and J. L. Walsh) *A bibliography of orthogonal polynomials,* Bull. Nat. Res. Council, no. 103, August 1940, ix + 204 pp.

[Ob. 1] *Thomas Hakon Gronwall—In Memoriam,* Bull. Amer. Math. Soc. vol. 38 (1932) pp. 775-786.

[Ob. 2] *Jacob David Tamarkin—his life and work,* Bull. Amer. Math. Soc. vol. 53 (1947) pp. 440-457.

[B 1] *Functional Analysis and Semi-groups,* Amer. Math. Soc. Colloquium Publications, vol. XXXI, New York, 1948, xii + 528 pp.

[B 2] (with R. S. Phillips) *Functional Analysis and Semi-groups,* Amer. Math. Soc. Colloquium Publications, vol. XXXI, rev. ed., Providence, 1957, xii + 808 pp. Revised reprint 1964.

[B 3] *Analytic Function Theory,* vol. I, Ginn and Co., Boston, 1959, xi + 308 pp.

[B 4] *Analytic Function Theory,* vol. II, Ginn and Co., Boston, 1962, xii + 496 pp.

[B 5] *Analysis,* Volume I, Blaisdell Publishing Co., New York, 1964, xii + 626 pp.

[B 6] *Analysis,* Volume II, Blaisdell Publishing Co., New York, 1966, xii + 672 pp.

[B 7] (with S. L. Salas) *First-Year Calculus,* Blaisdell Publishing Co., New York, 1968, xi + 415 pp.

[B 8] *Lectures on Ordinary Differential Equations,* Addison-Wesley Publishing Co., Reading, Massachusetts, 1969, xi + 723 pp.

[B 9] *Methods in Classical and Functional Analysis,* Addison-Wesley Publishing Co., Reading, Massachusetts, 1972, ix + 486 pp.

[B 10] (with S. L. Salas) *Calculus: One and Several Variables,* Xerox Publishing Co., Waltham, Massachusetts, 1971, vii + 800 pp.

Foreword

One of my first recollections of Einar Hille goes back to the return train trip to New York from the American Mathematical Society meeting in New Brunswick in 1945, when Hille, Feller, and I shared a double seat. The two of them were old friends and I was essentially unknown to them, yet they made me feel at ease in their company and we had a pleasant trip. This was the first of many kindnesses that I received from Hille. A couple of years later when I was trying to put together my mathematical career after four years at the MIT Radiation Lab and after I had started studying his book on *Functional Analysis and Semi-groups,* I pestered him with a continuous flow of letters; he answered my questions with a great deal of care and encouragement, and it was due in good part to this that I really got involved in semi-group theory. Finally in 1952 he asked me to collaborate with him on the second edition of his book. I found him a generous colleague, extremely patient, and perhaps a little too permissive for the good of the book. Although the new edition is one-and-a-half times the size of the original and contains several new developments of mine, it is largely in fact and entirely in spirit very much like the original edition.

Hille began the first edition of *Functional Analysis and Semi-groups* with a quote from Kipling: "And each man hears as the twilight nears, to the beat of his dying heart, the Devil drum on the darkened pane: 'You did it, but was it Art?' " There is no doubt in my mind that this book, which was written with great care and remarkable thoroughness, was indeed a work of art. It was both a textbook on functional analysis and a monograph on the theory of semi-groups of operators; in it was to be found for the first time functional analysis presented as a tool for classical analysis. In addition to the usual theory of function spaces and linear transformations, Hille organized into a unified whole the calculus and integration theory for vector-valued functions, function theory for vector-valued functions and functionals, Banach algebras, and the operational calculus. In the process of reorganizing this material, Hille made use of his extensive background in analysis to introduce many new vector-valued analogues of classical results. A notable instance of this is the chapter on the Laplace transform and binomial series. As a measure of his success, let me recall that this book was for many years one of the principal texts in functional analysis.

That Hille was able to present the theory of functional analysis in such a fresh and modern fashion at that time was no small feat. It is a little difficult to imagine from the vantage point of the 1970s what functional analysis was like in the thirties and early forties; it seems incredible now that in those days it had a rather feeble existence. Most established analysts dismissed it as merely generalization, and no doubt much of it was just that. Hille was one of the few analysts who recognized its possibilities and were willing to employ it in their investigations of classical problems; I suspect that Hille's association with Tamarkin was partly responsible for this. Actually Hille's first effort in semi-group theory arose out of his studies of the classical Gauss-Weierstrass and Poisson kernels in 1936 [63].* Within twelve years after this modest start, he developed a beautiful and substantial theory almost single-handedly.

Given Hille's background, it is not surprising that quite early in his investigations, in fact in 1938, he studied semi-groups holomorphic in a sector and that his three subsequent papers [73, 74, and 85] became the basis of the most beautiful chapter in the book. In this chapter he showed that, if a semi-group of operators, defined on the positive real axis, is also analytic in some interval $(0,\delta)$, then it has a holomorphic continuation into a semi-module and this extension also has the semi-group property. This is followed by a remarkable theorem characterizing the convex hull of the spectrum of the infinitesimal generator of a holomorphic semi-group of operators in terms of the exponential growth of the semi-group along the various rays in the sector of definition, a result suggested by the classical results of Phragmén-Lindelöf and Pólya. This chapter also contains a basic generation theorem that gives necessary and sufficient conditions for an operator to generate a semi-group holomorphic in a sector.

It was not until 1942, starting with paper [83], that Hille began an attack on the larger class of semi-groups of operators which were merely strongly continuous. It is a tribute to his excellent mathematical sense that the theory which he developed at that time is so well suited to the initial value problems of mathematical physics, since in fact these problems were not seriously studied from this point of view until Yosida began his researches on the diffusion equation in 1949. Here the basic result is the generation theorem, published independently by Hille and Yosida[1] in 1948, which gives necessary and sufficient conditions for a closed operator with dense domain to be the infinitesimal generator of a strongly continuous semi-group of contraction operators. This theorem opened up an entirely new approach to partial differential equations (perhaps hinted at earlier in work on the Schrödinger equation of

*Numbers in brackets refer to Hille's bibliography; raised numbers following an author's name refer to the list of references at the end of the foreword.

quantum mechanics) and established the theory of semi-groups of operators as one of the basic theories in analysis.

In the 1942 paper Hille discovered the representation of the resolvent of the generator as the Laplace transform of the semi-group. A standard approach to initial value problems in partial differential equations strongly suggests the use of the Laplace transform in this connection. It is therefore remarkable and significant to the understanding of Hille to note that he was mainly motivated in this discovery by the theory of analytic functions of exponential type as developed by Pólya.

The logarithm of the norm of a semi-group operator is a subadditive function on the half-line. Subadditive functions on integers had been previously studied, but Hille is responsible for developing the measure-theoretic properties of subadditive functions on the half-line. Likewise the fact that holomorphic semi-groups are defined on sectors, and more generally on semi-modules in the complex plane, led Hille and Max Zorn to an investigation of the theory of semi-modules [87] in 1943.

The chapter on ergodic theory also exploits the Laplace transform

$$R(\lambda) = \int_0^\infty e^{-\lambda \xi}\, T(\xi)\, d\xi\,,$$

since it is the principal ingredient of the Abel limit

$$(A) - \lim_{\xi \to 0,\infty} T(\xi) = \lim_{\lambda \to \infty, 0} \lambda R(\lambda).$$

Again Hille's knowledge of classical analysis stood him in good stead [89]. In this case he was able to make use of Tauberian theorems to show that, under certain auxiliary conditions, Abel summability implies the stronger Cesàro summability. Hille also was able to related the existence of Abel limits both at 0 and at ∞ to certain projection operators and decompositions of the space. These theorems extend to semi-groups of operators, results obtained by Dunford and others for iterates of a single operator.

The chapter on spectral theory is outstanding and has provided the basis for a great deal of subsequent research. Here for the first time one finds an operational calculus and a corresponding spectral mapping theorem explicitly constructed for the generators of semi-groups. Hille takes as the basic relation between the function $f(\lambda)$ and the corresponding operator $f(A)$ the formulas

$$f(\lambda) = \int_0^\infty e^{\lambda\xi} \, d\beta \quad \text{and} \quad f(A) = \int_0^\infty T(\xi) \, d\beta,$$

A being the generator of $T(\xi)$. Both $f(\lambda)$ and $f(A)$ are in a sense Laplace-Stieltjes transforms. In the earlier versions of operational calculus, it was always assumed that $f(\lambda)$ was holomorphic on the spectrum of A. In Hille's theory the spectral mapping theorem holds for functions of $f(\lambda)$ that are holomorphic in the interior of the spectrum of A but may be merely continuous at points of the spectrum that lie on the abscissa of convergence for the Laplace-Stieltjes transform of β. In order to relate the fine structure of the spectrum of A to that of $T(\xi)$ an ingenious, but somewhat different, approach is used, a very crude version of which is to be found in [85].

Part three of the book deals with a wide variety of examples of semi-groups of operators occurring in many fields of mathematics. The chapters on trigonometric semi-groups and translation semi-groups are related to the classical problem of factor sequences for Fourier series and factor functions for L_p spaces, and they make good use of Hille's great familiarity with this field [9, 22, 48, 56]. It is clear that this material served as one of the principal guides for his early work on semi-groups [63, 73, 85]. The chapter on semi-groups of self-adjoint operators on a Hilbert space also builds on an early paper [72]. Although this theorem is somewhat special, I suspect that it served to reinforce his faith in holomorphic semi-groups.

The chapter on partial differential equations provides a modest introduction to what has proved to be the most important application of the theory of semi-groups of operators. Here Hille exhibits the solutions of several problems from mathematical physics either in known form or by means of Fourier transforms and shows that they do indeed form semi-groups of operators. It is somewhat ironic to note that Hille overlooked the present-day approach to these problems which employs the then available Hille-Yosida theorem.

The concluding chapter in this part contains a rich variety of other examples taken from summability theory, Markoff chains, stochastic processes, and fractional integration. All of these examples are treated in a masterly fashion, and in each instance Hille is able to illuminate the subject with his theory of semi-groups of operators.

The book also contains an appendix giving some basic properties of Banach algebras without a unit, including the Gelfand theory. For the most part this material is not new; however it does contain a generalization of a theorem of

Gelfand on characters, proved by Hille in 1944 [88], and a new result on Lie semi-groups.

By 1948 Hille had gone much more deeply into the theory of Lie semi-groups. He spoke on this work [96] at the annual meeting of the American Mathematical Society in his Retiring Presidential address. It consisted of both an investigation of the underlying parameter semi-group $\pi \subset \overline{E}_n^{\,+}$ and a study of its representation $[T(p), p \in \pi]$ in the space of bounded linear operators $E(X)$ on a Banach space X. The basic law of composition was taken as

$$T(p)T(q) = T(p \circ q) ,$$

where $p \circ q$ denotes the product operation in π. The main problem with respect to π was to determine the canonical sub-semi-groups. Hille then showed that to every canonical sub-semi-group there corresponds an infinitesimal generator, that these generators form a positive cone, and that they satisfy the analogues of the three fundamental theorems of Lie. He was also able to prove strong continuity for $T(p)$ as a function of p, assuming only strong measurability. Although this work is a natural generalization of the theory of semi-groups of operators, it has remained dormant and without applications.

Work started on the second edition of *Functional Analysis and Semi-groups* early in 1952. Several chapters still remained unfinished by the fall of 1953 when I moved to New Haven to facilitate the completion of the book. It was during that year that I got to know Einar quite well, in spite of the fact that he was loaded down with his usual teaching duties and was then chairman of the graduate program in mathematics at Yale. I was very impressed by how much time Hille spent working at his desk on mathematics. His lectures were always very polished and were in fact written out in longhand the night before in his class notebook. Hille was a gracious host, more perhaps in the European tradition that the informal American fashion. He had a dry but ready sense of humor; an anecdote may serve to illustrate this. The high point of the academic year was a Christmas party for the graduate students and faculty held in Leet Oliver. It was a festive occasion and under the doorway was hung the traditional mistletoe. My wife Jean and Anne Rickart had decided to trap the every-shy Shizuo Kakutani under this doorway either by trickery or by force. Shizuo was not about to be tricked so the girls tried force, and in the ensuing struggle were themselves positioned under the mistletoe. Everyone else was completely absorbed in this life-and-death struggle, everyone that is except Einar, who seized the opportunity to kiss both of the girls.

The second edition consists of a thorough reworking of the first edition plus many new results. The Gelfand theory of commutative Banach algebras is introduced early in the book and now plays a major role in its development. The influence of Yosida and Feller is quite evident; and of course I exercised my prerogative as coauthor to include my own results on extended classes of semi-groups and the associated generation theorems, perturbation theory, the adjoint semi-group, the operational calculus, and spectral theory. There is also a new chapter on Hille's development of the Lie theory of semi-groups and an expanded section on the integration of the Kolmogoroff differential equations based in part on Hille's papers [116, 117]. The chapter on partial differential equations was omitted because by that time the subject required a treatise of its own. However we did include a discussion of the abstract Cauchy problem which had been initially developed by Hille and on which we had both worked.

After Yosida's 1949 paper[2] it was clear that semi-group theory was going to play an important part in the initial value problems of partial differential equations, and it is typical of Hille that he launched an attack on this problem on two different levels: the abstract and the concrete. In the abstract direction he formulated [103, 106-110, 115, 119] what he called the abstract Cauchy problem (ACP): Given a complex Banach space X and a linear operator U with domain $D(U)$ and given an element y_0 in X, find a function $y(t) = y(t; y_0)$ such that for $t > 0$

1. $y(t)$ is absolutely continuous and $y'(t)$ exists,
2. $y(t) \in D(U)$ and $y'(t) = U[y(t)]$,
3. $\lim_{t \to 0} y(t; y_0) = y_0$.

A solution is said to be of normal type ω if

$$\limsup_{t \to \infty} t^{-1} \log |y(t; y_0)| = \omega < \infty.$$

Hille showed that the ACP has at most one solution of normal type if U is a closed operator whose point spectrum is not dense in the right half-plane. On the other hand, if the spectrum of U covers the entire plane and X is an L-space, then the ACP may have explosive solutions [100, 109]. It is obvious that if U generates a semi-group of operators then the ACP is solvable for every y_0 in $D(U)$ and the solutions are of uniform normal type for all such y_0 of norm $\leqslant 1$. Hille was able to prove the converse assertion and in doing so demonstrated how basic the semi-group theory was to the ACP. After reading the manuscript of [109], I managed to prove a stronger form of the converse theorem and this is what appears in our book.

Another effect of the 1949 Yosida paper was to start Hille working on the forward diffusion equation

$$\frac{\partial f}{\partial t} = Lf \equiv \frac{\partial}{\partial x} \left\{ \frac{\partial}{\partial x} (b(x)f) - a(x)f \right\}, \quad \alpha < x < \beta,$$

in $L_1(\alpha,\beta)$, and soon he had found new proofs [98] for Yosida's results under less restrictive conditions. These results were communicated to Feller who suggested an attack on the backward diffusion equation

$$\frac{\partial g}{\partial t} = Cg \equiv b(x) \frac{\partial^2}{\partial x^2} g + a(x)g, \quad \alpha < x < \beta,$$

in $C[\alpha,\beta]$ by the same methods. This suggestion proved very fruitful and resulted in the note [99]. Another result of this interchange of ideas was to get Feller interested in the problem, and with the aid of Hille's preliminary results Feller[3] was able to find a fairly complete solution for both the forward and backward equations by 1952.

Hille's investigations proceeded more slowly and it was not until 1954 that he published a comprehensive account of them [109; see also 119]. He set himself the problem of finding necessary and sufficient conditions on the coefficients a and b in order that the maximally defined operators C and L generate semi-groups of positive contraction operators in $C[\alpha,\beta]$ and $L_1(\alpha,\beta)$, respectively. By posing the problem in this way, Hille, unlike Feller, rejected the use of boundary conditions. Introducing the auxiliary functions

$$W(x) = \exp \left\{ -\int_0^x \frac{a(s)}{b(s)} ds \right\}, \quad q(x) = \int_0^x [b(s)W(s)]^{-1} ds,$$

$$W_1(x) = \int_0^x W(s)ds, \quad W_2(x) = \int_0^x W_1(s)dq(s),$$

$$\Omega(x) = \int_0^x [W_1(x) - W_1(s)] \, dq(s),$$

Hille proved that C generates a positive contraction semi-group if and only if $\Omega(\alpha) = \infty = \Omega(\beta)$; that L generates a positive contraction semi-group if and only if $W_2(\alpha) = \infty = W_2(\beta)$; and that the L solution is isometric for positive initial data if and only if both conditions hold. Hille also studied the spectral properties of C and L, and the analytic representations and series expansions of the solutions. Although superseded in many ways by Feller, Hille's paper contains a wealth of additional information about the operators C and L.

In 1953 Hille also began working on the Kolmogoroff differential equations:

$$Y'(t) = AY(t) \quad \text{and} \quad Z'(t) = Z(t)A,$$

where the matrix $A = (a_{ij})$ is a Kolmogoroff matrix:

$$a_{ii} \leqslant 0, \quad a_{ij} \geqslant 0 \quad \text{for } i \neq j \text{ and } \sum_j a_{ij} = 0,$$

and both $Y(t)$ and $Z(t)$ belong to the Markoff algebra M of matrices $B = (b_{ij})$ with complex elements such that

$$|B| = \sup_i \sum_j |b_{ij}| < \infty.$$

The context in which Hille chose to treat these equations was that of Y and Z acting as operators on M, or more precisely on a subspace M_A of M; the solution semi-groups were then strongly continuous on M_A. The domain $D(A)$ is defined as

$$D(A) = [B \in M; AB \in M]$$

and M_A = closure of $D(A)$. A solution is said to consist of transition operators $\{P(t)\}$ if

$$0 \leqslant p_{ij}(t), \quad \sum_j P_{ij}(t) \equiv 1, \qquad \lim_{t \to 0} p_{ij}(t) = \delta_{ij}$$

and

$$P(s + t) = P(s)P(t).$$

One of Hille's main results is the following: Suppose that A is triangular, that is, $a_{ij} = 0$ for $j > i$, and define the restriction A_0 of A on

$$D(A_0) = [B \in D(A); AB \in M_A].$$

Then A_0 generates a strongly continuous semi-group of transition operators satisfying both Kolmogoroff differential equations. Another typical result has to do with null solutions: Suppose that $Z(t)$, satisfying the second Kolmogoroff differential equation, is differentiable in the norm topology for $t > 0$, $\lim_{t\to 0} |Z(t)| = 0$, and that $Z(t)$ is of normal type. Then for A triangular and arbitrary B in M, the equation $Q'(t) = Q(t)(A + B)$ has a solution with the same properties.

The above discussion covers only about a third of Hille's mathematical work which has been some 57 years in the making and is still going strong; the rest is for the most part outside of my field of competence.

I have had the good fortune to be one of his collaborators and I have learned much both from the man and the mathematician; for all this I am very grateful.

Ralph Phillips

References

1. K. Yosida, *On the differentiability and the representation of one-parameter semi-groups of linear operators,* J. Math. Soc. Japan vol. 1 (1948) pp. 15-21.
2. K. Yosida, *An operator-theoretic treatment of temporally homogeneous Markoff Processes,* J. Math. Soc. Japan vol. 1 (1949) pp. 244-253.
3. W. Feller, *The parabolic differential equations and the associated semi-group of transformations,* Ann. of Math. (2) vol. 55 (1952) pp. 468-519.

Chapter 1
Functional Equations

Papers included in this chapter are

Commentaries on the two papers in this chapter can be found starting on page 651.

A PYTHAGOREAN FUNCTIONAL EQUATION.*

By Einar Hille.

1. **Introduction.** It is well known that

$$|\sin(x+iy)|^2 = \sin^2 x + \sinh^2 y. \tag{1}$$

This relation can also be written

$$|\sin(x+iy)|^2 = |\sin x|^2 + |\sin iy|^2, \tag{2}$$

or, in other words, the function $\sin z$ $(z = x+iy)$ satisfies the functional equation

$$|f(z)|^2 = |f(x)|^2 + |f(iy)|^2, \tag{3}$$

a *Pythagorean Theorem* in the theory of functions.

It is easy to see that $\sin z$ is not the only solution of (3). Evidently $C\sin z$ will also satisfy where C is an arbitrary complex constant. A special solution is found to be given by Cz. It is not hard to verify that

$$C\frac{\sin az}{a} \tag{4}$$

is a solution of (3) where C is an arbitrary constant and a is either real or purely imaginary. We shall prove that this is the general solution, if we restrict ourselves to analytic solutions.

In § 2 we reduce the problem to the solving of a differential equation in two independent and two dependent variables. This equation we integrate by a specialization of one of the independent variables which is equivalent to assuming the solutions of (3) single-valued and analytic in the neighbourhood of the origin and at that point.

In § 3 this assumption is justified in different ways. This section also contains an *a priori* discussion of the singularities of solutions of the functional equation.

---

* Presented to the American Mathematical Society, October 28, 1922. The results of this paper may be considered trivial, but the author believes that the methods used are of some interest and may find applications to more general problems.

2. The reduction. Let us introduce

$$(5)\qquad [f(z)]^2 = F(z);$$
$$|F(z)| = R(z); \qquad |F(x)| = G(x); \qquad |F(iy)| = H(y),$$

which carries (3) over into the form

$$(6)\qquad R(z) = G(x) + H(y),$$

which is easy to solve. We can define $G(x)$ and $H(y)$ arbitrarily and $R(z)$ will be defined by (6). But when is the solution so obtained, the absolute value of an analytic function?

It is necessary that log $[R(z)]$ *is a harmonic function.* Hence

$$(7)\qquad \Delta\{\log[R(z)]\} = 0,$$

where Δ stands for the Laplacian. If we substitute for $R(z)$ its expression (6), form the Laplacian and simplify, we get

$$(8)\qquad [G(x)+H(y)]\,[G''(x)+H''(y)] - [G'(x)]^2 - [H'(y)]^2 = 0,$$

where the primes denote differentiation with respect to the argument of the function in question. Here we have a mixed differential equation containing two unknown functions $G(x)$ and $H(y)$. In order to obtain $G(x)$ we specialize y. The most natural specialization would certainly be to equate y to 0. This is allowable, at least if the solutions of (3) are single-valued and analytic at the origin.

We assume that this is the case and shall justify our assumption in the next section. Suppose a solution $f(z)$ of (3) is given by a power series

$$(9)\qquad f(z) = c_0 + c_1 z + c_2 z^2 + \cdots,$$

convergent for small values of z; $f(z)$ being analytic we can put $z = 0$ in equation (3) obtaining

$$(10)\qquad f(0) = c_0 = 0.$$

Further put $|c_1| = c$, then

$$(11)\qquad R(z) = c^2|z|^2\{1 + p(|z|^2)\},$$

where $p(|z|^2)$ is a power series in $|z|^2$ vanishing with z. Hence

$$H(0) = H'(0) = 0; \qquad H''(0) = 2c^2. \tag{12}$$

Substituting $y = 0$ in (8) we obtain

$$G(x)\, G''(x) - [G'(x)]^2 + 2c^2\, G(x) = 0. \tag{13}$$

If we put

$$G(x) = u^2, \tag{14}$$

the differential equation (13) is carried over into

$$uu'' - (u')^2 + c^2 = 0. \tag{15}$$

This equation is integrable by elementary methods and the general solution is found to be

$$u = c\,\frac{\sin a(x-\alpha)}{a}, \tag{16}$$

where a and α are arbitrary constants. The limiting value for $a \to 0$, namely $c(x-\alpha)$, is also a solution. We want to get the solution which vanishes at the origin. Hence $\alpha = 0$. Further $u^2 = G(x) > 0$ which shows that u must be a real function of x. c being real, we have to take a either real or purely imaginary. Thus

$$G(x) = |f(x)|^2 = c^2\,\frac{\sin^2 ax}{a^2}. \tag{17}$$

In a similar manner we show that

$$H(y) = |f(iy)|^2 = d^2\,\frac{\sin^2 by}{b^2}. \tag{18}$$

But $f(z)$ is analytic at the origin, thus

$$c = d, \qquad b = ia. \tag{19}$$

Hence

$$|f(z)|^2 = c^2\left[\frac{\sin^2 ax}{a^2} + \frac{\sinh^2 ay}{a^2}\right] = c^2\left|\frac{\sin az}{a}\right|^2. \tag{20}$$

If the absolute value of an analytic function is known in a simply connected region, then the argument is determined uniquely up to an additive real constant. Consequently

$$f(z) = C\frac{\sin az}{a}, \tag{21}$$

where C is an arbitrary complex constant, $|C| = c$, and a is a constant, either real or purely imaginary. When $a = 0$, formula (21) is understood to mean $f(z) = Cz$.

3. **The singularities.** We have assumed in § 2 that an arbitrary solution, $f(z)$, of (3) is single-valued and analytic at the origin. This assumption can be justified in different ways.

We can use equation (8) for this purpose. Suppose that every analytic solution of (3) is single-valued and analytic in the neighbourhood of some point on the axis of imaginaries. This point may, of course, differ from one solution to another. Suppose that for the solution $f(z)$ we may take $y = y_0$ and assume

$$H(y_0) = A, \quad H'(y_0) = B, \quad H''(y_0) = C.$$

Here $A > 0$, B and C are real. The equation (8) becomes

$$[G(x)+A][G''(x)+C]-[G'(x)]^2-B^2 = 0. \tag{22}$$

This equation is also integrable in terms of elementary functions. The general integral is of the form

$$G(x) = a\sin(bx+c)+d, \tag{23}$$

where two of the constants are expressible in terms of the other two and A, B, C. Their values are of no interest in this connection.

Supposing $f(z)$ single-valued and analytic even in the neighbourhood of $z = x_0$, we obtain in a similar manner

$$H(y) = \alpha\sin(\beta y+\gamma)+\delta \tag{24}$$

where, of course, α, β, γ, δ are not independent of a, b, c, d. Substitute the expressions (23) and (24) in equation (8) and we obtain

$$\begin{aligned}(25)\quad -a^2b^2-\alpha^2\beta^2-(d+\delta)\,[ab^2\sin(bx+c)+\alpha\beta^2\sin(\beta y+\gamma)]\\ -a\alpha(b^2+\beta^2)\sin(bx+c)\sin(\beta y+\gamma) = 0.\end{aligned}$$

From this relation we get

$1^\circ\ a^2b^2 + \alpha^2\beta^2 = 0.$ (Put $bx + c = 0$, $\beta y + \gamma = 0$).

$2^\circ\ d + \delta = 0.$ (Put $bx + c = 0$).

$3^\circ\ b^2 + \beta^2 = 0.$ (Put x and y arbitrary).

Thus

$$R(z) = a[\sin(bx + c) + \sin(biy + \gamma)]. \tag{26}$$

If we remember that $R(z)$ is positive for $x = 0$, y close to y_0, and for $y = 0$, x close to x_0, we conclude that ab and ac are real, and, moreover, a, b, c either all real or all purely imaginary. If we take b real, we find further $c = (4m + 3)\pi/2$, $\gamma = (4n + 1)\pi/2$. Then (26) takes the form

$$\begin{aligned} R(z) &= +|a|[-\cos bx + \cosh by] \\ &= 2|a|\left[\sin^2\left(\frac{b}{2}x\right) + \sinh^2\left(\frac{b}{2}y\right)\right], \end{aligned} \tag{27}$$

or the same result as in formula (20).

Finally, it is possible to gain some information concerning the possible singular points of analytic solutions of (3), supposed single-valued within their domain of existance, without solving the equation. The following principle will help us to discuss the singular points.

A solution of (3) *such that every simply connected region in the plane, no matter how small, contains an interior region, also simply connected, in which the solution is single-valued and analytic, can not have a finite singular point* $z_0 = x_0 + iy_0$, *such that it is possible to find a monotonic point-set* $x_1, x_2 \ldots x_n \ldots$ *or* $y_1, y_2, \ldots, y_n \ldots$ *with* $\lim_{n\to\infty} x_n = x_0$ *or* $\lim_{n\to\infty} y_n = y_0$ *and* $\lim_{n\to\infty} f(x_n) = \infty$ *or* $\lim_{n\to\infty} f(y_n) = \infty$. *In special no limit is possible at the origin other than* 0.

As the function $f(z)$ exists in the whole plane the line $x = x_0$ can not be a singular line throughout. Hence we can find a value $z = x_0 + iY_0$ in the neighbourhood of which $f(z)$ is analytic. But

$$\lim_{n\to\infty} |f(x_n + iY_0)|^2 = \lim_{n\to\infty} |f(x_n)|^2 + |f(iY_0)|^2,$$

which leads to a contradiction, no matter what value $f(iY_0)$ has. The special statement for the origin is proved in the same manner.

This principle excludes poles for the solutions. It also excludes isolated essentially singular points, because we can always find a Jordan curve along which $f(z)$ tends to the limit ∞.* We project the curve on the axes. $f(z)$ can not remain bounded on these projections, hence we can pick out a point set as required. The extension to non-isolated points is possible as long as $f(z)$ can not remain bounded on its singular points. Hence the principle is applicable if the set of singular points is enumerable but generally not if the set is perfect.†

* Proof by Valiron: Démonstration de l'existance pour les fonctions entières, de chemins de détermination infinie, Paris, C. R., 166 (1918), pp. 382–384, for entire functions. The extension to general essentially singular points is obvious. Also W. Gross: Über die Singularitäten analytischer Funktionen, Monatsh. Math. Phys. 29 (1918), pp. 3-47.

† Examples of single-valued and analytic functions whose singular points form a perfect discontinuous set but which remain less than a constant have been constructed by Denjoy and Pompéju.

STOCKHOLM,
August 15, 1922.

A CLASS OF FUNCTIONAL EQUATIONS.*

By Einar Hille.

1. The author has previously† studied the functional equation

$$(1) \qquad |F(x+iy)|^2 = |F(x)|^2 + |F(iy)|^2.$$

This equation is evidently a special case of the following general problem: Given an analytic function $R(u, v)$ of two variables u and v; is it possible to determine an analytic function $f(z)$ such that

$$(2) \qquad |f(x+iy)| = R\{|f(x)|, |f(iy)|\} ?$$

We shall assume that $R(u, v)$ is a rational function of u and v. The extension to the case in which $R(u, v)$ is an algebraic function does not introduce any additional difficulties, but will not be considered here. In §§ 4 and 5 of the present paper we indicate a general method of attacking this problem which will give the solution whenever a solution exists. The functions $f(z)$ which are defined in this manner are as a rule infinitely many-valued; assuming $R(u, v)$ to be symmetric, we determine in § 6 the possible single-valued solutions. The latter solutions are either of the form $\exp(c_1 z^2 + c_2 z + c_3 i)$ or are elliptic functions of simple type including degenerate cases of such functions. The analysis is carried through in a couple of illustrating cases in § 7.

2. It is obvious that the function $R(u, v)$ cannot be a perfectly arbitrary rational function. To begin with, $R(u, v)$ must have real coefficients and a region should exist in the real uv-plane in which $R(u, v) > 0$. Before we go any further let us dispose of the trivial case in which $R(u, v)$ depends upon only one variable, u for instance. Then we have

$$(3) \qquad w = R(u).$$

If we let $y \to 0$, $w \to u$ and $u = R(u)$. This equation may be satisfied identically or only for discrete values of u. In the former case the functional equation becomes

$$(4.1) \qquad w = u,$$

* Received October 11, 1927; presented to the American Mathematical Society, October 28, 1922.

† A Pythagorean Functional Equation, Annals of Math., (2), vol. 24, no. 2, pp. 175–180, 1922.

the solution of which is given by

$$f(z) = e^{c_1 z + c_2} \tag{4.2}$$

where c_1 and c_2 are constants, the former real the latter imaginary.

In the latter case the equation $u = R(u)$ must have a certain number of real positive roots $u_1, u_2, \cdots, u_n$, if the problem shall have a solution. The corresponding solutions are

$$f_\nu(z) = u_\nu e^{i\theta_\nu} \qquad (\nu = 1, 2, \cdots, n)$$

where θ_ν are arbitrary real constants.

3. From this point on we shall assume that $R(u, v)$ depends effectively upon two variables u and v. In order to simplify the discussion we shall make the restrictive assumption that $R(u, v)$ is a symmetric function of u and v.

Let us further assume that the solution $f(z)$ approaches definite limiting values when z tends towards the origin along either of the axes and let us put

$$\lim_{y\to 0} |f(iy)| = a, \qquad \lim_{x\to 0} |f(x)| = b \tag{5}$$

where $0 \leqq a, b \leqq +\infty$. Then

$$u = R(u, a) \quad \text{and} \quad v = R(b, v). \tag{6}$$

If a and b are finite numbers they satisfy the algebraic equation

$$t = R(t, t). \tag{7}$$

By considering all the different roots of this equation which are real and $\geqq 0$, we obtain all possible values for a and b. We may lose an infinite root but if such a root exists the equation

$$\frac{1}{s} = R\left(\frac{1}{s}, \frac{1}{s}\right)$$

has the root zero. Let a be a root of (7). There are three different possibilities to consider, namely

$$\text{(i) } R(u, a) \equiv u, \qquad \text{(ii) } R(u, a) \equiv a, \qquad \text{(iii) } R(u, a) \not\equiv u, a. \tag{8}$$

The first case is the interesting one for our purpose. We shall show that there exists only one such critical value. Suppose there were two values, a and b. Then

$$u = R(u, a), \qquad u = R(u, b). \tag{9}$$

Hence

$$a = \lim_{u \to a} R(u, b) = R(a, b),$$
$$b = \lim_{u \to b} R(u, a) = R(b, a).$$

In view of the symmetry of $R(u, v)$ in u and v these expressions are equal, consequently $a = b$.

We can always assume that the critical value a is either 0 or 1. In fact, if a has a finite value, c, different from 0 or 1 we can replace the given functional equation (2) by

$$(2^*) \qquad W = \frac{1}{c} R(cU, cV) = R_1(U, V)$$

which admits of $a = 1$ as critical value. If $f(z)$ is the general solution of the new functional equation, $c f(z)$ is the general solution of the old one. On the other hand, if the critical value is infinite we replace u, v, w by $\frac{1}{u}, \frac{1}{v}, \frac{1}{w}$ respectively. The resulting equation has 0 as critical value.

We notice that if $v = a$ then

$$(10) \qquad R(u, v) \equiv u, \quad \frac{\partial}{\partial u}[R(u, v)] = 1, \quad \frac{\partial^2}{\partial u^2}[R(u, v)] = 0.$$

The second set of values defined by (8) we call singular values. They correspond to singular constant solutions. In fact, suppose that

$$(11.1) \qquad R(u, s) \equiv s$$

where $s \geqq 0$ is some constant. Then the functional equation has the solution

$$(11.2) \qquad f(z) = s e^{i\theta}.$$

The third set of values has no particular connection with our problem.

4. We shall outline a general method of attacking the problem. We assume that $R(u, v)$ is positive in a certain region of the uv-plane, e. g. when $u \geqq c$, $v \geqq c$, and that a value $a \geqq c$ exists such that $R(u, a) \equiv u$. The hypothesis that $R(u, v)$ is symmetric in u and v is not necessary for the discussion immediately following but will be used later.

Suppose that $f(x + iy)$ is an analytic solution of (2), and consider any simply-connected region within which $f(z)$ is holomorphic and different from zero. In such a region $\log |f(x + iy)|$ is a harmonic function and satisfies Laplace's equation:

$$(12.1) \qquad \Delta \log |f(x + iy)| = 0$$

or

$$\left(\frac{\partial^2}{\partial x^2}+\frac{\partial^2}{\partial y^2}\right)\log R(u,v)=0. \tag{12.2}$$

The equation (12.2) is found to be a mixed differential equation in two independent variables x and y and two dependent variables u and v, namely

$$\begin{aligned} R\frac{\partial R}{\partial u}\frac{d^2u}{dx^2}+R\frac{\partial R}{\partial v}\frac{d^2v}{dy^2}+\left[R\frac{\partial^2 R}{\partial u^2}-\left(\frac{\partial R}{\partial u}\right)^2\right]\left(\frac{du}{dx}\right)^2 \\ +\left[R\frac{\partial^2 R}{\partial v^2}-\left(\frac{\partial R}{\partial v}\right)^2\right]\left(\frac{dv}{dy}\right)^2=0. \end{aligned} \tag{13}$$

If $u = g(x)$ and $v = h(y)$ form a solution of this differential equation, then $u = g(\alpha x+\beta)$ and $v = h(\alpha y+\gamma)$ also form a solution and that for arbitrary values of α, β and γ. Not all these solutions of the differential equation can be solutions of the functional equation. In fact, assuming that $R(u, v)$ is symmetric in u and v, we have $u = a$ and $v = a$ for x and y respectively equal to zero. This implies two conditions on the constants. Further, we have the requirement that $u \geqq 0$ and $v \geqq 0$ for real values of x and y respectively. This imposes still another restriction upon the constants.

It may or may not be possible to fulfill these conditions. Suppose for the moment that we can find a pair of functions

$$u = G(x), \qquad v = H(y),$$

satisfying equation (13) and such that

(1) $G(x) \geqq 0$, $H(y) \geqq 0$ when x and y are real;

(2) $G(0) = a$, $H(0) = a$;

(3) $R(G(x), H(y)) > 0$ when $-x_1 \leqq x \leqq +x_1$, $-y_1 \leqq y \leqq +y_1$.

Then

$$P(x, y) = \log R(G(x), H(y))$$

is a harmonic function in the rectangle indicated. The conjugate harmonic function $Q(x, y)$ is determined uniquely save for an arbitrary additive constant. Then form

$$f(x+iy) = e^{P(x,y)+iQ(x,y)}.$$

This function is a solution of the functional equation.

5. If we keep y (or x) constant in equation (13) we obtain an ordinary differential equation which can be integrated by means of 3 quadratures.

The most favorable choice which we can make for a constant value of y is 0. Then $v = |f(iy)| = a$. Suppose that

$$\frac{dv}{dy} \to s, \quad \frac{d^2 v}{dy^2} \to t \text{ when } y \to 0. \tag{14}$$

If a equals zero we have $s = 0$, but it may happen that $s = 0$ for other values of a as well. Then (13) becomes

$$u\frac{d^2 u}{dx^2} - \left(\frac{du}{dx}\right)^2 + P(u) = 0 \tag{15}$$

where

$$P(u) = tu\frac{\partial R}{\partial v}\bigg]_{v=a} + s^2 \left[u\frac{\partial^2 R}{\partial v^2} - \left(\frac{\partial R}{\partial v}\right)^2\right]_{v=a} \tag{16}$$

Let us introduce a new variable

$$U = \left(\frac{du}{dx}\right)^2. \tag{17.1}$$

Then

$$\frac{1}{2} u \frac{dU}{du} - U + P(u) = 0 \tag{17.2}$$

or

$$U = u^2 \left[C - \int_a^u \frac{P(u)}{u^3} du\right]. \tag{17.3}$$

This expression has to be modified if $a = 0$ owing to the singularity of the integrand at the origin. Hence

$$x = \int_a^u v^{-1} \left[C - \int_a^v \frac{P(\tau)}{\tau^3} d\tau\right]^{-\frac{1}{2}} dv. \tag{18}$$

Since $u = |f(x)|$ and $v = |f(iy)|$ we conclude that if $f(z)$ is holomorphic at the origin

$$u(0) = v(0) = a, \quad \left[\frac{du}{dx}\right]_{x=0} = \left[\frac{dv}{dy}\right]_{y=0} = s. \tag{19}$$

Hence if $a \neq 0$ we have $C = s^2 a^{-2}$. If $a = 0$ we have to make appropriate changes in the lower limit of integration in (17.3) and (18).

The expression for y is obtained by changing u into v in the formulas above. The expressions for u and v are obtained by inversion from formula (18) and the corresponding formula for y. If this process leads to a pair of functions having the properties specified at the end of § 4, we have obtained a particular solution of the problem from which the general solution is easily obtained.

6. We are now in position to consider the problem of determining when the functions $u(x)$ and $v(y)$ and consequently also $f(z)$ will be single-valued functions of their arguments. Painlevé* has investigated the differential equations of the second order

$$y'' = F(y', y, x)$$

which have fixed critical points. The right hand side of the equation is supposed to be rational in y', algebraic in y and analytic in x. In particular, the equation

$$y'' = \frac{y'^2}{y} + a(y), \tag{20.1}$$

where $a(y)$ is an algebraic function of y not depending upon x, has a single-valued solution only if $a(y)$ has either of the following two forms, namely

$$a(y) = \beta y, \tag{20.2}$$

or

$$a(y) = \alpha y^3 + \beta y^2 + \gamma + \frac{\delta}{y}. \tag{20.3}$$

In the first case the solutions have the form

$$y = Ke^Y, \qquad Y = \frac{\beta}{2}x^2 + K_1 x. \tag{21}$$

In the second case the solutions are elliptic functions of simple type or degenerate cases of such functions.

Hence, if a functional equation of the type (2) has a single-valued solution, this solution must be either an exponential function of the type represented by formula (21), or an elliptic function or a degenerate case of such a function. The elliptic functions which satisfy equation (20.1) with (20.3) and in addition are real upon the axes form a very restricted class. On account of the conditions of reality the lattice points $2n\omega_1 + 2m\omega_3$ where $2\omega_1$ and $2\omega_3$ are the primitive periods, must be symmetrically distributed with respect to the axes. Hence the primitive periods are either conjugate imaginary or one is real and the other purely imaginary.

* Acta Math., vol. 25, pp. 1–85, 1902. The results which we need are to be found in Tableau II, p. 24, compared with Tableau XI, equations (13) and (15), p. 55. The discussion of Painlevé was completed in one important point by Gambier, Acta Math., vol. 33, 1910, pp. 1–55. The latter's results are not needed for our purposes.

7. We shall study a couple of simple examples. We begin with the equation

$$w = uv. \tag{22}$$

There are two critical values namely 0 and 1. The former is a singular value, the latter not. The singular solution is identically zero. In order to find the other solution we form the differential equation (13) corresponding to the problem. This equation can be written in the form

$$\frac{d^2}{dx^2}[\log u] + \frac{d^2}{dy^2}[\log v] = 0. \tag{23}$$

The first term is a function of x alone, the second one of y alone. Hence each term separately must be equal to a constant. Integrating the equation

$$\frac{d^2}{dx^2}[\log u] = C, \tag{24}$$

we find that

$$u = e^{\frac{1}{2}Cx^2 + C_1x + C_2},$$

$$v = e^{-\frac{1}{2}Cy^2 + C_3y + C_4}.$$

Since $u(0) = v(0) = 1$ we have $C_2 = C_4 = 0$. Further, the remaining constants must be real. Hence

$$f(z) = e^{\frac{1}{2}Cz^2 + Kz + K_1 i} \tag{25}$$

where C and K_1 are real constants and K a complex one, otherwise arbitrary. This case corresponds to $a(y) = Cy$ the solution of which was given in formula (21).

Now let us take the equation

$$w = \frac{u+v}{1+uv}. \tag{26}$$

Here again we have the critical values 0 and 1 of which the latter is singular. To this latter value corresponds an arbitrary unit vector as a singular solution. The value $a = 0$ corresponds to an elliptic function. In this case the differential equation (13) reads

$$\begin{aligned} &(u+v)(1+uv)(1-v^2)u'' - [(1+uv)^2 + v^2(u+v)^2](u')^2 \\ &+ (u+v)(1+uv)(1-u^2)v'' - [(1+uv)^2 + u^2(u+v)^2](v')^2 = 0. \end{aligned} \tag{27}$$

If we put $y = 0$ we have $v_0 = 0$, $v_0' = 0$. Let us put $v''(0) = 2b^2$. Then the equation takes on the form

$$uu'' - (u')^2 + 2b^2 u(1 - u^2) = 0. \tag{28}$$

In order to integrate this equation we put $u = U^2$ and obtain

$$UU'' - (U')^2 + b^2(1 - U^4) = 0.$$

Multiplying by $2\frac{U'}{U^3}$ and integrating we obtain

$$(U')^2 = b^2(U^4 - cU^2 + 1). \tag{29}$$

Let us determine k from the relation

$$k + \frac{1}{k} = c \tag{30}$$

and put $U = \sqrt{k}\, t$. Then we find that

$$t = \sin \operatorname{am} \left(\frac{bx}{\sqrt{k}} \Big/ k\right)$$

or

$$u = k \sin \operatorname{am}^2 \left(\frac{bx}{\sqrt{k}} \Big/ k\right). \tag{31}$$

Finally

$$f(z) = e^{i\theta} k \sin \operatorname{am}^2 \left(\frac{bz}{\sqrt{k}} \Big/ k\right). \tag{32}$$

Since c is a real constant we find that k is either real or lies on the unit-circle; b is either real or purely imaginary; θ is a real arc.

The special case $k = \pm 1$ gives

$$u = \tan^2(bx) \tag{33}$$

and

$$f(z) = e^{i\theta} \tan^2(bz) \tag{34}$$

where b is either real or purely imaginary.

Chapter 2
Zero Point Problems and Asymptotic Integration of Second Order Linear Differential Equations

Papers included in this chapter are

[5] On the zeros of Sturm-Liouville functions

[12] An existence theorem

[15] On the zeros of the functions defined by linear differential equations of the second order

[23] Zero point problems for linear differential equations of the second order

An introduction to Hille's work on complex oscillation theory can be found starting on page 659.

ARKIV FÖR MATEMATIK, ASTRONOMI OCH FYSIK.
BAND 16. N:o 17.

On the Zeros of Sturm-Liouville Functions.

By

EINAR HILLE.

Communicated September 14th 1921 by I. BENDIXSON and H. VON KOCH.

1. Let $K(z)$ and $G(z)$ be single-valued and analytic functions of z in a simply connected region D in the z-plane where furthermore $K(z) \neq 0$ and consider the differential equation

$$\frac{d}{dz}\left[K(z)\frac{dw}{dz}\right] + G(z)w = 0. \tag{1.1}$$

A solution of (1.1) we agree to call a *Sturm-Liouville function* or, briefly, an *SL-function*. Such a function is uniquely determined by the values of

$$w(z_0) = c_0, \quad w'(z_0) = c_1 \tag{1.2}$$

at a point z_0 that is not singular to the differential equation. If we restrict ourselves to linearly independent solutions only, it is sufficient to know the value of the quotient

$$\frac{w'(z_0)}{w(z_0)} = \lambda. \tag{1.3}$$

This particular solution we denote by

$$w(z \mid z_0, \lambda). \tag{1.4}$$

In the present paper we are going to be concerned with the zeros of SL-functions. The problem we aim at is to give

a qualitative description of the distribution in the complex plane of the zeros of a given SL-function.

It is not the question of giving an exact determination of the position of the zeros; a problem of »quantitative» analysis that scarcely can be attacked at present. It is rather a sort of »analysis situs» that we aim at; to characterize the relative position of the zeros — supposed to exist — with as much precision as we can. The problem contains many special questions, *e. g.*

What restrictions do we put on the location of the zeros by fixing the position of one of them, or by assuming the solution to have specified (admissible) properties at a point or along a curve in the plane?

How are the zeros distributed in the neighbourhood of an irregular-singular point of the equation?

Is it possible to find a characterization of the type of distribution that holds for the whole family of solutions, save possibly for a certain set of exceptional solutions?

The second of the special problems, mentioned above, will get very little attention in the paper, except incidentally, as we have treated the question at some length in another place.[1] We try to solve the first problem by giving a method for assigning regions in the plane where neither the solution nor its derivative can vanish, a so called *zero-free region.* The form of the regions so obtained shows that the plane can be divided up into parts where the zeros form sequences of a regular type; the argument of the difference between two zeros being restricted within limits, determined by the variation of the arguments of the coefficients of the equation in the region.

The method we use is based upon a certain integral equality where the integration is performed in the complex plane. Similar equalities but with real limits of integration have been used frequently in the theory of boundary problems. HURWITZ was the first mathematician, as far as the

[1] Oscillation Theorems in the Complex Plane, Transactions of the American Mathematical Society, vol. 23, especially Section 4. This paper is quoted as O. T. below.

author knows, who applied complex integration to problems of the kind that we are dealing with.[1]

2. Let us put

$$\begin{cases} w_1 = w, \\ w_2 = K(z)\dfrac{dw}{dz}, \end{cases} \tag{2.1}$$

then the equation (1.1) is equivalent to the differential system

$$\begin{cases} \dfrac{dw_1}{dz} = \dfrac{w_2}{K(z)}, \\ \dfrac{dw_2}{dz} = -G(z)w_1. \end{cases} \tag{2.2}$$

Replace the first equation in (2.2) by its conjugate; multiply the new first equation by w_2 and the second one by $\overline{w}_1$ (where u stands for the conjugate of u), add the results and integrate between two limits z_1 and z_2. Thus we obtain

$$[\overline{w}_1 w_2]_{z_1}^{z_2} - \int_{z_1}^{z_2} |w_2|^2 \left[\overline{\frac{dz}{K(z)}}\right] + \int_{z_1}^{z_2} |w_1|^2 G(z)\, dz = 0. \tag{2.3}$$

Let us put

$$G(z)\,dz = d\Gamma, \quad \frac{dz}{K(z)} = d\mathrm{K}, \tag{2.4}$$

whereby (2.3) becomes

$$[\overline{w}_1 w_2]_{z_1}^{z_2} - \int_{z_1}^{z_2} |w_2|^2 \overline{d\mathrm{K}} + \int_{z_1}^{z_2} |w_1|^2 d\Gamma = 0. \tag{2.5}$$

This expression we call the *Green's transform* of the

[1] Über die Nullstellen der BESSEL'schen Funktionen, Mathematische Annalen, vol. 33, pp. 246—266, 1889, and Über die Nullstellen der hyper geometrischen Funktion, ibid., vol. 64, pp. 517—560, 1907.

differential equation. It is goirg to be the instrument with the help of which the following investigation will be carried through.

3. Let us first see how (2. 5) is changed by change of variables. Putting

$$z = f(\zeta), \tag{3.1}$$

we obtain

$$[\overline{\omega_1}\,\omega_2]_{\zeta_1}^{\zeta_2} - \int_{\zeta_1}^{\zeta_2} |\omega_2|^2 \,\overline{d\varkappa} + \int_{\zeta_1}^{\zeta_2} |\omega_1|^2 \,d\gamma = 0, \tag{3.2}$$

where

$$\left\{\begin{aligned} &\omega_1(\zeta) \equiv w_1(z); \\ &\omega_2(\zeta) = k(\zeta)\frac{d\omega_1}{d\zeta} \equiv K(z)\frac{dw}{dz}; \\ &k(\zeta) = \frac{K[f(\zeta)]}{f'(\zeta)}; \\ &g(\zeta) = G[f(\zeta)]f'(\zeta); \\ &d\varkappa = \frac{d\zeta}{k(\zeta)} \equiv d\mathbf{K}; \\ &d\gamma = g(\zeta)d\zeta \equiv d\mathbf{\Gamma}. \end{aligned}\right. \tag{3.3}$$

By special choice of the transformation the equation will take a simpler form. If we put

$$d\zeta = \frac{dz}{K(z)}, \tag{3.41}$$

the relation (3. 2) becomes

$$\left[\overline{\omega}\frac{d\omega}{d\zeta}\right]_{\zeta_1}^{\zeta_2} - \int_{\zeta_1}^{\zeta_2} \left|\frac{d\omega}{d\zeta}\right|^2 \overline{d\zeta} + \int_{\zeta_1}^{\zeta_2} |\omega|^2 d\gamma = 0, \tag{3.42}$$

corresponding to a differential equation

$$\frac{d^2\omega}{d\zeta^2} + g(\zeta)\,\omega = 0. \tag{3.43}$$

A symmetric form can be given the differential equation by putting

$$d\zeta = \sqrt{\frac{G(z)}{K(z)}}\,dz, \tag{3.51}$$

namely

$$\frac{d}{d\zeta}\left[g(\zeta)\frac{d\omega}{d\zeta}\right] + g(\zeta)\omega = 0, \tag{3.52}$$

where

$$g(\zeta) = \sqrt{G(z)\,K(z)}. \tag{3.53}$$

The GREEN's transform becomes

$$\left[\bar{\omega}\, g(\zeta)\frac{d\omega}{d\zeta}\right]_{\zeta_1}^{\zeta_2} - \int_{\zeta_2}^{\zeta_2}\left|\frac{d\omega}{d\zeta}\right|^2 g(\zeta)d\bar{\zeta} + \int_{\zeta_1}^{\zeta_2}|\omega|^2 g(\zeta)d\zeta = 0. \tag{3.54}$$

If we make a further transformation

$$W = \sqrt{g(\zeta)}\,w \tag{3.55}$$

the differential equation (3. 52) is carried over into

$$\frac{d^2 W}{d\zeta^2} + [1 + \Phi(\zeta)]W = 0, \tag{3.56}$$

where

$$\Phi(\zeta) = \frac{3}{4}\left(\frac{g'}{g}\right)^2 - \frac{1}{2}\frac{g''}{g}. \tag{3.57}$$

If $\Phi(\zeta)$ tends to zero in a sufficiently rapid manner when $\zeta \to \infty$ the form (3. 56) is very convenient for the study of the distribution of the zeros in the neighbourhood of an irregular-singular point.

There are of course other double transformations of the type

$$\begin{cases} w = h(\mathbf{Z})W \\ z = f(Z) \end{cases} \tag{3.6}$$

than the one used in formulae (3.51) and (3.55) that are useful for our purpose. To each form of the equation, supposed to be in self-adjoint form, there is a GREEN's transform.

4. Let z_1 be a point in a region, D, where $G(z)$ and $K(z)$ are analytic and $K(z) \neq 0$. Proceed from z_1 to a point z_2 in D along a path, C, that has a continuous tangent everywhere, except at a finite number of points at most. In a point z_0 on C where the slope-angle of the tangent is φ_0 the differentials $\overline{d\mathbf{K}}$ and $d\boldsymbol{\Gamma}$ have definite arguments, namely

$$\arg K(z_0) - \varphi_0 = \varkappa_0 \quad \text{and} \quad \arg G(z_0) + \varphi_2 = \gamma_0 \tag{4.1}$$

respectively. Let us mark in a separate plane the vectors

$$e^{i(\pi+\varkappa_0)} \quad \text{and} \quad e^{i\gamma_0} \tag{4.2}$$

and repeat this construction for all the points on C. Mark further the vector

$$-[\overline{w}_1 w_2]_{z=z_1} \tag{4.3}$$

the argument of which is

$$\arg K(z_1) + \arg \lambda + \pi . \tag{4.31}$$

The set of vectors so obtained we call *the vector field of the path with regard to the solution $w(z \mid z_1, \lambda)$.*

Now suppose we can find an angle θ having the following properties.

$$\text{I.} \qquad \theta \leq \arg\left[-(\bar{w}_1 w_2)_{z=z_1}\right] \leq \theta + \pi .$$

$$\text{II.} \qquad \begin{cases} \theta \leq \arg\left[\;\; d\overline{\mathbf{\Gamma}}\right] & \leq \theta + \pi, \\ \theta \leq \arg\left[-d\overline{\mathbf{K}}\right] & \leq \theta + \pi, \end{cases}$$

almost everywhere on C, *i. e.* excepting a point set of measure zero.

III. The signs of equality are *not* fulfilled, either in I, or, if such a sign is valid in I, not in II, in a point set on C of measure different from zero.

In this case the vector field is said to be *definite.*

If an angle θ can be found that fulfills the conditions I and II but not III, then the field is called *ambiguous.* Finally, the field is said to be *indefinite* in the remaining cases.

It is evidently sufficient in order that the field be definite that it is contained in an angle of opening less than π, the condition is evidently not necessary. Let us make the following

Definition 1: A curve C in the complex plane, made up of regular points of the differential equation (1. 1), and having a definite vector field with regard to a solution of (1. 1), $w(z|z_1, \lambda)$, where z_1 is a point on C, is called a line of influence with reference to the point z_1 and the solution $w(z|z_1, \lambda)$.

With this definition we obtain the following

Theorem I. A line of influence with reference to a point z_1 and the solution $w(z|z_1, \lambda)$ cannot contain a zero of this function or of its first derivative, the solution being continued analytically along the curve.

The proof is obvious. In view of formula (2. 5) and the previous definition we conclude that an angle θ can be found such that

$$(4.4) \qquad \begin{cases} \theta \leq \arg\left[-(\bar{w}_1 w_2)_{z=z_1}\right] \leq \theta + \pi, \\ \theta \leq \arg\left[-\displaystyle\int_{z_1}^{z_2} |w_2|^2 d\overline{\mathbf{K}}\right] \leq \theta + \pi, \\ \theta \leq \arg\left[\;\;\displaystyle\int_{z_1}^{z_2} |w_1|^2 d\overline{\mathbf{\Gamma}}\right] \leq \theta + \pi, \end{cases}$$

z_2 being a point on the curve C which we suppose to be a line of influence. In view of condition III we infer that the sum of the three vectors in (4. 4) has an argument α where $\theta < \alpha < \theta + \pi$, and is different from zero. Consequently

$$[\overline{w}_1 w_2]_{z=z_2} = \left[K(z)\,\overline{w(z)}\,\frac{dw}{dz}\right]_{z=z_2} \neq 0. \tag{4.5}$$

Q. e. d.

A point z_2 may belong to two different lines of influence, C' and C'', from z_1. If C'' can be deformed continuously into C' without sliding over a singular point of the differential equation, the determination of $w(z)$ in $z = z_2$ is the same whether we arrive at z_2 on C' or on C''. If the paths cannot be deformed into each other, the determinations are in general different. In order to avoid this difficulty we introduce the RIEMANN surface $\mathfrak{R}$, with branch-points at the singular points of the differential equation, on which $w(z \mid z_1, \lambda)$ is single-valued. We mark the lines of influence with reference to the point z_1 and the solution $w(z \mid z_1, \lambda)$ on $\mathfrak{R}$, observing that such a line must break off on arriving at a branch-point of $\mathfrak{R}$. Then we make the

Definition 2. The set of points on the RIEMANN *surface* $\mathfrak{R}$ *that are regular points of the differential equation (1. 1) and which belong to at least one of the lines of influence with reference to a point* z_1 *and a solution* $w(z \mid z_1, \lambda)$ *of (1. 1) is called the domain of influence with reference to the point* z_1 *and the solution* $w(z \mid z_1, \lambda)$ *and is designed by* $DI(z_1, \lambda)$.

From Theorem I we conclude:

Theorem II: A point of $DI(z_1, \lambda)$ *cannot be a zero of the corresponding solution* $w(z \mid z_1, \lambda)$ *or of its derivative, if these functions are continued analytically on* $\mathfrak{R}$ *along paths in* $DI(z_1, \lambda)$.

We have supposed z_1 to be a regular point of (1. 1). Certain conditions fulfilled, we can omit this restriction. All that is really required is, that the expressions in (2. 5) tend toward finite limits when $z \to z_1$. In case z_1 is an irregular-singular point, special caution is needed as in general there are only certain directions in which the singular point can be approached.

There are other generalisations possible. The form of the domain of influence depends essentially upon the value of the argument of $\lambda = \left[\frac{w'(z)}{w(z)}\right]_{z=z_1}$. This argument may be known along some curve C on the RIEMANN surface, mentioned above. The most important case is, of course, when $w(z)$ is real on a segment of the real axis, $\arg\lambda$ being 0 or π. If the argument is known on C, we can construct the domain of influence of every point z_0 on C with reference to the given solution. Thus

Definition 3: By the domain of influence, $DI(C, w)$, of a curve C with reference to a solution $w(z)$ for which $\arg\left[K(z)\,\overline{w(z)}\,\frac{dw}{dz}\right]$ is known on C, we understand the set of points on $\mathfrak{R}$ that belong to at least one of the domains of influence with reference to points on C and the given solution.

Consequently we have the

Theorem III: There are no zeros of $w(z)\frac{dw}{dz}$ in $DI(C, w)$, the functions being continued analytically in this region.

5. Let us map a region D of the z-plane conformally on a region $\varDelta$ of the ζ-plane by means of the function

$$\zeta = f(z) \quad \text{or} \quad z = F(\zeta). \tag{5.1}$$

The differential equation (1. 1) is carried over into

$$\frac{d}{d\zeta}\left\{\frac{K[F(\zeta)]}{F'(\zeta)}\frac{d\omega}{d\zeta}\right\} + F'(\zeta)\,G[F(\zeta)]\,\omega = 0, \tag{5.2}$$

the GREEN's transform of which is

$$[\overline{\omega_1}\,\omega_2]_{\zeta_1}^{\zeta_2} - \int_{\zeta_1}^{\zeta_2} |\omega_2|^2\,\overline{d\varkappa} + \int_{\zeta_1}^{\zeta_2} |\omega_1|^2\,d\gamma = 0, \tag{5.3}$$

with the same notation as in art. 3. Let z_1 be a point in D, corresponding to a point ζ_1 in $\varDelta$ and take a solution $w(z\,|\,z_1, \lambda)$ at $z = z_1$. To this function corresponds the solution $\omega(\zeta\,|\,\zeta_1, \varLambda)$ of (5. 2) at $\zeta = \zeta_1$, where

$$\varLambda = \lambda + F'(\zeta_1). \tag{5.4}$$

Let C be a line of influence in D from $z = z_1$ with reference to $w(z \mid z_1, \lambda)$. This curve is mapped on a curve Γ in $\varDelta$ in the ζ-plane. *Then Γ is a line of influence from $\zeta = \zeta_1$ with reference to $\omega(\zeta \mid \zeta_1, \varDelta)$.*

C contains regular points of the equation only, except possibly the further end-point. As $f'(z) \neq 0$ on C, all points on Γ, the further end-point possibly excepted, are regular points of the equation (5. 2). Further

$$\left\{K(z)\,\overline{w(z)}\,\frac{dw}{dz}\right\}_{z=z_1} \equiv \left\{\frac{K[F(\zeta)]}{F'(\zeta)}\,\overline{\omega(\zeta)}\,\frac{d\omega}{d\zeta}\right\}_{\zeta=\zeta_1} \tag{5.51}$$

$$\left.\begin{aligned} d\mathbf{K} &\equiv d\varkappa \\ d\mathbf{\Gamma} &\equiv d\gamma \end{aligned}\right\} \text{ in corresponding points.} \tag{5.52}$$

C being a line of influence with reference to the point z_1 and the solution $w(z \mid z_1, \lambda)$, it follows that Γ has the same property with reference to the point ζ_1 and the solution $\omega(\zeta \mid \zeta_1, \varDelta)$.

If D contains the whole domain of influence $DI(z_1, \lambda)$ $\varDelta$ will contain the conformal map thereof and this image is the domain of influence $DI(\zeta_1, \varDelta)$. We express this property by

Theorem IV: A domain of influence is covariant under conformal transformation.

This property does not hold if we make a simultaneous change of both variables of the type (3. 6).

6. Let us suppose the equation has been reduced to the form

$$\frac{d^2w}{dz^2} + G(z)w = 0 \tag{6.1}$$

with the GREEN's transform

$$\left[\bar{w}\frac{dw}{dz}\right]_{z_1}^{z_2} - \int_{z_1}^{z_2}\left|\frac{dw}{dz}\right|^2 \overline{dz} + \int_{z_1}^{z_2} |w|^2 G(z)\,dz = 0. \tag{6.2}$$

We assume $G(z)$ to be single-valued and construct the curves

$$(6.3) \qquad \begin{cases} \Re[G(z)] = G_1 = 0, \\ \Im\mathfrak{m}[G(z)] = G_2 = 0. \end{cases}$$

These curves divide the plane into a denumerable set of regions where G_1 and G_2 keep a constant sign. Such a region we agree to denote by B. If a region B has the property that lines parallel to the axes cut its boundary in two points at most, we put a subscript $_+$ on the B, *e. g.* B_+. In case the region is convex we denote it by B_*.

Let $z_1 = x_1 + iy_1$ be a point in a region B_+ and take the solutions $w(z \mid z_1, 0)$ or $w(z \mid z_1, \infty)$ which have the same domain of influence. Draw the vertical $\Re(z) = x_1$ and the horizontal $\Im\mathfrak{m}(z) = y_1$. These lines divide B_+ in four parts B_1, B_2, B_3 and B_4 where the indices are applied in such order that the regions are disposed as the four quadrants of the plane. Then we have

Theorem V: The domain of influence with reference to a point z_1 in a region B_+ and a solution such that $\left[w(z)\dfrac{dw}{dz}\right]_{z=z_1} = 0$, *contains B_1 and B_3 if $G_2 = \Im\mathfrak{m}[G(z)] > 0$ in B_+, but B_2 and B_4 if $G_2 < 0$.*

Assume $G_2 > 0$ for instance and let z_2 be a point in B_+. Then there is a path in B_1 from z_1 to z_2 such that at every point

$$\frac{\pi}{2} \leq \arg[-dz] \leq \pi,$$

$$\begin{cases} 0 < \arg[G(z)dz] < \pi, \text{ if } G_1 > 0 \text{ in } B_+, \text{ but} \\ \dfrac{\pi}{2} < \arg[G(z)dz] < \dfrac{3\pi}{2}, \text{ if } G_2 < 0 \text{ in } B_+. \end{cases}$$

This path is evidently a line of influence. The proof is conducted in a similar manner in the other cases.

Let a sequence of complex numbers be given

$$(6.41) \quad a_1 = \alpha_1 + i\beta_1, \; a_2 = \alpha_2 + i\beta_2, \; \ldots, \; a_n = \alpha_n + i\beta_n,$$

and put

$$\Delta \alpha_\nu = \alpha_{\nu+1} - \alpha_\nu, \quad \Delta \beta_\nu = \beta_{\nu+1} - \beta_\nu. \tag{6.42}$$

Definition 4: A set of complex numbers $a_1, a_2, \ldots, a_n$ is said to form a monotonic sequence if the differences $\Delta \alpha_\nu$ and $\Delta \beta_\nu$ each separately keep a constant sign throughout the sequence. This is said to be positively monotonic if $\Delta \alpha_\nu . \Delta \beta_\nu > 0$, and negatively monotonic if $\Delta \alpha_\nu . \Delta \beta_\nu < 0$.

From the preceding definition we conclude

Theorem VI: Let $w(z)$ be a solution of

$$\frac{d^2 w}{dz^2} + G(z) w = 0$$

such that $w(z) \dfrac{dw}{dz}$ vanishes at a number of points $a_1, a_2, \ldots, a_n$ in a region B_+. Then the $\{a\}$-sequence is monotonic; namely, positively monotonic if $\mathfrak{Im}\,[G(z)] < 0$ and negatively monotonic if $\mathfrak{Im}\,[G(z)] > 0$; the enumeration being properly chosen.

We shall be able to complete this result later. A sequence of zeros may evidently be monotonic and still have a very irregular distribution. Further, the theorem does not give any information about the fashion in which sequences of zeros in adjacent regions join together. Some information may be gathered by making a transformation

$$\zeta = e^{i\theta} z \tag{6.51}$$

which carries (6.1) into a similar equation

$$\frac{d^2 w}{d\zeta^2} + \tilde{G}(\zeta) w = 0, \tag{6.52}$$

to which corresponds a new division of the plane into B-regions. If θ is properly chosen, the zeros that we want to study will lie in the same region $\tilde{B}$. If this region happens to be a region B_+, the zeros will have to form a monotonic sequence with regard to the new set of axes. We can of course always, and in infinitely many ways, select an interior part of a region B that has the essential properties of a region B_+, namely that G_1 and G_2 are different from zero in the domain and the boundary of the same is cut in two points at most by any line parallel to either of the axes.

The following remarks may facilitate the study of zeros of solutions that are real on a segment of the real axis. Let (a, b) be a segment of the real axis where $G(z)$ is real, positive and analytic, except possibly in the end-points. Let B_0 and $\bar{B}_0$ be the adjacent B-regions above and below (a, b) respectively. From these regions we select a part, $\mathfrak{B}$, in the following way:

A. If $G_2 > 0$ in B_0 [above (a, b)] we take

1° the points in B_0 which can be reached by a path from an interior point on (a, b) along which $0 \leq \arg dz \leq \frac{\pi}{2}$;

2° the conjugate points in $\bar{B}_0$.

B. If $G_2 < 0$ in B_0, similarly, but $\frac{\pi}{2} \leq \arg dz \leq \pi$ instead.

Evidently the paths we have used are lines of influence with reference to a solution $w(z \mid x_1 \lambda)$ where $a < x_1 < b$ and λ is real. Hence

Theorem VII: If a solution $w(z)$ of (6.1) is real on a segment (a, b) of the real axis where $G(z) > 0$ and analytic, then $w(z)\frac{dw}{dz} \neq 0$ in the complex points of the corresponding region $\mathfrak{B}$.

7. In the preceding article we have determined parts of the domain of influence. To determine the complete domain is very laborious in general. There are however partial regions of great extent which are comparatively easy to construct and which often complete each other in a very fortunate manner, namely the *standard domain* and the *star*.

In order to obtain the standard domain we proceed as follows.[1] Put

$$(7.1) \qquad \begin{cases} d\mathbf{\Gamma} = d\mathbf{\Gamma}_1 + i d\mathbf{\Gamma}_2, \\ d\mathbf{K} = d\mathbf{K}_1 + i d\mathbf{K}_2. \end{cases}$$

If a curve C, starting from $z = z_1$ and made up of regular

[1] For the properties of the standard domain compare O. T., articles 3.6, 3.7 and 3.8.

points of the differential equation (1. 1), is such that one (always the same one) of the following four pairs of inequalities is fulfilled, namely

$$(7.2)\quad 1^\circ \begin{cases} d\mathbf{K}_1 \geq 0, \\ d\mathbf{\Gamma}_1 \leq 0; \end{cases} \Bigg| \; 2^\circ \begin{cases} d\mathbf{K}_1 \leq 0, \\ d\mathbf{\Gamma}_1 \geq 0; \end{cases} \Bigg| \; 3^\circ \begin{cases} d\mathbf{K}_2 \geq 0, \\ d\mathbf{\Gamma}_2 \geq 0; \end{cases} \Bigg| \; 4^\circ \begin{cases} d\mathbf{K}_2 \leq 0, \\ d\mathbf{\Gamma}_2 \leq 0; \end{cases}$$

then the curve is a line of influence with reference to z_1 and $w(z \mid z_1, 0)$ or $w(z \mid z_1, \infty)$. We can normalize our curves to consist of parts of the integral curves of the four differential equations

$$(7.3)\qquad d\mathbf{K}_1 = 0;\; d\mathbf{K}_2 = 0;\; d\mathbf{\Gamma}_1 = 0;\; d\mathbf{\Gamma}_2 = 0.$$

Definition 5: A curve, formed by arcs from the curve-nets defined by (7. 3), along which one of the inequalities in (7. 2) is fulfilled, is called a standard path. Such a curve is denoted by $S\mathbf{K}_1^+$, $S\mathbf{K}_1^-$, $S\mathbf{K}_2^+$ or $S\mathbf{K}_2^-$ according as the characteristic inequality is the 1st, 2nd, 3d or 4th. The totality of standard paths from $z = z_1$ on the RIEMANN *surface $\mathfrak{R}$ forms the standard domain of z_1, $D(z_1)$.*

As a corollary of theorem I we obtain

Theorem VIII: $w(z \mid z_1, 0)$ and $w(z \mid z_1, \infty)$ as well as the first derivatives of these functions do not vanish in the standard domain of z_1.

In a similar manner we can use standard paths when $\lambda \neq 0$ and ∞. If, for instance,

$$0 < \arg[K(z_0)] + \arg\lambda < \frac{\pi}{2},$$

we can use standard paths of the two types $S\mathbf{K}_1^+$ and $S\mathbf{K}_2^-$. The same remark applies to the domain of influence of a curve C with reference to a solution $w(z)$; such a domain can be approximated by standard paths.

Let us assume $K(z) \equiv 1$ and consider a region B_+ in the plane where *e. g.* $G_1 > 0$, $G_2 > 0$ Then according to theorem V the regions $B_1(z_1)$ and $B_2(z_1)$ are zero-free, provided $\left[w(z)\frac{dw}{dz}\right]_{z=z_1} = 0$; they also belong to the standard domain of z_1. All the curves $\mathbf{\Gamma}_2 = const.$ in B_+ have a negative slope; one of these curves, $\Gamma_2 z_1$, passes through z_1. We find that the part of the standard domain of z_1 that lies in B_+ consists of two parts, one above $\Im\mathfrak{m}(z) = y_1$ and to the right of $\Gamma_2 z_1$, the other below $y = y_1$ and to the left of $\Gamma_2 z_1$.

8. The star is also obtained by specializing the paths of integration in (6. 2), namely to the pencil of straight lines through $z = z_1$.[1] We assume $K(z) \equiv 1$ and put

$$z = z_1 + re^{i\theta}. \tag{8.1}$$

Then (6. 2) becomes

$$\left[\bar{w}\frac{dw}{dz}\right]_{z_1}^{z_2} - \int_{z_1}^{z_2}\left|\frac{dw}{dz}\right|^2 e^{-i\theta}dr + \int_{z_1}^{z_2}|w|^2 G(z)\, e^{i\theta}dr = 0. \tag{8.2}$$

Multiply by $e^{i\theta}$ and separate reals and imaginaries. Then we obtain

$$\Re\left\{\bar{w}\frac{dw}{dz}e^{i\theta}\right\}_{z_1}^{z_2} - \int_{z_1}^{z_2}\left|\frac{dw}{dz}\right|^2 dr + \int_{z_1}^{z_2}|w|^2 P(z_1)\, dr = 0, \tag{8.31}$$

$$\Im\mathfrak{m}\left\{\bar{w}\frac{dw}{dz}e^{i\theta}\right\}_{z_1}^{z_2} + \int_{z_1}^{z_2}|w|^2 Q(z_1)\, dr = 0, \tag{8.32}$$

where

$$P(z_1) + iQ(z_1) = e^{2i\theta}G(z). \tag{8.4}$$

We note that

$$r^2[P(z_1) + iQ(z_1)] = (z - z_1)^2 G(z), \tag{8.41}$$

which makes the construction of the curves

$$P(z_1) = 0, \quad Q(z_1) = 0 \tag{8.42}$$

an easy matter.

---

[1] For the properties of the zero-free star compare O. T., art. 3. 4.

A simple digestion shows that a segment of the line

$$z = z_1 + re^{i\theta},$$

starting from $z = z_1$ is a line of influence with reference to z_1 and $w(z|z_1, 0)$ or $w(z|z_1, \infty)$ if either $Q(z_1) \neq 0$ or $P(z_1) < 0$ throughout. Thus we obtain

Definition 6: The totality of linear segments, starting from $z = z_1$, and stopping, either at singular points of (6. 1), or at points where, 1° $Q(z_1)$ changes sign, 2° $P(z)$ changes sign from — to +, provided $P(z_1) < 0$ on the whole segment, forms the zero-free star with respect to z_1, $*(z_1)$.

Theorem IX: The solutions $w(z|z_1, 0)$ and $w(z|z_1, \infty)$ and their derivatives do not vanish in the zero-free star of z.

Take a region B_* (convex) and suppose $G_1 > 0$, $G_2 > 0$ in B_*. Then the star of z_1 contains at least the points of B_* for which

$$\begin{cases} 0 \leq \arg(z - z_1) \leq \frac{3\pi}{4}, \text{ or} \\ \pi \leq \arg(z - z_1) \leq \frac{7\pi}{4}. \end{cases}$$

This is obvious from the formulae

$$P(z_1) = \cos 2\theta\, G_1 - \sin 2\theta\, G_2. \tag{8.51}$$

$$Q(z_1) = \cos 2\theta\, G_2 + \sin 2\theta\, G_1. \tag{8.52}$$

We can sharpen the theorem VI considerably for a region B_* by means of the star. From the preceding result we conclude the correctness of

Theorem X: Let a region B be convex and contain a sequence of zeros of $w(z)\frac{dw}{dz}$, $a_1, a_2, \ldots, a_n$. Put $\Delta a_\nu = a_{\nu+1} - a_\nu$ and assume $\Re[\Delta a_\nu] > 0$. If

$$k\frac{\pi}{2} < \arg G(z) < (k+1)\frac{\pi}{2} \quad (k = -2, -1, 0, +1)$$

then

$$-\frac{k\pi}{4} > \arg \Delta a_\nu > -(k+1)\frac{\pi}{4}.$$

If the argument of $G(z)$ is restricted within closer limits in a region, we can obtain a similar restriction for the argument of Δa_ν, namely

Theorem XI: Let a region D in the plane be convex and let the argument of $G(z)$ in D satisfy

$$\vartheta_2 < \arg G(z) < \vartheta_1$$

where $\vartheta_1 - \vartheta_2 \leq \frac{\pi}{2}$. *If a_1 and a_2 are two zeros of $w(z)\frac{dw}{dz}$ in D, then either*

$$-\frac{\vartheta_1}{2} < \arg(a_1 - a_2) < -\frac{\vartheta_2}{2}$$

or

$$\pi - \frac{\vartheta_1}{2} < \arg(a_1 - a_2) < \pi - \frac{\vartheta_2}{2}.$$

In fact, in order to plot the zero-free star of a_1 we have to consider

$$\arg H(z) = \arg[(z - a_1)^2 G(z)] \tag{8.6}$$

on vectors from a_1. If $\arg(z - a_1)$ lies outside of the angles

$$k\frac{\pi}{2} - \frac{\vartheta_1}{2} \leq \theta \leq k\frac{\pi}{2} - \frac{\vartheta_2}{2} \quad (k = 0, 1, 2, 3)$$

$\arg H(z) \neq k\frac{\pi}{2}$, and in the angles corresponding to $k = 1$ and 3 we have $\frac{\pi}{2} < \arg H(z) < \frac{3\pi}{2}$. This digestion shows that $*(a_1)$ contains all points in D excepting those in the angles, mentioned in the theorem.

If $G(z_1) \neq 0$ the star of z_1 consists of two separate parts which however are tangent to each other. The boundary of the star in the neighbourhood of $z = z_1$ is made up of the branch of the curve $Q(z_1) = 0$ along which $P(z_1) > 0$, and the tangent of $Q(z_1) = 0$ at $z = z_1$. From (8.52) we conclude that the slope of the tangent is determined by the equation

$$\text{tg}\, 2\theta = -\frac{G_1(z_1)}{G_2(z_1)}. \tag{8.7}$$

For the slope-angle we obtain the value

$$\theta_a = -\frac{1}{2} \arg G(z_1). \tag{8.71}$$

As the slope of the tangent of $\Gamma_2 z_1$ (see above) is

$$-\frac{G_1(z_1)}{G_2(z_1)}$$

we conclude that the angle which the tangent of $\Gamma_2 z_1$ at $z = z_1$ makes with the line $y = y_1$ is bisected by the tangent of $Q(z_1) = 0$ at the point mentioned.

9. Let us pay some more attention to the lineal element that is defined by the point z_1 and the line $\arg(z - z_1) = \theta_a$. Replacing $\cos\theta$ and $\sin\theta$ by the proportional differentials dx and dy in (8.52) we find that the lineal element in question is a solution of the differential equation

$$[(dx)^2 - (dy)^2]\, G_1(z) + 2\, dx\, dy\, G_2(z) = 0 \tag{9.1}$$

or

$$\mathfrak{Im}[G(z)\,(dz)^2] = 0. \tag{9.11}$$

The condition $P(z) > 0$ is found to be equivalent to

$$\mathfrak{R}[G(z)\,(dz)^2] > 0. \tag{9.12}$$

Hence the lineal elements have to satisfy the differential equation

$$\mathfrak{Im}[\sqrt{G(z)}\, dz] = 0. \tag{9.2}$$

The integral curves of this equation we have studied in another place[1] and given the name *asymptotic zero-curves*

[1] Compare O. T., article 4. 5.

on account of their properties. Certain conditions fulfilled, these curves approximate the zeros of the solutions in the following sense. On a given integral curve, L', there is a set of points

$$A_1, A_2, \ldots, A_n, \ldots,$$

such that a solution $w(z)$ of (6.1) exists with zeros at

$$a_1, a_2, \ldots, a_n, \ldots,$$

where

$$|A_n - a_n| < \varepsilon$$

when $n > N_\varepsilon$ and there are no other zeros of $w(z)$ in a certain neighbourhood of L'. A sequence of zeros of this type we call a *string*. For the proof we carry the differential equation into the form (3.56). For the details see the paper quoted above.

If, for instance, $G(z)$ has a pole of order n at infinity, there are $(n+2)$ directions, making equal angles with each other, in which the asymptotic zero-curves tend toward infinity, and the general solution of the equation has $(n+2)$ strings of zeros in the neighbourhood of infinity.

It is perhaps worth while mentioning that along an asymptotic zero-curve

$$\text{(9.3)} \qquad \arg[-\overline{dz}] - \arg[d\Gamma] \equiv 0, \ [\text{mod } \pi].$$

Proof. From (9.1) it follows that

$$\text{(9.31)} \qquad dx\, d\Gamma_2 + dy\, d\Gamma_1 = 0$$

along a zero-curve. Hence

$$\frac{dy}{dx} = -\frac{d\Gamma_2}{d\Gamma_1}.$$

But

$$\arg[-\overline{dz}] = -\operatorname{arctg}\left(\frac{dy}{dx}\right)$$

$$\arg[d\Gamma] = \operatorname{arctg}\left(\frac{d\Gamma_2}{d\Gamma_1}\right)$$

which shows the truth of the statement.

10. Let us now sketch a general working scheme for the study of the zeros of a family of SL-functions. We suppose the defining differential equation to be reduced to the form (6.1). In most cases that are likely to arise in practice $G(z)$ is a rational function or some other elementary function. We proceed as follows.

1°. Mark the singular points of (6.1) in the plane.

2°. Draw the curves $G_1 = 0$ and $G_2 = 0$ and mark the properties of monotony in the resulting B-regions. For the convex regions note the limits of $\arg \Delta a$.

3°. Plot the curves $\Gamma_1 = const.$ and $\Gamma_2 = const.$

4°. If the equation has an irregular singular point, investigate the distribution of the zeros in the neighbourhood by some special method, for instance by making the transformations (3.51) and (3.55). Plot the asymptotic zero-curves. Pay special attention to solutions that miss one or more of the number of strings of zeros that characterizes the general solution. These exceptional solutions usually coincide with the solutions that tend toward zero when the variable approaches the singular point in a certain direction.

5°. Try to get hold of the »simplest» pair of solutions and investigate their zeros. Solutions, real on a segment of the real axis, are usually comparatively simple and deserve special attention.

6°. Note special properties of the differential equation as, for instance, transformations that leave the equation invariant.

7°. Approximate by means of standard paths and vectors the domain of influence with reference to some point z_1 and the general solution $w(z \mid z_1, \lambda)$. Place hypothetical zeros in the uncovered parts of the plane, observing information gathered from 2° and 4°; and construct standard domains and stars of these points.

8°. Make suitable change of the variables in order to sharpen the results.

9°. Vary the parameter λ continuously and observe the variation of the zeros. (This completes 5° and joins the results of 5° and 7°).

Tryckt den 28 mars 1922.

Uppsala 1922. Almqvist & Wiksells Boktryckeri-A.-B.

AN EXISTENCE THEOREM*

BY

EINAR HILLE

1. In an earlier paper† the author has considered a certain singular integral equation of Volterra's type, namely

$$w(z) = w_0(z) + \int_z^\infty \sin(t-z)\, \Phi(t)\, w(t)\, dt \tag{1}$$

where

$$w_0''(z) + w_0(z) = 0. \tag{2}$$

The path of integration is the ray $\arg(t-z) = 0$. The function $\Phi(t)$ is single-valued and analytic at every finite point of the sector S defined by

$$-\vartheta \leqq \arg z \leqq +\vartheta, \qquad |z| \geqq \varrho > 0 \tag{3}$$

and satisfies the inequality

$$|\Phi(z)| < \frac{M}{|z|^{1+\nu}} \tag{4}$$

in S, M and ν being positive constants. We shall take up the question of the existence of a solution of this integral equation for renewed consideration in some detail.‡

---

* Presented to the Society, March 1, 1924.

† *Oscillation theorems in the complex domain,* these Transactions, vol. 23, no. 4, pp. 350–385; June, 1922. The developments of the present paper are intended to complete the scanty discussion in § 4.2 of that paper.

‡ Integral equations of a similar type have been studied by Evans and Love for real variables. Love has used his results in researches concerning the behavior of solutions of linear differential equations for large positive values (see *On linear difference and differential equations,* American Journal of Mathematics, vol. 38 (1916), pp. 57–80, where further citations are to be found). Reference should also be made to the investigations of Horn (e. g. in Journal für die reine und angewandte Mathematik, vol. 133 (1908)) with the spirit of which the present paper has much in common.

2. We shall need approximate evaluations of the integral

$$I(z;\, a) = \int_z^\infty \frac{dt}{|t|^a} \tag{5}$$

where z is a complex number which is not real and negative; a is a real constant greater than $+1$, and the path of integration is $\arg(t-z) = 0$. Putting $t = z + u$ (u real) we obtain

$$I(z;\, a) = \int_0^\infty \frac{du}{|z+u|^a}. \tag{6}$$

Using the inequality

$$|re^{i\theta} + u| = \sqrt{(r+u)^2 \cos^2\frac{\theta}{2} + (r-u)^2 \sin^2\frac{\theta}{2}}$$

$$> (r+u)\cos\frac{\theta}{2},$$

we find that

$$I(re^{i\theta};\, a) < \left[\sec\frac{\theta}{2}\right]^a \int_0^\infty \frac{du}{(u+r)^a} = \left[\sec\frac{\theta}{2}\right]^a \frac{r^{1-a}}{a-1}. \tag{7}$$

This evaluation, however, is not very good when a is large. We can get a better one by actually computing the integral. We have

$$I(z;\, a) = r^{1-a} \int_0^\infty \frac{dv}{|v+e^{i\theta}|^a} = r^{1-a} J(\theta;\, a). \tag{8}$$

Further,

$$|v + e^{i\theta}|^{-a} = (1 + v^2 + 2v\cos\theta)^{-a/2}$$

$$= (1+v)^{-a}\left[1 - \frac{4\sin^2\frac{\theta}{2}\, v}{(1+v)^2}\right]^{-a/2}.$$

If we assume $|\theta| < \pi$, the second factor in this expression can be expanded by means of the binomial theorem in a series which is uniformly convergent when $0 \leqq v \leqq +\infty$. Integrating this series term-wise, using the known formula

$$\int_0^\infty \frac{v^k\,dv}{(1+v)^{a+2k}} = \frac{\Gamma(k+1)\,\Gamma(a+k-1)}{\Gamma(a+2k)}$$

we obtain

$$J(\theta;\,a) = \frac{1}{\Gamma\left(\frac{a}{2}\right)} \sum_{k=0}^{\infty} \frac{\Gamma\left(\frac{a}{2}+k\right)\,\Gamma(a+k-1)}{\Gamma(a+2k)} \left(4\sin^2\frac{\theta}{2}\right)^k.$$

This expression can be simplified with the aid of the multiplication theorem of the Γ-function and becomes

$$J(\theta;\,a) = \frac{1}{a-1}\,F\left(a-1,\,1,\,\frac{a+1}{2},\,\sin^2\frac{\theta}{2}\right).$$

Consequently,

$$I(z;\,a) = \frac{1}{(a-1)\,r^{a-1}}\,F\left(a-1,\,1,\,\frac{a+1}{2},\,\sin^2\frac{\theta}{2}\right). \tag{9}$$

A particularly important case is that in which $a = 2$; we have*

$$I(r\,e^{i\theta};\,2) = \frac{\theta}{r\sin\theta}. \tag{10}$$

In order to arrive at an approximate evaluation of $I(z;\,a)$ we use the expression of the hypergeometric series $F(\alpha,\,\beta,\,\gamma,\,x)$ in the neighborhood of $x = +1$. In the present case we find after some reduction

$$\begin{aligned} F\left(a-1,\,1,\,\frac{a+1}{2},\,\sin^2\frac{\theta}{2}\right) &= -F\left(a-1,\,1,\,\frac{a+1}{2},\,\cos^2\frac{\theta}{2}\right) \\ &\quad + 2\sqrt{\pi}\,\frac{\Gamma\left(\frac{a+1}{2}\right)}{\Gamma\left(\frac{a}{2}\right)}\,|\sin\theta|^{1-a}. \end{aligned} \tag{11}$$

* Cf. Gauss, *Disquisitiones generales circa seriem infinitam* etc., *Werke,* vol. III, p. 127, formula XIV.

Since $a > +1$ the coefficients in the hypergeometric series in (9) are positive; consequently $J(\theta; a)$ is an increasing function of $|\theta|$, $0 \leqq |\theta| < \pi$. If $|\theta| \leqq \pi/2$ we get an upper limit for our function in $J(\pi/2; +a)$; from formula (11) we find

$$J\left(\frac{\pi}{2}; a\right) = \frac{\sqrt{\pi}\, \Gamma\left(\frac{a+1}{2}\right)}{(a-1)\, \Gamma\left(\frac{a}{2}\right)}. \tag{12}$$

If $\pi/2 < |\theta| < \pi$, formula (11) tells us that

$$J(\theta; a) < 2 \frac{\sqrt{\pi}\, \Gamma\left(\frac{a+1}{2}\right)}{(a-1)\, \Gamma\left(\frac{a}{2}\right)} |\sin\theta|^{1-a}. \tag{13}$$

Hence if we restrict a by the assumption

$$a \geqq a_0 > 1$$

we can find a constant C independent of a and of θ such that

$$I(z; a) < C \frac{R^{1-a}}{\sqrt{a-1}} \tag{14}$$

where

$$R = \begin{cases} |z|, & \text{if } -\dfrac{\pi}{2} \leqq \arg z \leqq +\dfrac{\pi}{2}, \\ |y|, & \text{if } \dfrac{\pi}{2} < |\arg z| < \pi \end{cases} \tag{15}$$

with the understanding $z = x + iy$.

3. In order to show the existence of a solution of (1) we use the method of successive approximations. We put

$$K(z, t) = \sin(t - z)\, \Phi(t) \tag{16}$$

and define the sequence of functions

$$
(17)\qquad
\begin{aligned}
w_1(z) &= w_0(z) + \int_z^\infty K(z,t)\, w_0(t)\, dt,\\
w_2(z) &= w_0(z) + \int_z^\infty K(z,t)\, w_1(t)\, dt,\\
&\cdots\cdots\cdots\cdots\\
w_n(z) &= w_0(z) + \int_z^\infty K(z,t)\, w_{n-1}(t)\, dt,\\
&\cdots\cdots\cdots\cdots
\end{aligned}
$$

Let Δ_0 be a strip of finite width in S defined by the inequalities $x \geqq A \geqq 0$, $B_1 \geqq y \geqq B_2$, and let L be the maximum of the absolute value of $w_0(z)$ in Δ_0. We have

$$
\begin{aligned}
w_{n+1}(z) - w_n(z) &= \int_z^\infty K(z,t)\,[w_n(t) - w_{n-1}(t)]\, dt\\
&= \int_0^\infty K(z, z+u)\,[w_n(z+u) - w_{n-1}(z+u)]\, du.
\end{aligned}
$$

Suppose that we have shown that for some value of n

$$
(18)\qquad |w_n(z) - w_{n-1}(z)| < \left(\frac{CM}{\sqrt{\nu}}\right)^n \frac{L}{\sqrt{n!}\,|z|^{n\nu}}
$$

when z lies in Δ_0. Then, using (4) and (14), we find

$$
\begin{aligned}
|w_{n+1}(z) - w_n(z)| &< \left(\frac{CM}{\sqrt{\nu}}\right)^n \frac{L}{\sqrt{n!}} \int_0^\infty \frac{|K(z, z+u)|}{|z+u|^{n\nu}}\, du\\
&< \left(\frac{CM}{\sqrt{\nu}}\right)^n \frac{LM}{\sqrt{n!}} \int_0^\infty \frac{du}{|z+u|^{(n+1)\nu+1}}\\
&< \left(\frac{CM}{\sqrt{\nu}}\right)^{n+1} \frac{L}{\sqrt{(n+1)!}\,|z|^{(n+1)\nu}}.
\end{aligned}
$$

But for $n = 1$ we have

$$w_1(z) - w_0(z) = \int_0^\infty K(z, z+u)\, w_0(z+u)\, du$$

and

$$|w_1(z) - w_0(z)| \leq \int_0^\infty |K(z, z+u)\, w_0(z+u)|\, du$$

$$< LM \int_0^\infty \frac{du}{|z+u|^{1+\nu}} < \frac{CM}{\sqrt{\nu}} \cdot \frac{L}{|z|^\nu}.$$

Hence (18) follows by complete induction. Consequently $w_n(z)$ converges uniformly in Δ_0 toward a single-valued and analytic function. On account of the uniform convergence the limiting function $w(z) = \lim_{n\to\infty} w_n(z)$ is a solution of the integral equation.

This is the only bounded solution. In fact, if a second bounded solution should exist, the difference, $D(z)$, of the two solutions would satisfy the integral equation

$$D(z) = \int_z^\infty K(z, t)\, D(t)\, dt.$$

Let Δ_X be the part of Δ_0 in which $x \geqq X$ where X is to be determined later, and let μ_X stand for the maximum of $|D(z)|$ in Δ_X. Then using formula (14) we conclude that

$$\mu_X \leqq \frac{CM}{\sqrt{\nu}} \cdot \frac{1}{X^\nu} \mu_X.$$

But X is at our disposal; if we make $X^\nu > CM/\sqrt{\nu}$, this inequality leads to a contradiction provided $\mu^X > 0$. Hence $D(z) \equiv 0$.

Since the width of the strip Δ_0 is arbitrary we have shown that (1) has a unique analytic solution in that portion of S which lies in the right half-plane. If the angle ϑ in formula (3) exceeds $\pi/2$ we can show that the solution exists also in the left half-plane in the following manner. Let b be an arbitrarily large but fixed positive number; then we can find a positive constant M_b such that

$$|\Phi(z)| < \frac{M_b}{|z+b|^{1+\nu}}$$

in S. If we go over the calculations again with this new majorant for $\Phi(z)$ we find that $w_n(z)$ converges to $w(z)$ provided the point $z = x + iy$ lies in S, $x > -b$, and (if $x < 0$) $|y| \geqq \varrho$. The convergence is uniform in any portion of this region in which y is bounded.

Various generalizations suggest themselves in connection with this proof. The function $w_0(z)$ need not satisfy the condition (2); all we have used in the proof is the property of $w_0(z)$ of being bounded in a strip where y is bounded. We could also carry through the proof with a slightly more general majorant for $\Phi(z)$ than the one furnished by formula (4).

4. Let us consider a closed region D in S in which y is bounded and x is bounded below and the points whose abscissas are negative have ordinates which exceed ϱ in absolute value. Let K be the maximum of $|w(z)|$ in this region and let $z_1 = x_1 + iy_1$ be the point in D where this maximum is taken on. Using (1) and (14) we find that

$$K < L + KM \int_0^\infty \frac{du}{|z_1 + u|^{1+\nu}} < L + K \frac{CM}{\sqrt{\nu}\, R_1^\nu}$$

where $R_1 = |z_1|$ or $|y_1|$ according as $x > 0$ or < 0. Let us choose D in such a fashion that $R_1^\nu > 2CM/\sqrt{\nu}$; then $K < 2L$ and

$$|w(z) - w_0(z)| < \frac{2CLM}{\sqrt{\nu}\, R^\nu} \tag{19}$$

where $R = |z|$ or $|y|$ according as $x > 0$ or < 0. We can evidently drop the assumption that x shall be bounded below in D. It is enough that y shall be bounded in order that (19) shall be true. We notice that L stands for the maximum of $|w_0(z)|$ in D.

We can arrive at a similar expression for $w(z)$ in the part of S where $y > B_1$ by considering the integral equation

$$w^+(z) = w_0^+(z) + \int_z^\infty K^+(z, t)\, w^+(t)\, dt, \tag{20}$$

where

$$w_0^+(z) = e^{iz}\, w_0(z), \qquad K^+(z, t) = e^{i(z-t)}\, K(z, t) \tag{21}$$

which is satisfied by $w^+(z) = e^{iz}\, w(z)$.

It is an easy matter to show that $|w^+(z)|$ is bounded in the region $y > B_1$. If we choose B_1 properly we can make the maximum of $|w^+(z)|$

in the resulting region less than twice the maximum of $|w_0^+(z)|$ in the same region. Denoting the latter by L^+ we arrive at the expression

$$(22) \qquad |e^{iz}[w(z)-w_0(z)]| < \frac{2CL^+M}{\sqrt{\nu}R^\nu}$$

where $R = |z|$ or $|y|$ according as $x > 0$ or < 0. A similar formula can be obtained for the lower half-plane.

We have assumed that $w_0(z) = c_1 e^{iz} + c_2 e^{-iz}$. If either c_1 or $c_2 = 0$ we can continue the corresponding solution of (1) into a wider region. In order to fix ideas, let us assume $c_1 = 1$, $c_2 = 0$ and denote the solution of (1) by $T_1(z)$.

It can be shown* by a study of the integral equation

$$(23) \qquad u(z) = 1 + \frac{1}{2i}\int_z^\infty [e^{2i(t-z)} - 1]\, \Phi(t)\, u(t)\, dt,$$

which is satisfied by $e^{-iz} T_1(z)$, that $T_1(z)$ is analytic in the sector

$$-\pi + \varepsilon \leqq \arg z \leqq 2\pi - \varepsilon, \qquad |z| \geqq \varrho,$$

and satisfies the condition

$$(24) \qquad e^{-iz} T_1(z) = 1 + \frac{\Theta_1(z)}{z^\nu}$$

where $|\Theta_1(z)|$ is bounded in the sector in question. In fact,

$$(25) \qquad e^{-iz} T_1(z) \to 1$$

along any path in the sector $-\pi \leqq \arg z \leqq +2\pi$ whose distance from the bounding rays $\theta = -\pi$ and $\theta = 2\pi$ ultimately becomes infinite.

---

* For a proof valid in the case in which $\nu = 1$ see § 2.24 of *On the zeros of Mathieu functions,* Proceedings of the London Mathematical Society, vol. 23 (1924).

PRINCETON UNIVERSITY,
PRINCETON, N. J.

ON THE ZEROS OF THE FUNCTIONS DEFINED BY LINEAR DIFFERENTIAL EQUATIONS OF THE SECOND ORDER

BY PROFESSOR EINAR HILLE,
Princeton University, Princeton, New Jersey, U.S.A.

1. A linear homogeneous differential equation of the second order can always be written in the form

$$w''+G(z)w=0. \tag{1}$$

We assume $G(z)$ to be a single-valued analytic function of z, the singularities of which shall have only a finite number of limit points. Let $w(z)$ be an arbitrary solution of (1) which is not identically zero.

The problem of integrating a linear differential equation has been interpreted in many different ways. In modern mathematics the problem is usually taken to mean the determination of the group of monodromy of the equation, inasmuch as this group gives the connection between the different expansions in series which can be obtained at the several singular points. If this problem is supposed to be solved, it is possible to construct the Riemann surface over the z-plane on which the solution $w(z)$ mentioned above is single-valued. Then the problem arises of how the values of w are distributed upon the surface. It is true that the known series expansions throw some light upon this question but only locally and in a very imperfect manner. A small but important detail of this problem is the study of the distribution of the value zero on the surface, in other words, *the problem of finding the zeros of the solutions*. The present paper is intended to call attention to a couple of methods which can be used for the purpose of elucidating this *zero point problem.*

The literature concerned with our problem is not extensive. We notice interesting investigations of Hurwitz, Van Vleck, Schafheitlin and Boutroux on various special equations. Important sidelight has been thrown on the problem through the study of irregular singular points of finite rank made by Horn, Birkhoff, Garnier and others. Finally the present writer has devoted a series of papers to this problem, some to general questions, some to particular cases*.

2. All methods which have been used for the purpose of studying the problem in question make use of one or several of the following four devices, namely:

*A paper in Proc. London Math. Soc. Ser. 2, Vol. 23 (1924) pp. 185-237: *On the zeros of Mathieu functions*, gives an outline of the general theory together with an important application.

(i) *Asymptotic representation* of the solutions in the neighbourhood of a singular point.

(ii) *Integral equalities* of the type used in the Sturmian theory of oscillation problems.

(iii) *Variation of the parameters* entering into the solutions.

(iv) *Conformal mapping* by means of the quotient of two solutions.

In the following we shall give a short account of convenient normalizations of the methods based upon the first two devices together with some reflections upon the latter two methods. For further details we have to refer the reader to the paper mentioned in the footnote.

3. In order to obtain the asymptotic representation of the solutions in the neighbourhood of a singular point we use the transformation

$$Z = Z(z;\ z_0) = \int_{z_0}^{z} \sqrt{G(z)}\, dz, \quad W = [G(z)]^{\frac{1}{4}}\, w. \tag{2}$$

The result of the transformation is a new differential equation

$$\frac{Wd^2}{dZ^2} + G^*(Z) W = 0, \tag{3}$$

where

$$G^*(Z) = g(z) = 1 - \tfrac{1}{4}\, \frac{G''(z)}{[G(z)]^2} + \tfrac{5}{16}\, \frac{[G'(z)]^2}{[G(z)]^3}\,. \tag{4}$$

This transformation has been much used in the study of oscillation problems for real variables since the time of Liouville. The importance of the transformation in the complex domain does not seem to have been generally recognized, though its normalizing power is scarcely less striking in the complex case than in the real one.

Suppose that $z = a$ is a singular point of the differential equation (1). Excluding the case in which $z = a$ is a simple pole of $G(z)$, we can always map a certain partial neighbourhood of $z = a$ upon a partial neighbourhood of $Z = \infty$ by means of the function $Z = Z(z;\ z_0)$. Suppose for instance that $z = a$ is a pole of order $k > 2$, in which case the point is an irregular singular point of finite rank. Then it is possible to find $k - 2$ *principal regions* which together form a complete neighbourhood of $z = a$ and each of which can be mapped upon a half-plane

$$\mathfrak{R}(Z) = X \geqq A > 0.$$

Furthermore

$$Z^2[G^*(Z) - 1]$$

remains bounded when Z tends to infinity within the half-plane, *i.e.*, when z tends to a within the principal region in question. If the rank is not finite, we may still be able to find principal regions abutting upon $z = a$ in which $g(z)$ behaves in the same manner as in the finite case.

On the other hand, if $z=a$ is a regular singular point such that the product c of the roots of the corresponding indicial equation is positive, then

$$e^{\frac{Z}{\sqrt{c}}}\left[G^*(Z)-1+\frac{1}{4c}\right] \qquad (\sqrt{c}>0)$$

remains bounded when Z tends to infinity in the half-plane $X>A$. Such an exponential approach of $G^*(Z)$ to its limit may also occur in cases in which $z=a$ is not a regular singular point.

4. The application of the transformation (2) involves two separate steps: (i) the determination of the principal regions in the z-plane which arise at the conformal representation of the z-plane upon the Z-plane, (ii) the study of the transformed equation (3) in the corresponding half-planes over the Z-plane.

We turn our attention to the second problem. We assume that we have obtained a transformed equation (3) such that

$$F(Z)=1-G^*(Z)$$

is single-valued and analytic in any finite portion of the region D_R,

$$|Z|\geqq R \text{ if } X\geqq 0 \text{ and } |Y|\geqq R \text{ if } X<0,\ R>0,$$

and

$$|F(Z)|<\frac{M}{|Z|^2} \tag{5}$$

when Z lies in D_R.

Let us associate the following singular integral equation of the Volterra type with the transformed equation (3), namely

$$W(Z)=W_0(Z)+\int_Z^\infty \sin(T-Z)F(T)W(T)dT, \tag{6}$$

where

$$W_0''(Z)+W_0(Z)=0 \tag{7}$$

and the path of integration is the line $\arg\,(T-Z)=0$.

It can be shown that this equation possesses an analytic solution, holomorphic in any finite portion of D_R and bounded in any infinite strip $B_1\leqq Y\leqq B_2$ wholly in D_R. Moreover, this is the only solution of (6) the modulus of which admits of a finite upper bound in any finite portion of D_R. Further, the solution of the integral equation satisfies the differential equation (3) no matter how the function $W_0(Z)$ has been chosen, subject to condition (7). Conversely, every solution of the differential equation (3) satisfies an integral equation of the type (6) provided $W_0(Z)$ is properly selected from the solutions of the auxiliary differential equation (7).

Thus there exists a one-to-one correspondence between the solutions of the transformed differential equation (3) and the solutions of the sine equation. The correspondence is such that linearly independent solutions of (7) correspond to linearly independent solutions of (3). Further, the corresponding solutions are asymptotically equal in D_R in a sense upon which we shall dwell somewhat.

Let us consider that part, Δ_0, of the strip $B_1 \leqq Y \leqq B_2$ which lies in D_R. Suppose that

$$\text{Max}\,|W_0(Z)| = L;\ \text{Max}|W(Z)| = K \tag{8}$$

in the strip Δ_0. An easy calculation shows that

$$|W(Z)| < L + MK\frac{\theta}{Y},$$

where $\theta = \arg Z$. Suppose that the constant R which enters into the determination of the region D_R satisfies the condition

$$R \geqq 2\pi M. \tag{9}$$

Then $M\dfrac{\theta}{Y} \leqq \frac{1}{2}$ when Z lies in Δ_0 and consequently

$$K < 2L.$$

Hence

$$|W(Z) - W_0(Z)| < 2LM\frac{\theta}{\sin\theta}\cdot\frac{1}{|Z|}. \tag{10}$$

A similar formula holds when Z lies in D_R above Δ_0, namely

$$|e^{iZ}[W(Z) - W_0(Z)]| < 2L_+M\frac{\theta}{\sin\theta}\cdot\frac{1}{|Z|}, \tag{11}$$

where

$$L_+ = \text{Max}\,|e^{iZ}W_0(Z)|,\ Y \geqq B_2. \tag{12}$$

The formula for the lower half-plane is analogous.

5. These formulae express the asymptotic relationship between the corresponding functions $W(Z)$ and $W_0(Z)$. They enable us to read off the asymptotic properties of $W(Z)$ knowing those of $W_0(Z)$. Equation (7) has two different types of solutions, namely, infinitely many linearly independent oscillatory solutions of the form $\sin(Z - Z_0)$ and two linearly independent non-oscillatory solutions e^{iZ} and e^{-iZ}. To the former class of functions $W_0(Z)$ correspond oscillatory solutions of (3) which have infinitely many zeros in D_R; to the latter correspond two truncated solutions which do not vanish at all in D_R.

In the oscillatory case we can take

$$W_0(Z) = \sin(Z - Z_0).$$

Suppose that the strip Δ_0 is so chosen that $B_1 < Y_0 < B_2$ where $Z_0 = X_0 + iY_0$. Let us divide Δ_0 into rectangles R_n by means of the lines $X = X_0 + (n - \frac{1}{2})\pi$ where n is an arbitrary integer. With the aid of formula (11) and of its analogon in the lower half-plane, it is easy to choose B_1 and B_2 in such a manner that the part of D_R in which $Y \leqq B_1$ or $\geqq B_2$ cannot contain any zero of $W(Z)$. On the other hand, the theorem of Rouché shows that each of the rectangles R_n contains one and only one zero of $W(Z)$ provided either Y_0 is a sufficiently large positive or negative number, or else n is a sufficiently large positive number. In the former case there is no restriction on n, in the latter none on Y_0. When $n \to +\infty$ the zero of $W(Z)$ in R_n which we denote by A_n tends to the centre of R_n, in other words

$$A_n - Z_0 - n\pi \to 0. \tag{13}$$

Similarly the zeros of $W'(Z)$ which we call the *extrema* of $W(Z)$ are approximated by the zeros of $\cos(Z - Z_0)$. Thus we find that the zeros and extrema of $W(Z)$ in D_R form a linear set, a *string*, in Δ_0. This set is guided by the line $Y = Y_0$; the points $Z_0 + n\pi$ and $Z_0 + (n - \frac{1}{2})\pi$ on this line give the approximate location of the zeros and extrema respectively when n is large.

The analysis of §§ 4 and 5 applies to the case in which $G^*(Z)$ approaches unity as a limit in the manner indicated by formula (5). If $G^*(Z)$ would approach some other limit, exponentially for instance, a similar analysis is possible. The essential condition for the success of the method is that the function $G^*(Z)$ shall be single-valued and analytic in a half-plane $X \geqq A$ and approach a finite limit a^2 when $X \to +\infty$; further the integrals

$$\int_Z^{\infty} [a^2 - G^*(T)]\, dT \tag{14}$$

shall be absolutely convergent for every value of Z in the half-plane*.

In order to obtain the distribution of the zeros of $w(z)$ in the principal region which corresponds to the half-plane $X \geqq A$ we have only to find the images of the points A_n. These points are approximated by a set of points upon an arc of the curve

$$\mathfrak{J}\left(\int_{z_0}^{z} \sqrt{G(z)}\, dz\right) = Y_0. \tag{15}$$

The extrema require a special discussion owing to the presence of the factor $[G(z)]^{\frac{1}{4}}$ in formula (3). Provided we can cover the Riemann surface of $w(z)$ completely with principal regions, the problem of the distribution of the zeros on the surface can be solved in a qualitative manner with the aid of the method outlined above. This solution can be given a higher degree of precision if the relations are known which express the truncated solutions belonging to one principal region in terms of the truncated solutions belonging to any other principal region.

6. The Sturmian methods are largely based upon the use of integral equalities. In order to obtain such an equality which is suitable for the complex domain, we multiply equation (1) by $\bar{w}$, the conjugate of w. If the resulting expression is integrated, a simple reduction yields the following relation

$$\Big[\bar{w}\, w'\Big]_{z_1}^{z_2} - \int_{z_1}^{z_2} |w'|^2 \overline{dz} + \int_{z_1}^{z_2} |w|^2 G(z)\, dz = 0. \tag{16}$$

*See the author's note in Proc. Nat. Acad. Sci. vol. 10, No. 12 (1924) pp. 488-493: *A general type of singular point.*

We call this equality *Green's transform* of the differential equation (1). In the following z_1 and z_2 shall be non-singular points of the differential equation, and the path C which joins them shall not pass through any singular point. Further, C shall be made up of a finite number of analytic arcs. Consequently, C has a continuously varying tangent except at a finite number of points where unique semi-tangents exist.

Formula (16) tells us that the necessary and sufficient condition in order that

$$w(z_2)w'(z_2)=0$$

is that

$$-\bar{w}(z_1)w'(z_1)-\int_{z_1}^{z_2}|w'(z)|^2\overline{dz}+\int_{z_1}^{z_2}|w(z)|^2G(z)dz=0, \tag{17}$$

or shorter

$$I_0+I_1+I_2=0.$$

7. The condition

$$\frac{w'(z_1)}{w(z_1)}=\lambda, \tag{18}$$

where λ is an arbitrary complex number including ∞, defines a solution of (1) uniquely save for an arbitrary multiplicative constant. We shall determine a region in which this solution $w(z;\, z_1,\, \lambda)$ cannot admit of any zeros or extrema. Let us put

$$\arg\lambda=\omega,\ \arg G(z)=\gamma_z,\ \arg dz=\theta_z. \tag{19}$$

Then we choose an arbitrary angle θ subject to the condition

$$\theta\leqq\omega+\pi\leqq\theta+\pi. \tag{20}$$

Now suppose that we choose a path C leading from $z=z_1$ to $z=z_2$ subject to the conditions mentioned in § 6 and, in addition, to the following conditions:

$$\theta\leqq\pi-\theta_z\leqq\theta+\pi, \tag{21}$$

$$\theta\leqq\gamma_z+\theta_z\leqq\theta+\pi. \tag{22}$$

Conditions (21) and (22) shall be fulfilled at all but a finite number of points on C. Further, the inequality signs shall hold either in (20) or else in one of the conditions (21) and (22) for a set of points forming a non-vanishing arc of C. Such a curve C shall be said to possess a *definite vector field* and to be *a line of influence with respect to the point* z_1 *and the solution* $w(z;\, z_1,\, \lambda)$.

It follows from the choice of C that

$$\theta\leqq\arg I_n\leqq\theta+\pi,\ |I_n|\neq 0,$$

where the relation holds for $n=0$, 1 and 2 if $\lambda\neq 0$ and ∞, and for $n=1$ and 2 if $\lambda=0$ or ∞ and the inequality signs hold for at least one value of n. Hence in every case

$$I_0+I_1+I_2\neq 0.$$

Consequently, in view of the discussion in § 6,

$$w(z_2)w'(z_2)\neq 0.$$

Now let us determine all the lines of influence with respect to $z=z_1$ and the solution $w(z;\ z_1,\ \lambda)$ which correspond to a given value of θ and then repeat the process for every value of θ compatible with condition (20). The end-points of all the lines so obtained form *the domain of influence of the point* $z=z_1$ *with respect to the solution* $w(z;\ z_1,\ \lambda)$ which we denote by $DI(z_1,\ \omega)$. In general this domain is made up of several disconnected regions which may contain some or all of their boundary points; the domain may or may not be over-lapping, in the former case it has to be spread out upon the Riemann surface of the solution.

There are no zeros or extrema of the solution $w(z;\ z_1,\ \lambda)$ *in the domain of influence of the point* $z=z_1$ *with respect to the solution in question.*

The points of $DI(z_1;\ \omega)$ are in general contained in the region $DI(z_1)$, which corresponds to $\lambda=0$ or ∞, for every value of ω. It may happen, however, that the vector field of a path is definite only if $I_0\neq 0$; in that event the end-point of the path belongs to $DI(z_1;\ \omega)$ for certain values of ω but does not belong to $DI(z_1)$.

8. It is of importance to notice that the domain of influence $DI(z_1;\ \omega)$ does not depend upon the value of $|\lambda|$ save for the case in which this value is either 0 or ∞. Further, we do not change any of these domains, if we replace $G(z)$ in equation (1) by $k^2G(z)$, k being a real quantity. Thus the function $w_k(z;\ z_1,\ re^{i\omega})$ which satisfies the differential equation

$$w''+k^2G(z)w=0 \tag{23}$$

and the logarithmic derivative of which reduces to $re^{i\omega}$ for $z=z_1$, does not vanish in the domain $DI(z_1;\ \omega)$ for any real value of k and any value of r, $0<r<+\infty$. In particular, the functions $w_k(z;\ z_1,\ 0)$ and $w_k(z;\ z_1,\ \infty)$ do not vanish in $DI(z_1)$ for any real value of k. The same statements hold for the first derivative of the function in question.

This situation raises an interesting question. The function $w_k(z;\ z_1,\ re^{i\omega})$ does not vanish anywhere in $DI(z_1;\ \omega)$ for any real value of k and any positive value of r. *Supposing that* $z=z_2$ *is a point in the complementary region of* $DI(z_1;\omega)$, *is it possible to determine a real value of* k *together with a positive value of* r, κ *and* ρ *respectively, say, such that*

$$w_\kappa(z_2;\ z_1,\ \rho e^{i\omega})=0?$$

In other words, does the locus of the zeros of $w_k(z;\ z_1,\ re^{i\omega})$ for real values of k and positive values of r, coincide with *co*-$DI(z_1;\ \omega)$? Ordinary considerations of continuity indicate the plausibility of an affirmative answer to this question.

9. The complete determination in a finite number of operations of the domain of influence of a point with respect to a given solution is seldom possible. But it is always possible to reach a more or less extensive portion of this domain in a finite number of steps, if the curves which are used for the purpose of building up lines of influence are properly standardized. As proper curves we want to mention the following six types:

$$\Re\left[\int_{z_1}^{z} dz\right]=\text{const.};\quad \Im\left[\int_{z_1}^{z} dz\right]=\text{const.}; \tag{24}$$

$$\mathfrak{R}\left[\int_{z_1}^{z} G(z)\,dz\right]=\text{const.};\quad \mathfrak{J}\left[\int_{z_1}^{z} G(z)\,dz\right]=\text{const.}; \tag{25}$$

$$\arg\left[\int_{z_1}^{z} dz\right]=\text{const.};\qquad \arg\left[\int_{z_1}^{z} G(z)\,dz\right]=\text{const.} \tag{26}$$

None of the first four types can be used alone with any greater amount of profit; the best result is obtained by combining all four types. Each of the last two types can be used alone and does not combine easily with the other or with the first four types. For the study of truncated solutions and for purposes of ordering the zeros and extrema valuable service is rendered by the curves

$$\mathfrak{R}\left[\int_{z_0}^{z} \sqrt{G(z)}\,dz\right] = \text{const.}, \tag{27}$$

which are the orthogonal trajectories of the curves in formula (15).

10. Let us give some simple applications of this *method of zero-free regions.*

First, let us suppose that $G(z)$ is holomorphic and different from zero in a convex region B and that in addition

$$\theta_1<\arg G(z)<\theta_2,\ 0<\theta_2-\theta_1<\pi, \tag{28}$$

for every point in B. Let z_1 and z_2 be two arbitrary zeros or extrema in B of some solution $w(z)$ of the differential equation (1). Then

$$-\tfrac{1}{2}\theta_2<\arg\,(z_1-z_2)<-\tfrac{1}{2}\theta_1 \tag{29}$$

if the enumeration of the points has been properly chosen.

This theorem gives an easy means of arranging the zeros and extrema of a solution within a certain region into a linear sequence.

Secondly, let us assume that $G(z)$ is real positive on an interval (a, b) of the real axis. Let D be a simply-connected region symmetric to the real axis, lying between the lines $x=a$ and $x=b$, the boundary of which is cut in two points only by any line $x=c$, $a<c<b$. Finally let $G(z)$ be holomorphic in the closed region D where in addition $\arg G(z)$ shall be $\neq \pm\pi$. Then, a solution of (1) cannot have any complex zeros or extrema in D if its logarithmic derivative is real at a point of the interval (a, b). This theorem is evidently of great applicability, especially in the study of differential equations which arise in mathematical physics.

11. Finally let us say a few words about the methods which are based upon variation of parameters and upon conformal mapping. Both these methods can be attached to the Sturmian method of zero-free regions sketched in §§ 6-10.

Let us refer to the quantity λ defined by equation (18) as *the parameter of the solution at the point* z_1. It is comparatively easy to follow the effect of varying this parameter. First we notice that the zeros and extrema of the solution are analytic functions of λ. Thus, if λ is varied continuously, the zeros and extrema move in a continuous manner except if λ passes through certain singular points. Now for every value of λ we have a domain of influence $DI(z_1;\omega)$. Further every moving zero and extremum has its own variable domain of influence. No matter how the motion is carried out, the zeros and the extrema have to stay outside of $DI(z_1; \omega)$ and outside of each other's domains of influence. This simple rule gives a means of following the motion of the zeros and the extrema during the variation of λ.

The information so obtained can be brought to bear upon the problem of conformal representation by means of the quotient of two linearly independent solutions of (1). If we choose

$$Q(z) = -\frac{w(z; z_1, \infty)}{w(z; z_1, 0)} \tag{30}$$

as the quotient, and if $Q = \lambda$ when $z = z_2$ then z_2 is a zero of

$$w(z; z_1, \lambda) = w(z; z_1, \infty) + \lambda w(z; z_1, 0). \tag{31}$$

Thus, if a certain zero of $w(z; z_1, \lambda)$ describes a path S when the parameter λ is varied along a path L, then S is the map in the z-plane of the path L in the Q-plane by means of the inverse of the function $Q(z)$*.

*The author's paper *On the zeros of the functions of the parabolic cylinder*, Arkiv för Mat., Astr. o. Fys., vol. 18, No. 26, 1924, gives a simple example of the stepwise application of the four different methods mentioned in §2, beginning with the asymptotic representation and culminating in the conformal mapping.

Zero point problems for linear differential equations of the second order*).

By Einar Hille.

1. *Introduction.* Let there be given a linear differential equation of the second order with analytic coefficients. Such an equation can always be reduced to the form

$$w'' + G(z)w = 0. \tag{E}$$

We consider a region B in the plane where $G(z)$ is single-valued and analytic; we further select a solution $w(z)$ of (E) and raise the question: What can be said about the distribution of the zeros of $w(z)$ in B?

Perhaps this detached question comes too suddenly. Let us look for a moment at the back-ground of the problem. The question of integrating a differential equation has been interpreted rather differently. Now-a-days we all agree that for equations with analytic coefficients the problem is a function-theoretical one. An equation like (E) defines implicitly a two parameter family of analytic functions. The integration problem for this equation consists in studying the analytic properties of these functions. Such a study is generally taken to imply the obtaining of local analytic representations of the functions in regions which together cover the whole plane, followed by a determination of the monodromic group of the equation which gives the connections between the different functional elements. But the problem could and should mean much more; we want to know the properties of growth of the solutions in the neighborhood of the singular points as well as the distribution of the values taken on by a solution upon the Riemann surface which belongs to it. For these properties we need special investigations to complete the information (frequently rather meager) rendered by the local series and by the group.

When we raise the question of how the value zero is distributed upon the Riemann surface of the solution, this is of course only a special case of the general question: How is the

*) Translation of lecture delivered to the Matematisk Forening, Copenhagen, February 24, 1927.

value C distributed? But it is in a way the most important special case, and it is to be expected that a more or less satisfactory solution of this special problem will throw light upon the general question.

2. *Methods.* It is not my intention to trace the history of this zero point problem. I shall only give some general points of view and then proceed to outline some of my own investigations.

If we disregard a few special investigations, we can distinguish essentially six different types of methods which have been used in the literature to elucidate the zero point problem. These methods are based upon the following typical ideas or tools:

1) Algebraic theorems and their extensions.
2) The associated *Riccati* equation.
3) Asymptotic integration.
4) Integral equalities of *Green's* type.
5) Variation of parameters.
6) Conformal mapping.

The first two methods will not be considered in the present paper. Typical investigations with methods of the first kind are due to *Laguerre* who studied extensively the polynomials which satisfy linear differential equations of the second order. We can also mention *Hurwitz'* investigation of *Bessel's* functions which he approximated uniformly with the aid of polynomials with real roots. — An application of the second method is to be found in *Boutroux's* famous prize memoir on the transcendental functions of *Painlevé* in which he devoted a chapter to a special case of *Bessel's* equation.

Under (3) we notice an imposing sequence of memoirs by *Birkhoff, Garnier, Horn, Poincaré,* and others, which permit a complete solution of the zero point problem in the neighborhood of an irregular singular point of finite rank (polar singularity of $G(z)$ in our case). I shall give below an arrangement of this method which permits treating quite complicated singularities of $G(z)$. The methods based upon asymptotic integration have the advantage of giving information about questions of growth as well as of distribution of values. In a certain sense they give existence proofs for the zero point problem.

The classical example of an application of methods of type

(4) is furnished by ***Hurwitz'*** investigation mentioned above. This method can be developed into a highly flexible and efficient tool for investigations of this kind. I shall give a convenient standardization of the method below in § 4.

Schafheitlin has treated the zeros of hypergeometric functions using methods of type (5); the method is, however, not much known and is difficult to apply. It is convenient to combine this method with the previous one. — Methods of type (6) are chiefly known through the work of *Klein* and his pupils upon the zeros of hypergeometric functions and of solutions of more general equations belonging to the Fuchsian class. These two methods will be illustrated by an example in §§ 6-7.

3. *Asymptotic integration.* For the purpose of asymptotic integration of (E) we use the transformation of Liouville. This transformation is frequently employed in oscillation problems for real variables; not much use has been made of it in the complex domain, in spite of its remarkable uniformizing power. It reads as follows

$$\text{(L)} \qquad Z = Z(z,z_0) = \int_{z_0}^{z} \sqrt{G(z)}\,dz, \quad W = [G(z)]^{\frac{1}{4}} w,$$

and transforms (E) into

$$\text{(E*)} \qquad W'' + H(Z)\,W = 0,$$

where

$$H(Z) = h(z) = 1 + \tfrac{5}{16}\frac{[G'(z)]^2}{[G(z)]^3} - \tfrac{1}{4}\frac{G''(z)}{[G(z)]^2}.$$

The singular points of $Z(z, z_0)$ arise partly from the singularities and partly from the zeros of $G(z)$; to these points we have to add the point at infinity, at any rate if this point is a singular point of (E). To these singularities correspond singular points of (E*) of which we shall take into account only those which arise from the singularities of (E). Let $z = \alpha$ be a singular point of (E) and let us assume that a certain partial neighborhood of $z = \alpha$ is mapped conformally upon a partial neighborhood af $Z = \infty$ through (L). To be more precise, we assume the existence of a region D abutting upon α with the following properties:

(1) D is a simply-connected, open region upon $\mathfrak{R}$, a Riemann surface upon which the general solution of (E) is single-valued,

and is bounded by a denumerable set of arcs $X =$ const, $Y =$ const, where $Z = X+iY$.

(2) $z = \alpha$ is a boundary point of D and the only singular point of (E) in $D +$ boundary. $G(z)$ shall vanish at most a finite number of times on the boundary and not at all in D itself.

(3) There exists a determination of $Z(z, z_0)$ which maps D, conformally and without overlapping, upon a region Δ in the Z-plane which contains a half-plane $X>A$ and is contained in another half-plane $X>B$.

(4) When $Z\to\infty$ in the sector $|\arg Z| \leqq \frac{\pi}{2}-\varepsilon$ on Δ, $\lim H(Z)$ shall exist and be equal to a^2, where a is independent of the path and $|\arg a| < \frac{\pi}{2}$.

(5) $\int_Z^\infty [a^2-H(T)]\, dT$ shall be absolutely convergent for an arbitrary path of integration within the sector. If $a = 0$, we demand the existence of $\int_Z^\infty |T^2 H(T) dT|$ instead.

Such a region D will be referred to as a principal region belonging to $z = \alpha$. It turns out that such regions exist in a great number of cases which are of practical interest. In a principal region the asymptotic integration of (E) can be carried out with the greatest ease by applying the method of successive approximations to the transformed equation (E*). I shall restrict myself to indicating the analysis in the most important special case, that in which

$$a = 1, \quad |1-H(Z)| < \frac{M}{|Z|^2} \text{ in } \Delta.$$

This case arises, *e. g.* when $z = \alpha$ is a pole of $G(z)$ of order greater than 2, *i. e.* $z = \alpha$ is an irregular singular point of finite rank.

$H(Z)$ is holomorphic in Δ in view of assumption (2). It is advantageous to assume in the following discussion that $H(Z)$ is holomorphic in a larger region Υ where it shall satisfy the same inequality as in Δ. Here Υ is that part of the Z-plane whose distance from the negative real axis exceeds a given positive number r. This assumption is actually true in the special case just mentioned.

The differential equation (E*) is then asymptotic in Υ to the equation

$$(E_0) \qquad W''+W=0$$

in an obvious sense. It is thus to be expected that the solutions of the two equations shall be asymptotic to each other. This is indeed the case; corresponding to an arbitrary solution $W_0(Z)$ of (E_0) we can find a solution of (E^*)

$$W(Z)=\lim_{n\to\infty} W_n(Z),$$

where

$$W_n(Z)=W_0(Z)+\int_Z^{\infty}\sin(T-Z)\ [1-H(T)]\ W_{n-1}(T)\,dT,$$

and $\arg\,(T-Z)=0$. This solution is holomorphic in Y; in any part of Y where $Y=\mathfrak{J}(Z)$ remains bounded we have

$$\text{(I)} \qquad |W(Z)-W_0(Z)|<2LM\frac{\Theta}{\sin\Theta\ |Z|},\quad \Theta=\arg Z,$$

provided $r\geqq 2\pi M$. Here $L=\text{Max}\,|W_0(Z+\varrho\pi)|$ when Z remains in the region under consideration and the real number ϱ runs from 0 to 1. We can get similar inequalities when Y is merely bounded either above or below. These inequalities give the key to the study of the rate of growth and of the distribution of values of $W(Z)$ in Y. These properties are simply asymptotically the same as those of $W_0(Z)$.

It is not necessary to go into further detail, but there is one essential point which should be brought out. Equation (E_0) has two different types of solutions, namely solutions which have zeros and solutions which have not. There is a continuum of solutions $\sin(Z-Z_0)$ of the former type which have infinitely many equi-distant zeros on a line parallel to the real axis and which are not bounded in any half-plane. Of the latter type there are only two linearly independent solutions, e^{iZ} and e^{-iZ}; they are further characterized by the property of tending to zero when Z tends to infinity in the upper and lover half-planes, respectively. To the former type correspond solutions of (E^*) which have an infinite number of zeros in Y, asymptotic to the point set $Z_0+n\pi$, and which are not bounded in any half-plane. To the latter type corresponds two solutions, hereafter referred to as truncated solutions, which have no zeros in Y if $r\geqq 2\pi M$, and which tend to zero in the upper and lower half-planes, respectively.

Upon returning to the z-plane we find the following situation holding in the principal region D. There exist two and only two linearly independent solutions which have only a finite number of zeros in D and which are characterized by their asymptotic behavior when $z \to \alpha$ along a curve $X = \text{const}$ in D. All other linearly independent solutions have infinitely many zeros in D which are asymptotic in a certain sense to the image points of a set $Z_0 + n\pi$.

Thus in case we can cover the whole plane with the aid of principal regions we have to a certain extent solved the problem of the distribution of values in general, and of zeros in particular, as well as the problem of the rate of growth. We need additional information away from the singular points and we also need some method of following a solution from one principal region to another.

4. *The method of zero-free regions.* This method is based upon the following simple integral equality

$$\text{(G)} \quad \Big[\overline{w(z)}\ w'(z)\Big]_{z_1}^{z_2} - \int_{z_1}^{z_2} |\, w'(z)\,|^2\ \overline{dz} + \int_{z_1}^{z_2} |\, w(z)\,|^2\ G(z)dz = 0$$

which is satisfied by every solution of (E).

Here z_1 and z_2 are in general regular points of the differential equation; the path of integration is a rectifiable curve which does not pass through any singular point. (G) is obtained from (E) by multiplying by $\overline{w(z)}$, the conjugate of $w(z)$, and integrating by parts between z_1 and z_2.

It is perhaps desirable to point out in advance what sort of information pertaining to the zero point problem is obtainable from (G). The results will always be negative: such and such a solution can not vanish in such and such a region. No information is received about the behavior of the solution in the complementary region except possibly by a renewed application of the method.

We start with one of the fundamental solutions at $z = z_1$, i. e. a solution such that $w(z_1)\ w'(z_1) = 0$. We determine a path of integration C from z_1 in the following manner. Set

$$\arg\ G(z) = \gamma_z, \quad \arg\ dz = \delta_z,$$

where we can assume that $G(z) \neq 0$ along the path and that C has a continuously varying tangent. These assumptions can easily be discarded when desirable. Let ϑ be a fixed but arbitrary angle and assume that C is chosen in such a manner that

$$\vartheta < \pi - \delta_z < \vartheta + \pi, \quad \vartheta < \gamma_z + \delta_z < \vartheta + \pi$$

at every point. Here we can replace $<$ by $\leqq$ provided actually $<$ holds along an arc of C. Under these circumstances we say that the path C has a definite vector field and is a line of influence with respect to $z = z_1$ and the solution chosen. It is easily seen that the assumptions imply that the two integrals in (G) are separately different from zero as well as their sum. In fact we have

$$\vartheta < \arg\left[-\int_{z_1}^{z_2} |w'(z)|^2 \, \overline{dz}\right] < \vartheta + \pi, \quad \vartheta < \arg\left[\int_{z_1}^{z_2} |w(z)|^2 \, G(z) \, dz\right] < \vartheta + \pi.$$

But then it follows from (G) that $w(z_2) \; w'(z_2) \neq 0$. Thus we have the following theorem:

If $z = z_2$ is the end-point of a line of influence starting at $z = z_1$ then the fundamental solutions at $z = z_1$ cannot vanish at $z = z_2$, nor can their first derivatives do so.

Hence, if we construct all the lines of influence starting at $z = z_1$, then their end-points form a set of points in the plane or on a Riemann surface such that the fundamental solutions at z_1 as well as their first derivatives are different from zero on the set. This point set we refer to as the domain of influence with respect to z_1 and the corresponding fundamental solutions.

It is far from necessary to restrict oneself to the fundamental solutions at $z = z_1$. We can very well consider solutions for which we know $\arg \dfrac{w'(z_1)}{w(z_1)}$. Then we have to take into account three terms in (G); this puts some additional restrictions upon δ_z, but the principle employed in constructing lines of influence is the same as in the special case treated above.

It is of course seldom possible in practice to determine a domain of influence completely. Instead we have to be satisfied with more or less extensive portions of the domain. Such portions can always be found by using standardized paths of the simplest kind, *e. g.* straight lines parallel to the axes or rays emanating from a point etc.

All domains of influence remain invariant, if we replace

$G(z)$ by $a\,G(z)$ where $a>0$. Consider for a moment the domain of influence with respect to z_1 and a corresponding fundamental solution. If a is chosen sufficiently large, this solution will have zeros in an arbitrary neighborhood of z_1. Thus there must exist points distinct from z_1 which are as near as we please to this point and which do not belong to the domain of influence. This domain is not a connected region except in very special cases.

5. *Examples.* Instead of continuing the general discussion I shall take up a particular equation whose solutions can claim some intrinsic function-theoretical interest; a case in which it is possible to obtain satisfactory results with simple means. We choose the equation*)

$$w''-(1+z^2)w=0. \tag{P}$$

The most important solutions of this equation are $e^{\frac{z^2}{2}}$ and

$$e(z)=e^{\frac{z^2}{2}}\int_0^z e^{-t^2}dt,\quad E(z)=e^{\frac{z^2}{2}}\int_z^\infty e^{-t^2}dt.$$

The two integrals involved,

$$k(z)=\int_0^z e^{-t^2}dt \quad\text{and}\quad K(z)=\int_z^\infty e^{-t^2}dt,$$

are known under various names, the error function and *Kramp's* transcendent are probably the most common ones.

The asymptotic behavior of these solutions is so well known that it does not pay to go deeply into the question of asymptotic integration, — just a bare outline of the facts will suffice. The principal regions are in this case four in number; we can choose them as the four quadrants. Thus the solutions of (P) which are entire functions have in general four strings of zeros, one in each quadrant, the points of which are given in a first approximation by equations of the form

$$\frac{i}{2}z^2=Z_0+n\pi.$$

To each of the four principal regions correspond two truncated solutions, i. e. we have at most eight such solutions. This number reduces to three, however; if $G(z)$ had been a general polynomial of degree two, we should have obtained four such solu-

*) I have already touched upon this equation in an earlier paper devoted to the case in which $G(z)$ is a polynomial of degree two. See On the zeros of the functions of the parabolic cylinder, Arkiv för Mat., Astr. o. Fys., vol. 18, no. 26 (1924) pp. 55–56.

tions instead. This depends upon the fact that a solution which is truncated in one principal region will also be truncated in one of the adjacent regions*. In this particular case we have a solution, namely $e^{\frac{z^2}{2}}$, which is truncated in all four regions. The other two truncated solutions are $E(z)$ and $E(-z)$.

That $E(z)$ is actually truncated in the first and the fourth quadrants follows from the following well known asymptotic expressions

$$(A_1) \qquad K(z) \backsim \frac{e^{-z^2}}{2z}\left\{1-\frac{1}{2z^2}+\frac{1\cdot 3}{2^2 z^4}-\cdots\right\},$$

$$(A_2) \qquad K(z) \backsim \sqrt{\pi}+\frac{e^{-z^2}}{2z}\left\{1-\frac{1}{2z^2}+\frac{1\cdot 3}{2^2 z^4}-\cdots\right\},$$

which are valid, the former when $|\arg z|<\frac{3\pi}{4}$, the latter when $|\arg(-z)|<\frac{3\pi}{4}$. From the second of these expressions we can compute an asymptotic expression for the zeros of $K(z)$ in the second quadrant namely

$$(A_0) \quad z_k=\sqrt{(2k-1)\pi}\, e^{\frac{3\pi i}{4}}\left\{1-i\frac{\log 2\pi\sqrt{2k-1}+\frac{3\pi i}{4}}{2\pi(2k-1)}\right\}+O\left\{\frac{\log^2 k}{k^{\frac{3}{2}}}\right\}.$$

The zeros in the third quadrant are obtained asymptotically by changing i into $-i$ throughout in (A_0). It follows from (A_0) that

$$\sqrt{(2k-1)\pi}<|z_k|<\sqrt{2k\pi}$$

when k is sufficiently large. The asymptotic formula actually gives the kth zero in the following sense:

If k is sufficiently large, there are exactly $2k$ zeros of $K(z)$ inside the circle $|z|=\sqrt{2k\pi}$.

To see this we have only to show that the change in the argument of $K(z)$ amounts to $2k\pi+o(1)$ when we let z describe the semi-circle $z=\sqrt{2k\pi}\,e^{i\Theta}$, Θ going from 0 to π. It is convenient to divide this semi-circle into three arcs (*i*) $0\leqq\Theta\leqq\frac{3\pi}{4}-\varepsilon$, (*ii*) $\frac{3\pi}{4}-\varepsilon\leqq\Theta\leqq\frac{3\pi}{4}$, and (*iii*) $\frac{3\pi}{4}\leqq\Theta\leqq\pi$. Here

$$\varepsilon=\tfrac{1}{2}\arcsin\frac{\log 8k\pi}{2k\pi}.$$

*) This is not a perfectly general theorem. It holds, however, in some of the most important cases which come up for consideration.

On the first arc we use (A_1) and find that the argument of $K(z)$ increases by $2k\pi-\frac{3\pi}{4}+O\left(\frac{\log^2 k}{k}\right)$. On the second and third arcs we use (A_2) and find that the increase amounts to $\frac{3\pi}{4}+o(1)$ and $o(1)$, respectively. Thus the total increase amounts to $2k\pi+o(1)$ which should be proved.

We now pass over to the applications of the method of zero-free regions to equation (P) and begin with $E(z)$. The formula (G) takes the following form

$$(G_0)\quad \left[\overline{w(z)}\,w'(z)\right]_{z_1}^{z_2}-\int_{z_1}^{z_2}|w'(z)|^2\,\overline{dz}-\int_{z_1}^{z_2}|w(z)|^2\,(1+z^2)\,dz=0.$$

If $w(z)=E(z)$ we have the right to let $z_2\Rightarrow\infty$ parallel to the positive real axis in (G_0) and obtain

$$\overline{E(z_1)}\,E'(z_1)+\int_{x_1}^{\infty}|E'(z)|^2\,dx+\int_{x_1}^{\infty}|E(z)|^2\,(1+z^2)\,dx=0,$$

where $z_1=x_1+iy_1$, $z=x+iy_1$. Assuming $E(z_1)\,E'(z_1)=0$, and separating reals and imaginaries we get

$$(G_1)\qquad \int_{x_1}^{\infty}\left[|E'(z)|^2+(1+x^2-y_1^2)\,|E(z)|^2\right]dx=0,$$

$$(G_2)\qquad y_1\int_{x_1}^{\infty}|E(z)|^2\,x\,dx=0.$$

It is obvious that $y_1\neq 0$; hence (G_2) requires that $x_1<0$. Further, (G_1) cannot be satisfied unless $y_1^2>1$. Thus if $z_1=x_1+iy_1$ is a zero of $E(z)\,E'(z)$, then $x_1<0$ and $|y_1|>1$.

To obtain additional information we try to approximate the domain of influence with respect to $z=0$ and $w=E(z)$. Suitable lines of influence are formed by the pencil of straight lines through the origin. Thus we set

$$z_1=0,\ z=re^{i\Theta},\ E(z_2)\,E'(z_2)=0$$

in (G_0) obtaining, after multiplying by $e^{i\Theta}$,

$$(G_3)\qquad -\frac{\sqrt{\pi}}{2}e^{i\Theta}+\int_0^{r_2}|E'|^2dr+\int_0^{r_2}|E|^2z^2(1+z^2)\frac{dr}{r^2}=0.$$

We can restrict Θ to lie between $\frac{\pi}{2}$ and π. Then the argument of the first term lies in the fourth quadrant. Hence, as long as the imaginary part of $z^2(1+z^2)$ is not positive, (G_3) cannot be true. This imaginary part equals $2xy(1+2x^2-2y^2)$, which is negative in the part of the second quadrant below the hyperbola

$$(B_1) \qquad 1+2x^2-2y^2=0.$$

In this manner we get the zeros of $E(z)$ bounded below.

To obtain upper bounds it is natural to select a point on the imaginary axis and approximate to the corresponding domain of influence. This point should not be chosen below $z=i$ as it would then merely give another lower bound, nor should it be chosen too far above $z=i$. After some trials we find that $z=\sqrt{2}\,i$ is a suitable point. Thus we set

$$z_1=\sqrt{2}i,\ \overline{E(\sqrt{2}i)}E'(\sqrt{2}i)=Re^{i\omega},\ z=\sqrt{2}i+re^{i\Theta}, E(z_2)E'(z_2)=0.$$

Here ω is found to be 4.012 (almost 230°), i. e. between $\frac{5\pi}{4}$ and $\frac{3\pi}{2}$. Substituting in (G_0) we obtain

$$(G_4) \quad Re^{i\omega}+e^{-i\Theta}\int_0^{r_2}|E'|^2dr+\int_0^{r_2}|E|^2\left[r^2e^{3i\Theta}+i\,2\sqrt{2}\,r\,e^{2i\Theta}-e^{i\Theta}\right]dr=0.$$

$$(G_5) \quad Re^{i(\omega+\Theta)}+\int_0^{r_2}|E'|^2dr+\int_0^{r_2}|E|^2(z^2+1)(z-\sqrt{2}i)^2\frac{dr}{r^2}=0,$$

of which the second expression is obtained from the first one by multiplying by $e^{i\Theta}$. We first assume $\frac{\pi}{2}\leqq\Theta\leqq\frac{3\pi}{4}$ and use (G_4). A tedious but simple calculation shows that the arguments of all five terms in (G_4) lie between $\frac{5\pi}{4}$ and $\frac{9\pi}{4}$, the limits included, and there are vectors in the interior of this interval. Thus (G_4) cannot hold for such values of Θ. We next assume $\frac{3\pi}{4}\leqq\Theta\leqq\pi$ and use (G_5). Now $\omega+\Theta$ lies in the first quadrant. Hence (G_5) cannot be valid if the imaginary part of $(z^2+1)(z-\sqrt{2}i)^2$ remains positive along the path of integration. This expression turns out to be

$$2x\,[2y(x^2-y^2)-\sqrt{2}(x^2-3y^2)-y-\sqrt{2}].$$

The curve

$$(B_2) \qquad 2y(x^2-y^2)-\sqrt{2}(x^2-3y^2)-y-\sqrt{2}=0$$

is a cubic of genus one with three real asymptotes which intersect at $z = \frac{\sqrt{2}}{2} i$ and have the slopes $+1$, 0 and -1. It consists of three separate branches which intersect the imaginary axis at $z = \sqrt{2} i$, $\frac{1}{4}[\sqrt{2}+\sqrt{10}]i$ and $\frac{1}{4}[\sqrt{2}-\sqrt{10}]i$ respectively. Above the upper of these branches $\mathfrak{I}[(z^2+1)(z-\sqrt{2}i)^2]$ is positive in the second quadrant; all points in this region are accessible by rays from $z = \sqrt{2} i$. Hence $E(z)\, E'(z) \neq 0$ in this region.

To sum up, we find that the zeros of $E(z)\, E'(z)$ in the second quadrant lie in the shaded strip S_2 of the figure 1. This

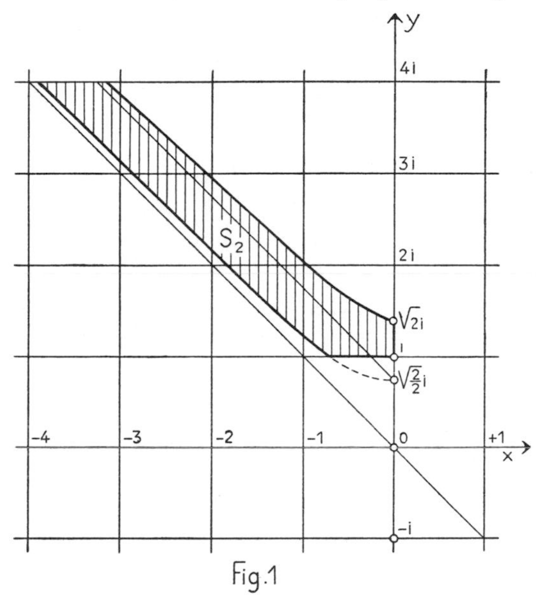

Fig.1

strip is bounded by arcs of the hyperbola (B_1) and of the cubic (B_2) plus two short segments of the lines $x = 0$ and $y = 1$.

Consider a line segment $x = \alpha$, $y \geqq 0$ or $y \leqq 0$. Such a segment can contain at most one zero of $w(z)\, w'(z)$ where $w(z)$ is an arbitrary solution of (P). This fact is easily seen with the aid of (G_0). The same holds for segments $y = \beta$, $x \geqq 0$ or $x \leqq 0$. Thus we can arrange the zeros of $E(z)\, E'(z)$ in the second quadrant according to increasing ordinates; this order is the same as that according to decreasing abscissae.

Having treated $E(z)$ so thoroughly we can be quite brief concerning the zeros of $e(z)$. This is an odd function, thus it vanishes at $z = 0$ and has a string of complex zeros in each quadrant. For these zeros we can obtain similar bounds as in the case of $E(z)$. In fact, the zeros of $e(z)$ in the second quadrant are enclosed in the strip S_2 of figure 1. It is however possible in this case to obtain a somewhat more favorable upper boundary than (B_2) by using the domain of influence with respect to $z = i$ instead of $z = \sqrt{2}i$. In this way we obtain the curve

$$(B_3) \qquad 2y(x^2 - y^2) - x^2 + 3y^2 - 1 = 0,$$

which is a rational cubic, double point at $z = i$ with tangents of slope $\pm\frac{\sqrt{3}}{3}$, having three asymptotes intersecting at $z = \frac{i}{2}$, of

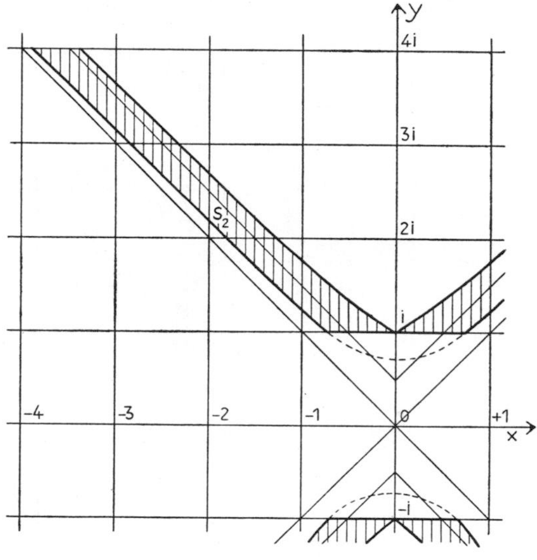

Fig. 2

slopes $+1$, 0 and -1. As a result we obtain that the zeros of $e(z)\, e'(z)$ in the second quadrant are enclosed in the strip s_2 of figure 2. This strip is bounded by arcs of (B_1) and (B_3) plus a segment of the line $y = 1$. The zeros in the other quadrants are of course enclosed in strips symmetric to s_2 with respect to the axes.

6. *The method of varying the parameters.* This method is somewhat special in its nature and range of applicability. But, given an equation with real coefficients and a solution the distribution of whose zeros is known in some detail, the method can be very useful when it comes to questions of enumeration and order. Perhaps I can best bring out these features of the method by using the equation (P) as an illustration.

Let the zeros of $E(z)$ in S_2 be denoted by z_j, those of $E'(z)$ by ζ_k where the enumeration is according to increasing ordinates $(j,k = 1,2,3,\cdots)$. It will be convenient in the following to refer to the points ζ_k as the extrema of $E(z)$, with similar notation for other functions. What is the relative order between the z_j and the ζ_k when we order the set $\{z_j, \zeta_k\}$ according to increasing ordinates? The method of varying the parameters permits proving that this order is $\zeta_1, z_1, \zeta_2, z_2, \cdots, \zeta_n, z_n, \cdots$ I shall merely bring out the essential steps in the proof.

The equation (P) is a special case of the equation

$$(\mathrm{P}_a) \qquad w'' - (a^2 + z^2) w = 0,$$

where we assume a to be real positive. We define two solutions $P_1(z,a)$ and $P_2(z,a)$ of this equation by the following conditions:

$$P_1(0,a) = 1, \quad P'_1(0,a) = 0;$$

$$P_2(0,a) = 0, \quad P'_2(0,a) = 1;$$

and consider the general solution

$$w(z,\lambda,a) = \lambda P_1(z,a) - P_2(z,a).$$

In particular, we have

$$w(z,0,1) = -e(z), \quad w\left(z, \frac{\sqrt{\pi}}{2}, 1\right) = E(z).$$

The method of zero-free regions now shows that $w(z,\lambda,a)$ has neither a real negative zero or extremum nor a purely imaginary one as long as $\lambda > 0$ and a is real. If $\lambda = 0$ we know that there is a zero at the origin, and if $a = 1$ there is no other purely imaginary zero. Further, $w(z,\lambda,a)$ has infinitely many zeros and extrema in the interior of the second quadrant when $\lambda > 0$ and a is real. Let these points be denoted by $z_j(\lambda,a)$ and $\zeta_k(\lambda,a)$ respectively. These functions are analytic in λ when λ is finite positive and in a when a is finite real. Moreover, the method of zero-free regions shows that for λ and a fixed, the line segment $y = \beta$, $x \leqq 0$, can contain at most one of the points $z_j(\lambda,a)$ and $\zeta_k(\lambda,a)$. Thus if we arrange these points after increasing ordinates for a fixed $\lambda > 0$ and a fixed real a, and choose the enumeration accordingly, this order will remain unchanged when λ and a vary in the manner indicated. We choose

$$z_j\left(\frac{\sqrt{\pi}}{2}, 1\right) = z_j, \quad \zeta_k\left(\frac{\sqrt{\pi}}{2}, 1\right) = \zeta_k.$$

Then we let $a \to 0$ and afterwards $\lambda \to 0$. The resulting solution $w(z, 0, 0)$ is easily shown to have all its zeros and extrema on the lines $y = \pm x$. The transformation $z = e^{\frac{3\pi i}{4}} t$ carries $y = -x$ over into the real axis of the t-plane, (P_0) becomes

$$\frac{d^2 w}{dt^2} + t^2 w = 0,$$

and $w(z, 0, 0)$ becomes $e^{\frac{3\pi i}{4}} \eta(t)$ where $\eta(t)$ is real when t is real. But then it follows from well known oscillation theorems that the zeros and extrema of $\eta(t)$ alternate on the real axis. This being the case, we conclude that the zeros and the extrema of $w(z, 0, 0)$ alternate on the line $y = -x$. Consequently we have

$$0 < \mathfrak{I}[\zeta_1(0, 0)] < \mathfrak{I}[z_1(0,0)] < \cdots < \mathfrak{I}[\zeta_n(0, 0)] < \mathfrak{I}[z_n(0, 0)] < \cdots$$

whence it follows that also

$$0 < \mathfrak{I}(\zeta_1) < \mathfrak{I}(z_1) < \cdots < \mathfrak{I}(\zeta_n) < \mathfrak{I}(z_n) < \cdots$$

This completes the proof.

7. *Conformal mapping.* Let $w_1(z)$ and $w_2(z)$ be two linearly independent solutions of (E) and consider the quotient

$$\lambda(z) = \frac{w_2(z)}{w_1(z)}.$$

Suppose that $\lambda = \lambda_1$ when $z = z_1$. This implies that $z = z_1$ is a zero of the solution $\lambda_1 w_1(z) - w_2(z)$ of (E). If we know that $\lambda = \lambda(z)$ maps a certain region R in the z-plane conformally upon a region P in the λ-plane, we know also that to every value of λ in P the solution $\lambda w_1(z) - w_2(z)$ will have one and only one zero in R. Thus the problem of conformal mapping through the quotient of two solutions of (E) and the zero-point problem for the solutions are merely different aspects of the same question. To illustrate this mode of attacking the zero point problem we again use the differential equation (P).

We put $w(z,\lambda, 1) = w(z,\lambda)$, i. e.

$$w(z,\lambda) = \lambda e^{\frac{z^2}{2}} - e^{\frac{z^2}{2}} \int_0^z e^{-t^2} dt,$$

and

$$\lambda(z) = \int_0^z e^{-t^2} dt = k(z).$$

Thus we are led to a study of the conformal mapping effected by the function $\lambda = k(z)$. Such a study has already been carried out by *F. Iversen**) to the paper of which we refer for further detail.

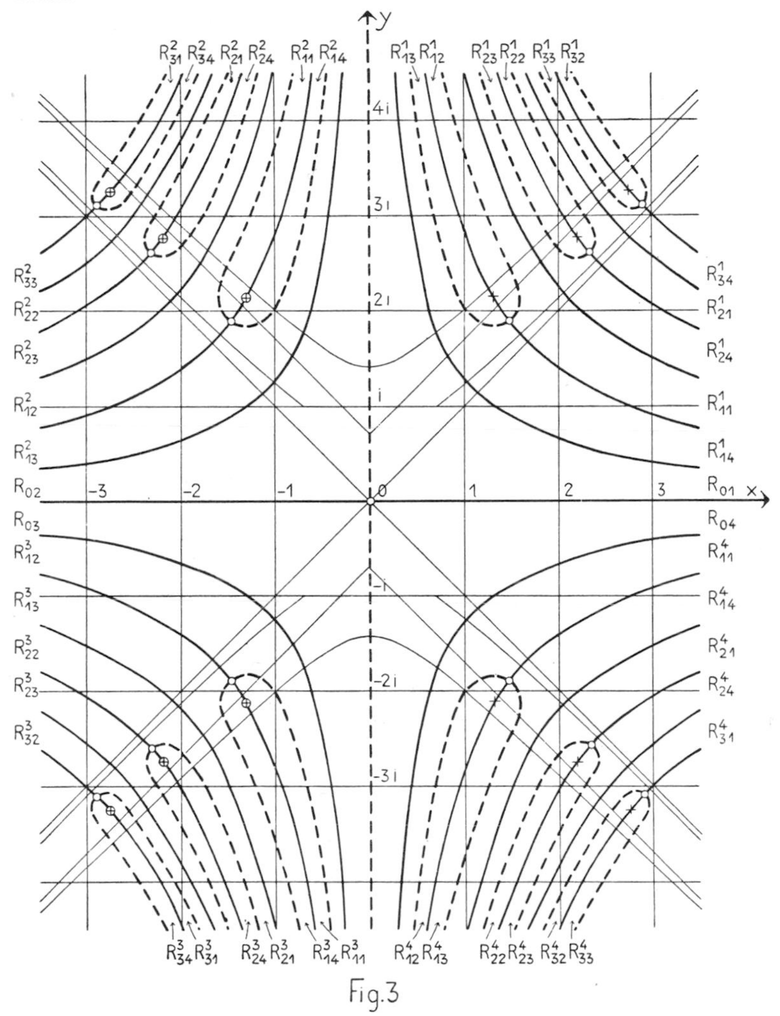

Fig. 3

We set $\lambda = u + iv$ and plot the curves $u = 0$, $v = 0$ in the z-plane. Fig. 3 gives a schematic representation of these curves, dotted lines indicate the first and lines drawn in full the second kind.

---

*) Sur une fonction entière dont la fonction inverse présente un ensemble de singularités de la puissance du continue, Öfversigt af Finska Vet.-Soc. Förhandlingar, Vol. 58, Sect. A, No. 3 (1915) 17 pp.

Iversen proves (l. c. p. 3) that there is one and only one curve $v=0$ between two consecutive hyperbolas of the family $xy=\pi(\frac{n}{2}+\frac{1}{4})$, $n=0,1,2,\cdots$ It is possible to find rather accurately where these curves intersect the line $y=x$. We have

$$k(re^{\frac{\pi i}{4}})=\frac{\sqrt{2}}{2}[c(r)+s(r)]+i\frac{\sqrt{2}}{2}[c(r)-s(r)],$$

where $c(r)$ and $s(r)$ are the integrals of *Fresnel,* namely

$$c(r)=\int_0^r \cos t^2\,dt,\quad s(r)=\int_0^r \sin t^2\,dt.$$

The points of intersection of $v=0$ with $y=x$ correspond to the real positive roots of the equation

$$c(r)-s(r)=0.$$

It can be shown that all the roots of this equation are of the form r_n, $-r_n$, ir_n and $-ir_n$ where $r_n>0$. The positive roots can be found approximately with the aid of the existing tables of Fresnel's integrals. The first six roots are*)

r_n	1.58	2.33	2.94	3.44	3.87	4.25
$k(r_n e^{\frac{\pi i}{4}})$	1.17	0.66	1.04	0.74	1.01	0.77

The error in r_n amounts to a couple of units in the last digit. For large values of n, $r_n \backsim \sqrt{(n-\frac{1}{4})\pi}$ and $k(r_n e^{\frac{\pi i}{4}}) \backsim \frac{\sqrt{\pi}}{2}$.

As $c(r)$ and $s(r)$ are both positive when $r>0$, it follows that the curves $u=0$ do not intersect the line $y=x$ (with the trivial exception of one curve, namely the imaginary axis), and consequently they have to stay above this line when $x,y>0$. We know that the intersections between the curves $u=0$ and $v=0$ must occur in the strip s_1 symmetric to the strip s_2 of Fig. 2.**) Thus we find by inspection that the three zeros of $k(z)$ in s_1 which are nearest to the origin are located approximately at the points

$$1.5+1.9i,\ 2.3+2.6i \text{ and } 2.9+3.1i,$$

with a probable error of a unit in the last digit of each coordinate.

---

*) These values are calculated by linear interpolation from the table on p. 26 in *Jahnke* & *Emde:* Funktionentafeln mit Formeln und Kurven, B. G. Teubner, Leipzig, 1907. The function $C(u)$ which occurs in this table is connected with our function $c(r)$ by the formula

$$C(u)=\sqrt{\frac{2}{\pi}}\,c\left(\sqrt{\frac{\pi}{2}}\,u\right).$$

**) In Fig. 3 the boundary lines of S_ν are marked, not those of s_ν.

The crosses in the right half-plane of Fig. 3 mark the approximate location of the points where $k(z) = -\frac{\sqrt{\pi}}{2}$, *i.e.* $K(-z) = 0$; similarly the ringed crosses in the left half-plane indicate where $k(z) = +\frac{\sqrt{\pi}}{2}$, *i.e.* $K(z) = 0$.

The curves $u = 0$ and $v = 0$ divide the z-plane into an infinite number of regions R_{on} and $R^{\nu}_{m,n}$, $n,\nu = 1, 2, 3, 4$; $m = 1, 2, 3, \cdots$ Each of these regions corresponds to a quadrant, namely the nth one, on the Riemann surface over the λ-plane. Four regions, R_{on} or $R^{\nu}_{m,n}$ where $n = 1, 2, 3$ and 4, which meet at a zero of $k(z)$ correspond to a full plane with a cut along a part of the real axis. In the case of R_{on} we need two cuts running from $+\frac{\sqrt{\pi}}{2}$ to $+\infty$ and from $-\frac{\sqrt{\pi}}{2}$ to $-\infty$; in the case of $R^{\nu}_{m,n}$ we need only the former cut when $\nu = 1$ or 4, but only the latter cut when $\nu = 2$ or 3. It is easy to see how these different cut planes are connected on the surface, but we refer the reader to Iversen's paper quoted above for this detail.

Once in possession of the conformal map effected by $\lambda = k(z)$ we can draw conclusions of interest to the zero point problem. If *e.g.* we give λ a fixed value in the first quadrant we conclude that the solution $w(z,\lambda)$ of (P) has one and only one zero in each of the regions R_{o1} and $R^{\nu}_{m,1}$, $\nu = 1, 2, 3, 4$ and $m = 1, 2, 3, \cdots$

One final remark of a slighly different nature! We have seen how the values $+\frac{\sqrt{\pi}}{2}$ and $-\frac{\sqrt{\pi}}{2}$ intervene in the preceding discussion. First, these together with ∞ are the asymptotic values of $k(z)$; $k(z) \to +\frac{\sqrt{\pi}}{2}$ in the sector $|\arg z| \leqq \frac{\pi}{4}$, to $-\frac{\sqrt{\pi}}{2}$ in the sector $|\arg(-z)| \leqq \frac{\pi}{4}$ and to ∞ in the remaining sectors. Consequently these three values are the only transcendental singular points of the inverse of $k(z)$. Secondly, the three values are the exceptional values in the sense of *R. Nevanlinna* of the entire function $k(z)$. Infinity is of course a *Picard* value of weight unity, the other two values have the weight one half.

In fact, the number of points where $k(z)$ equals $+\frac{\sqrt{\pi}}{2}\left(\text{or}-\frac{\sqrt{\pi}}{2}\right)$ inside the circle $|z|=\sqrt{2n\pi}$ is $2n+o(1)$, whereas the number of points where it equals any other fixed finite value is $4n+O(1)$. Thirdly, these three values of λ correspond to the three truncated solutions of the differential equation.

We have a similar situation if we consider the more general differential equation

$$w''+P_n(z)w=0,$$

where $P_n(z)$ is an arbitrary polynomial of degree n. Let $w_1(z)$ and $w_2(z)$ be two linearly independent solutions of this equation and consider

$$w(z,\lambda)=\lambda\, w_1(z)-w_2(z) \text{ and } \lambda(z)=\frac{w_2(z)}{w_1(z)}.$$

There exist in general $(n+2)$ values of λ with the following properties: (i) they form a complete set of asymptotic values of the meromorphic function $\lambda(z)$; (ii) they are the exceptional values of $\lambda(z)$ each having the same weight $\frac{2}{n+2}$; and (iii) they correspond to the truncated solutions of the differential equation. We see that the particular case which we have treated above differs in several respects from this normal situation; it is also clear why it does so.

Addendum.

Professor *N. E. Nørlund* has communicated to me the following observation concerning the curves $\Im[k(z)]=\Im(K(z)]=0$ which I publish with his kind permission. In the integral which defines $K(z)$ we use an arc of the hyperbola $\sigma\tau=xy$ $(t=\sigma+\tau i)$ as path of integration and obtain consequently the following expression for the imaginary part of $K(z)$, namely

$$v(x,y)=\int_x^{+\infty} e^{\left(\frac{xy}{\sigma}\right)^2-\sigma^2}\left[\sin 2xy-xy\frac{\cos 2xy}{\sigma^2}\right]d\sigma.\ (x>0)$$

In order that $v(x,y)$ shall be zero, it is necessary that the expression inside the bracket changes its sign at least once on the path of integration; obviously it cannot change sign more than once, namely at $\sigma=\sqrt{xy\cot 2xy}$. If we assume $xy>0$, this implies that $\cot 2xy$ must be positive along the curve $v(x,y)=0$

which again implies that $n\frac{\pi}{2} < xy < (n+\frac{1}{2})\frac{\pi}{2}$, $n = 1, 2, \cdots$ These limits are somewhat narrower than those found by Iversen. Moreover, they are the best possible limits. In fact, every hyperbola $xy = \alpha$ where $n\frac{\pi}{2} < \alpha < (n+\frac{1}{2})\frac{\pi}{2}$ intersects an arc of $v = 0$ once and only once. If we denote the x-coordinate of the point of intersection by x_α then $x_\alpha < \sqrt{\alpha \cot 2\alpha}$ and $x_\alpha \to 0$ when $\alpha \to (n+\frac{1}{2})\frac{\pi}{2}$ and $+\infty$ when $\alpha \to n\frac{\pi}{2}$, i. e. the arc in question lies entirely between the two hyperbolas $xy = n\frac{\pi}{2}$ and $xy = (n+\frac{1}{2})\frac{\pi}{2}$ and is tangent to these hyperbolas at infinity. A similar discussion can be given for the curves $\mathfrak{R}[K(z)] = 0$.

Chapter 3
Dirichlet Series

Papers included in this chapter are

A note on Hille's work on Dirichlet series can be found starting on page 662.

SOME REMARKS ON DIRICHLET'S SERIES

By EINAR HILLE.

[Received 3 May, 1925.—Read 14 May, 1925.]

[*Extracted from the Proceedings of the London Mathematical Society, Ser.* 2, *Vol.* 25, *Part* 3.]

1. The main purpose of the present paper is to derive a set of sufficient conditions that the sum of a Dirichlet's series shall be an entire function.

A necessary and sufficient condition for this can be read off from the analytic representation in the principal star given by M. Riesz.* But this test is rather laborious to handle, and supplementary results along different lines seem desirable. A sufficient condition that an ordinary Dirichlet's series shall represent an entire function was obtained by Hardy in connexion with his study of the application of Borel's method of summation to Dirichlet's series.† The present paper contains similar conditions for general Dirichlet's series obtained from the Cahen-Perron integral.

2. Let the Dirichlet's series

$$D(s;\ \lambda) = \sum_{n=1}^{\infty} a_n e^{-\lambda_n s} \quad (0 < \lambda_n < \lambda_{n+1}, \quad \lambda_n \to \infty) \tag{1}$$

be convergent for $\sigma > \sigma_0$ $(s = \sigma + it)$ where $0 \leqslant \sigma_0 < \infty$. Our starting point is the formula‡

$$D(s;\ \lambda) = \frac{1}{\Gamma(s/\nu)} \int_0^{\infty} x^{(s/\nu)-1} D(x;\ l^{\nu})\, dx \tag{2}$$

where $\nu > 0$ and

$$D(u;\ l^{\nu}) = \sum_{n=1}^{\infty} a_n e^{-l_n^{\nu} u} = \sum_{n=1}^{\infty} a_n e^{-e^{\nu\lambda_n} u}. \tag{3}$$

---

* *Acta Math.*, 35 (1911), 253–270 (266).

† *Proc. London Math. Soc.* (2), 8 (1910), 277–294 (290). See also Fekete, "Sur les séries de Dirichlet", *Comptes rendus*, 25 April, 1910.

‡ Perron, *Journal für Math.*, 134 (1908), 95–143 (138). The case $\nu = 1$ was considered by Cahen in his thesis. In the following the right-hand side of formula (2) will be referred to as the *Cahen-Perron integral.*

The representation is valid at least for $\sigma > \sigma_0$. The series (3) converges at least for $x > 0$, $u = x+iy$, and the function $D(u\,;\, l^\nu)$ is of finite order on the imaginary axis. In fact, Perron found that §

$$|D(x\,;\, l)| < 3C\,\Gamma(\alpha+1)\left(\frac{2}{x}\right)^{\alpha} e^{-\frac{1}{2}l_1 x}$$

where $\alpha = \sigma_0+\epsilon$ and

$$C = \text{Max}\,|C_n|, \quad C_n = \sum_{k=1}^{n} a_k\, e^{-\lambda_k \alpha}.$$

By exactly the same method as was used by Perron we find that

$$|D(u\,;\, l^\nu)| < C\Gamma(\tau+1)\left(1+2\frac{|u|}{x}\right)\left(\frac{2}{x}\right)^{\tau} e^{-\frac{1}{2}l_1^\nu x}, \tag{4}$$

where $\tau = \alpha/\nu$.

3. Let r denote a positive quantity, and write

$$\int_0^\infty x^{(s/\nu)-1} D(x\,;\, l^\nu)\, dx = \int_0^r + \int_r^\infty = F_\nu(s\,;\, r) + E_\nu(s\,;\, r). \tag{5}$$

The second integral converges uniformly as long as $|s|$ remains bounded, since

$$|E_\nu(s\,;\, r)| \leqslant |E_\nu(\sigma\,;\, r)| < 3C\Gamma(\tau+1)\, 2^\tau \int_r^\infty x^{(\sigma/\nu)-\tau-1} e^{-\frac{1}{2}l_1^\nu x}\, dx. \tag{6}$$

For $\sigma \geqslant \alpha+(\nu/4)$, say, the integral is less than

$$\Gamma\left(\frac{\sigma}{\nu}-\tau\right)\left(\frac{4}{l_1^\nu}\right)^{(\sigma/\nu)-\tau} e^{-\frac{1}{4}l_1^\nu r},$$

in view of Perron's lemma. For $\sigma < \alpha+(\nu/4)$, Schwarz's inequality shows that the integral is less than

$$\left[2\left(\frac{\sigma}{\nu}-\tau\right)-1\right]^{-\frac{1}{2}} r^{(\sigma/\nu)-\tau-\frac{1}{2}}\, l^{-(\nu/2)}\, e^{-\frac{1}{2}l_1^\nu r},$$

---

* *Loc. cit.*, 136, formula (56), which, however, contains an additional term $Cl^\alpha \exp(-l_1 x)$. This expression is an upper limit for the term $-C_N l_{N+1}^\alpha \exp(-l_{N+1} u)$, obtained by a partial summation. For $N=0$, which is the actual case, this term drops out since $C_0 = 0$ by definition.

which two inequalities together prove the statement. Hence $E_\nu(s\,;\,r)$ is an entire function of s.

Formula (4) shows that the first integral in (5) converges for $\sigma > \alpha$, and that the function $u^{-1} D(u\,;\,l^\nu)$, which is holomorphic in $|u-r| < r$, is of finite order, $\omega+1$ say, on the bounding circle. Hence $F_\nu(s\,;\,r)$ can be expanded in a factorial series in s/ν convergent for $\sigma > \nu\omega$*, namely

$$(7) \qquad F_\nu(s\,;\,r) = r^{s/\nu} \sum_{n=0}^{\infty} (-1)^n \frac{r^n D^{(n)}(r\,;\,l^\nu)}{s/\nu\{(s/\nu)+1\}\dots\{(s/\nu)+n\}}.$$

Thus we obtain the formula†

$$(8) \quad D(s\,;\,\lambda) = \frac{1}{\Gamma(s/\nu)} \int_r^\infty x^{(s/\nu)-1} D(x\,;\,l^\nu)\,dx + r^{(s/\nu)} \sum_{n=0}^{\infty} (-1)^n \frac{r^n D^{(n)}(r\,;\,l^\nu)}{\Gamma\{(s/\nu)+n+1\}},$$

valid for $\sigma > \nu\omega$.

4. In order to obtain an upper limit for ω we shall use a theorem due to Nörlund,‡ namely:

Let $f(z)$ be holomorphic in the interior of the unit circle on which (α, β) is an arc containing $z = 1$. Suppose a number k greater than or equal to 1 exists, such that

$$|f(z)|\,(1-|z|)^{k-1}\;\;|1-z|$$

remains less than a fixed constant in the region

$$0 \leqslant |z| < 1, \quad \alpha \leqslant \arg z \leqslant \beta.$$

Then the order of $f(z)$ on the arc (α, β) is less than or equal to k.

We shall apply this theorem to the function $u^{-1} D(u\,;\,l^\nu)$. We shall try to determine a k, such that

$$|D(u\,;\,l^\nu)|\,[r-|u-r|]^{k-1}$$

---

* Provided $\omega > 0$. See Nörlund, "Sur les séries de facultés", *Acta Math.*, 37 (1914), 327–387 (351).

† For ordinary binomial series a similar representation has been given by S. Wigert, *Arkiv för Mat.*, 7 (26) (1911), 1–12 (7). The reading of this paper gave me the idea for the present investigation in 1917. Formula (8) for $\nu = 1$, with the restrictive assumption that the series (1) should possess a half-plane of absolute convergence, was obtained at that time. The theorems **A–C** below in §§ 6–7 have obvious analogues for binomial series.

‡ *Loc. cit.*, 337, Theorem H.

remains bounded for $|u-r|<r$. Then $\omega+1 \leqslant k$, provided $k \geqslant 1$. In view of (4) we can determine a positive M_ν, independent of u, but not of ν, such that

$$|D(u;\ l^\nu)| < M_\nu |u| x^{-\tau-1} \quad (x>0).$$

If $|u-r|<r$, $|u|^2<2rx$; thus the inequality can be replaced by

$$(9) \qquad |D(u;\ l^\nu)| < M_\nu \sqrt{(2r)}\, x^{-\tau-\frac{1}{2}} = M_\nu \sqrt{(2r)}(r+\rho\cos\theta)^{-\tau-\frac{1}{2}}$$

on the circle $|u-r| = \rho < r$. Hence

$$|D(u;\ l^\nu)|(r-\rho)^{k-1} < M_\nu \sqrt{(2r)}(r+\rho\cos\theta)^{-\tau-\frac{1}{2}}(r-\rho)^{k-1}$$
$$\leqslant M_\nu \sqrt{(2r)}(r-\rho)^{k-\tau-\frac{3}{2}}.$$

Consequently, we can take $k = \tau+\frac{3}{2}$. Hence

$$\omega \leqslant \tau+\tfrac{1}{2} = \frac{\sigma_0+\epsilon}{\nu}+\tfrac{1}{2}$$

for every positive ϵ. We conclude that the expansion (8) holds for $\sigma>\sigma_0+(\nu/2)$. Thus, by taking ν sufficiently small, we obtain an expansion holding for every value of s in the half-plane $\sigma \geqslant \sigma_0+\delta$, δ fixed but arbitrarily small.

The factorial series (7) is absolutely convergent for $\sigma>\sigma_0+\nu$. This follows from a direct estimate of the growth of $|r^n D^{(n)}(r;\ l^\nu)|$ with n, which we shall give, assuming $r=1$, $\tau \geqslant \frac{1}{2}$; neither of these assumptions restricts the generality. We have, using (9),

$$|D^{(n)}(1;\ l^\nu)| = \left|\frac{n!}{2\pi i}\int_{|u-1|=\rho}\frac{D(u;\ l^\nu)}{(u-1)^{n+1}}\,du\right| \leqslant \frac{n!\ \sqrt{2}\,M_\nu}{2\pi\,\rho^n}\int_0^{2\pi}\frac{d\theta}{(1+\rho\cos\theta)^{\tau+\frac{1}{2}}}$$
$$\leqslant \frac{n!}{2\pi}\sqrt{2}\,M_\nu\,\rho^{-n}(1-\rho)^{-\tau+\frac{1}{2}}\int_0^{2\pi}\frac{d\theta}{1+\rho\cos\theta}$$
$$= n!\ \sqrt{2}\,M_\nu\,\rho^{-n}(1+\rho)^{-\frac{1}{2}}(1-\rho)^{-\tau} < n!\ \sqrt{2}\,M_\nu\,\rho^{-n}(1-\rho)^{-\tau}.$$

Putting $\rho = 1-(1/n)$, $n \geqslant 2$, we obtain finally

$$(10) \qquad |D^{(n)}(1;\ l^\nu)| < \sqrt{2}\,M_\nu\left(1-\frac{1}{n}\right)^{-n} n^\tau\, n! < 4\sqrt{2}\,M_\nu\, n^\tau\, n!.$$

Since

$$\frac{n^\tau\, n!}{|\Gamma\{(s/\nu)+n+1\}|} < \text{const}\, n^{\tau-(\sigma/\nu)},$$

the series (7) converges absolutely for $\sigma>\sigma_0+\nu+\epsilon$, $\epsilon>0$.

5. The assumption that the series (1) possesses a half-plane of convergence can be replaced by weaker assumptions, *e.g.* by :—

(i) The series (3) converges for $x > x_0 \geqslant 0$.

(ii) $D(u\,;\, l^\nu)$ is holomorphic on the positive real axis.

(iii) There exists an $r > 0$ such that $D(u\,;\, l^\nu)$ is holomorphic in $|u-r| < r$, and $u^{-1} D(u\,;\, l^\nu)$ is of finite order k on the bounding circle.

If these conditions are fulfilled for some positive ν, the divergent Dirichlet's series for $D(s\,;\,\lambda)$ is summable by the Cahen-Perron integral for $\sigma > \nu(k-1)$, at least, and the representation (8) is valid in the same half-plane. In the following we shall assume the Dirichlet's series, whether convergent or divergent, to be given by the Cahen-Perron integral, and the corresponding function $D(u\,;\, l^\nu)$ shall satisfy the conditions (i-iii).

It is obvious that if an $r = r_0$ exists for which condition (iii) is satisfied, then any positive $r < r_0$ will do just as well. It would be interesting to know whether the existence of a particular $\nu = \nu_0$, for which all the conditions are satisfied, implies the existence of other such ν values. The values $\nu > \nu_0$ seem promising.

6. Formula (8) is not of much interest if it does not yield an analytic continuation of $D(s\,;\,\lambda)$. The first question which arises is whether the series (7) can be convergent for all values of s in special cases. This will evidently happen if $D(u\,;\, l^\nu)$ is holomorphic at the origin.* In that event we can choose r so small that $D(u\,;\, l^\nu)$ is holomorphic in $|u-r| \leqslant R$ where $R > r$. Then

$$r^n |D^{(n)}(r\,;\, l^\nu)| < n!\left(\frac{r}{R}\right)^n,$$

and the series
$$\sum_{n=0}^{\infty} \frac{\Gamma(n+1)}{\Gamma\{(s/\nu)+n+1\}} \left(\frac{r}{R}\right)^n$$

is convergent for all values of s. Thus we obtain the theorem :—

A. *If $D(u\,;\, l^\nu)$ is holomorphic at $u = 0$ for some positive ν, then $D(s\,;\,\lambda)$*

---

* Cf. Pincherle, "Sur les fonctions déterminantes", *Annales Sci. de l'École Normale* (3), 22 (1905), 9-68 (50), where further references are given. See also Hurwitz: "Ueber die Anwendung eines functionentheoretischen Principes auf gewisse bestimmte Integrale", *Math. Annalen*, 53 (1900), 220–224. This paper should be consulted in connexion with Theorem **C** as well.

is an entire function which is represented by formula (8) *for all values of* s.

For $\lambda_n = \log n$, $\nu = 1$, this theorem gives Hardy's result quoted in the introduction. But the condition of Theorem **A** is far from being necessary in order that the factorial series (7) shall be convergent for all values of s. The abscissa of convergence of the series, if negative, is given by

$$(11)\qquad \sigma' = \overline{\lim}\, \frac{\log|R_{n+1}|}{\log n}, \qquad R_n = \sum_{m=n}^{\infty} (-1)^m \frac{D^{(m)}(r;\, l^\nu)}{m!} r^m.$$

In order that formula (8) shall be valid for all values of s we must have $\sigma' = -\infty$, which requires

$$(12)\qquad |R_n| = n^{-a_n} \text{ where } a_n \to +\infty \text{ with } n \quad (n \geqslant 2).$$

It is of some interest to formulate this condition differently.

B₁. *A necessary and sufficient condition in order that the formula* (8) *shall be valid for all values of s is that the series*

$$(13)\qquad \sum_{n=k}^{\infty} \frac{D^{(n)}(r;\, l^\nu)}{(n-k)!} (u-r)^{n-k} = D^{(k)}(u;\, l^\nu) \textit{ for } |u-r| < r$$

shall be convergent for $u = 0$, *for* $k = 0, 1, 2, \ldots$.

In fact, if $\sigma' = -\infty$ condition (12) shows that for $n > 2$

$$\left| \frac{D^{(n)}(r;\, l^\nu)}{n!} (-r)^n \right| \leqslant n^{-a_n} + (n-1)^{-a_{n-1}}$$

and

$$\frac{|D^{(n)}(r;\, l^\nu)|}{(n-k)!} r^{n-k} \leqslant \frac{n!}{(n-k)!} [n^{-a_n} + (n-1)^{-a_{n-1}}] r^{-k} < C_k (n-1)^{-a_{n-1}+k} r^{-k},$$

if we assume that $a_{n-1} \leqslant a_n$. Since $a_{n-1} > k+2$ from a certain point on, the series (13) is convergent for $u = 0$ and for every k. Conversely, if all the series (13) are convergent for $u = 0$ we have, for a fixed k, arbitrarily large

$$\frac{|D^{(n)}(r;\, l^\nu)|}{(n-k)!} r^{n-k} < 1,$$

if $n > n(k)$. Hence

$$\frac{|D^{(n)}(r;\, l^\nu)|}{n!} r^n < \frac{(n-k)!}{n!} r^k,$$

and
$$|R_m| < r^k \sum_{n=m}^{\infty} \frac{(n-k)!}{n!} < r^k c_k \sum_{n=m}^{\infty} \frac{1}{n^k} < \frac{2 r^k c_k}{(k-1) m^{k-1}},$$

if $m > m_0$. Consequently

$$\overline{\lim} \frac{\log |R_{m+1}|}{\log m} < -k+1.$$

But k is arbitrarily large, hence $\sigma' = -\infty$.

Using a terminology due to Hardy and Littlewood,* we can give the following paraphrase of Theorem $\mathbf{B}_1$, namely :—

$\mathbf{B}_2$. *A necessary and sufficient condition in order that formula* (8) *shall be valid for all values of s is that the series*

$$(14) \qquad \sum_{n=1}^{\infty} a_n l_n^{\nu k} \quad (k = 0, 1, 2, \ldots),$$

shall be summable by the circle method of type l^ν with radius r.

7. The results of § 6 show that the existence of derivatives of all orders of $D(u\,; l^\nu)$ at the origin is of importance for our problem. As a matter of fact, we have the following more general theorem :—

C. *A sufficient condition in order that the sum of $D(s\,;\lambda)$ shall be an entire function is that $D^{(n)}(x\,;\, l^\nu)$ shall tend to a finite limit when $x \to 0$ through positive values for $n = 0, 1, 2, \ldots$.*

The proof follows directly from results due to Pincherle.† Suppose that

$$(15) \qquad \lim_{x \to +0} D^{(n)}(x\,;\, l^\nu) = c_n \quad (n = 0, 1, 2, \ldots).$$

Then for $0 \leqslant x \leqslant r$ we have

$$(16) \qquad D(x\,;\, l^\nu) = \sum_{n=0}^{k} c_n x^n + \frac{x^{k+1}}{k\,!} \int_0^1 D^{(k+1)}(\theta x\,;\, l^\nu)(1-\theta)^k \, d\theta.$$

Assuming temporarily $\sigma > 0$, and substituting in the integral for $F_\nu(s\,;\, r)$, we obtain

(17)

$$F_\nu(s\,;\, r) = r^{s/\nu} \sum_{n=0}^{k} \frac{c_n r^n}{(s/\nu)+n} + \frac{1}{k\,!} \int_0^r x^{(s/\nu)+k} \left\{ \int_0^1 D^{(k+1)}(\theta x\,;\, l^\nu)(1-\theta)^k \, d\theta \right\} dx.$$

---

* *Rend. di Palermo*, 41 (1916), 36–53 (50). The notion of *type* in this connexion is not emphasized by these writers, who had no reason to do so in their investigation.

† *Loc. cit.*, 25–26.

The integral obviously converges for $\sigma > -(k+1)\nu$. In view of the factor $1/\Gamma(s/\nu)$ in the Cahen-Perron integral we conclude that $D(s;\lambda)$ is holomorphic in the half-plane $\sigma > -(k+1)\nu$. The integer k being arbitrarily large, it follows that $D(s;\lambda)$ is an entire function.

The same conclusion is evidently valid, if the condition (15) is fulfilled only for $n = 0, 1, 2, \ldots, k_\nu+1$, where $k_\nu \geqslant k$ for $\nu \geqslant \nu_0$.

If the condition (15) is fulfilled for all values of n, we can transform expression (5) by means of repeated integrations by parts operating on the integral for $F_\nu(s;r)$. After k such integrations we obtain the formula

$$(18)\quad D(s;\lambda) = \frac{1}{\Gamma(s/\nu)} \int_r^\infty x^{(s/\nu)-1} D(x;l^\nu)\,dx + r^{s/\nu} \sum_{n=0}^{k} (-1)^n \frac{r^n D^{(n)}(r;l^\nu)}{\Gamma\{(s/\nu)+n+1\}}$$
$$+(-1)^{k+1} \frac{1}{\Gamma\{(s/\nu)+k+1\}} \int_0^r x^{(s/\nu)+k} D^{(k+1)}(x;l^\nu)\,dx,$$

valid for $\sigma > -(k+1)\nu$, k being an arbitrary positive integer.* If it is permitted to modify suitably the path of integration in the remainder term in (18), this formula will yield an asymptotic representation of $D(s;\lambda)$ in a sector $-\alpha \leqslant \arg s \leqslant +\beta$, $\frac{1}{2}\pi \leqslant \alpha, \beta < \pi$.

* Cf. Pincherle, *loc. cit.*, 27.

Printed by C. F. Hodgson & Son, 2 Newton Street, Kingsway W.C. 2.

ON THE ABSOLUTE CONVERGENCE OF DIRICHLET SERIES.[1]

BY H. F. BOHNENBLUST AND EINAR HILLE.

The problem of the absolute convergence for Dirichlet series $\sum a_n e^{-\lambda_n s}$ deals with the relative position of the abscissa σ_a of absolute convergence and the abscissae σ_u of uniform convergence and σ_b of boundedness and regularity. For ordinary Dirichlet series $\sum a_n n^{-s}$ we have $\sigma_u = \sigma_b$.

In order to find an upper bound for the difference $\sigma_a - \sigma_u$, H. Bohr established in a paper published in 1913[2] a connection between the behavior of ordinary Dirichlet series and of power series in an infinite number of variables. Writing n as a product of prime numbers, the Dirichlet series

$$\sum_{n=1}^{\infty} a_n n^{-s} \tag{1}$$

can be written as a power series

$$c + \sum_{i_1=1}^{\infty} c_{i_1} x_{i_1} + \sum_{i_1 \leqq i_2 = 1}^{\infty} c_{i_1 i_2} x_{i_1} x_{i_2} + \cdots \tag{2}$$

in the variables $x_n = p_n^{-s}$, where p_n is the n-th prime number. The coefficient $c_{i_1 i_2 \cdots i_m}$ of (2) is equal to the coefficient a_n of (1), whose index n is equal to $p_{i_1} p_{i_2} \cdots p_{i_m}$. Bohr showed that, though actually functions of a single variable s, the variables $x_n = p_n^{-s}$ behave in many ways as if they were independent of one another. This is due to the linear independence of the quantities $\log p_n$.

The power series (2) will be said to be *bounded* in the domain $(G): |x_n| \leqq G_n$; where the G_n are non negative numbers, if

1° for every integer m, the m-truncated power series

$$c + \sum_{i_1=1}^{m} c_{i_1} x_{i_1} + \sum_{i_1 \leqq i_2 = 1}^{m} c_{i_1 i_2} x_{i_1} x_{i_2} + \cdots \tag{3}$$

obtained from (2) by putting $x_{m+1} = x_{m+2} = \cdots = 0$ is *absolutely convergent* in the domain (G), and

---

[1] Received March 10, 1931.—Presented to the American Mathematical Society, Dec. 31, 1930, (abstract nos. 37-1-91, 95). A short account of the main results of this paper appeared in the Comptes Rendus, t. **192**, pp. 30–32, séance du 5 janvier 1931.

[2] H. Bohr, Über die Bedeutung der Potenzreihen unendlich vieler Variablen in der Theorie der Dirichletschen Reihen $\sum a_n n^{-s}$, Gött. Nachr. (1913), p. 441–488.

2° if there exists an upper bound H, such that for every x_n of (G) and every m, the truncated power series (3) is in absolute value $\leqq H$.[3]

By means of this definition, Bohr was able to establish:[4]

THEOREM A: *Let σ_u be the abscissa of uniform convergence of the Dirichlet series* (1), *then the associated power series* (2) *is bounded in every domain* $|x_n| \leqq G_n = p_n^{-(\sigma_u+\delta)}$, *where δ is any arbitrarily small positive number,* and the converse

THEOREM B: *If the power series* (2) *is bounded in the domain* $|x_n| \leqq G_n = p_n^{-\sigma_0}$, *then* $\sigma_u \leqq \sigma_0$.

The problem of the absolute convergence of (1) is thus reduced to a problem on power series in an infinite number of variables.

If S is the least upper bound of all positive numbers α, such that every power series (2) bounded in (G) is absolutely convergent in (G'): $|x_n| \leqq G'_n = \varepsilon_n G_n$, whenever $0 < \varepsilon_n < 1$ and $\sum \varepsilon_n^\alpha$ converges, then

THEOREM C:[5] *The maximal width of the strip in which a Dirichlet series is uniformly, but non-absolutely convergent is* $= 1/S$.

Bohr showed that $S \geqq 2$, but could not prove the finiteness of S. In the same year, Toeplitz[6] settled this question by showing that $S \leqq 4$. There exist Dirichlet series for which the width $\sigma_a - \sigma_u$ of the strip of uniform but non-absolute convergence is arbitrarily close to $\frac{1}{4}$. His examples were constructed by means of quadratic forms in the x_n; the coefficients of which are essentially the elements of orthogonal (not normalized) matrices, and have the values ± 1.

For general Dirichlet series $\sum a_n e^{-\lambda_n s}$ we have to consider the abscissa σ_b of regularity and boundedness. Hardy, Carlson and Neder proved[7]

$$\sigma_a - \sigma_b \leqq \frac{D}{2},$$

where

$$D = \overline{\lim_{n\to\infty}} \frac{\log n}{\lambda_n}; \tag{4}$$

while Neder showed that this result is the best possible:

---

[3] D. Hilbert, Wesen und Ziele einer Analysis der unendlich vielen unabhängigen Variablen, Palermo Rend., **27**, (1909), p. 59–74.

[4] H. Bohr, loc. cit. Theorems VII and VIII, p. 472 and p. 475.

[5] H. Bohr, loc. cit. Theorem IX, p. 477.

[6] O. Toeplitz, Über eine bei den Dirichletschen Reihen auftretende Aufgabe aus der Theorie der Potenzreihen von unendlich vielen Veränderlichen, Göttinger Nachrichten (1913), p. 417–432.

[7] G. H. Hardy, The application of Abel's method of summation to Dirichlet series, Quarterly Journal, **47** (1916), p. 176–192.—F. Carlson, Sur les séries de Dirichlet, Comptes Rendus, t. **172** (1921), p. 838.—L. Neder, Zum Konvergenzproblem der Dirichletschen Reihen beschränkter Funktionen, Mathematische Zeitschrift, **14** (1922), p. 149–158.

For every $D \geqq 0$, there exist types satisfying (4) for which $\sigma_a - \sigma_b = \frac{D}{2}$. In the case $D = 1$ however, the type considered by Neder is not the type $\lambda_n = \log n$ of ordinary Dirichlet series.

In a different connection Littlewood[8] has considered bounded bilinear forms in an infinite number of variables. He obtained necessary conditions for boundedness and showed that they are the best of their kind.

In this paper we generalize in the two first sections the conditions of Littlewood to cover m-linear forms in an infinite number of variables and in the third section to cover symmetrical linear forms and m-ic forms. Then we apply the result to power series in an infinite number of variables (section 4) and to Dirichlet series (section 5). We shall prove that if σ_u is the abscissa of uniform convergence of the Dirichlet series (1), the series $\sum b_n n^{-s}$, where

$$b_n = \begin{cases} a_n & \text{when } n \text{ does not contain more than } m \text{ prime factors (the same or different),} \\ 0 & \text{otherwise,} \end{cases}$$

has an abscissa of absolute convergence $\leqq \sigma_u + \frac{m-1}{2m}$, and also that these inequalities are the best of their kind. Taking $m = 2$, we see that Toeplitz obtained the largest width $\sigma_a - \sigma_u$ possible for Dirichlet series associated with quadratic forms.

Then combining forms of different degrees, we shall prove in section 6 the main result

THEOREM VII. *For any given σ in the interval $0 \leqq \sigma \leqq \frac{1}{2}$, there exist ordinary Dirichlet series for which the width of the strip of uniform, but non-absolute convergence is exactly equal to σ.*

This theorem cannot be generalized to all types of Dirichlet series. However Bohr[9] has extended his results to cover certain types of Dirichlet series. Accordingly we are able to extend (section 7) our results for certain types and we obtain at the same time new examples proving Neder's result.

Throughout this paper we shall write the power series associated with (1) not in the form (2), where the summations are subject to the conditions $i_1 \leqq i_2 \leqq \cdots \leqq i_m$, but in the symmetrical form

$$c + \sum_{i_1=1}^{\infty} c_{i_1} x_{i_1} + \sum_{i_1 i_2=1}^{\infty} c_{i_1 i_2} x_{i_1} x_{i_2} + \cdots,$$

whose coefficients are the old ones divided by binomial factors.

[8] J. E. Littlewood, On bounded bilinear forms, Quarterly Journal of Mathematics (Oxford series), **1** (1930), p. 164–174.

[9] H. Bohr, Zur Theorie der allgemeinen Dirichletschen Reihen, Mathematische Annalen, **79** (1919), p. 136–156.

1. **Bounded m-linear forms in an infinite number of variables.** Let us consider the m-linear form

$$L(x^{(1)}, x^{(2)}, \cdots, x^{(m)}) \equiv \sum a_{i_1 i_2 \cdots i_m} x_{i_1}^{(1)} x_{i_2}^{(2)} \cdots x_{i_m}^{(m)}, \tag{1.1}$$

where the indices $i_1, i_2, \cdots, i_m$ independently assume all positive integral values. The coefficients $a_{i_1 \cdots i_m}$ and the variables $x_{i_\nu}^{(\nu)}$ are complex. The vector

$$x^{(\nu)} \equiv \{x_1^{(\nu)}, x_2^{(\nu)}, \cdots, x_n^{(\nu)}, \cdots\}$$

is said to belong to the domain (G_0), if for all values of n

$$|x_n^{(\nu)}| \leqq 1.$$

DEFINITION: *An m-linear form $L(x^{(1)}, \cdots, x^{(m)})$ is said to be bounded by H in the domain (G_0) if all the truncated forms*

$$\sum_{i_1=1}^{N_1} \sum_{i_2=1}^{N_2} \cdots \sum_{i_m=1}^{N_m} a_{i_1 i_2 \cdots i_m} x_{i_1}^{(1)} x_{i_2}^{(2)} \cdots x_{i_m}^{(m)}$$

are absolutely $\leqq H$ in (G_0).[10]

We can evidently suppose all the N_ν to be equal, say equal to N. In order to generalize the Littlewood inequalities we first introduce the following notations

$$\varrho = \frac{2m}{m+1},$$

$$S = \left[\sum_{i_1, \cdots, i_m=1}^{\infty} |a_{i_1 \cdots i_m}|^{\varrho}\right]^{\frac{1}{\varrho}}, \qquad T_{i_\nu}^{(\nu)} = [\sum |a_{i_1 \cdots i_m}|^2]^{\frac{1}{2}},$$

where the last sum is to be extended over $i_1, \cdots, i_{\nu-1}, i_{\nu+1}, \cdots, i_m$ from 1 to ∞; and finally

$$T^{(\nu)} = \sum_{i_\nu=1}^{\infty} T_{i_\nu}^{(\nu)}.$$

These values S, $T_{i_\nu}^{(\nu)}$, $T^{(\nu)}$ may of course be equal to ∞.

The first step in establishing the necessary conditions for boundedness is to prove the inequality

$$S^{\varrho} \leqq \sum_{\nu=1}^{m} T^{(\nu)\varrho}, \tag{1.2}$$

which shows in particular that S is finite if all the $T^{(\nu)}$ are finite. If one of the $T^{(\nu)}$ is divergent, there is nothing to prove. We may therefore suppose all the $T^{(\nu)}$ and hence all the $T_{i_\nu}^{(\nu)}$ to be finite. We can also assume for the proof that all the coefficients $a_{i_1 \cdots i_m}$ are real and non-

[10] This definition is equivalent to Hilbert's, since the truncated forms, containing only a finite number of terms, are absolutely convergent.

negative and that for every fixed ν, $T_{i_\nu}^{(\nu)}$ is a non-increasing function of i_ν. Should the last not be the case initially, the following permutation of the indices $i_1, \cdots, i_m$ will ensure it: Consider first the quantities $T_{i_1}^{(1)}$. A permutation of the i_1 interchanges the $T_{i_1}^{(1)}$, but does not alter $T^{(1)}$, nor S, nor any $T_{i_\nu}^{(\nu)}$ with a $\nu \geqq 2$. Since $T_{i_1}^{(1)} \to 0$, when $i_1 \to \infty$, we can rearrange the $T_{i_1}^{(1)}$ in non-increasing order of magnitude. Proceeding similarly for $\nu = 2, \cdots, m$ we obtain the desired order. Hence for all $\nu = 1, 2, \cdots, m$ and all $i_\nu = 1, 2, \cdots$

$$i_\nu T_{i_\nu}^{(\nu)} \leqq T^{(\nu)}. \tag{1.3}$$

Using the notation $\sum_{(i_\nu)}^{(\nu)}$ to indicate that ν and i_ν are fixed and that the summation is to be extended

1° over those values of $i_1, \cdots, i_{\nu-1}$, which do not exceed i_ν, and

2° over those values of $i_{\nu+1}, \cdots, i_m$ which are smaller than i_ν, we have by Hölder's inequality with $p = \dfrac{2}{2-\varrho} = m+1$, $q = \dfrac{2}{\varrho} = \dfrac{m+1}{m}$

$$\sum_{(i_\nu)}^{(\nu)} a_{i_1 \cdots i_m}^{\varrho} \leqq \left(\sum_{(i_\nu)}^{(\nu)} 1\right)^{\frac{1}{m+1}} \cdot \left(\sum_{(i_\nu)}^{(\nu)} a_{i_1 \cdots i_m}^{2}\right)^{\frac{\varrho}{2}}$$
$$\leqq i_\nu^{\frac{m-1}{m+1}} \cdot \left(\sum_{(i_\nu)}^{(\nu)} a_{i_1 \cdots i_m}^{2}\right)^{\frac{\varrho}{2}}.$$

Extending the summation in the last sum over the range $i_1, \cdots, i_{\nu-1}, i_{\nu+1}, \cdots, i_m = 1, 2, \cdots$ increases the right hand side; hence

$$\sum_{(i_\nu)}^{(\nu)} a_{i_1 \cdots i_m}^{\varrho} \leqq i_\nu^{\frac{m-1}{m+1}} \cdot T_{i_\nu}^{(\nu)\frac{\varrho}{2}}$$

or

$$\sum_{(i_\nu)}^{(\nu)} a_{i_1 \cdots i_m}^{\varrho} \leqq \left(i_\nu T_{i_\nu}^{(\nu)}\right)^{\frac{m-1}{m+1}} T_{i_\nu}^{(\nu)}.$$

Applying (1.3) we obtain

$$\sum_{(i_\nu)}^{(\nu)} a_{i_1 \cdots i_m}^{\varrho} \leqq \left(T^{(\nu)}\right)^{\frac{m-1}{m+1}} \cdot T_{i_\nu}^{(\nu)}. \tag{1.4}$$

On the other hand, as there is always a last largest number in a finite sequence of positive integers, every term in the infinite sum S^ϱ appears in one and only one $\sum_{(i_\nu)}^{(\nu)}$; hence

$$S^\varrho = \sum_{\nu=1}^{m}\left(\sum_{i_\nu=1}^{\infty}\left(\sum_{(i_\nu)}^{(\nu)} a_{i_1 \cdots i_m}^{\varrho}\right)\right)$$

and by (1.4)

$$S^\varrho \leqq \sum_{\nu=1}^{m}\left(\sum_{i_\nu=1}^{\infty} T^{(\nu)\frac{m-1}{m+1}} T_{(i_\nu)}^{(\nu)}\right) = \sum_{\nu=1}^{m} T^{(\nu)\varrho},$$

which is the inequality (1.2).

We now proceed to the second step. Supposing that H is an upper bound of the form (1.1), we are going to compare the series $T^{(\nu)}$ with H and prove the existence of a constant A (determined by m), such that

$$T^{(\nu)} \leqq A \cdot H, \tag{1.5}$$

for all forms (1.1). The proof of these inequalities is based upon the following lemma:[11]

Let $\lambda > 0$ and let $M_\lambda(f)$ denote the λ-th root of the mean value of the numbers $|f|^\lambda$ with respect to a set of discrete or continuous variables.

LEMMA 1: *Suppose that $M_2(f) \geqq T$, $M_4(f) \leqq A' T$.* (The A's are absolute constants, while T may depend on the function f). *Then will*

$$A'' \cdot T \leqq M_1(f) \leqq A''' \cdot T.$$

We shall apply this lemma, taking for f the function

$$f \equiv \sum_{i_1, \cdots, i_m = 1}^{n} a_{i_1 \cdots i_m} x_{i_1}^{(1)} \cdots x_{i_m}^{(m)}$$

with complex coefficients $a_{i_1 \cdots i_m}$, of m points $P_\nu(x_1^{(\nu)}, x_2^{(\nu)}, \cdots, x_n^{(\nu)})$. The mean value is the mean value of f, when each point P_ν runs independently through the vertices $(\pm 1, \pm 1, \cdots, \pm 1)$ of the unit cube in the n-dimensional euclidean space.

We first evaluate $M_2(f)$. Evidently

$$\begin{aligned} M_2^2(f) &= \frac{1}{2^{n \cdot m}} \sum_P \left(\sum_{i=1}^{n} a_{i_1 \cdots i_m} x_{i_1}^{(1)} \cdots x_{i_m}^{(m)} \right) \left(\sum_{j=1}^{n} \bar{a}_{j_1 \cdots j_m} x_{j_1}^{(1)} \cdots x_{j_m}^{(m)} \right) \\ &= \frac{1}{2^{n \cdot m}} \sum_{i,j} \left(a_{i_1 \cdots i_m} \bar{a}_{j_1 \cdots j_m} \sum_P x_{i_1}^{(1)} \cdots x_{j_m}^{(m)} \right). \end{aligned}$$

But since

$$\sum_{P_\nu} x_{i_\nu}^{(\nu)} x_{j_\nu}^{(\nu)} = 2^n \cdot \delta_{i_\nu, j_\nu},$$

it follows that

$$\sum_P x_{i_1}^{(1)} \cdots x_{j_m}^{(m)} = 2^{n \cdot m} \delta_{i_1 j_1} \cdots \delta_{i_m j_m},$$

and we obtain

$$M_2^2(f) = \sum_{i_1, \cdots, i_m = 1}^{n} | a_{i_1 \cdots i_m}^2 |,$$

i. e.

$$M_2(f) = \left\{ \sum_{i_1, \cdots, i_m = 1}^{n} | a_{i_1 \cdots i_m}^2 | \right\}^{\frac{1}{2}}. \tag{1.6}$$

[11] J. E. Littlewood, loc. cit. p. 169–170.

We next evaluate $M_4(f)$. We have

$$M_4^4(f) = \frac{1}{2^{n\cdot m}} \sum_{P,i,j,k,l} a_{i_1\cdots i_m} a_{j_1\cdots j_m} \bar{a}_{k_1\cdots k_m} \bar{a}_{l_1\cdots l_m} x_{i_1}^{(1)} \cdots x_{l_m}^{(m)}$$

$$= \frac{1}{2^{n\cdot m}} \sum_{i,j,k,l,} \left\{ a_{i_1\cdots i_m} \cdots \bar{a}_{l_1\cdots l_m} \sum_{P_1} x_{i_1}^{(1)} \cdots x_{l_1}^{(1)} \cdots \sum_{P_m} x_{i_m}^{(m)} \cdots x_{l_m}^{(m)} \right\}.$$

But the sum $\sum_{P_\nu} x_{i_\nu}^{(\nu)} \cdots x_{l_\nu}^{(\nu)}$ is equal to 2^n when either

$$\begin{aligned} &\alpha) \quad i_\nu = j_\nu \text{ and } k_\nu = l_\nu, \\ \text{or } &\beta) \quad i_\nu = k_\nu \text{ and } j_\nu = l_\nu, \\ \text{or } &\gamma) \quad i_\nu = l_\nu \text{ and } j_\nu = k_\nu, \end{aligned} \tag{1.7}$$

and equal to zero otherwise. Hence

$$M_4^4(f) = \sum a_{i_1\cdots i_m} \cdots \bar{a}_{l_1\cdots l_m} \leqq \sum |a_{i_1\cdots i_m}| \cdots |a_{l_1\cdots l_m}|,$$

where these sums are to be extended over all possible combinations (1.7) for all ν's. The combinations (1.7) are not mutually exclusive; but counting each of the overlapping terms separately on each occurence, we increase the right hand side and obtain

$$M_4^4(f) \leqq \sum_{\alpha,\cdots,\alpha,\alpha} + \sum_{\alpha,\cdots,\alpha,\beta} + \cdots + \sum_{\gamma,\cdots,\gamma,\gamma}. \tag{1.8}$$

These sums are to be understood as follows: for the indices $i_\nu, j_\nu, k_\nu, l_\nu$ we take the combination α, β, or γ in (1.7) according to the letter in the ν-th place under the sign Σ. Then we sum over all n^{2m} terms thus obtained. The right hand side of (1.8) contains exactly 3^m sums. Each of them is of the form

$$\sum_{\alpha,\cdots,\alpha,\beta,\cdots,\beta,\gamma,\cdots,\gamma} |a_{i_1\cdots i_m}| \, |a_{j_1\cdots j_m}| \, |a_{k_1\cdots k_m}| \, |a_{l_1\cdots l_m}|, \tag{1.9}$$

where the combination α occurs ω_1, the combination β ω_2 and the combination γ ω_3 times and $\omega_1 + \omega_2 + \omega_3 = m$. Let the indices $i_\alpha (= 1, 2, \cdots, n^{\omega_1})$, $i_\beta (= 1, 2, \cdots, n^{\omega_2})$ and $i_\gamma (= 1, 2, \cdots, n^{\omega_3})$ replace respectively the indices $i_1, i_2, \cdots, i_{\omega_1}$; $i_{\omega_1+1}, \cdots, i_{\omega_1+\omega_2}$; $i_{\omega_1+\omega_2+1}, \cdots, i_m$, then the sum (1.9) can be written in the form

$$\sum_{\substack{i_\alpha, i_\beta, i_\gamma \\ j_\alpha, j_\beta, j_\gamma}} |a_{i_\alpha i_\beta i_\gamma}| \, |a_{i_\alpha j_\beta j_\gamma}| \, |a_{j_\alpha i_\beta j_\gamma}| \, |a_{j_\alpha j_\beta i_\gamma}|$$

$$= \sum_{i_\alpha i_\beta j_\alpha j_\beta} \left(\sum_{i_\gamma} |a_{i_\alpha i_\beta i_\gamma}| \, |a_{j_\alpha j_\beta i_\gamma}| \right) \left(\sum_{i_\gamma} |a_{j_\alpha i_\beta i_\gamma}| \, |a_{i_\alpha j_\beta i_\gamma}| \right)$$

and using Cauchy's inequality

$$\leqq \sum_{i_\alpha i_\beta j_\alpha j_\beta} \left\{ \sum_{i_\gamma} | a^2_{i_\alpha i_\beta i_\gamma} | \right\}^{\frac{1}{2}} \left\{ \sum_{i_\gamma} | a^2_{j_\alpha j_\beta i_\gamma} | \right\}^{\frac{1}{2}} \left\{ \sum_{i_\gamma} | a^2_{j_\alpha i_\beta i_\gamma} | \right\}^{\frac{1}{2}} \left\{ \sum_{i_\gamma} | a^2_{i_\alpha j_\beta i_\gamma} | \right\}^{\frac{1}{2}}$$

$$= \sum_{i_\alpha j_\alpha} \left\{ \sum_{i_\beta} \left\{ \sum_{i_\gamma} | a^2_{i_\alpha i_\beta i_\gamma} | \right\}^{\frac{1}{2}} \left\{ \sum_{i_\gamma} | a^2_{j_\alpha i_\beta i_\gamma} | \right\}^{\frac{1}{2}} \right\}^2 .$$

Applying Cauchy's inequality a second time, we see that (1.4) is

$$\leqq \sum_{i_\alpha j_\alpha} \left(\sum_{i_\beta i_\gamma} | a^2_{i_\alpha i_\beta i_\gamma} | \right) \left(\sum_{i_\beta i_\gamma} | a^2_{j_\alpha i_\beta i_\gamma} | \right)$$

$$= \left\{ \sum_{i_\alpha i_\beta i_\gamma} | a^2_{i_\alpha i_\beta i_\gamma} | \right\}^2 = \left\{ \sum_{i_1 \cdots i_m} | a^2_{i_1 \cdots i_m} | \right\}^2 .$$

Since this result is independent of the choice of the combinations, the same evaluation applies for every term of (1.8). Hence

$$M_4^4(f) \leqq 3^m \left(\sum_{i_1 \cdots i_m} | a^2_{i_1 \cdots i_m} | \right)^2 ,$$

or

$$M_4(f) \leqq 3^{\frac{m}{4}} \left(\sum_{i_1 \cdots i_m} | a^2_{i_1 \cdots i_m} | \right)^{\frac{1}{2}} . \tag{1.10}$$

The integer m being fixed, $3^{\frac{m}{4}}$ is an absolute constant, and the hypotheses of Littlewood's lemma are satisfied with

$$T = \left\{ \sum_{i_1 \cdots i_m} | a^2_{i_1 \cdots i_m} | \right\}^{\frac{1}{2}} .$$

There exist therefore two constants A'' and A''' (depending only on m) such that

$$A'' \cdot (\sum | a |^2)^{\frac{1}{2}} \leqq M_1(f) \leqq A''' (\sum | a |^2)^{\frac{1}{2}} . \tag{1.11}$$

We can now prove (1.5) without difficulty. The maximum of every truncated form

$$\sum_{i_1, \cdots, i_m = 1}^{n} a_{i_1 \cdots i_m} x^{(1)}_{i_1} \cdots x^{(m)}_{i_m}$$

of a form (1.1) in the domain (G_0) is equal to the maximum of

$$\sum_{i_1=1}^{n} \left| \sum_{i_2 \cdots i_m = 1}^{n} a_{i_1 \cdots i_m} x^{(2)}_{i_2} \cdots x^{(m)}_{i_m} \right|$$

in the same domain. This maximum is larger or at least not less than any mean value taken with respect to vectors of the domain (G_0). We can apply (1.11) to every term,[12] and if H is an upper bound of (1.1) we obtain

$$H \geqq A^* \sum_{i_1=1}^{n} \left\{ \sum_{i_2\cdots i_m=1}^{n} | a^2_{i_1\cdots i_m} | \right\}^{\frac{1}{2}}$$

for every n, and therefore

$$H \geqq A^* \sum_{i_1=1}^{\infty} \left\{ \sum_{i_2\cdots i_m=1}^{\infty} | a^2_{i_1\cdots i_m} | \right\}^{\frac{1}{2}} = A^* \cdot T^{(1)},$$

and similarly for all other $T^{(\nu)}$:

(1.5) $$T^{(\nu)} \leqq A \cdot H$$

and using (1.2)

(1.12) $$S \leqq A_1 \cdot H.$$

This proves

THEOREM I. *In order that an m-linear form* (1.1) *be bounded by H in the domain* (G_0), *it is necessary that $T^{(1)}, T^{(2)}, \cdots, T^{(m)}$ and S should all be less than $A \cdot H$, where A is a constant depending only on m.*

2. Theorem I in the first section is the best result of its kind. As in the case $m = 2$, the proof is based upon the construction for every n of m-linear forms

(2.1) $$\sum_{i_1,\cdots,i_m=1}^{n} a_{i_1\cdots i_m} x^{(1)}_{i_1}\cdots x^{(m)}_{i_m},$$

for which

(2.2) $$\left\{ \sum_{i_1,\cdots,i_m=1}^{n} | a_{i_1\cdots i_m} |^{\frac{2m}{m+1}} \right\}^{\frac{m+1}{2m}} \geqq A_m \cdot H_n.$$

The value A_m is determined by m, independently of n; and H_n is the maximum of the absolute value of the form (2.1) in the domain (G_0). We can express these results by means of the series S, $T^{(\nu)}$:

THEOREM II. *Given any positive $t^{(\nu)}_{i_\nu}$ and $s_{i_1\cdots i_m}$, for which*

$$\lim_{i_\nu \to \infty} t^{(\nu)}_{i_\nu} = \lim_{i_1,\cdots,i_m \to \infty} s_{i_1\cdots i_m} = \infty,$$

there exist bounded forms for which

$$\sum t^{(\nu)}_{i_\nu} T^{(\nu)}_{i_\nu} \text{ and } \sum s_{i_1\cdots i_m} | a_{i_1\cdots i_m} |^{\frac{2m}{m+1}}$$

are divergent.

We turn first to the construction of the special forms (2.1). Let $\| a_{rs} \|$ be an n-rowed square matrix satisfying the conditions

[12] We use (1.11) for $(m-1)$ instead of m.

$$(2.3)\qquad \begin{cases} \sum_{t=1}^{n} a_{rt}\,\bar{a}_{st} = n\cdot\delta_{rs} \\ |a_{rs}| = 1. \end{cases}$$

Examples of such matrices have been given by Toeplitz[13] and Littlewood.[14] The simplest is

$$a_{rs} = e^{2\pi i\frac{r\cdot s}{n}} \qquad (r, s = 1, 2, \cdots, n).$$

From any two matrices satisfying (2.3) $\|a_{rs}\|$, $\|b_{pq}\|$ (with possibly not the same number of rows) we can construct a third by "substituting" the second into the first one:

$$\left\| \begin{matrix} a_{11}\,\|b_{pq}\| & \cdots & a_{1n}\,\|b_{pq}\| \\ \cdot & \cdot & \cdot \\ \cdot & \cdot & \cdot \\ a_{n1}\,\|b_{pq}\| & \cdots & a_{nn}\,\|b_{pq}\| \end{matrix} \right\|;$$

the resulting array we regard as a new matrix, whose number of rows will be equal to the product of the number of rows of $\|a_{rs}\|$ and $\|b_{pq}\|$.

Let $\|a_{rs}\|$ be any n-rowed matrix satisfying (2.3) and consider the form

$$(2.4)\qquad \sum_{i_1,\cdots i_m=1}^{n} a_{i_1 i_2}\cdot a_{i_2 i_3}\cdots a_{i_{m-1} i_m}\, x_{i_1}^{(1)}\cdots x_{i_m}^{(m)}.$$

The coefficients of this form are all equal to one in absolute value, hence

$$T^{(\nu)} = S = n^{\frac{m+1}{2}}.$$

We now proceed to evaluate the maximum H of this form in the domain (G_0). For any vectors $x^{(\nu)}$ in (G_0) we have

$$\left| \sum_{i_1,\cdots,i_m=1}^{n} a_{i_1 i_2}\cdots a_{i_{m-1} i_m}\, x_{i_1}^{(1)}\cdots x_{i_m}^{(m)} \right|$$

$$\leqq \sum_{i_m=1}^{n} \left| \sum_{i_1,\cdots,i_{m-1}=1}^{n} a_{i_1 i_2}\cdots a_{i_{m-1} i_m}\, x_{i_1}^{(1)}\cdots x_{i_{m-1}}^{(m-1)} \right|,$$

which by Cauchy's inequality is

$$\leqq n^{\frac{1}{2}} \left\{ \sum_{i_m=1}^{n} \sum_{\substack{i_1,\cdots,i_{m-1} \\ j_1,\cdots,j_{m-1}}=1}^{n} a_{i_1 i_2}\,\bar{a}_{j_1 j_2}\cdots a_{i_{m-1} i_m}\,\bar{a}_{j_{m-1} i_m}\, x_{i_1}^{(1)}\cdots \bar{x}_{j_{m-1}}^{(m-1)} \right\}^{\frac{1}{2}}$$

$$= n^{\frac{1}{2}} \left\{ \sum_{\substack{i_1,\cdots,i_{m-1} \\ j_1,\cdots,j_{m-1}}=1}^{n} a_{i_1 i_2}\cdots \bar{a}_{j_{m-2} j_{m-1}}\, x_{i_1}^{(1)}\cdots \bar{x}_{j_{m-1}}^{(m-1)} \sum_{i_m=1}^{n} a_{i_{m-1} i_m}\,\bar{a}_{j_{m-1} i_m} \right\}^{\frac{1}{2}}$$

---

[13] O. Toeplitz, loc. cit. p. 422–424.

[14] J. E. Littlewood, loc. cit. p. 172.

$$= n \left\{ \sum_{\substack{i_1,\cdots,i_{m-2} \\ j_1,\cdots,j_{m-2}}} \sum_{i_{m-1}} a_{i_1 i_2} \cdots \bar{a}_{j_{m-2} i_{m-1}} x_{i_1}^{(1)} \cdots \bar{x}_{i_{m-1}}^{(m-1)} \right\}^{\frac{1}{2}}$$

$$\leqq n \left\{ \sum_{\substack{i_1,\cdots,i_{m-2} \\ j_1,\cdots,j_{m-2}}} \sum_{i_{m-1}} a_{i_1 i_2} \cdots \bar{a}_{j_{m-2} i_{m-1}} x_{i_1}^{(1)} \cdots \bar{x}_{j_{m-2}}^{(m-2)} \right\}^{\frac{1}{2}},$$

since the coefficients of $|x_{i_{m-1}}^{(m-1)}|^2$ are $\geqq 0$. Repeating this evaluation for the summation over the other indices, we obtain by induction

$$H \leqq n^{\frac{m+1}{2}}$$

and therefore

$$S = T^{(\nu)} \geqq H,$$

that is (2.2). The proof of Theorem II proceeds now on exactly the same lines as Littlewood's proof for the case $m = 2$.

3. Symmetric m-linear forms and m-ic forms. In the application to Dirichlet series, we shall deal with forms

$$Q(x) \equiv \sum a_{i_1 \cdots i_m} x_{i_1} \cdots x_{i_m}$$

of the m-th degree instead of m-linear forms, where the coefficients $a_{i_1 \cdots i_m}$ are supposed to be symmetrical. To every such form we associate the m-linear form

$$L(x^{(1)}, x^{(2)}, \cdots, x^{(m)}) \equiv \sum a_{i_1 \cdots i_m} x_{i_1}^{(1)} \cdots x_{i_m}^{(m)}$$

and conversely, to every symmetrical linear form there corresponds an m-ic form, obtained by putting all $x^{(\nu)} = x$. Denoting by H, resp. $\mathfrak{H}$, the maxima of $Q(x)$, resp. $L(x^{(1)}, \cdots, x^{(m)})$ for the domain (G_0), we obviously have $H \leqq \mathfrak{H}$. On the other hand the identity

$$\begin{aligned} &L(x^{(1)}, \cdots, x^{(m)}) \\ (3.1)\quad &= \frac{1}{2^{m-1} \cdot m!} \sum_{\varepsilon} (-1)^{\varepsilon_2 + \varepsilon_3 + \cdots + \varepsilon_m} Q(x^{(1)} + (-1)^{\varepsilon_2} x^{(2)} + \cdots + (-1)^{\varepsilon_m} x^{(m)}), \end{aligned}$$

where the sum is to be extended over $\varepsilon_\nu = 0, 1$; $\nu = 2, 3, \cdots, m$, shows that

$$\mathfrak{H} \leqq \frac{m^m}{m!} H.$$

Hence:

A symmetrical m-linear form is bounded in the domain (G_0) if and only if the corresponding m-ic form is bounded in (G_0).

The identity (3.1) can be proved as follows. We have

$$\sum_{\varepsilon}(-1)^{\varepsilon_2+\cdots+\varepsilon_m} Q(x^{(1)}+(-1)^{\varepsilon_2} x^{(2)}+\cdots+(-1)^{\varepsilon_m} x^{(m)})$$
$$=\sum_{\varepsilon, i, \nu}(-1)^{\varepsilon_2+\cdots+\varepsilon_m+\varepsilon_{\nu_1}+\cdots+\varepsilon_{\nu_m}} \cdot a_{i_1\cdots i_m} x_{i_1}^{(\nu_1)} \cdots x_{i_m}^{(\nu_m)}$$
$$=\sum_{i, \nu} a_{i_1\cdots i_m} x_{i_1}^{(\nu_1)} \cdots x_{i_m}^{(\nu_m)} \sum_{\varepsilon}(-1)^{\varepsilon_2+\cdots+\varepsilon_m+\varepsilon_{\nu_1}+\cdots+\varepsilon_{\nu_m}}. \text{[15]}$$

Let α_r be the number of ν's equal to r $(r=1,2,\cdots,m)$ for a fixed set of values ν. We have $\alpha_1+\alpha_2+\cdots+\alpha_m=m$, where the α's are non-negative integers. Then

$$(3.2)\qquad \sum_{\varepsilon}(-1)^{\varepsilon_2+\cdots+\varepsilon_{\nu_m}}=\prod_{r=2}^{m}(1+(-1)^{\alpha_r+1})$$
$$=\begin{cases} 2^{m-1} \text{ when all } \alpha_r+1 \text{ are even,} \\ 0 \text{ otherwise.}\end{cases}$$

In the first case, since $\alpha_r \geqq 0$, all $\alpha_r\,(r=2,3,\cdots,m)$ are positive odd integers and therefore

$$\alpha_1+(m-1)+2x=m_1$$

where x is an integer larger than the number of $\alpha_r \geqq 3$. Hence $x=0$, $\alpha_1=\alpha_r=1$ and the sum (3.2) is zero except, when the $\nu_1, \nu_2, \cdots, \nu_m$ are a permutation of $1,2,\cdots,m$, and the proof of (3.1) is completed.

Theorem I therefore holds for m-ic forms; but as the examples we have constructed for Theorem II are not symmetrical, we must refine them in order to show that Theorem I is also the best result of its kind for the symmetrical case. We can still vary the matrix $\|a_{rs}\|$, subject only to conditions (2.3). Let p be a prime number $>m$. Starting from the matrix

$$M_1 \equiv \left\|e^{2\pi i \frac{r \cdot s}{p}}\right\|, \qquad (r, s=1,2,\cdots,p),$$

with p rows, we form successively for $\mu=2,3,\cdots$ the matrices M_μ, which are obtained by "substituting" the matrix $M_{\mu-1}$ into M_1:

$$M_\mu=\left\|e^{2\pi i \frac{r \cdot s}{p}} M_{\mu-1}\right\|.$$

The matrix M_μ contains $p^{2\mu}$ elements, each being a root of the equation $z^p-1=0$.

We put

$$\|a_{rs}\|=M_\mu \qquad (r, s=1,2,\cdots,p^\mu)$$

and form for this matrix, the form (2.5)

$$\sum_{i_1,\cdots,i_m=1}^{p^\mu} a_{i_1 i_2} a_{i_2 i_3} \cdots a_{i_{m-1} i_m} x_{i_1} \cdots x_{i_m}.$$

---

[15] Whenever ε_1 appears, it is put equal to 0, as the coefficient of $x^{(1)}$ is always $+1$.

The coefficients of this form are all roots of $z^p - 1 = 0$. It is not symmetrical, but by adding all the forms obtained by permuting the indices $i_1, i_2, \cdots, i_m$ we obtain a symmetrical one. Every coefficient of it is equal to

$$\sum_{k=0}^{p-1} \lambda_k \cdot \zeta_k,$$

where the ζ_k are the roots of $z^p = 1$, and where the λ_k are non negative integers, whose sum $\sum \lambda_k = m!$. Since p is a prime number, there are no relations of the form[16]

$$\sum_{k=0}^{p-1} \lambda_k \zeta_k = 0,$$

except possibly those in which all λ's are equal, say $= \lambda$. In such a case λ satisfies the equation

$$p \cdot \lambda = m!,$$

which is impossible, since p is a prime number larger than m. Thus

No coefficient of our symmetrical form is zero.

Every λ_k being $\leqq m!$, there exists only a finite number of values $\sum \lambda_k \zeta_k$, thus

There are only a finite number of coefficients different from each other.

There exists therefore an $\eta > 0$, such that all coefficients of the form are $\geqq \eta$ in absolute value. Then

$$S \geqq \eta \cdot n^{\frac{m+1}{2}}$$

and therefore

$$S \geqq A_m H_n, \qquad (n = p^\mu). \tag{3.3}$$

It is readily verified, that for the proof of Theorem II it suffices to have forms satisfying (3.3) for all $n = p^\mu$.

Theorem II *remains true for symmetrical and for m-ic forms.*

The coefficients of the form (3.1) possess a certain symmetry, since the permutation

$$\begin{pmatrix} i_1 & i_2 & \cdots & i_n \\ i_n & i_{n-1} & \cdots & i_1 \end{pmatrix}$$

leaves the coefficients $a_{i_1 \cdots i_m}$ invariant. In order to obtain a symmetrical form it suffices therefore to consider $m!/2$ permutations instead of $m!$. For $m = 3$, this enables us to take $p = 2$ and to construct "best possible" examples with real coefficients, these all have one of the values $\pm 1, \pm 3$.

[16] Such a relation with different λ's, would imply the existence of a polynomial $P(z)$ of degree $m - 1$ with rational coefficients, essentially different from the irreducible polynomial $z^{p-1} + z^{p-2} + \cdots + 1$ and having roots in common with it.

4. Application to power series in an infinite number of variables. In this section we take the first steps toward the application of the preceding results to ordinary Dirichlet series. Bohr proved the following theorem.[17]

Let G_n be a sequence of positive numbers and

$$P(x_1, x_2, \cdots, x_n, \cdots) \equiv c + \sum_{i=1}^{\infty} c_i x_i + \sum_{i_1, i_2 = 1}^{\infty} c_{i_1 i_2} x_{i_1} x_{i_2} + \cdots$$

a power series in an infinite number of variables, bounded in the domain $|x_n| \leqq G_n$. Let further ε_n be a sequence of positive number such that 1° $0 < \varepsilon_n < 1$ and 2° $\sum \varepsilon_n^2$ is convergent; then the power series P is absolutely convergent in the domain $|x_n| \leqq \varepsilon_n G_n$.

We now prove the following results, which complete Bohr's theorem.

THEOREM III. *If the power series P is bounded in $|x_n| \leqq G_n$: then its m-th polynomial P_m*

$$P_m \equiv c + \sum_{i=1}^{\infty} c_i x_i + \cdots + \sum_{i_1, \cdots, i_m = 1}^{\infty} c_{i_1 \cdots i_m} x_{i_1} \cdots x_{i_m}$$

is absolutely convergent in $|x_n| \leqq \varepsilon_n G_n$, when $\sum \varepsilon_n^{\sigma_m}$ converges, $\sigma_m = \dfrac{2m}{m-1}$.

This exponent σ_m is the best possible one:

THEOREM IV. *There exist polynomials of the m-th degree in an infinite number of variables bounded in $|x_n| \leqq 1$, such that for every $\delta > 0$, the polynomial is non-absolutely convergent for $x_n = \varepsilon_n$ although the series $\sum \varepsilon_n^{\sigma_m + \delta}$ converges.*

Since $\sigma_m \to 2$, when $m \to \infty$ it follows immediately from this theorem that the exponent 2 of Bohr's theorem cannot be increased.

REMARK: Condition 1° of Bohr's theorem, namely $0 < \varepsilon_n < 1$, is not essential to the truth of Theorem III.

Proof of Theorem III. It is obviously sufficient to prove this theorem for the domain (G_0). We show first the truth of

LEMMA 2. *If the power series $P(x_1, \cdots)$ is bounded by H in (G_0); then H is also an upper bound for the form*

$$\sum_{i_1, \cdots i_m = 1}^{\infty} c_{i_1 \cdots i_m} x_{i_1} \cdots x_{i_m}. \tag{4.1}$$

The proof follows the same lines as Bohr's proof in the case $m = 1$. (Theorem V of Bohr). We have to show, that every truncated form

$$\sum_{i_1, \cdots i_m = 1}^{N} c_{i_1 \cdots i_m} x_{i_1} \cdots x_{i_m}$$

[17] H. Bohr, loc. cit. p. 462.

is in absolute value $\leqq H$. Let x_n^* be the set of the variables for which the truncated form takes on a value H^*, which is absolutely the largest. This set exists since the truncated domain $(G_0): |x_n| \leqq 1, n \leqq N; x_n = 0, n > N$, is compact. Put $x_n = t \cdot x_n^*$, then the truncated power series is a power series in t, such that the coefficient of t^m equals this value H^*. By Cauchy's evaluation of the coefficients of a power series, we see that this value is absolutely $\leqq H$.

It suffices therefore to prove Theorem III for m-ic forms. Let $\varepsilon_n > 0$ be any sequence for which $\sum \varepsilon_n^{\sigma_m}$ converges. By Hölder's inequality with $p = \dfrac{2m}{m-1}$, $q = \dfrac{2m}{m+1}$

$$\sum_{i_1, \cdots, i_m = 1}^{N} \varepsilon_{i_1} \cdots \varepsilon_{i_m} |c_{i_1 \cdots i_m}| \leqq \left(\sum_1^N (\varepsilon_{i_1} \cdots \varepsilon_{i_m})^{\frac{2m}{m-1}} \right)^{\frac{m-1}{2m}} \cdot \left(\sum_1^N |c_{i_1 \cdots i_m}|^{\frac{2m}{m+1}} \right)^{\frac{m+1}{2m}}$$

$$\leqq E^{\frac{m-1}{2}} \cdot A_m \cdot H,$$

where E is the sum of the series $\sum \varepsilon_n^{\sigma_m}$, A_m a constant determined by m and H an upper bound of the form (4.1). Hence the sum

$$\sum_1^\infty |c_{i_1 \cdots i_m}| \varepsilon_{i_1} \cdots \varepsilon_{i_m}$$

is convergent, which proves Theorem III.

In the case $m = 2$, Theorem III shows that the examples given by Toeplitz were the best obtainable from quadratic forms.

Proof of Theorem IV. We use the examples constructed in section 3. The integers m and p being fixed, let $Q_\mu(x)$ be the m-ic form corresponding to $n = p^\mu$. The variables which enter in the different $Q_\mu(x)$ shall be independent of each other; in order to differentiate between them we use a superscript μ:

$$Q_\mu(x^{(\mu)}) \equiv \sum_{i_1, \cdots i_m = 1}^{p^\mu} c_{i_1 \cdots i_m} x_{i_1}^{(\mu)} \cdots x_{i_m}^{(\mu)}.$$

Let

$$Q(x) \equiv \sum_{\mu=1}^{\infty} \frac{1}{\mu^2} p^{-\mu \cdot \frac{m+1}{2}} Q_\mu(x^{(\mu)}). \tag{4.2}$$

This form $Q(x)$ is an m-ic form in the variables

$$\{x_n\} \equiv \{x_1^{(1)}, \cdots, x_p^{(1)}; x_1^{(2)}, \cdots, x_{p^2}^{(2)}; x_1^{(3)}, \cdots\}.$$

It is bounded in the domain (G_0), because the maxima of the forms Q_μ (truncated or not) are of order of magnitude $p^{\mu \cdot \frac{m+1}{2}}$, and the series $\sum \frac{1}{\mu^2}$ is convergent.

To study the absolute convergence, we use the fact, that there exists an $\eta > 0$, such that all coefficients of the forms Q_μ are in absolute value $\geqq \eta$. The m-ic form $Q(x)$ will therefore be non-absolutely convergent for a set of values $x_n^{(\mu)}$ if

$$\sum_{\mu=1}^{\infty} \frac{1}{\mu^2} p^{-\mu \cdot \frac{m+1}{2}} \left\{ \sum_{n=1}^{p^\mu} |x_n^{(\mu)}| \right\}^m \tag{4.3}$$

diverges.

Given an arbitrarily small number $\delta > 0$, p^δ is larger than one and we can choose $k < 1$, so that

$$h = p^\delta \cdot k^{\frac{m-1}{2} \cdot (1-\delta)} > 1.$$

We then take all the variables $x_n^{(\mu)}$ of the same form Q_μ equal to each other,

$$x_n^{(\mu)} = (k \cdot p^{-1})^{\mu \cdot \frac{m-1}{2m}(1-\delta)}.$$

The series $\sum x_n^{\frac{2m}{m-1} \frac{1}{1-\delta}}$ is thus equal to

$$\sum_{\mu=1}^{\infty} \sum_{n=1}^{p^\mu} (x_n^{(\mu)})^{\frac{2m}{m-1} \cdot \frac{1}{1-\delta}} = \sum_{\mu=1}^{\infty} k^\mu,$$

which converges since $k < 1$.

But for these special values $x_n^{(\mu)}$ the expression (4.3) diverges. It is equal to

$$\sum_{\mu=1}^{\infty} \frac{1}{\mu^2} \left(p^\delta k^{\frac{m-1}{2}(1-\delta)} \right)^\mu = \sum_{\mu=1}^{\infty} \frac{1}{\mu^2} h^\mu$$

and the last series is divergent, since $h > 1$. When $\delta \to 0$

$$\frac{2m}{m-1} \cdot \frac{1}{1-\delta} \to \frac{2m}{m-1};$$

we have therefore constructed a form in the variables x_n, bounded in (G_0), which for any given $\delta' > 0$, is non-absolutely convergent for a certain set of values x_n with convergent $\sum x_n^{\sigma_m + \delta'}$; completing thus the proof of Theorem IV.

5. **Applications to ordinary Dirichlet series.** We now apply the results of the preceeding section to ordinary Dirichlet series, with the help of Bohr's Theorems A, B, and C, mentioned in the introduction. We proved $S = 2$, hence by Theorem C:

There exist ordinary Dirichlet series for which the widths of their strips of uniform, non-absolute convergence are arbitrarily close to $\frac{1}{2}$.

Consider now Theorems III and IV; applying them to the theory of Dirichlet series we shall prove

THEOREM V. *Let σ_u be the abscissa of uniform convergence of the Dirichlet series*

$$\sum_{n=1}^{\infty} a_n n^{-s}. \tag{5.1}$$

If we replace by zero those terms for which n, decomposed into a product of prime numbers, contains more than m factors, then the new series is absolutely convergent in the half plane

$$\sigma > \sigma_u + \frac{m-1}{2m},$$

and

THEOREM VI. *There exist ordinary Dirichlet series with $a_n = 0$, when n contains more than m prime factors, which converge uniformly, but non-absolutely in strips whose widths are exactly equal to* $\frac{m-1}{2m}$:

$$\sigma_a - \sigma_u = \frac{m-1}{2m}.$$ [18]

When $m = 1$, we obtain a result proved by Bohr.[19]

The proof of these two theorems proceeds in the same way as Bohr's proof of Theorem C.

Let σ_u be the abscissa of uniform convergence of (5.1). Then by Theorem A of Bohr the power series corresponding to (5.1) is bounded in the domain $|x_n| \leqq p_n^{-\sigma_u-\delta}$, for every $\delta > 0$. Hence by Theorem III, and since

$$\sum_{n=1}^{\infty} p_n^{-1-\delta \cdot \frac{2m}{m-1}}$$

is convergent, the m-th polynomial is absolutely convergent in the domain $|x_n| \leqq p_n^{-\sigma_u - \frac{m-1}{2m} - 2\delta}$. By Theorem B of Bohr, the Dirichlet series associated with this m-th polynomial is absolutely convergent in

$$\sigma > \sigma_u + \frac{m-1}{2m} + 2\delta,$$

which proves Theorem V.[20]

---

[18] If the abscissa of uniform convergence is $\leqq 0$, then as Hardy and Carlson (cf. introduction) have proved, the series $\sum a_n^2$ is convergent. Theorem V states essentially that $\sum a_n^{2m/(m+1)}$ converges, if the summation is extended only over those indices n, which contain no more than m prime factors. Theorem VI shows that this exponent $2m/(m+1)$ is the best possible.

[19] H. Bohr, loc. cit. p. 468.

[20] It is interesting to interpret Lemma 2 as a result concerning Dirichlet series. Putting certain coefficients of a Dirichlet series equal to zero can change the position of the abscissa σ_u of uniform convergence. Lemma 2 states that σ_u is never shifted to the right, when all coefficients, whose indices contain more than m prime factors, are replaced by zero.

In order to prove Theorem VI, we consider the Dirichlet series associated with the example (4.2). By Theorem B, the abscissa of uniform convergence is $\leqq 0$. On the other hand, there exists for any $\delta' > 0$, at least one set of real, non-increasing values x_n with convergent

$$\sum_{n=1}^{\infty} x_n^{\frac{2m}{m-1}+\delta'},$$

such that the form (4.2) is non-absolutely convergent for these x_n. They are non-increasing, hence

$$x_n^{\frac{2m}{m-1}+\delta'} = O\left(\frac{1}{n}\right)$$

as $n \to \infty$, or

$$x_n = O\left(n^{-\frac{1}{\frac{2m}{m-1}+\delta'}}\right)$$

But p_n being the n-th prime number, we know that for any $\delta'' > 0$

$$p_n \cdot = O(n \cdot \log n) = o(n^{1+\delta''});$$

hence

$$n^{-(1+\delta'')} = o(p_n^{-1})$$

and

$$x_n = o\left(p_n^{-\frac{m-1}{2m}+\delta}\right)$$

for every $\delta > 0$. Thus there exist a constant A, such that

$$x_n \leqq A \cdot p_n^{-\frac{m-1}{2m}+\delta}$$

This shows that the Dirichlet series associated with (4.2) is non-absolutely convergent in any half-plane

$$\sigma \geqq \frac{m-1}{2\,m} - \delta \qquad (\delta > 0)$$

and therefore

$$\sigma_a \geqq \frac{m-1}{2\,m}.$$

Since $\sigma_u \leqq 0$

$$\sigma_a - \sigma_u \geqq \frac{m-1}{2\,m},$$

but by Theorem V, the left hand side cannot exceed $\frac{m-1}{2\,m}$, hence

---

This is not true in general. Given any Dirichlet series for which $\sigma_a - \sigma_u$ is different from zero, there exist coefficients, such that putting these equal to zero shifts σ_u to the right by the difference $\sigma_a - \sigma_u$.

$$\sigma_u = 0, \qquad \sigma_a = \frac{m-1}{2m}, \qquad \sigma_a - \sigma_u = \frac{m-1}{2m}.$$

6. Solution of the main problem. We give first an example of an ordinary Dirichlet series, for which $\sigma_a - \sigma_u = \frac{1}{2}$.

In the preceding sections we have only shown, that the inequality $\sigma_a - \sigma_u \leqq \frac{1}{2}$ cannot be replaced by a better one. But it is now easy to construct a Dirichlet series for which the width of the strip of uniform, but non-absolute convergence is exactly equal to $\frac{1}{2}$.

We start from the examples (4.2)

$$Q_m(x) \equiv \sum_{\mu=1}^{\infty} \frac{1}{\mu^2} p^{-\mu \cdot \frac{m+1}{2}} Q_\mu(x^{(\mu)}),$$

where the index m refers to the degree of the form. Dividing each of these forms by its upper bound, we obtain new forms $Q_m^*(x)$ bounded by 1 in the domain (G_0). Let q_m be any sequence of positive numbers, such that the series $\sum q_m$ is convergent. Put

$$P(x) \equiv \sum_{m=1}^{\infty} q_m Q_m^*(x).$$

The power series P is obviously bounded by $\sum q_m$ in (G_0), the abscissa of uniform convergence of the associated Dirichlet series is non-positive

$$\sigma_u \leqq 0.$$

On the other hand, the forms Q_m^* have no term in common, because of their different degrees. Hence the power series is non-absolutely convergent as soon as one of the forms Q_m^* is non-absolutely convergent. This implies

$$\sigma_a \geqq \frac{1}{2},$$

because, if $\sigma_a < \frac{1}{2}$, we can find an m large enough, such that the corresponding Q_m^* would be non-absolutely convergent for $x_n = p_n^{-\sigma_0}$, where $\sigma_a < \sigma_0 < \frac{1}{2}$. But since $\sigma_u \leqq 0$ and since the difference $\sigma_a - \sigma_u \leqq \frac{1}{2}$, it must be exactly $= \frac{1}{2}$:

$$\sigma_a = \frac{1}{2}; \qquad \sigma_u = 0.$$

We turn now to the proof of the main theorem:

THEOREM VII. *For any given σ, in the interval $0 \leqq \sigma \leqq \frac{1}{2}$, there exist ordinary Dirichlet series for which the width of the strip of uniform, but non-absolute convergence is exactly equal to σ.*[21]

The preceding example proved this theorem in the case $\sigma = \frac{1}{2}$, we therefore can suppose $\sigma < \frac{1}{2}$. We determine m such that

[21] Cf. Remark at the end of the paper.

$$\frac{m-1}{2m} \geqq \sigma$$

and Theorem VII will be proved, if we can show

THEOREM VIII. *There exist ordinary Dirichlet series associated with m-ic forms, for which* $\sigma_a - \sigma_u = \sigma$, *for every* σ *in the interval* $0 \leqq \sigma \leqq \dfrac{m-1}{2m}$.

We take up again the m-ic forms

$$Q_\mu(x^{(\mu)}) \equiv \sum_{i_1, \cdots i_m = 1}^{p^\mu} c_{i_1 \cdots i_m} x_{i_1}^{(\mu)} \cdots x_{i_m}^{(\mu)},$$

which were used to build up the example (4.2). By means of a parameter u, $0 \leqq u \leqq 1$, we are going to deform them into the worst examples, where all the coefficients are positive. We put

$$Q_\mu(x^{(\mu)}, u) \equiv \sum_1^{p^\mu} \varepsilon_{i_1 \cdots i_m}(u)\, c_{i_1 \cdots i_m} x_{i_1}^{(\mu)} \cdots x_{i_m}^{(\mu)},$$

and suppose that the functions $\varepsilon_{i_1 \cdots i_m}(u)$ are continuous in u; equal to one in absolute value, and satisfy the boundary conditions

$$\varepsilon_{i_1 \cdots i_m}(0) = 1 \text{ and } \varepsilon_{i_1 \cdots i_m}(1) = \frac{|c_{i_1 \cdots i_m}|}{c_{i_1 \cdots i_m}}.$$

Let $H_\mu(u)$ denote the maximum of the form $Q_\mu(u)$ in the domain (G_0). It is readily seen that for all μ, $H_\mu(u)$ is a continuous function of u with the following properties:

There exist absolute constants A_1, A_2, A_3 such that

1° for all values u and all $n = p^\mu$

$$H_\mu(u) \geqq A_1 \cdot n^{\frac{m+1}{2}}. \quad \text{(Theorem I)}$$

2° for $u = 0$ and all $n = p^\mu$

$$H_\mu(0) \leqq A_2\, n^{\frac{m+1}{2}}. \quad \text{(Theorem II)}$$

3° for $u = 1$ and all $n = p^\mu$

$$H_\mu(1) \geqq A_3\, n^m.\text{[22]}$$

We have $A_1 \leqq A_2$ and can suppose $A_1 \leqq A_3$.

LEMMA 3. *There exist two constants* B_1 *and* B_2 *such that for every* τ, $\dfrac{m+1}{2} \leqq \tau \leqq m$, *and every* μ, *there exists a* u_μ *for which*

$$B_1 \cdot n^\tau \leqq H_\mu(u_\mu) \leqq B_2 \cdot n^\tau.$$

---

[22] This follows from the fact that there exists an $\eta \geqq 0$, such that $|c_{i_1 \cdots i_m}| \geqq \eta$.

It is evidently sufficient to prove this lemma for large values of n. Let N be so large that

$$A_2 N^{\frac{m+1}{2}} \leqq A_3 N^m.$$

Then for all $n \geqq N$, we shall prove

$$A_1 n^\tau \leqq H_\mu(u_\mu) \leqq A_2 n^\tau. \tag{6.1}$$

Since (Condition 2)

$$H_\mu(0) \leqq A_2 n^{\frac{m+1}{2}} \leqq \begin{cases} A_3 n^m, \\ A_2 n^\tau, \end{cases}$$

and (Condition 3)

$$H_\mu(1) \geqq A_3 n^m,$$

we can find a u_μ, such that

$$H_\mu(u_\mu) = \text{Min}(A_3 n^m, A_2 n^\tau). \tag{6.2}$$

This $H_\mu(u_\mu)$ satisfies (6.1) because

$$A_1 n^\tau \leqq \text{Min}(A_3 n^m, A_2 n^\tau).$$

LEMMA 4. *For every ϱ in the interval $1 \leqq \varrho \leqq \frac{2m}{m+1}$, and every $n = p^\mu$, there exist m-ic forms in n dimensions, for which*

$$C_1 \cdot n^{\frac{m}{\varrho}} \leqq C_2 H_\mu \leqq \left\{ \sum_{i_1 \cdots i_m = 1}^{n} |C_{i_1 \cdots i_m}|^\varrho \right\}^{\frac{1}{\varrho}} \leqq C_3 H_\mu \leqq C_4 n^{\frac{m}{\varrho}},$$

where the constants C_1, C_2, C_3 and C_4 are independent of n.

It is easily verified that the forms $Q_\mu(x^{(\mu)}, u_\mu)$, where u_μ is determined by (6.2) satisfy these conditions.

We proceed from now on exactly as for the extreme case $\varrho = \frac{2m}{m+1}$. We put

$$Q(x) \equiv \sum_{\mu=1}^{\infty} \frac{1}{\mu^2} p^{-\mu \cdot \frac{m}{\varrho}} Q_\mu(x^{(\mu)}, u_\mu)$$

and prove that $Q(x)$ satisfies the following conditions:

1° *If $Q(x)$ is bounded in $|x_n| \leqq G_n$, then it is absolutely convergent in $|x_n| \leqq \varepsilon_n G_n$; when $\sum \varepsilon_n^\varkappa$ converges $\left(\varkappa = \frac{\varrho}{\varrho - 1}\right)$.*

2° *$Q(x)$ is bounded in $|x_n| \leqq 1$.*

3° *For every $\delta > 0$, there exist a set of real values x_n with convergent $\sum x_n^{\varkappa+\delta}$, for which $Q(x)$ is non-absolutely convergent.*

The associated Dirichlet series will be an example proving Theorem VIII, if we take $\varrho = \frac{1}{1-\sigma}$.

7. Generalization to more general types of Dirichlet series. A theorem similar to Theorem VII cannot be formulated for general Dirichlet series $\sum a_n e^{-\lambda_n s}$. The inequality

$$\sigma_a - \sigma_u \leqq \frac{D}{2}; \qquad D = \overline{\lim} \frac{\log n}{\lambda_n}, \tag{7.1}$$

is not necessarily the best possible one, when the type $\{\lambda_n\}$ is given and not merely the quantity D, as in the case considered by Neder. For the type determined by the Dirichlet series associated with the polynomials of the mth degree, (7.1) can be replaced (Theorem V) by

$$\sigma_a - \sigma_u \leqq \frac{m-1}{m} \cdot \frac{D}{2},$$

the upper limit D being equal to one for this type.

However Bohr[23] extended the results of his paper in the Göttinger Nachrichten to cover certain types of general Dirichlet series and for these types, with a further restriction to prevent the occurrence of examples similar to the one just mentioned, the inequality (7.1) is the best obtainable result.

Suppose that $\{b_n\}$ is a sequence of increasing positive numbers, which tend to ∞ and which are linearly independent in the field of rational numbers. Consider then the values

$$\sum_{\nu=1}^{n} r_\nu b_\nu,$$

for any positive n, and any non-negative integral coefficients r_ν. These values form a sequence, $\{\lambda_n\}$, which can be arranged in increasing order

$$\lambda_1 < \lambda_2 < \cdots < \lambda_n < \cdots,$$

and whose elements λ_n tend to ∞. This sequence can therefore be considered as the *type* of a Dirichlet series. The first sequence $\{b_n\}$ forms an integral base for $\{\lambda_n\}$; the associated function obtained by putting $x_n = e^{-b_n s}$ is a power series in x_n. The methods of proving Theorems A, B of Bohr and our results apply in this case:

For such types the inequality (7.1) *is the best possible result.*

In the construction of examples for which $\sigma_a - \sigma_u = \frac{D}{2}$ it is essential to observe that

$$D(b) = \overline{\lim} \frac{\log n}{b_n} = D(\lambda) = \overline{\lim} \frac{\log n}{\lambda_n},$$

[23] H. Bohr, Zur Theorie der allgemeinen Dirichletschen Reihen, Math. Annalen **79**, (1919), p. 136–156.

which is equivalent to the statement:

The exponent $D(b)$ *of convergence of the sequence* $\{e^{b_n}\}$ *is equal to the exponent* $D(\lambda)$ *of convergence of the sequence* $\{e^{\lambda_n}\}$.

We have obviously $\lambda_n \leqq b_n$ and therefore

$$D(b) \leqq D(\lambda),$$

and it remains only to show that $\sum e^{-\lambda_n \sigma}$ converges when $\sum e^{-b_n \sigma}$ converges. For the subsequence $\{\lambda_n'\}$, obtained by taking only the m first elements of the base, we have

$$\sum e^{-\lambda_n' \sigma} = \sum_{r_1, \cdots r_m = 0}^{\infty} e^{-\sigma \sum_1^m r_\nu b_\nu} = \prod_{\nu=1}^{m} \frac{1}{1 - e^{-\sigma b_\nu}}$$

$$\leqq \prod_{\nu=1}^{m} \left(1 + \frac{e^{-\sigma b_\nu}}{1 - e^{-\sigma b_1}}\right) \leqq \prod_{\nu=1}^{\infty} \left(1 + \frac{e^{-\sigma b_\nu}}{1 - e^{-\sigma b_1}}\right).$$

This last infinite product is convergent, proving thus

$$D(b) \geqq D(\lambda),$$

and therefore

$$D(b) = D(\lambda).$$

We are able now to construct new examples proving Neder's result. Given any non-negative D, there exist Dirichlet series for which

$$\sigma_a - \sigma_u = \frac{D}{2}$$

of all types $\{\lambda_n\}$, obtained from a base $\{b_n\}$, $0 < b_1 < \cdots b_n \cdots \to \infty$, satisfying

$$D = \overline{\lim} \frac{\log n}{b_n}.$$

The simplest example is

$$b_n = \frac{1}{D} \cdot \log n,$$

which gives Dirichlet series of the type $\sum a_n \cdot n^{-\frac{s}{D}}$, obtained from ordinary Dirichlet series by the substitution $\left(s \,|\, \frac{s}{D}\right)$.

PRINCETON UNIVERSITY.

REMARK (added in proof, May, 1931). As Prof. Bohr pointed out to us, Theorem VII can be proved more simply as follows. Let $f(s)$ be a Dirichlet series whose $\sigma_u = 0$ and whose $\sigma_a = \frac{1}{2}$. If $\zeta(s)$ denotes the Riemann Zeta-function ($\sigma_a = \sigma_u = 1$) and σ any real number $0 \leqq \sigma \leqq \frac{1}{2}$, then $f(s) + \zeta(s + 1 - \sigma)$ is a Dirichlet series for which $\sigma_a - \sigma_u = \sigma$. Similar examples prove Theorem VIII. It may be interesting however, to see that the method used to obtain best possible examples is flexible enough to give the whole range of possible values for the difference $\sigma_a - \sigma_u$.

Chapter 4
Integral Equations

The paper included in this chapter is

A comment on Hille's work on integral equations can be found starting on page 665.

Reprinted from the Proceedings of the NATIONAL ACADEMY OF SCIENCES.
Vol. 14, No. 12, pp. 911–914. December, 1928.

ON THE CHARACTERISTIC VALUES OF LINEAR INTEGRAL EQUATIONS

BY EINAR HILLE AND J. D. TAMARKIN

PRINCETON UNIVERSITY AND BROWN UNIVERSITY

Communicated November 7, 1928

1. We consider the integral equation

$$f(x) = \varphi(x) - \lambda \int_a^b K(x, s)\varphi(s)ds, \tag{1}$$

where the kernel is supposed to belong to the class (L^2), i.e., $K(x, s)$ and its square are Lebesgue integrable in $a \leqq x, s \leqq b$. The equation is known to possess a resolvent kernel which is the quotient of two entire functions of λ, defined by Fredholm's formulas with the modifications due to Hilbert.[1] The characteristic values of the equation are the zeros of the denominator $D_K^*(\lambda)$ in this quotient. The order ν of $D_K^*(\lambda)$ is $\leqq 2$, and if $\nu = 2$ the function belongs to the minimal type; its genus is at most unity. This result is due to Carleman.[2]

2. The proof given by Carleman is ingenious but also rather complicated. It is possible, however, to base the proof upon a simple and well-known method, namely, that of infinitely many equations in infinitely many unknowns. Let $\{\omega_i(x)\}$ be a complete orthonormal system for the interval (a, b). Then $\{\omega_i(x)\overline{\omega_j(s)}\}$ constitutes a complete orthonormal system for the square $a \leqq x, s \leqq b$. Put

$$f_i = \int_a^b f(s)\overline{\omega_i(s)}ds, \quad \varphi_i = \int_a^b \varphi(s)\overline{\omega_i(s)}ds,$$

$$a_{ij} = \int_a^b \int_a^b K(x, s)\overline{\omega_i(x)}\omega_j(s)\, dx\, ds, \tag{2}$$

$$k_i(x) = \int_a^b K(x, s)\omega_i(s)ds, \quad K^{(n)}(x, s) = \sum_{i=1}^{n} k_i(x)\overline{\omega_i(s)}.$$

Repeated application of Parseval's identity leads to the following system of linear equations

$$\sum_{j=1}^{\infty} (\delta_{ij} - \lambda a_{ij})\varphi_j = f_i \ (i = 1, 2, \ldots). \tag{3}$$

The infinite determinant $A_K(\lambda)$ of this system is absolutely convergent when $\sum|a_{ii}|$ converges, which is the case, for example, when

$$K(x, s) = \int_a^b K_1(x, t)K_2(t, s)dt \tag{4}$$

where K_1 and K_2 are two kernels belonging to (L^2). If $\sum|a_{ii}|$ diverges, $A_K(\lambda)$ may cease to exist. In this case we replace the system (3) by an equivalent system obtained by multiplying the ith equation in (3) by $e^{\lambda a_{ii}}(i = m + 1, m + 2, \ldots)$. The determinant $A_K^{(m)}(\lambda)$ of the transformed system is absolutely convergent. A fairly simple application of Hadamard's determinant theorem leads to the fundamental inequality

$$\log|A_K^{(m)}(\lambda)| < C + m \log|\lambda| + \frac{e}{2}|\lambda|^2 \sum_{i=m+1}^{\infty} \sum_{j=1}^{\infty} |a_{ij}|^2, \tag{5}$$

where C is independent of m and λ. It follows that for any given $\epsilon > 0$ we can find an m such that

$$\log|A_K^{(m)}(\lambda)| < \epsilon|\lambda|^2 \tag{6}$$

when $|\lambda|$ is large.

3. We put

$$A_K^*(\lambda) = e^{\lambda \sum_1^m a_{ii}} A_K^{(m)}(\lambda) = A_K^{(0)}(\lambda) \tag{7}$$

and notice that

$$A_K(\lambda) = \lim_{m \to \infty} A_K^{(m)}(\lambda) = A_K^{(\infty)}(\lambda)$$

in case the limit exists. Marty has proved that $A_K(\lambda)$ equals $D_K(\lambda)$, the denominator of Fredholm, when $A_K(\lambda)$ is a normal determinant and $K(x, s)$ is defined as the sum of its Fourier series in $\{\omega_i(x) . \overline{\omega_j(s)}\}$. Mollerup has also given a proof valid when (4) is satisfied.[3] The main difficulty in the proof is caused by the fact that $A_K(\lambda)$ remains unchanged when $K(x, s)$ be replaced by any equivalent function, whereas $D_K(\lambda)$ may be thrown out of existence by such a change.

It is possible, however, to prove that $A_K^*(\lambda) = D_K^*(\lambda)$ for every kernel belonging to (L^2). Indeed, we are dealing with two holomorphic functions of the point (a_{ij}) in a Hilbert space which are identical in a sub-space characterized by the convergence of the series $\sum_{ij}|a_{ij}|$. It follows that these functions are identical throughout the Hilbert space.[4] We can conclude also that $A_K(\lambda) = D_K(\lambda)$ whenever the series $\sum|a_{ii}|$ converges and the integral $\int_a^b K(s, s)ds$ exists and equals $\sum_1^\infty a_{ii}$.

Once the equality of $A_K^*(\lambda)$ and $D_K^*(\lambda)$ is proved, Carleman's theorem on the order of $D_K^*(\lambda)$ becomes a fairly simple consequence of formulas (6) and (7). Carleman has constructed examples showing that the upper limit two for the order is actually reached even in the case of continuous kernels.[5] Considerably simpler examples can be found with the aid of a class of power series recently studied by one of us.[6]

4. The characteristic values of (1) are the zeros of $D_K^*(\lambda)$ or of any one of the functions $A_K^{(m)}(\lambda)$ which differ merely by an exponential factor. Let $n(x)$ denote the number of characteristic values λ_n with $|\lambda_n| \leqq x$. By a well-known theorem

$$\int_0^r \frac{n(x)}{x} dx \leqq \operatorname*{Max}_{|\lambda| = r} |A_K^{(m)}(\lambda)| = M^{(m)}(r). \tag{8}$$

But we have already found an upper limit for $M^{(m)}(r)$ in (5). Since m is arbitrary we may choose its value so as to make $M^{(m)}(r)$ a minimum for any given value of r.

In this fashion we are lead to a remarkable connection between the problem of the frequency of the characteristic values and the theory of approximation of a function of a real variable by means of a given set of functions. In fact, we have

$$\sum_{i=m+1}^{\infty} \sum_{j=1}^{\infty} |a_{ij}|^2 = \int_a^b ds \int_a^b |K(s, t) - K^{(m)}(s, t)|^2 dt \tag{9}$$

where the interior integral represents the mean square error ρ_m of the approximation of $K(s, t)$ by the sum of the first m terms of its Fourier series in $\{\omega_i(t)\}$, s being regarded as a parameter. It is well known that ρ_m is the minimum of the mean square error when $K(s, t)$ is approximated by a linear combination of the functions $\omega_1(t), \omega_2(t), \ldots, \omega_m(t)$. The infinitesimal order of ρ_m is fairly well known in the case of approximation by ordinary or trigonometric polynomials. Utilizing the available results in this direction we are able to get better estimates for the order of the entire functions $D_K(\lambda)$ or $D_K^*(\lambda)$ when the kernel is subject to suitable restrictions.

5. Fredholm had found that the order of $D_K(\lambda)$ is at most $\frac{2}{2\alpha + 1}$ if $K(x, s)$ satisfies a Lipschitz condition of order α with respect to s uniformly in x.[7] We can prove the same result assuming $K^*(x, s)$ to belong to the class Lip$(\alpha, 2)$ with respect to s uniformly in x.[8] Here $K^*(x, s) = K(x, s)$ when $a \leqq s < b$ and $K^*(x, s + b - a) = K^*(x, s)$.

If we suppose that $K(x, s)$ possesses a partial derivative of order r with respect to s which satisfies a Lipschitz condition of order α with respect to s uniformly in x, we find that the order of $D_K(\lambda)$ cannot exceed $\frac{2}{2r + 2\alpha + 1}$.[9]

Suppose, finally, that $K(x, s)$ is a holomorphic function of s in a region T of the s-plane, containing the interval (a, b) in its interior and, moreover, independent of x. Then it can be shown that

$$\log|D_K(\lambda)| = O(\log^2|\lambda|), \; |\lambda_n| > e^{\beta n} \tag{10}$$

where λ_n is the nth characteristic value and $\beta(> 0)$ is a constant depending upon T. If $K(x, s)$ is an entire function of s we can even replace the "O" by "o" in (10) and β can be taken arbitrarily large. It is easy to construct examples showing that these estimates cannot be improved upon essentially; on the other hand, we can construct analytic kernels whose characteristic values increase as rapidly as we please.

Further developments and extensions of the above theory will be published in a later paper which will appear elsewhere.

[1] D. Hilbert, *Grundzüge einer allgemeinen Theorie der linearen Integralgleichungen*, B. G. Teubner, Leipzig, **1912,** p. 30, et seq. That the formulas in question apply to all kernels of (L^2) was first proved by T. Carleman, *Math. Z.*, **9,** 196–217, 1921.

[2] Loc. cit., p. 217.

[3] J. Marty, *Bull. Sci. Math. (Darboux)* (2), **33,** 1909 (296–300). J. Mollerup, *Ibid.*, **36,** 130–136, **1912,** reprinted in *Beretning om den anden skandinaviske Matematikerkongres i Kjöbenhavn* **1911,** Gyldendalske Boghandel, Copenhagen, 81–87, 1912.

[4] We are using the ideas of R. Gâteaux, posthumous publication in *Bulletin Soc. Math. France*, **47,** 70–96, 1919.

[5] T. Carleman, *Acta Math.*, **41,** 377–389, 1918.

[6] E. Hille, these PROCEEDINGS, **14,** 217–220, 1928. Similar use can be made of various series presented as *Gegenbeispiele* by Hardy and Littlewood, *Math. Z.*, **28,** 612–634, 1928.

[7] I. Fredholm, *Acta Math.*, **27,** 365–390, 1903.

[8] The terminology is that of Hardy and Littlewood, loc. cit., p. 612.

[9] This result has already been found by S. A. Gheorghiu, *Comptes Rendus*, **186,** 838–840, 1928, with the aid of a totally different method. We are indebted to Professor Hadamard for this reference. The results mentioned at the end of Mr. Gheorghiu's note concerning kernels which are continuous and of bounded variation (or have a derivative with these properties) follow immediately from our formulas. Two earlier notes by Gheorghiu, *Ibid.*, **184,** 864–865, 1309–1311, 1927, have a bearing on our general problem.

Chapter 5
Summability

The paper included in this chapter is

[48] (with J. D. Tamarkin) On the summability of Fourier series. II

A commentary on this paper can be found starting on page 667.

ON THE SUMMABILITY OF FOURIER SERIES. II.*

By Einar Hille and J. D. Tamarkin.

1. Introduction. The present paper is the second of a series of memoirs devoted to the general theory of summability of Fourier series. The first and third memoirs of the series[1] are dealing with the problem of summation of Fourier series with the aid of Nörlund's and Hausdorff's means respectively. Here we are concerned with general definitions of summation which satisfy the condition of being "effective" in a certain sense when used for summation of Fourier series. We establish relationships between definitions of summation possessing various effectiveness properties. Most of our results are valid not only for trigonometric Fourier series, but for general Fourier series associated with arbitrary orthonormal sets.

The following terminology and notation will be employed. Let

$$u_0 + u_1 + \cdots + u_n + \cdots \tag{1.01}$$

be an arbitrary series, convergent or not, and $\mathfrak{A} = (a_{mn})$ be a matrix of a definition of summation, which consists in constructing the "m-th transform"

$$t_m = \sum_{n=0}^{\infty} a_{mn} u_n \tag{1.02}$$

of (1.01) and evaluating the "generalized sum" of (1.01) as defined by

$$s = \lim_{m \to \infty} t_m. \tag{1.03}$$

For brevity we shall speak simply of a "transformation $\mathfrak{A}$".

Let $f(x)$ be an arbitrary function integrable over a given interval (a, b). If in addition

$$\int_a^b |f(x)|^p \, dx < \infty, \quad p \geqq 1,$$

we shall say that $f(x) \subset L_p$. It is convenient to allow $p = \infty$ and to introduce the corresponding class L_∞. It is customary to identify this class with the class B of functions measurable and essentially bounded on

* Received August 20, 1932. Revised during November 1932. Some of the results contained in this paper were presented to the Am. Math. Soc., March 26, 1932, and to the International Congress in Zürich, September 1932.

[1] Hille-Tamarkin [1, 2]. The numbers in brackets refer to the Bibliography at the end of the present paper.

(a, b) (when sets of values of x of measure zero can be disregarded). For reasons to be explained below, we shall depart from this usual notation and shall designate by L_∞ the class of all functions continuous on (a, b).

The class L_p becomes a linear vector metric complete space[2] if we introduce the norm of the element $f \subset L_p$,

$$(1.04)\qquad \begin{aligned} \|f\|_p &\equiv \left[\frac{1}{b-a}\int_a^b |f(x)|^p\,dx\right]^{1/p} \text{ if } p<\infty; \\ \|f\|_\infty &= \max_{a\leq x\leq b} |f(x)|, \end{aligned}$$

and define the distance between two "points" f, g of the space L_p as

$$(1.05)\qquad (f, g) \equiv \|f-g\|_p.$$

Let

$$(1.06)\qquad \varphi_1(x), \varphi_2(x), \cdots, \varphi_n(x), \cdots; \int_a^b \varphi_i(x)\overline{\varphi_j(x)}\,dx = \delta_{ij}$$

be an arbitrary orthonormal set for the interval (a, b). It will be assumed throughout this paper that the set $\{\varphi_n(x)\}$ satisfies the following conditions.

(α) The functions $\varphi_n(x)$ are continuous.

(β) The set $\{\varphi_n(x)\}$ is closed in L_∞, which means that an arbitrary $f(x) \subset L_\infty$ can be approximated in L_∞ as closely as we like by means of finite linear combinations of the functions $\varphi_n(x)$.

As a consequence of condition (β) we state

(γ) The set $\{\varphi_n(x)\}$ is complete in L_1, which means that the conditions

$$(1.07)\qquad \int_a^b f(x)\overline{\varphi_n(x)}\,dx = 0, \qquad n = 1, 2, \cdots; \quad f \subset L_1$$

imply $f(x) = 0$ almost everywhere.[3]

Let $f(x)$ be an arbitrary function $\subset L_p$. On setting

$$(1.08)\qquad f_n = \int_a^b f(x)\overline{\varphi_n(x)}\,dx, \qquad n = 1, 2, \cdots$$

we designate by $S(x; f)$ the formal Fourier series of f associated with the set $\{\varphi_n(x)\}$,

$$(1.09)\qquad f(x) \sim S(x; f) \equiv \sum_{n=1}^{\infty} f_n \varphi_n(x).$$

We shall say that this series $S(x; f) \subset L_p$ if its generating function $f(x) \subset L_p$.

Let us apply the transformation $\mathfrak{A}$ to the series (1.09). We shall be concerned with the problem of convergence in L_p of the m-th transform

---

[2] We refer to Hildebrandt [1] and Banach [1] concerning the terminology of the general theory of linear operations and transformations used in the present paper.

[3] Orlicz 1, pp. 4, 6.

of this series to the function $f(x)$. There is no difficulty in computing the m-th transform of (1.09) when $\mathfrak{A}$ is of finite reference. When $\mathfrak{A}$ is of infinite reference, the series

$$\sum_{n=1}^{\infty} a_{mn} f_n \varphi_n(x) \tag{1.10}$$

defining this m-th transform may not converge. Still it can be used for our purposes provided (1.10) is also a Fourier series $\subset L_p$. In this case we define as the m-th transform of (1.09) the function which generates (1.10), and put

$$\tau_m(x; f) \sim \sum_{n=1}^{\infty} a_{mn} f_n \varphi_n(x).$$

With this notation we can formulate the following

DEFINITION 1.1. *A transformation $\mathfrak{A}$ is said to be effective in the mean of order p, or simpler, (L_p)-effective, if it satisfies the conditions*

(E. 1) *The series*

$$\tau_m(x; f) \sim \sum_{n=1}^{\infty} a_{mn} f_n \varphi_n(x) \subset L_p, \qquad m = 1, 2, \cdots \tag{1.11}$$

whenever $f(x) \subset L_p$;

(E. 2) *The m-th transform $\tau_m(x; f)$ converges to $f(x)$ in L_p,*

$$\|\tau_m(x; f) - f(x)\|_p \to 0 \quad \text{as} \quad m \to \infty. \tag{1.12}$$

If conditions (E. 1), (E. 2) *are satisfied not for all $f(x) \subset L_p$, but only for those of a certain sub-class L_p^0 of L_p, we shall say that $\mathfrak{A}$ is (L_p)-effective on L_p^0.*

It is customary to designate by p' the conjugate exponent,

$$p' = \frac{p}{p-1};\ \infty;\ \text{or } 1 \text{ according as } 1 < p < \infty;\ p = 1;\ \text{or } p = \infty \tag{1.13}$$

respectively, so that always

$$\frac{1}{p} + \frac{1}{p'} = 1. \tag{1.14}$$

It will be assumed that $1 \leqq p \leqq \infty$.

Our main result can now be stated as follows:

(A) *The classes of (L_p)- and of $(L_{p'})$-effective transformations are identical.*

(B) *If $\mathfrak{A}$ is (L_p)-effective, then $\mathfrak{A}$ is also (L_q)-effective for an arbitrary q between p and p'.*

We see therefore that the property of being (L_p)-effective is symmetric with respect to $p = 2$, and that the class of (L_p)-effective transformations, considered as a function of p, increases or decreases, according as p increases from 1 to 2, or from 2 to ∞, reaching its possible maximum for $p = 2$.

There is a considerable number of articles in the literature dealing with (L_p)-effectiveness, and particularly, with (L_1)- and (L_∞)-effectiveness of various specified classes of transformations.[4] To the best of our knowledge, however, the relationship between the various classes of "effective" transformations has not been investigated.

The reader will notice that our Theorems (A) and (B) strongly resemble some well-known results of the theory of factor-sequence transformations of Fourier series.[5] It will be seen in the sequel that the resemblance is not merely formal, and that these results are fundamental for the proofs of our theorems.

Our discussion is considerably simplified by using some simple facts of the general theory of linear transformations in abstract spaces. In order not to interrupt the exposition we collect all necessary preliminaries, in the form of lemmas, in § 2. The general results concerning (L_p)-effective transformations are derived in § 3. The remainder of the paper deals with trigonometric Fourier series. In § 4 we give the general form of certain, fairly extended, subclasses of (L_1)-, and (L_∞)-effective transformations for trigonometric Fourier series. In the last § 5 we discuss the (L_1)-effectiveness of the "convergence" transformation for some special classes of sine and cosine Fourier series. From our present point of view there is an essential difference between the "identity" and the "convergence" transformations, whose matrices $\mathfrak{A}^i$ and $\mathfrak{A}^c$ are represented respectively by

$$(\mathfrak{A}^i) \qquad a_{mn} = 1, \qquad m, n = 1, 2, \cdots;$$

$$(\mathfrak{A}^c) \qquad a_{mn} = \begin{cases} 1, & n = 1, 2, \cdots, m, \\ 0, & n = m+1, \cdots. \end{cases}$$

[4] Without any pretense of being exhaustive we mention the following memoirs which have points of contact with the ideas or methods of the present paper: Gross [1]; Hahn [1]; Orlicz [1]; W. H. and G. C. Young [1]; Zygmund [1]. The definition of (L_∞)-effectiveness is closely related, but not identical, to the definition of (F)-effectiveness introduced in our memoirs [1, 2]. It should be observed that the property of being (L_∞)-effective implies the uniform convergence of the m-th transform $\tau_m(x; f)$ to the generating (continuous) function $f(x)$. This property is apparently more restrictive than (L_1)-effectiveness, so that the identity of these two classes of transformations is far from being obvious a priori. In fact, the present paper has originated in an attempt to construct a transformation $\mathfrak{A}$ which is (L_1)-effective without being (L_∞)-effective. As observed before, our definition of the class L_∞ deviates from the usual one where L_∞ designates the class B of all essentially bounded measurable functions $g(x)$ with the metric $\|g\|_B =$ upper measurable bound of $|g(x)|$. In such a space, however, even the simplest and best known transformations obviously are not effective since, in general, the m-th transform $\tau_m(x; f)$ does not converge uniformly to $f(x)$. That is why we are compelled to restrict the space L_∞ to contain only continuous functions. [Nevertheless, the problem of effectiveness of a transformation in the space B can be attacked successfully as will be shown in an addition to the present paper, to appear in the next number of these Annals. Added in the proof.]

[5] M. Riesz [1] pp. 487–490, where other references are found. See also Orlicz [1]. Our notation will differ slightly from that of Riesz.

The matrix $\mathfrak{A}^t$ is of course always (L_p)-effective, $1 \leqq p \leqq \infty$, while $\mathfrak{A}^c$ is (L_p)-effective only for $1 < p < \infty$.

2. **Auxiliary propositions.** I. We start with some indispensable notions and facts of the general theory of transformations in abstract spaces. Let $U(x)$ be a transformation which coördinates with each point x of a linear vector metric complete space $\mathfrak{E}$[5a] a unique point y of another linear vector metric complete space $\mathfrak{E}_1$. The transformation U will be designated as an operation, if $\mathfrak{E}_1$ reduces to the space of real or, more generally, of complex numbers; in this case $\|U(x)\| = |U(x)|$. Without any danger of confusion we shall use the same notation for the norms in the spaces $\mathfrak{E}$ and $\mathfrak{E}_1$. We say that

1. $U(x)$ is continuous at x if $\|U(x') - U(x)\| \to 0$ whenever $\|x' - x\| \to 0$, and continuous (in $\mathfrak{E}$) if it is continuous at each $x \subset \mathfrak{E}$.
2. $U(x)$ is limited, with the bound M_U if

$$M_U = \underset{0 \neq x \subset \mathfrak{E}}{\text{l. u. b.}} \frac{\|U(x)\|}{\|x\|} = \underset{\|x\|=1}{\text{l. u. b.}} \|U(x)\| < \infty. \tag{2.01}$$

3. $U(x)$ is linear if

$$U(c_1 x_1 + c_2 x_2) = c_1 U(x_1) + c_2 U(x_2). \tag{2.02}$$

4. A sequence of transformations $\{U_n(x)\}$ is bounded on a set $E \subset \mathfrak{E}$ if there exists a constant C_E depending on E such that

$$\|U_n(x)\| \leqq C_E, \qquad x \subset E. \tag{2.03}$$

The sequence $\{U_n(x)\}$ converges to $U_0(x)$ on E if

$$\|U_n(x) - U_0(x)\| \to 0 \quad \text{as} \quad n \to \infty; \qquad x \subset E. \tag{2.04}$$

Of course E may reduce to a single point x.

5. A sequence of linear limited transformations $\{U_n(x)\}$ is uniformly limited if

$$M_{U_n} \leqq M < \infty \tag{2.05}$$

which is equivalent to the condition

$$\|U_n(x)\| \leqq M\|x\|, \qquad n = 1, 2, \cdots \tag{2.06}$$

where M does not depend on x or n.

The following lemma is fundamental in the theory of sequences of linear limited transformations.

LEMMA 2.1. *If a sequence of linear limited transformations $\{U_n(x)\}$ converges at each point of $\mathfrak{E}$ then the sequence $\{U_n(x)\}$ is uniformly limited.*[6] *The sequence $\{U_n(x)\}$ then converges to a linear limited transformation $U_0(x)$.*

[5a] A space of type (B) in the terminology of Banach.

[6] Banach [1], p. 80, Theorem 5.

It is well-known that in order that a linear transformation $U(x)$ be limited it is necessary and sufficient that either (i) $U(x)$ be bounded on the boundary of the unit sphere $\|x\| = 1$, or (ii) $U(x)$ be bounded in an arbitrary fixed sphere $\|x - x_0\| \leqq r_0$, or else (iii) $U(x)$ be continuous at a single point x_0.

We note that the most general form of a linear limited operation $l(f)$ on L_p $(1 \leqq p < \infty)$ is given by

$$l(f) = \int_a^b f(t)\, g(t)\, dt \tag{2.07}$$

with

$$g(t) \subset L_{p'}, \qquad M_l = \|g\|_{p'} \tag{2.08}$$

where, in the case $p = 1$, $g(t)$ is an arbitrary *bounded* function.

In the space L_∞ we have

$$l(f) = \int_a^b f(s)\, d\alpha(s), \qquad M_l = \int_a^b |d\alpha(s)| \equiv V(\alpha), \tag{2.09}$$

where $\alpha(s)$ is a function of bounded variation, normalized according to the condition

$$\alpha(s_0) = \frac{1}{2}[\alpha(s_0 + 0) + \alpha(s_0 - 0)], \qquad a < s_0 < b. \tag{2.10}$$

II. We shall need some properties of linear limited transformations on L_p to a sub-set of L_p.

LEMMA 2.2. *Let $U(f)$ be a linear limited transformation on L_p to a sub-set of L_p and $W(g)$ a linear limited transformation on $L_{p'}$ to a sub-set of $L_{p'}$, $1 \leqq p \leqq \infty$. If the relation*

$$\int_a^b g\, U(f)\, dx = \int_a^b f\, W(g)\, dx \tag{2.11}$$

is satisfied for an arbitrary pair of functions $f \subset L_p$, $g \subset L_{p'}$, then the transformations U and W have the same bounds

$$M_U = M_W. \tag{2.12}$$

Proof. Let M be the bound of the bilinear operation

$$Q(f, g) \equiv \frac{1}{b-a} \int_a^b g\, U(f)\, dx = \frac{1}{b-a} \int_a^b f\, W(g)\, dx,$$

viz.,

$$M = \text{l. u. b.}\, |Q(f, g)|, \qquad \|f\|_p = \|g\|_{p'} = 1.$$

Since

$$\left| \frac{1}{b-a} \int_a^b \varphi(x)\, \psi(x)\, dx \right| \leqq \|\varphi\|_p \|\psi\|_{p'}; \quad \varphi \subset L_p, \quad \psi \subset L_{p'},$$

we have

$$|Q(f,g)| \leqq \begin{cases} \|g\|_{p'} \, \|U(f)\|_p \leqq M_U, \\ \|f\|_p \, \|W(g)\|_{p'} \leqq M_W, \end{cases}$$

whence

$$M \leqq M_U, \qquad M \leqq M_W.$$

On the other hand, an arbitrarily small $\varepsilon > 0$ being given, we can find two functions $f_\varepsilon \subset L_p$, $g_\varepsilon \subset L_{p'}$ such that

$$\|U(f_\varepsilon)\|_p = M_U - \varepsilon, \qquad \|W(g_\varepsilon)\|_{p'} = M_W - \varepsilon, \qquad \|f_\varepsilon\|_p = \|g_\varepsilon\|_{p'} = 1.$$

At this juncture we have to consider separately the cases $1 < p < \infty$, $p = 1$, $p = \infty$. Without loss of generality we may assume that all the functions concerned are real-valued. Let $1 < p < \infty$. Upon putting

$$\begin{aligned} f'(x) &= [\|W(g_\varepsilon)\|_{p'}]^{1-p'} \, |W(g_\varepsilon)|^{p'-1} \operatorname{sgn} W(g_\varepsilon), \qquad \|f'\|_p = 1, \\ g'(x) &= [\|U(f_\varepsilon)\|_p]^{1-p} \, |U(f_\varepsilon)|^{p-1} \operatorname{sgn} U(f_\varepsilon), \qquad \|g'\|_{p'} = 1, \end{aligned}$$

we find

$$\begin{aligned} Q(f', g_\varepsilon) &= \|W(g_\varepsilon)\|_{p'} = M_W - \varepsilon < M, \\ Q(f_\varepsilon, g') &= \|U(f_\varepsilon)\|_p = M_U - \varepsilon < M, \end{aligned}$$

whence it follows $M = M_U = M_W$.

Now let $p = 1$, $p' = \infty$. There exists a function $g'' \subset L_\infty$ such that

$$Q(f_\varepsilon, g'') \geqq \|U(f_\varepsilon)\|_1 - \varepsilon \geqq M_U - 2\varepsilon, \qquad \|g''\|_\infty = 1.$$

Indeed, on putting $g_0 = \operatorname{sgn} U(f_\varepsilon)$ we have

$$Q(f_\varepsilon, g_0) = \|U(f_\varepsilon)\|_1.$$

The function g_0 is measurable and bounded and $\|g_0\|_B = 1$. It is plain that there exists a sequence of continuous functions $\{g_\nu(x)\}$ which converges to $g_0(x)$ almost everywhere, while $\|g_\nu\|_\infty = 1$. We have then

$$\int_a^b g_\nu(x)\, U(f_\varepsilon)\, dx \to \int_a^b g_0(x)\, U(f_\varepsilon)\, dx \quad \text{as} \quad \nu \to \infty.$$

Hence, for ν sufficiently large,

$$|Q(f_\varepsilon, g_\nu) - Q(f_\varepsilon, g_0)| < \varepsilon,$$

and we can take $g''(x) = g_\nu(x)$.

There also exists a function $f'' \subset L$, such that

$$Q(f'', g_\varepsilon) \geqq \|W(g_\varepsilon)\|_\infty - \varepsilon \geqq M_W - 2\varepsilon, \qquad \|f''\|_1 = 1.$$

Indeed, $W(g_\varepsilon) \subset L_\infty$ is continuous; there will exist an interval $I \subset (a, b)$ in which

$$\| W(g_\varepsilon) \|_\infty - \varepsilon \leqq | W(g_\varepsilon) | \leqq \| W(g_\varepsilon) \|_\infty .$$

Now define

$$f''(x) = \begin{cases} \dfrac{(b-a) \operatorname{sgn} W(g_\varepsilon)}{mI}, & x \subset I, \qquad mI = \text{length of } I, \\ 0 \text{ elsewhere.} \end{cases}$$

Obviously

$$Q(f'', g_\varepsilon) = \frac{1}{mI} \int_I | W(g_\varepsilon) | \, dx \geqq \| W(g_\varepsilon) \|_\infty - \varepsilon, \qquad \| f'' \|_1 = 1.$$

Thus we conclude again that $M = M_U = M_W$. The case $p = \infty$, $p' = 1$ can be treated in an analogous fashion and Lemma 2.2 is now completely established.[7]

III. We now pass to the theory of factor-sequence transformations. A sequence of constants $\{a_n\}$ is designated as a sequence of class (p, p), $1 \leqq p \leqq \infty$, with respect to the orthonormal set $\{\varphi_n(x)\}$ if, when being applied to an arbitrary Fourier series

$$\sum_{n=1}^{\infty} f_n \varphi_n(x)$$

of class L_p, it leads to a Fourier series

$$\sum_{n=1}^{\infty} a_n f_n \varphi_n(x)$$

of class L_p. It is well-known[8] that the classes of factor-sequences (p, p) and (p', p') are identical.

LEMMA 2.3. *If a factor-sequence* $\{a_n\} \subset (p, p)$, $1 \leqq p \leqq \infty$, *then the transformation* $T_a(x; f) \equiv T_a(f)$ *defined by*

$$f(x) \sim \sum_{n=1}^{\infty} f_n \varphi_n(x) \subset L_p, \tag{2.13}$$

$$T_a(x; f) \sim \sum_{n=1}^{\infty} a_n f_n \varphi_n(x) \subset L_p \tag{2.14}$$

---

[7] In the case $1 < p < \infty$ Lemma 2.2 is also immediately derived from the fact that the transformations $U(f)$, $W(g)$ satisfying (2.11) are adjoint of each other, and so have the same bounds. Cf. Hildebrandt [1], p. 200; Banach [1], p. 100. In the case $p = 1$ (or $p = \infty$) a difficulty arises owing to the fact that the space L_∞ (or L_1) is only a non-dense part of the space where the adjoint of U (or of W) is defined. We prefer to give a direct proof instead of investigating this point, which can easily be done, however.

[8] M. Riesz [1], p. 489, in the case of trigonometric Fourier series. In the general case, Orlicz [1], pp. 22–24. It should be observed that our notation (∞, ∞) corresponds to (C, C) of Orlicz.

is a linear limited transformation on L_p to a sub-set of L_p. The corresponding transformation defined by

$$(2.15) \qquad g(x) \sim \sum_{n=1}^{\infty} g_n \varphi_n(x) \subset L_{p'},$$

$$(2.16) \qquad T_a(x; g) \sim \sum_{n=1}^{\infty} a_n g_n \varphi_n(x) \subset L_{p'},$$

is also a linear limited transformation on $L_{p'}$ to a sub-set of $L_{p'}$, and we have

$$(2.17) \qquad M_a^{(p)} \equiv \underset{\|f\|_p=1}{\text{l. u. b.}} \|T_a(f)\|_p = M_a^{(p')} \equiv \underset{\|g\|_{p'}=1}{\text{l. u. b.}} \|T_a(g)\|_{p'}.$$

Furthermore, the relation

$$(2.18) \qquad \int_a^b g(x)\, T_a(x; f)\, dx = \int_a^b f(x)\, T_a(x; f)\, dx$$

holds for an arbitrary pair of functions $f(x) \subset L_p$, $g(x) \subset L_{p'}$.

Proof. The facts stated in this lemma are partly known and partly are readily derived from various passages of the above-mentioned papers by M. Riesz and Orlicz. The fact that the transformations $T_a(f)$ and $T_a(g)$ are limited follows immediately from a general result of Banach.[9] When $1 \leqq p < \infty$ the relation (2.18) has been proved by Orlicz.[10] Being symmetric in f, g it holds also when $p = \infty$. Now it only remains to apply Lemma 2.2.

LEMMA 2.4. *If $\{a_n\} \subset (p, p) = (p', p')$, $1 \leqq p \leqq \infty$, then also $\{a_n\} \subset (q, q)$ for any q between p and p'. The transformation $T_a(x; h)$, $h \subset L_q$, is limited on every L_q and has the bound*

$$(2.19) \qquad M_a^{(q)} \leqq M_a^{(p)} = M_a^{(p')}.$$

The expression $\log M_a^{(q)}$ is a convex function of q.[11]

3. (L_p)-effective transformations. We return to the notation of § 1 and first investigate conditions under which a transformation $\mathfrak{A}$ is (L_p)-effective. We observe that, in view of the above discussion, the m-th transforms $\tau_m(x; f)$, $\tau_m(x; g)$ of the Fourier series

$$(3.01) \qquad f(x) \sim \sum_{n=1}^{\infty} f_n \varphi_n(x) \subset L_p,$$

$$(3.02) \qquad g(x) \sim \sum_{n=1}^{\infty} g_n \varphi_n(x) \subset L_{p'}$$

[9] [1], p. 113, Theorem 8, proof.

[10] [1], p. 23.

[11] This lemma is merely a special case of a fundamental result of M. Riesz, [1], p. 481, Theorem V.

determine linear limited transformations on L_p, $L_{p'}$, respectively, provided condition (E.1) is satisfied for a fixed value of p or p'. Hence the problem of (L_p)-effectiveness reduces to the investigation of the convergence of the sequence of linear limited transformations $\tau_m(x; f)$ in the space L_p.

LEMMA 3.1. *If a transformation* $\mathfrak{A}$ *is* (L_p)*-effective for a fixed* p, $1 \leqq p \leqq \infty$, *then*

$$a_{mn} \to 1 \quad as \quad m \to \infty, \qquad n = 1, 2, \cdots. \tag{3.03}$$

Proof. If $\mathfrak{A}$ is (L_p)-effective, we must have

$$\|\tau_m(x; f) - f(x)\|_p \to 0 \quad \text{as} \quad m \to \infty$$

for an arbitrary function $f(x) \subset L_p$. On setting here $f(x) = \varphi_n(x)$ we obtain

$$\|\tau_m(x; \varphi_n) - \varphi_n(x)\|_p = |a_{mn} - 1| \, \|\varphi_n\|_p \to 0$$

which yields the desired result.

THEOREM 3.1. *In order that a transformation* $\mathfrak{A}$ *be* (L_p)*-effective,* $1 \leqq p \leqq \infty$, *it is necessary and sufficient that* (i) $\mathfrak{A}$ *be* (L_p)*-effective on every sub-set* L_p^0 *which is dense in* L_p, *and* (ii) *that the sequence of the associated m-th transforms of the series* (3.01) *be a uniformly limited sequence of transformations on* L_p. *Insofar as the sufficiency is concerned, condition* (i) *can be replaced by a less restrictive one, viz.,* (i′) $\mathfrak{A}$ *is* (L_p)*-effective on a* (*single*) *sub-set* L_p^0 *dense in* L_p.

Proof. The necessity of the condition (i) is obvious. The necessity of (ii) follows immediately from Lemma 2.1, since the (L_p)-effectiveness of the transformation $\mathfrak{A}$ implies

$$\|\tau_m(x; f) - f(x)\|_p \to 0 \quad \text{as} \quad m \to \infty,$$

i. e. the convergence in L_p of the sequence of linear limited transformations $\tau_m(x; f)$ at each point f of the space L_p.

Assume now that conditions (i′) and (ii) are satisfied. Then an arbitrary function $f(x) \subset L_p$ can be approximated in L_p as closely as we like by a function of L_p^0. Hence, if $f(x) \subset L_p$ and an arbitrarily small $\varepsilon > 0$ are given, we can find a function $\psi_\varepsilon(x) \subset L_p^0$ such that

$$\|f - \psi_\varepsilon\|_p \leqq \frac{\varepsilon}{2(M+1)}, \qquad M_{\tau_m} \leqq M.$$

Since $\mathfrak{A}$ is assumed to be (L_p)-effective on L_p^0 there exists an $m_0 = m_0(\varepsilon)$ such that

$$\|\tau_m(x; \psi_\varepsilon) - \psi_\varepsilon(x)\|_p \leqq \frac{\varepsilon}{2} \quad \text{for} \quad m \geqq m_0.$$

Then

$$\|\tau_m(x;f)-f(x)\|_p \leqq \|\tau_m(x;\psi_\varepsilon)-\psi_\varepsilon(x)\|_p+\|\tau_m(f-\psi_\varepsilon)\|_p+\|f-\psi_\varepsilon\|_p$$
$$\leqq \frac{\varepsilon}{2}+M\|f-\psi_\varepsilon\|_p+\|f-\psi_\varepsilon\|_p \leqq \varepsilon$$

which proves the sufficiency of conditions (i′), (ii).[12]

Remark 3.1. If condition (E.1) is satisfied, $\tau_m(x;f)$ is a linear limited transformation on L_p. Let $M_m^{(p)}$ be the bound of $\tau_m(x;f)$. Condition (ii) of Theorem 3.1 states that the sequence $\{M_m^{(p)}\}$ must be bounded. We set

$$\overline{\lim_{m\to\infty}}\; M_m^{(p)} \equiv M_{\mathfrak{A}}^{(p)} \tag{3.04}$$

and designate this constant as the (L_p)-bound of the transformation $\mathfrak{A}$.

THEOREM 3.2. *The classes of (L_p)-effective and of $(L_{p'})$-effective transformations are identical, and have the same (L_p)- and $(L_{p'})$-bounds.*

Proof. Let $\mathfrak{A}$ be (L_p)-effective. First we see, by Lemma 3.1, that

$$a_{mn}\to 1 \quad \text{as} \quad m\to\infty; \qquad n=1,2,\cdots. \tag{3.03}$$

Now consider the set L^0 consisting of all finite linear combinations of the functions $\varphi_n(x)$. By property (γ) (p. 330) of our orthonormal set $\{\varphi_n(x)\}$ the set L^0 is dense in L_∞. Since $\|f\|_{r_1}\leqq\|f\|_{r_2}$ whenever $r_1\leqq r_2$, the set L^0 is also dense in any L_q, $1\leqq q\leqq\infty$. It is readily seen that $\mathfrak{A}$ is $(L_{p'})$-effective on L^0. Indeed, let

$$\psi(x)=\sum_{\nu=1}^{s} c_\nu\varphi_\nu(x)$$

be an arbitrary function $\subset L^0$. Then, by (3.03),

$$\|\tau_m(x;\psi)-\psi(x)\|_{p'}=\left\|\sum_{\nu=1}^{s}(a_{m\nu}-1)c_\nu\varphi_\nu(x)\right\|_{p'}$$
$$\leqq\sum_{\nu=1}^{s}|a_{m\nu}-1|\,|c_\nu|\,\|\varphi_\nu\|_{p'}\to 0 \quad \text{as} \quad m\to\infty.$$

Hence condition (i′) of Theorem 3.1 for the $(L_{p'})$-effectiveness of $\mathfrak{A}$ is satisfied. In view of the (L_p)-effectiveness of $\mathfrak{A}$ condition (E.1) is satisfied, and so the transformations $\tau_m(x;f)$, $\tau_m(x;g)$, $f\subset L_p$, $g\subset L_{p'}$, are limited, and, by Lemma 2.3, have the same bounds $M_m^{(p)}=M_m^{(p')}$. By Theorem 3.1 we have

$$M_{\mathfrak{A}}^{(p')}=\overline{\lim_{m\to\infty}}\,M_m^{(p')}=\overline{\lim_{m\to\infty}}\,M_m^{(p)}=M_{\mathfrak{A}}^{(p)}<\infty.$$

[12] The argument which we use here is analogous to an argument of Orlicz [1], pp. 13–14. Hahn [1] has considered a general class of representations of a function $f(x)$ by means of singular integrals which possess the property of converging in the space L_1 to $f(x)$. It is readily seen that in the case of Fourier series the singular integrals of Hahn reduce to transformations which satisfy the sufficient conditions of our Theorem 3.1.

It follows that condition (ii) of Theorem 3.1 for the $(L_{p'})$-effectiveness is also satisfied, and so $\mathfrak{A}$ is (L_p)-effective.

Remark 3.2. If the condition (α) (p. 330) of continuity of the functions $\varphi_n(x)$ is replaced by the condition: (α') The function $\varphi_n(x)$ are essentially bounded (not necessarily uniformly in n), then an analogous argument will show that *every matrix $\mathfrak{A}$ which is (L_∞)-effective with respect to the set $\{\varphi_n(x)\}$ is also (L_1)-effective.* We do not know whether the converse holds under these, more general, assumptions.

THEOREM 3.3. *If a transformation $\mathfrak{A}$ is (L_p)-effective, $1 \leq p \leq \infty$, then $\mathfrak{A}$ is also (L_q)-effective for any q between p and p', and*

$$M_{\mathfrak{A}}^{(q)} \leq M_{\mathfrak{A}}^{(p)} = M_{\mathfrak{A}}^{(p')}. \tag{3.05}$$

Proof. Using the same set L^0 as in the proof of Theorem 3.2 we see that condition (i') for the (L_q)-effectiveness is satisfied. Condition (ii) is also satisfied; indeed the (L_p)-effectiveness of $\mathfrak{A}$ implies, in view of Lemma 2.4, that $\tau_m(x; h)$, $h \subset L_q$, is a limited transformation on L_q, while

$$M_m^{(q)} \leq M_m^{(p)}, \qquad M_{\mathfrak{A}}^{(q)} \leq \overline{\lim_{m\to\infty}}\, M_m^{(p)} = M_{\mathfrak{A}}^{(p)}.$$

We close this section by proving

THEOREM 3.4. *The condition* (3.03) *and*

$$|a_{mn}| \leq A, \qquad m, n = 1, 2, \cdots \tag{3.06}$$

are necessary and sufficient in order that the matrix $\mathfrak{A}$ be (L_2)-effective. In particular every regular transformation is (L_2)-effective.[13]

Proof. Let (3.03), (3.06) be satisfied. If $f(x)$ is an arbitrary function $\subset L_2$, we have by Parseval's identity

$$\|\tau_m(x; f)\|_2 = \left[\sum_{n=1}^{\infty} |a_{mn}|^2 |f_n|^2\right]^{1/2} \leq A\left[\sum_{n=1}^{\infty} |f_n|^2\right]^{1/2} = A\,\|f\|_2.$$

Now let

$$s_n(x; f) \equiv s_n \equiv \sum_{\nu=1}^{n} f_\nu\, \varphi_\nu(x)$$

be the n-th partial sum of the series (1.09). Then

$$\begin{aligned}\|\tau_m(f)-f\|_2 &\leq \|\tau_m(f)-\tau_m(s_n)\|_2 + \|s_n - f\|_2 + \|\tau_m(s_n) - s_n\|_2\\ &\leq (A+1)\,\|s_n - f\|_2 + \|\tau_m(s_n) - s_n\|_2.\end{aligned}$$

Since $\|s_n - f\|_2 \to 0$ as $n\to\infty$ and, in view of (3.03), $\|\tau_m(s_n) - s_n\|_2 \to 0$ as $m\to\infty$, n being fixed, it follows that $\|\tau_m(f)-f\|_2 \to 0$, and so, conditions (3.03), (3.06) are sufficient.

[13] A regular transformation is one which transforms every convergent series into a convergent series with the same sum. See the discussion in § 5.

To prove the necessity of these conditions we observe that the transformation which carries $f(x) \subset L_2$ into $\tau_m(x; f) \subset L_2$ may be considered as a transformation which carries an arbitrary element $\xi = (x_1, x_2, \cdots)$ of the space l_2 with the metric $\|\xi\| = [\sum |x_\nu|^2]^{1/2}$ into the element

$$\tau_m(\xi) = \eta^{(m)} = (a_{m1} x_1, a_{m2} x_2, \cdots) \subset l_2.$$

It is well-known that a necessary and sufficient condition that this transformation be limited is that the sequence $\{a_{mn}\}$ (for fixed m) be bounded. Furthermore

$$M_{\tau_m} = \underset{n=1,2,\cdots}{\text{l. u. b.}} |a_{mn}|.$$

The condition of the (L_2)-effectiveness of $\mathfrak{A}$ can be interpreted by stating that $\|\tau_m(\xi) - \xi\| \to 0$ at each point $\xi \subset l_2$. By Lemma 2.1 the sequence of transformations $\{\tau_m(\xi)\}$ must be uniformly limited, and (3.06) follows at once. Necessity of (3.03) follows from Lemma 3.1.

4. **(L_1)- and (L_∞)-effectiveness for trigonometric Fourier series.** In the present section we are concerned with trigonometric Fourier series

$$f(x) \sim S(x; f) \equiv f_0 + \sum_{n=1}^{\infty} (f_n e^{inx} + f_{-n} e^{-inx}), \qquad f_n = \frac{1}{2\pi} \int_{-\pi}^{\pi} f(t) e^{-int} dt.$$

The interval (a, b) reduces here to $(-\pi, \pi)$ and $f(x)$ is assumed to be periodic. The transform of $S(x; f)$ by means of a factor sequence $\{a_0, a_1, \cdots, a_n, \cdots\}$ is defined at present by

$$T_a(x; f) \equiv T_a(f) \sim a_0 f_0 + \sum_{n=1}^{\infty} a_n (f_n e^{inx} + f_{-n} e^{-inx})$$

while the m-th transform of $S(x; f)$ by means of a matrix $\mathfrak{A}$ is

$$\tau_m(x; f) \sim a_{m0} f_0 + \sum_{n=1}^{\infty} a_{mn} (f_n e^{inx} + f_{-n} e^{-inx}).$$

It is obvious how the results of § 3 should be modified with the present notation.

As concerns the factor-sequences of class $(1, 1) = (\infty, \infty)$ we state

LEMMA 4.1. *A necessary and sufficient condition that a factor-sequence* $\{a_n\} \subset (1, 1) = (\infty, \infty)$ *is that*

$$a_n = \int_{-\pi}^{\pi} e^{int} d\alpha(t), \qquad n = 0, 1, \cdots \tag{4.01}$$

where $\alpha(t)$ *is an arbitrary function of bounded variation such that*

$$\alpha(-t) = -\alpha(t). \tag{4.02}$$

It can be assumed without loss of generality that $\alpha(t)$ is normalized according to

$$(4.03)\qquad \alpha(t) = \frac{1}{2}[\alpha(t+0)+\alpha(t-0)], \quad -\pi<t<\pi.$$

The transformation $T_a(f)$ is then represented by

$$(4.04)\qquad T_a(x;f) = \int_{-\pi}^{\pi} f(x+t)\,d\alpha(t).$$[14]

We observe that the integral of the right-hand member is taken over the closed interval $(-\pi,\pi)$ in the sense of Young-Lebesgue-Stieltjes, and reduces to the classical Riemann-Stieltjes integral when $f \subset L_\infty$. In the case $f \subset L_1$ this integral exists for almost all x and represents a function $\subset L_1$.

In the following discussion we shall consider the class $\mathfrak{S}$ of transformations $\mathfrak{A}$ which sum the Fourier series of an arbitrary step-function $\varphi(x)$ to the value $\varphi(x)$ at each point of continuity of $\varphi(x)$.

THEOREM 4.1. *The most general expression of the elements of a matrix $\mathfrak{A} \subset \mathfrak{S}$ which is (L_1)- or (L_∞)-effective is given by*

$$(4.05)\qquad a_{mn} = \int_{-\pi}^{\pi} e^{int}\,d\alpha_m(t); \qquad m, n = 0, 1, 2, \cdots.$$

Here $\alpha_m(t)$ are functions of bounded variation such that

$$(4.06)\qquad \begin{aligned} \alpha_m(t) &= \frac{1}{2}[\alpha_m(t+0)+\alpha_m(t-0)], \qquad -\pi<t<\pi,\\ \alpha_m(-t) &= -\alpha_m(t), \qquad \alpha_m(0) = 0,\end{aligned}$$

$$(4.07)\quad \alpha_m(x)\to\frac{1}{2}, \quad \alpha_m(x+0)-\alpha_m(x-0)\to 0 \quad as \quad m\to\infty, \quad 0<x\leqq\pi.$$

$$(4.08)\qquad V_m \equiv 2\int_0^{\pi} |d\alpha_m(t)| \leq A < \infty.$$

The (L_1)- and (L_∞)-bounds of $\mathfrak{A}$ are given by

$$(4.09)\qquad M_{\mathfrak{A}}^{(1)} = M_{\mathfrak{A}}^{(\infty)} = \overline{\lim_{m\to\infty}}\, V_m.$$

Proof. If $\mathfrak{A}$ is (L_1)- or (L_∞)-effective, conditions (E.1) and (E.2) are satisfied with $p=1$ and $p=\infty$. As in the proof of Theorems 3.1, 3.2 it follows that the sequence of transformations $\{\tau_m(x;f)\}$ is uniformly limited (in L_1 and in L_∞). Since

[14] M. Riesz [1], p. 488. The notation of Riesz is slightly changed here. The function $\alpha(t)$ is defined only for $-\pi\leqq t\leqq\pi$. We shall assume, without loss of generality, that outside of this interval $\alpha(t)$ equals $\alpha(\pm\pi)$ according as $t>\pi$ or $t<-\pi$. The same agreement will be made tacitly concerning the functions $\alpha_m(t)$ below.

(4.10) $$M_m^{(1)} = M_m^{(\infty)} = \int_{-\pi}^{\pi} |d\alpha_m(t)|$$

we conclude that (4.09) holds; (4.08) then follows.

By hypothesis $\mathfrak{A} \subset \mathfrak{S}$. If we choose for $\varphi(t)$ the periodic step-functions defined by

$$\varphi_1(t) = \begin{cases} 1 & \text{if } 0 \leqq |t| < x \leqq \pi, \\ 0 & \text{if } |t| \geqq x, \end{cases}$$

$$\varphi_2(t) = \begin{cases} 0 & \text{if } -\pi \leqq t < x, \quad 0 < x \leqq \pi, \\ 1 & \text{if } t = x, \\ 0 & \text{if } t > x, \end{cases}$$

we find respectively (Lemma 4.1) that

$$\tau_m(0;\varphi_1) = \int_{-x+0}^{x-0} d\alpha_m(t) = 2\alpha_m(x-0) \to \varphi_1(0) = 1,$$

$$\tau_m(0;\varphi_2) = \alpha_m(x+0) - \alpha_m(x-0) \to \varphi_2(0) = 0,$$

which in view of (4.06) establishes the necessity of (4.07). The necessity of (4.05) and (4.06) follows from Lemma 4.1.

Assume that conditions (4.05)–(4.08) are satisfied. It follows immediately from (4.07) that $\mathfrak{A} \subset \mathfrak{S}$. We now pass on to the proof that the transformation $\mathfrak{A}$ is (L_∞)- and (L_1)-effective. First let $f(x)$ be an arbitrary function $\subset L_\infty$, viz., periodic and continuous. In view of (4.05)–(4.07) we have, as $m \to \infty$,

$$\tau_m(x;f) - f(x) = \int_0^{\pi} \omega(x,t)\, d\alpha_m(t) + o(1),$$

where

$$\omega(x,t) \equiv f(x+t) + f(x-t) - 2f(x).$$

Let us investigate the integral

$$I \equiv \int_0^{\pi} \omega(x,t)\, d\alpha_m(t) = \int_0^{\delta} + \int_{\delta}^{\pi} \equiv I_1 + I_2.$$

Since $f(x)$ is continuous, being given an arbitrarily small $\varepsilon > 0$, we can fix a δ so small that

$$|I_1| \leqq A \max_{0 \leqq t \leqq \delta} |\omega(x,t)| \leqq \frac{\varepsilon}{3}.$$

Let $\varphi(x)$ be an arbitrary step-function and

$$\omega_1(x,t) \equiv \varphi(x+t) + \varphi(x-t) - 2\varphi(x).$$

We can write

$$I_2 = \int_{\delta}^{\pi} [\omega(x,t) - \omega_1(x,t)]\, d\alpha_m(t) + \int_{\delta}^{\pi} \omega_1(x,t)\, d\alpha_m(t) \equiv I_2' + I_2''.$$

The number δ being fixed, select the function $\varphi(x)$ in such a way that

$$|\omega(x, t) - \omega_1(x, t)| \leqq \eta, \qquad |I_2'| \leqq \eta A \leqq \frac{\varepsilon}{3}.$$

In view of (4.07), after the function $\varphi(x)$ has been fixed,

$$\int_\delta^\pi \omega_1(x, t)\, d\alpha_m(t) \to 0 \quad \text{as} \quad m \to \infty,$$

uniformly in x. Hence, for sufficiently large values of m and uniformly in x,

$$|I_2''| \leqq \frac{\varepsilon}{3}.$$

On combining these results we obtain

$$|\tau_m(x; f) - f(x)| \leqq \varepsilon,$$

uniformly in x, or

$$\|\tau_m(x; f) - f(x)\|_\infty \to 0 \quad \text{as} \quad m \to \infty$$

which establishes the (L_∞)-effectiveness of $\mathfrak{A}$.

The (L_1)-effectiveness of $\mathfrak{A}$ follows from Theorem 3.2.

The transformations $\mathfrak{A}$ which transform convergent series into convergent series with the same sum are called regular. We designate by $\mathfrak{R}$ the class of all regular transformations. It is plain then that $\mathfrak{R} \subset \mathfrak{S}$. Necessary and sufficient conditions for the regularity of a transformation $\mathfrak{A}$ are given by

(R.1) $$a_{mn} \to 1 \quad \text{as} \quad m \to \infty; \qquad n = 0, 1, 2, \cdots;$$

(R.2) $$\sum_{n=0}^{\infty} |a_{mn} - a_{m\,n+1}| \leqq A$$

where A is a fixed positive constant.[15]

THEOREM 4.2. *The most general expression of the elements of a regular matrix $\mathfrak{A}$ which is (L_1)- or (L_∞)-effective is given by* (4.05), *where the function $\alpha_m(t)$, in addition to conditions* (4.06)–(4.08), *satisfies also the condition*

(4.11) $$\alpha_m(t) = a_m \eta(t) + \int_0^t \beta_m(u)\, du,$$

where a_m is a constant and $\eta(t)$ is the function defined by

(4.12) $$\eta(t) = \begin{cases} \frac{1}{2} \text{ if } t > 0, \\ 0 \text{ if } t = 0, \\ -\frac{1}{2} \text{ if } t < 0, \end{cases} \quad |t| < \pi; \quad \eta(t + 2\pi) = \eta(t),$$

[15] Cf. Hahn [2], pp. 33–34.

and $\beta_m(t)$ *is a function* $\subset L_1$ *such that the series*

$$\sum_{n=0}^{\infty} |\gamma_{mn}|, \qquad \gamma_{mn} = \int_{-\pi}^{\pi} e^{nit}(1-e^{it})\,\beta_m(t)\,dt \tag{4.13}$$

converges for each $m = 0, 1, 2, \cdots$ *and has a sum which is uniformly bounded in* m.

Proof. We have only to investigate the additional conditions (4.11)–(4.13). Since $\mathfrak{A}$ is (L_∞)-effective, we have for $f(x) = e^{inx}$,

$$a_{mn} = \int_{-\pi}^{\pi} e^{int}\,d\alpha_m(t) = \tau_m(0; f) \to f(0) = 1$$

so that condition (R. 1) of the regularity of $\mathfrak{A}$ is automatically satisfied. For simplification we define a_{mn} for negative values of n by putting $a_{m,-n} = a_{mn}$. Then

$$a_{mn} - a_{mn+1} = \int_{-\pi}^{\pi} e^{int}(1-e^{it})\,d\alpha_m(t) = \int_{-\pi}^{\pi} e^{int}\,dA_m(t)$$

where

$$A_m(t) = \int_0^t (1-e^{iu})\,d\alpha_m(u).$$

We set

$$B(t) = ct + A_m(t)$$

and choose the constant c under the condition

$$B(\pi) = B(-\pi).$$

Then

$$a_{mn} - a_{mn+1} = \int_{-\pi}^{\pi} e^{int}\,dB(t) = -in\int_{-\pi}^{\pi} e^{int}\,B(t)\,dt \equiv \pi b_n, \quad n \neq 0,$$

$$b_0 \equiv 0.$$

Condition (R. 2) of the regularity of $\mathfrak{A}$ implies, and is implied by, the convergence of $\sum_{-\infty}^{\infty} |b_n|$. Hence $\sum_{-\infty}^{\infty} b_n e^{inx}$ is as an absolutely convergent Fourier series and represents a continuous function which equals

$$\frac{d}{dx}\sum_{n \neq 0} \frac{b_n}{-in}\,e^{-inx} = \frac{dB(x)}{dx}.$$

Thus

$$B(t) = ct + \alpha_m(t)(1-e^{it}) + \int_0^t e^{iu}\,\alpha_m(u)\,du$$

is absolutely continuous and has a continuous derivative, which shows that

$$\alpha_m(t)(1-e^{it})$$

is absolutely continuous on $(-\pi, \pi)$. Hence $\alpha_m(t)$ is itself absolutely continuous on $(-\pi, \pi)$, except perhaps at $t = 0$, where $\alpha_m(t)$ may have a jump, a_m say. Thus we have (4.11) and the proof of the last statement of Theorem 4.2 follows at once.

5. (L_1)-**effectiveness of convergence.** We have already mentioned in § 1 that the "convergence" transformation which is defined by

$$a_{mn} = 1, \text{ or } 0 \text{ according as } n \leqq m \text{ or } n > m, \quad m = 0, 1, 2, \cdots, \tag{5.01}$$

$$\tau_m(x; f) = s_m(x; f) \equiv s_m(x) \equiv s_m \equiv f_0 + \sum_{n=1}^{m} (f_n e^{inx} + f_{-n} e^{-inx}), \tag{5.02}$$

is not (L_1)-effective. We shall indicate two sub-sets of L_1 on which the convergence is (L_1)-effective. These sub-sets will be constituted by sine- or cosine-series

$$S(x) \equiv \sum_{n=1}^{\infty} q_n \sin nx, \tag{5.03}$$

$$C(x) \equiv \sum_{n=1}^{\infty} q_n \cos nx \tag{5.04}$$

where the sequence $\{q_n\}$, in addition to being positive and $\downarrow 0$, is subjected to certain other restrictions. Instead of saying that the convergence transformation is (L_1)-effective we shall say that the corresponding Fourier series converge in the mean to their generating functions.

THEOREM 5.1. *In order that the sine-series*

$$S(x) = \sum_{n=1}^{\infty} q_n \sin nx, \quad q_n \downarrow 0, \tag{5.05}$$

converge in the mean to a function $f(x) \subset L_1$, it is necessary and sufficient that the series

$$\sum_{n=1}^{n} \frac{q_n}{n} \tag{5.06}$$

converge.

Proof. The condition (5.06) is necessary and sufficient in order that (5.05) be a Fourier series, and hence that $S(x) \subset L_1$.[16] Now, if $\|f - s_n\|_1 \to 0$ then $S(x) \subset L_1$ which shows the necessity of (5.06) for the convergence of (5.05) in the mean.

To prove the sufficiency of (5.06) we first observe that (5.06) implies that

$$\sum_{n=1}^{\infty} (q_n - q_{n+1}) \log n < \infty \tag{5.07}$$

and that

$$q_n \log n \to 0 \quad \text{as} \quad n \to \infty. \tag{5.08}$$

[16] Young [1], p. 46.

This follows from

$$\sum_{\nu=1}^{n} \frac{q_\nu}{\nu} = \sum_{\nu=1}^{n-1} \sigma_\nu (q_\nu - q_{\nu+1}) + \sigma_n q_n,$$

$$\sum_{\nu=m}^{n} \frac{q_\nu}{\nu} > q_n (\sigma_n - \sigma_{m-1}) \sim q_n \log\left(\frac{n}{m}\right),$$

where

$$\sigma_n = 1 + \frac{1}{2} + \cdots + \frac{1}{n} \sim \log n.$$

Now we have

$$2 \sin\frac{x}{2} [s_n(x) - s_{m-1}(x)] = \sum_{\nu=m}^{n} q_\nu 2 \sin\frac{x}{2} \sin \nu x$$

$$= \sum_{\nu=m}^{n} q_\nu [t_{\nu+1}(x) - t_\nu(x)] = -q_m t_m(x) + q_n t_{n+1}(x) + \sum_{\nu=m}^{n-1} (q_\nu - q_{\nu+1}) t_{\nu+1}(x),$$

$$t_\nu(x) = 1 - \cos\left(\nu - \frac{1}{2}\right) x,$$

whence

$$\| s_n(x) - s_m(x) \|_1 \leqq q_m I_m + q_n I_{n+1} + \sum_{\nu=m}^{n-1} (q_\nu - q_{\nu+1}) I_{\nu+1},$$

where

$$I_\nu = \int_0^{\pi} \left[1 - \cos\left(\nu - \frac{1}{2}\right) x\right] \left[\sin\frac{x}{2}\right]^{-1} dx = O(\log \nu).$$

From (5.07), (5.08) it follows that the sequence $\{s_n(x)\}$ converges in the mean. There will exist then a function $f(x) \subset L_1$, to which $s_n(x)$ converges in the mean. Since $f(x)$ has the same Fourier coefficients as $S(x) \subset L_1$, these two functions must be equal almost everywhere. Hence $\| S - s_n \|_1 \to 0$, and Theorem 5.1 is established.

In the case of the cosine-series (5.04) the situation is less satisfactory owing to the fact that no simple necessary and sufficient conditions are known which have to be imposed on the sequence in order to make $C(x) \subset L_1$. It is known[17] that the condition $q_n \downarrow 0$ is not sufficient that (5.04) be a Fourier series. On the other hand simple sufficient conditions have been found by Young and Kolmogoroff.[18] They consist in the assumption that $q_n \to 0$ while the series

$$\sum_{n=1}^{\infty} n |\Delta^2 q_n|, \qquad \Delta q_n = q_n - q_{n+1}, \qquad \Delta^2 q_n = \Delta q_n - \Delta q_{n+1},$$

converges. These conditions are satisfied in the particular case where $q_n \downarrow 0$, $\Delta q_n \downarrow 0$ (the sequence $\{q_n\}$ is convex). For this class of cosine

[17] Szidon [1], p. 126.

[18] Young [1], pp. 44–45; Kolmogoroff [1]. The conditions of Kolmogoroff are more general than those of Young.

series it has been proved by Kolmogoroff that a necessary and sufficient condition that (5.04) converge in the mean to $f(x) \subset L$ is given by $q_n \log n \to 0$.

Bibliography.

S. Banach.

1. Théorie des opérations linéaires. Warsaw. 1932.

W. Gross.

1. Zur Poissonschen Summierung. Sitzungsber. Wiener Akademie, Abt. IIa, 124 (1915), pp. 1017–1037.

H. Hahn.

1. Über die Darstellung gegebener Funktionen durch singuläre Integrale. II. Mitt., Denkschr. Wiener Akademie, Math.-Naturw. Kl., 93 (1916), pp. 657–692.
2. Über Folgen linearer Operationen. Monatshefte Math. u. Ph., 32 (1922), pp. 1–88.

H. T. Hildebrandt.

1. Linear functional transformations in general spaces. Bull. Amer. Math. Soc., 37 (1931), pp. 185–212.

E. Hille and J. D. Tamarkin.

1. On the summability of Fourier series. I. Trans. Amer. Math. Soc., 34 (1932), pp. 757–783.
2. On the summability of Fourier series. III. To appear in the Mathemat. Annalen.

A. Kolmogoroff.

1. Sur l'ordre de grandeur des coefficients de la série de Fourier. Bull. Intern. de l'Acad. Polonaise, Classe des sc. math. 1923, pp. 83–86.

W. Orlicz.

1. Beiträge zur Theorie der Orthogonalentwickelungen. Studia Math., 1 (1929), pp. 1–39.

M. Riesz.

1. Sur les maxima des formes bilinéaires et sur les fonctionnelles linéaires. Acta Math. 49 (1926), pp. 465–497.

M. Steinhaus.

1. Sur les développements orthogonaux. Bull. Intern. Ac. Polon. Crac., 1926, pp. 11–39.

S. Szidon.

1. Reihentheoretische Sätze und ihre Anwendungen in der Theorie der Fourierschen Reihen. Math. Zeitschr. 10 (1921), pp. 121–127.

W. H. Young.

1. On the Fourier series of bounded functions. Proc. London Math. Soc., (2) 12 (1913), pp. 41–70.

A. Zygmund.

1. Sur les fonctions conjuguées. Fund. Math., 13 (1929), pp. 284–303.

PRINCETON UNIVERSITY, BROWN UNIVERSITY.

ADDITION TO THE PAPER "ON THE SUMMABILITY OF FOURIER SERIES II".[19] [20]

BY EINAR HILLE AND J. D. TAMARKIN.

6. **(B)-effective transformations.** In § 1 we explained the reasons which compelled us to denote by L_∞ the class of all continuous functions rather than the class B of all essentially bounded functions. It appears, however, that results of some interest may be obtained for transformations of Fourier series of functions $\subset B$, provided the limit concept in this space is defined not by means of the metric of B, but rather by means of *asymptotic convergence.*[21] We shall denote asymptotic convergence by the symbol $\xrightarrow{\text{as}}$. The following known property is essential for our discussion.

LEMMA 6.1. *If a sequence of functions $\{f_n(x)\} \subset L_1$ converges in L_1, it also converges asymptotically. Conversely, if $\{f_n(x)\}$ converges asymptotically and is bounded in B (that is, the functions $f_n(x)$ are uniformly essentially bounded), then $\{f_n(x)\}$ converges in L_1.*

We now introduce

DEFINITION 6.1. *A transformation $\mathfrak{A}$ is said to be (B)-effective if it satisfies the following conditions*

(E$'$.1) *The series*

$$\tau_m(x; f) \sim \sum_{n=1}^{\infty} a_{mn} f_n \varphi_n(x) \subset B \qquad m = 1, 2, \cdots, \tag{6.1}$$

whenever $f(x) \subset B$.

(E$'$.2) *The sequence of the m-th transforms converges asymptotically to $f(x)$,*

$$\tau_m(x; f) \xrightarrow{\text{as}} f(x) \text{ when } m \to \infty. \tag{6.2}$$

(E$'$.3) *For every fixed function $f(x) \subset B$ the sequence of the m-th transforms is bounded in B. In other words, there exists a constant $K(f)$ depending only on $f(x)$ such that*

$$\| \tau_m(x; f) \|_B \leqq K(f). \tag{6.3}$$

Our purpose is now to prove

THEOREM 6.1. *The classes of (B)-effective, (L_1)-effective and (L_∞)-effective transformations are identical.*

---

[19] Received March 15, 1933.

[20] The present note constitutes § 6 of our article in the preceding issue of these Annals, pp. 329–348.

[21] Banach [1], pp 3–4. We are indebted to Professor M. Riesz for calling our attention to such a possibility.

Proof. Let $\{a_n\}$ be a factor-sequence which transforms the Fourier series of an arbitrary function $f(x) \subset B$ into a Fourier series $\subset B$. The class of all such factor-sequences is denoted by (B, B). It is well-known (see footnote [8] on p. 336) that the classes (B, B) and $(1, 1) = (\infty, \infty)$ are identical. Hence, whenever a transformation $\mathfrak{A}$ satisfies condition (E. 1) of p. 331 with $p = 1$ or $p = \infty$, it also satisfies condition (E'.1), and vice versa. Now assume that $\mathfrak{A}$ is (L_1)-effective. Let $f(x)$ be an arbitrary function $\subset B$. Since (E.1), and therefore (E'.1), is satisfied, each transform $\tau_m(x; f) \subset B$. By hypothesis

$$(6.4) \quad \|\tau_m(x;f) - f(x)\|_1 \to 0, \quad \|\tau_m(x;f)\|_1 \leqq M_m^{(1)} \|f\|_1 \leqq M_m^{(1)} \|f\|_B.$$

Consequently, by Lemma 6.1,

$$\tau_m(x; f) \xrightarrow{\text{as}} f(x) \text{ when } m \to \infty,$$

and so condition (E'.2) is satisfied. To prove that (E'.3) is also satisfied we have to show that the sequence $\{\tau_m(x; f)\}$ is uniformly essentially bounded. Let $g(x)$ be an arbitrary function $\subset L_1$. We can apply relation (2.18) of p. 337 which holds for an arbitrary pair of functions $f(x) \subset B$, $g(x) \subset L_1$ (see footnote [10] on p. 337). Thus, upon putting

$$(6.5) \qquad U_m(g) = \frac{1}{b-a} \int_a^b g(x)\,\tau_m(x; f)\, dx,$$

we have

$$(6.6) \qquad U_m(g) = \frac{1}{b-a} \int_a^b f(x)\,\tau_m(x; g)\, dx,$$

and, by (6.4),

$$(6.7) \qquad |U_m(g)| \leqq M_m^{(1)} \|f\|_B \|g\|_1, \qquad m = 1, 2, \cdots.$$

If we consider $U_m(g)$ as a linear operation on L_1, we see that $\{U\ (g)\}$ is a sequence of linear limited operations bounded on L_1. By a theorem of Banach[22] the sequence of their bounds, $\{\|\tau_m(x; f)\|_B\}$, is bounded. Thus condition (E'.3) is satisfied and $\mathfrak{A}$ is (B)-effective.

Conversely, assume that $\mathfrak{A}$ is (B)-effective. Condition (E'.1), hence also condition (E.1) with $p = 1$ and $p = \infty$, is satisfied. Moreover, by (E'.2) and (E'.3), for an arbitrary function $f(x) \subset B$ we have

$$\tau_m(x; f) \xrightarrow{\text{as}} f(x), \qquad \|\tau_m(x; f)\|_B \leqq K(f).$$

In view of Lemma 6.1 this implies

$$\|\tau_m(x; f) - f(x)\|_1 \to 0 \quad \text{when} \quad m \to \infty.$$

The set of all essentially bounded functions is dense in L_1, and so the transformation $\mathfrak{A}$ is (L_1)-effective over a dense set of L_1. Consider the

[22] Banach [1], p. 80, Theorem 5.

sequence of the mth transforms $\{\tau_m(x; g)\}$, where $g(x)$ is an arbitrary function of L_1. To apply Theorem 3.1 of p. 338 we have only to show that this sequence is uniformly limited on L_1. We now write

$$(6.8)\quad V_m(f) = \frac{1}{b-a}\int_a^b f(x)\tau_m(x; g)\,dx = \frac{1}{b-a}\int_a^b g(x)\,\tau_m(x; f)\,dx.$$

This is a linear limited operation on B with the bound $\|\tau_m(x; g)\|_1$. For each element $f(x) \subset B$ we have

$$(6.9)\qquad |V_m(f)| \leqq \|\tau_m(x; f)\|_B \|g\|_1 \leqq K(f)\|g\|_1, \qquad m = 1, 2, \cdots,$$

whence the sequence of operations $\{V_m(f)\}$ is bounded on B. By the same theorem of Banach[22] the sequence of bounds $\{\|\tau_m(x; g)\|_1\}$ is bounded. Consequently the sequence $\{\tau_m(x; g)\}$ of linear limited transformations on L_1 is bounded on L_1. Another application of Banach's theorem will show that the sequence $\{\tau_m(x; g)\}$ is uniformly limited on L_1. Since all the conditions of Theorem 3.1 are satisfied, we conclude that $\mathfrak{A}$ is (L_1)-effective, and Theorem 6.1 is completely proved.

Remark 6.1. Let $f(x)$ be an arbitrary function $\subset B$. By referring to the papers mentioned in footnotes [9] and [10] on p. 337 it is readily found that Lemma 2.3 remains valid if the space L_∞ is replaced by B. Hence, for a fixed m, $\tau_m(x; f)$ defines a linear limited transformation on B to B, with the bound $M_m^{(B)} = M_m^{(1)}$. Thus

$$(6.10)\qquad \|\tau_m(x; f)\|_B \leqq M_m^{(1)}\|f\|_B.$$

The quantity

$$M_{\mathfrak{A}}^{(B)} = \overline{\lim_{m\to\infty}}\, M_m^{(B)} = \overline{\lim_{m\to\infty}}\, M_m^{(1)} = M_{\mathfrak{A}}^{(1)}$$

may be designated as the (B)-bound of the transformation $\mathfrak{A}$. We conclude that the (B)-bound of a (B)-effective transformation $\mathfrak{A}$ is equal to its (L_1)- and (L_∞)-bounds.

Remark 6.2. Let a sequence $\{f_k(x)\} \subset B$ be uniformly essentially bounded and let

$$f_k(x) \xrightarrow{\text{as}} f(x) \quad \text{when} \quad k \to \infty.$$

If $\{a_n\}$ is an arbitrary sequence $\subset (B, B)$, then

$$T_a(x; f) \sim \sum_{n=1}^{\infty} a_n f_n \varphi_n(x), \qquad f(x) \sim \sum_{n=1}^{\infty} f_n \varphi_n(x),$$

is a linear limited transformation on B to B. We shall prove that it is continuous in the sense that the sequence of functions $\{T_a(x; f_k)\}$ is uniformly essentially bounded, while

(6.11)
$$T_a(x; f_k) \xrightarrow{\text{as}} T_a(x; f)$$
whenever
(6.12)
$$f_k(x) \xrightarrow{\text{as}} f(x), \qquad \|f_k(x)\|_B \leqq K.$$
Indeed, since
$$\|T_a(x; f_k)\|_B \leqq M_a^{(1)} \|f_k(x)\|_B \leqq K M_a^{(1)},$$

the sequence of functions $\{T_a(x; f_k)\}$ is essentially uniformly bounded. On the other hand, by Lemma 6.1,
$$\|f_k(x) - f(x)\|_1 \to 0, \qquad \|T_a(x; f_k) - T_a(x; f)\|_1 \to 0,$$
whence
$$T_a(x; f_k) \xrightarrow{\text{as}} T_a(x; f) \quad \text{when} \quad k \to \infty.$$

Remark 6.3. We call a transformation $\mathfrak{A}$ which satisfies only conditions (E′.1) and (E′.2), but not (E′.3), (B')-effective. The argument of the first part of the proof of Theorem 6.1 shows that if $\mathfrak{A}$ is (L_1)-effective, it is also (B')-effective. It is an open question whether the class of (B')-effective transformations coincides with the class of (L_1)-effective transformations or is actually larger than this class.

Chapter 6
Fourier Transforms and Analytic Function Theory

Papers included in this chapter are

Commentaries on the four papers in this chapter can be found starting on page 669.

ON A THEOREM OF PALEY AND WIENER.*

By Einar Hille and J. D. Tamarkin.

1. In their recent work on the theory of quasi-analytic functions Paley and Wiener prove the following important theorem which is the basis of their discussion.

Let $\omega(x)$ be a real non-negative function not equivalent to zero defined for $-\infty < x < \infty$, and of integrable square in this range. A necessary and sufficient condition that there exist a real- or complex-valued function $F(x)$ defined in the same range, vanishing for $x < 0$, and such that the Fourier transform $G(x)$ of $F(x)$ satisfy the condition $|G(x)| = \omega(x)$ is that

$$\int_{-\infty}^{\infty} \frac{|\log \omega(x)|}{1+x^2} dx < \infty.$$ [1]

In the present note we prove a slightly more general theorem. This proof uses essentially the same idea as that of Paley and Wiener, but seems to throw some additional light on the situation.

2. We shall deal with two complex variables, $z = x + iy$, and $w = \varrho e^{i\theta}$ which are related by

$$w = \frac{1+iz}{1-iz}, \tag{1}$$

so that the half-plane $\Im(z) > 0$ is mapped on the interior of the unit circle $|w| < 1$. The variables $\xi = e^{i\varphi}$ and $\zeta = \tan\frac{\varphi}{2}$ will trace the circumference of the unit-circle in the w-plane and the real axis in the z-plane respectively.

Let p be any number such that $1 \leqq p < \infty$. The classes of functions $\Phi(\xi)$, $F(\zeta)$ for which the integrals

$$\int_{-\pi}^{\pi} |\Phi(\xi)|^p d\varphi, \quad \int_{-\infty}^{\infty} |F(\zeta)|^p d\zeta$$

exist will be designated by L_p, $\mathfrak{L}_p$ respectively. The formula

$$2\int_{-\infty}^{\infty} \frac{|F(\zeta)|^p}{1+\zeta^2} d\zeta = \int_{-\pi}^{\pi} |\Phi(\xi)|^p d\varphi, \quad \Phi(\xi) = F\left(\tan\frac{\varphi}{2}\right),$$

---

* Received March 16, 1933.

[1] R. E. A. C. Paley and N. Wiener, *Notes on the theory and application of Fourier transforms. I. On a theorem of Carleman.* Trans. Amer. Math. Soc., 35 (1933), pp. 348–355. Paley and Wiener write $x > 0$ instead of $x < 0$. This distinction is due to a different notation for the Fourier transform of a function.

shows that $\Phi(\xi) \subset L_p$ whenever $F(\zeta) \subset \mathfrak{L}_p$. We designate by H_p the class of functions $\gamma(w)$, analytic in $|w|<1$, such that the integral

$$\int_{-\pi}^{\pi} |\gamma(\varrho e^{i\theta})|^p \, d\theta$$

is bounded for $0 \leqq \varrho < 1$.

The integrals

$$I(z; F) = (2\pi i)^{-1} \int_{-\infty}^{\infty} \frac{F(\zeta)}{\zeta - z} \, d\zeta, \tag{2}$$

$$J(w; \Phi) = (2\pi i)^{-1} \int_{-\pi}^{\pi} \frac{\Phi(\xi)}{\xi - w} \, d\xi \tag{3}$$

will define the integrals of Cauchy type associated with the functions $F(\zeta)$, $\Phi(\xi)$, while the integrals

$$Q(z; F) = \frac{1}{\pi} \int_{-\infty}^{\infty} \frac{F(\zeta) y}{(\zeta - x)^2 + y^2} \, d\zeta = I(z; F) - I(\bar{z}; F), \quad \bar{z} = x - iy, \tag{4}$$

$$P(w; \Phi) = \frac{1}{2\pi} \int_{-\pi}^{\pi} \Phi(\xi) \frac{1-\varrho^2}{1+\varrho^2 - 2\varrho\cos(\varphi - \theta)} \, d\varphi = J(w; \Phi) - J(w^*; \Phi), \quad w^* = \bar{w}^{-1}, \tag{5}$$

will define the Poisson integrals associated with the same functions. It is plain that $I(z; F)$, $J(w; \Phi)$ are analytic in their respective variables whenever $\mathfrak{J}(z) \neq 0$, $|w| \neq 1$, $F(\zeta) \subset \mathfrak{L}_p$, $\Phi(\xi) \subset L_p$. It is well known that, for almost all ζ, ξ,

$$\lim_{z \to \zeta} Q(z; F) = F(\zeta),$$

$$\lim_{w \to \xi} P(w; \Phi) = \Phi(\xi),$$

where z and w tend to their respective limits along arbitrary non-tangential paths. If the variables z, w are related by (1), the variables ζ, ξ by $\xi = \dfrac{1+i\zeta}{1-i\zeta} = e^{i\varphi}$, $\zeta = \tan\dfrac{\varphi}{2}$, and the functions $F(\zeta)$, $\Phi(\xi)$ by $F(\zeta) = F\left(\tan\dfrac{\varphi}{2}\right) = \Phi(\xi)$, then a direct computation shows that

$$J(w; \Phi) = I(z; F) - C, \quad C = I(-i; F), \tag{6}$$

$$P(w; \Phi) = Q(z; F). \tag{7}$$

Let $g(z)$ be analytic in the half-plane $\mathfrak{J}(z) > 0$. For almost all ζ let $g(z)$ tend to a finite limit as $z \to \zeta$ along all non-tangential paths. These limiting values define the limit-function $G(\zeta)$ of $g(z)$. In an analogous fashion we define the limit-function $\Gamma(\xi)$ of the function $\gamma(w)$ analytic in the unit circle $|w|<1$. We shall be particularly concerned with the analytic functions which are represented by their own Cauchy integrals[2]

$$g(z) = I(z; G), \qquad \gamma(w) = J(w; \Gamma).$$

It will be assumed that $G(\zeta) \subset \mathfrak{L}_p$, $\Gamma(\xi) \subset L_p$ for some fixed p, $1 \leqq p < \infty$.

3. We now are in a position to state and prove our results.

THEOREM 1. *A necessary and sufficient condition that a function* $G(\zeta) \subset \mathfrak{L}_p$ *be the limit-function of a function* $g(z)$ *which is analytic in the half-plane* $\mathfrak{J}(z)>0$ *and which is represented by its Cauchy integral, is that* $G(\zeta)$ *satisfy the condition*

$$(8) \qquad I(\bar{z}; G) = (2\pi i)^{-1}\int_{-\infty}^{\infty} \frac{G(\zeta)\,d\zeta}{\zeta-\bar{z}} = 0, \qquad \bar{z} = x-iy, \quad y>0.$$

Proof.[3] If $g(z)$ is represented by its Cauchy integral then

$$g(z) = I(z; G)$$

and, in view of (4),

$$(9) \qquad g(z) = Q(z; G)+I(\bar{z}; G).$$

Since the limit-functions of $g(z)$ and of $Q(z; G)$ are the same $(= G(\zeta))$, the limit-function of $I(\bar{z}; G)$ is zero for almost all ζ. Hence $I(\bar{z}; G)$, being an analytic function of the variable $\bar{z}$, vanishes identically for $y>0$ by the uniqueness theorem of analytic functions. Conversely, if (8) is satisfied, then (9) shows that $g(z) \equiv Q(z; G)$ is the analytic function satisfying the conditions of Theorem 1.

THEOREM 2. *Let* $g(z)$ *be analytic in the half-plane* $\mathfrak{J}(z)>0$ *and let its limit-function* $G(\zeta) \subset \mathfrak{L}_p$. *If* $g(z)$ *is represented by its Cauchy integral it is also represented by its Poisson integral, and vice versa.*

Proof. We observe that the analogous theorem in the case of a function analytic in the unit circle is well known. If now $g(z)$ is represented by its Cauchy integral, then Theorem 1 and formula (9) show that

[2] We refer to the following two articles for the various properties of the Cauchy and Poisson integrals, and of the classes H_p, needed in the subsequent discussion: G. Fichtenholz, *Sur l'intégrale de Poisson et quelques questions qui s'y rattachent.* Fund. Math., 13 (1929), pp. 1–33; W. Smirnoff, *Sur les valeurs limites des fonctions regulières à l'intérieur d'un circle.* Jour. Soc. Phys.-Math. de Léningrade, 2 (1928–29), pp. 22–37. Other bibliographical references are found in these articles.

[3] Cf. Fichtenholz, loc. cit. [2], pp. 21–23.

$g(z) = Q(z;\,G)$ is also represented by its Poisson integral. Assume conversely that

(10) $$g(z) = Q(z;\,G) = I(z;\,G) - I(\bar{z};\,G).$$

If we pass to the variable w and set $\gamma(w) = g(z)$ we conclude from (7) that $\gamma(w) = P(w;\,\Gamma)$ is represented by its Poisson integral since $\Gamma(\xi) = G\left(\tan\frac{\varphi}{2}\right) = G(\zeta)$ obviously is the limit-function of $\gamma(w)$. Then $\gamma(w)$ is represented also by its Cauchy integral, and we have

$$\gamma(w) = J(w;\,\Gamma).$$

On returning to the variable z and using (6) we get

(11) $$g(z) = I(z;\,G) - C.$$

A comparison of (10) and (11) yields

$$I(\bar{z};\,G) = C, \qquad y > 0.$$

Let x be fixed and $y \to \infty$. It is readily seen that $I(\bar{z};\,G) \to 0$. This is obvious when $p = 1$, and the general case follows from the inequality

$$\left|\int_{-\infty}^{\infty} \frac{G(\zeta)}{\zeta - \bar{z}}\,d\zeta\right| \leqq \left[\int_{-\infty}^{\infty} |G(\zeta)|^p\,d\zeta\right]^{1/p} \left[\int_{-\infty}^{\infty} \frac{d\zeta}{\{(\zeta - x)^2 + y^2\}^{p'}}\right]^{1/p'},$$

$$p' = \frac{p}{p-1}, \quad p > 1.$$

It results that $I(\bar{z};\,G) = 0$, and (10) shows that $g(z)$ is represented by its Cauchy integral.

Remark. From the above argument it follows incidentally that if $G(\zeta) \subset \mathfrak{L}_p$ and $\Gamma(\xi) \subset L_p$ are the limit-functions of the analytic functions $g(z) = \gamma(w)$, where $w = \dfrac{1+iz}{1-iz}$, then the corresponding Cauchy integrals are transformed into one another,

$$I(z;\,G) = J(w;\,\Gamma).$$

4. In the subsequent discussion we shall employ Fourier transforms defined in a slightly more general manner than usual. Let $F(\zeta)$ be measurable over $(-\infty, \infty)$ and integrable over every finite range. Let

(12) $$\mathfrak{F}_N(u) \equiv \mathfrak{F}_N(u;\,F) \equiv (2\pi)^{-1/2} \int_{-N}^{N} e^{-iu\zeta}\,F(\zeta)\,d\zeta.$$

We assume that:

(i) For every fixed finite $A > 0$

(13) $$\int_{-A}^{A} |\mathfrak{F}_N(u)|\,du \leq M_A,$$

where the constant M_A depends only on A and on $F(\zeta)$, but not on N.

(ii) There exists a fixed number $\varepsilon_0 \geqq 0$ such that for each $\varepsilon > \varepsilon_0$

$$\int_{-\infty}^{-A} |\mathfrak{F}_N(u)| e^{\varepsilon u} \, du \to 0, \quad \int_{A}^{\infty} |\mathfrak{F}_N(u)| \, e^{-\varepsilon u} \, du \to 0 \tag{14}$$

as $A \to \infty$, the approach being uniform in N.

Under these assumptions it is readily proved that there exists a sequence of values $\{N_k\}$, $N_k \uparrow \infty$, and a function $\mathfrak{F}^0(u) \equiv \mathfrak{F}^0(u; F)$ such that

$$\int_0^u \mathfrak{F}_{N_k}(t) \, dt \to \mathfrak{F}^0(u) \text{ as } N_k \to \infty \tag{15}$$

for every finite range of u. This function $\mathfrak{F}^0(u)$ will be called a Fourier transform of $F(\zeta)$ of order 1. To prove these statements we have only to observe that, for a fixed A, the functions

$$\int_0^u \mathfrak{F}_N(t) \, dt, \qquad -A \leqq u \leqq A, \tag{16}$$

are of uniformly bounded variation. Hence, by a classical theorem of Helly, there exists a subsequence $\mathfrak{F}_i(A; u)$ of this set and a function of bounded variation $\mathfrak{F}(A; u)$ such that

$$\int_0^u \mathfrak{F}_i(A; t) \, dt \to \mathfrak{F}(A; u) \quad \text{as} \quad i \to \infty, \quad |u| \leqq A. \tag{17}$$

We now take a sequence $\{A_j\}$, $A_j \uparrow \infty$, of values of A and apply the "diagonal process" to the double sequence of functions $\{\mathfrak{F}_i(A_j; t)\}$. This process establishes the existence of the sequence $\{N_k\}$ and of the function $\mathfrak{F}^0(u)$ satisfying (15).

From the construction it follows that $\mathfrak{F}^0(u)$ is of bounded variation over every finite range, while

$$\int_\alpha^\beta \psi(t) \, \mathfrak{F}_{N_k}(t) \, dt \to \int_\alpha^\beta \psi(t) \, d\mathfrak{F}^0(t) \quad \text{as} \quad N_k \to \infty$$

holds for an arbitrary continuous function $\psi(t)$. On setting $\psi(t) = e^{-\varepsilon|t|}$ and using property (ii) we see at once that the integral

$$\int_{-\infty}^{\infty} e^{-\varepsilon|t|} \, |d\mathfrak{F}^0(t)| \tag{18}$$

exists (as an absolutely convergent improper Stieltjes integral). On the basis of these facts it is now readily seen that the relation

$$\int_{-\infty}^{\infty} e^{-\varepsilon|t|} \, \psi(t) \, \mathfrak{F}_{N_k}(t) \, dt \to \int_{-\infty}^{\infty} e^{-\varepsilon|t|} \, \psi(t) \, d\mathfrak{F}^0(t), \qquad N_k \to \infty, \tag{19}$$

holds for an arbitrary continuous function $\psi(t)$ bounded in $-\infty < u < \infty$. Finally we note that

$$(20) \qquad \mathfrak{F}^0(0) = 0, \qquad \mathfrak{F}^0(u) = \frac{1}{2}[\mathfrak{F}^0(u+0) + \mathfrak{F}^0(u-0)].$$

The first of these relations follows from (17) while the second may be assumed without modifying (19).

An important special case is obtained if we impose the additional restriction:

(iii) For every fixed A the set of integrals (16) is uniformly (with respect to N) absolutely continuous.

Under this assumption the function $\mathfrak{F}^0(u)$ is itself absolutely continuous. On setting

$$(21) \qquad \mathfrak{F}^0(u) = \int_0^u \mathfrak{F}(t)\, dt$$

we call the function $\mathfrak{F}(u) = \mathfrak{F}(u; F)$ the Fourier transform of $F(\zeta)$. According as the conditions (i–ii) or (i–iii) are satisfied we shall say that $F(\zeta)$ has a Fourier transform of order 1 or a Fourier transform respectively.

5. This notion of Fourier transform can be generalized in the direction of the Hahn-Wiener-Bochner theory of generalized trigonometric integrals. We do not intend to enter here into a discussion of such generalizations and their relationship with the theories mentioned. Neither do we investigate here the question of the uniqueness of the Fourier transform as defined above. We prove only the following

LEMMA. *The Fourier transform of order* 1 *of* $F(\zeta)$ *is uniquely determined at its points of continuity whenever* $F(\zeta) \subset \mathfrak{L}_p$, $1 \leqq p$. *Moreover, at all such points we have*

$$(22) \qquad \int_0^u \mathfrak{F}_N(t)\, dt \to \mathfrak{F}^0(u)$$

no matter how N *tends to infinity. The function* $\mathfrak{F}^0(u)$ *is uniquely determined everywhere if it is normalized according to* (20).

Proof. We start from the directly verifiable formulas

$$(23) \qquad \begin{aligned} \frac{1}{x-\zeta+iy} &= \frac{1}{i}\int_0^{\infty} e^{i(x-\zeta+iy)t}\, dt, \\ \frac{1}{x-\zeta-iy} &= -\frac{1}{i}\int_{-\infty}^{0} e^{i(x-\zeta-iy)t}\, dt, \qquad y>0, \\ \frac{2y}{(\zeta-x)^2+y^2} &= \int_{-\infty}^{\infty} e^{i(x-\zeta)t} e^{-y|t|}\, dt. \end{aligned}$$

Let $F(\zeta) \subset \mathfrak{L}_p$ and let conditions (i) and (ii) be satisfied. Thus we have

$$(24)\quad \begin{aligned} 2y\int_{-N}^{N} \frac{F(\zeta)\,d\zeta}{(\zeta-x)^2+y^2} &= \int_{-\infty}^{\infty} e^{ixt}\, e^{-y|t|}\,dt \int_{-N}^{N} e^{-i\zeta t}\, F(\zeta)\,d\zeta \\ &= (2\pi)^{1/2}\int_{-\infty}^{\infty} e^{ixt}\, e^{-y|t|}\,\mathfrak{F}_N(t)\,dt. \end{aligned}$$

On allowing $N = N_k \to \infty$ we get, in view of (19),

$$(25)\quad \Omega(x, y) = (2/\pi)^{1/2}\, y \int_{-\infty}^{\infty} \frac{F(\zeta)\,d\zeta}{(\zeta-x)^2+y^2} = \int_{-\infty}^{\infty} e^{ixt}\,d\,V(t),$$

$$(26)\quad V(u) = \int_0^u e^{-y|t|}\,d\mathfrak{F}_0(t), \qquad y > \varepsilon_0.$$

Here $V(u)$ is of bounded variation over the whole infinite range. Hence we can conclude that[4]

$$(27)\quad V(t) = \lim_{A\to\infty} (2\pi)^{-1}\int_{-A}^{A} \Omega(x, y)\, \frac{e^{-ixt}-1}{-ix}\,dx.$$

Thus $\mathfrak{F}^0(u)$ is uniquely determined by $F(\zeta)$ at all the points of continuity, and everywhere if $\mathfrak{F}^0(u)$ is normalized according to (20).

Remark 1. If all three conditions (i–iii) are satisfied then an analogous argument[5] will show that

$$(28)\quad \mathfrak{F}(u) = e^{y|u|} \lim_{A\to\infty} (2\pi)^{-1}\int_{-A}^{A} \Omega(x, y)\, e^{-ixu}\,dx.$$

Finally, if in condition (ii) $\varepsilon_0 = 0$, we can allow $y \to 0$ in (25) with the result

$$(29)\quad F(x) = \lim_{y\to 0} (2\pi)^{-1/2}\int_{-\infty}^{\infty} e^{ixt}\, e^{-y|t|}\,\mathfrak{F}(t)\,dt,$$

almost everywhere in $-\infty < x < \infty$. Thus we have obtained the classical Sommerfeld integral.

Remark 2. Conditions (i—iii) are satisfied when $1 \leqq p \leqq 2$. Then $F(\zeta)$ has a Fourier transform in $\mathfrak{L}_{p'}$, $p' = \dfrac{p}{p-1}$, that is

$$\mathfrak{F}(u) = \underset{N\to\infty}{\text{l.i.m.}}\ \mathfrak{F}_N(u),$$

[4] Cf. S. Bochner, *Vorlesungen über Fouriersche Integrale,* Leipzig, 1932, pp. 65–68. Bochner assumes that $V(u)$ is increasing, but, as he remarks himself, his argument is valid for any function of bounded variation in $(-\infty, \infty)$.

[5] Cf. Bochner, loc. cit. [4], p. 42.

where l.i.m. stands for the limit in the mean of order p' if $p' < \infty$, and for the uniform limit when $p' = \infty$[6]. In this case we have $\varepsilon_0 = 0$. But there exist functions $F(\zeta) \subset \mathfrak{L}_p$, $p > 2$, possessing no Fourier transform in any $\mathfrak{L}_q$, but such that the Fourier transform in the above more general sense exists, while $\varepsilon_0 > 0$ in condition (ii)[7].

6. We now return to our main subject matter.

THEOREM 3. *Let $G(\zeta) \subset \mathfrak{L}_p$, $p \geqq 1$, and suppose $G(\zeta)$ possesses the Fourier transform $\mathfrak{F}^0(u)$ of order* 1. *A necessary and sufficient condition that $G(\zeta)$ be a limit-function of a function $g(z)$ which is analytic in the half-plane $\mathfrak{J}(z) > 0$ and which is represented by its Cauchy integral, is that $\mathfrak{F}^0(u)$ vanish for $u < 0$.*[8]

Proof. As in § 5 it is readily seen that

$$\int_{-\infty}^{\infty} \frac{G(\zeta)\, d\zeta}{\zeta - x + iy} = -i(2\pi)^{1/2} \int_{-\infty}^{0} e^{ixt}\, e^{yt}\, d\mathfrak{F}^0(t), \qquad y > \varepsilon_0. \tag{30}$$

By the Lemma of § 5 the function $\mathfrak{F}^0(t)$ is uniquely determined if it is normalized according to (20). Consequently the vanishing of the integral (8) implies, and is implied by, the vanishing of $\mathfrak{F}^0(t)$ for $t < 0$. Now it remains only to apply Theorem 1.

THEOREM 4. *The condition $\mathfrak{F}^0(u; G) = 0$ for $u < 0$ is necessary and sufficient that the Poisson integral $Q(z; G)$ represent a function $g(z)$ analytic in the half-plane $\mathfrak{J}(z) > 0$.*

Proof. This theorem is a simple consequence of Theorems 1, 2 and 3.

THEOREM 5 (Paley and Wiener). *Let $0 \leqq F(\zeta) \subset \mathfrak{L}_p$, $1 \leqq p \leqq 2$. The condition*

$$\int_{-\infty}^{\infty} \frac{|\log F(\zeta)|\, d\zeta}{1 + \zeta^2} < \infty \tag{31}$$

is necessary and sufficient that there exist a function $G(\zeta) \subset \mathfrak{L}_p$ such that

$$|G(\zeta)| = F(\zeta), \tag{32}$$

while the Fourier transform $\mathfrak{F}(u; G)$ of $G(\zeta)$ vanishes for $u < 0$.

Proof. To prove the necessity of (31) assume that $G(\zeta)$ is a function satisfying the conditions of Theorem 5. By Theorem 4, the function

[6] E. C. Titchmarsh, *A contribution to the theory of Fourier transforms.* Proc. London Math. Soc., (2) 23 (1924–25), pp. 279–289.

[7] E. Hille and J. D. Tamarkin, *On the summability of Fourier Series.* III. To appear shortly in the Math. Annalen. It should be mentioned, however, that the method employed in § 18 of this paper could be used to construct functions $F(\zeta) \subset \mathfrak{L}_p$, $p > 2$, such that $\lim \mathfrak{F}_N(u)$ exists, but the theory developed in the present paper does not apply.

[8] For a special case of this theorem see N. Wiener, *The operational calculus,* Math. Ann., 95 (1926), pp. 557–584 (580).

$g(z) = Q(z; G)$ is analytic in the half-plane $\mathfrak{J}(z) > 0$ and is represented by its Cauchy, as well as by its Poisson, integral. The same will be true in the w-plane for the corresponding function $\gamma(w) = g(z)$. This implies, however, the integrability of $\log|\Gamma(\xi)|$, where $\Gamma(\xi) = G(\zeta)$ is the limit-function of $\gamma(w)$.[9] Hence

$$\int_{-\infty}^{\infty} \frac{|\log F(\zeta)|\, d\zeta}{1+\zeta^2} = \frac{1}{2}\int_{-\pi}^{\pi} \left|\log|\Gamma(\xi)|\right| d\varphi < \infty.$$

Conversely, let (31) be satisfied. Then $\Phi(\xi) = F\left(\tan\frac{\varphi}{2}\right) \subset L_p$, while $\log \Phi(\xi) \subset L_1$. Under these circumstances[9] there exists an analytic function $\gamma(w)$ such that the limit-function $\Gamma(\xi)$ satisfies the relation $|\Gamma(\xi)| = \Phi(\xi)$, and $\gamma(w)$ is represented by its Cauchy, as well as by its Poisson, integral. In view of the preceding theorems, the same will be true of the corresponding function $g(z) = \gamma(w)$, and the Fourier transform of the limit-function $G(\zeta)$ must vanish for the negative values of its argument. Finally (32) holds. This completes the proof of Theorem 5.

Remark. The conditions of Theorem 5 remain necessary when $p > 2$. It is an open question whether they will be also sufficient for such values of p.

---

[9] Cf. Smirnoff, loc. cit. [2].

ON THE THEORY OF FOURIER TRANSFORMS

BY EINAR HILLE AND J. D. TAMARKIN

1. *Introduction*. Let $g(s) \subset L_2$ over $(-\infty, \infty)$. Let

$$G(u; a) = (2\pi)^{-1/2} \int_{-a}^{a} e^{-ius} g(s) ds.$$

According to the classical result of the Plancherel theory of Fourier transforms, $G(u; a)$ tends in the mean of order 2 to a function $G(u) \subset L_2$ as $a \to \infty$. This function is designated as the Fourier transform (in L_2) of $g(s)$. We shall write

$$T\{u; g(s)\} = G(u) = \operatorname*{l.i.m}_{a\to\infty} G(u; a). \tag{1}$$

The functions $g(s)$ and $G(u)$ are reciprocal in the sense that

$$g(s) = T\{s; G(-u)\}, \tag{2}$$

which means that

$$g(s) = \operatorname*{l.i.m}_{a\to\infty} g(s; a); \quad g(s; a) = (2\pi)^{-1/2} \int_{-a}^{a} e^{ius} G(u) du. \tag{3}$$

As an immediate consequence of the convergence in the mean of $G(u; a)$ and $g(s; a)$, we have, almost everywhere,

$$G(u) = (2\pi)^{-1/2} \frac{d}{du} \int_{-\infty}^{\infty} g(s) \frac{1 - e^{-ist}}{is} ds, \tag{4}$$

$$g(s) = (2\pi)^{-1/2} \frac{d}{ds} \int_{-\infty}^{\infty} G(u) \frac{1 - e^{ius}}{-iu} du. \tag{5}$$

The reciprocity between g and G is expressed here in terms not involving the convergence in the mean.

Assume now that $g(s) \subset L_p$, $1 < p < 2$. Denote by p' the *conjugate* exponent, $p' = p/(p-1)$, $1/p + 1/p' = 1$. Titchmarsh* showed that Plancherel's theory can be extended, at least in part, to the present case. Indeed he proved that $G(u; a)$ con-

* E. C. Titchmarsh, *A contribution to the theory of Fourier transforms*, Proceedings of the London Mathematical Society, (2), vol. 23 (1925), pp. 279–289. We have slightly modified Titchmarsh's notation inasmuch as he deals with cosine- and sine-transforms, while we use exponential transforms.

verges in the mean of order p' to a function $G(u) \subset L_{p'}$, which satisfies the same reciprocity relations (4), (5). Titchmarsh did not prove, however, that, conversely, (3) holds. On the basis of the material developed by Titchmarsh we do not know whether $g(s; a)$ tends in the mean of order p to $g(s)$; hence we do not know whether $g(s)$ is the Fourier transform (in L_p) of $G(-u)$. A scrutiny of the literature reveals the unexpected fact that this question has never been investigated,* which, in our opinion, represents an undesirable gap in the theory of Fourier transforms. The purpose of the present note is to fill in this gap and to prove (3) in the case $1 < p \leqq 2$. An analogy between the theory of Fourier transforms and that of Fourier series should be pointed out here. We may consider the Fourier transform $G(u)$ of $g(s)$ as an analog of the *sequence* of Fourier coefficients $\{G_n\}$ of $g(s)$ (in the case where g is periodic and is being expanded in Fourier series). Then $g(s; a)$ appears as an analog of the nth partial sum $s_n(s; g)$ of $g(s)$ which is known to converge to $g(s)$ in the mean of order p. This analogy goes even further. It is known that if $g(s) \subset L_1$, over $(-\pi, \pi)$, $s_n(s; g)$ does not necessarily converge in the mean of order 1 to $g(s)$.† We shall prove that this is also the case for the Fourier transforms.

It will be essential for our discussion to consider the class of functions $\subset L_p$ over $(-\infty, \infty)$ as a linear vector metric complete space with the metric defined by

$$\|g\|_p = \left[\int_{-\infty}^{\infty} |g(s)|^p ds\right]^{1/p}.$$

By $\mathfrak{L}^p$ we shall mean the limit in the mean of order p. Some known properties of linear transformations of such spaces will be used in the sequel.‡

* Contrary to our unfounded statement in a recent paper. See E. Hille and J. D. Tamarkin, *On the summability of Fourier series*, III, Mathematische Annalen, vol. 108 (1933), pp. 525–577 (p. 530). According to results of the present note, however, this statement is correct.

† The case $p=1$ goes back to H. Hahn, *Ueber die Darstellung gegebener Funktionen durch singuläre Intergrale*, II, Denkschriften Wiener Akademie, Math.-Nat. Kl., vol. 93 (1916), p. 557–692 (p. 681). For $p>1$ see M. Riesz, *Sur les fonctions conjuguées*, Mathematische Zeitschrift, vol. 27 (1927), pp. 218–244 (p. 230). M. Riesz states at the end of his paper that to his results concerning Fourier series there correspond similar results for Fourier integrals.

‡ We refer for these properties to the recent book by S. Banach, *Théorie des Opérations Linéaires*, Warsaw, 1932.

2. *The Case* $1<p\leqq 2$. If $g(s)\subset L_p$, $1<p\leqq 2$, we have

$$G(u) = \underset{a\to\infty}{\mathfrak{L}^{p'}} (2\pi)^{-1/2}\int_{-a}^{a} g(t)e^{-iut}dt.$$

Here we may multiply by any function $\subset L_p$ and integrate under the integral sign. Thus

$$\begin{aligned} g(s;\, a) &= (2\pi)^{-1/2}\int_{-a}^{a} G(u)e^{isu}du \\ &= (2\pi)^{-1}\int_{-\infty}^{\infty} g(t)dt\int_{-a}^{a} e^{iu(s-t)}du \qquad (6)\\ &= \frac{1}{\pi}\int_{-\infty}^{\infty} g(t)\frac{\sin a(s-t)}{s-t}dt. \end{aligned}$$

Let $f(s)$ be any function $\subset L_p$. The integral

$$\tilde{f}(s) = \frac{1}{\pi}\int_{-\infty}^{\infty}\frac{f(t)}{s-t}dt$$

exists (in the sense of Cauchy's principal value at $t=s$) almost everywhere and is designated as the function conjugate to $f(s)$. Moreover, $\tilde{f}(s)\subset L_p$ whenever $f(s)\subset L_p$. Furthermore, there exists a positive constant M_p depending only on p and such that*

$$\|\tilde{f}\|_p \leqq M_p\|f\|_p. \tag{7}$$

Upon introducing the functions

$$g_a'(t) = g(t)\cos at, \qquad g_a''(t) = -\, g(t)\sin at$$

and their conjugates $\tilde{g}_a'(s)$, $\tilde{g}_a''(s)$,† we may rewrite (6) in the form

$$g(s;\, a) = \tilde{g}_a'(s)\sin as + \tilde{g}_a''(s)\cos as.$$

By (7),

$$\|\tilde{g}_a'\|_p \leqq M_p\|g(t)\cos at\|_p \leqq M_p\|g\|_p, \quad \|\tilde{g}_a''\|_p \leqq M_p\|g\|_p.$$

* M. Riesz, loc. cit., p. 234.

† For the analogous procedure in the case of Fourier series, see M. Riesz, loc. cit., p. 230, and A. Kolmogoroff, *Sur les fonctions harmoniques conjuguées et sur les séries de Fourier*, Fundamenta Mathematicae, vol. 7 (1925), pp. 23–28.

Hence

$$\|g(s;a)\|_p \leqq \|\tilde{g}_a'\|_p + \|\tilde{g}_a''\|_p \leqq 2M_p\|g\|_p. \tag{8}$$

For each a, the expression $g(s; a)$ constitutes as a linear transformation on L_p to L_p, and (8) shows that the family of these transformations, obtained when a varies, is uniformly limited.

Now assume that $g(s)$ is a step function of a finite number of steps, vanishing outside of a finite interval $(-N, N)$. For such a function we shall prove directly that

$$\|g(s;a) - g(s)\|_p \to 0 \text{ as } a \to \infty .$$

Since our function $g(s)$ is a linear combination of step functions defined by

$$g(s) = \begin{cases} 1 \text{ if } \alpha \leqq s \leqq \beta, \\ 0 \text{ outside } (\alpha, \beta), \end{cases} \qquad (-N \leqq \alpha < \beta \leqq N),$$

it will be sufficient to prove our assertion for a function of this type. For such a function, however,

$$g(s;a) = \int_\alpha^\beta \frac{\sin a(s-t)}{s-t} dt.$$

It is readily seen that

$$g(s;a) = O[(s-N)^{-1}a^{-1}], \qquad (s \geqq 2N).$$

Since $g(s)=0$ for $|s|>N$, we have

$$\int_{2N}^\infty |g(s) - g(s;a)|^p ds = O\left[a^{-p}\int_{2N}^\infty \frac{ds}{(s-N)^p}\right] = O(a^{-p})$$

as $a\to\infty$. Similarly

$$\int_{-\infty}^{-2N} |g(s) - g(s;a)|^p ds = O(a^{-p}).$$

On the other hand the integral

$$\frac{1}{\pi}\int_\alpha^\beta \frac{\sin a(s-t)}{s-t} dt = \frac{1}{\pi}\int_{-2N}^{2N} g^*(t)\frac{\sin a(s-t)}{s-t} dt$$

may be considered as the classical Dirichlet integral of the function $g^*(s)$ which is periodic, of period $4N$, and coincides with $g(s)$ in the interval $(-2N, 2N)$. It is well known from the theory of Fourier series, and also can be proved directly, that this Dirichlet integral converges to $g^*(s)$ in the mean of order p, over the interval $(-2N, 2N)$. Hence

$$\int_{-2N}^{2N} | g(s) - g(s; a) |^p ds$$
$$= \int_{-2N}^{2N} | g^*(s) - g(s; a) |^p ds \to 0 \text{ as } a \to \infty .$$

Consequently

$$\int_{-\infty}^{\infty} | g(s) - g(s; a) |^p ds = \int_{-2N}^{2N} \cdots + \int_{-\infty}^{-2N} \cdots + \int_{2N}^{\infty} \cdots \to 0$$

as $a \to \infty$, which is the desired result.

Since the family of transformations represented by $g(s; a)$ is uniformly limited, and since the step-functions constitute a dense sub-set of L_p, it follows immediately that

$$\|g(s) - g(s; a)\|_p \to 0 \text{ as } a \to \infty,$$

for an arbitrary function $g(s) \subset L_p$. We may therefore state the following addition to the results of Titchmarsh mentioned above.

THEOREM. *If* $g(s) \subset L_p$, $1 < p \leqq 2$, *and if*

$$G(u) = \underset{a \to \infty}{\mathfrak{L}^{p'}} (2\pi)^{-1/2} \int_{-a}^{a} g(s) e^{-ius} ds$$

is the Fourier transform (in $L_{p'}$*) of* $g(s)$*, then, conversely,* $g(s)$ *is the Fourier transform (in* L_p*) of* $G(-u)$*, and we have*

$$g(s) = \underset{a \to \infty}{\mathfrak{L}^{p}} (2\pi)^{-1/2} \int_{-a}^{a} G(u) e^{ius} du.$$

3. *The Case* $p=1$. We now proceed to show that our theorem fails in the case $p=1$. To do this we have to exhibit an example of a function $g(s) \subset L_1$, such that, if we set

$$G(u) = (2\pi)^{-1/2} \int_{-\infty}^{\infty} g(s) e^{-ius} ds,$$

$$g(s; a) = (2\pi)^{-1/2} \int_{-a}^{a} G(u) e^{ius} du,$$

the function $g(s; a)$ does not converge to $g(s)$ in the mean of order 1. The example which we are going to construct will show even more, namely, that the corresponding $g(s; a)$ does not even belong to L_1.

In constructing the example in question we are again guided by the analogy with Fourier series. It is well known* that the series

$$f(x) = \sum_{n=2}^{\infty} \frac{\cos nx}{\log n}$$

is a Fourier series of $f(x) \subset L_1$, but that its partial sums $s_p(x; f)$ do not converge to $f(x)$ in the mean of order 1 over $(-\pi, \pi)$. We now put

$$g(s) = (2\pi)^{-1/2} \int_0^{\infty} \frac{\cos us}{\log(2+u)} du. \tag{9}$$

It is plain that this integral converges uniformly over every interval, finite or not, which is separated from the point $s=0$. Moreover, it is easy to show that $g(s)>0$ for $s \neq 0$. It suffices to consider positive values of s only. Let $s>0$ be fixed. We put

$$q(u) = (2+u)\log^2(2+u), \quad \delta_\nu = \nu\pi/s, \quad (\nu = 0, 1, 2, \cdots).$$

Then, on integrating by parts in the right-hand member of (9), we see that

$$(2\pi)^{1/2} g(s) = \frac{1}{s} \int_0^{\infty} \frac{\sin us}{q(u)} du = \sum_{\nu=0}^{\infty} \frac{1}{s} \int_{\delta_\nu}^{\delta_{\nu+1}} \frac{\sin us}{q(u)} du. \tag{10}$$

Put

$$I_\nu(s) = \frac{1}{s} \int_{\delta_\nu}^{\delta_{\nu+1}} \frac{\sin us}{q(u)} du = (-1)^\nu s^{-2} \int_0^{\pi} \frac{\sin \tau d\tau}{q(\nu\pi + \tau/s)} \cdot$$

Hence

$$| I_\nu | < I_{\nu-1} | |, \qquad (\nu = 1, 2, \cdots),$$

and $\sum_0^\infty I_\nu$ appears as an alternating series in which the absolute value of the general term approaches zero and the first term is positive. The sum of such a series is positive. Hence $g(s)>0$. An integration by parts applied to the middle term in (10) shows that, for large values of $s, g(s) = O(s^{-2})$. This ensures the integrability of $g(s)$ over every interval (ϵ, ∞), $\epsilon>0$. Since $g(s)>0$, to

---

* A. Kolmogoroff, *Sur l'ordre de grandeur de coefficients de la série de Fourier*, Bullétin International de l'Académie Polonaise, Classe de Sciences mathématiques, 1923, pp. 83–86. See also E. Hille and J. D. Tamarkin, *On the summability of Fourier series*, II, Annals of Mathematics, (2), vol. 34 (1933), pp. 329–348 (pp. 347–348).

establish the integrability of $g(s)$ over $(0, \epsilon)$ it suffices to prove that, for any fixed positive α,

$$\int_\epsilon^\alpha g(s)ds$$

is bounded as $\epsilon \to 0$. In view of the uniform convergence of the integral in (9) over $(\epsilon, 1)$ we have

$$\int_\epsilon^1 g(s)ds = (2\pi)^{-1/2} \int_0^\infty \frac{du}{\log(2+u)} \int_\epsilon^1 \cos us\, ds$$

$$= (2\pi)^{-1/2} \left\{ \int_0^\infty \frac{\sin u\, du}{u \log(2+u)} - \int_0^\infty \frac{\sin u\, du}{u \log(2+u/\epsilon)} \right\}.$$

A simple application of the second law of the mean shows that the last integral tends to zero with ϵ. Thus we see that $g(s)$ is integrable over $(0, \infty)$, and, being even, it is integrable over $(-\infty, \infty)$. By the uniqueness theorem of Fourier integrals, it is plain that

$$G(u) = [\log(2 + |u|)]^{-1} = (2\pi)^{-1/2} \int_0^\infty g(s) \cos su\, ds$$

is the Fourier transform of $g(s)$ in L_∞, that is, $G(u; a)$ converges uniformly to $G(u)$ as $a \to \infty$. Now compute

$$g(s; a) = (2\pi)^{-1/2} \int_0^a G(u) \cos su\, du = (2\pi)^{-1/2} \int_0^a \frac{\cos su\, du}{\log(2+u)}.$$

An integration by parts shows that

$$(2\pi)^{1/2} g(s; a) = \frac{\sin sa}{s \log(2+a)} + \frac{1}{s} \int_0^a \frac{\sin su}{q(u)} du.$$

As before, we see that the second term of the right-hand member is $O(s^{-2})$ for $s \to \infty$, and since the first term is not absolutely integrable over $(0, \infty)$ it results that $g(s; a)$ does not belong to L_1.

BROWN UNIVERSITY AND YALE UNIVERSITY

A Remark on Fourier Transforms and Functions Analytic in a Half-Plane

by

Einar Hille and J. D. Tamarkin
New Haven (Conn.) Providence (R. I.)

Let $\mathfrak{L}_p$ be the class of functions $f(x)$ measurable over $(-\infty, \infty)$ such that the integral

$$\int_{-\infty}^{\infty} | f(x) |^p dx, \qquad p \text{ fixed, } p \geqq 1, \tag{1}$$

is finite. It is well known that $\mathfrak{L}_p$ becomes a complete linear metric vector space if we define its metric by

$$\| f(x) \|_p = \| f \|_p = \left[\int_{-\infty}^{\infty} | f(x) |^p dx \right]^{1/p}. \tag{2}$$

The value $p = \infty$ will also be admitted, with the agreement that $\mathfrak{L}_\infty$ is the class of functions $f(x)$ continuous and bounded over $(-\infty, \infty)$ with the metric

$$\| f \|_\infty = \sup_{-\infty < x < \infty} | f(x) |. \tag{3}$$

Let $g(x) \subset \mathfrak{L}_p$, $1 \leqq p < \infty$. Put

$$G(u\,;\ a) = (2\pi)^{-1/2} \int_{-a}^{a} g(x) e^{-iux} dx. \tag{4}$$

If there exists a function $G(u) \subset \mathfrak{L}_q$, $1 \leqq q \leqq \infty$, such that

$$\| G(u\,;\ a) - G(u) \|_q \to 0 \text{ as } a \to \infty, \tag{5}$$

then we say that $G(u)$ is the Fourier transform of $g(x)$ in $\mathfrak{L}_q$ and write

$$G(u) = \underset{a \to \infty}{L^q} G(u\,;\ a). \tag{6}$$

Such a function is known to exist when $1 \leqq p \leqq 2$ and $q = p' = \dfrac{p}{p-1}$.

Let $f(z)$ be a function of the complex variable $z = x + iy$, analytic in the upper half-plane $y > 0$. If, for almost all x,

$f(z)$ tends to a definite limit, $f(x)$, as $z \to x$ along all non-tangential paths, the function $f(x)$ so defined is said to be the limit-function of $f(z)$. We shall assume that $f(x) \subset \mathfrak{L}_p$, $1 \leqq p < \infty$.

In the present note we are concerned with the class $\mathfrak{A}_p$ of functions $f(z)$, analytic in the half-plane $y > 0$, each possessing a limit-function $f(x) \subset \mathfrak{L}_p$, $1 \leqq p < \infty$, and representable by its Cauchy integral,

$$(7) \qquad f(z) = \frac{1}{2\pi i} \int_{-\infty}^{\infty} \frac{f(\xi) d\xi}{\xi - z} \equiv I(z\,;\, f),$$

or, what is equivalent [1]), by its Poisson integral,

$$(8) \qquad f(z) = \frac{1}{\pi} \int_{-\infty}^{\infty} \frac{y f(\xi) d\xi}{(\xi - x)^2 + y^2} \equiv Q(z\,;\, f).$$

The following problem presents itself naturally: find necessary and sufficient conditions which must be satisfied by a given function $f(x) \subset \mathfrak{L}_p$ in order that $f(x)$ be the limit-function of a function $f(z) \subset \mathfrak{A}_p$. In a recent note [2]) we gave a solution of this problem under the assumption that $f(x)$ possesses a Fourier transform (in a certain generalized sense). It turns out that this transform must vanish for negative values of its argument. The purpose of the present note is to investigate the same problem under a different assumption, that $f(x)$ itself is the Fourier transform in $\mathfrak{L}_p$ of a function $g(u) \subset \mathfrak{L}_q$, $1 \leqq q \leqq \infty$.

Theorem. Let $f(x)$ be the Fourier transform in $\mathfrak{L}_p$ of a function $\varphi(u) \subset \mathfrak{L}_q$, $1 \leqq q \leqq \infty$. In order that $f(x)$ be the limit-function of a function $f(z) \subset \mathfrak{A}_p$ it is necessary and sufficient that $\varphi(u)$ vanish for $u > 0$.

Proof. For convenience we replace $\varphi(u)$ by $g(-u)$ and set

$$(9) \qquad f(\xi) = \underset{a \to \infty}{L^p} F(\xi;\, a),$$

where

$$(10) \qquad F(\xi\,;\, a) = \int_{-a}^{a} g(u) e^{i\xi u} du,$$

and $g(u) \subset \mathfrak{L}_q$. We first assume that $1 < q < \infty$, and proceed to the computation of the Poisson integral of $f(\xi)$,

---

[1]) Cf. Hille and Tamarkin, On a theorem of Paley and Wiener [Annals of Math. (2) **34** (1933), 606—614].

[2]) Loc. cit. [1]) For a special case see N. Wiener, The operational calculus [Math. Ann. **95** (1926), 557—584; (580)].

$$(11)\quad Q(z;f)=\frac{1}{\pi}\int_{-\infty}^{\infty}\frac{yf(\xi)d\xi}{(\xi-x)^2+y^2}=\lim_{N\to\infty}\frac{1}{\pi}\int_{-N}^{N}\frac{yf(\xi)d\xi}{(\xi-x)^2+y^2},$$

which obviously converges absolutely. In view of (9) we have

$$(12)\quad \int_{-N}^{N}\frac{yf(\xi)d\xi}{(\xi-x)^2+y^2}=\int_{-\infty}^{\infty}g(u)du\int_{-N}^{N}\frac{ye^{e\xi u}d\xi}{(\xi-x)^2+y^2}.$$

On the other hand, by a direct computation,

$$(13)\quad \frac{y}{(\xi-x)^2+y^2}=\frac{1}{2}\int_{-\infty}^{\infty}e^{i(x-\xi)t}e^{-y|t|}dt.$$

Hence

$$(14)\quad \begin{aligned}\int_{-N}^{N}\frac{ye^{i\xi u}d\xi}{(\xi-x)^2+y^2}&=\frac{1}{2}\int_{-\infty}^{\infty}e^{ixt-y|t|}\,dt\int_{-N}^{N}e^{i\xi(u-t)}d\xi=\\&=\int_{-\infty}^{\infty}e^{ixt-y|t|}\frac{\sin N(u-t)}{u-t}\,dt.\end{aligned}$$

On substituting into (12) and interchanging the order of integration, which is clearly permissible, we get

$$(15)\quad \begin{aligned}\frac{1}{\pi}\int_{-N}^{N}\frac{yf(\xi)d\xi}{(\xi-x)^2+y^2}&=\int_{-\infty}^{\infty}e^{ixt-y|t|}\,dt\int_{-\infty}^{\infty}g(u)\frac{1}{\pi}\frac{\sin N(u-t)}{u-t}du=\\&=\int_{-\infty}^{\infty}e^{ixt-y|t|}\,dt\int_{-\infty}^{\infty}g(u)\,\mathfrak{D}_N(u-t)\,du,\end{aligned}$$

where

$$\mathfrak{D}_N(u)=\frac{1}{\pi}\frac{\sin Nu}{u}$$

is the classical Dirichlet kernel. Now put

$$g_N(t)=\int_{-\infty}^{\infty}g(u)\,\mathfrak{D}_N(u-t)du.$$

It is known [1]) that

$$\|g_N(t)-g(t)\|_q\to 0\text{ as }N\to\infty.$$

Consequently

$$(16)\quad \begin{aligned}Q(z;\,f)&=\int_{-\infty}^{\infty}e^{ixt-y|t|}\,g(t)dt=\\&=\int_{0}^{\infty}e^{izt}\,g(t)dt+\int_{-\infty}^{0}e^{i\bar{z}t}g(t)dt\equiv\Omega_1(z)+\Omega_2(\bar{z}),\end{aligned}$$

[1]) See, for instance, HILLE and TAMARKIN, On the theory of Fourier transforms [Bulletin of the Amer. Math. Soc. 39 (1933)].

where $\bar{z} = x - iy$. Here $\Omega_1(z)$ is analytic in z and $\Omega_2(\bar{z})$ is analytic in $\bar{z}$, while

$$\Omega_1(z) \to 0, \ \Omega_2(\bar{z}) \to 0 \text{ as } |x| \to \infty.$$

Hence the vanishing of $\Omega_2(\bar{z})$ is a necessary and sufficient condition for the analyticity of $Q(z; f)$. Now, if $f(z) \subset \mathfrak{A}_p$, then $f(z)$ is represented by its Poisson integral $Q(z; f)$, and $\Omega_2(\bar{z}) \equiv 0$; conversely, if $\Omega_2(\bar{z}) \equiv 0$, $Q(z; f)$ is analytic and represents a function $f(z) \subset \mathfrak{A}_p$ whose limit-function is precisely $f(x)$. We see therefore that the condition $\Omega_2(\bar{z}) \equiv 0$ is necessary and sufficient in order that $f(x)$ be the limit-function of a function $f(z) \subset \mathfrak{A}_p$. In view of the uniqueness theorem for Fourier integrals, however, the condition $\Omega_2(\bar{z}) \equiv 0$ is equivalent to the condition $g(t) = 0$ for $t < 0$; which is the desired result.

The treatment of the cases $q = 1$, $q = \infty$ is slightly more complicated, but formula (16) and all the subsequent conclusions will still be valid. When $q = 1$ or $q = \infty$, we apply the method of arithmetic means to evaluate the integral (11). Thus

$$\frac{1}{N}\int_0^N dn \frac{1}{\pi}\int_{-n}^{n} \frac{yf(\xi)d\xi}{(\xi - x)^2 + y^2} = \lim_{a\to\infty}\int_{-a}^{a} g(u)du \frac{1}{N}\int_0^N dn \frac{1}{\pi}\int_{-n}^{n} \frac{ye^{i\xi u}d\xi}{(\xi - x)^2+y^2} =$$

$$= \lim_{a\to\infty}\int_{-a}^{a} g(u)du \frac{1}{N}\int_0^N dn \int_{-\infty}^{\infty} e^{ixt - y|t|}\frac{1}{\pi}\frac{\sin n(u-t)}{u-t}dt =$$

$$= \lim_{a\to\infty}\int_{-a}^{a} g(u)du \int_{-\infty}^{\infty} e^{ixt - y|t|}\frac{1}{2N\pi}\left[\frac{\sin\frac{N(u-t)}{2}}{\frac{u-t}{2}}\right]^2 dt = \tag{17}$$

$$= \lim_{a\to\infty}\int_{-\infty}^{\infty} e^{ixt - y|t|}dt \int_{-a}^{a} g(u)\mathfrak{F}_N(u-t)du = \int_{-\infty}^{\infty} e^{ixt - y|t|}dt \int_{-\infty}^{\infty} g(u)\mathfrak{F}_N(u-t)du$$

where

$$\mathfrak{F}_N(u) = \frac{1}{2N\pi}\left[\frac{\sin\frac{Nu}{2}}{\frac{u}{2}}\right]^2$$

is the Fejér kernel. Now put

$$g_N'(t) = \int_{-\infty}^{\infty} g(u)\mathfrak{F}_N(u-t)du.$$

When $N \to \infty$, the left-hand member of (17) tends to $Q(z; f)$

since the integral (11) converges. On the other hand, when $q = 1$,

$$\| g'_N(t) - g(t) \|_1 \to 0$$

while, when $q = \infty$, $g'_N(t) \to g(t)$ boundedly, and in fact uniformly over every finite range. Hence allowing $N \to \infty$, we obtain (16) again, and finish the proof as above. This method of course might have been used in the case $1 < q < \infty$ as well.

(Received, August 12th, 1933).

On the absolute integrability of Fourier transforms.

By

Einar Hille (New Haven) and J. D. Tamarkin (Providence).

1. Introduction. In their important paper [1] [1]) Hardy and Littlewood proved that if $\varphi(\theta)$ belongs to a Lebesgue class L_p over $(-\pi, \pi)$, $p > 1$, and if $\{c_m\}$, $m = \ldots, -2, -1, 0, 1, 2, \ldots$, is the sequence of complex Fourier coefficients of $\varphi(\theta)$ then

$$\sum_{m=-\infty}^{\infty} |c_m|^p (|m|+1)^{p-2} \leqq C_p \int_{-\pi}^{\pi} |\varphi(\theta)|^p \, d\theta,$$

where C_p is a constant depending only on p. This constant tends to ∞ as $p \to 1$, and the result does not hold when $p = 1$. It does hold however in the special case where $\varphi(\theta)$ ist the limit function of a function

$$\psi(w) = \sum_{n=0}^{\infty} a_n w^n$$

analytic in the unit circle $|w| < 1$, and such that

$$\int_{-\pi}^{\pi} |\psi(r e^{i\theta})| \, d\theta$$

is bounded for $0 \leqq r < 1$. In this case the result of Hardy and Littlewood can be stated in the form

$$\sum_{n=0}^{\infty} |a_n|/(n+1) \leqq C_1 \int_{-\pi}^{\pi} |\varphi(\theta)| \, d\theta$$

[1]) The numbers in brackets refer to the list at the end of this paper.

and the hypothesis can be replaced by an equivalent one, viz. that both functions $\varphi(\theta)$ and its conjugate $\tilde{\varphi}(\theta)$ are integrable, where $\tilde{\varphi}(\theta)$ is defined by

$$\begin{aligned}\tilde{\varphi}(\theta) &= -\lim_{\varepsilon\to 0}(2\pi)^{-1}\int_{\varepsilon}^{\pi}[\varphi(\theta+\tau)-\varphi(\theta-\tau)]\cot\frac{\tau}{2}\,d\tau \\ &= (2\pi)^{-1}\,P.V.\int_{-\pi}^{\pi}\varphi(\tau)\cot\frac{\theta-\tau}{2}\,d\tau,\end{aligned}\tag{1.1}$$

and *P. V.* stands for the Cauchy principal value of the integral in question. As a corollary of this result it follows that if both functions, $\psi(\theta)$ and its conjugate $\tilde{\psi}(\theta)$, are of bounded variation over $(-\pi, \pi)$, then their Fourier series converge absolutely. Almost simultaneously with Hardy and Littlewood, Fejér [1] showed that the best possible value of the constant C_1 is $1/2$. An extremely simple proof of Fejér's result was given by Zygmund in his recent book ([1], pp. 157—162). Fejér also gave the following elegant geometric interpretation of the preceding results. Let

$$\psi(w) = \sum_{n=0}^{\infty} b_n w^n$$

be analytic in the unit circle $|w| < 1$, and let $\zeta = \psi(w)$ map conformally the unit circle into a bounded domain of the ζ-plane, not necessarily simply covered, and bounded by a rectifiable curve of length l. Then

$$\sum_{n=0}^{\infty} |b_n| \leqq \tfrac{1}{2} l,$$

the coefficient $1/2$ being the best possible.

It is the purpose of the present note to investigate the analogues of the preceding results in the case where Fourier series are replaced by Fourier transforms and mapping of the unit circle is replaced by mapping of a half-plane. Hardy and Littlewood ([1], p. 203) state that if $f(x) \in L_p$ over $(0, \infty)$, $p > 1$, and if $F(x)$ is its cosine or sine Fourier transform then

$$\int_0^{\infty} |F(x)|^p x^{p-2}\,dx \leqq C_p \int_0^{\infty} |f(x)|^p\,dx.$$

This result in general is false when $p=1$. It holds true however in a case which is entirely analogous to that mentioned above in connection with functions analytic in the unit circle, viz. when $f(x)$ is the limit function of a function $f(z)$ analytic in the half-plane $\Im z > 0$ and such that

$$\int_{-\infty}^{\infty} |f(x+iy)|\, dx \quad \text{is bounded for } y > 0.$$

The proof of this result and of various other analogues of results of Hardy and Littlewood, and of Fejér, are found in the last § 4 of the present note [2]).

The following notation will be used throughout this paper. The class of functions $f(x)$ measurable over $(-\infty, \infty)$ and such that

$$\int_{-\infty}^{\infty} |f(x)|^p\, dx < \infty, \quad p \geqq 1,$$

will be designated by $\mathfrak{L}_p$, the notation L_p being reserved for the analogous class of functions $\varphi(\theta)$ defined over $(-\pi, \pi)$.

Let $f(x) \varepsilon \mathfrak{L}_p$. The conjugate function $\tilde{f}(x)$ is defined by

$$(1.2)\quad \tilde{f}(x) = -\frac{1}{\pi} \lim_{\varepsilon \to 0} \int_{\varepsilon}^{\infty} [f(x+t) - f(x-t)]/t\, dt = \frac{1}{\pi}\, P.\, V. \int_{-\infty}^{\infty} \frac{f(t)\, dt}{x-t}.$$

It is known that $\tilde{f}(x)$ exists for almost all x and, in case $p>1$, $\tilde{f}(x) \in \mathfrak{L}_p$. (See e. g. Zygmund [1], Ch. XII). Let $f(z)$, $z = x+iy$, be analytic in the half-plane $y>0$. If the limit

$$\lim_{y \to 0} f(x+iy) = f(x)$$

exists for almost all x, $f(x)$ will be called the limit function of $f(z)$.

By $\mathfrak{H}_p$ we denote the class of functions $f(z)$ analytic in the half-plane $y>0$ and such that the integral

$$J(y; f) = \int_{-\infty}^{\infty} |f(x+iy)|^p\, dx \leqq M^p,$$

where M is a positive constant which depends only on f and p.

[2]) One of these results was stated without proof in our note [1].

Let $g(x)$ be measurable over $(-\infty, \infty)$. We set

$$(1.3) \qquad I(z; g) = \frac{1}{2\pi i}\int_{-\infty}^{\infty} g(t)\frac{dt}{t-z},$$

$$(1.4) \qquad P(z; g) = \frac{1}{\pi}\int_{-\infty}^{\infty} g(t)\frac{y\,dt}{(t-x)^2+y^2} = \int_{-\infty}^{\infty} g(t)\,K(t; z)\,dt,$$

$$(1.5) \qquad \tilde{P}(z; g) = -\frac{1}{\pi}\int_{-\infty}^{\infty} g(t)\frac{(t-x)\,dt}{(t-x)^2+y^2} = \int_{-\infty}^{\infty} g(t)\,\tilde{K}(t; z)\,dt,$$

where

$$(1.6) \qquad K(t; z) = \frac{1}{\pi}\frac{y}{(t-x)^2+y^2} = \Re\frac{1}{\pi i(t-z)},$$

$$(1.7) \qquad \tilde{K}(t; z) = -\frac{1}{\pi}\frac{t-x}{(t-x)^2+y^2} = \Im\frac{1}{\pi i(t-z)}.$$

All these integrals converge absolutely whenever $g(t) \in \mathfrak{L}_p$, $p \geqq 1$, $y \neq 0$. We shall call $I(z; g)$ and $P(z; g)$ integrals of Cauchy type and of Poisson type respectively, associated with the function $g(t)$. We observe that, on setting $\bar{z} = x - iy$,

$$(1.8) \qquad P(z; g) = \frac{1}{2\pi i}\int_{-\infty}^{\infty} g(t)\left[\frac{1}{t-z} - \frac{1}{t-\bar{z}}\right]dt = I(z; g) - I(\bar{z}; g),$$

while

$$(1.9) \qquad 2I(z; g) = P(z; g) + i\tilde{P}(z; g).$$

If $g(x) = f(x)$ is the limit function of a function $f(z)$ analytic for $y > 0$, and such that

$$f(z) = I(z; f) \quad \text{or} \quad f(z) = P(z; f)$$

we shall say that $f(z)$ is represented by its proper Cauchy integral $I(z; f)$, or by its proper Poisson integral $P(z; f)$, omitting the adjective „proper“ if no confusion arises.

The linear transformation

$$(1.10) \qquad w = \frac{1+iz}{1-iz}, \qquad z = i\frac{1-w}{1+w}$$

maps the half-plane $y > 0$ into the interior of the unit circle $|w| < 1$, the correspondence between the boundaries being given by

$$x = \tan \theta/2. \tag{1.11}$$

If $\varphi(w)$ is analytic in $|w| < 1$, we define the limit function of $\varphi(w)$ by

$$\varphi(e^{i\theta}) = \lim_{r\to 1} \varphi(r\, e^{i\theta}),$$

whenever it exists for almost all θ. As usual [3]) H_p will denote the class of functions $\varphi(w)$ analytic in $|w| < 1$ and such that

$$\int_{-\pi}^{\pi} |\varphi(r\, e^{i\theta})|^p \, d\theta \leqq M^p, \quad w = r\, e^{i\theta}.$$

We shall also use properties of integrals of Cauchy type,

$$I_c(w; \gamma) = \frac{1}{2\pi i} \int_{-\pi}^{\pi} \gamma(\tau) \frac{d(e^{i\tau})}{e^{i\tau} - w} \tag{1.12}$$

and of Poisson type,

$$\begin{aligned} P_c(w; \gamma) &= \frac{1}{2\pi} \int_{-\pi}^{\pi} \gamma(\tau) \frac{(1 - r^2)\, d\tau}{1 - 2r\cos(\theta - \tau) + r^2} \\ &= \frac{1}{2\pi i} \int_{|\zeta|=1} \gamma(\tau) \left[\frac{1}{\zeta - w} - \frac{1}{\zeta - w^*}\right] d\zeta = I_c(w; \gamma) - I_c(w^*; \gamma), \\ &\qquad \zeta = e^{i\tau}, \quad w^* = 1/\bar{w}, \end{aligned} \tag{1.13}$$

associated with the given function $\gamma(\tau)$, as well as those of the proper Cauchy and Poisson integrals, $I_c(w; \varphi)$, $P_c(w; \varphi)$ associated with the given function $\varphi(w)$ analytic in $|w| < 1$.

The fundamental notions of the theory of Fourier transforms are of course indispensable for what follows. For these properties we refer to Zygmund [1], Ch. XII. Let $f(x)$ be integrable over every finite range and let

$$\mathfrak{f}_a(x) = (2\pi)^{-1/2} \int_{-a}^{a} f(t)\, e^{-ixt}\, dt.$$

[3]) We refer to the papers by F. Riesz [1], Fichtenholz [1], and Smirnoff [1], concerning various properties of classes H_p and of integrals of Cauchy type and Poisson type which will be used in the sequel.

If $\mathfrak{f}_a(x)$ converges in $\mathfrak{L}_q$, $1 \leqq q \leqq \infty$, to a function $\mathfrak{f}(x)$, so that as $a \to \infty$,

$$\int_{-\infty}^{\infty} |\mathfrak{f}(x) - \mathfrak{f}_a(x)|^q \, dx \to 0,$$

or

$$\mathfrak{f}_a(x) \to \mathfrak{f}(x) \text{ uniformly over } (-\infty, \infty),$$

according as $\infty > q$ or $q = \infty$, then we write

$$\mathfrak{f}(x) = T(x; f),$$

and call $\mathfrak{f}(x)$ the Fourier transform of $f(x)$ in $\mathfrak{L}_q$. It is well known that whenever $f(x) \in \mathfrak{L}_p$, $1 \leqq p \leqq 2$, it has a Fourier transform $\mathfrak{f}(x)$ in $\mathfrak{L}_{p'}$, $p' = p/(p-1)$.

An important tool in investigating various classes of functions analytic in the unit circle is furnished by the classical „factorization theorem" of F. Riesz-Ostrowski, according to which a function $\varphi(w)$ of a given class can be represented as a product of the „Blaschke function" $b_\varphi(w)$, associated with $\varphi(w)$, and of a function of the same class, which does not vanish in the unit circle. Such is for instance the case of functions of the class H_p. An analogous factorization theorem for functions analytic in the half-plane is indispensable for our discussion. A theorem of this kind is proved in the next § 2 for the functions of class $\mathfrak{H}_p$, $p \geqq 1$, together with some other properties of this important class of functions. Such factorization theorems are obtainable for much more general classes of functions analytic in a half-plane (Gabriel [1, 2]). This problem is of considerable importance in itself. We intend to return to this question in another paper, restricting our investigation at present to the class $\mathfrak{H}_p$, $p \geqq 1$, which will be sufficient for our immediate purposes. In this case the simplest procedure consists merely in showing that, under the transformation (1.10) the class $\mathfrak{H}_p$ is transformed into a sub-class of H_p, which is readily done by using an elegant method introduced by Gabriel. Our proof in § 4 appeals also to some properties of conjugate functions which are discussed in § 3.

2. Functions of class $\mathfrak{H}_p$. In this paragraph we shall discuss some general properties of functions of class $\mathfrak{H}_p$, which are important for the theory of Fourier transforms and also are interesting in themselves. The discussion will be based on a few preliminary lemmas. Unless explicitly stated to the contrary, it will be understood that $1 \leqq p < \infty$.

Lemma 2.1. *If $U(w)$ is $\geqq 0$ and subharmonic in the interior of a circle Γ, and is continuous in the closed area (Γ) bounded by Γ, then for any circle C in (Γ)*

$$\int_C U(w)\,|dw| \leqq 2 \int_\Gamma U(w)\,|dw|. \tag{2.1}$$

The proof is found in Gabriel [3].

Lemma 2.2. *Let $g(t) \in \mathfrak{L}_p$ over $(-\infty, \infty)$ and let*

$$\gamma(\tau) = g(t) = g\left(\tan \frac{\tau}{2}\right)$$

be its transform on the unit circle. The integrals of Cauchy type and of Poisson type associated with $g(t)$ are transformed under the transformation

$$w = \frac{1+iz}{1-iz}, \qquad z = i\,\frac{1-w}{1+w} \tag{2.2}$$

according to the formulas

$$I(z;g) = I_c(w;\gamma) + C_0, \qquad C_0 = I(-i;g), \tag{2.3}$$

$$P(z;g) = P_c(w;\gamma). \tag{2.4}$$

The constant C_0 vanishes when $I(z;g)$ is the proper Cauchy integral of a function of class $\mathfrak{H}_p$.

Lemma 2.3. *If $f(z)$ is analytic in the half-plane $y > 0$ and has a limit function $f(x) \in \mathfrak{L}_p$ then whenever $f(z)$ is represented by its proper Cauchy integral, it is also represented by its proper Poisson integral and vice versa.*

The proof of these lemmas is found in Hille and Tamarkin [2].

Lemma 2.4. *A function $f(z) \in \mathfrak{H}_p$ tends uniformly to zero when z tends to infinity in any closed half-plane $y \geqq \varepsilon > 0$, where ε is arbitrarily small but fixed.*

The proof is based on the representation of $f(z)$ by means of its Cauchy and Poisson integrals related to the half-plane $y \geqq y_0 > 0$. Such a representation was proved by Bochner [1] in the case $p = 1$ and by Paley and Wiener [1] in the case $p = 2$.

In our proof we follow the line of argument used by **Paley** and **Wiener**.

On applying Cauchy's formula to the rectangle with vertices at the points $(\pm T + iy_0)$, $(\pm T + iY)$ we have

$$2\pi i f(z) = \int_{-T}^{T} \frac{f(t+iy_0)}{t+iy_0-z}\,dt - \int_{-T}^{T} \frac{f(t+iY)}{t+iY-z}\,dt +$$

$$+ i\int_{y_0}^{Y} \frac{f(T+i\eta)}{T+i\eta-z}\,d\eta - i\int_{y_0}^{Y} \frac{f(-T+i\eta)}{-T+i\eta-z}\,d\eta =$$

$$= I_1(T, y_0) - I_1(T, Y) + I_2(T) - I_2(-T), \quad 0 < y_0 < y < Y.$$

Choose X such that $2|x| < X$. Then

$$\frac{1}{X}\int_{X}^{2X} |I_2(\pm T)|\,dT \leqq \frac{1}{X}\int_{X}^{2X} dT \int_{y_0}^{Y} \frac{|f(\pm T+i\eta)|}{|\pm T+i\eta-z|}\,d\eta \leqq$$

$$\leqq \frac{2}{X^2}\int_{y_0}^{Y} d\eta \int_{X}^{2X} |f(\pm T+i\eta)|\,dT \leqq$$

$$\leqq \frac{2}{X^2}\int_{y_0}^{Y} d\eta\, X^{1/p'}\left[\int_{X}^{2X} |f(\pm T+i\eta)|^p\,dT\right]^{1/p} \leqq 2MX^{1/p'-}\,(Y-y_0)$$

where

$$\int_{-\infty}^{\infty} |f(t+iy)|^p\,dt \leqq M^p,$$

and, in case $p = 1$, $p' = \infty$, the factor $X^{1/p'}$ is to be replaced by 1. On keeping y_0 and Y fixed and allowing $X \to \infty$ we get

$$2\pi i f(z) = \lim_{X\to\infty} \frac{1}{X}\int_{X}^{2X} [I_1(T, y_0) - I_1(T, Y)]\,dT.$$

Since the limits

$$I_1(\infty, y_0) = \lim_{T\to\infty} I_1(T, y_0) = \int_{-\infty}^{\infty} \frac{f(t+iy_0)}{t+iy_0-z}\,dt = I_1(y_0),$$

$$I_1(\infty, Y) = \lim_{T\to\infty} I_1(T, Y) = \int_{-\infty}^{\infty} \frac{f(t+iY)}{t+iY-z}\,dt = I_1(Y)$$

exist as absolutely convergent integrals it is seen that

$$2\pi i f(z) = I_1(y_0) - I_1(Y).$$

Now let $Y \to \infty$. In case $p = 1$ we have

$$|I_1(Y)| \leqq \int_{-\infty}^{\infty} \frac{|f(t+iY)|}{Y-y}\, dt \leqq M(Y-y)^{-1} = O(1/Y),$$

while, in case $p > 1$,

$$|I_1(Y)| \leqq \left[\int_{-\infty}^{\infty} |f(t+iY)|^p dt\right]^{1/p} \left[\int_{-\infty}^{\infty} \frac{dt}{\{(t-x)^2+(Y-y_0)^2\}^{p'/2}}\right]^{1/p'} =$$

$$= O(Y^{-1-\frac{1}{p'}}).$$

Hence $I_1(Y) \to 0$ in either case, and

$$(2.5) \qquad f(z) = (2\pi i)^{-1} \int_{-\infty}^{\infty} \frac{f(t+iy_0)}{t+iy_0-z}\, dt, \quad 0 < y_0 < y.$$

The same argument shows that

$$(2.6) \quad 0 = (2\pi i)^{-1} \int_{-\infty}^{\infty} \frac{f(t+iy_0)}{t+iy_0-z'}\, dt, \quad z' = x+iy', \quad y' < y_0.$$

Let $\varepsilon > 0$ be given. In formulas (2.5) and (2.6) put

$$y_0 = \frac{\varepsilon}{2}, \quad z' = x + i(2y_0 - y), \quad y > y_0$$

and subtract (2.6) from (2.5). Thus we obtain the desired representation

$$(2.7)\ f(z) = \frac{1}{\pi} \int_{-\infty}^{\infty} \frac{f(t+iy_0)(y-y_0)}{(t-x)^2+(y-y_0)^2}\, dt = \int_{-\infty}^{\infty} f(t+iy_0) K(t; z-iy_0)\, dt.$$

To prove our lemma assume $y \geqq \varepsilon = 2y_0$ and first consider the case $p = 1$. A positive δ being given choose T_0 so large that

$$\int_{-\infty}^{-T_0} |f(t+iy_0)|\, dt + \int_{T_0}^{\infty} |f(t+iy_0)|\, dt < \delta.$$

Since $K(t; z-y_0) \leqq (y-y_0)^{-1} < y_0^{-1}$ the contribution of the corresponding range of integration in (2.7) will not exceed δ/y_0. After

T_0 has been so fixed, the contribution of the remaining range $(-T_0, T_0)$ is obviously $O(1/z)$ uniformly in the half-plane $y \geqq \varepsilon$.

If $p > 1$ then, by the convexity property,

$$|f(z)|^p \leqq \int_{-\infty}^{\infty} |f(t+iy_0)|^p \, K(t; z - iy_0) dt \tag{2.8}$$

and the preceding argument can be applied without modifications to the integral of the right-hand member of (2.8).

Lemma 2.5. *Under the transformation* (2.2) *the class* $\mathfrak{H}_p$ *is transformed into a sub-class of* H_p.

Consider any half-plane $y > \varepsilon > 0$. Its boundary $y = \varepsilon$ is mapped by (2.2) into a circle Γ_ε in the w-plane, tangent from the inside to the circle $|w| = 1$ at $w = -1$. The half-plane itself is transformed into the interior of Γ_ε. Let $f(z) \in \mathfrak{H}_p$. The function $\varphi(w) = f(z)$ is analytic in $|w| < 1$, hence $|\varphi(w)|^p$ is subharmonic in the interior of Γ_ε and by Lemma 2.4, is continuous in the closed area (Γ_ε). Let C be any given circle $|w| = r < 1$. If ε is sufficiently small, C will be in (Γ_ε). Then, by Lemma 2.1,

$$\int_{-\pi}^{\pi} |\varphi(re^{i\theta})|^p \, d\theta = \frac{1}{r} \int_C |\varphi(w)|^p \, |dw| \leqq \frac{2}{r} \int_{\Gamma_\varepsilon} |\varphi(w)|^p \, |dw| =$$

$$= \frac{2}{r} \int_{-\infty}^{\infty} |f(t+i\varepsilon)|^p \, \frac{2\,dt}{(1+\varepsilon)^2 + t^2} < \frac{4}{r} M^p$$

which shows that $\varphi(w) \in H_p$.

We now pass on to the main theorems of this paragraph.

Theorem 2.1. (i) *A function* $f(z) \in \mathfrak{H}_p$ *for almost all* x *has a limit function* $f(x) \in \mathfrak{L}_p$ *to which it tends along any nontangential path.*

(ii) *Any* $f(z) \in \mathfrak{H}_p$ *is represented by its proper Cauchy and Poisson integrals. In terms of the real part of the limit function* $f(x)$ *we also have*

$$\begin{aligned} f(z) &= \frac{1}{\pi i} \int_{-\infty}^{\infty} \Re f(t) \, \frac{dt}{t-z} = 2\, I(z; \Re f) = \\ &= P(z; \Re f) + i\, \tilde{P}(z; \Re f). \end{aligned} \tag{2.9}$$

(iii) *Any* $f(z) \in \mathfrak{H}_p$ *tends to its limit function* $f(x)$ *in the mean of order* p,

$$\int_{-\infty}^{\infty} |f(x+iy)-f(x)|^p \, dx \to 0 \quad as \quad y \to 0.$$

Moreover, as $y \downarrow 0$,

$$T(y;f)=\int_{-\infty}^{\infty} |f(x+iy)|^p \, dx \uparrow T(0;f)=\int_{-\infty}^{\infty} |f(x)|^p \, dx.$$

(iv) *If* $f(x) \in \mathfrak{L}_p$ *and*

$$f(z)=P(z;f)$$

is analytic for $y>0$, *then* $f(z) \in \mathfrak{H}_p$ *and therefore is represented by its proper Poisson and Cauchy integrals* $P(z;f)$, $I(z;f)$ *respectively, as well as by* (2.9).

We observe that the analogous properties of functions of class H_p are well known. Let now $f(z) \in \mathfrak{H}_p$ and let $\varphi(w)$ be the transform of $f(z)$ under the transformation (2.2). By Lemma 2.5, $\varphi(w) \in H_p$, hence $\varphi(w)$ has a limit function $\varphi(e^{i\theta}) \in L_p$ to which it tends along any non-tangential path. Thus the limit function $f(x)$ of $f(z)$ exists almost everywhere along any nontangential path. Moreover, by Fatou's, theorem

$$\int_{-\infty}^{\infty} |f(x)|^p \, dx \leqq \lim_{y \to 0} \int_{-\infty}^{\infty} |f(x+iy)|^p \, dx \leqq M^p$$

so that $f(x) \in \mathfrak{L}_p$. This proves statement (i) of our theorem. To prove statement (ii) construct the integral $P(z;f)$. By Lemma 2.2 we have

$$P(z;f)=P_c(w;\varphi)=\varphi(w)=f(z)$$

since $\varphi(w) \in H_p$ is represented by its proper Poisson integral. Thus $f(z)$ is represented by its Poisson and by Lemma 2.3 by its Cauchy integral. To complete the proof of (ii) we observe that $f(z)=P(z;f)$ implies in view of (1.6)

$$\Re f(z)=P(z;\Re f)=\Re\{2I(z;\Re f)\}.$$

Since $f(z)$ and $2I(z;\Re f)$ are analytic for $y>0$, and tend to zero when $y \to \infty$, (2.9) follows. To prove statement (iii) we observe that, by (ii),

$$f(z)-f(x)=\int_{-\infty}^{\infty} [f(t)-f(x)] \, K(t;z) \, dt,$$

whence, by the convexity property,

$$|f(x+iy)-f(x)|^p \leqq \int_{-\infty}^{\infty} |f(t)-f(x)|^p K(t; z)\, dt =$$

$$= \int_{-\infty}^{\infty} |f(x+t)-f(x)|^p K(t; iy)\, dt.$$

and

$$\int_{-\infty}^{\infty} |f(x+iy)-f(x)|^p\, dx \leqq \int_{-\infty}^{\infty} dt\, K(t; iy) \int_{-\infty}^{\infty} |f(x+t)-f(x)|^p\, dx.$$

Since the function

$$F(t) = \int_{-\infty}^{\infty} |f(x+t)-f(x)|^p\, dx$$

is continuous at $t=0$, statement (iii) becomes a consequence of the classical property of the Poisson integral for

$$\int_{-\infty}^{\infty} F(t)\, K(t; iy)\, dt \to F(0) = 0 \quad \text{as} \quad y \to 0.$$

The fact that $T(y; f)$ increases when y decreases is well known and is readily derived from (2.8). The same applies to statement (iv).

Theorem 2.2. *A function $f(z) \in \mathfrak{H}_p$ can be represented as a product*

$$f(z) = b_f(z)\, h(z), \tag{2.10}$$

where

$$b_f(z) = \prod_{(\nu)} \frac{z - z_\nu}{z - \bar{z}_\nu} \frac{\bar{z}_\nu - i}{z_\nu + i}$$

is the Blaschke function associated with $f(z)$ and $h(z) \in \mathfrak{H}_p$, but does not vanish in the half-plane $y > 0$. Here $\{z_\nu\}$ is the sequence of zeros of $f(z)$ in the half-plane $y > 0$ and the condition

$$\sum_{(\nu)} y_\nu/(1 + x_\nu^2 + y_\nu^2) < \infty \tag{2.11}$$

must be satisfied. If $f(z)$ does not vanish for $y > 0$, $b_f(z) \equiv 1$, $h(z) = f(z)$. Otherwise

$$|b_f(z)| < 1, \quad y > 0, \quad |b_f(x)| = 1 \quad \textit{almost everywhere.}$$

The limit function $h(x)$ of $h(z)$ satisfies the condition

$$(2.12) \qquad |h(x)| = |f(x)| \quad \textit{almost everywhere}$$

so that

$$(2.13) \qquad \int_{-\infty}^{\infty} |h(z)|^p \, dx \leqq \int_{-\infty}^{\infty} |h(x)|^p \, dx = \int_{-\infty}^{\infty} |f(x)|^p \, dx \leqq M^p.$$

The analogous theorem for functions of class H_p is well known. Now, if $f(z) \in \mathfrak{H}_p$ and $\varphi(w) = f(z)$, then by Lemma 2.5, $\varphi(w) \in H_p$. Hence we have the representation

$$\varphi(w) = b_\varphi(w)\, \eta(w),$$

where $b_\varphi(w)$ is the Blaschke function relative to the unit circle associated with $\varphi(w)$, and $\eta(w) \in H_p$ but does not vanish in $|w| < 1$. We set $b_f(z) = b_\varphi(w)$, $h(z) = \eta(w)$. The properties of $b_f(z)$ stated in the theorem are readily derived from the known properties of $b_\varphi(w)$. As to $h(z)$, we have by Lemma 2.2,

$$P(z; h) = P_c(w; \eta) = \eta(w) = h(z)$$

and the properties of $h(z)$ now are proved by using Theorem 2.1. Finally, condition (2.11) is derived from the corresponding condition for the roots $\{w_\nu\}$ of $\varphi(w)$, viz. that the infinite product $\prod_{(\nu)} |w_\nu|$ must converge[4])

3. Conjugate functions. The properties of conjugate functions which we discuss in the present paragraph are partially known, at least under more restrictive assumptions. For the reader's convenience we state explicitly, in form of lemmas, those properties we need. It will be understood again that $1 \leqq p < \infty$.

Lemma 3.1. *Let $g(x) \in \mathfrak{L}_p$ and let its conjugate function*

$$\tilde{g}(x) = \frac{1}{\pi} \, P.V. \int_{-\infty}^{\infty} \frac{g(t)\, dt}{x - t}$$

[4]) The results stated in Theorem 2.2 are not essentially new. An analogous theorem was proved by Hardy, Ingham and Pólya [1], Theorem 1, under conditions which are not equivalent to ours. Formula (2.10) could also be derived from a theorem by Gabriel [1, 2]. It is not at all obvious, however, that the functions of class $\mathfrak{H}_p$ satisfy the conditions required by Gabriel; that they actually do so follows from our Lemma 2.4. The proof and application of this lemma is therefore the only essential novelty of our discussion. Our method obviously breaks down if $p < 1$.

exist for a given value of x, this being the case for almost all values of x. Then (*see* (1.5), (1.7))

$$\widetilde{P}(z; g) = \int_{-\infty}^{\infty} g(t)\, \widetilde{K}(t; z)\, dt \to \widetilde{g}(x) \quad \textit{as} \quad y \to 0. \tag{3.1}$$

There is no loss of generality in assuming $x = 0$. Then

$$\widetilde{P}(z; g) = \widetilde{P}(iy; g) = -\frac{1}{\pi}\int_{0}^{\infty} [g(t) - g(-t)] \frac{t\,dt}{t^2 + y^2}.$$

But

$$\frac{t}{t^2 + y^2} - \frac{1}{t} = \frac{y^2}{t^2 + y^2},$$

and by hypothesis

$$\widetilde{g}(0) = -\lim_{\varepsilon \to 0} \frac{1}{\pi} \int_{\varepsilon}^{\infty} [g(t) - g(-t)] \frac{dt}{t}$$

exists. Hence

$$\widetilde{P}(iy; g) = \widetilde{g}(0) + \lim_{\varepsilon \to 0} \frac{1}{\pi} \int_{\varepsilon}^{\infty} \frac{g(t) - g(-t)}{t} \frac{y^2}{t^2 + y^2}\, dt \equiv \widetilde{g}(0) + T(y).$$

We consider $T(y)$ as an improper integral and write

$$T(y) = \int_{0}^{\eta} + \int_{\eta}^{\infty} \equiv T_1(y) + T_2(y),$$

where η will be chosen sufficiently small but fixed. In the integral $T_2(y)$ the integrand is dominated by a fixed integrable function $|[g(t) - g(-t)]/t|$ and almost everywhere tends to 0 as $y \to 0$. Hence

$$T_2(y) \to 0 \quad \text{as} \quad y \to 0.$$

As to $T_1(y)$, on introducing the improper integral

$$G(t) = \frac{1}{\pi} \int_{0}^{t} [g(\tau) - g(-\tau)] \frac{d\tau}{\tau}$$

and integrating by parts, we have

$$T_1(y) = G(\eta) \frac{y^2}{y^2 + \eta^2} - \int_{0}^{\eta} G(t)\, d_t \frac{y^2}{t^2 + y^2}.$$

Since $G(\eta) \to 0$ as $\eta \to 0$, the integrated term here is $o(1)$ as $\eta \to 0$, and the second term is

$$o\left(-\int_0^\eta d_t \frac{y^2}{t^2+y^2}\right) = o(1), \text{ uniformly in } y.$$

This proves lemma 3.1. We observe that the relation

$$P(z; g) \to g(x) \quad \text{as} \quad y \to 0 \tag{3.2}$$

for almost all x is well known.

Theorem 3.1. *Let* $g(x)$ *and* $\tilde{g}(x) \in \mathfrak{L}_p$ [5]. *The function*

$$f(z) = 2I(z; g) = P(z; g) + i\tilde{P}(z; g) \tag{3.3}$$

is analytic in the half-plane $y > 0$ *and is representable by its proper Cauchy and Poisson integrals. Furthermore its limit function* $f(x)$ *is such that*

$$f(x) = g(x) + i\tilde{g}(x). \tag{3.4}$$

Conversely if $f(z)$ *is representable by its proper Cauchy or Poisson integral with the limit function* $f(x) \varepsilon \mathfrak{L}_p$, *then*

$$\mathfrak{I} f(x) = \tilde{\mathfrak{R}} f(x). \tag{3.5}$$

Let $f(z)$ be given by (3.3). By Lemma 3.1 its limit function $f(x) = g(x) + i\tilde{g}(x)$. On the other hand $f(z)$ is represented by an integral of Cauchy type; by Lemma 2.2. (2.3), its transform $\varphi(w)$ in the w-plane is represented also by an integral of Cauchy type; since, however, the limit function $\varphi(e^{i\theta}) \in L_p$, $\varphi(w)$ is also represented by its proper Cauchy or Poisson integrals; again by Lemma 2.2, (2.4), it follows that $f(z)$ is represented by its proper Poisson integral, and, by Lemma 2.3, by its proper Cauchy integral.

To prove the converse let

$$f(z) = I(z; f) = P(z; f), \qquad f(x) \in \mathfrak{L}_p.$$

On the other hand by Theorem 2.1, (iv), we also have

$$f(z) = P(z; \mathfrak{R} f) + i\tilde{P}(z; \mathfrak{R} f).$$

[5]) If $p > 1$ and $g(x) \in \mathfrak{L}_p$, then also $\tilde{g}(x) \in \mathfrak{L}_p$. In this case the results stated in Lemma 3.2 were obtained by M. Riesz [1].

Since $\mathfrak{R} f \in \mathfrak{L}_p$, Lemma 3.1 shows that for almost all x,

$$P(z; \mathfrak{R} f) \to \mathfrak{R} f(x), \qquad \tilde{P}(z; f) \to \tilde{\mathfrak{R}} f(x).$$

Consequently for almost x

$$f(x) = \mathfrak{R} f(x) + i \mathfrak{J} f(x) = \mathfrak{R} f(x) + i \tilde{\mathfrak{R}} f(x),$$

and (3.5) follows.

Corollary. *If $g(x)$ and $\tilde{g}(x)$ both $\in L_p$, $p \geqq 1$, then $-g(x)$ is the conjugate of $\tilde{g}(x)$ and the reciprocity relations*

$$(3.6) \quad \tilde{g}(x) = \frac{1}{\pi} P.\,V. \int_{-\infty}^{\infty} \frac{g(t)\,dt}{x-t}, \qquad g(x) = -\frac{1}{\pi} P.\,V. \int_{-\infty}^{\infty} \frac{\tilde{g}(t)\,dt}{x-t}$$

hold [6]).

Indeed, we have

$$f(z) = 2I(z; g) = I(z; g) + iI(z; \tilde{g}) = 2iI(z; \tilde{g}),$$

so that

$$P(z; g) + i\tilde{P}(z; g) = iP(z; \tilde{g}) - \tilde{P}(z; \tilde{g})$$

and we have only to apply Lemma 3.1. If $p > 1$ it is sufficient to assume only that either $g(x)$ or $\tilde{g}(x) \in \mathfrak{L}_p$.

Theorem 3.2. *Assume that the functions $g(x)$ and $\tilde{g}(x)$ both $\in \mathfrak{L}_p$ and that they both are of bounded variation over $(-\infty, \infty)$. Under these assumptions we have*

(i) *$g(x)$ and $\tilde{g}(x)$ are absolutely continuous so that the derivatives $g'(x)$ and $\tilde{g}'(x)$ both $\in \mathfrak{L}_1$.*

(ii) *The function $\tilde{g}'(x)$ is the conjugate of $g'(x)$.*

(iii) *If*

$$f(z) = 2I(z; g) = I(z; g + i\tilde{g}) = I(z; f), \qquad f(x) = g(x) + i\tilde{g}(x),$$

then the derivative $f'(z) \in \mathfrak{H}_1$ and thus is represented by its Cauchy and Poisson integrals. The limit function of $f'(z)$ is

$$f'(x) = g'(x) + i\tilde{g}'(x).$$

By Theorem 3.1 $f(z)$ is representable by its Cauchy and Poisson integrals and has the limit function $f(x) = g(x) + i\tilde{g}(x)$. By assump-

[6]) In case $p > 1$ a proof was given by M. Riesz [1].

tion this function is of bounded variation over $(-\infty, \infty)$. Hence its transform $\varphi(e^{i\theta})$ on the unit circle $|w|=1$, which is the limit function of the transform $\varphi(w)=f(z)$, is also of bounded variation over $(-\pi, \pi)$. It is well known that under these conditions $\varphi(e^{i\theta})$ is absolutely continuous in θ, hence $f(x)$ is absolutely continuous in x. Thus statement (i) is proved.

To prove statement (iii) we write

$$f(z)=P(z;f)=\int_{-\infty}^{\infty} f(t)\, K(t;z)\, dt,$$

$$f'(z)=\frac{\partial f(z)}{\partial x}=\int_{-\infty}^{\infty} f(t)\frac{\partial}{\partial x} K(t;z)\, dt$$

$$=\int_{-\infty}^{\infty} f(t)\frac{\partial}{\partial t} K(t;z)\, dt=\int_{-\infty}^{\infty} f'(t)\, K(t;z)\, dt.$$

The operations of differentiation under integral sign and at integrating by parts are obviously permissible; the integrated terms vanish since $K(t;z)$ vanishes for $t\to\pm\infty$, while $f(t)$ is bounded. It follows that $f'(x)$ is the limit function of $f'(z)$ and, by Theorem 2.1, iv, $f'(z)\in\mathfrak{H}_1$. Thus statement (iii) is proved. The proof of (ii) is now obtained by an easy application of Theorem 3.1.

4. Applications to the theory of Fourier transforms. We shall need some additional lemmas whose analogues in the theory of power series are quite trivial.

Lemma 4.1. *Let $f(x)\in\mathfrak{L}_p$, $1\leqq p<\infty$. If $f(x)$ has a Fourier transform $\mathfrak{f}(x)$ in some $\mathfrak{L}_q$, $1\leqq q\leqq\infty$, then in order that $f(x)$ be the limit function of a function $f(z)\in\mathfrak{H}_p$ (or, which is the same, of a function analytic in the half-plane $y>0$ and representable by its Cauchy or Poisson integrals) it is necessary and sufficient that $\mathfrak{f}(x)=0$ for $x<0$.*

Lemma 4.2. *If $f(x)\in\mathfrak{L}_p$, $1\leqq p\leqq\infty$ and has a Fourier transform $\mathfrak{f}(x)$ in $\mathfrak{L}_q$, $1\leqq q\leqq\infty$, then the Poisson integral associated with $f(x)$ can be written in the form*

$$(4.1)\quad P(z;g)=\frac{1}{\pi}\int_{-\infty}^{\infty} f(t)\frac{y\, dt}{(t-x)^2+y^2}=(2\pi)^{-1/2}\int_{-\infty}^{\infty} e^{ixt}e^{-y|t|}\mathfrak{f}(t)\, dt,\ y>0.$$

If $f(x)$ *is the limit function of a function* $f(z) \in \mathfrak{H}_p$, $1 \leqq p < \infty$, *then*

$$(4.2) \qquad P(z; f) = (2\pi)^{-1/2} \int_0^\infty e^{ixt} e^{-yt} \mathfrak{f}(t)\, dt, \quad y > 0 \text{ [7]}.$$

Lemma 4.3. *If* $f(x)$ *and* $g(x) \in \mathfrak{L}_2$ *and* $\mathfrak{f}(x)$, $\mathfrak{g}(x)$ *are their Fourier transforms in* $\mathfrak{L}_2$, *then*

$$(4.3) \qquad \int_0^\infty dy \int_0^\infty e^{-yt} |\mathfrak{f}(t)|\, dt \int_0^\infty e^{-ys} |\mathfrak{g}(s)|\, ds \leqq$$

$$\leqq \pi \left[\int_{-\infty}^\infty |f(x)|^2\, dx \int_{-\infty}^\infty |g(x)|^2\, dx \right]^{1/2}.$$

Indeed the left-hand member of (4.3) is equal to

$$\int_0^\infty \int_0^\infty |\mathfrak{f}(t)\, \mathfrak{g}(s)|/(t+s)\, dt\, ds$$

and by Hilbert's inequality (Hardy, Ingham and Pólya [1]) does not exceed

$$\pi \left[\int_0^\infty |\mathfrak{f}(t)|^2\, dt \int_0^\infty |\mathfrak{g}(s)|^2\, ds \right]^{1/2} \leqq \pi \left[\int_{-\infty}^\infty |\mathfrak{f}(t)|^2\, dt \int_{-\infty}^\infty |\mathfrak{g}(s)|^2\, ds \right]^{1/2} =$$

$$= \pi \left[\int_{-\infty}^\infty |f(x)|^2\, dx \int_{-\infty}^\infty |g(x)|^2\, dx \right]^{1/2}.$$

In passing to our main results we first give a new proof of a theorem due to M. Riesz [1].

Theorem 4.1. *If* $f(z) \in \mathfrak{H}_p$, $1 \leqq p < \infty$, *then for each* x,

$$(4.4) \qquad \int_0^\infty |f(x+iy)|^p\, dy \leqq \frac{1}{2} \int_{-\infty}^\infty |f(t)|^p\, dt \text{ [8]}.$$

[7]) The proof of these lemmas is found in Hille and Tamarkin [2]. The discussion given there was concerned with Fourier transforms in a certain generalized sense, but is valid without any modifications in the case of Fourier transforms in $\mathfrak{L}_q$.

[8]) This is an analogue of a classical theorem by Fejér and F. Riesz according to which if $\varphi(w) \in H_p$ and D is any diameter of the unit circle I, then

$$\int_D |\varphi(w)|^p\, |dw| \leqq \frac{1}{2} \int_I |\varphi(w)|^p\, |dw|.$$

We start with the case $p = 2$. Then $f(x)$ has a Fourier transform $\mathfrak{f}(x) \in \mathfrak{L}_2$ which vanishes for $x \leqslant 0$, so that

$$\int_{-\infty}^{\infty} |f(x)|^2 \, dx = \int_0^{\infty} |\mathfrak{f}(x)|^2 \, dx.$$

By Lemma 4.2 we have

$$f(z) = P(z; f) = (2\pi)^{-1/2} \int_0^{\infty} e^{ixt} e^{-yt} \mathfrak{f}(t) \, dt, \qquad y > 0,$$

whence, by Lemma 4.3,

$$\int_0^{\infty} |f(x + iy)|^2 \, dy \leqq (2\pi)^{-1} \int_0^{\infty} dy \left[\int_0^{\infty} e^{-yt} |\mathfrak{f}(t)| \, dt \right]^2$$

$$\leqq \frac{1}{2} \int_0^{\infty} |\mathfrak{f}(t)|^2 \, dt = \frac{1}{2} \int_{-\infty}^{\infty} |f(x)|^2 \, dx.$$

Thus (4.4) is proved in the case $p = 2$. In the general case, if $f(z)$ has no zeros in the half-plane $y > 0$, we apply the preceding result to the function $[f(z)]^{p/2} \in \mathfrak{H}_2$. Finally, if $f(z)$ has zeros in $y > 0$, we use Theorem 2.2 with the result

$$\int_0^{\infty} |f(x + iy)|^p \, dy \leqq \int_0^{\infty} |h(x + iy)|^p \, dy \leqq \frac{1}{2} \int_{-\infty}^{\infty} |h(x)|^2 \, dx =$$

$$= \frac{1}{2} \int_{-\infty}^{\infty} |f(x)|^2 \, dx.$$

On the basis of the preceding results we now can prove our main.

Theorem 4.2. *If both $g(x)$ and its conjugate $\tilde{g}(x) \in \mathfrak{L}_1$, and if $\mathfrak{f}(x)$ is the Fourier transform (in $\mathfrak{L}_\infty$) of the function*

$$f(x) = g(x) + i\tilde{g}(x)$$

then

$$\int_0^{\infty} |\mathfrak{f}(t)|/t \, dt \leqq \left(\frac{\pi}{2}\right)^{1/2} \int_{-\infty}^{\infty} |f(x)| \, dx, \tag{4.5}$$

the constant $\left(\frac{\pi}{2}\right)^{1/2}$ being the best possible.

We first observe that, by Theorem 3.1, $f(x)$ is the limit function of a function $f(z) \in \mathfrak{H}_1$. Hence by Lemma 4.1, $\mathfrak{f}(x) = 0$ for $x < 0$ If now $\mathfrak{f}(x)$ is real-valued and $\geqq 0$, the proof of Theorem 4.2 is almost trivial. Indeed under this assumption we have in view of Lemma 4.2 and Theorem 4.1,

$$(4.6) \qquad \int_0^\infty \mathfrak{f}(t)/t\,dt = \int_0^\infty dy \int_0^\infty e^{-yt}\mathfrak{f}(t)\,dt = (2\pi)^{1/2} \int_0^\infty f(iy)\,dy \leqq \leqq \left(\frac{\pi}{2}\right)^{1/2} \int_{-\infty}^\infty |f(x)|\,dx.$$

In general, however, this argument is not valid, and we have to use a device analogous to that used by Zygmund in the case of the power series ([1], pp. 158—159). If $f(z)$ has zeros in the half-plane $y > 0$, we use Theorem 2.2 to write

$$f(z) = b_f(z)\,h(z),$$

$h(z) \neq 0$ in $y > 0$, $h(z) \varepsilon \mathfrak{H}_1$, $|h(x)| = |f(x)|$ almost everywhere,

$$|b_f(z)| < 1 \quad \text{for} \quad y > 0.$$

We set

$$f(z) = f_1(z)\,f_2(z), \quad f_1(z) = b_f(z)\,[h(z)]^{1/2}, \quad f_2(z) = [h(z)]^{1/2},$$

so that $f_k(z) \in \mathfrak{H}_2$, $k = 1, 2$, while

$$\int_{-\infty}^\infty |f_k(x)|^2\,dx = \int_{-\infty}^\infty |h(x)|\,dx = \int_{-\infty}^\infty |f(x)|\,dx, \quad k = 1, 2.$$

If $f(z)$ has no zeros in $y > 0$ the situation is even simpler, but for the uniformity of notation we shall put

$$h(z) = f(z), \quad f_1(z) = f_2(z) = [f(z)]^{1/2}.$$

Now introduce the Fourier transforms $\mathfrak{f}_k(x)$ (in $\mathfrak{L}_2$) of $f_k(x)$. Again, since $\mathfrak{f}_k(x) = 0$ for $x < 0$, we have

$$\int_0^\infty |\mathfrak{f}_k(x)|^2\,dx = \int_{-\infty}^\infty |\mathfrak{f}_k(x)|^2\,dx = \int_{-\infty}^\infty |f_k(x)|^2\,dx = \int_{-\infty}^\infty |f(x)|\,dx.$$

The Fourier transform $\mathfrak{f}(x)$ of $f(x) = f_1(x) f_2(x)$ can be expressed in terms of $\mathfrak{f}_k(x)$ by means of the well known „Faltung" rule (e. g. Wiener [1], pp. 70–71),

$$(4.7)\quad \mathfrak{f}(t) = (2\pi)^{-1/2} \int_{-\infty}^{\infty} \mathfrak{f}_1(u)\, \mathfrak{f}(t-u)\, du = (2\pi)^{-1/2} \int_0^t \mathfrak{f}_1(u)\, \mathfrak{f}_2(t-u)\, du.$$

This yields

$$\int_0^\infty |\mathfrak{f}(t)|/t\, dt = \int_0^\infty dy \int_0^\infty e^{-yt} |\mathfrak{f}(t)|\, dt \leqq$$

$$\leqq (2\pi)^{-1/2} \int_0^\infty dy \int_0^\infty e^{-yt}\, dt \int_0^t |\mathfrak{f}_1(u)|\, |\mathfrak{f}_2(t-u)|\, du =$$

$$= (2\pi)^{-1/2} \int_0^\infty dy \int_0^\infty e^{-yu} |\mathfrak{f}_1(u)|\, du \int_0^\infty e^{-yv} |\mathfrak{f}_2(v)|\, dv \leqq$$

$$\leqq \left(\frac{\pi}{2}\right)^{1/2} \left[\int_0^\infty |\mathfrak{f}_1(u)|^2\, du \int_0^\infty |\mathfrak{f}_2(u)|^2\, du\right]^{1/2} = \left(\frac{\pi}{2}\right)^{1/2} \int_0^\infty |f(x)|\, dx,$$

which is the desired result.

It remains only to show that the coefficient $\left(\frac{\pi}{2}\right)^{1/2}$ in the right-hand member of (4.5) is the best possible. By considering the conformal mapping of an appropriate ellipse into the unit circle it is easy to construct a function $\psi(w)$, analytic in the closed unit circle $|w| \leqq 1$, real valued for real values of w, and such that

$$\int_0^1 \psi(w)\, dw > \left(\frac{1}{2} - \varepsilon\right) \int_{|w|=1} |\psi(w)|\, |dw|\ ^{9)},$$

where ε can be taken arbitrarily small but fixed. Using the transformation

$$w = \frac{1+iz}{1-iz}, \qquad z = i\,\frac{1-w}{1+w},$$

consider the function

$$f(z) = -i\,\psi(w)\,\frac{dw}{dz} = \psi(w)\,\frac{2}{(1-iz)^2}.$$

[9]) Cf. Fejér [1], p. 118, footnote.

Since $\psi(w)$ is bounded in $|w| \leqq 1$, $f(z) \in \mathfrak{H}_1$ and

$$\int_{|w|=1} |\psi(w)|\,|d w| = \int_{-\infty}^{\infty} |f(x)|\,dx.$$

Now the segment $(0, 1)$ of the w-plane is mapped into the segment $(0, i)$ of the y-axis in the z-plane and a simple computation shows that

$$\int_0^1 \psi(w)\,d w = \int_0^1 f(i y)\,dy > \left(\frac{1}{2} - \varepsilon\right) \int_{-\infty}^{\infty} |f(x)|\,dx.$$

Thus in view of (4.2) we have for this particular function $f(z)$

$$\int_0^\infty |\mathfrak{f}(t)|/t\,dt = \int_0^\infty dy \int_0^\infty e^{-yt}\,|\mathfrak{f}(t)|\,dt \geqq (2\pi)^{1/} \int_0^1 |P(iy; f)|\,dy \geqq$$

$$\geqq (2\pi)^{1/2} \int_0^1 P(iy; f)\,dy = (2\pi)^{1/2} \int_0^1 f(iy)\,dy > \left\{\left(\frac{\pi}{2}\right) - \delta\right\} \int_{-\infty}^{\infty} |f(x)|\,dx$$

where $\delta = \varepsilon (2\pi)^{1/2}$. Since δ can be made as small as we please the proof of Theorem 4.2 is completed.

The following theorem appears as in immediate corollary of Theorem 4.2.

Theorem 4.3. *Assume that $g(x)$ and its conjugate $\tilde{g}(x)$ are both of bounded variation over $(-\infty, \infty)$, and in addition $\in \mathfrak{L}_p$, $1 \leqq p < \infty$. Then $f(x) = g(x) + i\tilde{g}(x)$ is absolutely continuous and the Fourier transform $\mathfrak{f}(x)$ of $f(x)$, defined by*

$$\mathfrak{f}(x) = (2\pi)^{-1/2} \lim_{a \to \infty} \int_{-a}^{a} f(t)\,e^{-itx}\,dt, \tag{4.8}$$

exists for all $x \neq 0$ and is continuous for $x \neq 0$.

Furthermore $\mathfrak{f}(x) = 0$ for $x < 0$ and is absolutely integrable over $(0, \infty)$. More precisely

$$\int_0^\infty |\mathfrak{f}(t)|\,dt \leqq \left(\frac{\pi}{2}\right)^{1/2} \int_{-\infty}^{\infty} |f'(x)|\,dx,$$

the constant $\left(\frac{\pi}{2}\right)^{1/2}$ being the best possible.

By Theorem 3.2 $f(x)$ is absolutely continuous and is the limit function of a function $f(z)$ analytic in the half-plane $y > 0$ and such that $f'(z) \in \mathfrak{H}_1$, with the limit function $f'(x)$. By Theorem 4·2 then

$$\int_0^\infty |\mathfrak{f}_1(t)|/t\,dt \leqq \left(\frac{\pi}{2}\right)^{1/2} \int_{-\infty}^\infty |f'(x)|\,dx,$$

where $\mathfrak{f}_1(x)$ is the Fourier transform (in $\mathfrak{L}_\infty$) of $f'(x)$. On the other hand since under our assumptions $f(x) \to 0$ as $x \to \pm\infty$, we have

$$(2\pi)^{-1/2} \int_{-a}^{a} f(t)\,e^{-ixt}\,dt = (2\pi)^{-1/2} \left\{ \frac{e^{-ixt}}{-ix} f(t) \Big]_{-a}^{a} + \frac{1}{ix} \int_{-a}^{a} f'(t)\,e^{-ixt}\,dt \right\}$$

$$\to \frac{1}{ix}\mathfrak{f}_1(x),$$

whence

$$\mathfrak{f}(x) = \frac{1}{ix}\mathfrak{f}_1(x), \qquad x \neq 0.$$

It should be observed that whenever $f(x)$ has a Fourier transform in some $\mathfrak{L}_q$, this Fourier transform will coincide with $\mathfrak{f}(x)$ defined by (4.8). This remains true even when $f(x)$ has a Fourier transform in the sense of various more general definitions.

We may finally state the following geometric interpretation of Theorem 4.2.

Theorem 4.4. *Let $\zeta = F(z)$ map the half-plane $y > 0$ into a domain $\mathfrak{D}$ (not necessarily simply covered) of the ζ-plane, in such a way that the lengths L_{y_0} of the images of the lines $y = y_0$ are bounded. Then the boundary of $\mathfrak{D}$ is a rectifiable curve, of length $L = \lim_{y_0 \to 0} L_{y_0}$; the function $F'(z) = f(z) \in \mathfrak{H}_1$, its limit function $f(x) \in \mathfrak{L}_1$ and the Fourier transform $\mathfrak{f}(x)$ of $f(x)$ vanishes for $x < 0$ and is such that*

$$\int_0^\infty |\mathfrak{f}(t)|/t\,dt \leqq \left(\frac{\pi}{2}\right) L.$$

The proof of this theorem does not offer any difficulties and may be left to the reader.

List of references.

S. Bochner.

[1] Vorlesungen über Fouriersche Integrale. Leipzig, 1932.

L. Fejér.

[1] Über gewisse Minimumprobleme der Funktionentheorie. Mathematische Annalen, 97 (1927), 104—123.

Gr. Fichtenholz.

[1] Sur l'intégrale de Poisson et quelques questions qui s'y rattachent. Fundamenta Mathematicae, 13 (1929).

R. M. Gabriel.

[1] Concerning the zeros of a function regular in a half-plane. Journal of the London Mathematical Society, 4 (1929), 133—139.

[2] An improved result concerning the zeros of a function regular in a half-plane. Ibidem, 4 (1929), 307—309.

[3] An inequality concerning the integrals of positive subharmonic functions along certain curves. Ibidem, 5 (1930), 129—131.

G. H. Hardy, A. E. Ingham and G. Pólya.

[1] Theorems concerning mean values of analytic functions. Proceedings of the Royal Society, (A) 113 (1927), 542—569.

G. H. Hardy and J. E. Littlewood.

[1] Some new properties of Fourier constants. Mathematische Annalen 97 (1927), 159—20.

G. H. Hardy, J. E. Littlewood, G. Pólya.

[1] Inequalities. Cambridge, 1934.

E. Hille and J. D. Tamarkin.

[1] On the summability of Fourier Series. Third Note. Proceedings of the National Academy of Sciences, 16 (1930), 594—598.

[2] On a Theorem of Paley and Wiener. Annals of Mathematics (2) 34 (1933), 606—614.

R. E. A. C. Paley and N. Wiener.

[1] Notes on the theory and application of Fourier transforms. I—II. Transactions of the American Mathematical Society, 35 (1933), 348—355.

F. Riesz.

[1] Über die Randwerte einer analytischen Funktion. Mathematische Zeitschrift, 18 (1923), 87—95.

M. Riesz.

[1] Sur les fonctions conjuguées. Ibidem, 27 (1927), 218—244.

V. Smirnoff.

[1] Sur les valeurs limites des fonctions régulières à l'intérieur d'un circle. Journal de la Société Phys.-Math. de Leningrade, 2 (1930), 122—137.

N. Wiener.

[1] The Fourier integral and certain of its applications. Cambridge, 1933.

A. Zygmund.

[1] Trigonometrical Series. Monogr. Matem. Warsaw, 1935.

Chapter 7
Laplace Integrals

The paper included in this chapter is

[62] On Laplace integrals

A commentary on this paper can be found starting on page 672.

On Laplace integrals.

By

EINAR HILLE[1].

A number of integrals of rather different types are designated as Laplace integrals in the literature. In the present paper this term will be applied to Riemann-Stieltjes integrals of the form

$$(1) \qquad \int_0^\infty e^{-zu}\, dA(u).$$

$A(u)$ is supposed to be either of bounded variation in every finite interval $[0, \omega]$ or continuous in every such interval or, finally, the sum of two such functions. Further, $A(0) = 0$. Doubly infinite Laplace integrals are considered in the last paragraph, i. e., the limits of integration are $-\infty, \infty$ instead of $0, \infty$.

Two fundamental questions in the theory of Laplace integrals are the *problem of representation* and the *problem of analytic continuation,* neither of which is adequately treated in the literature. The questions involve a discussion of what analytic functions can be represented by a convergent Laplace integral in a given half-plane, and how the analytic continuation of such a function is obtained outside of the half-plane of convergence. The second problem can be regarded as one in the theory of summability in as much as a divergent integral is to be summed as far as possible. But, as will be shown in the present paper, it is advantageous to connect the problem with that of determining *what functions can be represented as the quotient of two absolutely convergent Laplace integrals.* The latter problem can be solved completely.

---

[1] Free use is made in the present paper of results announced by E. Hille and J. D. Tamarkin. See Proc. Nat. Acad. Sci., 19 (1933) 573—577, 902—908, 908—912, 20 (1934) 140—144.

The point of view and the notions of modern abstract algebra are helpful in dealing with the general theory of Laplace integrals. This depends upon the fact that the absolutely convergent Laplace integrals form a *domain of integrity.* It is natural to consider the algebraic properties of this domain, for instance, divisibility, theory of ideals, algebraic and transcendental extensions and so on. It turns out that these problems are really of interest to various branches of pure analysis. The present paper deals largely with the simplest and most trivial of these questions, viz. the extension of a domain of integrity to a field by the forming of quotients.

1. **Convergence and inversion.** It is necessary to recall the simplest notions in the theory of Laplace integrals. Let $f(z)$ be an analytic function, regular in a half-plane $x > R$, which is representable by a Laplace integral in some portion of the plane. With this integral are associated three *abscissas of convergence*, $\sigma_0[f]$, $\sigma_u[f]$ and $\sigma_a[f]$, referring to *ordinary*, *uniform* and *absolute convergence* respectively. These abscissas satisfy the inequalities.

$$(2) \qquad -\infty \leqq R \leqq \sigma_0 \leqq \sigma_u \leqq \sigma_a \leqq +\infty,$$

and are otherwise independent of each other. The mass function $A(u)$ is determined by the formula

$$(3) \qquad \frac{1}{2}[A(u+0)+A(u-0)] = \frac{1}{2\pi i} \lim_{T\to\infty} \int_{c-iT}^{c+iT} f(z)\, e^{uz} \frac{dz}{z},$$

where $c > \max(0, \sigma_0)$.

The following properties of Laplace integrals are of some interest as back ground for the subsequent considerations. For $x \geqq \sigma_0[f] + \varepsilon$, we have

$$(4) \qquad f(z) = O(1) + o(|y|),$$

whereas $f(z)$ is bounded for $x \geqq \sigma_u[f] + \varepsilon$. The values of $f(z)$ in the interior of the half-plane of absolute convergence show a remarkable degree of coherance since the total variation of $f(z)$ along any rectifiable path in this domain is bounded, provided merely the slope is bounded along the path. The same is true for the total variation of the product $z^n f^{(n)}(z)$ for any positive integer n. It is not necessarily true for paths of unbounded slope.

2. **The representation problem.** We pass now to the problem of what functions are representable by a convergent Laplace integral. A necessary and sufficient condition can be read off from formula (3). The function $f(z)$ in question must be regular in some half-plane and satisfy condition (4). For sufficiently large values of c the formula must define a function $A(u), 0 < u < \infty$, independent of c, such that the integral in formula (1) exists and has a domain of convergence. The existence of the integral between finite limits $0, \omega$ is ensured if $A(u)$ is of bounded variation or continuous or the sum of two such functions in $[0, \omega]$ — these being the cases to which we are limiting ourselves — and the convergence of the infinite integral implies that $A(u) = O(e^{Bu})$ for some fixed finite B. If a half-plane of absolute convergence is required, it is necessary and sufficient that $A(u)$ is of bounded variation, the total variation of $A(u)$ in $[0, \omega]$ being $O(e^{C\omega})$ for some fixed finite C.

This is a solution of the representation problem in principle, but the solution is clearly not of much value in this form. At the present stage of the question it would be desirable to get some sufficient conditions stated directly in terms of $f(z)$. Conditions of integrability suggest themselves in an obvious manner. We shall begin by introducing certain classes of functions $H_p(a)$.

Definition 1. *$f(z) \varepsilon H_p(a)$, where p is fixed, $1 \leqq p < \infty$, if it is regular for $x > a$, and*

$$\left\{ \int_{-\infty}^{\infty} |f(x+iy)|^p \, dy \right\}^{\frac{1}{p}} \leqq C, \; x > a. \tag{5}$$

If $f(z) \varepsilon H_p(a)$, then $\lim_{x \to a} f(x+iy)$ exists for almost all y, and the limit function belongs to $L_p(-\infty, \infty)$. It is convenient to regard $H_p(a)$ as a *linear vector space*, and to introduce a suitable *metric*, for example by taking the *distance* between $f_1(z)$ and $f_2(z)$ to be $\| f_1(z) - f_2(z) \|_p$, where the *norm* of $f(z)$, $\| f(z) \|_p$, is the least number C satisfying (5). The space is *complete* in terms of this metric, i. e., every fundamental sequence converges to a limit function which also belongs to the space. It should be observed that for every $f(z) \varepsilon H_p(a)$ formula (3) with $c > \max(0, a)$ defines $A(u)$ as a continuous function of u, $0 < u < \infty$, and that for fixed u, *$A(u)$ is a continuous functional of the element $f(z)$*, defined all over $H_p(a)$. For our present purposes any other metric would do just as well, provided it has the two properties (*i*) the space is complete, and (*ii*) $A(u)$ is a continuous functional of $f(z)$ for u fixed. We recall finally that the ordinary notions of the theory of point sets exist in a metric space. In particular, we understand by a *set of the first category* a point set which is the union of a denumerable or finite number of nowhere dense sets.

Definition 2. *$f(z)\,\varepsilon\,I(a)$, if $f(z)$ is representable by a Laplace integral, the mass function $A(u)$ being of bounded variation on every finite interval $[0, \omega]$, and $\sigma_a[f] \leqq a$.*

The following theorem holds.

THEOREM 1. *Every $f(z)\,\varepsilon\,H_p(a)$ is representable by a convergent Laplace integral for $x > a$, and*

$$A(u) = \frac{1}{2\pi}\int_0^u e^{av}\,d_v \int_{-\infty}^{\infty} f(a+it)\frac{e^{itv}-1}{it}\,dt. \tag{6}$$

If $1 \leqq p \leqq 2, A(u)$ is absolutely continuous, and $\sigma_a[f] \leqq a$, i. e., $H_p(a) \subset I(a)$. If $2 < p$, the cross section of $H_p(a)$ with the union of all classes $I(b)$ is a point set which is of the first category in $H_p(a)$, and, moreover, the set of elements $f(z)$ such that the corresponding function $A(u)$ is of bounded variation on any interval whatever is also of the first category in $H_p(a)$.

No proof can be given here, but some comments are in order to explain the remarkable difference between the two cases $1 \leqq p \leqq 2$ and $2 < p$. The existence of the representation for $x > a$ together with formula (6) is a fairly simple consequence of Cauchy's Theorem. If $1 \leqq p \leqq 2$ the boundary values $f(a+it)$ possess a Fourier transform in $L_{p'}(-\infty, \infty)$, and formula (6) reduces to

$$A(u) = \frac{1}{\sqrt{2\pi}}\int_0^u e^{av} F_a(v)\,dv, \tag{7}$$

where

$$F_a(u) = \frac{1}{\sqrt{2\pi}}\ \underset{T\to\infty}{\mathrm{l.\,i.\,m.}} \int_{-T}^{T} e^{iut} f(a+it)\,dt. \tag{8}$$

This proves that $A(u)$ is absolutely continuous, and a simple estimate of its total variation proves that $\sigma_a[f] \leqq a$. On the other hand, if $2 < p < \infty$, the boundary values need not have a Fourier transform in the ordinary sense. We find accordingly that $A(u)$, while being continuous, is ordinarily not absolutely continuous or even of bounded variation. As an example we may take the function

$$e^{-\omega_1 z}(1+z)^{-\frac{1}{2}} \prod_{n=2}^{\infty} \frac{n \log^2 n - z}{n \log^2 n + z}, \quad \omega_1 \geqq 0, \tag{9}$$

which belongs to every $H_p(a)$ with $p > 2, a > -2 \log^2 2$. Here $A(u) = 0$ for $0 \leqq u \leqq \omega_1$, and is an analytic function of u for $\omega_1 < u$. This function exists in the half-plane $\Re(u) > \omega_1$, but oscillates so violently as $u \to \omega_1$ along the real axis that it cannot be of bounded variation in any interval $[\omega_1, \omega_2]$. In order to pass from this isolated example to the sweeping statements of the theorem, we can use the tools of modern functional analysis. The total variation of $A(u)$ in $[\omega_1, \omega_2]$ is a functional defined, as a positive number or $+\infty$, for every element $f(z)$ of $H_p(a)$. This functional in its turn is the limit of a sequence of continuous quasi-linear functionals. By an important theorem of Banach-Saks-Steinhaus such a functional is either bounded in the whole space or else finite at most in a set of the first category. Example (9) shows that the latter case must hold when $p > 2$. It is not difficult to find sufficient conditions in order that $A(u)$ be absolutely continuous even when $p > 2$, but considerations of space forbid further excursions.

We have now carried the discussion of the representation problem as far as the scope of the present lecture permits.

3. **Multiplication theorems.** We shall need certain theorems on the multiplication of Laplace integrals in the subsequent discussion. The product of two Laplace integrals

$$f(z) = \int_0^\infty e^{-zu}\, dA(u), \quad g(z) = \int_0^\infty e^{-zu}\, dB(u), \tag{10}$$

can be written as a formal Laplace integral

$$h(z) \sim \int_0^\infty e^{-zu}\, dC(u), \tag{11}$$

where

$$C(u) = \int_0^u A(u-v)\, dB(v) = \int_0^u B(u-v)\, dA(v). \tag{12}$$

The convergence of the integrals in (10) for a given value of z does not imply the convergence of the integral in (11), but if the latter does converge, its value is $f(z)\, g(z)$. Formal integration by parts in (11) gives

$$h(z) = z \int_0^\infty e^{-zu}\, C(u)\, du,$$

where the integral converges to the sum $f(z)\,g(z)$ for

$$x > \max\,(0, \sigma_0[f], \sigma_0[g]).$$

A particularly interesting case is that in which one of the factors, say $g(z)$, is absolutely convergent, and $\sigma_a[g] < \sigma_0[f]$. In this case the product integral is always convergent when the integral for $f(z)$ converges. Similarly, uniform or absolute convergence of the integral for $f(z)$ induces the same type of convergence for the product integral. In particular

$$\sigma_0[fg] \leqq \sigma_0[f],\ \sigma_u[fg] \leqq \sigma_u[f],\ \sigma_a[fg] \leqq \sigma_a[f]. \tag{13}$$

Or, to put it roughly, *multiplication of a Laplace integral by an absolutely convergent Laplace integral leads to a new integral, the convergence properties of which are at least as good as those of the first integral.* In other words, *multiplication by an absolutely convergent Laplace integral is a convergency preserving transformation, or a regular definition of summability.* This simple observation is one of the basic ideas in our discussion of the continuation problem.

It is obvious that the sum and the difference of two Laplace integrals are Laplace integrals. We have just seen that the product of two absolutely convergent Laplace integrals is an absolutely convergent Laplace integral. Further, $f(z) = 1$ is representable by an absolutely convergent Laplace integral for $x > -\infty$. It follows that the class of all Laplace integrals whose abscissas of absolute convergence are less than a given real number a, i. e., the class $I(a)$, is a *commutative domain of integrity* having a unit element. This domain is not a field as is shown by simple examples. But we may expect to get a better notion of its properties by extending it to a field in the usual manner through the adjunction of the quotients.

4. **The continuation problem.** We are now ready to tackle the continuation problem. This may present itself in various forms. Thus a Laplace integral may be given, having a half-plane of convergence in which it defines an analytic function. It is desired to obtain the analytic continuation of this function in its domain of existence or at least in its half-plane of regularity. But we may also have a given function with an associated formal Laplace integral. We have to imagine that a mass function $A(u)$ has been computed from formula (3), but that the resulting integral does not exist in the accepted sense. Here we have to sum the formal integral so that it gives an acceptable representation

of the function in the largest possible domain. We shall show how these two phases of the problem can be handled by a formation of quotients.

Let $f(z)$ be an analytic function, having a half-plane of regularity, whose representation or analytic continuation is desired. Let $g(z)$ be another analytic function which is representable by a Laplace integral such that $\sigma_a[g] < \min\{\sigma_0[f], \infty\}$. Our point of departure is the trivial identity

$$f(z) = \frac{f(z)\,g(z)}{g(z)}. \tag{14}$$

In case $\sigma_0[f] < \infty$, we know that $f(z)\,g(z)$ is representable by a Laplace integral, the convergence properties of which are at least as good as those of the integral for $f(z)$. If $f(z)$ has only a formal representation, there is still the possibility that $g(z)$ can be so chosen that an absolutely convergent representation becomes available for the product $f(z)\,g(z)$. Theorem 1 shows that a sufficient condition for this to be the case is the existence of a multiplier $g(z)$ with the properties stated above such that $f(z)\,g(z)\,\varepsilon\,H_p(a)$ for some a and some $p, 1 \leqq p \leqq 2$. In either case we obtain a new representation of $f(z)$, viz. as the quotient of two absolutely convergent Laplace integrals. This method of summability for Laplace integrals might be referred to as the *quotient method.*

In order to estimate the effectiveness of this method, we must answer the following question: *What functions can be represented as the quotient of two absolutely convergent Laplace integrals in a given half-plane?* This question admits of a simple definite answer, which is perhaps surprising in view of the fact that we were unable to answer completely the question of what functions are representable as Laplace integrals. It turns out, however, that this new question is closely related to an old problem: *What functions are representable as the quotient of two bounded functions in a given domain?* This question has been solved by the brothers Nevanlinna.[1] It is easy to see the connection. Indeed, if

$$f(z) = L_1(z) : L_2(z), \tag{15}$$

where the Laplace integrals are absolutely convergent for $x \geqq a$, then we have already a representation of $f(z)$ as the quotient of two bounded functions in this half-plane. It follows that the corresponding necessary

[1] See F. und R. Nevanlinna, Über die Eigenschaften analytischer Funktionen in der Umgebung einer singulären Stelle oder Linie, Acta Sci. Fenn., 50: 5 (1922) 46 pp., especially pp. 23—26.

and sufficient conditions for the existence of such a representation must be satisfied. On the other hand, if they are satisfied, then we have

$$f(z) = B_1(z) : B_2(z), \tag{16}$$

where $B_\nu(z)$ are bounded for $x > a$. Using Theorem 1 we see that

$$(z + a + 1)^{-1} B_\nu(z) = L_\nu(z), \quad \nu = 1, 2, \tag{17}$$

are Laplace integrals, absolutely convergent for $x > a$, but possibly not on the line $x = a$. The quotient of these two integrals gives the required representation of $f(z)$ for $x > a$. The criteria of F. and R. Nevanlinna consequently lead to the following.

THEOREM 2. *In order that $f(z)$ shall be representable as the quotient of two Laplace integrals, absolutely convergent for $x \geqq a$, it is necessary that* (i) *$f(z)$ is meromorphic for $x \geqq a$, and those of its poles $\{b_n\}$ which lie outside of the circle $|z - a| = \delta$ satisfy the condition*

$$\sum \Re\{(b_n - a)^{-1}\} < \infty,$$

(ii)
$$\int_{-\infty}^{\infty} \overset{+}{\log} |f(a + iy)| \frac{dy}{1 + y^2} < \infty,$$

(iii)
$$\frac{1}{\varrho} \int_{-\frac{\pi}{2}}^{\frac{\pi}{2}} \overset{+}{\log} |f(a + \varrho e^{i\theta})| \cos\theta \, d\theta$$

is bounded for all large values of ϱ. Conversely, if these conditions are satisfied, then $f(z)$ is representable as the quotient of two Laplace integrals, absolutely convergent for $x > a$.

This theorem gives immediately

THEOREM 3. *A suitable application of the quotient method gives the analytic representation of a Laplace integral, convergent or formal, in the largest half-plane where the function in question satisfies the conditions of Theorem 2.*

Thus the quotient method solves the continuation problem for a large class of Laplace integrals, but it is a priori obvious that not every formal Laplace integral is associated with a function satisfying the conditions of Theorem 2 in its half-plane of meromorphism. For such a function the method gives the analytic continuation only in a portion of this domain. The possibility of modifying the quotient method so as to make it more powerful will be considered in § 6.

Other aspects of this method are also worthy of attention, and in particular the connections with the theory of summability. The object of the method is to seek the representation of the function by formula (15), i. e., as the quotient of two absolutely convergent Laplace integrals, $L_1(z)$ and $L_2(z)$, in some half-plane $x > a$. We shall refer to the denominator, $L_2(z)$, as a *multiplier* of $f(z)$ for $x > a$, the terminology being suggested by formula (14). To every given integral $L_2(z)$ and every real $a > \sigma_a[L_2]$ belongs a certain class of functions $f(z)$ having $L_2(z)$ as multiplier for $x > a$, i. e., $f(z)\, L_2(z)$ is a Laplace integral, and $\sigma_a[f L_2] \leq a$. Conversely, to every $f(z)$ and every real a corresponds a certain class of multipliers for $x > a$. This class may be vacuous, but if it contains one element it has the power of the continuum. A multiplier $\mu_1(z)$ is *at least as strong as* another multiplier $\mu_2(z)$ in $x > a$, if $\mu_1(z)$ is a multiplier of every $f(z)$ which admits $\mu_2(z)$ as a multiplier in $x > a$. A necessary and sufficient condition for this to be the case is that $\mu_1(z) : \mu_2(z)$ is representable by an absolutely convergent Laplace integral for $x > a$. Two multipliers are *equivalent* for $x > a$ if each is at least as strong as the other in this half-plane. We have similar notions in closed half-planes $x \geq a$. Those who are familiar with the Hausdorff theory of summability will recognize the origin of these notions.

5. **Special cases.** After these general considerations let us discuss the effectiveness of some special multipliers. We have first the functions in Hausdorff's logarithmic scale

$$(18) \quad (z - a_0)^{-\alpha_0}[\log(z - a_1)]^{-\alpha_1}[\log_2(z - a_2)]^{-\alpha_2} \ldots (\log_k(z - a_k)]^{-\alpha_k},$$

where the first exponent α_ν different from zero must have positive real part. The most interesting case here is that of pure powers, $(z - a)^{-\alpha}$, $\Re(\alpha) > 0$. In the simplest case $a = 0$ we obtain the representation

$$(19) \qquad f(z) = z^\alpha \int_0^\infty e^{-zu}\, dA_\alpha(u),$$

$$(20) \qquad A_\alpha(u) = \frac{1}{\Gamma(\alpha + 1)} \int_0^u (u - v)^\alpha\, dA(v).$$

These formulas are well-known from the theory of M. Riesz' typical means of the first kind. Indeed, if a Laplace integral is summable

$(R, \lambda, \varkappa)$ at $z = z_0 = x_0 + iy_0$, $x_0 > 0$, then the generalized value of the integral is given by formula (19) with $z = z_0$, $a = \varkappa$. On the other hand, if (19) is valid in the half-plane $x > a \geqq 0$, then the integral (1) is summable (R, λ, a) for $x > a$, provided $A(u)$ is defined and has the properties stated in the introduction. It is only on the line of summability that any difference can occur between summability by typical means and summability by power multipliers. The effectiveness of the latter method appears from the following theorem which is a generalization of a well-known theorem of M. Riesz for Dirichlet series. It is a consequence of Theorem 1.

THEOREM 4. *If $f(z)$ is regular for $x > a \geqq 0$, and if its Lindelöf mu-function is $\leqq \mu_0$ in this half-plane, then $f(z)$ is representable by formula (19) for $x > a$, $\alpha > \mu_0$. The corresponding mass function $A_\alpha(u)$ is continuous, but not necessarily of bounded variation, unless $\alpha > \mu_0 + \frac{1}{2}$, in which case it is absolutely continuous, and the integral is absolutely convergent for $x > a$.*

The function $A_\alpha(u)$ referred to in this theorem is given by formula (20) when this formula has a sense which is not always the case, however.

Replacing the function $z^{-\alpha}$ by the Gamma quotient

$$\frac{\Gamma(\alpha + 1)\,\Gamma(z)}{\Gamma(\alpha + z)}, \tag{21}$$

we arrive at a method of summation which is essentially equivalent with the typical means of the second kind. The two multipliers are equivalent for $x > 0$, but not for $x \geqq 0$. Finally, if it is desired to enter the left half-plane, it is enough to make a suitable change of variable $z \mid z + b$ in the multipliers.

Theorem 4 shows that powers and gamma quotients are effective multipliers in that portion of the half-plane of regularity where the function is of finite order with respect to the ordinate, i. e., where

$$\overline{\lim_{|y| \to \infty}} \log |f(x + iy)| : \log |y| < \infty. \tag{22}$$

If the rate of growth of the function is still stronger, we may resort to one of the following multipliers

$$\exp\left[- A\,[\log(z - a)]^{\alpha}\right], \tag{23}$$

$$\exp\left[- A(z - a)^{\beta}\right], \tag{24}$$

where $A > 0, a > 0, 1 > \beta > 0$. The first of these multipliers is stronger than any power in $x > a$, provided $a > 1$, and its strength grows with a. The second multiplier is stronger than the first in $x > a$, and its strength grows with β. Still stronger multipliers are available, e. g.,

$$\exp\left[-A(z-a)\log^{-1}(z-a+1)\right]. \tag{25}$$

On the other hand, the Phragmén-Lindelöf theorem shows that $g(z) \equiv 0$ is the only analytic function which is bounded in the half-plane $x > a$, and which is $O[\exp(-A\,|\,y\,|^{\beta})]$ with $\beta \geqq 1$ on $x = a$. Thus the strength of available multipliers is strictly limited which of course also follows from Theorem 2.

6. **The doubly infinite case.** It is obvious that the ideas of the present paper are capable of further applications. In particular, we can apply a similar analysis to Laplace integrals with doubly infinite limits

$$\int_{-\infty}^{\infty} e^{-zu}\, dA(u), \tag{26}$$

where the integral may be regarded as being defined by its principal value. A rapid survey of the main results will not be devoid of interest. Theorem 1 has the following analogue.

THEOREM 5. *If $f(z)$ is regular in the strip $a < x < b$, $b - a < \infty$, and if for a fixed p, $1 \leqq p < \infty$,*

$$\left\{\int_{-\infty}^{\infty} |f(x+iy)|^p\, dy\right\}^{\frac{1}{p}} \leqq C, \quad a < x < b, \tag{27}$$

then $f(z)$ is representable by a doubly infinite Laplace integral in $a < x < b$, and $A(u)$ is given by formula (6) for $u \geqq 0$ and by the same formula with a replaced by b for $u < 0$. If $1 \leqq p \leqq 2$, $A(u)$ is absolutely continuous, and the integral is absolutely convergent in $a < x < b$. If $2 < p < \infty$, $A(u)$ is continuous, but as a rule not of bounded variation.

The last statement can be made precise as in Theorem 2 by making a metric space out of the class of functions in question. It is sufficient that the metric makes the space complete, and that $A(u)$ becomes a continuous functional of $f(z)$ for u fixed. Then the exceptional set of functions $f(z)$, whose mass functions $A(u)$ are of bounded variation in any interval, is a set of the first category in the space.

The multiplication theorems hold, mutatis mutandis, for doubly infinite Laplace integrals. The necessary modifications do not prevent us from being able to use the idea of forming quotients in order to obtain the analytic continuation of such integrals. The analogue of Theorem 2 reads:

THEOREM 6. *A necessary condition for $f(z)$ to be representable as the quotient of two doubly infinite Laplace integrals, absolutely convergent for $a \leqq x \leqq b$, is that* (i) *$f(z)$ is meromorphic in this strip, and, putting $\frac{\pi}{b-a} = \lambda$, the series*

$$\sum e^{-\lambda |y_n|} \sin \lambda (x_n - a),$$

extended over the poles $\{x_n + iy_n\}$, is convergent,

$$\text{(ii)} \qquad \int_{-\infty}^{\infty} [\overset{+}{\log} |f(a+iy)| + \overset{+}{\log} |f(b-iy)|] \, e^{-\lambda |y|} \, dy < \infty,$$

$$\text{(iii)} \qquad e^{-\lambda t} \int_{a}^{b} [\overset{+}{\log} |f(x+it)| + \overset{+}{\log} |f(x-it)|] \sin \lambda (x-a) \, dx$$

is bounded for $t > 0$. Conversely, if these conditions hold, $f(z)$ can be written as the quotient of two doubly infinite Laplace integrals, absolutely convergent for $a < x < b$.

In the present case we have at our disposal a much richer variety of multipliers. A few examples must suffice. In the strip $a < x < b$ we can use $\exp[-\sin \lambda (z-a)]$, $\lambda = \frac{\pi}{b-a}$. The function $\Gamma[\varrho(z-a)]$ has been used by Hardy as a multiplier in the theory of Dirichlet series. The exponential functions

$$\text{(28)} \qquad \exp[(-z)^n], \; n > 1,$$

are available as multipliers in the whole plane. As a consequence we get

THEOREM 7. *If $f(z)$ is meromorphic and of finite order in the half-plane $x > a$, then it can be written as the quotient of two absolutely convergent doubly infinite Laplace integrals in this half-plane.*

This theorem implies, in particular, that *a simply infinite Laplace integral can be continued analytically with the aid of the modified quotient method in the largest half-plane where it is meromorphic and of finite order.* Whether or not the applications of the quotient method to the continuation problem can be pushed still further has to be left as an open question for the present.

Chapter 8
Factorisatio Numerorum and Möbius Inversion

Papers included in this chapter are

A note on these papers can be found on page 674.

A problem in "Factorisatio Numerorum".

By

Einar Hille (New Haven, Conn.)

1. Introduction. This note is devoted to a study of the number theoretic function $f(n)$ which gives the number of representations of the natural number n as a product of factors greater than one. Here two representations are considered identical if and only if they contain the same factors written in the same order. We define $f(1) = 1$. Some generalizations are indicated at the end of the note.

This function $f(n)$ does not seem to have attracted much attention. It is intimately connected with the algorithm of Möbius

$$(1.1) \qquad \begin{cases} a_1 b_1 = 1, \\ \sum_{d \mid n} a_d b_{n/d} = 0, \qquad n = 2, 3, \ldots, \end{cases}$$

which arises in a number of analytical problems, for instance, in the expansion of the reciprocal of an ordinary Dirichlet series into a series of the same type. The relationship is simply

$$(1.2) \qquad D(s) = \sum_{n=1}^{\infty} a_n n^{-s}, \frac{1}{D(s)} = \sum_{n=1}^{\infty} b_n n^{-s}, a_1 \neq 0,$$

where the coefficients are connected by (1.1). Taking $a_1 = 1$, $a_n = -1$, $n = 2, 3, \ldots$, we get $b_n = f(n)$ as will be shown below.

The only papers on this function which are known to the author are those of L. Kalmár. [1]) These contain a study of the summatory function

$$(1.3)\qquad F(n) = \sum_{m=1}^{n} f(m).$$

Denoting by ρ the positive root of the equation

$$(1.4)\qquad \zeta(s) = 2,$$

Kalmár proved that

$$(1.5)\qquad F(n) = -\frac{n^{\rho}}{\rho\,\zeta'(\rho)}\{1 + o(1)\},$$

and gave various estimates of the remainder.

It is obvious that $f(n)$ itself is a very irregular function, and next to nothing seems to be known about its behavior under different assumptions regarding the number of prime factors in n. A study of this problem does not call for particularly complicated machinery, and the results are not quite devoid of interest. They were found as a by-product in an investigation of the algorithm of Möbius which will be published elsewhere. [2])

2. Elementary properties. Let

$$(2.1)\qquad n = p_{i_1}^{\alpha_1} p_{i_2}^{\alpha_2} \ldots p_{i_\nu}^{\alpha_\nu}$$

be the representation of n as a product of prime factors. It is clear that $f(n)$ depends only upon the divisibility properties of n and not upon the actual numerical values of the prime factors. It follows that $f(n)$ is a symmetric function of the ν variables $\alpha_1, \alpha_2, \ldots, \alpha_\nu$.

Let $d \mid n$ and put $n = m\,d, m > 1$. All the factorizations of n which contain m as first factor are obtained by considering all factorizations of d. They are $f(d)$ in number. It is clear that these particular factorizations can be obtained in no other way. Hence we have

$$(2.2)\qquad f(n) = \sum_{d \mid n}{}' f(d),$$

---

[1]) A „factorisatio numerorum" problémájáról, Matematikai és Fizikai Lapok, **38** (1931) 1—15, and Über die mittlere Anzahl der Produktdarstellungen der Zahlen. (Erste Mitteilung) Acta Litterarum ac Scientiarum, Szeged, **5** (1931) 95—107. The second part of the latter paper does not seem to have appeared. I am indebted to Prof. O. Szász for this reference.

[2]) See also E. Hille and O. Szász, On the completeness of Lambert functions, Part I, Bulletin Amer. Math. Soc., **42** (1936), and Part II, Annals of Math., (2) **37** (1936)

where the summation extends over all divisors d of n which are $<n$ This functional equation together with the initial conditions

$$f(1)=f(p)=1 \tag{2.3}$$

determine $f(n)$ completely.

Formula (2.2) recalls Dedekind's inversion formula. If

$$h(n)=\sum_{d\mid n} g(d)$$

for all n, then

$$g(n)=h(n)-\sum h\left(\frac{n}{p_{i_1}}\right)+\sum h\left(\frac{n}{p_{i_1}p_{i_2}}\right)-\ldots,$$

where the summations extend over all the combinations 1, 2, 3, ... at a time of the distinct prime factors $p_{i_1}, p_{i_2}, \ldots, p_{i_\nu}$ of n. Putting $h(n)=2f(n)$, $g(n)=f(n)$ we get after simplification

$$f(n)=2\left\{\sum f\left(\frac{n}{p_{i_1}}\right)-\sum f\left(\frac{n}{p_{i_1}p_{i_2}}\right)+\ldots\right\}, \tag{2.4}$$

which is more suitable for numerical computation than (2.2). In particular [3])

$$f(p^\alpha)=2f(p^{\alpha-1})=2^{\alpha-1}, \tag{2.51}$$

$$f(p^\alpha q^\beta)=2\,[f(p^\alpha q^{\beta-1})+f(p^{\alpha-1}q^\beta)-f(p^{\alpha-1}q^{\beta-1})]. \tag{2.52}$$

Let us return to (2.2). Sum both sides of this equation with respect to all values of n for which $\alpha_1+\alpha_2+\ldots+\alpha_\nu=k$, a given integer. Here $\alpha_\mu\geqq 0$ and the basis $p_{i_1}, p_{i_2}, \ldots, p_{i_\nu}$ is kept fixed. Then

$$S_{k,\nu}=\sum_{(n)} f(n)=\sum_{(n)}\sum_{d\mid n}{}' f(d). \tag{2.6}$$

The values of d which occur in the last member are all of the form $d=p_{i_1}^{\beta_1}p_{i_2}^{\beta_2}\ldots p_{i_\nu}^{\beta_\nu}$ where $\beta_1+\beta_2+\ldots+\beta_\nu=\varkappa$, $0\leqq\varkappa\leqq k-1$. Moreover, every such integer occurs in the sum. Let us collect all terms which have the same value of $\varkappa$. A divisor d with $\varkappa=k-1$ divides exactly ν different values of n. It follows that every such d occurs ν times, and $\Sigma f(d)$ extended over these values equals $\nu S_{k-1,\nu}$. A d with $\varkappa=k-2$ is a divisor of exactly $\binom{\nu}{1}+\binom{\nu}{2}=\binom{\nu+1}{2}$ different values of n and these

[3]) Formula (2.52) was communicated to me by Dr. Marshall Hall who had found it by a different method.

terms contribute $\binom{\nu+1}{2} S_{k-2,\nu}$. We can prove by complete induction that

$$(2.7) \qquad S_{k,\nu} = \binom{\nu}{1} S_{k-1,\nu} + \binom{\nu+1}{2} S_{k-2,\nu} + \ldots + \binom{\nu+k-1}{k} S_{0,\nu}$$

Let us now put

$$(2.8) \qquad S_\nu(z) = \sum_{k=0}^{\infty} S_{k,\nu} z^k, \quad \nu = 1, 2, 3, \ldots$$

Using (2.7) we see that

$$1 + (1-z)^{-\nu} S_\nu(z) = 2 S_\nu(z),$$

whence

$$(2.9) \qquad S_\nu(z) = \frac{(1-z)^\nu}{2(1-z)^\nu - 1}.$$

$S_\nu(z)$ has ν simple poles at $z = z_{\mu,\nu} = 1 - 2^{-1/\nu}\omega^\mu$ where $\omega = e^{2\pi i/\nu}$, $\mu = 0, 1, 2, \ldots, \nu-1$. Expanding the corresponding principal parts in geometric series and adding, we get

$$(2.10) \qquad \begin{cases} S_{0,\nu} = 1, \\ S_{k,\nu} = \dfrac{1}{\nu} 2^{-1-1/\nu} \displaystyle\sum_{\mu=0}^{\nu-1} \omega^\mu (1 - 2^{-1/\nu}\omega^\mu)^{-k-1}, \quad k \geqq 1. \end{cases}$$

In this sum the term corresponding to $\mu = 0$ dominates all the rest. Indeed, putting $2^{-1/\nu} = R$, we have for $\nu > 1$, $0 < \mu < \frac{\nu}{2}$,

$$|1 - R\omega^\mu| > R\,|1 - \omega^\mu| = 2R \sin\frac{\mu\pi}{\nu} > 4R\frac{\mu}{\nu} > 2\sqrt{2}\,\frac{\mu}{\nu},$$

$$1 - R < \frac{\log 2}{\nu},$$

whence

$$(2.11) \qquad \begin{cases} S_{k,\nu} = \dfrac{1}{\nu} 2^{-1-1/\nu} (1 - 2^{-1/\nu})^{-k-1} [1 + \eta_k], \\ |\eta_k| < 4^{-k}. \end{cases} \qquad (k \geqq 1, \nu > 1)$$

It follows in particular that

$$(2.12) \qquad \begin{cases} f(p_{i_1}^{\alpha_1} p_{i_2}^{\alpha_2} \ldots p_{i_\nu}^{\alpha_\nu}) < \dfrac{1}{\nu}(1 - 2^{-1/\nu})^{-k-1}, \\ \alpha_1 + \alpha_2 + \ldots + \alpha_\nu = k, \quad k = 0, 1, 2, \ldots, \nu = 1, 2, 3, \ldots \end{cases}$$

It seems likely that this estimate is not a very close one for large values of ν.

3. Generating Dirichlet series. Let $f_k(n)$ denote the number of representations of n as the product of k factors, each greater than one when $n > 1$, the order of the factors being essential. It is well known that

$$\sum_{n=2}^{\infty} f_k(n)\, n^{-s} = [\zeta(s) - 1]^k, \quad k = 1, 2, 3, \ldots$$

for $\Re(s) = \sigma > 1$. Since

$$f(n) = \sum_{k=1}^{\infty} f_k(n), \quad f(1) = 1,$$

we have

$$\sum_{n=1}^{\infty} f(n)\, n^{-s} = \{2 - \zeta(s)\}^{-1} \tag{3.1}$$

for $\Re(s) > \rho$, where ρ is the positive root the equation (1.4).

Let us now consider ν distinct primes $p_{i_1} < p_{i_2} < \ldots < p_{i_\nu}$, and let P denote the multiplicative system of all integers of the form $p_{i_1}^{\alpha_1} p_{i_2}^{\alpha_2} \ldots p_{i_\nu}^{\alpha_\nu}$, where the exponents are non-negative integers. We refer to $p_{i_1}, p_{i_2}, \ldots, p_{i_\nu}$ as the basis of P. Put

$$\zeta(s;\, P) = \prod_{\mu=1}^{\nu} [1 - p_{i_\mu}^{-s}]^{-1}, \quad \Re(s) > 0. \tag{3.2}$$

Then

$$\sum_{(P)} f(n)\, n^{-s} = \{2 - \zeta(s;\, P)\}^{-1} \equiv F(s;\, P), \tag{3.3}$$

where on the left the summation extends over all integers in P By a well-known theorem of Landau on Dirichlet series with positive coefficients, the series converges for $\Re(s) > \rho(P)$, where $\rho(P)$ is the positive root of the equation

$$\zeta(s,\, P) = 2. \tag{3.4}$$

Since $\zeta(\sigma;\, P)$ is monotone decreasing from $+\infty$ to 1 as σ goes from 0 to $+\infty$, this equation has one and only one positive root. We have clearly $\zeta(\sigma;\, P) < \zeta(\sigma)$ for $\sigma > 1$. Hence the former function reaches the value 2 before the latter does, i. e.,

$$0 < \rho(P) < \rho. \tag{3.5}$$

If $p_{i_1} = 2$ and $\nu > 1$, $\zeta(1; P) > 2$, so that the lower bound 0 can be replaced by 1 in (3.5).

Taking a sequence of multiplicative systems of the type described above, such that $P_1 \subset P_2 \subset \ldots \subset P_\mu$, where $\subset$ indicates a proper subset, we have obviously

$$\zeta(\sigma; P_1) < \zeta(\sigma; P_2) < \ldots < \zeta(\sigma; P_\mu), \qquad \sigma > 0,$$

so that

$$\rho(P_1) < \rho(P_2) < \ldots < \rho(P_\mu) < \rho. \tag{3.6}$$

If the sequence is infinite and $P_\mu \subset P_{\mu+1}$ for all μ, there exists a unique limiting system P_∞ which contains all systems P_μ and is the smallest multiplicative system of positive integers having this property. Let N be the system of all natural numbers, i. e., the multiplicative system based on all primes. Then $P_\infty \subseteqq N$. Put

$$\zeta(s; P_\infty) = \prod_{\nu=1}^{\infty} [1 - p_{i_\nu}^{-s}]^{-1}, \tag{3.7}$$

where the product is extended over the basis of P_∞. Let the abscissa of (absolute) convergence of the product be σ_0, $0 \leqq \sigma_0 \leqq 1$. For $\sigma_0 + \varepsilon \leqq \sigma \leqq 1/\varepsilon$, $\zeta(\sigma; P_\mu)$ converges uniformly to $\zeta(\sigma; P_\infty)$ as $\mu \to \infty$. Moreover, $\zeta(\sigma; P_\mu) < \zeta(\sigma; P_\infty)$ and for $\sigma > 1$ the latter is $\leqq \zeta(\sigma; N) = \zeta(\sigma)$ where the first sign of equality holds if and only if it holds identically. Consequently

$$\rho(P_\mu) \uparrow \rho(P_\infty) \leqq \rho, \tag{3.8}$$

with obvious notation.

In (3.6) it was assumed that the systems involved had finite bases. But it is clear that $P_1 \subset P_2 \subset N$ implies

$$\rho(P_1) < \rho(P_2) < \rho(N) = \rho, \tag{3.9}$$

whether or not the bases are finite.

4. The summatory functions. Let P be a multiplicative system of the type described above, the basis being finite or infinite. The well known relation in the theory of Dirichlet series betwen the partial sums of the coefficients and the abscissa of convergence shows that

$$\sum_{(P)}^{m \leqq n} f(m) = O[n^{\rho(P)+\varepsilon}] \tag{4.1}$$

for every positive ε. Here the summation is extended over all integers m in P which are $\leqq n$. We can get much more precise results, however.

Suppose first that the basis consists of a single prime p. Then by (2.51)

$$\sum_{p^\alpha \leqq n} f(p^\alpha) = B(n)\, n^{\rho(P)}, \quad \rho(P) = \frac{\log 2}{\log p}, \tag{4.2}$$

where $\frac{1}{2} \leqq B(n) \leqq 1$. We note that the corresponding generating Dirichlet series $F(s; P)$ has infinitely many poles on the line $\sigma = \rho(P)$.

If $1 < \nu \leqq \infty$, the situation is different. In order that $\zeta(\sigma + it; P) = 2$ for $\sigma = \rho(P)$ it is necessary as well as sufficient that $p_{i_\mu}^{-s}$ is positive for all values of μ. For a $t \neq 0$, this condition contradicts the linear independence of the logarithms of the prime numbers. Hence $F(s, P)$ has a single pole on the line $\sigma = \rho(P)$ at $s = \rho(P)$. This pole is simple and the residue obviously equals $-[\zeta'(\rho(P); P)]^{-1}$. There are of course infinitely many poles in any strip $\rho(P) - \delta \leqq \sigma \leqq \rho(P)$ for every $\delta > 0$ owing to the almost periodic character of $\zeta(s; P)$ in such a strip, but the only thing we need to know about these poles is the fact that there is only one of them on the right boundary of the strip. It follows that the hypotheses of the Ikehara-Wiener theorem[4]) are satisfied so that

$$\sum_{(P)}^{m \leqq n} f(m) = -\frac{n^{\rho(P)}}{\rho(P)\, \zeta'(\rho(P); P)} \{1 + o(1)\}. \tag{4.3}$$

Here P is any multiplicative system whose basis contains at least two primes. In particular, we might take $P = N$ in which case we obtain Kalmár's formula (1.5).

5. Estimates of $f(n)$ in terms of n. It follows from (4.3) that there exists a positive $C_1(P)$ such that

$$f(n) < C_1(P)\, n^{\rho(P)}, \quad n \in P. \tag{5.1}$$

If ν is finite we can also get a converse inequality. The number of terms on the left hand side of (4.3) is then equal to the number of solutions of the inequality

$$\alpha_1 \log p_{i_1} + \alpha_2 \log p_{i_2} + \ldots + \alpha_\nu \log p_{i_\nu} \leqq \log n \tag{5.2}$$

in non-negative integers. This number is obviously $O[(\log n)^\nu]$. Suppose that n is so chosen that $f(n)$ is the largest term of the sum in (4.3). This will happen for infinitely many values of n Hence there exists a positive $C_2(P)$ such that

[4]) See, e. g., N. Wiener, The Fourier integral, Cambridge, 1933, pp. 127—130

$$f(n) > C_2(P)\ (\log n)^{-\nu}\, n^{\rho(P)},\quad n \in P, \tag{5.3}$$

for infinitely many values of n in P. But here we can replace ν by $\nu-1$ and if $\nu>1$ we can take $C_2(P)$ as large as we please. The case $\nu=1$ is already settled by formula (2.51) so we can take $\nu>1$. Suppose that it has been shown that for some choice of μ and τ

$$f(n) = o\,[\,(\log n)^{-\mu}\, n^{\tau}\,]$$

for n in P, $n \to \infty$. We have then also

$$\sum_{(P)}^{m\leq n}{}' f(m) = o\left\{\sum_{(P)}^{m\leq n}{}' (\log m)^{-\mu}\, m^{\tau}\right\}$$

$$= o\left\{\sum [\,\alpha_1 \log p_{i_1} + \ldots + \alpha_\nu \log p_{i_\nu}\,]^{-\mu}\,[\,p_{i_1}^{\alpha_1} \ldots p_{i_\nu}^{\alpha_\nu}\,]^{\tau}\right\},$$

where the summation here and below extends over all positive α's satisfying (5.2). If μ is taken to be positive, the relation between the arithmetic and the geometric means shows that the last sum is bounded above by some multiple of

$$\sum (\alpha_1 \ldots \alpha_\nu)^{-\frac{\mu}{\nu}}\,[p_{i_1}^{\alpha_1} \ldots p_{i_\nu}^{\alpha_\nu}]^{\tau}\,.$$

Summing for α_ν we get an expression which is of the order of magnitude of

$$n^{\tau} \sum (\alpha_1 \ldots \alpha_{\nu-1})^{-\frac{\mu}{\nu}}\,(\log n - \alpha_1 \log p_{i_1} - \ldots - \alpha_\nu \log p_{i_\nu})^{-\frac{\mu}{\nu}},$$

and if $\mu<\nu$ the sum is of the order of magnitude of the integral

$$\int_S [x_1 \ldots x_{\nu-1}\ (\log n - x_1 \log p_{i_1} - \ldots - x_{\nu-1} \log p_{i_{\nu-1}})]^{-\frac{\mu}{\nu}}\, dS$$

taken over that portion of the $(\nu-1)$-dimensional space in which all factors of the bracket are positive. A simple calculation shows that this integral is $O\,[(\log n)^{\nu-1-\mu}]$. Thus, taking $\tau=\rho(P)$ we get

$$\sum_{(P)}^{m\leq n} f(m) = o\,[(\log n)^{\nu-1-\mu}\, n^{\rho(P)}]\,. \tag{5.4}$$

Strictly speaking, the summation should extend over only those integers in P which have exactly ν distinct prime factors, but formula

(3.6) shows that the added terms do not disturb the estimate. But (5.4) contradicts (4.3) unless $\mu < \nu - 1$. It follows that

$$f(n) > M(\log n)^{-\nu+1} n^{\rho(P)} \tag{5 5}$$

for an arbitrarily large M and for infinitely many values of n in P, provided $\nu > 1$. It is possible that this estimate could be still further improved. But it is good enough to show that if ρ is the root of equation (1.4) and δ is a fixed, arbitrarily small positive number then the inequality

$$f(n) > n^{\rho-\delta} \tag{5.6}$$

holds for infinitely many values of n.

6. Generalizations. The previous discussion admits of very considerable extensions. In § 2 we considered a number theoretical function $f(n)$ satisfying the functional equation (2.2) with the initial conditions (2.3). In the discussion we have used only the divisibility properties of the natural numbers. It is clear that the results will remain unchanged if we replace the natural numbers by any other system having similar divisibility properties. Consider the set of integral ideals $\mathfrak{a}$ in a commutative ring R without divisors of zero. We suppose that the finite chain condition is satisfied, that the prime ideals, except the null ideal, have no proper divisors, and that R is integrally closed with respect to its quotient field. In this case every ideal in R has a unique representation as product of prime ideals $\mathfrak{p}$, the conditions mentioned being necessary and sufficient for unique factorization. We can then define $\mathfrak{f}(\mathfrak{a})$ as the number of representations of $\mathfrak{a}$ as product of ideals, omitting powers of the unit ideal $\mathfrak{o}$, i. e., R considered as an ideal, two representations being considered equal if and only if they involve the same factors written in the same order. We have $\mathfrak{f}(\mathfrak{o}) = \mathfrak{f}(\mathfrak{p}) = 1$. Further (2.2) is satisfied, and it is clear that

$$\mathfrak{f}(\mathfrak{p}_{i_1}^{e_1} \mathfrak{p}_{i_2}^{e_2} \ldots \mathfrak{p}_{i_\nu}^{e_\nu}) = f(2^{e_1} 3^{e_2} \ldots p_\nu^{e_\nu}). \tag{6.1}$$

In order to extend the discussion of §§ 3, 4, and 5 we need a definition of the absolute value of an ideal, $|\mathfrak{a}|$, such that

$$|\mathfrak{a}\mathfrak{b}| = |\mathfrak{a}|\,|\mathfrak{b}|, \tag{6.2}$$

and that it is possible to form an analog of the zeta function

$$\zeta_R(s) = \sum_{(\mathfrak{a})} |\mathfrak{a}|^{-s}. \tag{6.3}$$

The convergence of such a series imposes certain restrictions on R and on the choice of the absolute value. Thus, only a finite number of ideals in R can have the same absolute value, and $|\mathfrak{a}| > 1 + \delta$ for some fixed $\delta > 0$ if $\mathfrak{a} \neq \mathfrak{o}$.

These conditions are of course satisfied in the classical case of integral ideals in an algebraic field if we take $|\mathfrak{a}| = N\mathfrak{a}$, i. e., the number of residue classes in R modulo $\mathfrak{a}$. (6.3) is then simply the Dedekind zeta function. It is known, however, that in a ring having unique factorization of ideals it is always possible to introduce an evaluation (= Bewertung), for the elements as well as for the ideals, based upon a norm $\|\alpha\|$ which satisfies the conditions of Kürschák, one of which is (6.2). But the existence of a zeta function is in general not ensured in an evaluation ring. A noteworthy exception is given by the fields obtained by a finite algebraic extension of the field of all rational functions of an indeterminate with coefficients modulo a prime p. The number theory of such fields, which goes back to Dedekind, has been developed in recent yeurs by Artin, Hasse, F. K. Schmidt and others.[5])

In these two cases we can obtain estimates of the summatory function $\Sigma f(\mathfrak{a})$ where the summation extends over all ideals in the ring the absolute values of which do not exceed a given integer. This follows from the formula

$$\sum_{(\mathfrak{a})} f(\mathfrak{a})\, |\mathfrak{a}|^{-s} = \{2 - \zeta_R(s)\}^{-1}. \tag{6.4}$$

In the case of algebraic fields the Ikehara-Wiener theorem applies and gives the analog of Kalmár's estimate; in the case of characteristic p we are dealing with a rational function of p^s. We can also obtain estimates of $f(\mathfrak{a})$ itself by restricting the summation to those ideals of absolute value less than a given integer which involve a given sub-set of prime ideals. The necessary generating Dirichlet series can be formed as in § 3 and the discussion of §§ 4 and 5 is easily carried over. In the congruence case we are still dealing with rational functions of an exponential function (at least as long as the basis is finite), and the same is true in the Dedekind case if the norms of the prime ideals in the basis are all powers of the same rational prime. In the general Dedekind case the Ikehara-Wiener theorem still gives the estimates.

---

[5]) See, e. g., F. K. Schmidt, Analytische Zahlentheorie in Körpern der Charakteristik p, Math. Zeitschrift, 33 (1931) 1—32. I am indebted to Profs. Ö. Ore and M. Zorn for calling my attention to this possibility and for explications.

It should be noted, however, that the existence of a zeta function for a sub-set of finite basis is independent of the existence of such a function for the whole set. In order to define an analog of $\zeta(s; P)$ we only need to have $|\mathfrak{a}| > 1$ for $\mathfrak{a} \neq \mathfrak{o}$ in the sub-set. It follows that it is possible to obtain estimates for $\mathfrak{f}(\mathfrak{a})$ in terms of $|\mathfrak{a}|$ in other cases than those mentioned above.

Finally it is possible to extend some of these considerations to the case of non-commutative rings.

(Received June 4, 1936.)

Reprinted from DUKE MATHEMATICAL JOURNAL
Vol. 3, No. 4, December, 1937

THE INVERSION PROBLEM OF MÖBIUS

BY EINAR HILLE

1. **Introduction.** The present paper represents an attempt to give a rigorous treatment of certain inversion problems which have their origin in a little-known paper by A. F. Möbius.[1]

As a typical, though not the oldest, example of these inversion problems we might take the linear functional equation with constant coefficients

$$\sum_{n=1}^{\infty} a_n f(nz) = g(z), \tag{1.1}$$

a formal solution of which has the form

$$\sum_{n=1}^{\infty} b_n g(nz) = f(z). \tag{1.2}$$

These problems all lead to the same infinite system of bilinear equations

$$a_1 b_1 = 1, \qquad \sum_{d \mid n} a_d b_{n/d} = 0, \quad n > 1, \tag{1.3}$$

for which the *algorithm of Möbius* seems a fitting name.

This algorithm is perhaps best known from the problem of finding the reciprocal of an ordinary Dirichlet series, i.e., a solution of the problem

$$\sum_{n=1}^{\infty} a_n n^{-s} \sum_{n=1}^{\infty} b_n n^{-s} = 1. \tag{1.4}$$

We shall see that the properties of these series are fundamental in all these inversion problems.

This observation suggests that there is a class of inversion problems associated with the problem of expressing the reciprocal of a general Dirichlet series or, still more generally, of a Laplace-Stieltjes integral as a function of the same class. In general the reciprocal is not so expressible, but whenever it is, certain functional equations of the type

$$\int_1^{\infty} f(uz)\, dA(u) = g(z) \tag{1.5}$$

have solutions of the form

$$\int_1^{\infty} g(uz)\, dB(u) = f(z), \tag{1.6}$$

Received April 24, 1937; presented to the American Mathematical Society, March 26, 1937.

[1] *Ueber eine besondere Art von Umkehrung der Reihen*, Journal f. Math., vol. 9 (1832), pp. 105–123; *Gesammelte Werke*, vol. IV, 1887, pp. 589–612.

where

$$(1.7)\qquad \int_1^u A\left(\frac{u}{v}\right) dB(v) = 1, \quad 1 < u.$$

The last equation is the transcendental analogue of the algorithm of Möbius.

In §2 of the present paper there is a discussion of the original problem of Möbius, of the algorithm of Möbius and of the problem of finding the reciprocal of an ordinary Dirichlet series. While there is comparatively little that is strictly new in this paragraph, the results are necessary for the rest of the paper and do not appear to be well known. In §3 we discuss equation (1.1) and various connected problems. In §4 we discuss equation (1.5) and the problem of expressing the reciprocal of a Laplace-Stieltjes integral as an integral of the same kind.

2. Some classical problems.

2.1. *The algorithm of Möbius.* Let $\mathfrak{A} = \{a_n\}$ be a given infinite sequence of real or complex numbers. The sequence is *proper* or *improper* according as $a_1 \neq 0$ or $= 0$, and in the former case it is *normalized* if $a_1 = 1$. Following O. Hölder,[2] we call $\mathfrak{B} = \{b_n\}$ *the reciprocal sequence* of $\mathfrak{A}$ if the latter is proper and $\mathfrak{A}$ and $\mathfrak{B}$ satisfy the algorithm of Möbius

$$(2.1.1)\qquad a_1 b_1 = 1, \qquad \sum_{d \mid n} a_d b_{n/d} = 0, \quad n > 1,$$

or symbolically $\mathfrak{A}\mathfrak{B} = 1$. The underlying product definition is that of *Dirichlet multiplication*, i.e., in general $\mathfrak{A}\mathfrak{B} = \mathfrak{C}$, where the sequence $\mathfrak{C} = \{c_n\}$ is defined by

$$(2.1.2)\qquad c_n = \sum_{d \mid n} a_d b_{n/d}.$$

Thus, formally,

$$\sum_{n=1}^{\infty} a_n n^{-s} \cdot \sum_{n=1}^{\infty} b_n n^{-s} = \sum_{n=1}^{\infty} c_n n^{-s}.$$

In case of the reciprocal sequence, $\mathfrak{C}$ is simply the unit sequence 1, 0, 0, $\cdots$.

The system (2.1.1) determines $\mathfrak{B}$ uniquely. We have

$$(2.1.3)\qquad b_n = \sum (-1)^{\alpha_1+\alpha_2+\cdots} C_{\alpha_1\alpha_2\cdots} (a_{d_1})^{\alpha_1} (a_{d_2})^{\alpha_2} \cdots,$$

where the summation extends over all factorizations of $n = d_1^{\alpha_1} d_2^{\alpha_2} \cdots$, and $C_{\alpha_1\alpha_2\cdots}$ is the combinatorial function which gives the number of possible arrangements of a set consisting of α_1 objects of one kind, α_2 objects of a second, etc. We have

$$(2.1.4)\qquad \sum (-1)^{\alpha_1+\alpha_2+\cdots} C_{\alpha_1\alpha_2\cdots} = \mu(n),$$

[2] *Über gewisse der Möbiusschen Funktion $\mu(n)$ verwandte zahlentheoretische Funktionen, die Dirichletsche Multiplikation und eine Verallgemeinerung der Umkehrungsformeln,* Berichte d. Sächs. Akad. d. Wiss., Math.-phys. Kl., vol. 85 (1933), pp. 1–28.

the Möbius' μ-function, whereas

$$\sum C_{\alpha_1 \alpha_2 \cdots} = \pi(n), \tag{2.1.5}$$

the number of factorizations of n into factors $\neq 1$ ($n \neq 1$, $\pi(1) = 1$) when attention is paid to the order of the factors. $\pi(n)$ is a highly irregular function.[3] For the following it is enough to note that

$$\sum_{n=1}^{\infty} \pi(n) n^{-s} = [2 - \zeta(s)]^{-1} \tag{2.1.6}$$

for $\Re(s) > \rho$, $\zeta(\rho) = 2$, and that

$$\pi(n) < C_1 n^{\rho} \tag{2.1.7}$$

for all values of n, whereas for every $\epsilon > 0$ there are infinitely many values of n for which

$$\pi(n) > C_2 n^{\rho - \epsilon}. \tag{2.1.8}$$

2.2. *The reciprocation problem for ordinary Dirichlet series.* That the reciprocal of an ordinary Dirichlet series with $a_1 \neq 0$ can be represented by a series of the same kind is well known. The best theorem in this connection is one due to E. Landau.[4]

THEOREM 2.2.1. *Let*

$$D(s; a_n) = \sum_{n=1}^{\infty} a_n n^{-s}, \quad a_1 \neq 0 \tag{2.2.1}$$

have a domain of convergence, and let the function represented by the series be holomorphic and different from zero for $\sigma > \alpha$. *Then*

$$[D(s; a_n)]^{-1} \equiv \sum_{n=1}^{\infty} b_n n^{-s} \tag{2.2.2}$$

is convergent for $\sigma > \alpha$.

This theorem lies quite deep. It naturally brings up the question whether it is possible to assign upper bounds for the real parts of the possible zeros of $D(s; a_n)$ and thus also for the abscissa of convergence of the reciprocal. The answer is in the affirmative and is fairly trivial.

THEOREM 2.2.2. *Let* $\{r_n\}$ *be a given sequence of positive numbers,* $r_1 = 1$, $r_n = O(n^{\kappa})$ *for some fixed real* κ. *Consider the class* $\mathfrak{D}$ *of all Dirichlet series* $D(s; a_n)$ *with* $a_1 = 1$, $|a_n| = r_n$, $n \geqq 2$. *Let* S *be the abscissa of convergence of* $D(s; r_n)$ *and put* $D(S + 0; r_n) = R \leqq \infty$. *If* $R > 2$, *the equation*

$$D(\sigma; r_n) = 2 \tag{2.2.3}$$

[3] See E. Hille, *A problem in "factorisatio numerorum"*, Acta Arithmetica, vol. 2 (1936), pp. 134–144. $\pi(n)$ is denoted by $f(n)$ in this paper.

[4] *Über den Wertevorrat von* $\zeta(s)$ *in der Halbebene* $\sigma > 1$, Göttinger Nachrichten, 1933, pp. 81–91, p. 90.

has a real root $\rho = \rho(\mathfrak{D}) > S$. *No series of* $\mathfrak{D}$ *has any zeros in the half-plane* $\sigma > \rho$, *whereas there exist series in* $\mathfrak{D}$ *having infinitely many zeros in the strip* $\rho - \epsilon < \sigma < \rho$ *for every* $\epsilon > 0$. *If, on the other hand,* $1 < R \leqq 2$, *there are no zeros of any series in* $\mathfrak{D}$ *for* $\sigma > S$ *and there are series having either zeros or singular points in every strip* $S - \epsilon < \sigma < S$.

Proof. The verification of the fact that the series in $\mathfrak{D}$ cannot have zeros in the half-planes $\sigma > \rho(\mathfrak{D})$ and $\sigma > S$, respectively, is elementary and may be left to the reader. Further, the series $[D(s; a_n)]^{-1} \equiv D(s; b_n)$ are easily shown to be absolutely convergent in the same half-planes.

If $R > 2$, we note that the series

$$1 - \sum_{n=1}^{\infty} r_n n^{-s}$$

is a member of the class $\mathfrak{D}$ and vanishes at $s = \rho(\mathfrak{D})$. It further has infinitely many zeros in any strip $\rho(\mathfrak{D}) - \epsilon < \sigma < \rho(\mathfrak{D})$. If $1 < R \leqq 2$, the point $s = S$ is a non-polar singularity of $D(s; r_n)$ and consequently also a singularity of $[D(s; r_n)]^{-1}$. It follows that in either case the estimates given are the best possible valid for the whole class $\mathfrak{D}$.

It would be of some interest to know if these estimates for the upper bound of the real parts of the zeros are imposed upon us by a relatively small set of elements in $\mathfrak{D}$ or if they represent the rule rather than the exception. A discussion of this question in general calls for an interpretation of $\mathfrak{D}$ as a topological space, i.e., a definition of closure, possibly based upon a definition of distance or of measure.

But there is one very special case in which a complete answer is available without any topology. Suppose that $a_n = 0$ unless n is a prime, and put

$$a_{p_k} = \alpha_k, \qquad r_{p_k} = \rho_k, \qquad D(s; a_n) = P(s; \alpha_k), \qquad \mathfrak{D} = \mathfrak{P}.$$

Let us suppose that $R > 2$. Using a classical theorem of H. Bohr on the relation between the set of values of a Dirichlet series and of the associated power series in infinitely many unknowns,[5] we conclude that every series $P(s; \alpha_k)$ has infinitely many zeros in every strip $\rho(\mathfrak{P}) - \epsilon < \sigma < \rho(\mathfrak{P})$. Indeed, the associated power series is simply the linear form

$$L(x) = 1 + \sum_1^{\infty} \alpha_k x_k \equiv 1 + \sum_1^{\infty} \rho_k e^{i\theta_k} x_k,$$

and putting $x_k = -e^{-i\theta_k} p_k^{-\rho}$, we get $L(x) = 1 - \sum_1^{\infty} \rho_k p_k^{-\rho} = 0$. By Bohr's theorem the value 0 is taken on infinitely often by $P(s; \alpha_k)$ in every strip $\rho - \epsilon < \sigma < \rho$. Hence in this case all the reciprocal series in $\mathfrak{P}$ have the same abscissa of convergence, viz., $\rho(\mathfrak{P})$.

[5] *Über die Bedeutung der Potenzreihen unendlich vieler Variabeln in der Theorie der Dirichletschen Reihen* $\Sigma a_n n^{-s}$, Göttinger Nachrichten, 1913, pp. 441–488, p. 451.

2.3. *Möbius' problem.* Möbius[6] raised the following question: given a power series

$$f(z) = \sum_{n=1}^{\infty} a_n z^n, \quad a_1 \neq 0, \tag{2.3.1}$$

find the expansion of z in terms of the functions $f(z^n)$, $n = 1, 2, 3, \cdots$. Let it be

$$z = \sum_{n=1}^{\infty} b_n f(z^n). \tag{2.3.2}$$

A straightforward calculation shows that the b's must satisfy the algorithm of Möbius. Further, it is clear that if

$$F(z) = \sum_{n=1}^{\infty} A_n z^n, \tag{2.3.3}$$

then

$$F(z) = \sum_{n=1}^{\infty} B_n f(z^n), \tag{2.3.4}$$

where

$$B_n = \sum_{d \mid n} b_d A_{n/d}. \tag{2.3.5}$$

All this is highly formal and an analyst naturally wants to know the range of validity of the formulas, conditions for convergence, etc.

As a preliminary step in this study, let us suppose that the power series in (2.3.1) has a circle of convergence, and form the adjoint power series

$$\varphi(z) = \sum_{n=1}^{\infty} b_n z^n. \tag{2.3.6}$$

We call $\varphi(z)$ the Möbius transform of $f(z)$,

$$\varphi(z) = \mathfrak{M}[f(z)]. \tag{2.3.7}$$

The Möbius algorithm shows that conversely $f(z)$ is the Möbius transform of $\varphi(z)$,

$$\mathfrak{M}[\mathfrak{M}[f(z)]] = f(z), \tag{2.3.8}$$

i.e., the Möbius transformation is an involution.

We must show that the power series in (2.3.6) is also convergent. This is established in

THEOREM 2.3.1. *Let the radii of convergence of the power series in* (2.3.1) *and* (2.3.6) *be* R_1 *and* R_2 *respectively. If* $0 < R_1 < 1$, *then* $R_1 = R_2$; *if* $1 \leqq R_1$, *then also* $1 \leqq R_2$.

[6] Loc. cit., *Werke*, vol. IV, p. 591.

Proof. Choose $R_0 < R_1$. Then there exists an M such that $|a_n| \leqq MR_0^{-n}$ for every n. Hence by (2.1.3)

$$|b_n| \leqq \sum C_{\alpha_1\alpha_2\cdots} M^{\alpha_1+\alpha_2+\cdots} R_0^{-(\alpha_1 d_1+\alpha_2 d_2+\cdots)}.$$

If $R_1 \leqq 1$, then $R_0 < 1$. We can assume $M \geqq 1$ without restricting the generality. Further,

$$\alpha_1 d_1 + \alpha_2 d_2 + \cdots \leqq n, \qquad \alpha_1 + \alpha_2 + \cdots \leqq p(n),$$

where $p(n)$ denotes the total number of prime factors of n. Hence by (2.1.5)

$$|b_n| \leqq \pi(n) M^{p(n)} R_0^{-n}. \tag{2.3.9}$$

Since $p(n) = O(\log n)$ and $\pi(n) = O(n^\rho)$, we conclude that $R_0 \leqq R_2$ or $R_1 \leqq R_2$.

Suppose next that $R_1 > 1$. We can then choose $R_0 > 1$. Further,

$$\alpha_1 d_1 + \alpha_2 d_2 + \cdots \geqq \nu_1 p_{i_1} + \nu_2 p_{i_2} + \cdots \equiv P(n)$$

if $n = p_{i_1}^{\nu_1} p_{i_2}^{\nu_2} \cdots$, where the p_{i_k} are the distinct prime factors of n. Hence

$$|b_n| \leqq \pi(n) M^{p(n)} R_0^{-P(n)}. \tag{2.3.10}$$

But $P(n)$ is infinitely often $o(n)$. It follows that

$$\overline{\lim_{n\to\infty}} |b_n|^{1/n} \leqq 1,$$

or $R_2 \geqq 1$.

In order to complete the proof for the case $R_1 < 1$, we use the involutory character of the transformation. We have shown that $R_1 \leqq R_2$. If $R_2 \leqq 1$, we can conclude that $R_2 \leqq R_1$, i.e., $R_1 = R_2$, simply by noticing that $f(z)$ is the Möbius transform of $\varphi(z)$. On the other hand, the assumption $R_2 > 1$ implies by the same argument that $R_1 \geqq 1$, and this contradicts the original assumption. Hence $R_1 < 1$ implies $R_1 = R_2$.

If $R_1 = 1$, we may well have $R_2 > 1$. The situation becomes clearer by introducing the associated Dirichlet series

$$D(s; a_n) = \sum_{n=1}^{\infty} a_n n^{-s}, \qquad D(s; b_n) = \sum_{n=1}^{\infty} b_n n^{-s}.$$

The assumption $R_2 > 1$ implies that $D(s; b_n)$ converges for all s and is an entire function of s. Hence $D(s; a_n)$ has a half-plane of absolute convergence and is not merely a formal Dirichlet series; moreover, $a_n = O(n^\kappa)$ for some finite value of κ.

Thus if $R_1 = 1$, we have $R_2 = R_1$ unless $D(s; b_n)$ is an entire function, and then $R_2 \geqq R_1$. This case can arise only when $a_n = O(n^\kappa)$.

If $R_1 > 1$, $D(s; a_n)$ is an entire function of s, and normally $R_2 = 1$, unless $D(s; a_n) \not\equiv 0$, in which case $R_2 \geqq 1$.

After this discussion it is easy to discuss the validity of Möbius' inversion formula.

THEOREM 2.3.2. *If R_1 is the radius of convergence of the power series for $f(z)$,* *the inversion formula* (2.3.2) *is valid for* $|z| < \min(R_1, 1)$. *If* $R_1 < 1$, *the series diverges for* $R_1 < |z| < 1$. *The series may converge outside of the unit circle, but normally it does not represent* z *for such values. If the radius of convergence of* $F(z)$ *in* (2.3.3) *is* R, *formula* (2.3.4) *is valid for* $|z| < \min(R, R_1, 1)$.

Proof. Let $0 < R_1 \leqq 1$. Then all the terms in (2.3.2) are regular analytic functions of z in $|z| < R_1$. By (2.3.9)

$$\begin{aligned}\sum_{n=1}^{\infty} |b_n f(z^n)| &\leqq \sum_{n=1}^{\infty} |b_n| \sum_{m=1}^{\infty} |a_m| \cdot |z|^{mn} \\ &\leqq \sum_{n=1}^{\infty} \pi(n) M^{p(n)+1} R_0^{-n} \frac{|z|^n}{1 - |z|^n/R_0},\end{aligned} \tag{2.3.11}$$

and this is clearly convergent for $|z| < R_0$. Here $R_0 < R_1$ and as near to R_1 as we please, i.e., the series in (2.3.2) is absolutely convergent for $|z| < R_1$. Moreover, the double series obtained by substituting the power series for $f(z^n)$ on the right side of (2.3.2) is absolutely convergent, as we have just seen. It can consequently be rearranged at liberty. Collecting powers of equal degree and reducing with the aid of the algorithm of Möbius, the double series reduces to its first term z. This completes the proof for the case $R_1 \leqq 1$.

If $R_1 > 1$, the terms of the series (2.3.2) are holomorphic for $|z| \leqq 1$ and in no larger region. For such values formula (2.3.10) shows that the double series is dominated by

$$\sum_{n=1}^{\infty} \pi(n) M^{p(n)+1} R_0^{-P(n)} \frac{|z|^n}{1 - |z|^n/R_0}. \tag{2.3.12}$$

It follows that the inversion formula is valid for $|z| < 1$.

Let $|z| < 1$. Then

$$\overline{\lim_{n\to\infty}} |b_n f(z^n)|^{1/n} = |z| \overline{\lim_{n\to\infty}} |b_n|^{1/n} = |z|/R_2.$$

It follows that if $R_1 < 1$, so that $R_2 = R_1$, the series (2.3.2) diverges in the annulus $R_1 < |z| < 1$.

Outside of the unit circle the situation may differ considerably in different cases. Thus if

$$f(z) = \frac{z}{1-z}, \quad \text{then} \quad z = \sum_{n=1}^{\infty} \mu(n) \frac{z^n}{1-z^n},$$

which clearly diverges for $|z| > 1$. But if

$$f(z) = \frac{z}{1-z^2},$$

then

$$b_n = \begin{cases} \mu(2k+1), & n = 2k+1, \\ 0, & n = 2k, \end{cases}$$

so that the series

$$\sum_{n=1}^{\infty} b_n \frac{z^n}{1 + z^{2n}}$$

converges also outside of the unit circle, but to $-1/z$ instead of to z. Finally, if

$$f(z) = ze^{-z^k},$$

where k is a positive integer, then the series

$$\sum_{n=1}^{\infty} b_n z^n e^{-z^{nk}}$$

converges on the rays $|z| > 1$, $\arg z = \nu 2\pi/k$, $\nu = 0, 1, \cdots, k - 1$, and nowhere else outside of the unit circle. The sum of the series tends to zero as $z \to \infty$ along the rays in question; thus, the sum of the series cannot be z, but I am unable to determine its actual value.

The reader will have no difficulty in verifying the statements concerning $F(z)$ in Theorem 2.3.2 on the basis of the estimates for the coefficients, and this part of the proof will be omitted.

3. A class of linear functional equations.

3.1. *The Möbius $\mathfrak{A}$-transform.* Let $\mathfrak{A} = \{a_n\}$ be a given proper, normalized sequence, $\mathfrak{B} = \{b_n\}$ the reciprocal sequence in the sense of §2.1. Möbius[7] observed that the same algorithm enters in the study of the functional equation

$$G(z) = \sum_{n=1}^{\infty} a_n F(z^n) \tag{3.1.1}$$

for which he proposed the solution

$$F(z) = \sum_{n=1}^{\infty} b_n G(z^n). \tag{3.1.2}$$

Conversely, (3.1.1) is a solution of (3.1.2) if $F(z)$ is the given function. Though Möbius claims that these relations hold for arbitrary given functions, he has presumably only had power series in mind, and there is no indication that he looked into convergence questions at all.

Putting

$$G(e^z) = g(z), \qquad F(e^z) = f(z),$$

we can rewrite the functional equations as follows:

$$g(z) = \sum_{n=1}^{\infty} a_n f(nz), \tag{3.1.3}$$

$$f(z) = \sum_{n=1}^{\infty} b_n g(nz). \tag{3.1.4}$$

[7] Loc. cit., *Werke*, vol. IV, p. 593.

These forms are more convenient to handle than the original ones, and will serve as the basis of the discussion in the present paragraph. Special cases have long been in the literature. Thus the case $a_n = 1$, $b_n = \mu(n)$, $z = k$, gives the inversion formulas

$$g(k) = \sum_{n=1}^{\infty} f(nk), \tag{3.1.5}$$

$$f(k) = \sum_{n=1}^{\infty} \mu(n)g(nk), \tag{3.1.6}$$

which are used extensively in analytical number theory.[8]

We proceed to an analytical discussion of equation (3.1.3). Let E be a set of points in the complex plane such that if E contains the point z_0 it also contains all multiples nz_0 of z_0, $n = 2, 3, \cdots$. Let $f(z)$ be given in E and such that

$$\mathfrak{M}[f(z); \mathfrak{A}] \equiv \sum_{n=1}^{\infty} a_n f(nz) \tag{3.1.7}$$

converges in E. We call this function the Möbius $\mathfrak{A}$-transform of $f(z)$ with similar notation and terminology for $\mathfrak{B}$ and for other sequences. The reciprocity of the two sequences $\mathfrak{A}$ and $\mathfrak{B}$ is reflected in the property

$$\mathfrak{M}[\mathfrak{M}[f(z); \mathfrak{A}]; \mathfrak{B}] = \mathfrak{M}[\mathfrak{M}[f(z); \mathfrak{B}]; \mathfrak{A}] = f(z), \tag{3.1.8}$$

valid for sufficiently restricted classes of functions $f(z)$.

THEOREM 3.1.1. *A sufficient condition for the validity of* (3.1.8) *is the absolute convergence of the series*

$$S[f] \equiv \sum_{m=1}^{\infty} \sum_{n=1}^{\infty} a_m b_n f(mnz). \tag{3.1.9}$$

Proof. The series, being absolutely convergent, can be rearranged arbitrarily. If summed by columns its sum is $\mathfrak{M}[\mathfrak{M}[f(z); \mathfrak{A}]; \mathfrak{B}]$, if summed by rows, $\mathfrak{M}[\mathfrak{M}[f(z); \mathfrak{B}]; \mathfrak{A}]$, whereas summation over constant values of the product mn gives simply $f(z)$, by virtue of Möbius' algorithm.

3.2. *Inversion of the $\mathfrak{A}$-transform.* We now turn to the question of finding the inverse of the $\mathfrak{A}$-transform, i.e., the resolution of the equation

$$\mathfrak{M}[f(z); \mathfrak{A}] = g(z) \tag{3.2.1}$$

for $f(z)$ in terms of $g(z)$.

THEOREM 3.2.1. *A sufficient condition that*

$$f(z) = \mathfrak{M}[g(z); \mathfrak{B}] \tag{3.2.2}$$

[8] See, e.g., P. Bachmann, *Die analytische Zahlentheorie*, Leipzig, 1894, p. 310 et seq. Bachmann does not seem to have been aware of Möbius' paper. Thus he credits the introduction of the function $\mu(n)$ to F. Mertens, *Ueber einige asymptotische Gesetze der Zahlentheorie*; Journal f. Math., vol. 77 (1874), pp. 289–338.

be a solution of (3.2.1) *which is absolutely convergent in* E *is that the series* $S[g]$ *be absolutely convergent in* E. *On the other hand, there can be at most one solution of* (3.2.1) *which renders the series* $S[f]$ *absolutely convergent, and whenever it exists this solution is given by* (3.2.2).

Proof. The assumption that $S[g]$ is absolutely convergent implies that $\mathfrak{M}[\mathfrak{M}[g(z); \mathfrak{B}]; \mathfrak{A}] = g(z)$ by Theorem 3.1.1. Hence (3.2.2) gives a solution under these circumstances, and the solution is evidently absolutely convergent. Conversely, if $f(z)$ is a solution of (3.2.1) such that $S[f]$ is absolutely convergent, then for the same reason

$$S[f] \equiv f(z) = \mathfrak{M}[\mathfrak{M}[f(z); \mathfrak{A}]; \mathfrak{B}] = \mathfrak{M}[g(z); \mathfrak{B}],$$

so the solution in question is uniquely determined and given by (3.2.2).

It must be granted that Theorem 3.2.1 is of a rather restrictive character. It should be pointed out, however, that the mere existence of $\mathfrak{M}[f(z); \mathfrak{A}]$ is not enough to insure that this function is a solution of (3.2.1). Thus, for example, if there exists a non-vanishing function $g(z)$ such that $\mathfrak{M}[g(z); \mathfrak{B}] \equiv 0$, then formula (3.2.2) certainly does not give a solution of (3.2.1). See further §3.3.

The following theorem is of a somewhat different character.

THEOREM 3.2.2. *Let* $g(z)$ *be holomorphic in the sector* S, $\theta_1 < \arg z < \theta_2$, $|z| > R > 0$, *and let*

$$g(z) = z^{-\alpha}\left[c_0 + O\left(\frac{1}{|z|}\right)\right] \quad \text{as } z \to \infty \text{ in } S. \tag{3.2.3}$$

Further, suppose that $D(s; a_n) \equiv \sum_1^\infty a_n n^{-s}$ *is convergent and different from zero for* $\Re(s) > \Re(\alpha) - \epsilon$, $\epsilon > 0$. *Then* (3.2.2) *defines a solution of* (3.2.1), *holomorphic in* S, *and*

$$f(z) = z^{-\alpha}\left[c_0[D(\alpha; a_n)]^{-1} + O\left(\frac{1}{|z|}\right)\right] \quad \text{as } z \to \infty \text{ in } S, \tag{3.2.4}$$

and this is the only solution of such asymptotic character.

Proof. We have to show that $\mathfrak{M}[\mathfrak{M}[g(z); \mathfrak{B}]; \mathfrak{A}] = g(z)$ under the given assumptions. We start by observing that

$$\mathfrak{M}[z^{-\alpha}; \mathfrak{A}] = D(\alpha; a_n)z^{-\alpha}, \tag{3.2.5}$$

$$\mathfrak{M}[z^{-\alpha}; \mathfrak{B}] = D(\alpha; b_n)z^{-\alpha}, \tag{3.2.6}$$

the convergence of $D(\alpha; b_n)$ being a consequence of Landau's Theorem 2.2.1. Let us put

$$g(z) = c_0 z^{-\alpha} + g_1(z), \tag{3.2.7}$$

$$f(z) = c_0 D(\alpha; b_n)z^{-\alpha} + f_1(z). \tag{3.2.8}$$

Then if $f(z)$ is a solution of (3.2.1), $f_1(z)$ is a solution of

$$\mathfrak{M}[f_1(z); \mathfrak{A}] = g_1(z). \tag{3.2.9}$$

But to this equation we can apply Theorem 3.2.1. Indeed, by assumption $g_1(z) = O(|z|^{-1-\gamma})$, $\gamma = \Re(\alpha)$, and the series $D(s; a_n)$ and $D(s; b_n)$ are absolutely convergent for $s = 1 + \gamma$, since they converge for $s = \gamma - \epsilon/2$. Forming $S[g_1]$ and replacing each term by its absolute value, we find that the series is dominated by a constant multiple of

$$|z|^{-1-\gamma} \sum_{m=1}^{\infty} \sum_{n=1}^{\infty} |a_m b_n| (mn)^{-1-\gamma}$$

convergent in S. Hence $f_1(z) = \mathfrak{M}[g_1(z); \mathfrak{B}]$ is a solution of (3.2.9), and $\mathfrak{M}[g(z); \mathfrak{B}]$ is a solution of (3.2.1).

Further,

$$f_1(z) = O\left\{|z|^{-1-\gamma} \sum_{n=1}^{\infty} |b_n| n^{-1-\gamma}\right\},$$

or

$$|f_1(z)| \leqq M |z|^{-1-\gamma}, \tag{3.2.10}$$

whence it follows that the double series $S[f_1]$ is absolutely convergent in S. Hence, by Theorem 3.2.1, $f_1(z)$ is the only solution of (3.2.9) having this property; and, a fortiori, the only solution satisfying (3.2.10). It follows that $\mathfrak{M}[g(z); \mathfrak{B}]$ is the only solution of (3.2.1) of the form (3.2.8), where $f_1(z)$ satisfies (3.2.10).

It is obvious that the solution breaks down if $D(\alpha; a_n) = 0$. In general, it also breaks down if $D(s; a_n) = 0$ for an s with $\Re(s) > \Re(\alpha)$. It should be noted, however, that

$$[D(\alpha; a_n)]^{-1} z^{-\alpha}$$

is a solution of

$$\mathfrak{M}[f(z); \mathfrak{A}] = z^{-\alpha}$$

under the sole assumption that $D(\alpha; a_n)$ is convergent and different from zero. This observation may sometimes be used in order to extend the validity of our solution.

Thus, for example, Theorem 3.2.2 does not apply to the case $a_n = (-1)^{n-1}$ if $0 < \alpha < \frac{1}{2}$, but formula (3.2.2), nevertheless, gives a solution of the corresponding equation. This is readily seen by running over the proof again with this particular choice of the parameters.

Further, it should be noted that the assumption on the remainder in (3.2.3) is chosen merely with the view of insuring that the Dirichlet series $D(s; a_n)$ and $D(s; b_n)$ be absolutely convergent for $s = 1 + \gamma$. If there should exist a δ, $0 < \delta < 1$, such that these series converge absolutely for $s = \delta + \gamma$, it is sufficient for our purposes to assume that $z^{\alpha} g(z) = c_0 + O(|z|^{-\delta})$.

3.3. *Additional remarks.* We are dealing with two adjoint equations

$$\mathfrak{M}[x(z), \mathfrak{A}] = g(z), \tag{3.3.1}$$

$$\mathfrak{M}[y(z), \mathfrak{B}] = h(z), \tag{3.3.2}$$

and the corresponding homogeneous equations

$$\mathfrak{M}[u(z), \mathfrak{A}] = 0, \tag{3.3.3}$$

$$\mathfrak{M}[v(z), \mathfrak{B}] = 0. \tag{3.3.4}$$

We have already observed that

$$\mathfrak{M}[z^{-\alpha}, \mathfrak{A}] = D(\alpha; a_n)z^{-\alpha}, \tag{3.3.5}$$

provided $\Re(\alpha) > \sigma_0$, the abscissa of convergence of the series for $D(s; a_n)$. In the same domain we have

$$\mathfrak{M}\left[\frac{\partial^n}{\partial\alpha^n} z^{-\alpha}, \mathfrak{A}\right] = \frac{\partial^n}{\partial\alpha^n}[D(\alpha; a_n)z^{-\alpha}]. \tag{3.3.6}$$

From this we conclude that if the equation

$$D(s; a_n) = 0$$

has a k-fold zero at $s = \alpha$ with $\Re(\alpha) > \sigma_0$, then

$$z^{-\alpha}, z^{-\alpha} \log z, \cdots, z^{-\alpha} (\log z)^{k-1} \tag{3.3.7}$$

are solutions of the homogeneous equation (3.3.3).

It is clear that, if there are infinitely many zeros of $D(s; a_n)$ in the domain of convergence of the Dirichlet series, then any function of the form

$$\sum_{n=1}^{\infty} c_n z^{-\alpha_n} \tag{3.3.8}$$

satisfies (3.3.3), provided the double series

$$\sum_{m=1}^{\infty} a_m \sum_{n=1}^{\infty} c_n(mz)^{-\alpha_n}$$

can be rearranged so as to interchange the order of the summations.

It is perhaps worth while remarking that the series (3.3.8) do not form a dense set in any of the function spaces usually considered, such as $C[1, \infty]$ or $L_p(1, \infty)$, $1 \leqq p < \infty$. Indeed, the set $\{z^{-\alpha_n}\}$ will be closed in the space in question only if the series

$$\sum \frac{a + b\Re(\alpha_n)}{1 + |\alpha_n|^2}$$

diverges, where a and b are constants depending upon the space. But in our case $\Re(\alpha_n)$ is bounded; hence we are demanding the divergence of the series $\sum |\alpha_n|^{-2}$. But according to Landau[9] the number of zeros of an ordinary Dirichlet series in the domain $\sigma \geqq \sigma_0 + \epsilon$, $|t| \leqq T$, is $O(T \log T)$ and this frequency clearly does not permit the divergence of $\sum |\alpha_n|^{-2}$.

[9] E. Landau, *Über die Nullstellen der Dirichletschen Reihen*, Berliner Sitzungsberichte, vol. 14 (1913), pp. 897–907.

Our next remark concerns the existence of a solution of (3.3.1) when $g(z)$ satisfies (3.3.4). It is clear that the inversion formula of Möbius cannot give a solution in this case. It seems very plausible that no solution can exist under these circumstances. In certain simple cases it is possible to verify this surmise. Take, for example,

$$\sum_{k=0}^{\infty} f(2^k z) = 1.$$

Here $D(s; b_n) = 1 - 2^{-s}$ and $\mathfrak{M}[1, \mathfrak{B}] = 0$. Consider any domain E of the type described in §3.1, i.e., if $z_0 \in E$, so do nz_0 for $n = 2, 3, \cdots$. If the equation holds for $z = z_0$ and for $z = 2z_0$, then we get by subtraction $f(z_0) = 0$. If it is true for $z = 2^k z_0$ for every integer k, we get by the same argument that $f(2^k z_0) = 0$ for every k. But this clearly contradicts the assumption that the equation holds for $z = z_0$. Thus the equation in question cannot have any solution in E or even in a point set S which contains $2z_0$ whenever it contains z_0.

The final remark of this paragraph concerns the solution of (3.3.1) when $g(z)$ is a solution of (3.3.3). It is enough to consider the case $g(z) = z^{-\alpha}$, where $s = \alpha$ is a k-fold root of the equation $D(s; a_n) = 0$ in the half-plane of convergence of the Dirichlet series. Equation (3.3.6) then shows that

$$x(z) = z^{-\alpha}\{(-1)^k [D^{(k)}(\alpha; a_n)]^{-1} (\log z)^k + \sum_{\nu=0}^{k-1} c_\nu (\log z)^\nu\}$$

is a solution of (3.3.1). This result shows a further analogy between the formal theory of the equations here considered and that of linear differential equations.

3.4. *Further equations with the same algorithm.* It was known to Möbius[10] that his algorithm entered in the study of other functional equations. The following example is slightly more general than the situation which Möbius had in mind.

We consider two multiplicative systems, i.e., we give two sequences $\{\epsilon(p)\}$ and $\{\omega(p)\}$, where p runs through the primes, we take $\epsilon(1) = \omega(1) = 1$ and define $\epsilon(n)$ and $\omega(n)$ by the equations

$$\epsilon(mn) = \epsilon(m)\epsilon(n), \qquad \omega(mn) = \omega(m)\omega(n). \tag{3.4.1}$$

Let us form the functional equation

$$\sum_{n=1}^{\infty} a_n \epsilon(n) f(\omega(n)z) = g(z). \tag{3.4.2}$$

It is not difficult to see that a formal solution is given by

$$f(z) = \sum_{n=1}^{\infty} b_n \epsilon(n) g(\omega(n)z), \tag{3.4.3}$$

[10] Möbius, loc. cit., *Werke*, vol. IV, p. 594.

where $\{b_n\}$ as usual is the reciprocal sequence of $\{a_n\}$. The elementary methods of §3.2 can be used to develop sufficient conditions for the validity of this inversion formula.[11] The details can be left to the reader.

4. General algorithms.

4.1. *The Möbius $A(u)$-transform.* We can write the Möbius $\mathfrak{A}$-transform as a Stieltjes integral, viz.,

$$\mathfrak{M}[f(z), \mathfrak{A}] = \int_1^\infty f(uz)\, dA(u),$$

where

$$A(u) = \sum_{n<u} a_n .$$

This suggests a generalization of the inversion problem to arbitrary functions $A(u)$ of bounded variation.

Let $A(u)$ be given for $u \geqq 1$, $A(1) = 0$, of bounded variation in every finite interval, and such that

$$A(u) = O(u^{\omega+\epsilon}) \tag{4.1.1}$$

for every $\epsilon > 0$, where $\omega \geqq 0$ is a constant.

Suppose now that $f(z)$ is an analytic function satisfying the following conditions: (i) $f(z)$ is holomorphic in a sectorial domain S such that if z is in S, so is uz for every $u \geqq 1$; (ii) the integral

$$\mathfrak{M}[f(z), A(u)] \equiv \int_1^\infty f(uz)\, dA(u) \tag{4.1.2}$$

exists in some domain $S_0 \subset S$.

We shall as a rule use the abbreviated notation $\mathfrak{M}[f, A]$ and refer to this function as the Möbius $A(u)$-transform of $f(z)$. To every fixed function $A(u)$ satisfying the above conditions there is a class $\mathfrak{F}[A]$ of functions $f(z)$ which admit $A(u)$-transforms in the sense of the definition.

The effective determination of $\mathfrak{F}[A]$ may be quite laborious except in the simplest cases, but it is easy to find a subclass of $\mathfrak{F}[A]$. Let $f(z)$ be holomorphic in a sectorial domain S and

$$|f'(z)| < M|z|^{-\gamma-1}, \qquad \gamma > \omega. \tag{4.1.3}$$

It is easy to see that $\mathfrak{M}[f, A]$ exists in S; thus every such function belongs to $\mathfrak{F}[A]$.

4.2. *The reciprocation problem for Laplace integrals.* In the case of the $\mathfrak{A}$-transform the inverse or reciprocal transform was given a priori and had a sense for a sufficiently restricted but not vacuous class of functions. For the

[11] Some instances of this inversion formula figured in the papers of E. Hille and O. Szász, *On the completeness of Lambert functions,* of which the first part appeared in the Bulletin of the American Mathematical Society, vol. 42 (1936), pp. 411–418, and the second in the Annals of Mathematics, vol. 37 (1936), pp. 800–815.

$A(u)$-transform the situation is different, and the inverse transform need not exist at all. By analogy with the sequence case we should consider the Laplace-Stieltjes integral

$$D(s) = \int_1^\infty u^{-s}\,dA(u). \tag{4.2.1}$$

The hypothesis (4.1.1) insures the convergence of this integral for $\sigma > \omega$. We should then take the reciprocal of $D(s)$ and find its representation as a Laplace-Stieltjes integral. But it is well known that $[D(s)]^{-1}$ ordinarily is not representable in this manner.

Necessary and sufficient conditions in order that $[D(s)]^{-1}$ shall be representable by a convergent Laplace-Stieltjes integral do not seem to be known. In the following I shall give two sets of two conditions each. One condition is common to the two sets; the first set is necessary, but perhaps not sufficient, the second set is sufficient, but certainly not necessary.[12]

THEOREM 4.2.1. *Let $D(s)$ be a function representable by a convergent Laplace-Stieltjes integral. In order that $[D(s)]^{-1}$ shall also admit such a representation, it is necessary that* (i) $\lim_{\sigma\to+\infty} D(\sigma) \neq 0$, *and* (ii) *there exist a half-plane $\sigma > \sigma_0$ in which $D(s) \neq 0$.*

Proof. That the conditions are necessary is obvious. Since

$$\lim_{\sigma\to+\infty} D(\sigma) = \lim_{u\to 1+} A(u),$$

we can replace condition (i) by the equivalent condition (i′) $A(u)$ is discontinuous at $u = 1$.

THEOREM 4.2.2. *If $A(u)$ is discontinuous at $u = 1$, $A(1+0) = a \neq 0$, and if the Laplace-Stieltjes integral representing $D(s)$ is absolutely convergent for $\sigma > \alpha$, then $[D(s)]^{-1}$ is representable by a Laplace-Stieltjes integral absolutely convergent for $\sigma >$* max (α, β), *where β is the root of the equation*

$$|a| = \int_1^\infty u^{-\sigma}\,dV_1^u[A(v) - a],^{13} \tag{4.2.3}$$

if it exists; otherwise $\beta = -\infty$.

Proof. Let us put

$$A_1(v) = A(v) - a, \qquad A_1(1) = 0,$$

$$V_1^u A_1(v) = A_0(u),$$

$$A_1(u, s) = \int_1^u v^{-s}\,dA_1(v),$$

$$A_0(u, s) = \int_1^u v^{-s}\,dA_0(v).$$

[12] It is to be hoped that the investigations by R. H. Cameron and N. Wiener, now in progress, will throw further light on this question.

[13] Here and in the following, $V_a^b f(t)$ denotes the total variation of $f(t)$ in $a \leq t \leq b$.

Then $A_0(u)$, $A_1(u)$, $A_0(u, s)$, and $A_1(u, s)$ are continuous at $u = 1$ and tend to zero as $u \to 1$. A simple consideration shows that

$$V_1^u A_1(v, s) \leqq A_0(u, \sigma). \tag{4.2.4}$$

By assumption

$$A_0(u) = O(u^{\alpha+\epsilon})$$

for every $\epsilon > 0$. It follows that for $\sigma > \alpha$ the increasing function $A_0(u, \sigma)$ tends to the finite limit $A_0(\infty, \sigma)$ as $u \to \infty$. Further, it is obvious that $A_0(\infty, \sigma)$ is monotone decreasing when σ increases and tends to zero as $\sigma \to \infty$. The latter conclusion follows from the fact that $A_0(u) \to 0$ monotonically from above as $u \to 0+$. Hence the equation

$$A_0(\infty, \sigma) = |a| \tag{4.2.5}$$

has at most one root $\geqq \alpha$. We define β to be equal to this root if it exists; otherwise we take $\beta = -\infty$.

Now let $\sigma > \max(\alpha, \beta)$. Then

$$D(s) = a + A_1(\infty, s),$$

and

$$|A_1(\infty, s)| \leqq A_0(\infty, \sigma) < |a|.$$

Hence

$$[D(s)]^{-1} = \sum_{n=0}^{\infty} (-1)^n a^{-n-1} [A_1(\infty, s)]^n, \tag{4.2.6}$$

and the series is absolutely convergent. We shall rewrite this series as a Laplace-Stieltjes integral. We have for $\sigma > \gamma > \max(\alpha, \beta)$

$$A_1(\infty, s) = \int_1^{\infty} u^{-(s-\gamma)} \, d_u A_1(u, \gamma). \tag{4.2.7}$$

Hence

$$[A_1(\infty, s)]^n = \int_1^{\infty} u^{-(s-\gamma)} \, d_u A_n(u, \gamma), \tag{4.2.8}$$

where

$$A_n(u, \gamma) = \int_1^u A_{n-1}(u/v, \gamma) \, d_v A_1(v, \gamma) \qquad (n = 2, 3, \cdots). \tag{4.2.9}$$

Here (4.2.8) is absolutely convergent, being the product of absolutely convergent Laplace-Stieltjes integrals.[14] This fact also follows from the subsequent estimates of $A_n(u, \gamma)$.

[14] For the properties of Laplace-Stieltjes integrals used in this paper, consult D. V. Widder, Trans. Amer. Math. Soc., vol. 31 (1929), pp. 694–743, and E. Hille and J. D. Tamarkin, Proc. Nat. Acad. Sci., vol. 19 (1933), pp. 573–577, 902–912; vol. 20 (1934), pp. 140–144.

We shall prove the inequality

$$(4.2.10) \qquad | A_n(u, \gamma) | \leqq V_1^u A_n(v, \gamma) \leqq [A_0(u, \gamma)]^n.$$

The inequality is obviously true for $n = 1$. Suppose that it has been proved for $n = k$. It follows in particular that $A_k(u, \gamma)$ is continuous at $u = 1$ and tends to zero as $u \to 1$. Using (4.2.9) with $n = k + 1$, we see that $A_{k+1}(u, \gamma)$ has the same property, whence it follows that it is sufficient to prove the second half of the inequality. But

$$\begin{aligned} V_1^u A_{k+1}(v, \gamma) &= \int_1^u \left| d_v \int_1^v A_k(v/t, \gamma)\, d_t A_1(t, \gamma) \right| \\ &\leqq \int_1^u \left| d_v \int_1^v | A_k(v/t, \gamma) | \cdot | d_t A_1(t, \gamma) | \right| \\ &\leqq \int_1^u d_v \int_1^v [A_0(v/t, \gamma)]^k\, d_t A_0(t, \gamma) \\ &= \int_1^u [A_0(u/t, \gamma)]^k\, d_t A_0(t, \gamma) \\ &\leqq [A_0(u, \gamma)]^k \int_1^u d_t A_0(t, \gamma) \\ &= [A_0(u, \gamma)]^{k+1}. \end{aligned}$$

This completes the proof of the inequality.

Let us put

$$(4.2.11) \qquad \begin{cases} B(u. \gamma) = a^{-1} + \sum_{n=1}^{\infty} (-1)^n a^{-n-1} A_n(u, \gamma), & u > 1, \\ B(1, \gamma) = 0. \end{cases}$$

Then by (4.2.10)

$$(4.2.12) \qquad | B(u, \gamma) | \leqq V_1^u B(v, \gamma) \leqq \sum_{n=0}^{\infty} | a |^{-n-1} [A_0(u, \gamma)]^n,$$

the series being absolutely convergent and uniformly bounded for $1 \leqq u \leqq \infty$.

Hence $B(u, \gamma)$ is of bounded variation in $[1, \infty]$ and

$$(4.2.13) \qquad [D(s)]^{-1} = \int_1^{\infty} u^{-(s-\gamma)}\, d_u B(u, \gamma) = \int_1^{\infty} u^{-s}\, dB(u),$$

where

$$(4.2.14) \qquad B(u) = \int_1^u v^{\gamma}\, d_v B(v, \gamma).$$

The second integral in (4.2.13) is clearly absolutely convergent for $\sigma > \max(\alpha, \beta)$. This completes the proof of the theorem.

The assumption that $D(s)$ has a half-plane of absolute convergence is obviously unnecessarily restrictive. But simple convergence is not enough to insure the existence of even a formal Laplace integral for the reciprocal, much less of a half-plane of convergence.

4.3. *Inversion of the $A(u)$-transform.* In the present paragraph we shall assume that the reciprocal of $D(s)$ admits a representation by means of a convergent Laplace-Stieltjes integral. The functions $A(u)$ and $B(u)$ are then joined by the relation

$$\int_1^u B(u/v)\,dA(v) = \int_1^u A(u/v)\,dB(v) = 1, \qquad u > 1, \tag{4.3.1}$$

for almost all values of u. This is the transcendental analogue of the Möbius algorithm to which it reduces when $A(u)$ is a step function with jumps at the integers.

We can now expect that for a sufficiently restricted class of functions $f(z)$ we shall have

$$\mathfrak{M}\{\mathfrak{M}[f, A], B\} = \mathfrak{M}\{\mathfrak{M}[f, B], A\} = f(z). \tag{4.3.2}$$

We have the following analogue of Theorem 3.2.1.

THEOREM 4.3.1. *A sufficient condition that* (4.3.2) *shall hold is that the double integral*

$$I[f] \equiv \int_1^\infty \int_1^\infty f(uvz)\,dA(u)\,dB(v) \tag{4.3.3}$$

be absolutely convergent.

Proof. Let

$$V_1^u A(t) = A_0(u), \qquad V_1^u B(t) = B_0(u).$$

The condition of the theorem is then that the integral

$$\int_1^\infty \int_1^\infty |f(uvz)|\,dA_0(u)\,dB_0(v)$$

be convergent. It is then permissible to regard (4.3.3) as a repeated Stieltjes integral and the order in which the integrations are performed is immaterial. Now

$$\mathfrak{M}\{\mathfrak{M}[f, A], B\} = \int_1^\infty \left\{\int_1^\infty f(uvz)\,dA(u)\right\} dB(v),$$

$$\mathfrak{M}\{\mathfrak{M}[f, B], A\} = \int_1^\infty \left\{\int_1^\infty f(uvz)\,dB(v)\right\} dA(u).$$

Hence these two operations exist and give the same result. On the other hand, going back to the definition of the Stieltjes integral as a double sum and "summing by hyperbolas" uv = const. before passing to the limit, we can show that the double integral can be written in the form

$$\int_1^\infty f(wz)\,d_w \int_1^w B(w/u)\,dA(u).$$

Formula (4.3.1) shows that this expression reduces to $f(z)$. This completes the proof of the theorem.

We can now pass to the question of inversion.

THEOREM 4.3.2. *Let $B(u)$ exist as a function of bounded variation and satisfy* (4.3.1). *A sufficient condition that*

$$f(z) = \mathfrak{M}[g, B] \tag{4.3.4}$$

be a solution of

$$g(z) = \mathfrak{M}[f, A] \tag{4.3.5}$$

in the domain S is that the double integral $I[g]$ be absolutely convergent in S. On the other hand, there cannot be more than one solution $f(z)$ of (4.3.5) *which renders $I[f]$ absolutely convergent, and whenever it exists this solution is given by* (4.3.4).

Proof. The assumption that $I[g]$ is absolutely convergent implies that

$$\mathfrak{M}\{\mathfrak{M}[g, B], A\} = g(z)$$

by Theorem 4.3.1. Hence (4.3.4) gives a solution of (4.3.5) and the integral is obviously absolutely convergent.

Conversely, if $f(z)$ is a solution of (4.3.5) such that $I[f]$ is absolutely convergent, then for the same reason

$$f(z) = I[f] = \mathfrak{M}\{\mathfrak{M}[f, A], B\} = \mathfrak{M}[g, B],$$

so that the solution in question is uniquely determined and given by (4.3.4).

Again it is necessary to remark that the mere existence of $\mathfrak{M}[g, B]$ in a domain S is not sufficient to insure that this function be a solution of (4.3.5). Indeed, suppose that the Laplace-Stieltjes integral

$$[D(s)]^{-1} = \int_1^\infty u^{-s}\, dB(u)$$

has a zero $s = \alpha$ in the half-plane of convergence. Then

$$\mathfrak{M}[z^{-\alpha}, B] = z^{-\alpha} \int_1^\infty u^{-\alpha}\, dB(u) \equiv 0$$

and is certainly not a solution of the equation

$$\mathfrak{M}[f(z), A(u)] = z^{-\alpha}.$$

4.4. *Concluding remarks.* In the previous discussion I have perhaps overemphasized the rôle of the associated Laplace-Stieltjes integral

$$D(s) = \int_1^\infty u^{-s}\, dA(u).$$

It should be observed that this function and its reciprocal are mainly tools in the discussion, and the decisive rôle is really played by the reciprocal functions $A(u)$ and $B(u)$ which are supposed to satisfy the algorithm (4.3.1). We know from the sequence case that these functions may very well exist without the associated Laplace-Stieltjes integrals having any domain of convergence or even any a priori obvious significance. The theorems of §4.3 really presuppose merely the existence of a pair of functions satisfying (4.3.1) and not the existence of the associated Laplace-Stieltjes integrals.

But it must be admitted that when the integrals do not exist, the problem of finding the function $B(u)$ reciprocal to a given function $A(u)$ is in general not a very promising one. Sometimes we may circumvent the difficulties by a preliminary application of a suitable method of summation. Thus it may happen that $s^{-n}[D(s)]^{-1}$ is representable by a Laplace-Stieltjes integral even though $[D(s)]^{-1}$ is not. In this case the corresponding function $B_n(u)$ can be used to find an n-fold integral of the formal solution of (4.3.5) from which the solution itself may be found by solving an integral equation of the Abel type.

There are of course other functional equations which may be treated by the methods of this paper; for instance, the equation

$$\int_{-\infty}^{\infty} F(z - u)\, da(u) = G(z),$$

which is associated with Laplace-Stieltjes integrals having 0 instead of 1 as the lower limit of integration.

Finally, it should be remarked that there are various relations, some obvious, others less so, between the functional equations treated in this paper on one hand, and the theory of Watson transforms and the Karamata-Wiener Tauberian theory on the other. The author hopes to return to these questions at a later opportunity.

YALE UNIVERSITY.

Chapter 9
Hermitian Series and Differential Operators of Infinite Order Which Are Entire Functions of Finite Operators

Papers included in this chapter are

An introduction to Hille's papers on Hermitian series can be found starting on page 675.

ANALYSE MATHÉMATIQUE. — *Sur les séries associées à une série d'Hermite.*

Note de M. EINAR HILLE.

1 Soit $h_n(z) = e^{-\frac{1}{2}z^2} H_n(z) = (-1)^n e^{\frac{1}{2}z^2} \frac{d^n}{dz^n}(e^{-z^2})$ la $n^{\text{ième}}$ fonction d'Hermite. Nous posons

$$A_{2k} = |H_{2k}(0)|, \qquad A_{2k+1} = |H'_{2k+1}(0)|(4k+3)^{-\frac{1}{2}},$$

$$c_n(z) = A_n \cos\left[(2n+1)^{\frac{1}{2}} z - n\frac{\pi}{2}\right], \qquad s_n(z) = A_n \sin\left[(2n+1)^{\frac{1}{2}} z - n\frac{\pi}{2}\right],$$

$$e_n^+(z) = c_n(z) + i s_n(z), \qquad e_n^-(z) = c_n(z) - i s_n(z).$$

2. Il est facile de voir que $h_n(z)$ satisfait à l'équation intégrale

$$h_n(z) = c_n(z) + (2n+1)^{-\frac{1}{2}} \int_0^z t^2 \sin\left[(2n+1)^{\frac{1}{2}}(z-t)\right] h_n(t)\, dt.$$

La solution obtenue par la méthode des approximations successives peut s'écrire

$$h_n(z) = \frac{1}{2}[e_n^+(z) + e_n^-(z)] + \frac{1}{2}\sum_{k=1}^{\infty}\left[2i(2n+1)^{\frac{1}{2}}\right]^{-k} z^{3k}$$
$$\times \sum_{m=1}^{2^k} (-1)^{\mu} \int_0^1 \int_0^1 \cdots \int_0^1 R_k(r)[e_n^+(R_{k,m}^+ z) + e_n^-(R_{k,m}^- z)]\, dr_1\, dr_2 \ldots dr_k.$$

Ici on a posé $R_k(r) = r_1^{3k-1} r_2^{3k-4} \ldots r_k^2$ et les $R_{k,m}^+$ et $R_{k,m}^-$ sont des sommes de la forme

$$R_{k,m}^+ = \sum_{\alpha=0}^{k-1} \pm r_0 r_1 \ldots r_\alpha (1 - r_{\alpha+1}) + r_0 r_1 \ldots r_k \qquad (r_0 \doteq 1),$$

$$R_{k,m}^- = \sum_{\alpha=0}^{k-1} \pm r_0 r_1 \ldots r_\alpha (1 - r_{\alpha+1}) - r_0 r_1 \ldots r_k,$$

où les signes $\pm$ sont choisis d'une manière arbitraire et la sommation par rapport à m s'étend sur toutes les 2^k combinaisons de signes possibles,

μ étant le nombre des signes négatifs dans $R^{+}_{k,m}$. Cette représentation de $h_n(z)$ est valable dans tout le plan.

On démontre sans difficulté que le premier terme de cette série, c'est-à-dire la fonction $c_n(z)$, est aussi le plus grand terme quand $z = o\left(n^{\frac{1}{6}}\right)$, z étant non réel.

3. Soit maintenant $f(z) = \sum_{0}^{\infty} f_n h_n(z)$ une *série d'Hermite* donnée et introduisons les quatre *séries associées*

$$C(z) = \sum_{0}^{\infty} f_n c_n(z), \qquad S(z) = \sum_{0}^{\infty} f_n s_n(z),$$

$$E^{+}(z) = \sum_{0}^{\infty} f_n e^{+}_n(z), \qquad E^{-}(z) = \sum_{0}^{\infty} f_n e^{-}_n(z).$$

Pour une valeur non réelle de z, *on peut démontrer que la série de* $f(z)$ *converge ou diverge selon que la série de* $C(z)$ [*ou de* $S(z)$] *converge ou diverge. Il suit que le domaine de convergence absolue est une bande* $-\tau < y < \tau$ et que les séries de $f(z)$, $C(z)$ et $S(z)$ divergent pour $|y| > \tau$. On trouve que $\tau = -\lim\sup(2n+1)^{-\frac{1}{2}} \log[A_n |f_n|]$.

La série de $E^{+}(z)$ converge dans le demi-plan $y > -\tau$ et celle de $E^{-}(z)$ dans $y < \tau$.

4. En introduisant les séries intégrées

$$E^{+}_{k}(z) = \sum_{n=0}^{\infty} f_n (2n+1)^{-\frac{1}{2}k} e^{+}_n(z), \quad E^{-}_{k}(z) = \sum_{n=0}^{\infty} f_n (2n+1)^{-\frac{1}{2}k} e^{-}_n(z),$$

on démontre *la relation fondamentale entre la série d'Hermite et ses séries de Dirichlet associées*

$$f(z) = \frac{1}{2}[E^{+}(z) + E^{-}(z)] + \frac{1}{2} \sum_{k=1}^{\infty} (2i)^{-k} z^{3k}$$
$$\times \sum_{m=1}^{2k} (-1)^{\mu} \int_0^1 \int_0^1 \cdots \int_0^1 R_k(r) [E^{+}_{k}(R^{+}_{k,m} z) + E^{-}_{k}(R^{-}_{k,m} z)]\, dr_1\, dr_2 \ldots dr_k,$$

valable tout d'abord pour $-\tau < y < \tau$.

Mais il y a plus. *Soient* $\mathcal{E}^{+}$ *et* $\mathcal{E}^{-}$ *les étoiles principales des fonctions* $E^{+}(z)$ *et* $E^{-}(z)$. *Soit de plus* $\mathcal{C}$ *le domaine commun des quatre étoiles* $\mathcal{E}^{+}$, $-\mathcal{E}^{+}$, $\mathcal{E}^{-}$

et $-\mathcal{E}^-$. *On démontre sans peine que* $f(z)$ *est une fonction holomorphe de* z *dans* $\mathcal{C}$ *où elle est représentée par la série convergente donnée ci-dessus.*

5. Il y a donc évidemment une relation étroite entre la série d'Hermite et ses séries associées $E^+(z)$ et $E^-(z)$, qui sont des séries de Dirichlet avec les exposants $\pm(2n+1)^{\frac{1}{2}}$. On peut se demander si cette relation donne quelque renseignement sur les singularités d'une fonction donnée par une série d'Hermite si les points singuliers des séries de Dirichlet associées sont connus. Il n'est pas évident en général qu'une singularité de $E^+(z)$ ou de $E^-(z)$ donne naissance à une singularité de $f(z)$. Néanmoins on peut utiliser la relation d'une façon indirecte dans des cas étendus. Soit, par exemple, $i^{-n}f_n \geqq 0$ pour $n > n_0$. Par le théorème de Pringsheim-Vitali-Landau le point $z = \tau i$ est singulier pour $E^-(z)$, et c'est aussi une singularité de $f(z)$ comme on s'en convainc par une démonstration analogue à celle du théorème classique. On peut donc formuler des théorèmes pour les séries d'Hermite en utilisant l'analogie avec les séries de Dirichlet aux exposants $\pm(2n+1)^{\frac{1}{2}}$, et dans un grand nombre de cas les méthodes familières dans les séries de Dirichlet en donnent les démonstrations. A titre d'exemple nous ajoutons seulement le théorème suivant, analogue du théorème de Fabry-Carlson-Landau-Szàsz : *la fonction définie par la série* $\sum_{k=1}^{\infty} f_{n_k} h_{n_k}(z)$ *ne peut pas être prolongée au delà de sa bande de convergence si la suite* $\{n_k\}$ *satisfait à une quelconque des conditions suivantes :* (1) $n_k(k\log k)^{-2} \to \infty$; (2) $n_k k^{-2} \to \infty$, $\inf\,(n_{k+1}-n_k)\,n_k^{-\frac{1}{2}} > 0$; *ou* (3) $n_k k^{-2} \to \infty$, $(n_{k+1}-n_k)\,n_k^{-\frac{1}{2}} > e^{-\varepsilon\varpi(n)}$, $n > n(\varepsilon)$, $\varpi(n) = \inf_{m \geqq n} n_m m^{-2}$.

(Extrait des *Comptes rendus des séances de l'Académie des Sciences*, t. 209, p. 714, séance du 13 novembre 1939.)

GAUTHIER-VILLARS, IMPRIMEUR-LIBRAIRE DES COMPTES RENDUS DES SÉANCES DE L'ACADÉMIE DES SCIENCES.
113482-39 Paris. — Quai des Grands-Augustins, 55.

CONTRIBUTIONS TO THE THEORY OF HERMITIAN SERIES

II. THE REPRESENTATION PROBLEM*

BY

EINAR HILLE

1. **Introduction.** In the first note of this series [2]† the author laid a broad foundation for the theory of Hermitian series in the complex domain. A number of questions encountered during this investigation were merely mentioned and detailed discussion had to be postponed till later communications. The *representation* or *expansion problem* was such a question, that is, *the problem of finding necessary and sufficient conditions in order that an analytic function shall be representable by an Hermitian series for complex values of the variable.* This problem is solved in the present note.

Let $H_n(z)$ denote the nth normalized orthogonal function of Hermite

$$H_n(z) = [\pi^{1/2}2^n n!]^{-1/2}(-1)^n e^{z^2/2}\frac{d^n}{dz^n}[e^{-z^2}]. \tag{1.1}$$

If $f(x)$ is a measurable function such that $x^n \exp[-x^2/2]f(x) \in L_1(-\infty, \infty)$ for $n=0, 1, 2, \cdots$, then $f(x)$ has an associated Fourier-Hermite series

$$f(x) \sim \sum_{n=0}^{\infty} f_n H_n(x) \tag{1.2}$$

where

$$f_n = \int_{-\infty}^{\infty} f(t)H_n(t)dt.$$

This series is ordinarily not convergent, but if we assume for instance that $\exp[-\alpha x^2]f(x) \in L_1(-\infty, \infty)$ for some $\alpha < 1/2$, then the series (1.2) can be summed to the sum $f(x)$ for almost all x by a generalization of the Abel method of summation.‡ Under this assumption the series

$$f(x; s) = \sum_{n=0}^{\infty} f_n H_n(x)s^n$$

converges for

---

* Presented to the Society, October 28, 1939; received by the editors November 6, 1939.

† Numbers in square brackets refer to the references at the end of this paper.

‡ See E. Hille [1, pp. 448–450, 453].

$$|s| < \left(\frac{1-2\alpha}{1+2\alpha}\right)^{1/2} \equiv r_\alpha.$$

Further $f(x; s)$ is a holomorphic function of s in the interior of the ellipse in the complex s-plane having its vertices at the points $s = \pm 1$, $\pm ir_\alpha$. As $s \to +1$ along the real axis, $f(x; s) \to f(x)$ for almost all values of x. It follows in particular that if the series (1.2) converges in a set E of positive measure, its sum equals $f(x)$ almost everywhere in E provided $\exp[-\alpha x^2] f(x) \in L_1(-\infty, \infty)$ for some $\alpha < 1/2$.

It may happen that the series (1.2) converges also for complex values of the variable. If so, it is known that the domain of absolute convergence is a strip S_τ: $-\tau < y < \tau$, $z = x + iy$, where

$$\tau = -\limsup_{n\to\infty} (2n+1)^{-1/2} \log |f_n|. \tag{1.3}$$

For a proof, see E. Hille [2, chap. 2] and G. Szegö [7, p. 246]. The series may converge on the lines of convergence $y = \pm\tau$, but it diverges outside of S_τ. In this case the series

$$\sum_{n=0}^{\infty} f_n \boldsymbol{H}_n(z) \equiv f(z) \tag{1.4}$$

defines an analytic function which is holomorphic in S_τ.

Such a function $f(z)$ is of course characterized by the fact that its Fourier-Hermite coefficients satisfy the inequality

$$|f_n| \leqq M(\epsilon) \exp[-(\tau - \epsilon)(2n+1)^{1/2}] \tag{1.5}$$

for every positive ϵ, but a direct characterization of $f(z)$ in terms of function-theoretical properties would seem desirable. A complete solution of this problem is given by

THEOREM 1. *Let $f(z)$ be an analytic function. A necessary and sufficient condition in order that the Fourier-Hermite series*

$$\sum_{n=0}^{\infty} f_n \boldsymbol{H}_n(z), \qquad f_n = \int_{-\infty}^{\infty} f(t) \boldsymbol{H}_n(t)\,dt, \tag{1.6}$$

shall exist and converge to the sum $f(z)$ in the strip S_τ: $-\tau < y < \tau$, is that $f(z)$ is holomorphic in S_τ and that to every given β, $0 \leqq \beta < \tau$, there exists a finite positive $B(\beta)$ such that

$$|f(x+iy)| \leqq B(\beta) \exp[-|x|(\beta^2 - y^2)^{1/2}] \tag{1.7}$$

for $-\infty < x < \infty$, $-\beta \leqq y \leqq \beta$.

The proof that this condition is necessary will be given in §2. The sufficiency proof occurs in §5 and is preceded by material relating to the asymptotic behavior of $H_n(z)$ for large values of n based on the investigations of R. E. Langer [3]. An application to functions meromorphic in an upper or a lower half-plane is given in §6.

As far as I know there have only been two previous contributions to the representation problem for Hermitian series, those of G. N. Watson [9, pp. 417–421] and O. Volk [8]. Their equivalent conditions require that $f(z)$ shall be holomorphic in $-\tau \leqq y \leqq \tau$ and that the function $g(z) \equiv z^{-1} \exp(z^2/2) f(z)$ shall have the properties (i) $g(x+iy) \to 0$ as $|x| \to \infty$, $-\tau \leqq y \leqq \tau$, and (ii) $g(x \pm i\tau) \in L_1(-\infty, \infty)$. It is clear that these conditions while sufficient are far from necessary.

2. **Proof of necessity.** The reader will be supposed to have some knowledge of the properties of Hermitian polynomials and of solutions of the Weber-Hermite differential equations. For convenient summaries of the former theory see E. Hille [1] and G. Szegö [7], for the latter see Whittaker-Watson [12].

The only property which will be used in the present paragraph is the generating function*

$$\sum_{n=0}^{\infty} H_n(u) H_n(v) s^n = \pi^{-1/2}(1 - s^2)^{-1/2} \exp\left\{-\frac{(1+s^2)(u^2+v^2) - 4suv}{2(1-s^2)}\right\}. \tag{2.1}$$

The series converges for arbitrary complex values of u and v when $|s| < 1$. Putting $u = z = x + iy$, $v = \bar{z} = x - iy$, we get

$$\sum_{n=0}^{\infty} |H_n(z)|^2 s^n = \pi^{-1/2}(1 - s^2)^{-1/2} \exp\left\{-\frac{1-s}{1+s} x^2 + \frac{1+s}{1-s} y^2\right\}. \tag{2.2}$$

We shall also need information concerning integrals of the form

$$J(p, q, \nu) = \int_0^{\infty} \exp\{-pt - q/t\} t^{-\nu} dt,$$

where p and q are positive and $\nu = 3/2$, 2 or 5/2. Now

$$J(p, q, 3/2) = \pi^{1/2} q^{-1/2} \exp[-2(pq)^{1/2}], \tag{2.3}$$

as is easily verified. Since $J(p, q, 5/2) = -J_q'(p, q, 3/2)$,

$$J(p, q, 5/2) = \pi^{1/2} q^{-3/2} [1/2 + (pq)^{1/2}] \exp[-2(pq)^{1/2}]. \tag{2.4}$$

* See, for instance, E. Hille [1, pp. 439–440].

Finally, by Schwarz' inequality

$$J(p, q, 2) \leqq \pi^{1/2}q^{-1}[1/2 + (pq)^{1/2}]^{1/2} \exp [-2(pq)^{1/2}]. \tag{2.5}$$

Suppose now that the series (1.6) exists and converges to $f(z)$ in the strip $-\tau<y<\tau$. It follows immediately that $f(z)$ is holomorphic in the strip and it remains merely to estimate $f(z)$. Let β be given, $0<\beta<\tau$, and put $\alpha=(\beta+\tau)/2$. In view of (1.5), the series

$$\sum_{n=0}^{\infty} | f_n |^2 \exp (2\alpha N) \equiv A^2(\beta) \tag{2.6}$$

is convergent. Here and in the following we shall write

$$N = (2n + 1)^{1/2}. \tag{2.7}$$

Let us put

$$H(x, y; \alpha) = \sum_{n=0}^{\infty} | H_n(x + iy) |^2 \exp (-2\alpha N). \tag{2.8}$$

Cauchy's inequality then gives

$$| f(x + iy) | \leqq A(\beta)[H(x, y; \alpha)]^{1/2}. \tag{2.9}$$

Let us first obtain $H(x, y; \alpha)$ in closed form. Forming

$$\sum_{n=0}^{\infty} | H_n(z) |^2 J(2n + 1, \alpha^2, 3/2)$$

and using (2.2) and (2.3) we get after some simplifications

$$H(x, y; \alpha) = \frac{\alpha}{\pi}\int_0^{\infty} (1 - e^{-4t})^{-1/2} \exp \{-\alpha^2/t - t - x^2 \tanh t + y^2 \coth t\} t^{-3/2} dt, \tag{2.10}$$

where $\tanh t$ denotes the hyperbolic tangent of t.

Elementary considerations show that

$$\coth t \leqq (1 + t)/t, \qquad -\tanh t \leqq -t/(1 + t), \qquad t > 0.$$

Substituting these dominants for the hyperbolic functions in (2.10) we obtain

$$\begin{aligned} (\pi/\alpha)H(x, y; \alpha) &\leqq H_0(x, y; \alpha) \\ &\equiv e^{y^2}\int_0^{\infty} (1 - e^{-4t})^{-1/2} \exp \left\{-\frac{\alpha^2 - y^2}{t} - t - \frac{x^2 t}{1 + t}\right\} t^{-3/2} dt \\ &\equiv e^{y^2} H_1(x, y; \alpha), \end{aligned} \tag{2.11}$$

and it remains to estimate $H_1(x, y; \alpha)$.

Let $\xi=4\alpha|x|^{-1}$ and put

$$H_1 = H_{11} + H_{12}, \quad H_{11} = \int_0^{\xi} \cdots, \quad H_{12} = \int_{\xi}^{\infty} \cdots \tag{2.12}$$

with obvious notation.

We note that $t/(1+t)$ is an increasing function. Hence its least value in the interval (ξ, ∞) is taken on at $t=\xi$. The inequality $1+u \leqq e^u$ gives

$$(1+u)^{-1} \geqq e^{-u}, \qquad (1-e^{-u})^{-1} \leqq 1/u + 1$$

and finally

$$(1-e^{-4t})^{-1/2} \leqq (1/(4t)+1)^{1/2} \leqq t^{-1/2}/2 + 1.$$

Hence

$$H_{12} \leqq \exp\left[-\frac{4\alpha x^2}{4\alpha+|x|}\right]\{J(1, \alpha^2-y^2, 3/2) + (1/2)J(1, \alpha^2-y^2, 2)\}.$$

Supposing $|x| \geqq 4\alpha$, which is no essential restriction, and using (2.3) and (2.5), we get

$$H_{12} \leqq \pi^{1/2} \exp\left[-2\alpha|x| - 2(\alpha^2-y^2)^{1/2}\right](\alpha^2-y^2)^{-1} \cdot \{(1/2)[1/2+(\alpha^2-y^2)^{1/2}]^{1/2} + (\alpha^2-y^2)^{1/2}\}.$$

If $-\beta \leqq y \leqq \beta$, the right-hand side does not exceed

$$\frac{10\beta+7}{(\tau+3\beta)(\tau-\beta)} \exp\left[-2\alpha|x|\right],$$

that is,

$$H_{12}(x, y; \alpha) \leqq B_1(\beta) \exp\left[-2\alpha|x|\right] \tag{2.13}$$

for $|x| \geqq 4\alpha$, $-\beta \leqq y \leqq \beta$, and by a suitable modification of $B_1(\beta)$ we can of course gain that the inequality is true for all x. Here and in the following $B(\beta)$, with or without subscripts, denotes a monotone increasing function of β, defined and continuous for $0 \leqq \beta < \tau$, but conceivably tending to infinity as $\beta \rightarrow \tau$.

For $t>0$ we have

$$-t/(1+t) = -t + t^2/(1+t) < -t + t^2.$$

Further, for $0 < t \leqq \xi \leqq 1$, that is, for $|x| \geqq 4\alpha$,

$$(1-e^{-4t})^{-1/2} \leqq (1-e^{-4})^{-1/2} t^{-1/2} < 2t^{-1/2}.$$

Hence

$$
\begin{aligned}
H_{11} &< 2e^{\xi^2 x^2}\int_0^{\xi} \exp\{-(\alpha^2-y^2)/t - x^2 t\} t^{-2}dt \\
&< 2e^{16\alpha^2} J(x^2, \alpha^2 - y^2, 2) \\
&< 2\pi^{1/2} e^{16\tau^2}(\alpha^2-y^2)^{-1}[1/2 + |x|(\alpha^2-y^2)^{1/2}]^{1/2} \exp[-2|x|(\alpha^2-y^2)^{1/2}] \\
&< 4e^{16\tau^2}(\alpha^2-y^2)^{-1}[1+|x|^{1/2}(\alpha^2-y^2)^{1/4}]\exp[-2|x|(\alpha^2-y^2)^{1/2}].
\end{aligned}
$$

Suppose again that $|y| \leqq \beta$. Then $(\beta^2-y^2)^{1/2} < (\alpha^2-y^2)^{1/2}$ and the absolute maximum of

$$|x|^{1/2}(\alpha^2-y^2)^{1/4}\exp\{-2|x|[(\alpha^2-y^2)^{1/2}-(\beta^2-y^2)^{1/2}]\}$$

for all values of x equals

$$\frac{1}{2}\left\{\frac{\alpha}{e(\alpha-\beta)}\right\}^{1/2}.$$

Consequently

$$H_{11} < 4e^{16\tau^2}(\alpha^2-\beta^2)^{-1}[1+\alpha^{1/2}(\alpha-\beta)^{-1/2}]\exp[-2|x|(\beta^2-y^2)^{1/2}]$$

or

$$H_{11}(x, y; \alpha) < B_2(\beta)\exp[-2|x|(\beta^2-y^2)^{1/2}] \tag{2.14}$$

for $-\beta \leqq y \leqq \beta$ and all values of x, since clearly the restriction $|x| \geqq 4\alpha$ can be removed.

Combining the inequalities in (2.9) to (2.14) we get the desired estimate

$$|f(x+iy)| < B(\beta)\exp[-|x|(\beta^2-y^2)^{1/2}],$$

valid for every z in the strip $-\beta \leqq y \leqq \beta$ and for every $\beta < \tau$. This completes the proof of the necessity of condition (1.7).

3. **Some asymptotic formulas.** The sufficiency proof requires accurate information concerning the behavior of $\boldsymbol{H}_n(z)$ for large values of n when the distance of z from the line segment $(-N, N)$ remains bounded. This question will be studied in this and the next following sections.

The Weber-Hermite differential equation

$$w'' + (2\kappa + 1 - z^2)w = 0, \tag{3.1}$$

where κ is an arbitrary real or complex parameter, has a solution $h_\kappa(z)$ which is uniquely characterized by the asymptotic property

$$\lim e^{z^2/2}(2z)^{-\kappa}h_\kappa(z) = 1 \tag{3.2}$$

when $|z| \to \infty$ and $|\arg z| < 3\pi/4$. Here the power has its principal determination when z and κ are real positive. In the notation of Whittaker and Watson

$$h_\kappa(z) = 2^{\kappa/2}D_\kappa(2^{1/2}z).$$

Since $\boldsymbol{H}_n(z)$ satisfies (3.1) when $\kappa=n$ and

$$\boldsymbol{H}_n(z) = (\pi^{1/2}2^n n!)^{-1/2}e^{-z^2/2}[(2z)^n + \cdots],$$

we conclude that

$$\boldsymbol{H}_n(z) = (\pi^{1/2}2^n n!)^{-1/2}h_n(z). \tag{3.3}$$

It was shown by Whittaker [11] that $h_{-\kappa-1}(iz)$ and $h_{-\kappa-1}(-iz)$ are also solutions of (3.1) and Watson [9, p. 395] proved that

$$h_\kappa(z) = \pi^{-1/2}\Gamma(\kappa + 1)2^\kappa\{e^{\kappa\pi i/2}h_{-\kappa-1}(iz) + e^{-\kappa\pi i/2}h_{-\kappa-1}(- iz)\}. \tag{3.4}$$

The asymptotic behavior of $h_{-\kappa-1}(-iz)$ was studied by Watson on two occasions. The result of his earlier paper [9, p. 416] is not good enough for our purposes, but that of his second paper [10, p. 142] more than meets our needs. Watson's analysis is based upon special representations of the functions involved by means of definite integrals to which he applies the method of steepest descent.

I prefer a different approach to the problem based upon the work of Langer [3] on the asymptotic behavior of solutions of certain classes of second order linear differential equations for large values of a parameter. This is a much more general approach and brings out the reason why different expansions are to be expected in different parts of the plane—the so-called phenomenon of Stokes. N. Schwid [6] has proved that Langer's method applies to the Weber-Hermite equation. Schwid does not discuss $h_{-\kappa-1}(\pm iz)$ explicitly, however, and in that part of the plane where we require information his results are not valid.* It is not necessary to give a detailed presentation of Langer's method. We shall merely state the results and refer to Langer and Schwid for all details.

In the following, κ shall be real positive. We write

$$\mathrm{K} = (2\kappa + 1)^{1/2} \tag{3.5}$$

and†

$$\xi(z) = \int_{\mathrm{K}}^{z} [\mathrm{K}^2 - t^2]^{1/2}dt. \tag{3.6}$$

* His estimate of the order of the remainder term in formula (26), p. 353, is valid for bounded values of z only.

† Strictly speaking we should take $-\mathrm{K}$ as the lower limit of the integral when z is in the left half-plane. Since $\xi(-\mathrm{K})$ is a real quantity, the imaginary part of $\xi(z)$ is not affected by this change and the only place where it matters is in the remainder terms of formulas (3.7), (3.9), and (3.11) below. Here we should write $\xi-\xi(-\mathrm{K})$ instead of ξ in the left half-plane. The correction is of no importance, however, except in a small neighborhood of $z=-\mathrm{K}$ which is excluded from consideration anyway.

The Langer theory shows that for any value of $z \neq \pm K$ there are two linearly independent local solutions of (3.1) which for large values of κ are of the form

$$[\xi'(z)]^{-1/2} \exp [\pm i\xi(z)]\{1 + O(1/\xi) + O(1/\kappa)\} \tag{3.7}$$

and with the aid of which any solution can be expressed linearly. The coefficients will of course depend upon the determination chosen for the infinitely many-valued function $\xi(z)$. In the present case the situation is simplified by the fact that the imaginary part of $\xi(z)$ is only two-valued and the arcs $\mathfrak{I}[\xi(z)]=0$ are independent of the determination of $\xi(z)$. These arcs form a configuration symmetric with respect to the coordinate axes and divide the plane into four sectors S_j. The sector S_2, of main importance to us, is bounded by the line segment $(-K, K)$ and two transcendental curves Γ_1 and Γ_2 going from $\pm K$ to ∞. The angle of inclination of the tangent of Γ_1 with the positive real axis starts with the value $\pi/3$ at $z=K$ and tends to $\pi/4$ as $|z| \to \infty$.

Suppose the determination of $\xi(z)$ to be chosen at some point z_0 in S_2 and let a particular solution $w(z)$ of (3.1) be expressed linearly in terms of the two local solutions (3.7) in some neighborhood of $z=z_0$. Both solutions can be continued analytically in S_2, the continuations agreeing everywhere with the local fundamental system. It follows that the expression of $w(z)$ remains valid at all interior points of S_2. It is only when we cross the boundary of S_2 that the situation may change and a different substitution may have to be used to express $w(z)$ in terms of the local fundamental system. Thus to every sector S_j and to every given solution $w(z)$ there are corresponding coefficients C_{jk}, $j=1, 2, 3, 4$; $k=1, 2$, joining $w(z)$ with the fundamental system in S_j.

After these generalities let us choose $\xi(z)$ so that on the positive imaginary axis we have

$$\begin{aligned} \xi(iy) = & -\pi K^2/4 + (i/2)\{y(K^2 + y^2)^{1/2} \\ & + K^2 \log [y/K + (1 + y^2/K^2)^{1/2}]\}, \end{aligned} \tag{3.8}$$

where the square roots and the logarithm have their principal values. We can then take as the fundamental system in S_2 two solutions $w_{2,k}(z)$ which for large values of κ have the form

$$w_{2,k}(z) = \{1 - z^2/K^2\}^{-1/4} \exp [\pm i\xi(z)]\{1 + O(1/\xi) + O(1/\kappa)\}. \tag{3.9}$$

Here the plus sign goes with $k=1$ and the minus sign with $k=2$ and the fourth root is real positive on the positive imaginary axis. The parameter κ is supposed to be large and this will make $|\xi(z)|$ large if we omit small neighborhoods of the points $z= \pm K$ of radius $O(\kappa^{-1/6})$ inside of which (3.9) is not valid.*

* Compare the second footnote on page 86.

We have consequently

$$h_{-\kappa-1}(-iz) = A(\kappa)w_{2,1}(z) + B(\kappa)w_{2,2}(z) \tag{3.10}$$

in S_2. In order to determine $A(\kappa)$ and $B(\kappa)$ we use (3.2). Multiply both sides of (3.10) by $e^{-z^2/2}(-2iz)^{1+\kappa}$ and let $z\to\infty$, say along the positive imaginary axis. The left-hand side tends to the limit 1, whereas the right-hand side tends to a limit if and only if $B(\kappa)=0$. Setting $B(\kappa)=0$ and using (3.8) and (3.9), we can compute the value of $A(\kappa)$. The result is

$$\begin{aligned} h_{-\kappa-1}(-iz) &= 2^{-1/2}\mathrm{K}^{-\kappa-1}\exp\left[(1/4)\mathrm{K}^2(1+\pi i)\right] \\ &\quad\cdot\left\{1 - z^2/\mathrm{K}^2\right\}^{-1/4}\exp\left[i\xi(z)\right]\left\{1+O(1/\xi)+O(1/\kappa)\right\}, \end{aligned} \tag{3.11}$$

valid in S_2 except in neighborhoods of $z=\pm\mathrm{K}$ of radius $O(\kappa^{-1/6})$. This is the required basic asymptotic formula. It can be shown to be in agreement with Watson's formula quoted above.

4. **Estimates of the imaginary part of** $\xi(z)$. We shall need a couple of lemmas for the sufficiency proof.

LEMMA 1. *Let* $\xi(x)$ *be defined by* (3.6) *and let* $-\mathrm{K}<x<\mathrm{K}$, $0\leqq y\leqq \mathrm{K}$. *Then*

$$\begin{aligned} \Im[\xi(x+iy)] &= y(\mathrm{K}^2-x^2)^{1/2} + (1/6)y^3(\mathrm{K}^2-x^2)^{-1/2} \\ &\quad + \rho\mathrm{K}^2y^3(\mathrm{K}^2-x^2)^{-3/2}, \end{aligned} \tag{4.1}$$

where $|\rho|\leqq 5/24$.

Proof. We have

$$\begin{aligned} \Im[\xi(x+iy)] &= \Im\left\{\int_x^{x+iy}(\mathrm{K}^2-t^2)^{1/2}dt\right\} \\ &= \Re\left\{\int_0^y[\mathrm{K}^2-x^2-2ixv+v^2]^{1/2}dv\right\}, \end{aligned}$$

where the value of the square root is real positive when $v=0$. But by elementary transformations

$$(a^2+b)^{1/2} - a - \frac{b}{2a} = -\frac{b^2}{4a^2}\left[(a^2+b)^{1/2}+a+\frac{b}{2a}\right]^{-1}.$$

Substituting $a^2=\mathrm{K}^2-x^2$, $b=-2ixv+v^2$, we find that the right-hand side becomes equal to P/Q where

$$\begin{aligned} P &= v^2(v-2ix)^2, \\ Q &= 4(\mathrm{K}^2-x^2)\left\{[\mathrm{K}^2-x^2-2ixv+v^2]^{1/2}+(\mathrm{K}^2-x^2)^{1/2}\right. \\ &\quad \left.+(v^2-2ixv)(\mathrm{K}^2-x^2)^{-1/2}\right\}. \end{aligned}$$

Here $0 \leqq v \leqq y \leqq K$. The least value of $|Q|$ is obtained for $v=0$. On the other hand, $|v-2ix|^2 \leqq 5K^2$. Hence $|P/Q|$ does not exceed

$$(5/8)K^2(K^2 - x^2)^{-3/2}v^2,$$

the integral of which from $v=0$ to $v=y$ equals

$$(5/24)K^2(K^2 - x^2)^{-3/2}y^3.$$

This is the remainder term of formula (4.1). The other two terms are the result of substituting $a^2=K^2-x^2$, $b=-2ixv+v^2$ into $a+b/(2a)$, integrating with respect to v from 0 to y and finally taking the real part.

LEMMA 2. *For a fixed x, where $-K<x<K$, the maximum value of*

$$\Im[\xi(x+iy)] + |x|(\beta^2 - y^2)^{1/2}, \tag{4.2}$$

when $0 \leqq y \leqq \beta$, is not less than

$$\beta K + \frac{\beta^3}{6K^3}(K^2 - x^2) - \frac{5}{24}\frac{\beta^3}{K}. \tag{4.3}$$

In particular, this value is reached or exceeded on the ellipse

$$E(K, \beta):\quad \beta^2x^2 + K^2y^2 = \beta^2K^2. \tag{4.4}$$

Proof. By Lemma 1

$$\Im[\xi(x+iy)] \geqq y(K^2 - x^2)^{1/2} + (1/6)y^3(K^2 - x^2)^{-1/2} - (5/24)K^2y^3(K^2 - x^2)^{-3/2}.$$

The maximum value of

$$y(K^2 - x^2)^{1/2} + |x|(\beta^2 - y^2)^{1/2}$$

for a fixed x is reached when $y=(\beta/K)(K^2-x^2)^{1/2}$, that is, on the ellipse $E(K, \beta)$, and equals βK. Substitution of this value of y gives formula (4.3).

5. **The sufficiency proof.** After all these preparations we can attack the sufficiency proof. It is assumed that $f(z)$ is holomorphic in the strip S_τ: $-\tau<y<\tau$, and that to every β, $0<\beta<\tau$, there exists a finite positive $B(\beta)$ for which condition (1.7) holds.* This condition is evidently sufficient to ensure the existence of the Fourier-Hermite coefficients f_n defined by the second half of formula (1.6). Thus the Fourier-Hermite series of $f(z)$ has a sense, and condition (1.7) also ensures that this series is Abel summable to the sum $f(x)$ everywhere on the real axis. In order to complete the proof of the theorem it

* It is of course only values of β close to τ which are of interest, but it is no restriction to assume that $B(\beta)$ satisfies the conventions of §2 for all β, $0 \leqq \beta < \tau$. The same conventions can be imposed on all B-functions introduced below.

is then enough to prove that the series is convergent in the strip S_τ, and this will be proved if we are able to show the existence, for every β, $0<\beta<\tau$, of a finite positive $B^*(\beta)$ such that

$$|f_n| < B^*(\beta)e^{-\beta N}, \qquad N = (2n+1)^{1/2}, \tag{5.1}$$

for all n.

It is known that there exists a constant A such that†

$$|\boldsymbol{H}_n(t)| \leqq A \tag{5.2}$$

for all integers n and all real values of t. This inequality shows that

$$\left\{\int_{-\infty}^{-N+1} + \int_{N-1}^{\infty}\right\} |f(t)|\,|\boldsymbol{H}_n(t)|\,dt \leqq 2AB(\beta)\int_{N-1}^{\infty} e^{-\beta t}dt < B_1(\beta)e^{-\beta N}. \tag{5.3}$$

On the interval $(-N+1, N-1)$ we use formula (3.4) obtaining

$$\int_{-N+1}^{N-1} f(t)\boldsymbol{H}_n(t)dt = \pi^{-3/4}(2^n n!)^{1/2}\left\{i^n\int_{-N+1}^{N-1} f(t)h_{-n-1}(it)dt + (-i)^n\int_{-N+1}^{N+1} f(t)h_{-n-1}(-it)dt\right\}. \tag{5.4}$$

In these integrals we can choose the path of integration joining $-N+1$ with $N-1$ in an arbitrary manner provided it stays in the strip S_τ. Moreover, we can choose a different path for each of the integrals involved. Let $E(N, \beta)$ be the ellipse of Lemma 2 and denote the points of intersection of the ellipse with the lines $x = \pm(N-1)$ by P, Q, R, and S, where Q is in the first and P in the second quadrant.

The path of integration C_2 for the integral involving $h_{-n-1}(-it)$ shall consist of the line segments joining $-N+1$ with P and $N-1$ with Q connected by the arc PQ of the ellipse. The path of integration C_1 for the integral involving $h_{-n-1}(it)$ will be the image of C_2 in the real axis. Let us put

$$J_2 = \int_{C_2} |f(z)|\,|h_{-n-1}(-iz)|\,|dz|. \tag{5.5}$$

Along C_2 we are at a safe distance from the critical points $z = \pm N$ so that formula (3.11) is valid, if we replace κ and K by n and N respectively. It follows that there exists a constant M independent of z and n such that on $E(N, \beta)$ we have

† See E. Hille [1, pp. 435–436] and G. Szegö [7, p. 236].

$$(5.6)\qquad |h_{-n-1}(-iz)| < MA_n|1 - z^2/N^2|^{-1/4}\exp\{-\Im[\xi(z)]\},$$

where

$$(5.7)\qquad A_n = N^{-n-1}e^{n/2} < B(2^n n!)^{-1/2}n^{-1/4},$$

and B is an absolute constant.

On the vertical parts of C_2 it is enough to know that $\Im[\xi(z)]>0$. The length of the vertical parts is $O(n^{-3/4})$ and the factor $|1-z^2/N^2|^{-1/4}$ amounts at most to $O(n^{1/8})$. Their contributions to J_2 are consequently of the form

$$(5.8)\qquad B_2(\beta)[2^n n!]^{-1/2}n^{-7/8}e^{-\beta N}.$$

On the arc PQ of the ellipse, Lemma 2 shows that

$$\int_P^Q |f(z)||h_{-n-1}(z)||dz| < B\,B(\beta)[2^n n!]^{-1/2}n^{-1/4}$$

$$\int_P^Q \exp\{-|x|(\beta^2 - y^2)^{1/2} - \Im[\xi(x+iy)]\}|1 - z^2/N^2|^{-1/4}|dz|$$

$$< B_3(\beta)[2^n n!]^{-1/2}n^{-1/4}e^{-\beta N}\int_P^Q |1 - z^2/N^2|^{-1/4}|dz|.$$

The integral being $O(N)$, this expression is of the form

$$(5.9)\qquad B_4(\beta)[2^n n!]^{-1/2}n^{1/4}e^{-\beta N}.$$

Similar estimates hold on C_1. We have then

$$(5.10)\qquad \left|\int_{-N+1}^{N-1} f(t)\boldsymbol{H}_n(t)dt\right| < B_5(\beta)n^{1/4}e^{-\beta N},$$

and combining (5.3) and (5.10) we get finally

$$(5.11)\qquad |f_n| < B_6(\beta)n^{1/4}e^{-\beta N}.$$

This being true for every $\beta<\tau$, it is clear that the estimate implies the validity of (5.1). This completes the proof of Theorem 1.

6. **Functions holomorphic or meromorphic in a half-plane.** Suppose that $f(z)$ is holomorphic in the half-plane $y>-\alpha$, $\alpha>0$, and that $f(z)$ admits of a formal Fourier-Hermite series

$$(6.1)\qquad f(z) \sim \sum_{n=0}^{\infty} f_n\boldsymbol{H}_n(z), \qquad f_n = \int_{-\infty}^{\infty} f(t)\boldsymbol{H}_n(t)dt.$$

We shall show that if the series has a strip of convergence, then $f(z)$ cannot be of *exponential type* in the half-plane $y\geqq 0$. In other words, the existence of a strip of convergence implies a lower limit on the rate of growth of the function in its supposed half-plane of holomorphism. The precise statement is as follows.

THEOREM 2. *Let $f(z)$ be holomorphic in the half-plane $y>-\alpha$, $\alpha>0$. Further let*

$$\limsup_{r\to\infty} (1/r) \log |f(re^{i\theta})| = h(\theta) \leqq M, \qquad 0 \leqq \theta \leqq \pi. \tag{6.2}$$

*The Fourier-Hermite expansion of $f(z)$ can never converge outside of the real axis.**

Proof. For the proof we need the following

LEMMA 3. *If $f(z)$ is holomorphic in the half-plane $y\geqq 0$ and if* $\log |f(x)| < -\beta|x|$, *where* $\beta>0$, *and* $\log |f(re^{i\theta})| \leqq Mr$ *when* $0\leqq\theta\leqq\pi$, *then* $f(z)\equiv 0$.

This Lemma is implicitly contained in some results of F. Nevanlinna [4, pp. 11–12] and is explicitly stated by F. and R. Nevanlinna [5, p. 38] as a special case of a more general theorem. It is easily proved by forming the Laplace transform of $f(z)$ and rotating the line of integration through an angle of π in the upper half-plane.

This Lemma applies directly to the present situation since the assumption that the Fourier-Hermite series of $f(z)$ has a strip of convergence implies the inequality $\log |f(x)| < -\beta|x|$ for some positive β by virtue of Theorem 1 and this combined with the inequality (6.2) would force $f(z)$ to vanish identically.

We see in particular that *the Fourier-Hermite series of an entire function of order ρ will never converge outside of the real axis unless $\rho>1$ or $\rho=1$ and the function is of the maximal type.* Some further results of F. and R. Nevanlinna [5, pp. 40–42] throw light upon the possible width of the strip of convergence of the Fourier-Hermite series of an entire function of order one and particular maximal type. We state the result for functions meromorphic in a half-plane.

THEOREM 3. *Let $f(z)$ be meromorphic in the half-plane $y>-\alpha$, $\alpha>0$. Let its zeros and poles in the half-plane $y>-\alpha$ be $a_1, a_2, a_3, \cdots$ and $b_1, b_2, b_3, \cdots$ respectively, where in addition $\Im(b_n)\geqq\beta>0$. Put* $\arg a_n=\alpha_n$, $\arg b_n=\beta_n$, *and let*

$$d = \limsup_{r\to\infty} \frac{1}{r}\, [b(r) - a(r)],$$

where

$$a(r) = \sum \sin \alpha_n, \qquad b(r) = \sum \sin \beta_n$$

* The same conclusion is valid if the hypotheses refer to a lower half-plane instead. This remark also applies to Theorem 3 below.

and the summations extend over all zeros and poles respectively of absolute value between β *and* r. *Finally let*

$$q = \limsup_{r\to\infty} [r \log r]^{-1} \int_0^\pi \log | f(re^{i\theta}) | \sin \theta \, d\theta.$$

If $f(z)$ *has a Fourier-Hermite series whose ordinate of convergence equals* τ, *then* $\tau \leqq \min (\alpha, \beta, q + \pi d)$.

In order to connect this theorem with the Nevanlinna theory we have merely to observe that by Theorem 1

$$\limsup_{r\to\infty} \frac{1}{2r} \{\log | f(-r) | + \log | f(r) | \} \leqq -\tau$$

and substitution in the inequality of F. and R. Nevanlinna gives $\tau \leqq q + \pi d$. The other inequalities are self-evident.

Theorem 3 shows that the width of the strip of convergence is affected by the frequency of zeros and poles as well as by the rate of growth of the function in parts of the plane distant from the real axis, since all these factors influence the rate of growth of the function near the real axis. That Theorem 3 is the best of its kind is brought out by the following examples.

We first consider $f(z) = \Gamma(a - iz)$, $a > 0$. By Stirling's formula and Theorem 1 we have $\tau = \min (a, \pi/2)$. Applying Theorem 3 to $\Gamma(a - iz)$ in the half-plane $y > -a$, we find $\beta = +\infty$, $d = 0$, and, again by Stirling's formula, $q = \pi/2$, so that the inequality $\tau \leqq \min (\alpha, \beta, q + \pi d) = \min (a, \pi/2)$ becomes exact. We may of course also apply the theorem to an arbitrary lower half-plane. Then $\beta = a$, $d = 1$ and $q = -\pi/2$, so that the same inequality results.

Our second example is $f(z) = \operatorname{sech} az$, $a > 0$. Here we can take $\alpha = \beta = \pi/(2a)$ and find $\pi d = a$, $q = 0$ and $\tau \leqq \min (a, \pi/(2a))$, which is exact.

References

1. E. Hille, *A class of reciprocal functions*, Annals of Mathematics, (2), vol. 27 (1926), pp. 427–464.

2. ———, *Contributions to the theory of Hermitian series*, Duke Mathematical Journal, vol. 5 (1939), pp. 875–936.

3. R. E. Langer, *On the asymptotic solutions of differential equations, with an application to the Bessel functions of large complex order*, these Transactions, vol. 34 (1932), pp. 447–480.

4. F. Nevanlinna, *Zur Theorie der asymptotischen Potenzreihen*, Annales Academiae Scientiarum Fennicae, (A), vol. 12, no. 3 (1918), vii+81 pp.

5. F. and R. Nevanlinna, *Über die Eigenschaften analytischer Funktionen in der Umgebung einer singulären Stelle oder Linie*, Acta Societatis Scientiarum Fennicae, vol. 50, no. 5 (1922), 46 pp.

6. N. Schwid, *The asymptotic forms of the Hermite and Weber functions*, these Transactions, vol. 37 (1935), pp. 339–362.

7. G. Szegö, *Orthogonal Polynomials*, American Mathematical Society Colloquium Publications, vol. 23, New York, 1939, ix+401 pp.

8. O. Volk, *Über die Entwicklung von Funktionen einer komplexen Veränderlichen nach Funktionen, die einer linearen Differentialgleichung zweiter Ordnung genügen*, Mathematische Annalen, vol. 86 (1922), pp. 296–316.

9. G. N. Watson, *The harmonic functions associated with the parabolic cylinder*, Proceedings of the London Mathematical Society, (2), vol. 8 (1910), pp. 393–421.

10. ———, *The harmonic functions associated with the parabolic cylinder*, Proceedings of the London Mathematical Society, (2), vol. 17 (1918), pp. 116–148.

11. E. T. Whittaker, *On the functions associated with the parabolic cylinder in harmonic analysis*. Proceedings of the London Mathematical Society, vol. 35 (1903), pp. 417–427.

12. E. T. Whittaker and G. N. Watson, *Modern Analysis*, 4th edition, Cambridge, 1927, x+608 pp.

YALE UNIVERSITY,
NEW HAVEN, CONN.

Reprinted from DUKE MATHEMATICAL JOURNAL
Vol. 7, December, 1940

A CLASS OF DIFFERENTIAL OPERATORS OF INFINITE ORDER, I

BY EINAR HILLE

Introduction. The present paper is the first part of an investigation devoted to the theory of differential operators of infinite order[1] of the form

$$G(\delta_z) = \sum_{k=0}^{\infty} g_k \delta_z^k. \tag{1}$$

Here

$$G(w) = \sum_{k=0}^{\infty} g_k w^k$$

is supposed to be an entire function,[2] the order and type of which will be subjected to various restrictions;

$$\delta_z = z^2 - \frac{d^2}{dz^2} \tag{2}$$

is the differential operator of Hermite-Weber; and $\delta_z^k = \delta_z \cdot \delta_z^{k-1}$. Putting

$$h_n(z) = (-1)^n e^{\frac{1}{2}z^2} \frac{d^n}{dz^n} (e^{-z^2}) \equiv e^{-\frac{1}{2}z^2} H_n(z), \tag{3}$$

where $H_n(z)$ is the n-th polynomial of Hermite, we find that

$$\delta_z h_n(z) = (2n + 1)h_n(z). \tag{4}$$

The author has shown the importance of the differential operator $G(\delta_z)$ in the theory of Hermite series (see E. Hille [4]). There only those features of the theory were discussed which were of immediate use for Hermite series. In the present paper and its continuation we shall consider various questions omitted in the earlier discussion.

The basic notion of applicability of a differential operator was given on page 897 of the paper quoted above. Let $G(w)$ be a given entire function and let $\mathfrak{F}$ be a given class of analytic functions $\{f(z)\}$. We say that *the differential operator* $G(\delta_z)$ *applies to or is applicable to the class* $\mathfrak{F}$ *if the series*

$$G(\delta_z) \cdot f(z) = \sum_{k=0}^{\infty} g_k \delta_z^k f(z) \tag{5}$$

Received September 16, 1940; presented to the American Mathematical Society, September 10, 1940.

[1] For a survey of the general field of differential operators of infinite order see R. D. Carmichael [1] and H. T. Davis [2]. The latter has an extensive bibliography. Numbers in brackets refer to the bibliography at the end of this paper.

[2] For the theory of entire functions used in this paper consult the treatise of G. Valiron [9].

converges at every point where $f(z)$ is holomorphic, the sum of the series being a holomorphic function of z in any domain in which $f(z)$ is holomorphic, no matter what element $f(z)$ of $\mathfrak{F}$ we substitute in the series.

It was shown on page 898 of the same paper that if $\mathfrak{F}$ is the class of all analytic functions, a necessary and sufficient condition in order that $G(\delta_z)$ shall apply to $\mathfrak{F}$ is that the entire function $G(w)$ be of order $\sigma \leqq \frac{1}{2}$ and of *minimal type* if $\sigma = \frac{1}{2}$. A new proof of this theorem will be given in §4 below; and in part II of this paper we shall show that the condition is still sufficient if we replace δ_z by an arbitrary second-order differential operator, the coefficients of which are entire functions. The result also extends to differential operators of order n if we replace $\frac{1}{2}$ by n^{-1}.

The present paper is devoted mainly to the applicability of the differential operator $G(\delta_z)$ to classes of entire functions. §§1–3 contain preliminary material such as formal representations of the operators, fundamental estimates of $\delta_z^k f(z)$ for large values of k under different assumptions on $f(z)$, and a brief discussion of the continuity properties of the functional

$$F_\sigma(z; f) = \limsup_{k\to\infty} \left| \frac{\delta_z^k f(z)}{\Gamma(1 + k/\sigma)} \right|^{1/(2k)}, \tag{6}$$

which is analogous to the functional

$$\Phi_\sigma(z; f) = \limsup_{k\to\infty} \left| \frac{f^{(k)}(z)}{\Gamma(1 + k/\sigma)} \right|^{1/k} \tag{7}$$

in the theory of the differential operator $G(d/dz)$. Most of the material in these sections is new either in form or in substance, but there is a certain amount of unavoidable overlapping with §3.3 of the previous paper. Whenever possible we refer the reader to this paper for further details, however.

The main theme occupies §§4–8. Certain results bearing on this problem ([4], Theorems 3.3 and 3.4) have already been announced without proofs. These results are restated in amplified form and proved in the present paper. The fundamental notions are the *order relation*, defining the *conjugate order*, the *conjugate type* and the *critical ρ-order*, the conjugate of which is the *maximal σ-order*. The order relation is

$$\frac{1}{\rho} + \frac{1}{2\sigma} = 1, \tag{8}$$

valid for $\rho \geqq 2$ which is the critical ρ-order with $\sigma = 1$ as the corresponding maximal σ-order. The order relation can be made more precise by the introduction of the conjugate type. Let it be known that[3]

$$\limsup_{r\to\infty} r^{-\rho} \log M(r; f) \leqq \alpha, \qquad \rho > 2, \tag{9}$$

[3] $M(r; f)$ denotes the maximum modulus of $f(z)$ on the circle $|z| = r$. Similar notation is used for other functions of z or w.

and let β be the conjugate type of α determined by the equation

$$(\rho\alpha)^{1/\rho}(2\sigma\beta)^{1/(2\sigma)} = 1, \tag{10}$$

where σ is given by (8); then $G(\delta_z)$ applies to $f(z)$ whenever

$$\limsup_{r\to\infty} r^{-\sigma} \log M(r; G) < \beta, \tag{11}$$

and this limit is the best of its kind. If $\rho = 2$ there is still a conjugate type depending upon α, though we cannot determine its exact value. For $\rho < 2$ there exists a maximal type $\beta(0)$ such that $G(\delta_z)$ applies whenever (11) holds with σ replaced by 1 and β by $\beta(0)$. Again, we can determine fairly narrow limits for $\beta(0)$ but not its exact value. We also investigate how the order and type of the transform $G(\delta_z)\cdot f(z)$ depend upon those of $f(z)$ and $G(w)$. To prove that the various limits obtained are the best possible and to prove the existence of various phenomena, we need a large number of counter examples which are assembled in §8.

In §9 we have joined various loosely connected remarks on $G(\delta_z)$ and related operators. The applicability theory extends to operators which are entire functions in δ_z with coefficients which are polynomials in z of limited degree. The greater part of the section contains general reflections on the outstanding differences between the theories of the operators $G(\delta_z)$ and $G(d/dz)$. The presence of a finite maximal σ-order is such a difference, the irregular behavior of the functional $F_\sigma(z, f)$, defined by (6), is another. The fact that the functions $f(z)$ for which $F_\infty(z, f)$ is bounded are highly specialized compared with the functions for which $\Phi_\infty(z, f)$ is bounded is also noteworthy. This implies that the classical method of solving non-homogeneous differential equations of infinite order by operator series or their equivalents is of very limited interest in case of the equation $G(\delta_z)\cdot W = F(z)$.[4] Finally, we call attention to the fact that the operator $G(\delta_z)$ seems to be oriented in the complex plane, and that the values of $G(w)$ on the fixed sets $\{2n + 1\}$ and $\{-2n - 1\}$ are of fundamental importance for the behavior of the operator. There is nothing corresponding to this situation in the theory of the operator $G(d/dz)$.

In later papers we shall study the equation

$$G(\delta_z)\cdot W = F(z) \tag{12}$$

for given functions $F(z)$ and $G(w)$. We shall also extend parts of the applicability theory to operators of the form $G(D_z)$, where D_z is a given second-order differential operator.

1. **Basic formulas.** In the following $f(z)$ is an analytic function holomorphic within a domain D. We shall obtain various expressions for the operators

[4] A postulational treatment of operational equations which also applies to the operator δ_z is due to F. Schürer [8].

$\delta_z^k f(z)$ with the aid of Cauchy's formulas. For z in D we have obviously

$$\delta_z f(z) = \frac{1}{2\pi i} \oint \frac{z^2(t-z)^2 - 2}{(t-z)^3} f(t)\, dt.$$

Here we can replace z^2 by t^2 in the numerator without changing the value of the integral. As contour of integration we can choose a small circle $|t - z| = R$. By iteration we get

$$\delta_z^k f(z) = (2\pi i)^{-k} \oint \underset{(k)}{\cdots} \oint \left\{ \prod_{\nu=1}^{k} \frac{t_{\nu+1}^2 (t_\nu - t_{\nu+1})^2 - 2}{(t_\nu - t_{\nu+1})^3} \right\} f(t_1)\, dt_1 \cdots dt_k, \tag{1.1}$$

where t_{k+1} is to be replaced by z. The contours of integration can be taken as sufficiently small circles $|t_\nu - t_{\nu+1}| = R_\nu$, the sum of the radii of which is less than the distance from z to the boundary of D. The integrations are to be carried out in the natural order of the subscripts.

This formula is highly condensed and can be used for estimates but does not readily yield the best possible appraisals. In order to get fairly sharp estimates of the behavior of $\delta_z^k f(z)$ as $k \to \infty$ we had better resolve $\delta_z^k f(z)$ into components. The expression on the right side of (1.1) can evidently be written as the sum of the 2^k integrals

$$I_{i_1 \cdots i_k} = (-2)^m (2\pi i)^{-k} \oint \underset{(k)}{\cdots} \oint \left\{ \prod_{\nu=1}^{k} \frac{t_\nu^{2i_\nu}}{(t_\nu - t_{\nu+1})^{3-2i_\nu}} \right\} f(t_1)\, dt_1 \cdots dt_k. \tag{1.2}$$

Here the i_ν's are either zero or one and all combinations are permitted. Further $m = k - \sum i_\nu$.

As a matter of fact, the k-fold integrals in (1.2) can be replaced by m-fold ones if we use an obvious contraction process based upon Cauchy's integral formula. Each of the integrations with respect to a variable t_ν such that $i_\nu = 1$ can be suppressed if appropriate modifications are made in the remaining integrations. The result is an expression of the form

$$\begin{aligned} J_{j_0 j_1 \cdots j_m} &= (-\pi i)^{-m} z^{2j_0} \\ &\cdot \oint \underset{(m)}{\cdots} \oint \frac{s_1^{2j_1} s_2^{2j_2} \cdots s_m^{2j_m}}{(s_1 - s_2)^3 (s_2 - s_3)^3 \cdots (s_m - z)^3} f(s_1)\, ds_1 \cdots ds_m, \end{aligned} \tag{1.3}$$

where $j_0, j_1, \cdots, j_m$ are non-negative integers such that

$$j_0 + j_1 + \cdots + j_m = k - m. \tag{1.4}$$

We have then

$$\delta_z^k f(z) = \sum J_{j_0 j_1 \cdots j_m}. \tag{1.5}$$

Here the summation extends over all integers m $(0 \leqq m \leqq k)$ and all integers $j_0, j_1, \cdots, j_m$ subject to (1.4). The contours of integration are subject to the obvious restrictions.

We have shown ([4], p. 892) that the reduction can be carried still further.

For the details we refer to the quoted passage. The result is an expression of the form

$$
(1.6) \quad \begin{aligned} J_{j_0 j_1 \cdots j_m} &= (2\pi i)^{-\mu-1} \left\{ \prod_{\alpha=1}^{\mu+1} (d_\alpha)! \right\} z^{2j_0} \\ &\cdot \oint \cdots_{(\mu+1)} \oint \frac{u_1^{2\nu_1} u_2^{2\nu_2} \cdots u_\mu^{2\nu_\mu} f(u_1)\, du_1 \cdots du_{\mu+1}}{(u_1 - u_2)^{d_1+1} (u_2 - u_3)^{d_2+1} \cdots (u_{\mu+1} - z)^{d_{\mu+1}+1}}. \end{aligned}
$$

Here it is supposed that the subscripts $j_1, j_2, \cdots, j_m$ vanish except those in the places $i_1, i_2, \cdots, i_\mu$. Further

$$
j_{i_\alpha} = \nu_\alpha, \quad i_0 = 0, \quad i_{\mu+1} = m, \quad 2(i_\alpha - i_{\alpha-1}) = d_\alpha, \quad s_{i_\alpha} = u_\alpha .
$$

We note that

$$
d_1 + d_2 + \cdots + d_\mu + d_{\mu+1} = 2m.
$$

The formula remains valid if all $j_i = 0$. We have then $\mu = 0$, $d_1 = 2m$, $j_0 = k - m$, and a single integral. It is also valid if $i_\mu = m$. We have then $d_{\mu+1} = 0$ and take $(d_{\mu+1})! = 0! = 1$.

For the purpose of making estimates we have found it convenient to resolve $\delta_z^k f(z)$ into its 2^k component integrals. But when it comes to getting representations of differential operators of the type $G(\delta_z)$, it is advantageous to reassemble these components into groups.

Let us return to formula (1.3) and choose contours of integration which may depend upon k and m but not upon the subscripts $j_0, j_1, \cdots, j_m$. Summing all m-fold integrals, we get

$$
(-\pi i)^{-m} \oint \cdots \oint \frac{L_{k-m}(z, t_1, t_2, \cdots, t_m)}{(t_1 - t_2)^3 (t_2 - t_3)^3 \cdots (t_m - z)^3} f(t_1)\, dt_1 \cdots dt_m .
$$

Here

$$
(1.7) \qquad L_\nu(t_0, t_1, \cdots, t_m) = \sum t_0^{2j_0} t_1^{2j_1} \cdots t_m^{2j_m},
$$

where the summation extends over all non-negative integers $j_0, j_1, \cdots, j_m$ such that

$$
(1.8) \qquad j_0 + j_1 + \cdots + j_m = \nu.
$$

We have consequently

$$
(1.9) \quad \delta_z^k f(z) = \sum_{m=0}^{k} (-\pi i)^{-m} \oint \cdots_{(m)} \oint \frac{L_{k-m}(z, t_1, \cdots, t_m)}{(t_1 - t_2)^3 \cdots (t_m - z)^3} f(t_1)\, dt_1 \cdots dt_m ,
$$

where the contours of integration can still be disposed within obvious bounds. The first term in the sum corresponding to $m = 0$ is understood to be

$$
L_k(z) f(z) \equiv z^{2k} f(z).
$$

As an application of formula (1.9), let us compute $\delta_z^k z^n$, where n is a non-negative integer. The integrals are easily worked out and give

$$\delta_z^k z^n = \sum (-1)^m A_{n,k,m} z^{n+2k-4m}, \tag{1.10}$$

where

$$\begin{aligned} A_{n,k,m} = \sum & (n + 2j_1)(n + 2j_1 - 1)(n + 2j_1 + 2j_2 - 2) \\ & \cdot (n + 2j_1 + 2j_2 - 3) \cdots (n + 2j_1 + 2j_2 + \cdots + 2j_m - 2m + 2) \\ & \cdot (n + 2j_1 + 2j_2 + \cdots + 2j_m - 2m + 1). \end{aligned} \tag{1.11}$$

Here the summation extends over all non-negative integers $j_1, j_2, \cdots, j_m$ subject to the condition $0 \leqq j_1 + j_2 + \cdots + j_m \leqq k - m$. In (1.10) the summation extends over all non-negative integers m not exceeding the smaller of the numbers k and $\frac{1}{4}(2k + n)$. Evidently $A_{n,k,m}$ is a positive integer, and taking merely the term corresponding to $j_1 = k - m$, $j_2 = \cdots = j_m = 0$ we get the trivial estimate

$$A_{n,k,m} > \frac{(n + 2k - 2m)!}{(n + 2k - 4m)!} \tag{1.12}$$

which gives some idea of the rate of growth of these coefficients. We see that $\delta_z^k z^n$ is a polynomial of degree $n + 2k$ which reaches its maximum for fixed values of $|z|$ on the lines $y = \pm x$, $z = x + iy$.

On the basis of these formulas we can get expansions of the δ-transforms of an arbitrary analytic function holomorphic at the origin. Let

$$f(z) = \sum_{n=0}^{\infty} a_n z^n, \quad |z| < R; \qquad G(w) = \sum_{k=0}^{\infty} g_k w^k, \quad |w| < \infty. \tag{1.13}$$

Then

$$\delta_z^k f(z) = \sum_{n=0}^{\infty} a_n \sum_m (-1)^m A_{n,k,m} z^{n+2k-4m}, \tag{1.14}$$

$$G(\delta_z) \cdot f(z) = \sum_{k=0}^{\infty} g_k \sum_{n=0}^{\infty} a_n \sum_m (-1)^m A_{n,k,m} z^{n+2k-4m}. \tag{1.15}$$

Here formula (1.14) is easily justified and the double series is absolutely convergent for $|z| < R$. Formula (1.15), on the other hand, is of more problematic nature. If, however, the operator $G(\delta_z)$ is known to apply to all analytic functions, then (1.15) is a valid representation of the transform for $|z| < R$ if the triple series is summed in the order indicated, that is, first with respect to m, then n and finally k. The same conclusion is of course valid if $f(z)$ belongs to some more restricted class $\mathfrak{F}$ of analytic functions and it is known a priori that the operator $G(\delta_z)$ applies to $\mathfrak{F}$.

Formula (1.15) represents $G(\delta_z) \cdot f(z)$ as a double series in the polynomials $\delta_z^k z^n$. The domain of convergence of a polynomial series of course need not be a circle, but we have no idea of what domains are possible in the present case. Sometimes it is possible to rearrange the polynomial series into a power series within the circle $|z| < R$. The following is a particularly important case of which we shall make an application below in §8.6.

Suppose that[5]

$$\omega^{2k} g_k \geqq 0, \qquad \omega = e^{\frac{1}{4}\pi i} \quad \text{or} \quad e^{-\frac{1}{4}\pi i}, \tag{1.16}$$

and that $G(\delta_z)$ is known to apply to a class $\mathfrak{F}$ of the following structure. All functions $f(z)$ of $\mathfrak{F}$ are holomorphic at $z = 0$; and if

$$f(z) = \sum_{n=0}^{\infty} a_n z^n$$

belongs to $\mathfrak{F}$, then

$$f^*(z) = \sum_{n=0}^{\infty} a_n^* z^n$$

also belongs to $\mathfrak{F}$ if $|a_n^*| = |a_n|$ for all n. We can then assert that the triple series in (1.15) is absolutely convergent within the circle of holomorphism of $f(z)$ for every $f(z)$ in $\mathfrak{F}$. Indeed, the series converges for every $f(z)$ in $\mathfrak{F}$ by assumption. The series is known to remain convergent if we replace every a_n by $a_n^* = \omega^{-n} |a_n|$. If in this convergent series we put $z = \omega r$ $(0 < r < R)$ and observe (1.16) and the definition of a_n^*, we obtain a triple series all the terms of which are non-negative. This series is simply the series obtained when we replace every term in (1.15) by its absolute value. Consequently the series (1.15) is absolutely convergent within the circle of holomorphism of $f(z)$ for any $f(z)$ in $\mathfrak{F}$. Such a series can of course be rearranged as a power series in z. We do not insist further on the properties of the series (1.15).

Representations of $G(\delta_z) \cdot f(z)$ of more general usefulness can be obtained directly from formula (1.9). We have

$$G(\delta_z) \cdot f(z) = \sum_{k=0}^{\infty} g_k \sum_{m=0}^{k} (-\pi i)^{-m} \underbrace{\oint \cdots \oint}_{(m)} \frac{L_{k-m}(z, t_1, \cdots, t_m)}{(t_1 - t_2)^3 \cdots (t_m - z)^3} f(t_1)\, dt_1 \cdots dt_m. \tag{1.17}$$

Let us suppose that in this formula all m-fold integrals are taken along the same paths of integration, regardless of the value of k, and that the double series is absolutely convergent in such a manner that summation and integration can be interchanged. These hypotheses will be critically examined in §4 below. Introducing the entire functions

$$G_0(z) = \sum_{k=0}^{\infty} g_k L_k(z) \equiv G(z^2), \tag{1.18}$$

$$G_m(z, t_1, \cdots, t_m) = \sum_{k=0}^{\infty} g_k L_{k-m}(z, t_1, \cdots, t_m), \tag{1.19}$$

[5] For the sake of simplicity we have assumed that (1.16) is valid for all k but all large k would be sufficient for our purposes.

we can write

$$G(\delta_z)\cdot f(z) = G_0(z)\cdot f(z) + \sum_{m=1}^{\infty} (-\pi i)^{-m} \oint \cdots_{(m)} \oint \frac{G_m(z, t_1, \cdots, t_m)}{(t_1 - t_2)^3 \cdots (t_m - z)^3} f(t_1)\, dt_1 \cdots dt_m . \tag{1.20}$$

The right side may have a meaning even if the left has none. In this case we use the symbol $G^*(\delta_z)\cdot f(z)$ and regard $G^*(\delta_z)$ as an extension of $G(\delta_z)$. In all cases considered below, however, formula (1.20) will be found to represent $G(\delta_z)\cdot f(z)$ proper.

2. **Estimates of $\delta_z^k f(z)$.** We shall investigate how the iterates of $\delta_z f(z)$ grow in absolute value with k. Various estimates will be obtained from the different formulas of the preceding section.

We start with formula (1.1). Let us choose as contours of integration the circles

$$| t_\nu - t_{\nu+1} | = \frac{p}{k} \qquad (\nu = 1, 2, \cdots, k;\ t_{k+1} = z),$$

where p is any positive number less than $R(z)$, the radius of holomorphism of $f(z)$ at z. Let $M_z(p)$ denote the maximum modulus of $f(t)$ on the circle $| t - z | = p$ and put $| z | = r$. Then

$$\begin{aligned} | \delta_z^k f(z) | &\leq M_z(p)\left(\frac{k}{p}\right)^{2k} \prod_{\nu=1}^{k} \left\{\left(r + \frac{\nu}{k} p\right)^2 \frac{p^2}{k^2} + 2\right\} \\ &= M_z(p) 2^k \left(\frac{k}{p}\right)^{2k} \prod_{\nu=1}^{k} \left\{1 + \frac{p^2}{2k^2}\left(r + \frac{\nu}{k} p\right)^2\right\} \\ &< M_z(p) 2^k \left(\frac{k}{p}\right)^{2k} \exp\left\{\frac{p^2}{2k^2} \sum_{\nu=1}^{k} \left(r + \frac{\nu}{k} p\right)^2\right\} \end{aligned}$$

and finally

$$| \delta_z^k f(z) | < M_z(p) 2^k \left(\frac{k}{p}\right)^{2k} \cdot \exp\left\{\frac{1}{2}\left(\frac{pr}{k}\right)^2 + \frac{1}{2}\frac{p^3 r}{k^2}(k + 1) + \frac{1}{12}\frac{p^4}{k^3}(k + 1)(2k + 1)\right\}. \tag{2.1}$$

This estimate is not particularly good but is nevertheless quite useful. As an example, suppose that $f(z)$ is an entire function such that

$$\lim_{r\to\infty} r^{-2} \log M(r; f) = 0, \qquad M(r; f) = \max_{0\leq\theta<2\pi} | f(re^{i\theta}) |. \tag{2.2}$$

Then $M_z(p) \leq M(r + p; f)$. Let us choose $p = ak^{\frac{1}{2}}$, where a is independent of k and will be disposed of later. Formula (2.2) implies that the k-th root of

$M(r + ak^{\frac{1}{2}}; f)$ tends to unity as $k \to \infty$. It follows that

$$\limsup_{k\to\infty} k^{-1} | \delta_z^k f(z) |^{1/k} \leqq 2a^{-2} \exp\left(\tfrac{1}{6}a^4\right),$$

and this expression reaches its minimum for $a^4 = 3$. Hence

$$\limsup_{k\to\infty} k^{-1} | \delta_z^k f(z) |^{1/k} \leqq 2\left(\frac{e}{3}\right)^{\frac{1}{2}} \tag{2.3}$$

for any entire function satisfying (2.2). This means that the order ρ of the function is at most 2; and if $\rho = 2$ then the function is of minimal type. We are not able to improve on (2.3), but we are fairly sure that the constant on the right is not the best possible.

In passing we notice the estimate

$$| \delta_z^k f(z) | \leqq M_z(p)\left[(r + p)^2 + 2\left(\frac{k}{p}\right)^2 \right]^k \tag{2.4}$$

which is more favorable than (2.1) for small values of k.

Suppose now that $f(z)$ is an entire function of order $\rho > 2$ and type α, i.e.,

$$\limsup_{r\to\infty} r^{-\rho} \log M(r; f) = \alpha. \tag{2.5}$$

We then use formulas (1.3)–(1.6) to get our estimates.[6] Formula (1.5) expresses $\delta_z^k f(z)$ as the sum of 2^k integrals. We separate these into two groups according as $2m \leqq k$ or $> k$. The integrals of the first group we take in the form given by formula (1.3). We let p be a positive quantity, to be disposed of later, and in the integral for $J_{j_0, j_1, \cdots, j_m}$ we use as contours of integration the circles

$$| s_\nu - s_{\nu+1} | = \frac{p}{m} \qquad (\nu = 1, 2, \cdots, m; s_{m+1} = z).$$

We note that $| s_\nu | \leqq r + p$ for all ν. Hence

$$| J_{j_0, j_1, \cdots, j_m} | \leqq M(r + p; f) 2^m (r + p)^{2(k-m)} \left(\frac{m}{p}\right)^{2m}. \tag{2.6}$$

This estimate is also true for $m = 0$ if we replace the meaningless 0^0 by 1. Let the sum of the terms $J_{j_0, j_1, \cdots, j_m}$ of the first group in which $2m \leqq k$ be denoted by S_k^1. We have then

$$| S_k^1 | \leqq M(r + p; f)(r + p)^{2k} \sum_{2m \leqq k} \binom{k}{m} 2^m \left[\frac{m}{p(r + p)}\right]^{2m},$$

[6] For the subsequent discussion cf. E. Hille [4], pp. 892–894. The estimate of S_k^1 is new and the factor $2e^{\frac{1}{2}}$ in the exponent in (2.10) is better than the factor 4 previously found ([4], formula (3.3.11)), but it is still not the best possible factor. This affects formula (2.16) below adversely.

since there are $\binom{k}{m}$ m-fold integrals. The function $(x/A)^x$ is less than or equal to 1 for $0 < x \leqq A$. Further

$$\sum_{2m \leqq k} \binom{k}{m} 2^m < 2^{\frac{1}{2}k} \sum_0^k \binom{k}{m} = 2^{\frac{3}{2}k}.$$

Hence

$$(2.7) \qquad | S_k^1 | \leqq M(r + p; f)[2^{\frac{3}{4}}(r + p)]^{2k} \max \left\{ 1, \left[\frac{k}{2p(r + p)} \right]^k \right\}.$$

Let us denote the sum of the integrals $J_{j_0, j_1, \cdots, j_m}$ with $2m > k$ by S_k^2. Here we use the reduced form (1.6). Let q be another positive quantity to be disposed of later and choose as contours of integration the circles

$$| u_\alpha - u_{\alpha+1} | = \frac{d_\alpha q}{2m} \qquad (\alpha = 1, 2, \cdots, \mu + 1;\ u_{\mu+2} = z).$$

By the usual methods we get

$$(2.8) \qquad | J_{j_0, j_1, \cdots, j_m} | \leqq M(r + q; f)(r + q)^{2(k-m)} \left(\frac{2m}{q} \right)^{2m} \prod_\alpha \left\{ \frac{(d_\alpha)!}{d_\alpha^{d_\alpha}} \right\}.$$

It was shown ([4], pp. 893–894) that

$$(2.9) \qquad | S_k^2 | \leqq M(r + q; f) 4 \left(\frac{2k}{eq} \right)^{2k} \left\{ k^{\frac{1}{2}} + \sum_{j=1}^{\frac{1}{2}k} \frac{[eq(r + q)]^{2j}}{j^j j!} \right\}.$$

The expression within the braces is dominated by a suitably chosen exponential function, and a simple calculation shows that

$$(2.10) \qquad | S_k^2 | \leqq 8M(r + q; f) k^{\frac{1}{2}} \left(\frac{2k}{eq} \right)^{2k} \exp \{ 2e^{\frac{1}{2}} q(r + q) \}.$$

Combining (2.7) and (2.10) we get

$$(2.11) \qquad \begin{aligned} | \delta_z^k f(z) | \leqq {} & M(r + p; f)[2^{\frac{3}{4}}(r + p)]^{2k} \max \left\{ 1, \left[\frac{k}{2p(r + p)} \right]^k \right\} \\ & + 8M(r + q; f) k^{\frac{1}{2}} \left(\frac{2k}{eq} \right)^{2k} \exp \{ 2e^{\frac{1}{2}} q(r + q) \}, \end{aligned}$$

where p and q are arbitrary positive numbers.

It is clear that the first term in (2.11) cannot be made essentially less than k^k for large values of k, $| z |$ being fixed, no matter how p is chosen. The second term is much more affected by the choice of q.

So far we have made no use of hypothesis (2.5) so that formula (2.11) is valid for any entire function without restrictions on p and q. Let us now use (2.5).

We choose $p = 1$, $q = bk^{1/\rho}$, where b will be disposed of later. For $k \geqq 2(r + 1)$ the first term in (2.11) becomes

$$M(r + 1; f)[2^{\frac{1}{2}}(r + 1)k]^k. \tag{2.12}$$

Let us now define σ by the order relation

$$\frac{1}{\rho} + \frac{1}{2\sigma} = 1. \tag{2.13}$$

Then the second term in (2.11) may be written

$$8M(r + bk^{1/\rho}; f)k^{\frac{1}{2}+k/\sigma}\left(\frac{2}{be}\right)^{2k} \exp\,[2e^{\frac{1}{2}}bk^{1/\rho}(r + bk^{1/\rho})]. \tag{2.14}$$

Since $\sigma < 1$, the expression (2.14) evidently completely dominates (2.12). By (2.5)

$$\limsup_{k\to\infty}\,[M(r + bk^{1/\rho}; f)]^{1/k} \leqq \exp\,[\alpha b^\rho].$$

Consequently

$$\limsup_{k\to\infty}\, k^{-1/\sigma}\,|\,\delta_z^k f(z)\,|^{1/k} \leqq \left(\frac{2}{be}\right)^2 \exp\,[\alpha b^\rho]$$

for every positive b. Minimizing the right side, we get

$$\limsup_{k\to\infty}\, k^{-1/\sigma}\,|\,\delta_z^k f(z)\,|^{1/k} \leqq \left(\frac{2}{e}\right)^{1/\sigma} (\alpha\rho)^{2/\rho}. \tag{2.15}$$

We shall see in §8.1 that this estimate is the best of its kind in a certain sense.

The case $\rho = 2$, $\alpha > 0$, remains. Here the two terms of (2.11) become essentially of the same order of magnitude and the exponential factor in the second term also affects the estimate. We now choose $p = ak^{\frac{1}{2}}$, $q = bk^{\frac{1}{2}}$, where a and b are to be disposed of later. Assuming $2a^2 < 1$, we have

$$|\,\delta_z^k f(z)\,| \leqq M(r + ak^{\frac{1}{2}}; f)2^{\frac{1}{2}k}\left\{1 + \frac{r}{ak^{\frac{1}{2}}}\right\}^k k^k$$
$$+ 8M(r + bk^{\frac{1}{2}}; f)\left(\frac{2}{be}\right)^{2k} \exp\,\{2e^{\frac{1}{2}}bk^{\frac{1}{2}}(r + bk^{\frac{1}{2}})\}\, k^{k+\frac{1}{2}}.$$

Since $(A + B)^\gamma < A^\gamma + B^\gamma$ when $\gamma < 1$, we get

$$k^{-1}|\,\delta_z^k f(z)\,|^{1/k} \leqq [M(r + ak^{\frac{1}{2}}; f)]^{1/k}2^{\frac{1}{2}}\left\{1 + \frac{r}{ak^{\frac{1}{2}}}\right\}$$
$$+ [8M(r + bk^{\frac{1}{2}}; f)]^{1/k}\left(\frac{2}{be}\right)^2 \exp\left\{2e^{\frac{1}{2}}b^2\left(1 + \frac{r}{bk^{\frac{1}{2}}}\right)\right\} k^{1/(2k)}$$
$$\to 2^{\frac{1}{2}} \exp\,(\alpha a^2) + \left(\frac{2}{be}\right)^2 \exp\,[b^2(\alpha + 2e^{\frac{1}{2}})]$$

as $k \to \infty$. This is true for all $a > 0$ so we can let $a \to 0$, and it is also true for all values of $b > 0$ so we can choose $b^2 = (\alpha + 2e^{\frac{1}{2}})^{-1}$ which minimizes. Hence we get

$$\limsup_{k\to\infty} k^{-1} | \delta_z^k f(z) |^{1/k} \leqq \frac{4}{e} \alpha + \frac{8}{e^{\frac{1}{2}}} + 2^{\frac{1}{2}}. \tag{2.16}$$

No claim is made that this estimate is the best possible, but we shall prove in §8.1 that the factor $4/e$ in front of α cannot be replaced by any smaller quantity.

For small values of α formula (2.16) gives far too high an estimate. For such values we get a much better result by following the method used in deriving formula (2.3). This method gives

$$\limsup_{k\to\infty} k^{-1} | \delta_z^k f(z) |^{1/k} \leqq 2a^{-2} \exp [\alpha a^2 + \tfrac{1}{6} a^4] \tag{2.17}$$

for all real values of a. The minimum of the right side is reached for

$$a^2 = -\tfrac{3}{2}\alpha + (\tfrac{9}{4}\alpha^2 + 3)^{\frac{1}{2}}. \tag{2.18}$$

For small values of α this estimate gives a limit of the form

$$2\left(\frac{e}{3}\right)^{\frac{1}{2}} \{1 + 3^{\frac{1}{2}}\alpha + O(\alpha^2)\}. \tag{2.19}$$

For large values of α it gives

$$2e\alpha + O\left(\frac{1}{\alpha^2}\right). \tag{2.20}$$

Here (2.19) is much better than (2.16), while (2.20) is not so good as (2.16). We do not insist on further refinements of these estimates. The results obtained so far can be summarized as follows. We recall that statements regarding "best possible" estimates will be proved in §8.1.

THEOREM 2.1. *If $f(z)$ is an entire function of order ρ and type α, then*

(i) $$\limsup_{k\to\infty} k^{-1/\sigma} | \delta_z^k f(z) |^{1/k} \leqq \left(\frac{2}{e}\right)^{1/\sigma} (\alpha\rho)^{2/\rho}$$

when $\rho > 2$, where the conjugate exponent σ is determined by the order relation (2.13). This estimate is the best of its kind.

(ii) $$\limsup_{k\to\infty} k^{-1} | \delta_z^k f(z) |^{1/k} \leqq \min \left\{\frac{4}{e} \alpha + \frac{8}{e^{\frac{1}{2}}} + 2^{\frac{1}{2}}, 2a^{-2} \exp [\alpha a^2 + \tfrac{1}{6} a^4]\right\}$$

when $\rho = 2$, where a^2 is defined by (2.18). For large values of α the factor $4/e$ cannot be replaced by any smaller number.

(iii) $$\limsup_{k\to\infty} k^{-1} | \delta_z^k f(z) |^{1/k} \leqq 2\left(\frac{e}{3}\right)^{\frac{1}{2}}$$

when $\rho < 2$ or $\rho = 2$ and $\alpha = 0$.

In the discussion above we have restricted ourselves to entire functions of normal type or minimal type. More precise information could be obtained by the introduction of the so-called *proximate orders* of Boutroux and Lindelöf in both these cases and also in the case of functions of maximal type.[7] We leave such extensions to the interested reader.

3. **On a class of functionals.** The results obtained in the preceding section can be formulated in a slightly different manner which is of some interest. Let $f(z)$ be an analytic function holomorphic in some domain D. Let z be a fixed point of D, λ a real number $0 < \lambda \leqq 1$, and form

$$(3.1) \qquad F_\lambda(z;f) \equiv \limsup_{k\to\infty}\left[\frac{|\delta_z^k f(z)|}{\Gamma\left(1+\dfrac{k}{\lambda}\right)}\right]^{1/(2k)},$$

where we admit $+\infty$ as a possible value. This defines a non-negative function of z in D. *As a function of z it is not necessarily continuous in D even if it is bounded.* An example showing this will be found in §8.2.

But we can also regard $F_\lambda(z;f)$ for fixed z in D as a *functional* of the second argument defined for all functions holomorphic at the point in question. As such it is evidently *non-linear*, but it is *quasi-additive*

$$(3.2) \qquad F_\lambda(z;f_1+f_2) \leqq F_\lambda(z;f_1) + F_\lambda(z;f_2),$$

and for every constant $C \neq 0$

$$(3.3) \qquad F_\lambda(z;Cf) = F_\lambda(z;f).$$

These functionals are closely related to various types of *grades*[8] which have been considered in the theory of linear operations.

Now let $\mathfrak{F}_{\rho,\alpha}$ denote the class of all entire functions $f(z)$ such that

$$(3.4) \qquad \limsup_{r\to\infty} r^{-\rho}\log M(r;f) \leqq \alpha.$$

We can then reformulate Theorem 2.1 as follows:

THEOREM 3.1. *The functional $F_\lambda(z;f)$ has the following properties when $f(z)$ ranges over $\mathfrak{F}_{\rho,\alpha}$. $F_\lambda(z;f) \equiv 0$ for $\lambda < \sigma$ when $\rho > 2$, and for $\lambda < 1$ when $\rho \leqq 2$. For $\rho > 2$*

[7] See G. Valiron [9], §III, 6.

[8] This is the term used by H. T. Davis ([2], Chapter V, *Grades defined by special operators*) for the superior limit of the k-th root of the absolute value of the transform of the k-th power of the operator. This corresponds to the case $\sigma = \infty$ in (3.1) which seems to be of limited interest to us. I. M. Sheffer used the term exponential value; the German term "Stufe" was introduced by O. Perron. The notion itself in one form or another goes back to C. Bourlet and S. Pincherle. If the grade is infinite, H. T. Davis discusses how fast the sequence in question tends to infinity under different assumptions on the function. The introduction of the Gamma function in the functional puts the investigation on a systematic basis in our case. More general grades have been considered by P. Flamant [3].

$$F_\sigma(z; f) \leqq (\alpha\rho)^{1/\rho}(2\sigma)^{1/(2\sigma)}, \tag{3.5}$$

where σ is the conjugate order of ρ, and no better estimate is valid for the class $\mathfrak{F}_{\rho,\alpha}$. $F_1(z; f)$ is uniformly bounded on $\mathfrak{F}_{\rho,\alpha}$ when $\rho \leqq 2$, and bounds can be read off from (ii) *and* (iii) *of Theorem* 2.1. *Finally, $F_\lambda(z; f)$ is unbounded on $\mathfrak{F}_{\rho,\alpha}$ for $\lambda > \sigma$ when $\rho > 2$, and for $\lambda > 1$ when $\rho \leqq 2$.*

Let us now investigate the continuity properties of $F_\lambda(z; f)$ as a functional of $f(z)$ on $\mathfrak{F}_{\rho,\alpha}$. We shall say that $f_n(z)$ converges to $f(z)$ in $\mathfrak{F}_{\rho,\alpha}$ if $f_n(z) \in \mathfrak{F}_{\rho,\alpha}$ for all n, $f(z) \in \mathfrak{F}_{\rho,\alpha}$, and $f_n(z)$ converges uniformly to $f(z)$ in every fixed circle $|z| \leqq R$ of the z-plane. We then say as usual that $F_\lambda(z; f)$ is continuous at $f = f_0$, if $f_n \to f_0$ implies $F_\lambda(z; f_n) \to F_\lambda(z; f_0)$. Since $F_\lambda(z; f)$ is a non-negative quasi-additive functional which vanishes for $f = 0$, continuity anywhere implies continuity everywhere. Starting with this remark, we shall prove

THEOREM 3.2. *$F_\lambda(z; f)$ is not continuous anywhere on $\mathfrak{F}_{\rho,\alpha}$ for $\lambda = \sigma$ when $\rho > 2$ and for $\lambda = 1$ when $\rho \leqq 2$.*

For the proof it is enough to exhibit a sequence f_n converging to zero in $\mathfrak{F}_{\rho,\alpha}$, such that $F_\lambda(z; f_n)$ does not converge to zero at least for some values of z, where λ is σ or 1. The existence of such functions is an immediate consequence of formula (3.3). We have merely to choose a function $f(z)$ such that $F_\lambda(z; f) \not\equiv 0$ for $\lambda = \sigma$ or 1 respectively and then take $f_n(z) = C_n f(z)$, where $C_n \to 0$. The possibility of finding such a function $f(z)$ will be established in §8.1.

We have proved elsewhere and it will be proved again in §4 that $F_{\frac{1}{2}}(z; f)$ is definable over the class of all analytic functions. More precisely, it was shown that

$$F_{\frac{1}{2}}(z; f) \equiv \limsup_{k\to\infty} \left[\frac{|\delta_z^k f(z)|}{(2k)!}\right]^{1/(2k)} \leqq \frac{1}{R(z)}, \tag{3.6}$$

where $R(z)$ is the distance from z to the nearest singular point of $f(z)$, and that moreover this inequality is the best of its kind.[9] Judging from the analogy with the functional

$$\Phi_1(z; f) \equiv \limsup_{k\to\infty} \left[\frac{|f^{(k)}(z)|}{k!}\right]^{1/k} = \frac{1}{R(z)}, \tag{3.7}$$

one would imagine that the sign of equality should always hold in (3.6) for every f and every point z. Unfortunately this is not the case and it is even obvious that it could not be so.[10] Indeed, if $f^{(2n)}(0) = 0$ for all n, then $F_{\frac{1}{2}}(0; f) = 0$ regardless of the value of $R(0)$. Moreover, *$F_{\frac{1}{2}}(z; f)$ can be a discontinuous function of z inside the domain of holomorphism of $f(z)$ and it is an everywhere discontinuous functional over the class of analytic functions $f(z)$.* An example to prove the first point will be found in §8.2. The second statement is proved as Theorem 3.2.

[9] See [4], formula (3.3.13) and Theorem 3.1. Our present formula (3.6) gives a more pregnant formulation of the result. That the inequality is the best possible also follows from the example in §8.2, especially formula (8.2.3).

[10] This does not exclude the possibility that equality may hold almost everywhere, for instance, or that the origin is the only exceptional point.

4. **Entire functions of δ_z.** Let

$$G(w) = \sum_{k=0}^{\infty} g_k w^k$$

be an entire function of w, the order of which will be restricted below. We proceed now to a systematic study of the differential operator $G(\delta_z)$. We recall the definition of applicability given in the introduction: *$G(\delta_z)$ applies to the class $\mathfrak{F}$ of analytic functions $f(z)$ if the series*

$$G(\delta_z)\cdot f(z) = \sum_{k=0}^{\infty} g_k \delta_z^k f(z) \tag{4.1}$$

converges to a holomorphic function at every point where $f(z)$ is holomorphic, regardless of what function $f(z) \in \mathfrak{F}$ we take.

In §1 formula (1.20) was given as a formal representation of the transform

$$\begin{aligned} G(\delta_z)\cdot f(z) &= G_0(z)\cdot f(z) \\ &+ \sum_{m=1}^{\infty} (-\pi i)^{-m} \oint \cdots \oint \frac{G_m(z, t_1, t_2, \cdots, t_m)}{(t_1 - t_2)^3 \cdots (t_m - z)^3} f(t_1)\, dt_1 \cdots dt_m, \end{aligned} \tag{4.2}$$

where the contours of integration in the m-th term are circles $|t_\nu - t_{\nu+1}| = R_{m,\nu}$ such that $\sum_{\nu=1}^{m} R_{m,\nu} < R(z)$, the radius of holomorphism of $f(z)$ at z. The entire functions $G_m(z, t_1, \cdots, t_m)$ are defined by (1.7), (1.18), and (1.19). Now formula (4.2) is obtained by substituting the expression for $\delta_z^k f(z)$ furnished by formula (1.9) into (4.1) and rearranging terms. This is evidently permitted if (4.2) remains convergent when every quantity is replaced by its absolute value and, in addition, $|G_m(z, t_1, \cdots, t_m)|$ by $\sum |g_k| |L_{k-m}(z, t_1, \cdots, t_m)|$. All our convergence proofs will involve such replacements so we can rest assured that if the resulting series are convergent, they do give valid representations of the transform and the operator $G(\delta_z)$ does apply to the class of functions under consideration.

We shall begin the discussion by an investigation of the entire functions $G_m(z, t_1, \cdots, t_m)$. We shall suppose that

$$\limsup_{r\to\infty} r^{-\sigma} \log \max_{0\leq\theta<2\pi} |G(re^{i\theta})| \leq \beta, \tag{4.3}$$

restricting ourselves for reasons which will become apparent below to the case in which $0 < \sigma \leq 1$.[11] Let $\mathfrak{G}_{\sigma,\beta}$ denote the class of all such functions $G(w)$.[12]

[11] The case $\sigma = 0$ would require the introduction of proximate orders or some such device and is excluded for convenience. Only the range $\frac{1}{2} \leq \sigma \leq 1$ is of interest below.

[12] The reader should observe that the classes $\mathfrak{F}_{\rho,\alpha}$ and $\mathfrak{G}_{\sigma,\beta}$ become identical if $\rho = \sigma$, $\alpha = \beta$, $z = w$. It is convenient, however, to distinguish between a *space of operands* and a *space of operators* and our notation is chosen accordingly. Thus the letters f, z, $\mathfrak{F}$, ρ, α always refer to the operand space and G, w, $\mathfrak{G}$, σ, β always to the operator space.

LEMMA 4.1. *If* $G(w) \in \mathfrak{G}_{\sigma,\beta}$ *and* $|z|, |t_1|, \cdots, |t_m| \leqq u$, *then for every* $\epsilon > 0$

$$(4.4) \qquad |G_m(z, t_1, \cdots, t_m)| \leqq C[\sigma e(\beta + \epsilon)/m]^{m/\sigma} \exp[(\beta + \epsilon)u^{2\sigma}],$$

where C *depends upon* β, ϵ, *and* σ, *but not upon* m *or* u.

Proof. By assumption[13]

$$|g_k| \leqq C(\epsilon) \frac{(\beta + \epsilon)^{k/\sigma}}{\Gamma(1 + k/\sigma)}.$$

Since

$$|L_{k-m}(z, t_1, \cdots, t_m)| \leqq \binom{k}{m} u^{2(k-m)},$$

we have

$$(4.5) \qquad \begin{aligned} |G_m(z, t_1, \cdots, t_m)| &\leqq C(\epsilon) \sum_{k=m}^{\infty} \binom{k}{m} \frac{(\beta + \epsilon)^{k/\sigma}}{\Gamma(1 + k/\sigma)} u^{2(k-m)} \\ &= \frac{1}{m!} C(\epsilon)(\beta + \epsilon)^{m/\sigma} E_{1/\sigma}^{(m)}((\beta + \epsilon)^{1/\sigma} u^2), \end{aligned}$$

where

$$E_\alpha(z) = \sum_{k=0}^{\infty} \frac{z^k}{\Gamma(1 + \alpha k)}$$

is the Mittag-Leffler E_α-function. For large positive x it is known that

$$(4.6) \qquad E_{1/\sigma}(x) = \sigma \exp(x^\sigma) + O\left(\frac{1}{x}\right).$$

We can consequently find a quantity $B(\sigma)$ which is bounded for $0 < \epsilon \leqq \sigma \leqq 1$, such that for all $x \geqq 0$

$$E_{1/\sigma}(x) \leqq B(\sigma) \exp(x^\sigma).$$

By Cauchy's formula

$$\frac{1}{m!} E_{1/\sigma}^{(m)}(x) = \frac{1}{2\pi i} \int_{|z-x|=R} \frac{E_{1/\sigma}(z)\, dz}{(z - x)^{m+1}},$$

the right side of which does not exceed

$$R^{-m} E_{1/\sigma}(x + R) \leqq B(\sigma) R^{-m} \exp[(x + R)^\sigma] < B(\sigma) \exp(x^\sigma) R^{-m} \exp(R^\sigma)$$

since $\sigma \leqq 1$. The least value of the last expression is obtained for $R^\sigma = m/\sigma$.

[13] See, for instance, G. Valiron [9], p. 41. The constant $C(\epsilon)$ depends upon $G(w)$ and the choice of ϵ. Similarly, with other constants below. If it is of some importance that the constants are independent of the functions, attention will be called to the fact.

Hence

$$\frac{1}{m!}E_{1/\sigma}^{(m)}(x) < B(\sigma)\left(\frac{\sigma e}{m}\right)^{m/\sigma} \exp(x^\sigma).$$

If we combine this estimate with (4.5), formula (4.4) results and Lemma 4.1 is proved.

Formula (4.4) is the basic inequality in most of the subsequent discussion. It does not lead to quite as sharp estimates as those based on §2. This is in the main because of our replacing all variables $z, t_1, \cdots, t_m$ by the absolute value of the largest among them. In our applications it is known that $|t_j| \leqq |z| + (m-j)q(m)$, where $q(m) \to 0$ and $mq(m) \to \infty$ as $m \to \infty$. We consequently get $u = |z| + mq(m)$ which, when substituted into (4.4), leads to an unnecessarily high estimate. The excess becomes appreciable only when the orders of $f(z)$ and $G(w)$ are conjugate in the sense defined in the introduction. Since this case can be adequately handled by the methods of §2, we shall not attempt to improve upon (4.4).

We are now ready to start with applicability questions. When $\mathfrak{F}$ is the class of all analytic functions, the following theorem has already been stated and proved by the author.[14]

THEOREM 4.1. *A necessary and sufficient condition that $G(\delta_z)$ shall be applicable to the class of all analytic functions is that $G(w) \in \mathfrak{G}_{\frac{1}{2},0}$.*

It is easy to give a proof of the sufficiency of this condition on the basis of Lemma 4.1. Let $f(z)$ be an analytic function holomorphic at the point z_0, where the radius of holomorphism is to be $R(z_0)$. In the m-fold integral of formula (4.2) we choose as contours of integration the circles $|t_\nu - t_{\nu+1}| = p/m$, where p is any number less than $R(z_0)$. We take $z = z_0$, put $r = |z_0|$, and choose $u = r + p$, $\beta = 0$, and $\sigma = \frac{1}{2}$ in Lemma 4.1. We then obtain the following majorant of the series (4.2):

$$|G(z_0^2)||f(z_0)| + C(\epsilon)M(p)\exp[\epsilon(r+p)]\sum_{m=1}^{\infty}\frac{1}{2^m}\left(\frac{\epsilon e}{p}\right)^{2m}, \tag{4.7}$$

where $M(p)$ is the maximum modulus of $f(z)$ on the circle $|z - z_0| = p$. Since ϵ is at our disposal and can be chosen less than $2^{\frac{1}{2}}p/e$ a priori, we see that the majorant series is convergent. Moreover, the convergence is evidently uniform if z_0 is restricted to a bounded domain in which $f(z)$ is holomorphic and $R(z_0)$ has a positive infimum. Hence the transform is holomorphic at every finite point where $f(z)$ is holomorphic. This completes the proof of the sufficiency of the condition. For the necessity we refer to the passage quoted in footnote 14. The argument just given evidently also proves the following result.[15]

[14] [4], Theorem 3.2. Our previous proof was based upon a formula of type (2.11).

[15] This theorem can be generalized. Assuming the distance of D_2 to the complement of D_1 to be $\beta + \eta$, we can allow $G(w) \in \mathfrak{G}_{\frac{1}{2},\beta}$, the conclusion being unchanged. The proof

THEOREM 4.2. *Let $D_1 \supset D_2$ be bounded domains in the complex plane such that D_2 has a positive distance η from the complement of D_1. Let $\mathfrak{F}(D_1) \subset \mathfrak{F}(D_2)$ be the classes of all functions holomorphic and bounded in D_1 and D_2 respectively. Let $G(w) \in \mathfrak{G}_{\frac{1}{2},0}$. Then $G(\delta_z)$ defines a linear, bounded and consequently continuous transformation on $\mathfrak{F}(D_1)$ to $\mathfrak{F}(D_2)$.*

Indeed, choosing $p = \eta$ and $\epsilon = \eta/e$ in (4.7) we get

$$(4.8) \qquad \max_{z \in D_2} | G(\delta_z) \cdot f(z) | \leqq \max_{z \in D_2} \{| G(z^2) | + C_1(\eta) \exp [\eta/e | z |]\} \cdot \max_{z \in D_1} | f(z) |.$$

This inequality shows that we are dealing with a bounded transformation of one normed linear vector space upon another.

$G(\delta_z) \cdot f(z)$ is a bilinear transformation involving two function spaces $\mathfrak{F}$ and $\mathfrak{G}$. That it is continuous on $\mathfrak{F}$ when $G(w)$ is fixed in $\mathfrak{G}_{\frac{1}{2},0}$ is expressed by Theorem 4.2. We have also continuity on $\mathfrak{G}_{\frac{1}{2},0}$ for fixed $f(z)$ in $\mathfrak{F}$. This does not follow directly from (4.8), and we shall prove continuity only for a particular type of convergence which will be referred to as *dominated convergence in* $\mathfrak{G}_{\frac{1}{2},0}$. We shall prove

THEOREM 4.3. *Let $\{\Gamma_n(w)\}$ be a sequence of functions in $\mathfrak{G}_{\frac{1}{2},0}$ which converges to a function $G(w)$ and is such that there exists a fixed function $\Gamma_0(w)$ in $\mathfrak{G}_{\frac{1}{2},0}$ with $M(r; \Gamma_n) \leqq M(r; \Gamma_0)$ for all n and r. Let $f(z)$ be holomorphic in a domain D. Then $\Gamma_n(\delta_z) \cdot f(z)$ converges to $G(\delta_z) \cdot f(z)$ in D and the convergence is uniform in any domain $D_0 \subset D$ which is bounded and has a positive distance from the complement of D.*

The dominated convergence of $\Gamma_n(w)$ to $G(w)$ evidently implies uniform convergence in any finite domain and also implies that $G(w) \in \mathfrak{G}_{\frac{1}{2},0}$.[16] Hence we have also

$$\lim_{n \to \infty} \Gamma_{n,m}(z, t_1, \cdots, t_m) = G_m(z, t_1, \cdots, t_m)$$

for every fixed m, and the convergence is uniform when the variables are bounded. Further, the existence of a common dominant of the sequence of maximal moduli implies that the estimates of Lemma 4.1 hold uniformly with respect to n for

cannot be based upon formula (4.2) and Lemma 4.1, which would give a weaker result, but requires the more powerful machinery of §2. Compare G. Pólya [6], p. 600, for the operator $G(d/dz)$. Pólya assumes that D_2 has no points in common with the point-set obtained by adding the conjugate indicator diagram of $G(w)$, in his case a function of exponential type, to the complement of D_1. I do not know if the indicator diagram can be worked into the theory of the operator $G(\delta_z)$.

[16] The first statement follows from Vitali's theorem.—Uniform convergence relative to the function $M(r; \Gamma_0)$ in the sense of E. H. Moore implies but is not implied by dominated convergence. Relatively uniform convergence with respect to an "étalonnage" has been used by P. Flamant [3] in operator theory, but dominated convergence is possibly new in this connection. Cf., however, J. F. Ritt [7], p. 30 et seq.—Theorem 4.3 extends to operators in $\mathfrak{G}_{\sigma,\beta}$ ($\frac{1}{2} < \sigma \leqq 1$) and entire functions $f(z)$.

all the functions $\Gamma_{n,m}(z, t_1, \cdots, t_m)$. It follows that for $z \in D_0$, $p = 2^{-\frac{1}{2}}\eta$, $\epsilon = \frac{1}{2}\eta e^{-1}$ we have

$$\Bigg| \Gamma_n(\delta_z)\cdot f(z) - \Gamma_n(z^2)\cdot f(z) - \sum_{m=1}^{N} (-\pi i)^{-m}$$

$$\cdot \oint \cdots \oint \frac{\Gamma_{n,m}(z, t_1, \cdots, t_m)}{(t_1 - t_2)^3 \cdots (t_m - z)^3} f(t_1)\, dt_1 \cdots dt_m \Bigg|$$

$$\leqq 4^{-N} C_2(\eta) \exp\left(\frac{\eta}{2e}\,|\, z \,|\right) \max_{z \in D_1} |\, f(z) \,|,$$

where D_1 is the subset of D the points of which have a distance $\leqq 2^{-\frac{1}{2}}\eta$ from some point of D_0. This implies that

$$\limsup_{n\to\infty} |\, \Gamma_n(\delta_z)\cdot f(z) - G(\delta_z)\cdot f(z) \,| \leqq 2\cdot 4^{-N} C_2(\eta) \exp\left(\tfrac{1}{2}\eta e^{-1} |\, z \,|\right) \max_{z \in D_1} |\, f(z) \,|$$

for every N, whence the theorem follows.

THEOREM 4.4. *A necessary and sufficient condition that* $G(\delta_z)$ *shall apply to the class of all entire functions is that* $G(w) \in \mathfrak{G}_{\frac{1}{2},\beta}$ *for some finite* β.[17]

The necessity will follow from the example in §8.3. For the sufficiency proof, assume that (4.3) holds with $\sigma = \frac{1}{2}$ and a finite β. In formula (4.2) we choose as contours of integration in the m-fold integral the circles $|\, t_\nu - t_{\nu+1} \,| = q(m)$, where $q(m)$ is to be chosen suitably. Taking $u = r + mq(m)$, $\sigma = \frac{1}{2}$ in Lemma 4.1 we obtain the following dominant of the series (4.2)

$$|\, G(z^2) \,||\, f(z) \,| + C(\epsilon) \sum_{m=1}^{\infty} \left[\frac{e(\beta + \epsilon)}{2^{\frac{1}{2}} mq(m)}\right]^{2m} \exp\left[(\beta + \epsilon)(r + mq(m))\right] M(r + mq(m); f),$$

where $r = |\, z \,|$ and $M(r; f) = \max |\, f(z) \,|$ when $|\, z \,| = r$.

No matter how fast $M(r; f)$ tends to infinity with r, we can choose $q(m)$ subject to the following conditions:

(1) $q(m) \to 0$ as $m \to \infty$,

(2) $mq(m) \to \infty$,

(3) $\limsup_{m\to\infty} [M(r + mq(m); f)]^{1/m} \leqq 2$.

We see then that the series (4.8) is uniformly convergent in any fixed circle $|\, z \,| \leqq R$, since the m-th root of the m-th term tends to zero. It follows that $G(\delta_z)\cdot f(z)$ exists and is an entire function. This completes the proof of the sufficiency of the condition of the theorem.

5. The order relation, conjugate orders and types. The classes $\mathfrak{F}_{\rho,\alpha}$ and $\mathfrak{G}_{\sigma,\beta}$ are characterized by the inequalities

$$\text{(5.1)} \qquad \limsup_{r\to\infty} r^{-\rho} \log \max_{0 \leqq \theta < 2\pi} |\, f(re^{i\theta}) \,| \leqq \alpha,$$

[17] This theorem was announced in [4], footnote 10, p. 899.

(5.2) $$\limsup_{r\to\infty} r^{-\sigma} \log \max_{0\leqq\theta<2\pi} | G(re^{i\theta}) | \leqq \beta.$$

We say that ρ and σ are *conjugate orders* if

(5.3) $$\sigma = 1, \qquad \rho < 2,$$

(5.4) $$\frac{1}{\rho} + \frac{1}{2\sigma} = 1, \qquad \rho \geqq 2.$$

We call (5.4) the *order relation.*[18] The value $\rho = 2$ below which the order relation ceases to hold is the *critical ρ-order*, and the corresponding value $\sigma = 1$ is the *maximal σ-order*.

Suppose that $\rho > 2$ and σ is the conjugate order of ρ. We then say that α and β are *conjugate types* if

(5.5) $$(\alpha\rho)^{1/\rho}(2\beta\sigma)^{1/(2\sigma)} = 1.$$

The importance of these notions will become evident in the following. We shall begin by proving

THEOREM 5.1. *A necessary and sufficient condition that $G(\delta_z)$ shall be applicable to all classes $\mathfrak{F}_{\rho,\alpha}$ with ρ fixed $\geqq 2$ $(0 \leqq \alpha < \infty)$ is that $G(w) \in \mathfrak{G}_{\sigma,0}$, where σ is the conjugate order of ρ. If $\rho < 2$, the condition is merely sufficient.*[19]

At this point we shall prove only the sufficiency of the condition. The necessity will follow from the examples in §8.4. Let us assume then that $\rho \geqq 2$, σ is the conjugate of ρ, and $G(w) \in \mathfrak{G}_{\sigma,0}$. We use formula (1.20) = (4.2), choosing as contours of integration in the m-fold integral the circles

(5.6) $$| t_\nu - t_{\nu+1} | = m^{-1/(2\sigma)} \qquad (\nu = 1, 2, \cdots, m; t_{m+1} = z).$$

We can then substitute $\beta = 0$,

(5.7) $$u = r + m^{1/\rho}, \qquad | z | = r,$$

in Lemma 4.1. It follows that the series for $G(\delta_z)\cdot f(z)$ is dominated by the series

$$|G(z^2)||f(z)| + C(\epsilon, \sigma) \sum_{m=1}^{\infty} 2^m \left(\frac{\epsilon\sigma e}{m}\right)^{m/\sigma} \cdot m^{m/\sigma} \cdot \exp\,[\epsilon(r + m^{1/\rho})^{2\sigma}] M(r + m^{1/\rho}; f).$$

By assumption (5.1) holds with a finite α. We can then find finite quantities A and B such that

$$M(r; f) < B \exp\,(Ar^\rho)$$

[18] The terminology is mine. Cf. H. Muggli, [5], p. 152, for the corresponding relation in the case of the operator $G(d/dz)$. The basic facts appear to be due to G. Valiron [10], pp. 52–53. See also formula (9.3) below.

[19] If we demand instead that $G(\delta_z)$ shall apply to all entire functions of order $\leqq \rho$, including functions of maximal type of order ρ, then it is necessary and sufficient that the order of $G(w)$ be $< \sigma$. See H. Muggli, [5], p. 152, for $G(d/dz)$.

for $r \geqq 1$. The series is consequently dominated by

$$|G(z^2)||f(z)| + C(\beta, \epsilon, \sigma) \sum_{m=1}^{\infty} 2^m (\epsilon\sigma e)^{m/\sigma} \tag{5.8}$$
$$\cdot \exp \{\epsilon(r + m^{1/\rho})^{2\sigma} + A(r + m^{1/\rho})^{\rho}\}.$$

This series converges for every finite value of z because the m-th root of the m-th term has a limit superior less than or equal to

$$2(\epsilon\sigma e)^{1/\sigma} e^A,$$

and this quantity can be made as small as we please since ϵ is at our disposal. If $\rho = 2$ we have to replace A in the exponent by $A + \epsilon$, but the conclusion is the same. It follows that $G(\delta_z) \cdot f(z)$ exists as an entire function.

If $\rho < 2$ we have $\sigma = 1$ and (5.7) is to be replaced by $u = r + m^{\frac{1}{2}}$ and (5.8) by

$$|G(z^2)||f(z)| + C(\epsilon, \sigma) \sum_{m=1}^{\infty} (2\epsilon e)^m \exp \{\epsilon(r + m^{\frac{1}{2}})^2 + A(r + m^{\frac{1}{2}})^{\rho}\}. \tag{5.9}$$

This series is clearly convergent for every finite value of z. This completes the proof of the theorem.

6. **Closer estimates in the conjugate case.** We shall now utilize the machinery built up in §§2 and 3 for the purpose of discussing in more detail the case in which $f(z)$ and $G(w)$ are of conjugate orders. We start by proving

THEOREM 6.1. *A necessary and sufficient condition that $G(\delta_z)$ shall be applicable to the class $\mathfrak{F}_{\rho,\alpha}$, ρ fixed > 2, α fixed $(0 < \alpha < \infty)$, is that $G(w) \in \mathfrak{G}_{\sigma,\gamma}$ where $\gamma < \beta$. Here ρ and σ are conjugate orders, α and β conjugate types.*

We shall prove merely the sufficiency here; the necessity of the condition follows from the example in §8.4. By definition

$$G(\delta_z) \cdot f(z) = \sum_{k=0}^{\infty} g_k \delta_z^k f(z).$$

Since $G(w) \in \mathfrak{G}_{\sigma,\gamma}$

$$\limsup_{k\to\infty} \left[\Gamma\left(1 + \frac{k}{\sigma}\right) | g_k |\right]^{1/k} \leqq \gamma^{1/\sigma},$$

while

$$\limsup_{k\to\infty} \left[\frac{| \delta_z^k f(z) |}{\Gamma(1 + k/\sigma)}\right]^{1/k} = [F_\sigma(z; f)]^2 \leqq (2\sigma)^{1/\sigma} (\alpha\rho)^{2/\rho}$$

by formula (3.5). It follows that

$$\limsup_{k\to\infty} | g_k \delta_z^k f(z) |^{1/k} \leqq (2\sigma\gamma)^{1/\sigma} (\alpha\rho)^{2/\rho}.$$

Since $\gamma < \beta$, the conjugate of α, the product is less than 1. Hence $G(\delta_z) \cdot f(z)$ exists as an entire function when the condition of the theorem is satisfied.

If $\rho \leqq 2$, the results are less precise. We start with the case $\rho = 2$.

THEOREM 6.2. *There exists a positive decreasing function* $\beta(\alpha)$ *such that* $G(\delta_z)$ *applies to the class* $\mathfrak{F}_{2,\alpha}$, α *fixed* $(0 < \alpha < \infty)$, *whenever* $G(w) \in \mathfrak{G}_{1,\gamma}$ *if and only if* $\gamma < \beta(\alpha)$. *As* $\alpha \to \infty$, $\alpha\beta(\alpha) \to \frac{1}{4}$.

The existence of such a $\beta(\alpha)$ is proved by a Dedekind cut argument. If $\alpha_1 < \alpha_2$ then $\mathfrak{F}_{2,\alpha_1} \subset \mathfrak{F}_{2,\alpha_2}$, so that if $G(\delta_z)$ applies to $\mathfrak{F}_{2,\alpha_2}$ it also applies to $\mathfrak{F}_{2,\alpha_1}$. Hence the condition $\gamma < \beta(\alpha_2)$ must imply $\gamma < \beta(\alpha_1)$ whence $\beta(\alpha_2) \leqq \beta(\alpha_1)$. For $\beta(\alpha)$ we have the following inequalities

$$\beta_1(\alpha) \leqq \beta(\alpha) \leqq \beta_2(\alpha), \tag{6.1}$$

where

$$\frac{1}{\beta_1(\alpha)} = \sup_f \sup_z \, [F_1(z; f)]^2, \tag{6.2}$$

$f(z)$ ranging over $\mathfrak{F}_{2,\alpha}$, and

$$\frac{1}{\beta_2(\alpha)} = \sup_f \sup_z \, [F_1^*(z; f)]^2, \tag{6.3}$$

where

$$F_1^*(z; f) = \lim_{k \to \infty} \left[\frac{| \delta_z^k f(z) |}{\Gamma(1 + k)} \right]^{1/(2k)}, \tag{6.4}$$

and $f(z)$ ranges over the subset of functions in $\mathfrak{F}_{2,\alpha}$ for which such a limit exists at some points z. It is quite likely that $\beta_1(\alpha) = \beta_2(\alpha)$, but not being able to get good estimates of either quantity, the author must leave this question unsolved. An example in §8.5 shows that

$$\limsup_{\alpha \to \infty} \alpha\beta_2(\alpha) \leqq \tfrac{1}{4},$$

whence it follows in particular that $\beta(\alpha) \to 0$ as $\alpha \to \infty$. On the other hand, Theorem 2.1 (ii) shows that

$$\liminf_{\alpha \to \infty} \alpha\beta_1(\alpha) \geqq \tfrac{1}{4}.$$

Hence

$$\lim_{\alpha \to \infty} \alpha\beta(\alpha) = \tfrac{1}{4}$$

as claimed above. Formula (2.19) shows that

$$\beta(0) = \lim_{\alpha \to 0} \beta(\alpha) \geqq \frac{1}{2} \left(\frac{3}{e} \right)^{\frac{1}{2}}. \tag{6.5}$$

In case $\rho < 2$ we also have an unsatisfactory situation.

THEOREM 6.3. *If $G(w) \in \mathfrak{G}_{1,\gamma}$ with $\gamma < \beta(0)$, then $G(\delta_z)$ applies to the class $\mathfrak{F}_{2,0}$ and a fortiori to every class $\mathfrak{F}_{\rho,\alpha}$ with $\rho < 2$. There exists a quantity β_0 ($\beta(0) \leqq \beta_0 \leqq 1$) such that $G(\delta_z)$ applies to the class of all polynomials if and only if $G(w) \in \mathfrak{G}_{1,\gamma}$ with $\gamma < \beta_0$.*

The existence of β_0 follows again by a Dedekind cut argument; and since the class of all polynomials is a subset of $\mathfrak{F}_{2,0}$, we have $\beta(0) \leqq \beta_0$. That $\beta_0 \leqq 1$ follows from the example in §8.5 below. It seems quite likely that $\beta_0 = \beta(0)$, but we are unable to prove it.

7. On the order and type of $G(\delta_z)$-transforms.

We shall study the relations between orders and types of the three functions $f(z)$, $G(w)$, and $G(\delta_z)\cdot f(z)$. We restrict ourselves to the simplest case.

THEOREM 7.1. *Let $f(z)$ be an entire function of finite order ρ and finite type α. Let $G(w)$ be an entire function of order σ and finite type β. Let ρ' be the conjugate order of ρ, and suppose that $\sigma \leqq \rho'$ and that $\beta = 0$ if $\sigma = \rho'$. Then the entire function $G(\delta_z)\cdot f(z)$ is of order $\mathrm{P} \leqq \max(\rho, 2\sigma)$. If $\mathrm{P} = \rho$ and $\rho > 2\sigma$, the type is at most α; it is at most $\alpha + \beta$ if $\mathrm{P} = \rho = 2\sigma$, and at most β if $\mathrm{P} = 2\sigma$ and $\rho < 2\sigma$. These limits are the best possible.*[20]

We know that $G(\delta_z)\cdot f(z)$ exists as an entire function by Theorem 5.1. In order to get the required estimates we have to modify the analysis which led to formulas (5.8) and (5.9). Suppose first that $\rho > 2\sigma$ and go back to formula (1.20) = (4.2) once more. We choose as contours of integration the circles

$$|t_\nu - t_{\nu+1}| = m^{1/\rho - 1} \qquad (\nu = 1, 2, \cdots, m; t_{m+1} = z),$$

and in Lemma 4.1 we choose $\sigma = \sigma$, $\beta = \beta$, and

$$u = r + m^{1/\rho}.$$

Recalling that

$$M(r; f) < B(\epsilon) \exp[(\alpha + \epsilon)r^\rho],$$

we see that the difference $G(\delta_z)\cdot f(z) - G(z^2)\cdot f(z)$ is dominated by a quantity $C(\epsilon)$ multiplied into the infinite series

$$\sum_{m=1}^{\infty} 2^m [(\beta + \epsilon)\sigma e]^{m/\sigma} m^{am} \exp\{(\alpha + \epsilon)(r + m^{1/\rho})^\rho + (\beta + \epsilon)(r + m^{1/\rho})^{2\sigma}\},$$

where

$$a = 2\left\{1 - \frac{1}{\rho} - \frac{1}{2\sigma}\right\} \leqq 0.$$

The function $G(z^2)\cdot f(z)$ evidently has all the properties claimed for $G(\delta_z)\cdot f(z)$ in the theorem, i.e., its order is at most $\max(\rho, 2\sigma)$ and its type is at most

[20] See H. Muggli [5], p. 153, for $G(d/dz)$. Muggli does not discuss the type except when $\rho = 1$. See also G. Valiron [10], p. 53.

α, $\alpha + \beta$, and β according as $\rho > 2\sigma$, $\rho = 2\sigma$, and $\rho < 2\sigma$. Moreover, it is only when $\rho = 2\sigma$ that any lowering of order or type can occur for the product. It remains to discuss the infinite series.

We write $\sum = \sum_1 + \sum_2$, where $\sum_1$ contains all terms with $m < r^\rho$ and $\sum_2$ the rest. In $\sum_2$ we have $r + m^{1/\rho} < 2m^{1/\rho}$. Hence

$$\sum\nolimits_2 < \sum_{m > r^\rho} 2^m[(\beta + \epsilon)\sigma e]^{m/\sigma} m^{am} \exp [(\alpha + \beta + 2\epsilon)2^\rho m].$$

We recall that if $a = 0$, $\sigma = \rho'$, and $\beta = 0$; further, ϵ is at our disposal. It follows that $\sum_2 < A(r, \epsilon)$, a finite quantity which tends to zero as $r \to \infty$. In $\sum_1$ we have

$$(r + m^{1/\rho})^\rho < r^\rho + \rho(r + m^{1/\rho})^{\rho-1} m^{1/\rho} < r^\rho + \rho m^{1/\rho}(2r)^{\rho-1}$$

if $\rho > 1$, and

$$(r + m^{1/\rho})^\rho < r^\rho + m$$

if $\rho \leqq 1$. Similarly

$$(r + m^{1/\rho})^{2\sigma} < r^{2\sigma} + 2\sigma m^{1/\rho}(2r)^{2\sigma-1}, \qquad \text{or} \qquad < r^{2\sigma} + m^{2\sigma/\rho} < r^{2\sigma} + m$$

according as $\sigma > \frac{1}{2}$ or $\sigma \leqq \frac{1}{2}$. Hence

$$\sum\nolimits_1 < \exp \{(\alpha + \epsilon)r^\rho + (\beta + \epsilon)r^{2\sigma}\} \sum_{m < r^\rho} 2^m[(\beta + \epsilon)\sigma e]^{m/\sigma} m^{am} e^E$$

$$\equiv \exp \{(\alpha + \epsilon)r^\rho + (\beta + \epsilon)r^{2\sigma}\} \sum\nolimits_3 .$$

If both $\rho \leqq 1$ and $\sigma \leqq \frac{1}{2}$ we have $E < (\alpha + \beta + 2\epsilon)m$ and $\sum_3$ evidently remains bounded while $r \to \infty$. The most unfavorable case is that in which $\rho > 1$ and $\sigma > \frac{1}{2}$ and we can restrict ourselves to a detailed discussion of this case.

Here we have

$$E < (\alpha + \epsilon)\rho m^{1/\rho}(2r)^{\rho-1} + (\beta + \epsilon)2\sigma m^{1/\rho}(2r)^{2\sigma-1}$$

$$< [(\alpha + \epsilon)\rho + (\beta + \epsilon)2\sigma]m^{1/\rho}(2r)^{\rho-1} \equiv Br^{\rho-1}m^{1/\rho}.$$

$\sum_3$ is then seen to be dominated by an expression of the form

$$\sum_{m < r^\rho} a_m \exp [Br^{\rho-1} m^{1/\rho}],$$

where

$$0 < a_m < A(\Delta) \exp [-\Delta m],$$

Δ being an arbitrarily large fixed quantity. The latter inequality is obvious if $a < 0$, and if $a = 0$ we recall once more that $\beta = 0$ and that ϵ is at our disposal and can be made as small as we please in advance. We note that B is independent of m and r. Now the maximum of

$$\exp \{Br^{\rho-1}m^{1/\rho} - \Delta m\}$$

when m is a continuous variable equals

$$\exp \{(B\rho)^{\rho/(\rho-1)}(\rho - 1)\Delta^{-1/(\rho-1)}r^{\rho}\}.$$

It follows that

$$\sum\nolimits_3 < B(\Delta)r^{\rho} \exp \{C\Delta^{-1/(\rho-1)}r^{\rho}\}.$$

Since Δ is as large as we please, we conclude that

$$\sum\nolimits_1 < D(\epsilon) \exp [(\alpha + 2\epsilon)r^{\rho}].$$

Combining the estimates of $\sum_1$ and $\sum_2$ we see that

$$\sum < D_1(\epsilon) \exp [(\alpha + 2\epsilon)r^{\rho}]$$

for every positive ϵ. Consequently, if $\rho > 2\sigma$ the transform $G(\delta_z)\cdot f(z)$ is an entire function of order ρ and type α at most.

If $\rho = 2\sigma$ the same argument gives instead

$$\sum < D_1(\epsilon) \exp [(\alpha + \beta + 2\epsilon)r^{\rho}]$$

for every positive ϵ. We conclude that the order of $G(\delta_z)\cdot f(z)$ is at most $\rho = 2\sigma$ and if this is its true order, the type is at most $\alpha + \beta$.

If $\rho < 2\sigma$ we choose the radii of the circles of integration equal to

$$m^{1/(2\sigma)-1}$$

instead. This gives as majorant for the difference $G(\delta_z)\cdot f(z) - G(z^2)\cdot f(z)$ a constant multiple of the series

$$\sum_1^{\infty} 2^m[(\beta + \epsilon)\sigma e]^{m/\sigma} m^{am} \exp \{(\alpha + \epsilon)(r + m^{1/(2\sigma)})^{\rho} + (\beta + \epsilon)(r + m^{1/(2\sigma)})^{2\sigma}\}.$$

This series is discussed by the same method as above and shows that $G(\delta_z)\cdot f(z)$ now is an entire function of order at most 2σ and that, if this be the true order, the type is at most β.

That the results are the best possible will be proved by examples in §8.6.

Suppose that $\frac{1}{2} \leqq \sigma < \min (\frac{1}{2}\rho, \rho')$ and that $G(w) \in \mathfrak{G}_{\sigma,\beta}$, where β is an arbitrary fixed non-negative real number. Then $G(\delta_z)$ *defines a linear transformation on the class* $\mathfrak{F}_{\rho,\alpha}$ *to itself* by Theorem 7.1. *This transformation appears,* however, *to be neither bounded nor continuous.*[21] An example proving this for the case in which

$$\limsup_{n\to\infty} n^{-\frac{1}{2}} \log | C(2n + 1) | = +\infty$$

will be given in §8.7.

8. **Counter examples.**

In this section we shall give the various examples which will prove our statements in §§2–7 concerning best possible results or

[21] H. Muggli [5], p. 153, showed that $\exp [d^2/dz^2]$ does not define a continuous transformation on the classes to which it applies.

lack of continuity, etc. We shall give as few details as possible since many examples employ the same principle, but the first time the principle is used a fuller treatment will be given.

Many of our examples employ properties of the Hermite functions. We shall list what properties we need here for later reference. We refer the reader to E. Hille [4] for proofs.[22] The function $h_n(z)$ is defined by formula (3) of the introduction. For $r > 0$, $|z| = r$, we have

$$|h_n(z)| \leqq (-i)^n h_n(ir). \tag{8.1}$$

Further,

$$\begin{aligned} 0 &< (-1)^n h_{2n}(ir) - \frac{(2n)!}{n!} \cosh (4n+1)^{\frac{1}{2}} r \\ &< \frac{(2n)!}{n!} \exp [(4n+1)^{\frac{1}{2}} r] \{\exp [\tfrac{1}{6} r^3 (4n+1)^{-\frac{1}{2}}] - 1\}. \end{aligned} \tag{8.2}$$

There is a similar formula for the functions of odd order in which $4n + 1$ is replaced by $4n + 3$, the cosh by sinh, and the factor $(2n)!/n!$ by $2(2n + 1)!/(4n + 3)^{\frac{1}{2}} n!$. The formula

$$(-1)^n \frac{n!}{(2n)!} h_{2n}(ir) = 1 + \frac{n}{2!}(2r)^2 + \frac{n(n-1)}{4!}(2r)^4 + \cdots + \frac{n!}{(2n)!}(2r)^{2n} \tag{8.3}$$

proves that for fixed r the left side is an increasing function of n. A similar result holds for the functions of odd order. If $z = x + iy$, $y > 0$, we have

$$h_n(x+iy) = h_n(iy) \exp [-ix(2n+1)^{\frac{1}{2}}]\{1 + n^{-\frac{1}{2}} \eta_n(x, y)\}, \tag{8.4}$$

where $\eta_n(x, y)$ is bounded for $-1/\epsilon \leqq x \leqq 1/\epsilon$, $\epsilon \leqq y \leqq 1/\epsilon$. Finally we note that if in a Hermitian series the coefficients satisfy the condition $i^n c_n \geqq 0$ for all large n, then the point of intersection of the upper line of convergence with the imaginary axis is a singular point of the function defined by the series.

8.1. **Examples for Theorems 2.1 and 3.1.** For $\rho > 2$ we shall show the existence of entire functions $f(z)$ of order ρ and type α, such that

$$\lim_{k \to \infty} k^{-1/\sigma} |\delta_z^k f(z)|^{1/k} = \left(\frac{2}{e}\right)^{1/\sigma} (\alpha\rho)^{2/\rho} \tag{8.1.1}$$

for every value of z on a given line $x = x_0$ in the complex plane with the possible exception of the point $z = x_0$. For this purpose we consider the series[23]

$$f(z; x_0, a, b) = \sum_{n=0}^{\infty} (-1)^n \exp \{-a(4n+1)^b + ix_0(4n+1)^{\frac{1}{2}}\} \frac{n!}{(2n)!} h_{2n}(z), \tag{8.1.2}$$

where $x_0 \geqq 0$, $a > 0$, $\frac{1}{2} < b < 1$. We have obviously

$$|f(z; x_0, a, b)| \leqq f(ir; 0, a, b).$$

[22] See, in particular, Theorems 1.1, 1.4, and 5.1.

[23] The case $b = \frac{1}{2}$ is discussed in [4], pp. 895-896.

Let us choose $N = [\frac{1}{24}r^2]$ and break up the series in (8.1.2) into two parts $\sum_1$ and $\sum_2$, where $\sum_1$ contains all the terms with $n \leqq N$ and $\sum_2$ all the rest. Then using (8.2) and (8.3) we get

$$
\begin{aligned}
\left|\sum\nolimits_1\right| &\leqq \sum_0^N \exp\left[-a(4n+1)^b\right] \frac{n!}{(2n)!} \left| h_{2n}(ir) \right| \\
&< \frac{N!}{(2N)!} \left| h_{2N}(ir) \right| \sum_0^N \exp\left[-a(4n+1)^b\right] \\
&< C_1 \exp\left\{(4N+1)^{\frac{1}{2}} r + \tfrac{1}{6}(4N+1)^{-\frac{1}{2}} r^3\right\} \\
&< C_2 \exp\left\{\tfrac{1}{3}6^{\frac{1}{2}} r^2\right\}.
\end{aligned}
$$

Similarly

$$
\begin{aligned}
\left|\sum\nolimits_2\right| &< \sum_{N+1}^{\infty} \exp\left\{-a(4n+1)^b + r(4n+1)^{\frac{1}{2}} + \tfrac{1}{6}r^3(4n+1)^{-\frac{1}{2}}\right\} \\
&< \exp\left\{\tfrac{1}{6}6^{\frac{1}{2}} r^2\right\} \sum_{N+1}^{\infty} \exp\left\{-a(4n+1)^b + r(4n+1)^{\frac{1}{2}}\right\}.
\end{aligned}
$$

If r is very large, the exponent in the last formula has a single maximum for $n > N$. It follows by classical arguments that the infinite series is of the same order of magnitude as the integral

$$\int_0^{\infty} \exp\left[-au^{2b} + ru\right] u\, du$$

which by the method of Laplace is found to be less than a constant $A(a, b)$ times

$$r^{(2-b)/(2b-1)} \exp\left\{a(2b-1)\left(\frac{r}{2ab}\right)^{2b/(2b-1)}\right\}.$$

Consequently, $f(z; x_0, a, b)$ is an entire function whose order ρ and type α satisfy the inequalities

$$\text{(8.1.3)} \qquad \rho \leqq \frac{2b}{2b-1} \equiv b', \qquad \alpha \leqq a(2b-1)(2ab)^{-2b/(2b-1)} \equiv a'.$$

We observe that the quantities which figure on the right sides of these inequalities are the conjugates of b and a respectively in the sense of relations (5.4) and (5.5). We shall prove that equality holds in both places.

Now take $z = z_0 = x_0 + iy_0$, where y_0 is arbitrary but fixed and positive. Then

$$\delta_z^k f(z_0; x_0, a, b) = \sum_{n=0}^{\infty} (-1)^n (4n+1)^k \exp\left\{-a(4n+1)^b + ix_0(4n+1)^{\frac{1}{2}}\right\} \cdot \frac{n!}{(2n)!} h_{2n}(z_0),$$

the absolute value of which exceeds the real part which equals

$$\sum_{n=0}^{\infty} (4n+1)^k \exp[-a(4n+1)^b] \frac{n!}{(2n)!} \,|\, h_{2n}(iy_0) \,|\, \{1 + n^{-\frac{1}{2}} \Re[\eta_n(x_0, y_0)]\}.$$

Since $\eta_n(x, y)$ is uniformly bounded in $-1/\epsilon \leqq x \leqq 1/\epsilon$, $\epsilon \leqq y \leqq 1/\epsilon$, we can find an integer m such that $n^{-\frac{1}{2}}\Re[\eta_n(x_0, y_0)] > -\frac{1}{2}$ for $n \geqq m$ and any point z_0 in the rectangle mentioned. The terms of the series corresponding to $n < m$ can evidently be neglected for large values of k. The remainder exceeds

$$\begin{aligned} \frac{1}{2}\sum_{m}^{\infty} (4n+1)^k \exp[-a(4n+1)^b] \frac{n!}{(2n)!} \,|\, h_{2n}(iy_0) \,| \\ > \frac{1}{2}\sum_{m}^{\infty} (4n+1)^k \exp[-a(4n+1)^b] \\ > \tfrac{1}{2} \max_n \{(4n+1)^k \exp[-a(4n+1)^b]\} \\ > C\left(\frac{k}{abe}\right)^{k/b}. \end{aligned}$$

It follows that

$$\liminf_{k\to\infty} k^{-1/b} \,|\, \delta_z^k f(z_0; x_0, a, b) \,|^{1/k} \geqq (abe)^{-1/b}. \tag{8.1.4}$$

From this inequality we easily get the required results. We first notice that the inequality can be replaced by the equality

$$\lim_{k\to\infty} k^{-1/b} \,|\, \delta_z^k f(z_0; x_0, a, b) \,|^{1/k} = (abe)^{-1/b}. \tag{8.1.5}$$

Indeed, formula (2.15) is valid not merely for functions of order ρ and type α, but for functions satisfying (3.4), i.e., whose order is at most ρ and type at most α. Using this remark and the estimates (8.1.3), we obtain the inequality

$$\limsup_{k\to\infty} k^{-1/b} \,|\, \delta_z^k f(z_0; x_0, a, b) \,|^{1/k} \leqq (abe)^{-1/b},$$

which combined with (8.1.4) gives (8.1.5).

These formulas show that the functional $F_\lambda(z_0; f)$ is 0, $(abe)^{-1/b}$, or $+\infty$ according as λ is less than, equal to or greater than b. On the other hand, Theorem 3.1 tells us that $F_\lambda(z; f)$ is always 0 for $\lambda \leqq \sigma$, the conjugate of the order of $f(z)$. It follows that $\sigma \leqq b$ and hence that $\rho \geqq b'$. But this is precisely the opposite to the first inequality in (8.1.3). This inequality then must be an equality. From the inequality

$$(abe)^{-1/b} \leqq \left(\frac{2}{e}\right)^{1/b} (\alpha b')^{2/b'},$$

we then get the opposite of the second inequality under (8.1.3) which then also must be an equality.

We have consequently proved that formula (2.15) is the best possible in the sense that equality may actually hold for any preassigned value of z even if the limit superior is replaced by an ordinary limit.[24]

If $\rho = 2$ we can use the function $f(z; x_0, a, 1)$ and proceed as above. We obtain

$$\rho = 2, \qquad \alpha \leqq \tfrac{1}{3}6^{\frac{1}{2}} + \frac{1}{4a}, \tag{8.1.6}$$

$$\liminf_{k\to\infty} k^{-1} | \delta_z^k f(z_0; x_0, a, 1) |^{1/k} \geqq \frac{1}{ae}, \tag{8.1.7}$$

and finally

$$\liminf_{k\to\infty} k^{-1} | \delta_z^k f(z_0; x_0, a, 1) |^{1/k} \geqq \frac{4}{e}\alpha - \frac{4}{3}6^{\frac{1}{2}}. \tag{8.1.8}$$

This inequality proves that the factor $4/e$ in formula (2.16) cannot be replaced by any smaller quantity.

For $\rho < 2$ or, more precisely, for functions satisfying the condition (2.2) we proved that inequality (2.3) holds. That this inequality is not capable of very considerable improvement is shown by the fact that

$$\limsup_{k\to\infty} k^{-1} | \delta_z^k 1 |^{1/k} \geqq \frac{1}{e} \tag{8.1.9}$$

on the lines $y = \pm x$. Indeed, formulas (1.10) and (1.12) show that if $y = \pm x$, then

$$(-1)^\nu \delta_z^{2\nu} 1 \geqq A_{0,2\nu,\nu} > (2\nu)!,$$

and this obviously implies (8.1.9).

8.2. **Discontinuities of the functional** $F_\lambda(z;f)$. It was stated in §3 that $F_\lambda(z;f)$ may be a discontinuous function of z for a fixed $f(z)$. This is proved by considering the series

$$\varphi(z; a, b) = \sum_{n=0}^{\infty} (-1)^n \exp[-a(4n+3)^b] \frac{n!(4n+3)^{\frac{1}{4}}}{(2n+1)!} h_{2n+1}(z), \tag{8.2.1}$$

where $a > 0$, $\frac{1}{2} \leqq b \leqq 1$. This is a function of the same type as $f(z; 0, a, b)$ of formula (8.1.2) and it can be discussed by the same methods.

When $b = \frac{1}{2}$, the series is convergent in the strip $-a < y < a$ and the points $z = \pm ai$ are singular. It is obvious that

$$F_{\frac{1}{2}}(0; \varphi) = 0, \tag{8.2.2}$$

[24] The excluded case in which z_0 is real $\neq 0$ can be handled by a modification of the series. Cf. a similar argument in [4], pp. 896-897.

and it is an easy matter to show that

$$F_{\frac{1}{2}}(iy;\varphi) = \frac{1}{a-|y|} = \frac{1}{R(iy)} \qquad (-a < y < a,\ y \neq 0). \tag{8.2.3}$$

Here $R(z)$ denotes the radius of holomorphism of $\varphi(z)$ at the point z. This example then shows that $F_{\frac{1}{2}}(z;f)$ may be a discontinuous function of z. It also shows, incidentally, that (3.6) is the best possible estimate.

For $\frac{1}{2} < b < 1$ we are dealing with an entire function of order b' and type a' (see formulas (8.1.3)). We have

$$F_b(0;\varphi) = 0, \qquad F_b(iy;\varphi) = a^{-1/(2b)} \qquad (y \neq 0). \tag{8.2.4}$$

Thus $F_b(z;\varphi)$ is discontinuous at $z = 0$. The same result is true when $b = 1$.

8.3. **Example for Theorem 4.4.** We shall prove that if $G(w)$ is any entire function of order $\frac{1}{2}$ and maximal type, then there exists an entire function $f(z)$, usually of infinite order, such that $G(\delta_z)\cdot f(z)$ does not exist anywhere on the imaginary axis. By assumption we can find a monotone increasing function $\lambda(u)$ tending to infinity with u such that

$$(2k)!\,|\,g_k\,| > [\lambda(k)]^{2k} \tag{8.3.1}$$

for infinitely many values of k. Let us then choose a monotone increasing function $\mu(u)$, tending to infinity with u, such that

$$\lambda(u)\exp\{-\mu(4u^2)\} \to \infty \text{ with } u. \tag{8.3.2}$$

Then form

$$f(z) = \sum_{n=0}^{\infty}(-1)^n \exp\{-(4n+1)^{\frac{1}{2}}\mu(4n+1)\}\frac{n!}{(2n)!}h_{2n}(z). \tag{8.3.3}$$

It is easily seen that this is an entire function which is ordinarily of infinite order. Further

$$\delta_z^k f(iy) > \delta_z^k f(0) > C(2k)^{2k}\exp\{-2k\mu(4k^2)\}. \tag{8.3.4}$$

It follows that the series

$$\sum_0^{\infty} g_k\,\delta_z^k f(z)$$

cannot converge anywhere on the imaginary axis since there is a subsequence of the terms tending to $+\infty$. On the other hand, it may very well happen that the series

$$\sum_{n=0}^{\infty}(-1)^n \exp\{-(4n+1)^{\frac{1}{2}}\mu(4n+1)\}G(4n+1)\frac{n!}{(2n)!}h_{2n}(z), \tag{8.3.5}$$

obtained by termwise performance of the operation $G(\delta_z)$ on (8.3.3), converges for all values of z. This depends exclusively upon the asymptotic behavior of $G(4n+1)$ for large values of n.

8.4. **Example for Theorems 5.1 and 6.1.** The following example really refers to Theorem 6.1 but it is also a counter example for that part of Theorem 5.1 in which $\rho > 2$. We suppose then that $\rho > 2$ and let ρ and σ be conjugate orders and α and β conjugate types. Corresponding to a given entire function $G(w)$ of order σ and type β we shall show the existence of an entire function $f(z)$ of order ρ and type α, such that $G(\delta_z) \cdot f(z)$ does not exist. By assumption

$$|g_k| = \frac{\beta^{k/\sigma}}{\Gamma(k/\sigma)} \lambda_k \quad \text{where} \quad \frac{1}{k} \log \lambda_k \to 0 \tag{8.4.1}$$

as $k \to \infty$. The unfavorable case for us is that in which $\liminf \lambda_k = 0$. It is then possible to find a steadily decreasing function $\lambda(u)$ such that $\lambda(u) \to 0$, $u^{-1} \log \lambda(u) \to 0$ as $u \to \infty$ but $\lambda_k \geqq \lambda(k)$ for infinitely many values of k. We then form the following modification of our first counter example in formula (8.1.2):

$$f(z) = \sum_{n=0}^{\infty} (-1)^n \frac{\exp[-\beta(4n+1)^\sigma]}{\lambda[\sigma\beta(4n+1)^\sigma]} \frac{n!}{(2n)!} h_{2n}(z). \tag{8.4.2}$$

It is not difficult to show that this is an entire function of order ρ and type α. On the imaginary axis

$$\delta_z^k f(iy) > \frac{C}{\lambda(k)} \left(\frac{k}{\beta\sigma e}\right)^{k/\sigma}, \tag{8.4.3}$$

whence it follows that the series

$$\sum_{k=0}^{\infty} g_k \delta_z^k f(z) \tag{8.4.4}$$

diverges everywhere on the imaginary axis, because it contains infinitely many terms the absolute values of which exceed a positive constant.

The converse problem: *Given any function $f(z)$ of order ρ and type σ, construct an entire function $G(w)$ of conjugate order σ and conjugate type β, such that $G(\delta_z) \cdot f(z)$ does not exist,* appears to be much more difficult. It would be easily solved if it were known that for every entire function of order ρ and type α there exists at least one point z_0 where

$$F_\sigma(z_0 ; f) = (2\sigma)^{1/(2\sigma)} (\alpha\rho)^{1/\rho}.$$

Whether or not this is actually true is one of the open questions which we cannot answer.

The function defined by (8.4.2) shows that the condition of Theorem 6.1 is necessary for the truth of that theorem. It also shows that the condition of Theorem 5.1 is necessary when $\rho > 2$. If we set $\sigma = 1$ in (8.4.2), we obtain an entire function of order 2 for which the series (8.4.4) diverges on the imaginary axis. Thus the condition of Theorem 5.1 is also necessary when $\rho = 2$. That it is not necessary when $\rho < 2$ follows already from Theorem 6.3.

8.5. **Examples for Theorems 6.2 and 6.3.** We have first to prove formula (6.5). For this purpose we consider the function $f(z; 0, a, 1)$, a special case of the function defined by formula (8.1.2). Here we can sharpen (8.1.6) and determine the exact type of the function. We have[25]

$$\begin{aligned} f(iy; 0, a, 1) &= \sum_{n=0}^{\infty} (-1)^n \frac{n!}{(2n)!} h_{2n}(iy) e^{-(4n+1)a} \\ &> e^{-a} \sum_{n=0}^{\infty} (-1)^n \frac{1}{4^n n!} h_{2n}(iy) e^{-4na} \\ &= \left\{\frac{\pi}{2 \sinh 2a}\right\}^{\frac{1}{2}} \exp\left[\tfrac{1}{2} \coth 2a\, y^2\right]. \end{aligned}$$

In the opposite direction we get, for instance, by a suitable use of Cauchy's inequality, that

$$f(iy; 0, a, 1) < C(a) \mid y \mid \exp\left[\tfrac{1}{2} \coth 2a\, y^2\right],$$

where the exact value of $C(a)$ is immaterial. It follows that

$$\alpha = \tfrac{1}{2} \coth 2a. \tag{8.5.1}$$

Further, a more elaborate analysis shows that we can sharpen (8.1.7) in the present case to

$$\lim_{k \to \infty} k^{-1} \mid \delta_z^k f(iy; 0, a, 1) \mid^{1/k} = \frac{1}{ae}. \tag{8.5.2}$$

Thus the functional $F_1^*(z; f)$ defined by formula (6.4) has a sense for $f(z; 0, a, 1)$ at least on the imaginary axis. By virtue of (6.3) this leads to the simple inequality

$$\beta_2(\alpha) < a = \tfrac{1}{4} \log \frac{2\alpha + 1}{2\alpha - 1}, \qquad \alpha > \tfrac{1}{2}, \tag{8.5.3}$$

of which (6.5) is an immediate consequence. We notice that $\beta(\alpha)$ also satisfies the inequality (8.5.3).

We next have to prove that the quantity β_0 introduced in Theorem 6.3 is less than or equal to one. This follows from the fact that

$$\exp(-\beta\delta_z) \cdot 1$$

does not exist at the origin for any $\beta \geqq 1$ since $(-1)^\nu (\delta_z^{2\nu} 1)_{z=0} > (2\nu)!$.

8.6. **Examples for Theorem 7.1.** We shall show that if $\rho > 2\sigma$, the order and type of $G(\delta_z) \cdot f(z)$ may actually coincide with those of $f(z)$. We restrict ourselves to the case $\rho \geqq 2$. The function $f(z; 0, a, b)$ of formula (8.1.2) is an

[25] The second equality follows from the Abel-Hermite kernel also known as Mehler's generating function for Hermite polynomials.

entire function of order b' and type a' (see formula (8.1.3)). We take $G(w) = E_{1/\sigma}(w)$, the Mittag-Leffler E-function, where $\sigma < b$. Then

$$E_{1/\sigma}(\delta_z)\cdot f(z; 0, a, b) = \sum_{n=0}^{\infty} (-1)^n E_{1/\sigma}(4n+1) \cdot \exp[-a(4n+1)^b] \frac{n!}{(2n)!} h_{2n}(z), \tag{8.6.1}$$

and with the aid of the methods of §8.1 it is a simple matter to prove that this is also an entire function of order b' and type a' as long as $\sigma < b$. If $\sigma = b$ but $a > 1$, the order is still b' but the type increases to infinity as $a \to 1$. The transform does not exist for $a \leqq 1$.

Finally, we shall give an example to show that if $\rho < 2\sigma$, the transform may be of order 2σ and have the same type as $G(w)$. We take $f(z) = 1$ and

$$G(w) = \sum_{k=0}^{\infty} \frac{(aw)^{4k}}{\Gamma(1 + 4k/\sigma)}. \tag{8.6.2}$$

If $a > 0$, this is an entire function of order σ and type a^σ.[26] We assume $\sigma \leqq 1$, and apply formula (1.15), obtaining

$$G(\delta_z)\cdot 1 = \sum_{k=0}^{\infty} \frac{a^{4k}}{\Gamma(1 + 4k/\sigma)} \sum_{m=0}^{2k} (-1)^m A_{0,4k,m} z^{8k-4m}. \tag{8.6.3}$$

Here the operator $G(\delta_z)$ satisfies condition (1.16), so we can use the remarks made in connection with this condition. As the class $\mathfrak{F}$ we can take either $\mathfrak{F}_{2,0}$ or simply $\mathfrak{P}$, the class of all polynomials. In either case the class is left invariant by transformations which affect merely the arguments of the derivatives at the origin, leaving their absolute values unchanged. If $\sigma < 1$ it is known that the operator $G(\delta_z)$ applies to the class $\mathfrak{F}$; if $\sigma = 1$ the assumption of applicability imposes a condition on a. If $G(\delta_z)$ does apply, we know that the series (8.6.3) is absolutely convergent for all z. Hence we have

$$G(\delta_z)\cdot 1 = \sum_{\nu=0}^{\infty} (-1)^\nu z^{4\nu} \sum_{2k \geqq \nu} \frac{A_{0,4k,2k-\nu}}{\Gamma(1 + 4k/\sigma)} a^{4k}. \tag{8.6.4}$$

This expression reaches its maximum on the lines $y = \pm x$. Substituting $z = \omega r$, where ω is a primitive eighth root of unity, and noting that the A's are positive integers, we find readily that

$$G(\delta_z)\cdot 1 \geqq G(r^2) > C(\sigma) \exp(a^\sigma r^{2\sigma}). \tag{8.6.5}$$

It follows that $G(\delta_z)\cdot 1$ is an entire function whose order is exactly 2σ and whose type is a^σ. This conclusion is valid when $\sigma < 1$. If $\sigma = 1$ the existence of the transform is ensured only for sufficiently small values of a. For such values the order of the transform equals $2\sigma = 2$, but the type exceeds a and tends to

[26] See formula (4.6). $G(w)$ is evidently a linear combination of Mittag-Leffler E-functions.

infinity as a approaches the maximal value beyond which the transform does not exist.

8.7. **Discontinuity of** $G(\delta_z)$ **on** $\mathfrak{F}_{\rho,\alpha}$. Let us suppose that $G(w)$ is an entire function of order σ ($\frac{1}{2} \leqq \sigma \leqq 1$) such that

$$\limsup_{n\to\infty} n^{-\frac{1}{2}} \log | G(2n+1) | = +\infty. \tag{8.7.1}$$

If $2\sigma < \rho < \sigma'$, the conjugate of σ, the operator $G(\delta_z)$ defines a linear transformation on $\mathfrak{F}_{\rho,\alpha}$ to itself. Suppose in addition that $\rho \geqq 2$. I say that *the transformation* $G(\delta_z)$ *is not continuous and a fortiori not bounded on* $\mathfrak{F}_{\rho,\alpha}$.

By virtue of (8.7.1) we can find a positive monotone increasing function $\lambda(u)$ tending to infinity with u, such that

$$| G(2n+1) | > \exp [(2n+1)^{\frac{1}{2}}\lambda(n)] \tag{8.7.2}$$

for infinitely many values of n. We can assume, without restriction of the generality, that there are infinitely many even values of n for which (8.7.2) is true. Put $\mu(u) = [\lambda(u)]^{\frac{1}{2}}$, and define

$$f_n(z) = \frac{n!}{(2n)!} \exp [-(4n+1)^{\frac{1}{2}}\mu(2n)] h_{2n}(z). \tag{8.7.3}$$

Since $\rho \geqq 2$, these functions belong to $\mathfrak{F}_{\rho,\alpha}$,[27] and by virtue of formula (8.2) and the properties of $\mu(u)$ the sequence $\{f_n(z)\}$ converges to zero as $n \to \infty$, uniformly in any fixed circle $| z | \leqq R$. On the other hand,

$$G(\delta_z)\cdot f_n(z) = G(4n+1) f_n(z) \tag{8.7.4}$$

obviously does not converge to zero anywhere and $\limsup | G(\delta_z)\cdot f_n(z) | = \infty$ everywhere outside of the real axis. This proves that $G(\delta_z)$ is not continuous at $f = 0$ and hence nowhere in $\mathfrak{F}_{\rho,\alpha}$.

9. **Additional comments on** $G(\delta_z)$ **and related operators.** Let us first point out that the investigation given here of the operator $G(\delta_z)$ also extends to the more general operator

$$G(\delta_z, z, m) = \sum_{\nu=0}^{m} a_\nu z^\nu G_\nu(\delta_z), \tag{9.1}$$

where the a's are given constants and the $G_\nu(w)$ given entire functions. This operator is equivalent to

$$\sum_{k=0}^{\infty} P_k(z)\delta_z^k, \tag{9.2}$$

where $\{P_k(z)\}$ is a given sequence of polynomials of degree $\leqq m$.

[27] If $\rho = 2$, this requires $\alpha \geqq \frac{1}{2}$.

We find that the operator $G(\delta_z, z, m)$ applies to all analytic functions if and only if all $G_\nu(w) \in \mathfrak{G}_{\frac{1}{2},0}$. The operator applies to all functions of $\mathfrak{F}_{\rho,\alpha}$, ρ fixed $\geqq 2$, if and only if all $G_\nu(w) \in \mathfrak{G}_{\sigma,0}$, where σ is the conjugate of ρ, and so on.

Our remaining remarks are devoted to a comparison between the two operators $G(d/dz)$ and $G(\delta_z)$. The former operator is fairly well known, having been the object of much research in the past. There is much similarity between the two theories and we have called attention to such features in several places above. But there are also considerable differences. The applicability questions are much easier to solve for the operator $G(d/dz)$ than for $G(\delta_z)$. This is of course essentially because of the fact that it is much easier to discuss the rate of growth of $f^{(k)}(z)$ than of $\delta_z^k f(z)$. But the difference is not merely a difference in degree of accessibility to customary analytical technique. This would not be so interesting if the general situation were fundamentally the same in both cases. Actually there seem to exist differences of more profound nature.

One such difference is reflected in the different character of the order relations which govern the applicability of these operators to entire functions. These relations are[28]

$$(9.3) \qquad \frac{1}{\rho} + \frac{1}{\sigma} = 1 \quad \text{and} \quad \frac{1}{\rho} + \frac{1}{2\sigma} = 1$$

for d/dz and δ_z respectively. The difference in the coefficients of the formulas is immaterial. More essential is the fact that the first formula is valid whenever it has a sense, i.e., for $1 \leqq \sigma \leqq \infty$, while the second one holds only for $\frac{1}{2} \leqq \sigma \leqq 1$. Thus there is always a class $\mathfrak{F}_{\rho,\alpha}$ of entire functions $f(z)$ to which the operator $G(d/dz)$ applies when $G(w)$ is an entire function. Not so with $G(\delta_z)$. Here we can find a class $\mathfrak{F}_{\rho,\alpha}$ to which it applies only if $G(w) \in \mathfrak{G}_{\sigma,\beta}$ for $0 \leqq \sigma \leqq 1$ and not for any $\sigma > 1$. Moreover, if $G(w)$ is merely holomorphic in a finite circle $|w| < R$, the operator $G(d/dz)$ always applies to the class $\mathfrak{F}_{1,0}$, while the class of functions to which $G(\delta_z)$ applies appears to be highly special and certainly does not contain any class $\mathfrak{F}_{\rho,\alpha}$ as a core. Thus the phenomenon of a critical ρ-order and a maximal σ-order affects the operator $G(\delta_z)$ profoundly and there is no correspondence in the theory of $G(d/dz)$.

This phenomenon is intimately connected with the difference in behavior between the two basic functionals

$$(9.4) \qquad \Phi_\sigma(z;f) = \limsup_{k\to\infty} \left| \frac{f^{(k)}(z)}{\Gamma(1+k/\sigma)} \right|^{1/k},$$

$$(9.5) \qquad F_\sigma(z;f) = \limsup_{k\to\infty} \left| \frac{\delta_z^k f(z)}{\Gamma(1+k/\sigma)} \right|^{1/(2k)}.$$

We observed in §3 that if $f(z)$ is holomorphic at a point z whose distance from the nearest singularity is $R(z)$, then

$$\Phi_1(z;f) = \frac{1}{R(z)}, \qquad F_{\frac{1}{2}}(z;f) \leqq \frac{1}{R(z)},$$

[28] See H. Muggli [5], p. 152, for the first operator.

and the former functional is a continuous function of z within the domain of holomorphism while the latter need not be. If $f(z) \in \mathfrak{F}_{\rho,\alpha}$ $(1 < \rho)$, and σ is determined from the first formula under (9.3), then $\Phi_\sigma(z; f)$ exists and is a function of ρ and α, independent of z and $f(z)$. On the other hand, if $f(z) \in \mathfrak{F}_{\rho,\alpha}$ $(2 < \rho)$, and σ is determined from the second formula under (9.3), then $F_\sigma(z; f)$ exists as a finite quantity but depends upon $f(z)$ and may be a discontinuous function of z. No matter what value σ has $(0 < \sigma \leqq \infty)$, there is always a class of entire functions for which $\Phi_\sigma(z; f)$ is bounded everywhere. In particular, for $\sigma = \infty$, $\Phi_\infty(z; f) \leqq \alpha$ whenever $f(z) \in \mathfrak{F}_{1,\alpha}$, i.e., for every function of *exponential type* in Pólya's terminology, the type being less than or equal to α.

$F_\sigma(z; f)$ shows an entirely different behavior. In particular, the class of entire functions for which $F_\infty(z; f) \leqq \alpha$ seems to be quite special and connected with the more intricate part of the theory of the Hermite-Weber equation. Any solution of the equation

$$w'' + (2\kappa + 1 - z^2)w = 0, \qquad |2\kappa + 1| \leqq \alpha^2,$$

belongs to this class and other functions of the class can be generated by customary analytical devices from such solutions.

The determination of the class of functions for which $F_\infty(z; f) \leqq \alpha$ would seem to be of some importance. Indeed, the classical theory of the differential equation $G(d/dz) \cdot W = F(z)$ has been largely concerned with the class of functions for which $\Phi_\infty(z; f) \leqq \alpha$, i.e., the class $\mathfrak{F}_{1,\alpha}$. It is essentially this class which serves as the basis of the investigations of R. D. Carmichael, H. T. Davis, E. Hilb, H. von Koch, O. Perron, G. Pólya, and I. M. Sheffer, to mention only a few.[29] Moreover, F. Schürer [8] has developed a general theory of L-operations, satisfying certain postulates, which includes also the theory of the functional equation $G(L) \cdot W = F(z)$. He determines all solutions of this equation within the class of functions for which

$$\limsup_{k \to \infty} |L^k f(z)|^{1/k} \leqq q. \tag{9.6}$$

It is easy to formulate conditions under which Schürer's postulates are satisfied for $L = \delta_z$, but as long as the class of functions for which $F_\infty(z; f) \leqq q^2$ is not well defined, the existence theorems given by the Schürer theory are not of much interest in the present case.

Let us finally call attention to one more feature that makes for a distinct difference between the operators $G(d/dz)$ and $G(\delta_z)$. It goes back to the characteristic functions of the operators d/dz and δ_z. Since

$$\frac{d}{dz} e^{\lambda z} = \lambda e^{\lambda z}, \tag{9.7}$$

we can say that every complex number λ is a characteristic value of the operator d/dz with $e^{\lambda z}$ as the corresponding characteristic function. The growth prop-

[29] See R. D. Carmichael [1] and H. T. Davis [2] for further references and an outline of the results obtained by these writers.

erties of the characteristic functions are determined entirely by the characteristic value in question and the operator d/dz does not single out any particular value of λ or any particular direction in the z-plane.

The operator δ_z shows a different behavior. The equation[30]

$$\delta_z f(z) = (2\kappa + 1) f(z) \tag{9.8}$$

is satisfied by the Hermite-Weber functions of order κ. Again all complex numbers are characteristic values. But now all odd integers are exceptional in the sense that some of the corresponding solutions have exceptional growth properties, namely, $h_n(z)$ when $\kappa = n$ and $h_n(iz)$ when $\kappa = -n - 1$. Moreover, the growth properties are governed by κ only when it comes to the fine structure. In the first approximation it is much more important to know which particular solution is considered than the value of κ. The lines $y = \pm x$ divide the plane into four sectors in each of which there is a solution of (9.8) which tends exponentially to zero as $z \to \infty$ regardless of the value of κ which contributes only a minor correction. As a rule these four subdominant solutions are pair-wise linearly independent and dependence occurs if and only if κ is an integer.

Thus the operator δ_z is strongly oriented in the complex z-plane and gives a certain preference to the odd integers among the characteristic values. This orientation and preference shows itself in many ways in the properties of the operator $G(\delta_z)$ and becomes particularly important in the theory of the differential equation $G(\delta_z) \cdot W = F(z)$ which will be taken up for study in a later paper.

[The prototype of Theorem 4.2 for the operator $G(d/dz)$ is due to J. F. Ritt [7], pp. 34–35. I use this opportunity to amend some statements in my paper [4]. In footnote 4 credit for the first application of finite order differential operators to analytic continuation should not have been assigned to H. Cramér as the publication of Ritt's thesis preceded that of Cramér's by five months. Ritt also has the honor of having proved the first general gap theorem for Dirichlet series, preceding both Carlson-Landau and Szász, and this should have been pointed out in footnote 28. I apologize for these unintentional oversights. It should be observed, however, that my Theorems 4.3 and 5.7 are analogues of theorems due to Cramér and Carlson-Landau-Szász and not of related theorems due to Ritt. My methods, aside from the basic differential operator approach, have very little in common with those of Ritt. I expect to make extensive use of Ritt's methods in later papers, however. **Added November 15, 1940.**]

Bibliography

1. R. D. Carmichael, *Linear differential equations of infinite order*, Bulletin Amer. Math. Soc., vol. 42(1936), pp. 193-218.
2. H. T. Davis, *The Theory of Linear Operators from the Standpoint of Differential Equations of Infinite Order*, Bloomington, Ind., 1936, xiv + 628 pp.
3. P. Flamant, *La notion de continuité dans l'étude des transmutations distributives des fonctions d'une variable complexe et ses applications*, Bulletin des Sci. Math., (2), vol. 52(1928), pp. 26-48, 77-96, 104-128.

[30] For the properties of the solutions of the Weber-Hermite equation see E. T. Whittaker and G. N. Watson [11], Chapter XVI. They refer to them as the *functions of the parabolic cylinder.*

4. E. Hille, *Contributions to the theory of Hermitian series*, this Journal, vol. 5(1939), pp. 875-936.
5. H. Muggli, *Differentialgleichungen unendlich hoher Ordnung mit konstanten Koeffizienten*, Commentarii Math. Helvetici, vol. 11(1938), pp. 151-179.
6. G. Pólya, *Untersuchungen über Lücken und Singularitäten von Potenzreihen*, Math. Zeitschrift, vol. 29(1929), pp. 549-640.
7. J. F. Ritt, *On a general class of linear homogeneous differential equations of infinite order with constant coefficients*, Trans. Amer. Math. Soc., vol. 18(1917), pp. 27-49.
8. F. Schürer, *Eine gemeinsame Methode zur Behandlung gewisser Funktionalgleichungsprobleme*, Berichte d. Sächsichen Ges. d. Wiss. zu Leipzig, Math.-Phys. Kl., vol. 70(1918), pp. 185-240.
9. G. Valiron, *Lectures on the General Theory of Integral Function*, Toulouse, 1923, xi + 208 pp.
10. G. Valiron, *Sur les solutions des équations différentielles linéaires d'ordre infini et à coefficients constants*, Annales de l'École Norm. Sup., (3), vol. 46(1929), pp. 25-53.
11. E. T. Whittaker and G. N. Watson, *Modern Analysis*, 4th edition, Cambridge University Press, 1927, x + 608 pp.

Yale University.

Sur les fonctions analytiques définies par des séries d'Hermite [1];

Par Einar HILLE.

(New Haven, Connecticut, U. S. A.).

---

1. Introduction. — Soit

$$h_n(z) = e^{-\frac{1}{2}z^2} H_n(z) = (-1)^n e^{\frac{1}{2}z^2} \frac{d^n}{dz^n}(e^{-z^2}) \tag{1.1}$$

et posons

$$A_n = +\{[h_n(0)]^2 + (2n+1)^{-1}[h'_n(0)]^2\}^{\frac{1}{2}} \tag{1.2}$$

de sorte que

$$A_{2k} = \frac{(2k)!}{k!}, \qquad A_{2k+1} = 2\frac{(2k+1)!}{(4k+3)^{\frac{1}{2}} k!}.$$

Soit $\{f_n\}$ une suite de nombres tels que

$$-\limsup_{n\to\infty} (2n+1)^{-\frac{1}{2}} \log(A_n|f_n|) \equiv \tau > 0. \tag{1.3}$$

Alors, la série d'Hermite

$$\sum_{n=0}^{\infty} f_n h_n(z) \equiv f(z) \tag{1.4}$$

converge absolument dans la bande

$$-\tau < y < \tau, \qquad z = x + iy, \tag{1.5}$$

et y définit une fonction holomorphe.

---

[1] Recherche supportée par l'Air Research and Development Command, U. S. Air Force [Contrat SAR/AF 49 (638)-224].

Dans des travaux parus il y a vingt ans, j'ai signalé quelques propriétés des fonctions définies par des séries d'Hermite. On trouve là des relations entre une série d'Hermite et ses séries associées, des analogues de théorèmes classiques sur les séries entières attachés aux noms de Pringsheim et Fabry entre autres, une classe de multiplicateurs qui gardent l'holomorphie et ainsi de suite. A ces résultats j'ajoute maintenant deux théorèmes inédits, sur les relations entre coefficients et singularités, qui datent de la même époque.

La discussion repose sur l'observation suivante (*voir* la bibliographie [2], p. 930). Soient

$$(1.6)\quad \begin{cases} E^{+}(z;f)=\sum_{n=0}^{\infty}(-i)^{n}A_{n}f_{n}\exp\left[i(2n+1)^{\frac{1}{2}}z\right], \\ E^{-}(z;f)=\sum_{n=0}^{\infty}i^{n}A_{n}f_{n}\exp\left[-i(2n+1)^{\frac{1}{2}}z\right]. \end{cases}$$

Notons par $A[f]$, $A[E^{+}]$ et $A[E^{-}]$ les étoiles de Mittag-Leffler, par rapport à l'origine, des fonctions $f(z)$, $E^{+}(z;f)$ et $E^{-}(z;f)$ respectivement. Alors on a

$$(1.7)\quad A[f]\supset A[E^{+}]\cap\{-A[E^{+}]\}\cap A[E^{-}]\cap\{-A[E^{-}]\}.$$

2. Fonctions entières. — Nous commençons avec le

Théorème 1. — *Soit* $G(t)$ *holomorphe dans le demi-plan droit et soit*

$$(2.1)\quad \limsup_{r\to\infty} r^{-\frac{1}{2}}\log|G(re^{i\theta})|\equiv\eta(\theta)\qquad \left(-\frac{1}{2}\pi<\theta<\frac{1}{2}\pi\right),$$

où

$$(2.2)\quad \eta(\theta)<A(\operatorname{tg}|\theta|)^{\frac{1}{2}}-B$$

et où $A>0$, $B>0$. *Alors la série*

$$(2.3)\quad \sum_{n=1}^{\infty}A_{n}^{-1}\,G(n)\,h_{n}(z)\equiv f(z)$$

converge dans la bande

$$(2.4)\quad -2^{-\frac{1}{2}}B<y<2^{-\frac{1}{2}}B$$

et $f(z)$ *est une fonction entière.*

Pour démontrer le théorème il suffit d'établir que $E^{+}(z;f)$ et $E^{-}(z;f)$ sont des fonctions entières. On peut montrer cela en utilisant le Calcul des résidus. Considérons l'expression

$$(2.5) \qquad -\int G(t)\exp\left\{i\left[(2t+1)^{\frac{1}{2}}z-\frac{1}{2}\pi t\right]\right\}\frac{dt}{e^{-2\pi i t}-1}.$$

Si l'intégrale est prise le long du rectangle de sommets

$$k-\frac{1}{2}-iq, \quad m+\frac{1}{2}-iq, \quad m+\frac{1}{2}+iq, \quad k-\frac{1}{2}+iq$$

$$(1<k<m,\ 0<q),$$

elle a pour valeur

$$\sum_{n=k}^{m}(-i)^{n}G(n)\exp\left[i(2n+1)^{\frac{1}{2}}z\right].$$

Soit $0<B_1<B$. Grâce à la relation (2.2) nous pouvons choisir l'entier k de sorte que

$$(2.6) \quad \log|G(u+iv)|<2A\left\{v^{2}+\frac{v^{4}}{u^{2}}\right\}^{\frac{1}{4}}-B_1(u^{2}+v^{2})^{\frac{1}{4}}, \qquad k-\frac{1}{2}\leqq u.$$

Si l'on a aussi

$$(2.7) \qquad k>\frac{1}{2}+\left(\frac{4A}{\pi}\right)^{2},$$

un calcul facile montre que les intégrales prises le long des côtés horizontaux du rectangle tendent vers zéro quand q tend vers l'infini. Alors, en notant

$$(2.8) \qquad I_n(z)\equiv\int_{n-\frac{1}{2}-i\infty}^{n-\frac{1}{2}+i\infty}G(t)\exp i\left[(2t+1)^{\frac{1}{2}}z-\frac{1}{2}\pi t\right]\frac{dt}{e^{-2\pi i t}-1},$$

on obtient

$$\sum_{n=k}^{m}(-i)^{n}G(n)\exp\left[i(2n+1)^{\frac{1}{2}}z\right]=I_{m+1}(z)-I_k(z).$$

Montrons maintenant que

$$(2.9) \qquad \lim_{m\to\infty}I_{m+1}(z)=0$$

uniformément, par rapport à z, dans le secteur

$$S: \quad \frac{1}{4}\pi \leqq \arg z \leqq \frac{3}{4}\pi.$$

En effet, la valeur absolue de l'intégrande est majorée par

$$\exp\left[P(z, u, v, m) - \frac{1}{2}\pi|v|\right],$$

où

$$P(z, u, v, m) < 2A\left\{v^2 + \frac{v^4}{\left(m+\frac{1}{2}\right)^2}\right\}^{\frac{1}{4}}$$
$$- B_1\left[\left(m+\frac{1}{2}\right)^2 + v^2\right]^{\frac{1}{4}} + \mathcal{R}\left[iz(2m+2+iv)^{\frac{1}{2}}\right]$$

et où le dernier terme est toujours négatif dans le secteur S. Mais

$$|I_{m+1}(z)| < \int_{-\infty}^{\infty} \exp\left\{2A|v|^{\frac{1}{2}} + \left[\frac{2A}{\left(m+\frac{1}{2}\right)^{\frac{1}{2}}} - \frac{1}{2}\pi\right]|v|\right.$$
$$\left. - B_1\left[\left(m+\frac{1}{2}\right)^2 + v^2\right]^{\frac{1}{4}}\right\} dv;$$

donc $I_{m+1}(z)$ tend vers zéro quand m tend vers l'infini, uniformément par rapport à z dans S.

Nous avons donc

$$(2.10) \quad E^+(z; f) = \sum_{n=1}^{k-1} (-i)^n G(n) \exp\left[i(2n+1)^{\frac{1}{2}} z\right] - I_k(z) \qquad (z \in S).$$

Mais l'intégrale $I_k(z)$ converge pour chaque valeur finie de z, car

$$P(z, u, v, k-1) < 2A\left\{|v|^{\frac{1}{2}} + \frac{|v|}{\left(k-\frac{1}{2}\right)^{\frac{1}{2}}}\right\} + |z|\left[(2k)^{\frac{1}{2}} + |v|^{\frac{1}{2}}\right]$$

et nous supposons que k vérifie (2.7).

Cela étant, nous voyons que $E^{+}(z; f)$ est une fonction entière. La même méthode montre aussi que $E^{-}(z; f)$ est entière. Le théorème est donc démontré.

COROLLAIRE. — *Les séries*

$$\sum_{n=1}^{\infty} A_{2n}^{-1} G(2n) h_{2n}(z) \quad \text{et} \quad \sum_{n=0}^{\infty} A_{2n+1}^{-1} G(2n+1) h_{2n+1}(z) \tag{2.11}$$

représentent des fonctions entières.

En effet, ce sont les séries des fonctions

$$\frac{1}{2}[f(z)+f(-z)] \quad \text{et} \quad \frac{1}{2}[f(z)-f(-z)],$$

où $f(z)$ est définie par (2.3).

3. FONCTIONS AUX SINGULARITÉS FINIES. — L'introduction du facteur i^n dans le $n^{\text{ième}}$ terme de la série (2.3) change la situation d'une manière frappante. En général, la série

$$F(z) \equiv \sum_{n=1}^{\infty} i^n A_n^{-1} G(n) h_n(z), \tag{3.1}$$

qui a la même bande de convergence que (2.3), définit une fonction avec des singularités finies. Si $G(t)$ est réel positif avec t, alors le point $z = \tau i$ est singulier (*voir* [2], p. 918); ce résultat reste valable même si la suite $\{G(n)\}$ n'a pas un signe constant à condition que les changements de signe se présentent assez lentement. Dans le cas où $G(t)$ est une fonction analytique on peut obtenir quelques renseignements sur le domaine d'holomorphie de $F(z)$ en utilisant la méthode employée ci-dessus.

THÉORÈME 2. — *Soit* $G(t)$ *une fonction holomorphe dans le secteur*

$$-\pi \leqq -\vartheta_1 < \arg t < \vartheta_2 \leqq \pi,$$

telle que

$$\eta(\theta) \equiv \limsup_{r\to\infty} r^{-\frac{1}{2}} \log |G(re^{i\theta})| \qquad (-\vartheta_1 < \theta < \vartheta_2) \tag{3.2}$$

soit une fonction continue et bornée. Soit $\eta(0) < 0$. *Alors la série* (3.1) *converge dans la bande*

$$2^{-\frac{1}{2}}\eta(0) < y < -2^{-\frac{1}{2}}\eta(0) \tag{3.3}$$

et la fonction $F(z)$ *est holomorphe dans le domaine* Δ *obtenu par la construction suivante. Soit*

$$\Gamma = \left[z \mid \mathfrak{I}\left(z e^{\frac{1}{2}i\theta}\right) \leq \eta(\theta),\ -\mathfrak{S}_1 < \theta < \mathfrak{S}_2\right] \tag{3.4}$$

et soit Δ *ce qui reste du plan complexe si l'on néglige les ensembles* $\overline{\Gamma}$ *et* $-\overline{\Gamma}$.

Ici nous avons

$$\begin{cases} E^{+}(z; f) = \sum_{n=1}^{\infty} G(n) \exp\left[i(2n+1)^{\frac{1}{2}} z\right], \\ E^{-}(z; f) = \sum_{n=1}^{\infty} (-1)^n G(n) \exp\left[-i(2n+1)^{\frac{1}{2}} z\right]. \end{cases} \tag{3.5}$$

Considérons l'intégrale

$$\int G(t) \exp\left[i(2t+1)^{\frac{1}{2}} z\right] \frac{dt}{e^{2\pi i t} - 1} \tag{3.6}$$

prise le long du rectangle de sommets

$$\frac{1}{2} - iq, \quad m + \frac{1}{2} - iq, \quad m + \frac{1}{2} + iq, \quad \frac{1}{2} + iq.$$

La valeur de l'intégrale est

$$\sum_{n=1}^{m} G(n) \exp\left[i(2n+1)^{\frac{1}{2}} z\right].$$

Pour $y > 2^{-\frac{1}{2}}\eta(0)$ l'intégrale tend vers $E^{+}(z; F)$ quand m tend vers l'infini. Nous pouvons remplacer les lignes horizontales par les lignes obliques

$$\arg(2t+1) = \theta, \qquad u > \frac{1}{2}, \qquad 0 < \theta < \mathfrak{S}_2,$$

$$\arg(2t+1) = \theta_1, \qquad u > \frac{1}{2}, \qquad -\mathfrak{S}_1 < \theta_1 < 0.$$

L'intégrale prise le long du premier rayon converge si

$$\mathcal{J}\left(z\, e^{\frac{1}{2}i\theta}\right) > \eta(\theta), \tag{3.7}$$

tandis que la deuxième intégrale converge pour chaque valeur de z grâce au facteur $[e^{2\pi i t} - 1]^{-1}$. Alors pour chaque θ avec $0 < \theta < \vartheta_2$ nous obtenons le demi-plan (3.7) dans lequel $E^+(z; f)$ est holomorphe.

De la même manière on traite les valeurs de θ dans $(-\vartheta_1, 0)$. Il faut seulement remplacer l'intégrale (3.5) par

$$-\int G(t) \exp\left[i(2t+1)^{\frac{1}{2}} z\right] \frac{dt}{e^{-2\pi i t} - 1}. \tag{3.8}$$

Dans ce cas, le rayon $\arg(2t+1) = \theta_2$ est arbitraire dans le demi-plan supérieur sauf pour la condition $0 < \theta_2 < \vartheta_2$ et l'intégrale converge pour chaque z tandis que l'intégrale dans le demi-plan inférieur suivant le rayon $\arg(2t+1) = \theta$ converge si z vérifie la condition (3.7).

On voit donc que les singularités de $E^+(z; f)$ appartiennent à l'ensemble $\bar{\Gamma}$.

Quant à $E^-(z; f)$ on montre que c'est une fonction entière de z. Cela découle de la représentation

$$E^-(z; f) = \int G(t) \exp\left[-i(2t+1)^{\frac{1}{2}} z\right] \frac{e^{\pi i t}\, dt}{e^{2\pi i t} - 1}, \tag{3.9}$$

où l'intégrale est prise le long des deux rayons $\arg(2t+1) = \theta_k$, $-\vartheta_1 < \theta_1 < 0 < \theta_2 < \vartheta_2$ joints par un segment de la ligne $u = \frac{1}{2}$. Évidemment l'intégrale converge pour chaque z. Le théorème est donc démontré.

Il n'est pas nécessaire de supposer que $\eta(\theta)$ reste borné quand θ tend vers $-\vartheta_1$ ou ϑ_2. On peut remplacer cette hypothèse par une inégalité de la forme

$$\eta(\theta) < h(\theta) = \begin{cases} A\left(\operatorname{tg}\dfrac{\pi|\theta|}{2\vartheta_1}\right)^{\frac{1}{2}} - B, & -\vartheta_1 < \theta < 0, \\ A\left(\operatorname{tg}\dfrac{\pi\theta}{2\vartheta_2}\right)^{\frac{1}{2}} - B, & 0 \leq \theta < \vartheta_2. \end{cases} \tag{3.10}$$

Soient $\vartheta_1 = \vartheta_2 = \pi$ et $\eta(\theta) \leqq 0$ pour chaque θ. Dans ce cas on trouve que $\bar{\Gamma}$ se réduit à un segment $(-i\infty, -i\alpha)$ de l'axe imaginaire et Δ est tout le plan sauf les segments $(-i\infty, -i\alpha)$ et $(i\alpha, i\infty)$.

BIBLIOGRAPHIE.

[1] E. Hille, *Sur les séries associées à une série d'Hermite* (*C. R. Acad. Sc.*, t. **209**, 1939, p. 714-716).

[2] E. Hille, *Contributions to the theory of Hermitian series* (*Duke Math. J.*, t. 5, 1939, p. 875-936).

[3] E. Hille, *Contributions to the theory of Hermitian series*. II. *The representation problem* (*Trans. Amer. Math. Soc.*, t. 47, 1940, p. 80-94).

Chapter 10
Differential Transforms

The paper included in this chapter is

[84] On the oscillation of differential transforms and the characteristic series of boundary-value problems

Reprinted from SEMINAR REPORTS IN MATHEMATICS (Los Angeles)
UNIVERSITY OF CALIFORNIA PUBLICATIONS IN MATHEMATICS, N.S. VOLUME 2, NO. 1 (1944)

ON THE OSCILLATION OF DIFFERENTIAL TRANSFORMS AND THE CHARACTERISTIC SERIES OF BOUNDARY-VALUE PROBLEMS *

BY EINAR HILLE

[Received May 17, 1942]

1. The problem of S. Bernstein. The following is an old problem which is still very active: *Given a real function* f(x) *in a finite or infinite interval* (a, b), *find when* f(x) *is analytic in* (a, b). It is obvious that $f(x)$ must have derivatives of all orders, but this condition is merely necessary and not sufficient, so that further restrictions have to be imposed on the derivatives. A. Pringsheim has shown that *a necessary and sufficient condition that* f(x) *be holomorphic in the closed interval* [a, b], $(b-a<\infty)$, *is the existence of two quantities* M *and* R *such that* $|f^{(n)}(x)| \leqq MR^n n!$ *for* $a \leqq x \leqq b$ *and all* n.

In 1914 S. Bernstein attacked the problem from entirely different angles. One mode of attack was through the theory of polynomial approximation. If $f(x)$ is continuous in $[a, b]$, ($b - a$ finite), let $E_n[f(x)]$ denote the greatest lower bound of the maximum in $[a, b]$ of $|f(x) - P_n(x)|$, where $P_n(x)$ is an arbitrary polynomial of degree n. Then $f(x)$ is holomorphic in $[a, b]$ if and only if $E_n[f(x)] \leqq M\rho^n$ for all n, where $0 < \rho < 1$. This basic fact is intimately connected with his second line of attack which involves the signs of the derivatives. He proved that *a necessary and sufficient condition in order that* f(x) *shall be holomorphic in the circle* $|z - a| < b - a$ *is that* f(x) *be the difference of two functions* g(x) *and* h(x) *which are positive in the interval* (a, b) *together with their derivatives of all orders.* A much deeper result is the following: f(x) *is certainly analytic in* (a, b) *if no derivative changes its sign in* (a, b). The domain of holomorphism is now bounded by a circle with center at the mid-point of the segment and diameter one-fourth of its length plus the four tangents drawn to this circle from the end points of the segment. Bernstein could even allow some sign changes of the derivatives, provided they were of sufficiently low frequency.

The question of the relationship between the analytical character of f(x) *in* (a, b) *and the oscillatory properties of its derivatives is the problem of S. Bernstein.* It remained essentially untouched until a couple of years ago when interest in the problem suddenly flared up in this country. As evidence of this interest, mention may be made of the investigations by R. P. Boas, G. Pólya, A. C. Schaeffer, I. J. Schoenberg, G. Szegö, J. D. Tamarkin, D. V. Widder, Norbert Wiener, and the present writer. The results of most of this work are now in the course of publication.

Before passing over to the main subject of this paper, it should be noted that Schaeffer has succeeded in showing that *if* N_k, *the number of sign changes of* $f^{(k)}(x)$ *in* (a, b), *is bounded,* ($N_k \leqq N$ *for all* k), *then* f(x) *is analytic in* (a, b), the domain of holomorphism being a rhomboid with the given interval as one diagonal, the length of the other diagonal depending upon N. This result appears to lie very deep.

2. The investigations of Pólya-Wiener. For the case of periodic functions, Pólya

* Presented before the Mathematics Club of the University of California, Los Angeles, on April 30, 1942.

and Wiener devised an entirely new line of attack on the Bernstein problem which has led to remarkable results. I shall describe the results first and then analyze the method, which is essential for my own work.

Suppose that $f(x)$ is real, $f(x+2\pi) = f(x)$, and $f(x) \in C^{(\infty)}(-\infty, \infty)$. Suppose further that $f^{(k)}(x)$ has N_k sign changes in the period. Here N_k is necessarily an even number. The first and perhaps most striking result is: *If* $N_k \leqq N$ *for all* k, *then* f(x) *is a trigonometric polynomial of degree not greater than* N/2. *Conversely, if* f(x) *is a real trigonometric polynomial of degree* d, *then* $N_k = 2d$ *for all large* k. If N_k is unbounded, the results of Pólya and Wiener are less complete and, as it turned out, not the best possible. They derived no conditions for mere analyticity, but showed that if $N_k = o(k^{1/2})$ then $f(x)$ is an entire function; and if the order of N_k is still lower, they could assign upper bounds for the order of the entire function. Szegö has found new independent proofs of these results, and he could show that $N_k < 2k/(\log k)$ *implies that* f(x) *is entire and this condition could not be replaced by* $N_k = O(k)$, *whereas* $N_k = O(k^\alpha)$, $(0 < \alpha < 1)$, *implies that the order of* f(x) *is at most* $1/(1-\alpha)$ *and this order could actually be reached.*

After outlining the method used by Pólya and Wiener, I shall present the method in such a manner that the possibility of generalizing the problem becomes apparent. The class of functions considered by them is the class of Fourier series

$$f(x) = \frac{1}{2}a_0 + \sum_1^\infty (a_n \cos nx + b_n \sin nx)\,, \tag{2.1}$$

where the coefficients are real and

$$\sum_1^\infty n^k(|a_n| + |b_n|) \tag{2.2}$$

converges for every k. This merely expresses that the Fourier series of $f^{(k)}(x)$ can be obtained by termwise differentiation and that the resulting series are absolutely convergent for all k. We note that the differential operator D^2 takes $\cos nx$ into $-n^2 \cos nx$ and $\sin nx$ into $-n^2 \sin nx$, i.e., *the terms of* (2.1) *are characteristic functions of the operator* $L = D^2$ *corresponding to the boundary-value problem*

$$(D^2 + \mu)u = 0\,, \qquad u(-\pi) = u(\pi)\,, \qquad u'(-\pi) = u'(\pi)\,, \tag{2.3}$$

the characteristic values being $\mu = n^2$, $(n = 0, 1, 2, \cdots)$. We shall refer to (2.1) as *the characteristic series* of this boundary-value problem.

The basic observation of Pólya and Wiener is that the operator $D^2 - \lambda$, $(\lambda \geqq 0)$, is *oscillation preserving with respect to periodic functions* in the following sense: *If* f(x) *is periodic and has continuous first- and second-order derivatives and if* f(x) *has* N *sign changes in the period, then* $(D^2 - \lambda)f(x) = f''(x) - \lambda f(x)$ *has at least* N *sign changes in the period.* This follows from the representation

$$(D^2 - a^2)f(x) = e^{-ax}D\{e^{2ax}D[e^{-ax}f(x)]\}\,, \tag{2.4}$$

where $\alpha^2 = \lambda$. By assumption, $f(x)$ has an even number N of sign changes in the period. By Rolle's theorem, $D[e^{-\alpha x}f(x)] = e^{-\alpha x}[f'(x) - \alpha f(x)]$ has at least $N - 1$ sign changes in the period. But the oscillatory factor is periodic and must have an even number of sign changes in the period. Hence no sign change can be lost by the first differentiation and similarly for the second one, so that the left-hand side of (2.4) must have at least N sign changes in the period.

This observation is used as follows: Pólya and Wiener construct an auxiliary function which has at least as many sign changes in the period as $f^{(2k)}(x)$ has. They take

$$F(x, m, k) = \sum_{1}^{\infty} \left[\frac{4m^2 n^2}{(m^2 + n^2)^2}\right]^k (a_n \cos nx + b_n \sin nx) . \tag{2.5}$$

This function has the property

$$(D^2 - m^2)^{2k} F(x, m, k) = (4m^2)^k f^{(2k)}(x) , \tag{2.6}$$

whence it follows that its number of sign changes in the period is at least N_{2k}, the number of sign changes of $f^{(2k)}(x)$. Here m is an arbitrary positive integer. Now the "root eating factor"

$$\frac{4m^2n^2}{(m^2 + n^2)^2} \tag{2.7}$$

equals one if and only if $n = m$, and for all other values of n it is less than one. Choosing k sufficiently large, we can thus obtain that the mth term in the expansion is stressed at the expense of all other terms. Suppose that the mth term is actually present, i.e., that $|a_m| + |b_m| \neq 0$. Then $a_m \cos mx + b_m \sin mx$ has $2m$ sign changes in the period and they are separated by alternating positive maxima and negative minima. At such a maximum the mth term in $F(x, m, k)$ dominates the sum of the series, k being sufficiently large, so that $F(x, m, k)$ is also positive; similarly at the minima of the mth term. It follows that $F(x, m, k)$ has at least $2m$ sign changes in the period. By the oscillation-preserving property of the transformation $D^2 - m^2$, we conclude that for all large values of k, $f^{(2k)}(x)$ also has at least $2m$ sign changes in the period. But, by assumption, $f^{(2k)}(x)$ has N_{2k} sign changes in the period and $N_{2k} \leqq N$ for all k. Hence we must have $2m \leqq N$. In other words, $a_m = b_m = 0$ for $2m > N$ and we have proved that $f(x)$ is a trigonometric polynomial of degree not exceeding $N/2$. It is obvious that we can draw the same conclusion if we know merely that $N_{2k} \leqq N$ for infinitely many values of k. The converse is easily proved.

3. Exploration of generalizations. It should be reasonably obvious from this exposition of the leading ideas of the method of Pólya and Wiener that the problem can be generalized considerably. *The method is evidently designed for the discussion of the oscillatory properties of functions represented by expansions in terms of solutions of boundary-value problems associated with linear second-order differential equations.* Let us run over the ground more or less formally to start with. Afterward we shall see to what extent the indicated research program can be carried out rigorously.

We replace the operator D^2 by a more general linear second-order differential operator L such that

$$Ly = p_0(x)y + p_1(x)y' + p_2(x)y'' \tag{3.1}$$

with analytic coefficients. We take a finite or infinite interval (a, b) in which the coefficients are holomorphic and an associated boundary-value problem

$$(L + \mu)u = 0\,, \qquad \text{B.C.}(u, a, b) = 0\,, \tag{3.2}$$

where the nature of the boundary condition will be left open for the time being.

We assume the existence of an infinite number of characteristic values $\{\mu_n\}$ with associated characteristic functions $\{u_n(x)\}$, normalized in weighted mean square. We then consider the corresponding characteristic series

$$f(x) \sim \sum_1^\infty a_n u_n(x) \tag{3.3}$$

with real coefficients. Formally we have

$$L^k f(x) \sim \sum_1^\infty (-\mu_n)^k a_n u_n(x)\,, \tag{3.4}$$

since $Lu_n(x) = -\mu_n u_n(x)$. We shall assume that the right-hand side converges in weighted mean square to the left for every k. This amounts essentially to assuming that

$$\sum_1^\infty |\mu_n{}^k a_n| < \infty \tag{3.5}$$

for all k. We denote by F the class of functions defined by (3.3), the coefficients being real and satisfying (3.5). We can evidently expect that $f(x)$ and all its differential transforms $L^k f(x)$ shall satisfy the same boundary conditions as the $u_n(x)$.

The problem of Pólya and Wiener or the S. Bernstein problem has now been transformed into the following: *If* $f(x) \in F$, *what are the relations between the oscillatory properties of the differential transforms* $L^k f(x)$ *in* (a, b) *and the analytical properties of* $f(x)$*? If the number of sign changes of* $L^k f(x)$ *in* (a, b) *be denoted by* N_k, *what relations hold between the rate of growth of* N_k *and the analytical character of* $f(x)$*?* In this paper I shall restrict the discussion to the case in which N_k is bounded or, more generally, $\liminf_{k\to\infty} N_k = N < \infty$. In this case we obtain, in perfect analogy with the periodic case, that $f(x)$ is a finite linear combination of characteristic functions $u_n(x)$. Extensions to the case in which N_k is unbounded are possible, but much more difficult, and in the few cases in which the method of Szegö applies, the latter gives better results than the method of Pólya and Wiener.

The first step in the attack on the problem will be to show that $L - \lambda$ is oscillation preserving in the interval (a, b) with respect to the class F when $\lambda > 0$.

Here an analogue of formula (2.4) would be useful. The required formula has long been in the literature and reads

$$(L - \lambda)f(x) = \frac{p_2(x)W(x)}{Y(x)} D\left\{\frac{Y^2(x)}{W(x)} D\left[\frac{f(x)}{Y(x)}\right]\right\}. \tag{3.6}$$

Here $Y(x)$ is a solution of the auxiliary equation $(L - \lambda)y = 0$, and $W(x)$ is the Wronskian of $Y(x)$ and a second linearly independent solution of the same equation. This formula reduces to (2.4) when $L = D^2$ if we take $Y(x) = e^{ax}$ and remember that $W(x) =$ const. The choice of $Y(x)$ is highly arbitrary and the only obvious requirement is that $Y(x)$ shall be different from zero in (a, b). This, however, imposes a condition on the differential operator L: *the auxiliary equation* $(L - \lambda)y = 0$ *must possess non-oscillatory solutions in the interval* (a, b). We shall return to this condition below.

The proof of Pólya and Wiener for the oscillation-preserving character of the operator $D^2 - \lambda$ would seem to be based upon rather special considerations. If we keep in mind, however, that they base their conclusions on the fact that the class of functions to which the operator is applied satisfies the boundary conditions of the associated boundary-value problem, the mode of generalization becomes obvious. We have to utilize effectively the fact that (3.3) satisfies the boundary conditions of (3.2) and to choose $Y(x)$ in an appropriate manner. It is clear on the other hand that the requirement that $L - \lambda$ shall be oscillation preserving in (a, b) with respect to the class F will impose certain restrictions upon the boundary conditions defining F.

If these difficulties have been overcome, we arrive at the question of generalizing the auxiliary function $F(x, m, k)$ of formula (2.5) This evidently requires finding an analogue of the "root eating factor" (2.7). The latter, however, is constructed in a perfectly obvious manner in terms of characteristic numbers of the boundary-value problem, so it is obvious that we should try

$$\frac{4\mu_m\mu_n}{(\mu_m + \mu_n)^2} \tag{3.7}$$

and form

$$F(x, m, k) = \sum_1^\infty \left[\frac{4\mu_m\mu_n}{(\mu_m + \mu_n)^2}\right]^k a_n u_n(x). \tag{3.8}$$

The factor (3.7) has all the desirable properties, provided the characteristic values μ_n are all real non-negative, as is usually true. In order to be able to utilize the stressing effect of the multiplier, we have to have information regarding the oscillatory properties of the characteristic functions $u_n(x)$, as well as some information concerning their general behavior in (a, b). Now it is well known since the investigations of A. Haar and E. W. Hobson that the solutions of reasonably simple boundary-value problems behave essentially like simple harmonic oscillations in the interior of the interval, so this phase of the problem should not cause serious difficulties.

4. Sufficient conditions. We have now the question of putting these heuristic considerations on a firm basis. We begin by imposing the following conditions:

A_1. $p_m(x)$ *shall be holomorphic in* (a, b), (m = 0, 1, 2).
A_2. $p_0(x) \leqq 0$, $p_2(x) > 0$ *in* (a, b).

The first of these conditions ensures that the operation L can be repeated as often as we please on functions $f(x)$ having derivatives of all orders. Together the two conditions ensure that there is no singular point of the differential equation $(L + \mu)u = 0$ in the open interval (a, b) and that for negative values of μ the solutions are non-oscillatory, i.e., $u(x)u'(x)$ can vanish at most once in the interval. The behavior of the coefficients at the end points of the interval is left open, but will have to be adjusted to the needs of the boundary-value problem.

We shall now list the properties of the characteristic values and functions which will be desirable for the discussion. Let us put

$$(4.1)\quad P(x) = [p_2(x)]^{-1} \exp\left\{\int^x \frac{p_1(t)}{p_2(t)}\,dt\right\}, \quad K(x) = p_2(x)P(x), \quad G(x) = -p_0(x)P(x),$$

so that the self-adjoint form of the equation $Ly = 0$ becomes

$$(4.2)\qquad D[K(x)Dy] - G(x)y = 0.$$

We now require:

A_3. *The functions* $\{[P(x)]^{1/2}u_n(x)\}$ *form a real orthonormal system, complete in* $L_2(a, b)$.

A_4. $0 < \mu_n \leqq \mu_{n+1}$. *The series* $\sum \mu_n^{-a}$ *converges for some* $a > 0$.

A_5. *There exist constants* β *and* γ *and a non-negative function* $U(x)$, *continuous in* (a, b), *such that*

$$|u_n(x)| \leqq \mu_n^{\beta} U(x), \qquad |u_n'(x)| \leqq \mu_n^{\gamma} U(x).$$

A_6. *In every fixed interval* (c, d), $(a \leqq c < d \leqq b)$, *the number of zeros,* $Z_n(c, d)$ *say, of* $u_n(x)$ *in* (c, d) *tends to infinity with* n. $Z_n(a, b)$ *is finite and a never decreasing function of* n.

A system $S\{L, u_n(x), \mu_n; (a, b)\}$ satisfying these six assumptions will be called *admissible*. If the system is admissible, we call the characteristic series (3.3) admissible, provided the coefficients are real and (3.5) holds, and the function represented by the series is also called admissible. The assumptions are such that every admissible series will have to converge absolutely in (a, b) and uniformly with respect to x in every fixed interior interval. Moreover, the operation L can be carried out termwise and the resulting series is also admissible.

The boundary conditions have been left open so far. We consider four different types of boundary-value problems, referred to as the *Sturm-Liouville*, the *periodic*, the *singular*, and the *semi-singular* cases. Each problem is characterized by an end-point condition on the coefficients, which permits the formulation of the problem, and a boundary condition which sets up the problem. These conditions read as follows in the four cases:

CASE	END-POINT CONDITION	BOUNDARY CONDITION
1	$p_m(x)$ regular at $x = a$ and b and $p_2(a)p_2(b) \neq 0$	$u(a) = 0, \quad u(b) = 0$
2	Same as case 1 plus $K(a) = K(b)$	$u(a) = u(b), \quad u'(a) = u'(b)$
3	$R(x, x_0; \lambda) \to \infty$ when $x \to a$ and when $x \to b$	$u(x)/Y(x, \lambda) \to 0$ when $x \to a$ and when $x \to b$
4	Same as case 3 at $x = a$, same as case 1 at $x = b$	Same as case 3 at $x = a$, $C_1u(b) + C_2u'(b) = 0$

Here cases 1 and 2 need very little further explanation. They are simple cases of classical boundary-value problems. In both cases conditions A_3 through A_6 are trivially satisfied and can be omitted from the formulation of the problem. In case 1 other boundary conditions could be allowed. Cases 3 and 4 need explanation. Here

$$R(x, x_0; \lambda) = \int_{x_0}^{x} \frac{dt}{K(t)} \int_{x_0}^{t} [G(s) + \lambda P(s)]\,ds\,, \tag{4.3}$$

with x_0 an arbitrary interior point of (a, b) and $\lambda > 0$. The end-point condition is to hold for all $\lambda > 0$; but if it holds for one, it holds for all. Likewise the choice of x_0 is immaterial. The condition is necessary and sufficient in order that the equation $(L - \lambda)y = 0$ shall have a solution $Y(x, \lambda)$ which becomes infinite at both ends of the interval and has a positive minimum at $x = x_0$. This is the $Y(x, \lambda)$ which appears in the boundary condition. In case 4 we have to add that $C_1 \geqq 0$, $C_2 \geqq 0$, $C_1 + C_2 > 0$.

In cases 3 and 4 we need an additional condition to the effect that the function $U(x)$ entering in A_5 is dominated by some multiple of $Y(x, \lambda)$, where the multiplier may depend upon λ and x_0 but not upon x.

An admissible system which corresponds to any one of these four types of boundary-value problems will be called *conservative*. For a conservative system we have, first, *the transformation* $L - \lambda$ *is oscillation preserving for every* $\lambda > 0$ *in the interval* (a, b) *with respect to the class of admissible series.* Secondly, the Pólya-Wiener theorem holds in the following form:

If $f(x)$ *is admissible and* $\liminf_{k \to \infty} N_k = N$, *where* N_k *is the number of sign changes of* $L^k f(x)$ *in* (a, b), *then there exists an integer* $M = M(N)$ *such that*

$$f(x) = \sum_{1}^{M} a_n u_n(x)\,.$$

In the case of simple characteristic values, $M(N)$ *is the largest integer for which* $u_n(x)$ *has at most* N *sign changes in* (a, b). *In the case of double characteristic values* $M(N)$ *can exceed this number by at most one unit.*

The proof follows the same general pattern as in the Pólya-Wiener case but is naturally more complicated. In the Sturm-Liouville case all characteristic values are simple and $M(N) = N$. In the periodic case the results are equally simple.

It is in order to add some remarks concerning case 3, the singular boundary-value problem. This type of problem appears to be new, but among the orthogonal systems which arise out of such problems there occur some of the most important systems in analysis. It is enough to mention the Legendre, Jacobi, Hermite, and Laguerre polynomials, which all correspond to singular boundary-value problems.

The Legendre case is sufficiently simple to be considered here. We have

$$Ly=D[(1-x^2)Dy], \quad a=-1, \quad b=+1, \quad \mu_n=n(n+1), \quad u_n(x)=\left(n+\frac{1}{2}\right)^{1/2} P_n(x).$$

It is an easy matter to verify that the A-conditions are satisfied so that the system is admissible. In order to show that it is conservative of type 3, we form

$$R(x,0;\lambda)=\lambda\int_0^x \frac{t}{1-t^2}\,dt=\frac{1}{2}\lambda \log[1/(1-x^2)]\,.$$

Since this expression becomes infinite when $x\to\pm1$, the appropriate end-point condition is satisfied, and we have merely to verify that the Legendre polynomials are the solutions of the singular boundary-value problem

$$D[(1-x^2)Du]+\mu u=0\,, \qquad u(x)/[\log(1-x^2)]\to 0\,, \qquad x\to\pm1\,.$$

Now it is well known that in the neighborhood of the point $x = 1$ Legendre's equation has a regular solution and also a solution which becomes infinite as a constant times $\log 1/(1 - x)$, and at the point $x = -1$ we have also a regular solution and a solution which becomes infinite as a constant times $\log 1/(1 + x)$. The boundary condition then requires that we choose the solution which is regular at $x = 1$ and then choose μ so that this solution becomes regular at $x = -1$ as well. It is well known that this gives $\mu_n = n(n + 1)$ and $u_n(x)$ is a constant multiple of the nth Legendre polynomial $P_n(x)$. It is not difficult to show that the corresponding class of admissible functions is simply $C^{(\infty)}[-1, 1]$, so we arrive at the following result:

If $f(x) \in C^{(\infty)}[-1,1]$ *and* $\liminf_{k\to\infty} N_k = N$, *where* N_k *is the number of sign changes of* $[D(1-x^2)D]^k f(x)$ *in* $(-1, 1)$, *then* $f(x)$ *is a polynomial of degree* N. *Conversely, if* $f(x)$ *is a real polynomial of degree* N, *then* $N_k = N$ *for all large* k.

We have thus a characterization of ordinary polynomials in the interval $(-1, 1)$ by means of the Legendre operator analogous to the characterization of trigonometric polynomials given by Pólya and Wiener. The same type of characterization can be given in $(0, \infty)$ with the aid of the Laguerre operator $xD^2 + (1 - x)D$ and in $(-\infty, \infty)$ with the aid of the Hermite operator $D^2 - 2xD$.

[*Bibliographical Note*.—The reader who would like to know more about the questions discussed in the present paper may find the following references helpful. The basic paper of S. Bernstein appeared in volume 75 (1914) of the Mathematische Annalen; later the material was reproduced for the most part in his book *Leçons sur les propriétés extrémales, etc.* (Paris, 1926), pp. 162–197. For the modern development of the S. Bernstein problem, see a paper by G. Pólya, entitled *On the zeros of the derivatives of a function and its analytic character*, which was delivered before the American Mathematical Society on April 3, 1942; this appeared in volume 49 (1943) of the Bulletin of the Society. The paper by G. Pólya and N. Wiener, *On the oscillation of the derivatives of a periodic function*, appeared in volume 52 (1942) of the Transactions. This paper is to be followed by a series of papers on the oscillation of differential transforms of which the first four, two by G. Szegö, one by A. C. Schaeffer, and one by the present author, have appeared in volumes 52, 53, and 54 of the Transactions.]

Chapter 11
Ergodic Theory

The paper included in this chapter is

[89] Remarks on ergodic theorems

A commentary on this paper and a note on Hille's work on abstract summability in connection with orthogonal expansions, multipliers, and approximation theory can be found starting on page 677.

REMARKS ON ERGODIC THEOREMS

BY
EINAR HILLE

1. **Introduction.** In recent papers Nelson Dunford [7, 7a](1) has made a study of the spectral theory of linear bounded transformations on a Banach space to itself with a view to finding necessary and sufficient conditions for the convergence in various topologies of a sequence of linear transformations to a projection. This led him in particular to a systematic theory of ergodic theorems.

It is well known that in "discrete" ergodic theory one is concerned with the sequence of powers T^n of a given transformation T. Ordinarily this sequence does not converge to a limit in the topology in question, but the sequence may be summable to a generalized limit. Classical ergodic theorems involve the use of $(C, 1)$ means, but arbitrary methods of summation have been considered by L. W. Cohen [6] and others.

The assertion that the sequence $\{T^n\}$ is summable by the method of summation (S) is a statement that the series(2)

$$(1.1)\quad I + (T - I)\lambda^{-1} + (T^2 - T)\lambda^{-2} + \cdots + (T^n - T^{n-1})\lambda^{-n} + \cdots$$

is summable (S) for $\lambda = 1$. When the series converges, its sum is $(\lambda - 1)R(\lambda)$ where $R(\lambda)$ is the resolvent of T,

$$(1.2)\qquad R(\lambda) = \sum_{n=0}^{\infty} T^n \lambda^{-n-1}.$$

Both series converge outside the least circle $|\lambda| = r$ which contains the spectrum of T.

The fundamental importance of the resolvent for all these questions has been brought out very clearly by Dunford. The present paper, which is an outgrowth of discussions between Professor Dunford and the author, is concerned with two aspects of the relationship between the properties of $R(\lambda)$ and the validity of an ergodic theorem for T.

The first aspect is the question of relations between different types of summability and corresponding ergodic theorems, in particular (C, k) and Abel summability. By analogy with numerically valued series one would expect that the existence of

Presented to the Society, February 27, 1943, jointly with G. Szegö; received by the editors June 24, 1944.

(1) Numbers in brackets refer to the Bibliography at the end of the paper.

(2) In formulas (1.1) to (1.4) convergence and limits are taken in the sense of a fixed unspecified topology.

$$(1.3) \qquad (C,\, k)\text{-lim } T^n$$

should imply the existence of

$$(1.4) \qquad \lim_{\lambda \to 1+0} (\lambda - 1)R(\lambda) = (A)\text{-lim } T^n$$

and the equality of the two limits. Conversely, one might expect that the existence of the Abel limit should imply the existence of the (C, k) limit provided appropriate Tauberian conditions are satisfied.

In the latter direction Dunford (oral communication, unpublished) has already extended the Fatou-Riesz theorem (see M. Riesz [13]). If $(\lambda-1)R(\lambda)$ is holomorphic at $\lambda=1$ where its value is P and $\|T^n\| \leqq M$, then $(C, k)\text{-lim } T^n = P$ for $k>0$. In the present paper we extend N. Wiener's general Tauberian theorem (see [15, Theorem VIII]) and prove as a consequence that $(A)\text{-lim } T^n = P$ plus $\|T^n\| \leqq M$ implies $(C, k)\text{-lim } T^n = P$ for every $k>0$.

The second question concerns the character of $\lambda=1$ for the resolvent $R(\lambda)$ when an ergodic theorem holds for T. Dunford had shown (see [7, Theorem 3.16]) that if $(C, 1)\text{-lim } T^n = P$ in the uniform topology, then $\|T^n\| = o(n)$ and $(\lambda-1)R(\lambda)$ is holomorphic and equals P at $\lambda=1$. At the time when this investigation was started, Dunford had not yet considered the corresponding question in the strong topology and the first results (the case $\alpha=1$ of Theorems 10 and 11 below) showed that his conditions were no longer valid in the strong case. Though in the mean time Dunford [7, Theorem 3.19] has also settled the strong case, our results would still seem to be of some interest since they prove the existence of ergodic theorems of specified type for transformations which are of considerable interest in analysis, namely fractional integration.

In this paper a careful discussion is made of the transformations

$$(1.5) \qquad T_\alpha f = f(t) - \frac{1}{\Gamma(\alpha)} \int_0^t (t-u)^{\alpha-1} f(u)\, du$$

on the spaces $C_0[0, 1]$ and $L_1(0, 1)$. Here the resolvent $R_\alpha(\lambda)$ has $\lambda=1$ as its only singularity, but it is an essential singular point and not a pole. We prove the existence of $(A)\text{-lim } T_\alpha^n$ for $0<\alpha<2$, of $(C, k)\text{-lim } T_1^n$ for $k>\frac{1}{2}$, and of $\lim T_\alpha^n$ for $0<\alpha<1$, all in the strong topologies of the two spaces, and corresponding results for almost everywhere convergence when $f(t) \in L_1(0, 1)$. The limits found for the parameters α and k are sharp. Since, in particular, $(C, 1)\text{-lim } T_1^n$ exists and $\|T_1^n\|$ is of the order of $n^{1/4}$, an example is given of a strong ergodic theorem in which the usual condition $\|T^n\| \leqq M$ for all n is not satisfied. It is still a far cry from $O(n^{1/4})$ to $o(n)$, however. For $\alpha=1$ the results follow from known properties of Laguerre polynomials, but for $\alpha \neq 1$ they require a detailed investigation of a class of polynomials which apparently has not been studied before. I am indebted to Professor G. Szegö for the asymptotic expressions of these polynomials presented in §7 below.

The subject matter is distributed as follows. Abstract Abelian and Tauberian theorems are given in §2 with applications to ergodic theorems in §3. That $\lambda=1$ need not be an isolated singularity of the resolvent in the strong topology is shown in §4. The discussion of T_α occupies §§5, 6, and 7.

The present paper discusses relations between Tauberian theorems and spectral theory on one hand and ergodic theorems of the "discrete" type on the other. The "continuous" case may be subjected to a similar analysis, but here the discussion is more intimately connected with the analytic theory of semi-groups of linear transformations and will be given in another connection.

2. **Some abstract Abelian and Tauberian theorems.** In the present paragraph X is a complex Banach space of elements $x, y, \cdots$ and convergence of a sequence $\{z_n\}$ to the limit z_0 is meant in the strong sense, that is $\|z_n - z_0\| \to 0$. We shall extend a few theorems of Abelian or Tauberian character to abstract spaces mainly with a view to relating (C, k) and Abel summability.

THEOREM 1. *Let $\{x_n\} \in X$ and let k be fixed positive. Then*

$$(C, k)\text{-}\lim x_n = y_0 \tag{2.1}$$

implies that[3]

$$\|x_n\| = o(n^k), \tag{2.2}$$

$$(A)\text{-}\lim x_n = y_0. \tag{2.3}$$

We have to recall some formulas. Define $C_n^{(k)}$ by the generating function

$$(1-\mu)^{-1-k} = \sum_{n=0}^{\infty} C_n^{(k)} \mu^n,$$

so that

$$C_n^{(k)} = \binom{n+k}{n} = \frac{\Gamma(n+k+1)}{\Gamma(n+1)\Gamma(k+1)}, \tag{2.4}$$

and put

$$C_n^{(k)} x_n^{(k)} = \sum_{\nu=0}^{n} C_{n-\nu}^{(k-1)} x_\nu. \tag{2.5}$$

Then (2.1) asserts that $x_n^{(k)} \to y_0$ when $n \to \infty$. Put

$$\xi(\mu) = (1-\mu)\sum_{n=0}^{\infty} x_n \mu^n = (1-\mu)^{1+k} \sum_{n=0}^{\infty} C_n^{(k)} x_n^{(k)} \mu^n. \tag{2.6}$$

Then (2.3) asserts that $\xi(\mu) \to y_0$ when $\mu \to 1-0$. For the proof of (2.3) it is enough to observe that

(3) Let us note in passing that (2.1) implies (C, κ)-lim $x_n = y$ for every $\kappa > k$. Proof as in the numerically valued case. This fact is used repeatedly below.

$$\|\xi(\mu) - y_0\| \leqq (1-\mu)^{1+k} \sum_{n=0}^{\infty} C_n^{(k)} \|x_n^{(k)} - y_0\| \mu^n$$

and that $\{(1-\mu)^{1+k} C_n^{(k)} \mu^n\}$ is a limit preserving sequence of convergence factors. Hence from $\|x_n^{(k)} - y_0\| \to 0$ when $n \to \infty$ it follows that $\|\xi(\mu) - y_0\| \to 0$ when $\mu \to 1$. The proof given by S. Chapman [5, p. 379] for (2.2) in the case of numerically valued sequences can be followed step by step in the abstract case, merely replacing absolute values everywhere by norms.

We need the abstract form of Wiener's general Tauberian theorem [15, Theorem VIII].

THEOREM 2. *Let $K_1(u) \in L_1(-\infty, \infty)$ and suppose that*

$$(2.7) \qquad (2\pi)^{-1/2} \int_{-\infty}^{\infty} K_1(u) e^{-itu} du \neq 0$$

for all real t. Let $z(u)$ be a bounded measurable(4) *function on $(-\infty, \infty)$ to X and suppose that*

$$(2.8) \qquad \lim_{s \to \infty} \int_{-\infty}^{\infty} K_1(s-u) z(u) du = z \int_{-\infty}^{\infty} K_1(u) du,$$

where the integral on the left is taken in the sense of Bochner [4] *and $z \in X$. Let $K_2(u) \in L_1(-\infty, \infty)$. Then*

$$(2.9) \qquad \lim_{s \to \infty} \int_{-\infty}^{\infty} K_2(s-u) z(u) du = z \int_{-\infty}^{\infty} K_2(u) du.$$

We can follow Wiener's argument. It is clear that (2.8) implies

$$\lim_{s \to \infty} \int_{-\infty}^{\infty} \left[\sum_{n=1}^{N} a_n K_1(s + \lambda_n - u) \right] z(u) du = z \int_{-\infty}^{\infty} \left[\sum_{n=1}^{N} a_n K_1(u + \lambda_n) \right] du$$

for any choice of the complex constants a_n and the real constants λ_n. But Wiener's basic theorem [15, Theorem II] asserts that (2.7) is a necessary and sufficient condition in order that the linear combinations of the translations of $K_1(u)$ be dense in $L_1(-\infty, \infty)$. We may consequently choose the constants so that $\|K_2(u) - \sum_{n=1}^{N} a_n K_1(u+\lambda_n)\|_1 < \epsilon$. If, however, $\|K_2(u) - K_3(u)\|_1 < \epsilon$ then for all real s

$$\left\| \int_{-\infty}^{\infty} [K_2(s-u) - K_3(s-u)] z(u) du \right\| < \epsilon \operatorname*{ess\,sup}_{-\infty < u < \infty} \|z(u)\|.$$

Formula (2.9) is a direct consequence of these estimates.

---

(4) That is, $z(u)$ is the limit for almost all u of a sequence of finitely valued functions $z_n(u)$ such that the sets where $z_n(u) = x$ are (void or) measurable for all $x \in X$.

We shall give only one application of this theorem, namely to the $[O\text{-}A\rightarrow(C, k)]$ theorem(5).

THEOREM 3. *Let* $\|x_n\|\leqq M$ *for all* n *and* (A)-$\lim x_n=y_0$. *Then* (C, k)-$\lim x_n=y_0$ *for every positive* k.

Let us define

$$(2.10)\qquad x_k(u) = x_n^{(k)} \quad\text{for}\quad n \leqq u < n+1, \qquad n = 0, 1, 2, \cdots,$$

$$(2.11)\qquad \Xi(\lambda) = \frac{\lambda^{1+k}}{\Gamma(1+k)}\int_0^\infty u^k e^{-\lambda u} x_k(u)du,$$

where the integral may be taken in the sense of Riemann-Graves. Since $\|x_n\|\leqq M$ we have also $\|x_k(u)\|\leqq M$ for all u. From (2.4) it follows that $\Gamma(k+1)C_n^{(k)}-n^k=O(n^{k-1})$. It is then a simple matter to show that $\|\Xi(\lambda)-\xi(e^{-\lambda})\|\rightarrow 0$ when $\lambda\rightarrow+0$. Hence

$$(2.12)\qquad \lim_{\lambda\rightarrow+0} \lambda^{1+k}\int_0^\infty u^k e^{-\lambda u} x_k(u)du = y_0\Gamma(k+1).$$

This is a limit of Wiener's type if we take $\lambda=e^{-s}$, $u=e^v$, $K_1(v)=\exp[(1+k)v-e^v]$. Since

$$\int_{-\infty}^\infty K_1(v)e^{-itv}dv = \int_0^\infty e^{-u}u^{k-it}du = \Gamma(1+k-it) \neq 0,$$

condition (2.7) is satisfied. Let $h>0$ be fixed and define $K_2(v)$ to be e^{-v} for $0\leqq v\leqq\log(1+h)$ and zero outside of this interval. Formula (2.9) then gives after simplification

$$(2.13)\qquad \lim_{\omega\rightarrow\infty}\frac{1}{\omega h}\int_\omega^{\omega(1+h)} x_k(u)du = y_0$$

or

$$y_0 = \lim_{\omega\rightarrow\infty}\left\{x_k(\omega) + \frac{1}{\omega h}\int_\omega^{\omega(1+h)}[x_k(u)-x_k(\omega)]du\right\}.$$

In order to complete the proof we shall estimate the difference $[x_k(u)-x_k(\omega)]$ for $\omega\leqq u\leqq\omega(1+h)$. Using (2.5) we see that the required difference is of the form

$$S = [C_b^{(k)}]^{-1}\sum_{\nu=0}^{b} C_{b-\nu}^{(k-1)}x_\nu - [C_a^{(k)}]^{-1}\sum_{\nu=0}^{a} C_{a-\nu}^{(k-1)}x_\nu.$$

There is no restriction in assuming that ω tends to infinity through integral

(5) Cf. A. F. Andersen [1, p. 80] and the remarks after the proof of Theorem 3. We give a direct proof based on the ideas of Wiener.

values. We have then $\omega=[\omega]=a$ and $a<b<(1+h)a$, b being also an integer. Hence $S=S_1+S_2$ where

$$S_1 = \sum_{\nu=0}^{a} \{C_{b-\nu}^{(k-1)}[C_b^{(k)}]^{-1} - C_{a-\nu}^{(k-1)}[C_a^{(k)}]^{-1}\} x_\nu,$$

$$S_2 = [C_b^{(k)}]^{-1} \sum_{\nu=a+1}^{b} C_{b-\nu}^{(k-1)} x_\nu.$$

It follows that

$$\|S_2\| \leqq [C_b^{(k)}]^{-1} C_{b-a-1}^{(k)} M \leqq AMh^k,$$

where A is a constant, possibly depending upon k but not upon a or h. In the case of S_1 we note that

$$\frac{C_{n-\nu}^{(k-1)}}{C_n^{(k)}} = k \frac{\Gamma(n+1)}{\Gamma(n+1+k)} \frac{\Gamma(n-\nu+k)}{\Gamma(n-\nu+1)}.$$

Here the first fraction on the right is a decreasing function of n and if $k<1$, as we may suppose without restricting the generality(6), the second fraction is decreasing for $n>\nu$. Thus for $k<1$ the numerical coefficient of x_ν in S_1 is negative for $\nu=0, 1, \cdots, a$. Hence

$$\|S_1\| \leqq \{1 - [C_b^{(k)} - C_{b-a-1}^{(k)}][C_b^{(k)}]^{-1}\} M \leqq AMh^k$$

and

$$\|x_k(u) - x_k(\omega)\| \leqq 2AMh^k, \qquad \omega \leqq u \leqq \omega(1+h).$$

Thus, given any $\epsilon>0$ we can find an $\omega(\epsilon, h)$ such that for $\omega \geqq \omega(\epsilon, h)$

$$\|x_k(\omega) - y_0\| \leqq \epsilon + 2AMh^k.$$

From this we conclude that $x_k(\omega)\to y_0$ when $\omega\to\infty$ and the theorem is proved.

At the end of his paper on the Abel→Cesàro problem, J. E. Littlewood [10, p. 448] remarks that a series is $(C, 1)$ summable if its partial sums are bounded and it is summable Abel. That such a series is actually summable (C, k) for every $k>0$ was first proved by A. F. Andersen [1, p. 80] as a special case of a more general theorem. The latter can also be extended to Banach spaces. We state the result without proof.

THEOREM 4. *If* $\{x_n\} \in X$ *and* (A)-lim $x_n=y_0$ *and, in addition,* $\|x_n^{(k)}\| \leqq M$, *then* $(C, k+\delta)$-lim $x_n=y_0$ *for every* $\delta>0$.

The theorem of Andersen is also an extension of an earlier result of Littlewood (loc. cit.) according to which a series which is finite (C, r) and summable Abel is summable $(C, r+1)$.

(6) Cf. footnote 3.

On the other hand, from the assumption that (A)-lim x_n exists and $\|x_n\| \leqq Mn^\gamma$ for some fixed $\gamma > 0$ it is not possible to conclude that (C, k)-lim x_n exists for any k. Indeed, Littlewood (loc. cit.) gives an example of an Abel summable series, the partial sums of which are $O[(\log n)^{1+\epsilon}]$, which is not summable (C).

The above theorems can be extended in various other directions, however. An important case is that in which the elements x_n are functions of a numerical parameter ξ defined on a measurable set S, $x_n = x_n(\xi)$, $\xi \in S$. The assumptions regarding boundedness and convergence in the theorems may then depend upon ξ and as a consequence the conclusions will also depend upon ξ. We note three cases of some interest.

(1) If the assumptions hold uniformly with respect to ξ, so do the conclusions.

(2) If the assumptions hold for almost all ξ, so do the conclusions.

(3) If in the assumptions the bounds are integrable functions of ξ and convergence holds in measure with respect to ξ, the same is true in the conclusions.

We recall that $y_n(\xi)$ converges to $y_0(\xi)$ in measure (or asymptotically) with respect to ξ in S if to every given $\epsilon > 0$ the measure of the subset of S where $\|y_n(\xi) - y_0(\xi)\| > \epsilon$ tends to zero when $n \to \infty$.

3. **Applications to ergodic theorems**. In the following X is the basic and Z an auxiliary Banach space which will be identified with X or related spaces later. The space of all linear bounded operators on X to X is denoted by $\mathfrak{X}$. Finally E_2 is the space of complex numbers with the usual metric. A combination of Theorems 1 and 3 gives the following theorem.

THEOREM 5. *Let* $\{z_n\}$ *be a sequence of elements in a complex Banach space* Z. *Let*

$$R(\lambda; z) = \sum_{n=0}^{\infty} z_n \lambda^{-n-1}, \tag{3.1}$$

when the series converges(7). *A necessary condition in order that*(8)

$$\lim_{n\to\infty} [C_n^{(k)}]^{-1} \sum_{\nu=0}^{n} C_{n-\nu}^{(k-1)} z_\nu = z \tag{3.2}$$

for some fixed positive k *is that*

$$\lim_{\lambda\to 1+0} (\lambda - 1)R(\lambda; z) = z, \tag{3.3}$$

$$\lim_{n\to\infty} n^{-k} z_n = 0. \tag{3.4}$$

---

(7) The series converges for $|\lambda| > 1$ if (3.2) holds.

(8) More concisely expressed: *a* (C, k)*-ergodic sequence* $\{z_n\}$ *is* (A)*-ergodic and a bounded* (A)*-ergodic sequence is* (C, k)*-ergodic for every* $k > 0$. It is obvious that similar results hold for much more general definitions of summation than that of Cesàro.

Conversely, if (3.3) *holds and* (3.4) *is replaced by* $\|z_n\| \leqq M$ *for all* n, *then* (3.2) *holds for every positive* k.

Ergodic theorems are obtained by suitable specialization. Let us first take $Z = \mathfrak{X}$, $z_0 = I$, $z_1 = T, \cdots, z_n = T^n, \cdots$, where $T \in \mathfrak{X}$. Then

$$R(\lambda; z) = R(\lambda; T) = \sum_{n=0}^{\infty} T^n \lambda^{-n-1} \tag{3.5}$$

is the resolvent of T and we obtain a "uniform" ergodic theorem.

THEOREM 6. *A necessary condition for the existence of a* $P \in \mathfrak{X}$ *such that for some fixed* $k > 0$

$$\lim_{n \to \infty} [C_n^{(k)}]^{-1} \sum_{\nu=0}^{n} C_{n-\nu}^{(k-1)} T^{\nu} = P \tag{3.6}$$

is that

$$\lim_{\lambda \to 1+0} (\lambda - 1) R(\lambda; T) = P, \tag{3.7}$$

$$\lim_{n \to \infty} n^{-k} T^n = 0. \tag{3.8}$$

Conversely, if (3.7) *holds but* (3.8) *be replaced by* $\|T^n\| \leqq M$ *for all* n, *then* (3.6) *holds for every* $k > 0$. *Here convergence is taken in the sense of the uniform topology of* $\mathfrak{X}$.

The same theorem holds if $\mathfrak{X}$ is an arbitrary commutative normed ring.

We get a "strong" or "mean" ergodic theorem by taking $Z = X$, $z_0 = x$, $z_1 = Tx, \cdots, z_n = T^n x, \cdots, x \in X$, $T \in \mathfrak{X}$.

THEOREM 7. *A necessary condition for the existence of a* $P \in \mathfrak{X}$ *such that for some fixed* $k > 0$ *and all* $x \in X$

$$\lim_{n \to \infty} [C_n^{(k)}]^{-1} \sum_{\nu=0}^{n} C_{n-\nu}^{(k-1)} T^{\nu} x = Px \tag{3.9}$$

is that for all x

$$\lim_{\lambda \to 1+0} (\lambda - 1) R(\lambda; T) x = Px \tag{3.10}$$

and

$$\lim_{n \to \infty} n^{-k} T^n x = 0. \tag{3.11}$$

Conversely, if (3.10) *holds and if* (3.11) *be replaced by* $\|T^n\| \leqq M$ *for all* n, *then* (3.9) *holds for every* $k > 0$. *Convergence in* (3.9) *and* (3.10) *is taken in the sense of the strong topology of* $\mathfrak{X}$.

We omit the case of "weak" ergodic theorems and pass over to theorems of "almost everywhere" type ("strong" theorems in the terminology of G. D. Birkhoff [3]). Here we disregard the case contemplated at the end of §2 where the norm is a function of ξ. Let X be a complex Banach space the elements of which are complex-valued measurable functions $x(\xi)$ on a set S in a euclidean space. We take $Z=E_2$, $z_0=x(\xi)$, $z_1=Tx(\xi)$, $\cdots$, $z_n=T^n x(\xi)$, $\cdots$, where $T\in\mathfrak{X}$.

THEOREM 8. *A necessary condition for the existence of a transformation* $P\in\mathfrak{X}$ *such that for a fixed* $k>0$

$$\lim_{n\to\infty} \sigma_n^{(k)}[T]x(\xi) \equiv \lim_{n\to\infty} [C_n^{(k)}]^{-1} \sum_{\nu=0}^{n} C_{n-\nu}^{(k-1)} T^\nu x(\xi) = Px(\xi) \tag{3.12}$$

for all $x(.)\in X$ *and for almost all* ξ (where the exceptional set may depend upon $x(.)$) *is that*

$$\lim_{\lambda\to 1+0} (\lambda - 1)R(\lambda; T)x(\xi) = Px(\xi), \tag{3.13}$$

$$\lim_{n\to\infty} n^{-k}T^n x(\xi) = 0 \tag{3.14}$$

for all $x(.)$ *and almost all* ξ. *If* (3.13) *holds but* (3.14) *be replaced by* $|\sigma_n^{(\beta)}[T]x(\xi)| \leq M(\xi)$ *for all* n, *where* $\beta\geq 0$ *and* $M(\xi)$ *is a measurable function finite almost everywhere, then* (3.12) *holds for all* $k>\beta$. *Convergence is taken in the classical, pointwise sense*(9).

In the converse part we have used Theorem 4 instead of Theorem 3 to obtain greater generality. Similar extensions hold for Theorems 5, 6, and 7.

4. The non-polar character of $\lambda=1$ for "strong" resolvents. A sufficient condition for the validity of (3.3) in Theorem 5 is that $R(\lambda; z)$ have a simple pole at $\lambda=1$ of residue z. Dunford [7, Theorem 3.16] could show in the "uniform" case that (3.6) implies not merely (3.7) but also the much stronger conclusion that $\lambda=1$ is a simple pole of $R(\lambda; T)$ of residue P(10). We shall show that this is no longer true in the "strong" case.

Our first example is taken from the theory of semi-groups of linear self-adjoint bounded transformations on a Hilbert space to itself (see E. Hille [8, §4.1], where further references to the literature are to be found). The transformation T_s is defined for $s>0$ and the semi-group property $T_sT_t=T_{s+t}$ implies that T_s is positive definite. Consequently its spectrum $S(T_s)$ is real positive. If we impose the restriction

(9) We are tacitly assuming that the definitions of addition, scalar multiplication, and convergence for the elements $x(.)$ of X are consistent with those holding in E_2 for the point functions $x(\sigma)$. It is not necessary that S be a subset of a euclidean space or that measurability be taken in the sense of Lebesgue. We do not insist further on such refinements, however.

(10) If $\lambda=1$ should belong to the resolvent set instead of to the spectrum, $P=0$. Assumption (3.7) does not figure explicitly in Dunford's theorem, but it is usually implied by his assumption that the operator T is ergodic in a specified sense.

$$\|T_s\| \leqq 1, \tag{4.1}$$

then $S(T_s)$ will be restricted to the interval (0, 1) of the real axis in the λ-plane and $T_s x$ is represented by a Laplace-Stieltjes integral

$$T_s x = \int_0^\infty e^{-su} dE(u)x. \tag{4.2}$$

Here the integral converges for $\Re(s) \geqq 0$ and defines a semi-group of linear bounded transformations, not merely for real positive s, but in the whole half-plane of convergence. The self-adjoint character is of course lost for complex s, but T_s and $T_{\bar{s}}$ are adjoint transformations. The transformation $E(u)$ is the resolution of the identity of a positive definite self-adjoint transformation

$$B = \lim_{h \to \infty} \frac{1}{h} [I - T_h],$$

where the limit is taken in the strong sense and the domain of B is dense in the space. We normalize $E(u)$ by assuming it to be strongly left-continuous(11), that is $E(u-h)x \to E(u)x$ when $h \to +0$, u being fixed positive. Further $E(u)E(v) = E(w)$, $w = \min(u, v)$, $E(u) = 0$ for $u \leqq 0$ and $E(u)x \to x$ when $u \to +\infty$. From (4.2) we can compute the resolvent of T_s and find

$$R(\lambda, T_s)x = \int_0^\infty \frac{dE(u)x}{\lambda - e^{-su}} \tag{4.3}$$

for all λ not in $S(T_s)$. A point λ_0 of the interval (0, 1) belongs to the resolvent set of T_s if and only if $u_0 = (1/s)\log(1/\lambda_0)$ lies in an interval of constancy of $E(u)$. On the other hand, λ_0 belongs to the point spectrum of T_s if and only if $[E(u_0+0) - E(u_0)]x$ is not identically zero. A simple calculation shows that

$$(\lambda - \lambda_0)R(\lambda; T_s)x \to [E(u_0 + 0) - E(u_0)]x, \qquad u_0 = \frac{1}{s}\log\frac{1}{\lambda_0},$$

when $\lambda \to \lambda_0$ in the sectors $0 < \epsilon \leqq |\arg(\lambda - \lambda_0)| \leqq \pi - \epsilon$. In particular,

$$(\lambda - 1)R(\lambda; T_s)x \to E(+0)x \tag{4.4}$$

and here λ may approach 1 throughout the sector $|\arg(\lambda - 1)| \leqq \pi - \epsilon$.

Formulas (4.1) and (4.4) show that the assumptions of Theorem 7 are satisfied. Since $T_s^n = T_{ns}$, we have consequently

$$(C, k)\text{-lim } T_{ns}x = E(+0)x, \qquad k > 0, \tag{4.5}$$

(11) The customary normalization requires strong right semi-continuity. If this convention should be adopted here, it would become necessary to replace the lower limit 0 in formula (4.2) by an arbitrary negative quantity and to make similar adjustments in the other formulas of this paragraph.

for all x. On the other hand, $\lambda=1$ is a pole of $R(\lambda; T_s)$ if and only if $E(u)$ is constant for small positive u and has a saltus at $u=0$, that is, there exists a $\delta>0$ such that $E(u)=E(+0)\neq E(0)=0$ for $0<u\leqq\delta$. If there is no such interval of constancy of $E(u)$, then $\lambda=1$ is not an isolated singularity. This is to be regarded as the general case, since the semi-group T_s is defined as soon as $E(u)$ is given subject to the conditions stated above which do not involve any assumptions of constancy on an interval $(0, \delta)$ and do not imply any such restriction.

Actually (4.5) holds also for $k=0$, that is,

$$\lim_{n\to\infty} T_{ns}x = \lim_{\omega\to\infty} T_{\omega}x = E(+0)x. \tag{4.6}$$

This follows from a well known property of Laplace-Stieltjes integrals according to which the limit of the integral for $s\to+\infty$ is the saltus of the integrator at the origin. This property also holds for vector-valued integrals of the type considered here. We have consequently proved the following "mean" ergodic theorem.

THEOREM 9. *If* $\{T_s\}$, $s>0$, *is a semi-group of linear self-adjoint transformations on a Hilbert space satisfying* (4.1), *then* (4.6) *holds for all* x.

It should be observed that the corresponding theorem for groups of unitary transformations is false. This can be seen by considering, for instance, the group of translations in an L_2-space such that $U_s x(t)=x(t+s)$ where $x(t)\in L_2(a, b)$ and is continued periodically outside (a, b) if $b-a$ is finite and by symmetry if (a, b) extends to infinity one way only. For groups of unitary transformations on Hilbert space we have the quasi-ergodic theorem of J. von Neumann [11] which asserts, among other things, that $(C, 1)$-lim U_α^n exists in the strong sense and is a projection P_α. Our Theorem 7 shows that the stronger assertion(12) (C, k)-lim $U_\alpha^n=P_\alpha$ holds for every $k>0$, and here, as we have just remarked, k cannot be allowed to be zero.

The rest of the paper will be devoted to a discussion of ergodic theorems relating to the operation of fractional integration in the Riemann-Liouville sense.

5. **(A)-ergodic theorems relating to fractional integration.** We consider simultaneously the two function spaces $C_0[0, 1]$ and $L_1(0, 1)$. The first is the space of functions $f(t)$ continuous for $0\leqq t\leqq 1$ such that $f(0)=0$ with $\|f\|=\max|f(t)|$, the latter the space of integrable functions on $(0, 1)$ with the usual metric. We note that the elements of $C_0[0, 1]$ are dense in $L_1(0, 1)$.

Let α be fixed positive and define

$$J_\alpha[f] = \frac{1}{\Gamma(\alpha)}\int_0^t (t-u)^{\alpha-1}f(u)du, \qquad 0\leqq t\leqq 1, \tag{5.1}$$

(12) This is included in results of L. W. Cohen [6, p. 508], however, since his assumptions are obviously satisfied in the present case.

$$T_\alpha[f] = (I - J_\alpha)f. \tag{5.2}$$

These are obviously linear bounded transformations on the spaces in question. The norm of J_α does not exceed $[\Gamma(\alpha+1)]^{-1}$ in either space. We recall that $J_\alpha J_\beta = J_{\alpha+\beta}$, in particular, $J_\alpha^n = J_{n\alpha}$. We shall study the iterates of T_α. A simple calculation gives

$$T_\alpha^n[f] = f(t) - \int_0^t P_n(t-u, \alpha) f(u)du, \qquad n \geqq 1, \tag{5.3}$$

$$P_n(w, \alpha) = \sum_{\nu=1}^{n} (-1)^{\nu-1} \binom{n}{\nu} \frac{w^{\nu\alpha-1}}{\Gamma(\nu\alpha)}. \tag{5.4}$$

For $\alpha=1$ the latter expression is a Laguerre polynomial in w (for notation and properties see G. Szegö [14, chapters V, VIII])

$$P_n(w, 1) = L_{n-1}^{(1)}(w). \tag{5.5}$$

In the present paragraph we shall be concerned with (A)-ergodic properties of T_α of the strong or almost everywhere types, that is relations like (3.10) and (3.13) above, and start by computing the resolvent $R(\lambda, T_\alpha) \equiv R_\alpha(\lambda)$. For the following compare E. Hille and J. D. Tamarkin [9, pp. 524–525] and E. Hille [8, p. 43]. For $|\lambda-1| > \|J_\alpha\|$ we have

$$\begin{aligned} R_\alpha(\lambda) &= [\lambda I - T_\alpha]^{-1} = [(\lambda-1)I + J_\alpha]^{-1} \\ &= (\lambda-1)^{-1}[I + (\lambda-1)^{-1}J_\alpha]^{-1} \\ &= (\lambda-1)^{-1}I + (\lambda-1)^{-1}\sum_{n=1}^{\infty}(1-\lambda)^{-n}J_{n\alpha}. \end{aligned}$$

Hence

$$R_\alpha(\lambda)f = (\lambda-1)^{-1}f(t) + (\lambda-1)^{-1}\int_0^t \sum_{n=1}^{\infty} \frac{(t-u)^{n\alpha-1}}{\Gamma(n\alpha)(1-\lambda)^n} f(u)du. \tag{5.6}$$

This representation is evidently valid for all values of $\lambda \neq 1$ and not merely for $|\lambda-1| > \|J_\alpha\|$. Thus $R_\alpha(\lambda)f$ is an entire function of $1/(\lambda-1)$ [13]. The kernel

$$K_\alpha(w, \lambda) = \sum_{n=1}^{\infty} \frac{w^{n\alpha-1}}{\Gamma(n\alpha)(1-\lambda)^n} \tag{5.7}$$

is evidently expressible in terms of Mittag-Leffler's function $E_\alpha(z)$ as

$$K_\alpha(w, \lambda) = \frac{d}{dw} E_\alpha\left(\frac{w^\alpha}{1-\lambda}\right). \tag{5.8}$$

[13] The spectrum of T_α consists of one single point, $\lambda=1$, which, nevertheless, makes up the "continuous" spectrum. To see this, we note that the functions $f_n = t^n$ have norm 1 in $C_0[0, 1]$ but $(I-T_\alpha)f_n = J_\alpha f_n$ converges strongly to zero when $n \to \infty$. In $L_1(0, 1)$ we take $f_n = (n+1)t^n$ instead.

In discussing the ergodic properties of T_α we shall need a couple of lemmas concerning singular integrals. The proofs follow classical lines and are omitted here.

LEMMA 1. *Let $K_\nu(w) \in L_1(0, 1)$ for all ν and be such that*

$$\text{(i)} \qquad \int_0^t K_\nu(w)dw \to 1, \qquad K_\nu(t) \to 0,\ t > 0,$$

when $\nu \to \infty$, both relations holding uniformly with respect to t in $0 < \epsilon \leqq t \leqq 1$ for every $\epsilon > 0$, and

$$\text{(ii)} \qquad \int_0^1 |K_\nu(w)| dw \leqq M \quad \text{for all} \quad \nu.$$

Let $f(t) \in C_0[0, 1]$. Then

$$f_\nu(t) \equiv \int_0^t K_\nu(t-u)f(u)du \to f(t)$$

when $\nu \to \infty$, uniformly in t, $0 \leqq t \leqq 1$, that is, $f_\nu(t)$ converges strongly to $f(t)$.

LEMMA 2. *Let $K_\nu(w) \in L_1(0, 1)$ for all ν and satisfy the conditions of Lemma 1, but not necessarily uniformly with respect to t in* (i). *In addition, let there exist a dominant $G_\nu(w)$ such that $|K_\nu(w)| \leqq G_\nu(w)$ for all ν and w. Here $G_\nu(w)$ shall be a positive, monotone decreasing function of w for fixed ν which is absolutely continuous on $(\epsilon, 1)$ for every $\epsilon > 0$. Further, $G_\nu(w)$ shall satisfy condition* (ii) *and for fixed $w > 0$, $G_\nu(w) \to 0$ when $\nu \to \infty$. If $f(t) \in L_1(0, 1)$ then $f_\nu(t)$ exists for almost all t, $f_\nu(t) \in L_1(0, 1)$, and $f_\nu(t) \to f(t)$ when $\nu \to \infty$ for almost all t.*

We shall now prove the following theorem.

THEOREM 10. *Let α be fixed, $0 < \alpha < 2$. Then*

$$(5.9) \qquad (\lambda - 1)R_\alpha(\lambda)f \to 0 \quad \text{when} \quad \lambda \to 1$$

for every $f \in C_0[0, 1]$ or $L_1(0, 1)$ in the sense of strong convergence in the space in question. Here λ may tend to 1 radially(14) *in the sector $|\arg(\lambda - 1)| < (2 - \alpha)\pi/2$. If $f(t) \in L_1(0, 1)$, (5.9) also holds in the sense of pointwise convergence for almost all t. The theorem becomes false for $\alpha \geqq 2$.*

By virtue of formulas (5.6) and (5.7) we have

$$(5.10) \qquad (\lambda - 1)R_\alpha(\lambda)f = f(t) + \int_0^t K_\alpha(t-u, \lambda)f(u)du,$$

and we shall prove that the kernel $-K_\alpha(w, \lambda)$ satisfies the conditions of Lemmas 1 and 2. For this purpose we shall need the following classical for-

(14) More generally, along any Stolz' path.

mulas[15] from the theory of the function $E_\alpha(z)$ which are valid when $0<\alpha<2$:

$$(5.11)\quad E_\alpha(z) = \frac{1}{\alpha}\exp\left[z^{1/\alpha}\right] + o(1), \qquad \left|\arg z\right| < \alpha\frac{\pi}{2},$$

$$\left.\begin{aligned}(5.12)\quad E_\alpha(z) &= -\frac{1}{z\Gamma(1-\alpha)} + O\left(\frac{1}{z^2}\right)\\ (5.13)\quad E'_\alpha(z) &= \frac{1}{z^2\Gamma(1-\alpha)} + O\left(\frac{1}{z^3}\right)\end{aligned}\right\}, \qquad \alpha\frac{\pi}{2} < \arg z < (4-\alpha)\frac{\pi}{2}.$$

Using (5.8) and (5.12) we see that

$$(5.14)\qquad -\int_0^t K_\alpha(w, \lambda)dw = 1 - E_\alpha\left(\frac{t^\alpha}{1-\lambda}\right) \to 1$$

when $\lambda \to 1$ radially in the sector $\left|\arg(\lambda-1)\right| < (2-\alpha)\pi/2$. The convergence is obviously uniform with respect to t in any interval $(\epsilon, 1)$, $\epsilon>0$, and also uniform with respect to λ in any fixed interior sector. Formulas (5.8) and (5.13) show that $\lim_{\lambda\to 1} K_\alpha(w, \lambda)=0$, uniformly in $0<\epsilon\leqq w\leqq 1$. Thus condition (i) of Lemma 1 is satisfied.

Let $\epsilon>0$ be fixed and $(\alpha+\epsilon)\pi/2\leqq\theta\leqq(4-\alpha-\epsilon)\pi/2$. Formula (5.13) shows that $E_\alpha(z)$ is of bounded variation on every ray $\arg z=\theta$ and the total variation is less than some fixed quantity B_ϵ, that is

$$\int_0^\infty \left| E'_\alpha(re^{i\theta})\right| dr \leqq B_\epsilon.$$

But if $\arg(\lambda-1)=\pi-\theta$ then

$$(5.15)\qquad \int_0^1 \left| K_\alpha(w, \lambda)\right| dw = \int_0^{A(\lambda)} \left| E'_\alpha(re^{i\theta})\right| dr \leqq B_\epsilon,$$

where $A(\lambda)=\left|\lambda-1\right|^{-1}$. Hence condition (ii) is also satisfied and uniformly with respect to λ in the sector $\left|\arg(\lambda-1)\right|\leqq(2-\alpha-\epsilon)\pi/2$. It follows that the singular integral in (5.10) satisfies the conditions of Lemma 1 and consequently (5.9) holds in the sense of strong convergence in $C_0[0, 1]$. Moreover, the convergence is uniform with respect to λ in any interior sector[16].

(15) See, for example, A. Wiman [16, formula (5)] from which (5.11) and (5.12) can be read off. Differentiation of (5) gives (5.13).

(16) On the other hand, (5.9) cannot hold in the uniform topology. This is not difficult to show directly, but for $0<\alpha\leqq 1$ the contrary assumption would lead to a contradiction with Dunford [7, Theorem 3.16]. Indeed, this assumption by virtue of formula (6.5) below together with our Theorem 6 would show that the sequence T_α^n had to be (C)-ergodic in the uniform topology and this would imply that $\lambda=1$ were at most a simple pole of $R_\alpha(\lambda)$ which is obviously false.

Formula (5.15) shows that for $|\arg(\lambda-1)| \leqq (2-\alpha-\epsilon)\pi/2$

$$\|(\lambda - 1)R_\alpha(\lambda)\| \leqq 1 + B_\epsilon, \tag{5.16}$$

where it is immaterial if the norm is taken in $C_0[0, 1]$ or in $L_1(0, 1)$. But the elements of the former space are dense in the latter, and strong convergence of a sequence of elements in $C_0[0, 1]$ implies strong convergence of the same sequence in $L_1(0, 1)$. Formula (5.16) then implies that (5.9) holds for every $f \in L_1$, in the sense of strong (=mean) convergence, by the Hahn-Banach-Steinhaus theorem (see [2, p. 79]).

We now proceed to the question of pointwise almost everywhere convergence when $f \in L_1$. In order to apply Lemma 2 we have merely to determine a suitable dominant $G_\alpha(w, \lambda)$ of $K_\alpha(w, \lambda)$. Formula (5.13) shows, however, that we can determine a $C = C(\alpha)$ such that if $|\lambda - 1| < 1$, $|\arg(\lambda-1)| \leqq (2-\alpha-\epsilon)\pi/2$, and

$$G_\alpha(w, \lambda) = \begin{cases} C(\alpha)\,|\lambda - 1|^{-1/\alpha}, & 0 \leqq w < |\lambda - 1|^{1/\alpha}, \\ C(\alpha)\,|\lambda - 1|\, w^{-\alpha-1}, & |\lambda - 1|^{1/\alpha} \leqq w \leqq 1, \end{cases}$$

then $|K_\alpha(w, \lambda)| \leqq G_\alpha(w, \lambda)$ and all the conditions of Lemma 2 are satisfied. Hence $(\lambda-1)R_\alpha(\lambda)f(t) \to 0$ for almost all t when $\lambda \to 1$ in the sector indicated.

It remains to settle the case $\alpha \geqq 2$. For $\alpha = 2$ we have

$$(\lambda - 1)R_2(\lambda)f = f(t) - (\lambda - 1)^{-1/2} \int_0^t \sin\,[(\lambda - 1)^{-1/2}(t - u)]f(u)du$$

and for $\lambda > 1$

$$\int_0^1 |K_2(w, \lambda)|\, dw = \frac{2}{\pi}(\lambda - 1)^{-1/2} + O(1).$$

It is not difficult to see that the norm of $(\lambda-1)R_2(\lambda)$ in $C_0[0, 1]$ as well as in $L_1(0, 1)$ is of the same order of magnitude, that is, tends to infinity when $\lambda \to 1$. This excludes the possibility of (5.9) being true in the sense of strong convergence when $\alpha = 2$. The situation is still more unfavorable when $\alpha > 2$, because then $E_\alpha(z)$ and $E_\alpha'(z)$ are unbounded on every ray[17]. Both functions are oscillatory on the negative real axis in such a manner that amplitudes as well as frequencies tend to infinity with $|z|$. From this we conclude that $\limsup \|(\lambda-1)R_\alpha(\lambda)\| = \infty$ when $\lambda \to 1$ and strong convergence is out of question.

Pointwise convergence still remains as a possibility. We note, however, that (5.14) may be rewritten as

$$(\lambda - 1)R_\alpha(\lambda) \cdot 1 = E_\alpha\left(\frac{t^\alpha}{1 - \lambda}\right). \tag{5.17}$$

---

[17] See A. Wiman [16, pp. 220–221].

If $t=0$, the right-hand side always equals one, but if $0<t\leqq 1$ and $2\leqq\alpha$, the right-hand side never tends to a limit when $\lambda\to 1+0$. Thus there exists at least one function of $L_1(0, 1)$ for which $(\lambda-1)R_\alpha(\lambda)f(t)$ does not converge pointwise when $\lambda\to 1+0$ for any value of $\alpha\geqq 2$. This completes the proof of Theorem 10.

6. **(C)-ergodic theorems relating to fractional integration.** Here our main result is the following.

THEOREM 11. *The transformation T_α of formula* (5.2) *is strongly (C)-ergodic in $C_0[0, 1]$ and $L_1(0, 1)$ when $0<\alpha\leqq 1$ but never when $\alpha>1$. If $0<\alpha<1$, then $\|T_\alpha^n\|$ is bounded for all n and $T_\alpha^n f$ tends strongly to zero for every $f(t)$ in $C_0[0, 1]$ or in $L_1(0, 1)$. Further, $T_\alpha^n f(t)\to 0$ for almost all t for every $f(t)\in L_1(0, 1)$. If $\alpha=1$, $\|T_\alpha^n\|=O(n^{1/4})$ and (C, k)-lim $T_\alpha^n f=0$ for every $k>\frac{1}{2}$ but for no $k\leqq\frac{1}{2}$, convergence holding in the strong sense in both spaces. For $k>\frac{1}{2}$ we have also pointwise convergence to zero for almost all t. When $\alpha>1$ we have* $\log\|T_\alpha^n\|>C(\alpha)n^{1/(\alpha+1)}$ *and no (C)-ergodic theorems are possible.*

For the proof we refer back to formula (5.3). Let α be fixed, $0<\alpha<1$. It is sufficient to prove that the kernel $P_n(w, \alpha)$ satisfies the conditions of Lemmas 1 and 2. We start by proving that for $0<t\leqq 1$

$$\int_0^t P_n(w, \alpha)dw \to 1 \quad \text{when} \quad n\to\infty \tag{6.1}$$

uniformly with respect to t in any interval $(\epsilon, 1)$. For this purpose we evaluate $(\lambda-1)R_\alpha(\lambda)\cdot 1$ anew. One expression of this function was given in formula (5.17). Another follows from computing the sum of the series $\sum_0^\infty \lambda^{-n-1}T_\alpha^n\cdot 1$. Equating the two expressions we obtain the identity

$$(\lambda-1)\sum_{n=0}^{\infty}\lambda^{-n-1}\int_0^t P_n(w, \alpha)dw = 1 - E_\alpha\left(\frac{t^\alpha}{1-\lambda}\right)$$

and

$$\int_0^t P_n(w, \alpha)dw = 1 - \frac{1}{2\pi i}\int_C \lambda^n(\lambda-1)^{-1}E_\alpha\left(\frac{t^\alpha}{1-\lambda}\right)d\lambda, \tag{6.2}$$

where to start with the contour of integration may be taken as the circle $|\lambda|=r>1$. This contour, however, may be deformed into one inside the unit circle. Let Γ_n be the circle $|\lambda|=r_n<1$ and draw the two tangents to Γ_n from $\lambda=1$. If r_n is sufficiently near to 1 the tangents will make angles with the positive real axis which are numerically less than $(2-\alpha)\pi/2$. The new contour of integration C_n will then consist of the two tangents plus the major arc of the circle Γ_n between the points of tangency. By virtue of formula (5.12) the integrand is uniformly bounded on C_n; the contributions to the value of the integral from the rectilinear and the circular parts of the path are

$O[(1-r_n)^{1/2}]$ and $O(r_n^n)$ respectively. Choosing $r_n=1-n^{-1}\log n$ we see that the integral tends to zero when $n\to\infty$. This proves (6.2).

Formulas (7.4) and (7.11) below show that for a fixed α, $0<\alpha<1$, positive constants $A=A(\alpha)$ and $C=C(\alpha)$ can be found such that

$$(6.3)\qquad |P_n(w,\alpha)| \leqq \begin{cases} Anw^{\alpha-1}, & 0<nw^\alpha\leqq 1,\\ n^{1/\alpha}\exp\left[-C(nw^\alpha)^{1/(\alpha+1)}\right], & 1<nw^\alpha,\ w \text{ bounded}.\end{cases}$$

A simple calculation shows that this estimate implies

$$(6.4)\qquad \int_0^1 |P_n(w,\alpha)|\,dw \leqq B(\alpha)$$

and

$$(6.5)\qquad \|T_\alpha^n\| \leqq 1+B(\alpha)$$

in both spaces. Formulas (6.1), (6.3), and (6.4) show that Lemma 1 applies to the singular integral in (5.3). Hence for every $f(t)\in C_0[0,1]$ the sequence $T_\alpha^n f$ will converge strongly to zero. Since (6.5) holds in both spaces, we conclude that $T_\alpha^n f$ also converges strongly in $L_1(0,1)$ for every $f(t)\in L_1(0,1)$ (18).

In order to settle the question of pointwise convergence in $L_1(0,1)$ we note that a dominant $G_n(w,\alpha)$ of $|P_n(w,\alpha)|$, having the properties required in Lemma 2, can be obtained from (6.3) by the simple device of increasing the right-hand sides by multiplication by suitable constants, depending upon α but not upon n or w, so that the resulting estimates agree for $w=n^{-1/\alpha}$. The new right-hand sides then form a suitable dominant having all the required properties. This completes the argument when $0<\alpha<1$.

When $\alpha=1$ several formulas from the theory of Laguerre polynomials are needed. We have first (see G. Szegö [14, p. 193, formulas (8.22.4) and (8.22.5)]) for $\beta>-1$

$$(6.6)\qquad |L_n^{(\beta)}(w)| \leqq \begin{cases} C_1(\beta)n^\beta, & \text{if } 0<nw\leqq 1,\\ C_2(\beta)n^{\beta/2-1/4}w^{-\beta/2-1/4}, & \text{if } 1<nw<n.\end{cases}$$

These estimates give

$$(6.7)\qquad \int_0^1 |L_n^{(\beta)}(w)|\,dw \leqq C(\beta)\begin{cases} n^{\beta/2-1/4}, & -1<\beta<\tfrac{3}{2},\\ n^{1/2}\log n, & \beta=\tfrac{3}{2},\\ n^{\beta-1}, & \tfrac{3}{2}<\beta,\end{cases}$$

and here also the reversed inequalities are valid for a suitable positive $C(\beta)$, since the inequalities in (6.6) may be reversed on subsets of measure exceeding $1/(2n)$ in the first case and $1/2$ in the second.

(18) Uniform convergence is excluded by the argument given in footnote 16. The same remark applies to formula (6.14) below.

Next we note that (see [14, p. 97, (5.1.7) and (5.1.14)])

$$(6.8) \qquad L_n^{(\beta)}(0) = \binom{n+\beta}{n} = C_n^{(\beta)}, \qquad \frac{d}{dw} L_{n-1}^{(\beta-1)}(w) = - L_n^{(\beta)}(w)$$

whence

$$[L_n^{(\beta-1)}(0)]^{-1} \int_0^t L_n^{(\beta)}(w)dw = 1 - [L_{n-1}^{(\beta-1)}(0)]^{-1} L_{n-1}^{(\beta-1)}(t).$$

Combining (6.6) and (6.8) we see that for $\beta > \frac{1}{2}$

$$(6.9) \qquad [L_n^{(\beta-1)}(0)]^{-1} \int_0^t L_n^{(\beta)}(w)dw \to 1$$

uniformly for $0 < \epsilon \leqq t \leqq 1$.

Finally we shall need the formula

$$(6.10) \qquad L_n^{(\beta+k)}(w) = \sum_{\nu=0}^{n} C_{n-\nu}^{(k-1)} L_\nu^{(\beta)}(w)$$

which is an immediate consequence of the structure of the generating function of the Laguerre polynomials (see [14, formula (5.1.9)]).

Combining formula (5.3) for $\alpha = 1$ with (6.7) for $\beta = 1$ we see that

$$(6.11) \qquad \| T_1^n \| = O(n^{1/4})$$

and it is not difficult to show that this estimate cannot be improved. Hence strong convergence of $T_\alpha^n f$ when $n \to \infty$ is out of the question for $\alpha = 1$. Let us consider (C, k) summability instead. We form

$$(6.12) \qquad \sigma_n^{(k)}[f] = [C_n^{(k)}]^{-1} \sum_{\nu=0}^{n} C_{n-\nu}^{(k-1)} T_1^\nu[f].$$

We recall that

$$(6.13) \qquad T_1^\nu[f] = f(t) - \int_0^t L_{\nu-1}^{(1)}(t-u)f(u)du.$$

Substituting (6.13) in (6.12) and reducing with the aid of (6.10), where we take $\beta = 1$ and replace n by $n-1$, we obtain finally

$$(6.14) \qquad \sigma_n^{(k)}[f] = f(t) - [C_n^{(k)}]^{-1} \int_0^t L_{n-1}^{(k+1)}(t-u)f(u)du.$$

Formula (6.7) then shows that the norm of this transformation is bounded if and only if $k > \frac{1}{2}$. For such values of k the kernel $[C_n^{(k)}]^{-1} L_{n-1}^{(k+1)}(w)$ then satisfies condition (ii) of Lemma 1. But condition (i) is also satisfied by virtue of (6.6), (6.8), and (6.9)([19]). Hence $\sigma_n^{(k)}[f]$ converges strongly to zero for every $f(t)\in C_0[0, 1]$ when $k>\frac{1}{2}$ but not when $k\leqq\frac{1}{2}$. The extension to strong convergence in $L_1(0, 1)$ is made as above. Pointwise convergence is settled by observing that formula (6.6) gives an admissible dominant of the kernel by increasing the factors $C_1(\beta)$ and $C_2(\beta)$ so that the two resulting estimates agree for $w=1/n$. This completes the argument when $\alpha=1$.

The case $\alpha>1$ is entirely different. Here formula (7.11) shows that positive quantities $a(\alpha)$ and $b(\alpha)$ can be found such that in a subset of $[0, 1]$ of measure greater than 1/2

$$(6.15)\qquad |P_n(w, \alpha)| > a(\alpha)n^{1/2(\alpha+1)}w^{-(\alpha+2)/2(\alpha+1)} \exp\left[b(\alpha)n^{1/(\alpha+1)}w^{\alpha/(\alpha+1)}\right]$$

whence follows

$$(6.16)\qquad \int_0^1 |P_n(w, \alpha)|\, dw > a(\alpha) \exp\left[\tfrac{1}{2}b(\alpha)n^{1/(\alpha+1)}\right]$$

for all large n. This implies that $\|T_\alpha^n\|$ is of the same order of magnitude for large n, that is, condition (3.11) of Theorem 7, which is necessary for (C, k) summability, is violated for all k as soon as $\alpha>1$. We know already from Theorem 10 that the other necessary condition (3.10) is violated for $\alpha\geqq 2$. Thus the sequence T_α^n is not (C)-ergodic for $\alpha>1$ and the proof of Theorem 11 is complete.

7. **Asymptotic behavior of $P_n(w, \alpha)$ and related functions**([20]). By definition

$$(7.1)\qquad P_n(w, \alpha) = \sum_{\nu=1}^{n} (-1)^{\nu-1}\binom{n}{\nu}\frac{w^{\nu\alpha-1}}{\Gamma(\nu\alpha)}.$$

We introduce the entire function

$$(7.2)\qquad H_\alpha(z) = \sum_{\nu=1}^{\infty} \frac{(-1)^{\nu-1}}{\nu!\Gamma(\nu\alpha)} z^\nu,$$

which is an obvious generalization of the Bessel function J_1. Indeed

$$H_1(z) = z^{1/2}J_1(2z^{1/2}).$$

A simple argument shows that if $n\to\infty$ and $w\to 0$ in such a manner that $w^\alpha n\to z\geqq 0$ then

$$(7.3)\qquad wP_n(w, \alpha) \to H_\alpha(z), \qquad w^\alpha n \to z.$$

Hence for $w^\alpha n=O(1)$ we have $wP_n(w, \alpha)=O(w^\alpha n)$ or

$$(7.4)\qquad |P_n(w, \alpha)| \leqq A(\alpha)w^{\alpha-1}n \quad \text{when} \quad 0 < w^\alpha n \leqq 1.$$

([19]) In formula (6.9) we may obviously replace one of the subscripts n by $n-1$ without affecting the conclusion.

([20]) This section of the paper is entirely due to Professor G. Szegö, who has also contributed to the results of §6.

This is the first of the inequalities in (6.3) above. The second inequality requires a rather elaborate argument.

We assume next $w=O(1)$ and put

$$\eta = \left(\frac{w}{n}\right)^{\alpha/(\alpha+1)} \tag{7.5}$$

with the understanding that $\eta n \to \infty$. Using Hankel's integral for the Gamma function, we get after some simplifications

$$P_n(w, \alpha) = -\frac{1}{2\pi i w}\int_{-\infty}^{(0+)} e^t\left\{1-\left(\frac{w}{t}\right)^{\alpha}\right\}^n dt. \tag{7.6}$$

Here the path of integration is a loop surrounding the negative real axis and $t=0$ counterclockwise and t^α is real positive when t is so. Substitution of $t=\eta n\tau$ gives

$$P_n(w, \alpha) = -\frac{\eta n}{2\pi i w}\int_{-\infty}^{(0+)} [e^{\eta\tau}(1-\eta\tau^{-\alpha})]^n d\tau. \tag{7.7}$$

Putting

$$f_\eta(\tau) = \eta\tau + \log(1-\eta\tau^{-\alpha}), \tag{7.8}$$

we have

$$f_\eta'(\tau) = \eta + \frac{\eta\alpha\tau^{-\alpha-1}}{1-\eta\tau^{-\alpha}},$$

$$f_\eta''(\tau) = -\eta\alpha\frac{(\alpha+1)\tau^{-\alpha-2}(1-\eta\tau^{-\alpha})+\eta\alpha\tau^{-2\alpha-2}}{(1-\eta\tau^{-\alpha})^2},$$

where the primes denote differentiation with respect to τ.

Let S be the root of the equation $f_\eta'(\tau)=0$ which tends to

$$S_0 = \alpha^{1/(\alpha+1)}e^{\pi i/(\alpha+1)}$$

when $\eta \to 0$. This root can be expanded in powers of η for small η,

$$S = S_0 + S_1\eta + S_2\eta^2 + \cdots$$

We have then

$$f_\eta''(S) = \eta[(\alpha+1)S^{-1}-\eta] = (\alpha+1)\alpha^{-1/(\alpha+1)}e^{-\pi i/(\alpha+1)}\eta + O(\eta^2),$$

$$f_\eta(\tau) = f_\eta(S) + \tfrac{1}{2}\eta(\tau-S)^2[(\alpha+1)S^{-1}+\Delta(\tau,\eta)],$$

where $\Delta(\tau,\eta)$ is holomorphic in a neighborhood of $\tau=S$ and $|\Delta|$ is uniformly small provided η and $\max|\tau-S|=M$ are sufficiently small, M being independent of η.

Let $0<\alpha<1$. Then $\pi/2<\pi/(\alpha+1)<\pi$. We deform the loop into a contour consisting of the two rays $\arg\tau=\pm\arg S$, $|\tau|\geqq\delta$, plus the major arc of the circle $|\tau|=\delta$ joining the end points of the rays. We have then along the ray $\arg\tau=\arg S$

$$\tau - S = (\eta n)^{-1/2}(\alpha+1)^{-1/2} i S^{1/2}\sigma,$$

where

$$\arg\sigma = \arg(\pm iS^{1/2}) = \frac{\pi}{2(\alpha+1)} \pm \frac{\pi}{2} + \eta'$$

and $\eta'\to 0$ when $\eta\to 0$.

The customary argument in the method of steepest descents shows that the main contribution to the value of $P_n(w,\alpha)$ comes from the neighborhoods of $\tau=S$ and $\tau=\bar{S}$ in the integral (7.7) after the path has been deformed as indicated above and δ is chosen sufficiently small. Hence, neglecting terms of lower order, $P_n(w,\alpha)$ is found to be twice the real part of

$$(7.9)\qquad -\frac{\eta n}{2\pi i w}(\eta n)^{-1/2}(\alpha+1)^{-1/2} i S^{1/2}\left[e^{\eta S}(1-\eta S^{-\alpha})\right]^n \int \exp\left[-\tfrac{1}{2}\sigma^2(1+\Delta')\right]d\sigma.$$

Here Δ' is a quantity of the same character as Δ. The integration with respect to σ can be taken along the real axis and the integral tends to $(2\pi)^{1/2}$ when $\eta n\to\infty$.

Expanding into powers of η we get

$$f_\eta(S) \equiv \eta S + \log(1-\eta S^{-\alpha}) = \sum_1^\infty V_m\eta^m,$$

where the V_m depend only upon α and m. In particular,

$$V_1 = S_0 - S_0^{-\alpha} = \left(1+\frac{1}{\alpha}\right)\alpha^{1/(\alpha+1)}e^{\pi i/(\alpha+1)}.$$

The series is convergent for small values of η. We note that $n\eta^m\to 0$ when $n\to\infty$ as soon as $m>k$ where

$$k = \left[\frac{\alpha+1}{\alpha}\right].$$

Hence

$$(7.10)\qquad \left[e^{\eta S}(1-\eta S^{-\alpha})\right]^n = \exp\left\{n\sum_1^k V_m\eta^m\right\}\cdot(1+\delta_n)$$

where $\delta_n\to 0$ when $n\to\infty$. The first term in the finite sum gives the main contribution to the value.

In replacing $P_n(w,\alpha)$ by the contributions to the integral from the neighborhoods of $\tau=S$ and $\bar{S}$ we have neglected the distant parts of the path of integration and the neighborhood of $\tau=0$. Suppose that $\tau=\rho \exp \pi i/(\alpha+1)$ where $\rho \geqq (1+\epsilon)|S_0|$, ϵ a fixed positive number independent of n. The contribution of a circular arc connecting $(1+\epsilon)S$ and $(1+\epsilon)S_0$ may be neglected. We can then replace $\log(1-\eta\tau^{-\alpha})$ by $-\eta\tau^{-\alpha}$ with sufficient degree of accuracy and $\Re[f_\eta(\tau)]$ by $\eta\Re[\tau-\tau^{-\alpha}]=\eta C[\rho+\rho^{-\alpha}]$, $C=\cos(\pi/(\alpha+1))<0$. The minimum of $\rho+\rho^{-\alpha}$ is reached for $\rho=\alpha^{1/(\alpha+1)}=|S_0|=|S|+O(\eta)$. Using these observations it is found that the corresponding part of the integral in (7.7)

$$\left| \int_{(1+\epsilon)S_0}^{\infty} \right| \leqq \exp\{\eta n[\Re(V_1)-\epsilon']\}$$

with a fixed positive ϵ' independent of n. Now let $\tau=\delta e^{i\phi}$, $0\leqq\phi\leqq\pi/(\alpha+1)$, where both δ and $\eta\delta^{-\alpha}$ are small. We can replace $\Re[f_\eta(\tau)]$ by $\eta\Re[\tau-\tau^{-\alpha}]$ $=\eta(\delta\cos\phi-\delta^{-\alpha}\cos\alpha\phi)$. Here the last parenthesis is an increasing function of ϕ which is negative for $\phi=\pi/(\alpha+1)$ and numerically as large as we please if δ has been chosen properly. The contribution from the small circle is consequently negligible in comparison with the main term. This completes the argument for $0<\alpha<1$.

The case $\alpha=1$ is already in the literature and need not be considered here. The method used above also applies to the case $1<\alpha$. Here, however, $\pi/(\alpha+1)<\pi/2$ and we deform the loop in (7.7) into a contour consisting of the circle $|\tau|=|S|$ plus the negative real axis from $-\infty$ to $-|S|$ and back again. The main contribution still comes from the neighborhoods of $\tau=S$ and $\bar{S}$ and is given by formulas (7.9) and (7.10), where, however, k is now equal to one. The discussion of the neglected parts of the path follows similar lines as above. Combining the various estimates we obtain the following result.

THEOREM 12. *For $n^{-1/\alpha}<w<\omega$, $0<\alpha$,*

$$P_n(w,\alpha) = -\left(\frac{2}{\pi}\right)^{1/2}(\alpha+1)^{-1/2}\alpha^{1/2(\alpha+1)}w^{-(\alpha+2)/2(\alpha+1)}n^{1/2(\alpha+1)}$$
$$\times \Re\left\{(1+\delta_n)\exp\left[\frac{\pi i}{2(\alpha+1)}+n(V_1\eta+\cdots+V_k\eta^k)\right]\right\}, \tag{7.11}$$

where

$$\eta=\left(\frac{w}{n}\right)^{\alpha/(\alpha+1)}, \qquad V_1=\left(1+\frac{1}{\alpha}\right)\alpha^{1/(\alpha+1)}e^{i\pi/(\alpha+1)}, \qquad k=\left[1+\frac{1}{\alpha}\right],$$

$V_1,\cdots,V_n$ depend on α only, and $\delta_n\to 0$ when $n\to\infty$.

The difference between the three cases $\alpha<1$, $\alpha=1$, $\alpha>1$ is rather striking. For $\alpha<1$ the exponential factor gives rise to oscillations with a strong damping factor. For $\alpha=1$, we have $k=2$ and $nV_1\eta=2(wn)^{1/2}i$, $nV_2\eta^2=\frac{1}{2}w$, so that the change in amplitude depends upon w but not upon n, except for the trivial factor $n^{1/4}$. For $\alpha>1$ the exponential factor is still oscillatory but now the amplitudes tend very rapidly to infinity with n. Formulas (6.3), second line, and (6.15) are immediate consequences of (7.11).

The same method can be applied to the function $H_\alpha(z)$ of (7.2) and leads to the estimate

$$\begin{aligned} H_\alpha(z) &= -\left(\frac{2}{\pi}\right)^{1/2}(\alpha+1)^{-1/2}(\alpha z)^{1/2(\alpha+1)} \\ &\times \Re\left\{[1+\delta(z)]\exp\left[\frac{\pi i}{2(\alpha+1)}+\left(1+\frac{1}{\alpha}\right)(\alpha z)^{1/(\alpha+1)}e^{\pi i/(\alpha+1)}\right]\right\} \end{aligned} \tag{7.12}$$

where $\delta(z)\to 0$ when $z\to+\infty$. In particular, if $0<\alpha<1$,

$$|H_\alpha(z)| < \exp\left[-cz^{1/(\alpha+1)}\right], \qquad c>0, \qquad z \geqq 1. \tag{7.13}$$

For $\alpha>0$, $H_\alpha(z)$ has only real non-negative zeros and the same is true in the case of the polynomials

$$\Phi_n(z,\alpha) = \sum_{\nu=1}^{n}(-1)^{\nu-1}\binom{n}{\nu}\frac{z^\nu}{\Gamma(\nu\alpha)}. \tag{7.14}$$

(See G. Pólya and G. Szegö [12, vol. 2, p. 68, problem 167].) We note that

$$P_n(w,\alpha) = w^{-1}\Phi_n(w^\alpha,\alpha)$$

and

$$\Phi_n(z,1) = zL_{n-1}^{(1)}(z).$$

Bibliography

1. A. F. Andersen, *Studier over Cesàro's Summabilitetsmetode*, Thesis, Copenhagen, 1921, 100 pp.
2. S. Banach, *Théorie des opérations linéaires*, Warsaw, 1932.
3. G. D. Birkhoff, *Proof of the ergodic theorem*, Proc. Nat. Acad. Sci. U.S.A. vol. 17 (1931) pp. 656–660.
4. S. Bochner, *Integration von Funktionen, deren Werte die Elemente eines Vektorraumes sind*, Fund. Math. vol. 20 (1933) pp. 262–276.
5. S. Chapman, *On non-integral orders of summability of series and integrals*, Proc. London Math. Soc. (2) vol. 9 (1911) pp. 369–409.
6. L. W. Cohen, *On the mean ergodic theorem*, Ann. of Math. (2) vol. 41 (1940) pp. 505–509.
7. N. Dunford, *Spectral theory.* I. *Convergence to projections*, Trans. Amer. Math. Soc. vol. 54 (1943) pp. 185–217.
7a. ———, *Spectral theory*, Bull. Amer. Math. Soc. vol. 49 (1943) pp. 637–651.
8. E. Hille, *Notes on linear transformations.* II. *Analyticity of semi-groups*, Ann. of Math. (2) vol. 40 (1939) pp. 1–47.
9. E. Hille and J. D. Tamarkin, *On the theory of linear integral equations.* I. Ann. of Math. (2) vol. 31 (1930) pp. 479–528.
10. J. E. Littlewood, *The converse of Abel's theorem on power series*, Proc. London Math. Soc. (2) vol. 9 (1911) pp. 434–448.

11. J. von Neumann, *Proof of the quasi-ergodic hypothesis*, Proc. Nat. Acad. Sci. U.S.A. vol. 18 (1932) pp. 70–82.

12. G. Pólya and G. Szegö. *Aufgaben und Lehrsätze aus der Analysis*, 2 vols., Berlin, 1925.

13. M. Riesz, *Über einen Satz des Herrn Fatou*, Journal für Mathematik vol. 140 (1911) pp. 89–99.

14. G. Szegö, *Orthogonal polynomials*, Amer. Math. Soc. Colloquium Publications, vol. 23, New York, 1939.

15. N. Wiener, *Tauberian theorems*, Ann. of Math. (2) vol. 33 (1932) pp. 1–100.

16. A. Wiman, *Über die Nullstellen der Funktionen* $E_\alpha(x)$, Acta Math. vol. 29 (1905) pp. 217–234.

YALE UNIVERSITY,
NEW HAVEN, CONN.

Chapter 12
Semi-groups

The papers included in this chapter are

[63] Notes on linear transformations. I

[72] On semi-groups of transformations in Hilbert space

[73] Notes on linear transformations. II. Analyticity of semi-groups

[83] Representation of one-parameter semi-groups of linear transformations

[85] On the analytical theory of semi-groups

[88] On the theory of characters of groups and semi-groups in normed vector rings

[91] (with N. Dunford) The differentiability and uniqueness of continuous solutions of addition formulas

[96] Lie theory of semi-groups of linear transformations

An introduction to Hille's earlier papers on semi-groups and a note on Hille's work on semi-group theory in connection with approximation theory can be found starting on page 684.

NOTES ON LINEAR TRANSFORMATIONS, I*

BY
EINAR HILLE

Under the above title the author intends to publish some investigations on the properties of linear transformations in abstract spaces. In the present note the space is a suitable subset of the set of all measurable functions defined for $-\infty < x < \infty$, and the transformations are of the form

$$K_\alpha[f] \equiv \alpha \int_{-\infty}^{\infty} K(\alpha t) f(x+t) dt, \qquad \alpha > 0. \tag{1}$$

The results, which are somewhat loosely knit together, cluster around four problems. (i) *The originators of zero*, i.e., the solutions of the equation

$$K_\alpha[f] = 0. \tag{2}$$

(ii) *The invariant elements*, i.e., the solutions of the equation

$$K_\alpha[f] = f. \tag{3}$$

(iii) *The functional equations* satisfied by $K_\alpha[f]$ for special choices of the kernel. (iv) *The metric properties* of the transformation $K_\alpha[f]$, including *properties of contraction*, and *degree of approximation* of f by $K_\alpha[f]$ for large values of α. The material is grouped as follows. §1 gives a survey of problems (i), (ii) and (iv) for a general kernel $K(u) \epsilon L_1(-\infty, \infty)$, $K(u) \geqq 0$. It lies in the nature of things that the results for this case are rather incomplete. They probably do not offer much of any novelty to the workers in the field, but serve as background for the discussion in §§3–4. The existence of functional equations obtained by superposition is established in §2, and the equations are given for four particular kernels which may be associated with the names of Dirichlet, Picard, Poisson, and Weierstrass. A closer study of the last two kernels, which satisfy the same functional equation, is given in §3, whereas the kernel of Picard is treated in §4. It turns out that the study of problems (i), (ii) and (iv) for these special kernels is much simplified by the corresponding functional equations. Some results on the Dirichlet kernel occur in §5, but lack the same degree of completeness, sharpness and simplicity.†

---

* Presented to the Society, April 20, 1935; received by the editors March 14, 1935.
† The author is indebted to Professor J. D. Tamarkin for helpful criticism.

1. Non-negative kernels in $L_1(-\infty, \infty)$

1.1. We shall be concerned with kernels $K(u)$ satisfying the following conditions:

(K_1) *$K(u)$ is defined as a measurable non-negative function in* $(-\infty, \infty)$.

(K_2) *$\int_{-\infty}^{\infty} K(u)du$ exists and equals unity.*

Let $S=S(K)$ be the set of all functions $f(x)$ satisfying the two conditions

(S_1) *$f(x)$ is defined as a measurable function in $(-\infty, \infty)$, and*

$$K_\alpha[f] \equiv \alpha \int_{-\infty}^{\infty} K(\alpha t) f(x+t)dt \tag{1.11}$$

exists as an ordinary Lebesgue integral for almost all x and all $\alpha>0$.

(S_2) *$K_\alpha[f] \epsilon S(K)$ for all $\alpha>0$ whenever $f \epsilon S(K)$.*

It is obvious from these definitions that $S(K)$ is a linear vector space closed under the transformations K_α. We note that if $f(x) \epsilon S(K)$ then all translations of $f(x)$, i.e., the functions $f(x+h)$, also belong to $S(K)$, and that the two operations K_α and translation by h commute.

1.2. Problem (i) calls for the solution of the equation

$$K_\alpha[f] = 0. \tag{1.21}$$

A solution is clearly $f \sim 0$. But is this the only solution? Not always, as we shall see.

Let us denote the Fourier transform of $g(x)$ by $T[x; g]$. Suppose that $K(u) \epsilon L_2(-\infty, \infty)$. It then has a Fourier transform in the same space. Suppose that $f(x)$ is a solution of (1.21) in L_2. Then by a well known formula

$$T[x; f] T[-x/\alpha; K] = 0. \tag{1.22}$$

Here we have two possibilities. (1) $T[-x/\alpha; K]$ vanishes only in a null set. In this case (1.22) implies that $T[x; f] \sim 0$, and consequently also $f \sim 0$, so that $f \sim 0$ is the only solution of (1.21) in L_2. (2) $T[-x/\alpha; K]$ vanishes on a set S of positive measure. We can assume S to be bounded. Let $g(x)$ be a measurable function which is bounded in S and vanishes outside $\bar{S}$, and put $f(x) = T[-x; g]$. This function $f(x)$ is in L_2 and is a solution of (1.21) which is not equivalent to zero. That this case can actually arise is shown by the kernel $K(u) = \pi^{-1}u^{-2}(1-\cos u)$ whose Fourier transform vanishes for $|x|>1$.

It is obvious that this method is capable of some extension, but it suffers from the usual limitations due to the severe restrictions which must be imposed upon the function in order that it shall have a Fourier transform. The special kernels considered below in §§3–4 have Fourier transforms nowhere equal to zero, and the particular properties of the kernels will enable us to prove that $f=0$ is the only solution of problem (i) in the corresponding space $S(K)$.

1.3. Problem (ii) calls for the fixed points of $S(K)$, i.e., the solutions of the equation*

$$K_\alpha[f] = f. \tag{1.31}$$

Condition (K_2) shows that $f=1$ is a solution. In many important cases $K(u)$ is an even function of u. If this is so, and $x \epsilon S(K)$, then $f(x)=x$ is a solution of (1.31) for every $\alpha>0$. Consequently every linear function is an invariant. This case is realized for instance for the kernels of Picard and Weierstrass, treated below, but not for that of Poisson, because $f=x$ does not belong to the corresponding space $S(K)$.

The method of Fourier transforms leads to the equation

$$(2\pi)^{1/2}T[x; f]T[-x/\alpha; K] = T[x; f], \tag{1.32}$$

if we assume for the sake of simplicity that $K(u)$ and $f(x)$ are in L_2. We have again two cases. (1) If $T[-x/\alpha; K]=(2\pi)^{-1/2}$ only on a null set, $T[x; f]$ must vanish almost everywhere, i.e., $f\sim 0$ is the only solution of (1.31) in L_2. (2) If, on the other hand, $T[-x/\alpha; K]=(2\pi)^{-1/2}$ on a set of positive measure, a construction similar to that of §1.2 will lead to an invariant manifold in L_2.

1.4. Let us now consider a metric space $M(K)$ which is a sub-set of $S(K)$. We shall suppose that $M(K)$ has the following properties.

(M_1) *It is a normed linear vector space in the sense of Banach, complete with respect to its metric.*

(M_2) $f(x)\epsilon M(K)$ *implies* $K_\alpha[f]\epsilon M(K)$ *for every* $\alpha>0$.

(M_3) $\|K_\alpha[f]\| \leqq \|f\|$.

We shall first consider the possibilities of finding such spaces $M(K)$ in $S(K)$. It is a simple matter to see that every Lebesgue space $L_p(-\infty, \infty)$, $1\leqq p\leqq\infty$, is a sub-space of every $S(K)$, and the same is true of the space $C[-\infty, \infty]$ of the functions which are continuous for $-\infty\leqq x\leqq\infty$. That the customary metrics of these spaces satisfy condition (M_1) is well known, and

* There are some passing remarks on this problem by N. Wiener and E. Hopf in the introduction to their paper *Ueber eine Klasse singulärer Integralgleichungen*, Sitzungsberichte der Preussischen Akademie der Wissenschaften, Mathematisch-Physikalische Klasse, 1931, pp. 696–706. They assume that the kernel $K(u)$ vanishes exponentially for large values of $|u|$. In this case the method of bilateral Laplace transforms applies and shows that the solutions are essentially exponential functions. The discussion of the invariant elements of the Weierstrass kernel in §3.4 could have been made somewhat shorter with the aid of this method. [Added in proof, November 2, 1935.]

in order to see that (M_2) and (M_3) are satisfied it is enough to recall the following inequalities:

$$(1.41) \qquad \int_{-\infty}^{\infty} | K_\alpha[f] | \, dx \leqq \alpha \int_{-\infty}^{\infty} K(\alpha t) dt \int_{-\infty}^{\infty} | f(x+t) | \, dx,$$

$$(1.42) \qquad \int_{-\infty}^{\infty} | K_\alpha[f] |^p dx \leqq \alpha \int_{-\infty}^{\infty} K(\alpha t) dt \int_{-\infty}^{\infty} | f(x+t) |^p dx,$$

$$(1.43) \qquad \text{e.l.u.b.} \, | K_\alpha[f] | \leqq \text{e.l.u.b.} \, | f |.$$

The first inequality refers to the case in which $f \epsilon L_1$, and is immediate. The second inequality presupposes $f \epsilon L_p$, $1 < p < \infty$; it follows from Jensen's inequality for convex functions. In (1.43) $f \epsilon L_\infty$, but we have merely to replace the *essential least upper bound* by the *maximum* in order to get the corresponding inequality for $f \epsilon C$.

There is consequently no lack of sub-spaces of $S(K)$ which satisfy our conditions. It is perhaps also possible to find a metric satisfying these conditions which applies to the whole of $S(K)$. Various metrics valid for the space of measurable functions come to mind in this connection, but these metrics normally fail to satisfy the condition $\|\alpha f\| = |\alpha| \, \|f\|$ which is a part of (M_1). This condition is used extensively below, especially in §3.6. But this is actually the only part of our conditions which it seems difficult to impose on $S(K)$; in particular, (M_2) and (M_3) do not cause any trouble.

Condition (M_3) is consequently a natural assumption to make in the study of these kernels. Its geometric significance is that the transformation $K_\alpha[f]$ defines a contraction of the space $M(K)$ for every fixed $\alpha > 0$. In special cases this contraction will be continuous and monotone with respect to α; this is the case with the particular kernels discussed below.

1.5. It is an easy matter to show that (K_1) and (K_2) imply that

$$(1.51) \qquad \lim_{\alpha \to \infty} K_\alpha[f] = f(x)$$

at every point of continuity of $f(x)$. Under certain circumstances we can also show convergence of $K_\alpha[f]$ to f in the sense of the metric in $M(K)$.

For this purpose let us introduce the *modulus of continuity* of $f(x)$ defined as

$$(1.52) \qquad \omega(h; f) = \| f(x+h) - f(x) \|,$$

where h is fixed, and the norm is taken with respect to x. Further, let $P(u)$ and $Q(u)$ be even continuous functions of u, monotone increasing for $u > 0$, and vanishing for $u = 0$. We have then the following

THEOREM 1.5. *A sufficient condition that*

$$\lim_{\alpha\to\infty} \|K_\alpha[f] - f\| = 0 \tag{1.53}$$

for every $f(x)\epsilon M(K)$ *is that the following assumptions hold*:

(C_1) $f(x)\epsilon M(K)$ *implies* $f(x+h)\epsilon M(K)$, *and* $\|f(x+h)\| = \|f(x)\|$ *for every real* h.

(C_2) *There shall exist two functions* $P(u)$ *and* $Q(u)$ *with the properties stated above such that*

$$\|K_\alpha[f] - f\| \leqq P\{K_\alpha[Q(\omega(t; f))]\}$$

for every $f\epsilon M(K)$.

(C_3) $\lim_{h\to 0} \omega(h; f) = 0$ *for every* $f\epsilon M(K)$.

The proof of this theorem follows standard lines, and can be omitted here. Let us instead consider the justification of imposing such conditions. Our assumptions are satisfied in $L_p(-\infty, \infty)$ for $1 \leqq p < \infty$, but not for $p = \infty$. Indeed, in L_1

$$\int_{-\infty}^{\infty} | K_\alpha[f] - f | \, dx \leqq \alpha \int_{-\infty}^{\infty} K(\alpha t) dt \int_{-\infty}^{\infty} | f(x+t) - f(x) | \, dx$$
$$= K_\alpha[\omega(t; f)],$$

so that (C_2) is satisfied with $P(u) = Q(u) = |u|$. Conditions (C_1) and (C_3) are evidently also satisfied. If $1 < p < \infty$, we have instead

$$\int_{-\infty}^{\infty} | K_\alpha[f] - f |^p dx \leqq \alpha \int_{-\infty}^{\infty} K(\alpha t) dt \int_{-\infty}^{\infty} | f(x+t) - f(x) |^p dx$$
$$= K_\alpha[(\omega(t; f))^p],$$

so that (C_2) is satisfied with $P(u) = |u|^{1/p}$, $Q(u) = |u|^p$. The other conditions are also known to hold. If $p = \infty$, conditions (C_1) and (C_2) still hold, but not (C_3). Moreover, formula (1.53) cannot hold for every $f(x)\epsilon L_\infty$. Indeed, convergence in this space is essentially uniform convergence, and a sequence of continuous functions converges essentially uniformly if and only if it converges uniformly. This of course implies that the limit function is continuous. Since $K_\alpha[f]$ is always a continuous function of x, (1.53) cannot hold when $f(x)$ is discontinuous. In the case of $C[-\infty, \infty]$ we have

$$\max_x | K_\alpha[f] - f | \leqq \alpha \int_{-\infty}^{\infty} K(\alpha t) \max_x | f(x+t) - f(x) | \, dt$$
$$= K_\alpha[\omega(t; f)],$$

so that all three conditions hold.

The assumptions of Theorem 1.5 are clearly not necessary, and various modifications of these assumptions could be given which would preserve their sufficient character. The reader who reconstructs the omitted proof of the theorem will find that the convergence of $K_\alpha[f]$ to f as $\alpha\to\infty$ is uniform in any family of uniformly bounded, equi-continuous functions. He will also get some idea of what degree of approximation is to be expected. In the special cases treated in §§3–4 it is possible to find a best degree of approximation valid for all elements of $M(K)$ which are not invariant.

2. Some functional equations

2.1. For the work of the present paragraph it is convenient to add the following postulate:

$$(K_3) \qquad K(u)\epsilon L_2(-\infty, \infty).$$

We shall also need (K_1), (K_2), (S_1) and (S_2).

For every function $f(x)\epsilon S(K)$ we can form the iterated transformations $K_\alpha[K_\beta[f]]$ and $K_\beta[K_\alpha[f]]$, and they are also elements of $S(K)$. We are particularly interested in those cases in which these superposed transforms are expressible in terms of simple transforms $K_\gamma[f]$, where γ is some function of α and β. Such cases are revealed by the method of Fourier transforms.

Proceeding formally, let us write

$$K_\alpha[K_\beta[f]] = \int_{-\infty}^{\infty} K(u;\alpha,\beta)f(u+x)du, \tag{2.11}$$

where

$$K(u;\alpha,\beta) = \alpha\beta\int_{-\infty}^{\infty} K(\alpha s)K(\beta(u-s))ds. \tag{2.12}$$

Then

$$\begin{aligned} T[x; K(u;\alpha,\beta)] &= \alpha\beta(2\pi)^{1/2}T[x; K(\alpha s)]T[x; K(\beta s)] \\ &= (2\pi)^{1/2}T[x/\alpha; K(u)]T[x/\beta; K(u)], \end{aligned} \tag{2.13}$$

so that

$$K(u;\alpha,\beta) = (2\pi)^{1/2}T^{-1}\{u; T[x/\alpha; K(v)]T[x/\beta; K(v)]\}, \tag{2.14}$$

which can be used for the computation of the composed kernel. This formula is the basis of all the functional equations in the following.

2.2. Let us consider some important special cases.

I. Weierstrass's singular integral. Here

$$K(u) = \pi^{-1/2}e^{-u^2},$$

and

$$T[x; K(u)] = (2\pi)^{-1/2}e^{-x^2/4}.$$

It follows that the Fourier transform of the composed kernel is

$$\exp\left[-\left(\frac{1}{\alpha^2}+\frac{1}{\beta^2}\right)\frac{x^2}{4}\right],$$

so that the kernel itself becomes

$$\gamma\pi^{-1/2}e^{-\gamma^2u^2}, \qquad \frac{1}{\gamma^2}=\frac{1}{\alpha^2}+\frac{1}{\beta^2}.$$

Hence putting

$$\overline{W}_\alpha[f] = \alpha\pi^{-1/2}\int_{-\infty}^{\infty} e^{-\alpha^2u^2}f(u+x)du, \tag{2.21}$$

we obtain

$$\overline{W}_\alpha[\overline{W}_\beta[f]] = \overline{W}_\gamma[f], \qquad \frac{1}{\gamma^2}=\frac{1}{\alpha^2}+\frac{1}{\beta^2}. \tag{2.22}$$

II. Poisson's integral for the half-plane. Here

$$K(u) = \pi^{-1}(1+u^2)^{-1},$$

and putting

$$\overline{P}_\alpha[f] = \alpha\pi^{-1}\int_{-\infty}^{\infty}\frac{f(u+x)}{1+\alpha^2u^2}du, \tag{2.23}$$

we get

$$\overline{P}_\alpha[\overline{P}_\beta[f]] = \overline{P}_\gamma[f], \qquad \frac{1}{\gamma}=\frac{1}{\alpha}+\frac{1}{\beta}. \tag{2.24}$$

We note that this is essentially the same functional equation as that of the Weierstrass kernel.

III. Picard's singular integral. Here

$$K(u) = \tfrac{1}{2}e^{-|u|},$$

and putting

$$\Pi_\alpha[f] = \frac{\alpha}{2}\int_{-\infty}^{\infty} e^{-\alpha|u|}f(u+x)du, \tag{2.25}$$

we obtain

$$(2.26)\qquad (\alpha^2 - \beta^2)\Pi_\alpha[\Pi_\beta[f]] = \alpha^2\Pi_\beta[f] - \beta^2\Pi_\alpha[f].$$

IV. **Dirichlet's singular integral.** Here

$$K(u) = \frac{\sin u}{\pi u}.$$

Putting

$$(2.27)\qquad D_\alpha[f] = \frac{1}{\pi}\int_{-\infty}^{\infty}\frac{\sin \alpha u}{u} f(u+x)du,$$

we get

$$(2.28)\qquad D_\alpha[D_\beta[f]] = D_\gamma[f], \qquad \gamma = \min(\alpha, \beta).$$

We note that this kernel does not satisfy either (K_1) or (K_2). This fact makes the investigation of the corresponding transformation much more complicated.

Other examples of simple functional equations could undoubtedly be found in this connection. The importance of these four transformations is such, however, that a special investigation of their properties as revealed by the functional equations is warranted. This will be done below.

3. The Poisson-Weierstrass case

3.1. Equations (2.22) and (2.24) reduce to the common form

$$(3.11)\qquad F_\lambda[F_\mu[f]] = F_{\lambda+\mu}[f]$$

by an obvious change of parameters. This equation is consequently satisfied by the two transformations

$$(3.12)\qquad P_\lambda[f] = \frac{\lambda}{\pi}\int_{-\infty}^{\infty}\frac{f(u+x)}{u^2+\lambda^2}du,$$

$$(3.13)\qquad W_\lambda[f] = (\pi\lambda)^{-1/2}\int_{-\infty}^{\infty} e^{-u^2/\lambda} f(u+x)du.$$

This fact is undoubtedly well known to mathematical physicists.* $P_\lambda[f]$ is

* Several writers on the theory of the equation of heat conduction have observed such functional equations. P. Appell gave equation (3.11) for $W_\lambda[f]$ in Journal de Mathématiques, (4), vol. 8 (1892), pp. 187–216, p. 201. Cesàro, Académie Royale de Belgique, Bulletin de la Classe des Sciences, 1902, pp. 387–407, p. 392, noted that certain solutions form the elements of an Abelian group. G. Doetsch has produced a number of related transcendental addition theorems; see especially Mathematische Zeitschrift, vol. 25 (1926), pp. 608–626, p. 615. I am not aware of similar considerations having been made for Poisson's integral.

the solution of Dirichlet's problem for the upper half-plane corresponding to the boundary values $f(x)$ on the x-axis, whereas $W_\lambda[f]$ is a solution of the equation of heat conduction in one dimension corresponding to a given initial temperature $f(x)$. These interpretations make equation (3.11) intuitively obvious.

We choose for $S(P)$ and $S(W)$ the classes of measurable functions defined on $(-\infty, \infty)$ for which (3.12) and (3.13) respectively exist as proper Lebesgue integrals for every $\lambda > 0$. This choice is evidently in agreement with (S_1), and a moment's consideration will show that (S_2) is also fulfilled, and that (3.11) holds for any such function $f(x)$. $S(P)$ is simply the class of all $f(x)$ such that $f(x)/(1+x^2)\epsilon L_1(-\infty, \infty)$. $S(W)$ cannot be characterized in such simple terms.

The transforms $P_\lambda[f]$ and $W_\lambda[f]$ are analytic functions of x and of λ. For a fixed real x, $P_\lambda[f]$ defines one analytic function of λ in the right half-plane and another in the left, which are holomorphic in the half-planes in question, whereas $W_\lambda[f]$ is holomorphic in the right half-plane and ordinarily does not exist in the left one. For a fixed positive λ, $P_\lambda[f]$ is an analytic function of x, holomorphic in the strip $-\lambda < \mathfrak{I}(x) < \lambda$,* whereas $W_\lambda[f]$ is an entire function of x.

In the present case formula (1.51) holds in a sharper form, viz.,

$$\lim_{\lambda \to 0} P_\lambda[f] = f(x), \qquad \lim_{\lambda \to 0} W_\lambda[f] = f(x), \tag{3.14}$$

for almost all x whenever $f(x)\epsilon S(P)$ or $S(W)$.

3.2. Problem (i) has a very simple solution in this case:

THEOREM 3.2. *If* $P_\alpha[f]=0$ *or* $W_\alpha[f]=0$ *for a fixed* α, *and* $f\epsilon S(P)$ *or* $S(W)$ *respectively, then* $f(x)\sim 0$.

This is pretty well known. A proof is obtained by observing that $F_\alpha[f]=0$ implies $F_{\alpha+\beta}[f]=0$ for every $\beta>0$ by (3.11). $F_\lambda[f]$ being analytic in λ must then vanish identically, and (3.14) shows that this implies $f(x)\sim 0$.

The same argument shows that $F_\alpha[f_1] \neq F_\alpha[f_2]$ unless $f_1(x)\sim f_2(x)$.

3.3. Let us now consider problem (ii). It is required to find whether, for a fixed α, the equation

$$F_\alpha[f] = f \tag{3.31}$$

can have any solution in S other than the trivial one, $f=$ constant. If there exists such a solution $f(x)$ then (3.11) shows that the corresponding transform $F_\mu[f]$ satisfies the equation

* $P_\lambda[f]$ also defines two other analytic functions of x, one holomorphic above this strip, the other one below it.

$$F_{\lambda+\alpha}[f] = F_\lambda[f]$$

for every λ. Hence $F_\lambda[f]$ is an analytic function of λ with period α. From this point onwards the two cases must be treated separately.

In the Poisson case we note that if $f(x)\epsilon S(P)$ and x is fixed, then $P_\lambda[f] = o(|\lambda|)$. Hence, if $P_\lambda[f]$ is periodic in λ with period α, it must be a constant with respect to λ, i.e., $P_\lambda[f]=f$ identically in λ. But $P_\lambda[f]$ is a potential function for $\lambda>0$, i.e.,

$$\left\{\frac{\partial^2}{\partial x^2}+\frac{\partial^2}{\partial \lambda^2}\right\}P_\lambda[f] = 0.$$

Here

$$\frac{\partial^2}{\partial \lambda^2}P_\lambda[f] = 0,$$

since $P_\lambda[f]$ is independent of λ. Hence

$$\frac{\partial^2}{\partial x^2}P_\lambda[f] = 0,$$

so that $P_\lambda[f]=f$ is a linear function of x. But x is clearly not in $S(P)$, hence f is a constant. Thus we have proved

THEOREM 3.3. *The only function in $S(P)$ which is invariant under a Poisson transformation P_α is $f(x)=$const., and this function is invariant under all such transformations.*

3.4. The Weierstrass case is rather different. We have seen that $W_\lambda[f]$ must be an analytic function of λ with period α. Being holomorphic in the right half-plane, $W_\lambda[f]$ must then be an entire function of λ as well as of x. We have consequently

$$W_\lambda[f] = \sum_{-\infty}^{\infty} A_n(x)e^{2\pi i n\lambda/\alpha}. \tag{3.41}$$

Here the coefficients are entire functions of x which tend to zero faster than $\exp[-B|n|]$, B arbitrary, as $n\to\infty$, x being fixed. But $W_\lambda[f]$ is a solution of the partial differential equation

$$\frac{\partial^2 W}{\partial x^2} = 4\frac{\partial W}{\partial \lambda}, \tag{3.42}$$

and we are clearly entitled to differentiate term by term in (3.41). It follows that

$$A_n''(x) = 8\pi i n\alpha^{-1}A_n(x) \qquad (n = 0, \pm 1, \pm 2, \cdots). \tag{3.43}$$

Consequently $A_0(x)$ is a linear combination of

$$1 \text{ and } x, \tag{3.44}$$

$A_n(x)$ is a linear combination of

$$\exp\left[(2\pi in/\alpha)^{1/2}(1+i)x\right] \quad \text{and} \exp\left[(2\pi in/\alpha)^{1/2}(-1-i)x\right] \tag{3.45}$$

if $n>0$, and of

$$\exp\left[(2\pi i|n|/\alpha)^{1/2}(1-i)x\right] \text{ and } \exp\left[(2\pi i|n|/\alpha)^{1/2}(-1+i)x\right] \tag{3.46}$$

if $n<0$. It follows that the equation (3.31) has a continuum of solutions in the Weierstrass case. These solutions have a denumerable basis, viz., the functions of (3.44)–(3.46). Any linear combination of these functions, the coefficients of which satisfy the restriction of tending to zero faster than any function of n of the form $\exp[-B|n|]$ as $n\to\infty$, assuming that there are infinitely many terms, is a solution of (3.31).

These solutions are entire functions of x. Their rate of growth is subject to rather interesting limitations. Suppose that x is real and $|f(x)|\leqq Ae^{k|x|}$, where k is a positive constant. A simple calculation shows that for $\lambda=\sigma+i\tau$

$$|W_\lambda[f]| \leqq 2Ae^{k|x|}(\sigma^2+\tau^2)^{1/4}\sigma^{-1/2}\exp\left[k^2(\sigma^2+\tau^2)/(4\sigma)\right]. \tag{3.47}$$

Suppose now that it is known that $W_\lambda[f]$ has the period α. Then we have the same estimate if we replace λ by $\lambda+n\alpha$. Here we can choose n so as to minimize $((\sigma+n\alpha)^2+\tau^2)/(\sigma+n\alpha)$. This minimum lies arbitrarily close to $2|\tau|$ if $|\tau|$ is large. Hence for an $f(x)$ which produces a periodic solution we can replace (3.47) by

$$|W_\lambda[f]| \leqq 2^{3/2}A\exp\left[k|x|+k^2|\tau|/2\right]. \tag{3.48}$$

But this estimate implies that $W_\lambda[f]$ is a rational function of $w=\exp[2\pi i\lambda/\alpha]$ with singularities only at 0 and ∞, or more precisely, $W_\lambda[f]=w^{-n}P_{2n}(w)$ where $P_{2n}(w)$ is a polynomial in w of degree $\leqq 2n$ and $n=[k^2\alpha/(4\pi)]$. This result gives us additional information about the solutions of (3.31). It follows that any solution which involves infinitely many functions of the basis must occasionally grow faster than any function of the form $e^{k|x|}$ on the real axis. On the other hand, a simple calculation shows that if such a solution is an entire function of order two, it is of the minimal type of that order. Suitably chosen "lacunary series" in terms of the basis functions show that this estimate cannot be essentially improved upon. In the other direction we notice that *the only solution which is at most of the minimal type of order one is* $Ax+B$, *and this is the only invariant common to all Weierstrass transformations.*

It should be added that the preceding results also permit a complete determination of the solutions of the equation

$$F_\alpha[f] = F_\beta[f]$$

in the two cases under consideration. The reader will have no difficulties in supplying the details.

3.5. We shall now study the character of the deformation defined by $F_\lambda[f]$ in metric sub-spaces of $S(K)$. We consider two sub-sets $M(P)$ and $M(W)$ of $S(P)$ and $S(W)$ respectively which we suppose satisfy postulates (M_1), (M_2) and (M_3). In addition we shall require

(M_4) *$f(x)\epsilon M(K)$ implies $|f(x)|\epsilon M(K)$, and the inequality $|f(x)| \leqq |g(x)|$ for almost all x implies $\|f\| \leqq \|g\|$.*

A particular consequence of (M_4) is that $f(x)$ and $|f(x)|$ have the same norm since $|f(x)| \leqq ||f(x)||$ and vice versa.

An immediate consequence of (M_3) together with the functional equation (3.11) is that

$$\|F_\beta[f]\| \leqq \|F_\alpha[f]\| \text{ for } 0 < \alpha < \beta, \tag{3.51}$$

so that the transformation $F_\lambda[f]$ is a steady contraction of the space M, and

$$F_\beta[M] \subset F_\alpha[M] \subset M. \tag{3.52}$$

It follows that $\lim_{\alpha\to\infty} \|F_\alpha[f]\|$ exists and is $\geqq 0$. If $f(x)\epsilon L_p(-\infty, \infty)$, $1 \leqq p < \infty$, or more generally, if

$$\lim_{T\to\infty} \frac{1}{2T}\int_{-T}^{T} f(t)dt = 0, \qquad \frac{1}{2T}\int_{-T}^{T} |f(t)|\, dt \leqq A, \tag{3.53}$$

then

$$\lim_{\alpha\to\infty} F_\alpha[f] = 0 \tag{3.54}$$

for all x by a theorem of N. Wiener.*

$\|F_\lambda[f]\|$ is a functional of $f(x)$ and a function of λ. For a fixed λ it is clearly a continuous functional of $f(x)$ in M by virtue of (M_3). Let us now consider its properties as a function of the real positive variable λ for fixed $f(x)$.

Formula (3.51) expresses that $\|F_\lambda[f]\|$ is a monotone decreasing function of λ. We have for $0<h<\lambda$,

$$\begin{aligned} 0 &\leqq \|F_\lambda[f]\| - \|F_{\lambda+h}[f]\| \leqq \|F_{\lambda+h}[f] - F_\lambda[f]\| \\ &= \|F_\lambda[F_h[f] - f]\| \leqq \|F_h[f] - f\|, \end{aligned}$$

and

* See S. Bochner, *Fourierscher Integrale*, p. 30.

$$0 \leqq \|F_{\lambda-h}[f]\| - \|F_\lambda[f]\| \leqq \|F_\lambda[f] - F_{\lambda-h}[f]\|$$
$$= \|F_{\lambda-h}[F_h[f] - f]\| \leqq \|F_h[f] - f\|.$$

Hence if

$$\lim_{h\to 0} \|F_h[f] - f\| = 0, \tag{3.55}$$

then $\|F_\lambda[f]\|$ is continuous for every $\lambda \geqq 0$, and the elements $F_\lambda[f]$ form a continuous curve in M having f as one of its end points. On the other hand, if $\|F_\lambda[f]\|$ is not continuous at $\lambda = \lambda_0$, but has a jump j at this point, then

$$\lim_{h\to 0} \|F_{\lambda+h}[f] - F_{\lambda-h}[f]\| \geqq j$$

for every λ, $0 < \lambda \leqq \lambda_0$, and the distance from $F_\alpha[f]$ to $F_\beta[f]$ would be at least j if either α or β belongs to the range $[0, \lambda_0]$. In particular, the distance between $f(x)$ and any one of its transforms must be at least j. It seems difficult to exclude this possibility a priori, but we shall show that it cannot occur if F_λ be interpreted as P_λ or W_λ.

3.6. We have

$$P_{\lambda+h}[f] - P_\lambda[f] = \frac{h}{\lambda} P_\lambda[f]$$
$$- h(\lambda + h)(2\lambda + h) \frac{1}{\pi} \int_{-\infty}^{\infty} \frac{f(u + x)du}{(u^2 + \lambda^2)(u^2 + (\lambda + h)^2)}.$$

Here we take norms on both sides, noting that the norm of a sum is not greater than the sum of the norms. In the second term we note that $u^2 + (\lambda+h)^2 \geqq (\lambda+h)^2$, and apply hypothesis (M_4). Combining these steps we get for $h > 0$

$$\|P_{\lambda+h}[f] - P_\lambda[f]\| \leqq \frac{h}{\lambda}\left\{\|P_\lambda[f]\| + \frac{2\lambda + h}{\lambda + h} \|P_\lambda[\,|f|\,]\|\right\}$$
$$< \frac{h}{\lambda}\{\|P_\lambda[f]\| + 2\|P_\lambda[\,|f|\,]\|\},$$

or

$$\|P_{\lambda+h}[f] - P_\lambda[f]\| < 3\frac{h}{\lambda}\|P_\lambda[\,|f|\,]\| \leqq 3\frac{h}{\lambda}\|f\|. \tag{3.61}$$

If $-\lambda < h < 0$ we get instead

$$\|P_{\lambda+h}[f] - P_\lambda[f]\| < 3\frac{|h|}{\lambda}\|P_{\lambda+h}[\,|f|\,]\| \leqq 3\frac{|h|}{\lambda}\|f\|. \tag{3.62}$$

In the case of $W_\lambda[f]$ we have

$$W_{\lambda+h}[f] - W_\lambda[f] = \frac{h}{(\lambda+h)^{1/2}[\lambda^{1/2}+(\lambda+h)^{1/2}]} W_\lambda[f] + \frac{1}{(\pi(\lambda+h))^{1/2}} \int_{-\infty}^{\infty} [e^{-u^2/(\lambda+h)} - e^{-u^2/\lambda}] f(u+x)du.$$

A simple calculation shows that

$$\left| e^{-u^2/(\lambda+h)} - e^{-u^2/\lambda} \right| \leqq \begin{cases} \dfrac{h}{e\lambda} e^{-u^2/(2(\lambda+h))}, & 0 < h, \\ \dfrac{|h|}{e(\lambda+h)} e^{-u^2/(2\lambda)}, & -\lambda < h < 0. \end{cases}$$

Hence we get for $h>0$

$$\| W_{\lambda+h}[f] - W_\lambda[f] \| \leqq \frac{h}{\lambda}\left\{ \tfrac{1}{2}\| W_\lambda[f]\| + \frac{2^{1/2}}{e} \| W_{2(\lambda+h)}[|f|]\| \right\},$$

or

$$\| W_{\lambda+h}[f] - W_\lambda[f] \| \leqq 2\frac{h}{\lambda} \| W_\lambda[|f|]\| \leqq \frac{2h}{\lambda}\|f\|, \tag{3.63}$$

where we have used formula (3.51) in addition to hypothesis (M_4). For $-\lambda<h<0$ we get instead

$$\| W_{\lambda+h}[f] - W_\lambda[f] \| < \frac{2|h|}{\lambda+h} \| W_{\lambda+h}[|f|]\| \leqq \frac{2|h|}{\lambda+h}\|f\|. \tag{3.64}$$

These formulas show that $\|P_\lambda[f]\|$ and $\|W_\lambda[f]\|$ are continuous families of continuous transformations in $M(P)$ and $M(W)$ respectively for $\lambda>0$. If $\lambda \geqq \lambda_0 > 0$, $\|f\| \leqq B$, these families satisfy a Lipschitz condition of order one with respect to λ, uniformly in λ and in f. It follows in particular that the monotone decreasing functions $\|P_\lambda[f]\|$ and $\|W_\lambda[f]\|$ are continuous for $\lambda>0$.

It is not possible to prove continuity at $\lambda=0$ by these considerations. As a matter of fact we recall from the result of the discussion in §1.5 that (3.55) is not true for all metric sub-spaces of S. In particular, it was shown to be false in $L_\infty(-\infty, \infty)$.

3.7. Let us now consider the transformation

$$E_\lambda = I - F_\lambda, \tag{3.71}$$

where I is the identity. As a consequence of (3.11) we get

$$E_{\lambda+\mu}[f] = E_\lambda[f] + E_\mu[f] - E_\lambda[E_\mu[f]]. \tag{3.72}$$

It is easy to see that the operations E_λ and F_μ commute. Formula (3.72) is less useful to us than the mixed equation

$$E_{\lambda+\mu}[f] = E_\lambda[f] + E_\mu[F_\lambda[f]] = E_\lambda[f] + F_\lambda[E_\mu[f]]. \tag{3.73}$$

Using the first and the last member of this equation we get

$$\|E_{\lambda+\mu}[f]\| \leqq \|E_\lambda[f]\| + \|F_\lambda[E_\mu[f]]\|,$$

whence, by virtue of (M_3),

$$\|E_{\lambda+\mu}[f]\| \leqq \|E_\lambda[f]\| + \|E_\mu[f]\|. \tag{3.74}$$

A particular consequence of this relation is that

$$\|E_{\alpha 2^{-n}}[f]\| \geqq 2^{-n}\|E_\alpha[f]\|,$$

and this leads to the important conclusion that

$$\limsup_{h\to 0} \frac{1}{h}\|E_h[f]\| \geqq \frac{1}{\alpha}\|E_\alpha[f]\| \tag{3.75}$$

for every fixed positive α. It follows that the degree of approximation of a function $f(x)$ by its Poisson or Weierstrass transform is definitely limited to be of the first order at best. Indeed, if the limit on the left-hand side is zero, then $\|E_\alpha[f]\|=0$ for every α, i.e., $f(x)$ is an invariant element of the space M under all transformations F_α. These were determined in §3.3 for the Poisson case and in §3.4 for that of Weierstrass. We have consequently proved

THEOREM 3.7. *If $f(x)\epsilon M(P)$ and*

$$\lim_{h\to 0} \frac{1}{h}\|P_h[f] - f\| = 0, \tag{3.76}$$

then $f(x)=\text{const.}$ *If $f(x)\epsilon M(W)$ and*

$$\lim_{h\to 0} \frac{1}{h}\|W_h[f] - f\| = 0, \tag{3.77}$$

then $f(x)=Ax+B$.

It follows in particular that if $M(P)$ or $M(W)$ coincides with $L_p(-\infty, \infty)$, $1\leqq p<\infty$, then (3.76) or (3.77) implies $f(x)=0$. The theorem shows that an inequality of the form

$$\|F_h[f] - f\| > Ch \tag{3.78}$$

holds for every $f \epsilon M$ and for infinitely many values of $h \to 0$. Here C is a non-negative constant depending only upon f which equals zero if and only if f is invariant under all transformations F_α. The estimates of §3.6 show on the other hand that the inequality (3.78) can be reversed for all those functions of the space M which are themselves transforms, i.e., which can be written as $f = F_\alpha[g]$ with $g \epsilon M$. It follows that in a space M whose metric satisfies the conditions stated in §3.5 the degree of approximation of a function $f(x)$ by its Poisson or Weierstrass transform is at best of the first order with respect to α, except for the fixed elements, and that this order is actually reached for an infinite subclass of the space, namely by all the transforms.

4. The Picard case

4.1. We shall now take up for discussion Picard's equation

$$(\alpha^2 - \beta^2)\Pi_\alpha[\Pi_\beta[f]] = \alpha^2\Pi_\beta[f] - \beta^2\Pi_\alpha[f], \qquad \alpha \neq \beta. \tag{4.11}$$

$S(\Pi)$ is the class of all measurable functions $f(x)$ such that (2.25) exists as a proper Lebesgue integral for every $\alpha > 0$. This assumption means that (S_1) is satisfied, and it is easy to see that (S_2) is then also satisfied, and that (4.11) holds for any such function.

The transform $\Pi_\alpha[f]$ is an analytic function of α, regular in the right half-plane. It can be shown that $\Pi_\alpha[f]$ is absolutely continuous and possesses a second-order partial derivative with respect to x for almost all x, and satisfies the differential equation

$$\frac{\partial^2}{\partial x^2}\Pi_\alpha[f] = \alpha^2\{\Pi_\alpha[f] - f\} \tag{4.12}$$

almost everywhere.

It is well known that

$$\lim_{\alpha \to \infty} \Pi_\alpha[f] = f(x) \tag{4.13}$$

for almost all x when $f(x) \epsilon S(\Pi)$.

4.2. Formula (4.12) gives us the following complete solution of problem (i).

THEOREM 4.2. *If* $\Pi_\alpha[f] = 0$ *for some* $\alpha > 0$, *then* $f(x) \sim 0$.

The same conclusion can be drawn from (4.11) combined with (4.13). The same argument shows that $\Pi_\alpha[f_1] = \Pi_\alpha[f_2]$ implies $f_1(x) \sim f_2(x)$.

4.3. The question of invariant elements is also easily answered. Suppose that for some $\alpha > 0$

$$\Pi_\alpha[f] = f. \tag{4.31}$$

Formula (4.32) then shows that

$$\frac{\partial^2}{\partial x^2}\Pi_\alpha[f] = 0.$$

Hence we have proved

THEOREM 4.3. *The only functions in $S(\Pi)$ which are invariant under a Picard transformation Π_α are the linear functions, $Ax+B$, and these functions are invariant under all such transformations.*

We recall that it was shown in §1.3 that the linear functions are left invariant by every transformation $K_\alpha[f]$ whose kernel is an even function and which satisfies (K_1) and (K_2). The Picard transformation has consequently no other invariant elements than those common to this class of transformations.

The equation

$$\Pi_\alpha[f] = \Pi_\beta[f] \tag{4.32}$$

can be treated in the same manner. Together with (4.11) it implies

$$\Pi_\alpha[\Pi_\beta[f]-f]=0,$$

whence $\Pi_\beta[f]-f=0$, and $f(x)=Ax+B$.

4.4. Let us now consider a linear sub-space $M(\Pi)$ of $S(\Pi)$ in which we introduce a metric subject to postulates (M_1), (M_2) and (M_3). Note that (M_4) is not assumed. We shall show that (M_3), i.e., the assumption

$$\|\Pi_\alpha[f]\| \leqq \|f\| \tag{4.41}$$

for every $\alpha>0$, implies that

$$\|\Pi_\alpha[f]\| \leqq \|\Pi_\beta[f]\|, \qquad \alpha<\beta, \tag{4.42}$$

i.e., the analogue of (3.51). We can write (4.11)

$$\Pi_\alpha[f] = \frac{1}{\beta^2}\{\alpha^2\Pi_\beta[f] + (\beta^2-\alpha^2)\Pi_\alpha[\Pi_\beta[f]]\}.$$

This gives

$$\begin{aligned}\|\Pi_\alpha[f]\| &\leqq \frac{1}{\beta^2}\{\alpha^2\|\Pi_\beta[f]\| + (\beta^2-\alpha^2)\|\Pi_\alpha[\Pi_\beta[f]]\|\}\\ &\leqq \frac{1}{\beta^2}(\alpha^2+\beta^2-\alpha^2)\|\Pi_\beta[f]\| = \|\Pi_\beta[f]\|.\end{aligned}$$

It follows that $\|\Pi_\alpha[f]\|$ is a monotone increasing function of α, and

$$(4.43)\qquad \Pi_\alpha[M] \subset \Pi_\beta[M] \subset M.$$

In particular, $\|\Pi_\alpha[f]\|$ tends to a finite limit $\geqq 0$ as $\alpha \to 0$. The transformation $\Pi_\alpha[f]$ is ordinarily not defined for $\alpha = 0$, and need not tend to any finite limit as $\alpha \to 0$, as is shown by the simple example $\Pi_\alpha[x^2] = x^2 + 2\alpha^{-2}$. On the other hand, if the mean value of $f(x)$ over the range $(-T, T)$ is uniformly bounded with respect to T, and tends to a finite limit $\mathfrak{M}[f]$ as $T \to \infty$, then by Wiener's theorem

$$(4.44)\qquad \lim_{\alpha \to 0} \Pi_\alpha[f] = \mathfrak{M}[f]$$

for all x, uniformly over any fixed finite interval. But it is obvious a priori that this result does not enable us to draw any conclusion regarding the numerical value of $\lim_{\alpha \to 0}\|\Pi_\alpha[f]\|$.

4.5. The continuity properties of the Picard transform are on the whole simpler than in the Poisson-Weierstrass case. We can rewrite (4.11) in the form

$$(4.51)\qquad \Pi_\alpha[f] - \Pi_\beta[f] = \frac{\alpha^2 - \beta^2}{\beta^2} \Pi_\beta[f - \Pi_\alpha[f]].$$

Putting

$$(4.52)\qquad \mathrm{H}_\alpha[f] = f - \Pi_\alpha[f],$$

we get

$$(4.53)\qquad \Pi_\alpha[f] - \Pi_\beta[f] = \frac{\alpha^2 - \beta^2}{\beta^2} \Pi_\beta[\mathrm{H}_\alpha[f]].$$

This relation leads to the inequalities

$$(4.54)\qquad \begin{aligned} \|\Pi_\alpha[f] - \Pi_\beta[f]\| &\leqq \left|\left(\frac{\alpha}{\beta}\right)^2 - 1\right| \|\Pi_\beta[\mathrm{H}_\alpha[f]]\| \\ &\leqq \left|\left(\frac{\alpha}{\beta}\right)^2 - 1\right| \|\mathrm{H}_\alpha[f]\| \\ &\leqq 2\left|\left(\frac{\alpha}{\beta}\right)^2 - 1\right| \|f\|, \end{aligned}$$

since obviously

$$(4.55)\qquad \|\mathrm{H}_\alpha[f]\| \leqq 2\|f\|.$$

Since Π_β and H_α commute, we have also

$$\|\Pi_\alpha[f] - \Pi_\beta[f]\| \leqq 2\left|\left(\frac{\alpha}{\beta}\right)^2 - 1\right| \|\Pi_\beta[f]\|. \tag{4.56}$$

It follows from these inequalities that $\Pi_\alpha[f]$ regarded an an element of $M(\Pi)$ is continuous with respect to α, $0<\alpha<\infty$, and that $\|\Pi_\alpha[f]\|$ is a continuous function of α in the same range. We do not have continuity at either zero or infinity except in special cases. Formula (4.42) expresses the fact that $\|\Pi_\alpha[f]\|$ is an increasing function of α. The rate of growth is limited by the inequality

$$\|\Pi_\beta[f]\| \leqq \left[2\left(\frac{\beta}{\alpha}\right)^2 - 1\right] \|\Pi_\alpha[f]\|, \qquad \alpha < \beta, \tag{4.57}$$

which is a consequence of (4.54).

4.6. Let us now consider the transformation $H_\alpha[f]$ in more detail. It satisfies the functional equation

$$(\alpha^2 - \beta^2)H_\alpha[H_\beta[f]] = \alpha^2 H_\alpha[f] - \beta^2 H_\beta[f], \tag{4.61}$$

and the mixed equations

$$\alpha^2(H_\alpha[f] - H_\beta[f]) = (\beta^2 - \alpha^2)\Pi_\alpha[H_\beta[f]], \tag{4.62}$$

$$\alpha^2\Pi_\beta[H_\alpha[f]] = \beta^2\Pi_\alpha[H_\beta[f]]. \tag{4.63}$$

Suppose that $\alpha<\beta$. Then (4.62) gives

$$\begin{aligned}\alpha^2\|H_\alpha[f]\| &\leqq \alpha^2\|H_\beta[f]\| + (\beta^2 - \alpha^2)\|\Pi_\alpha[H_\beta[f]]\| \\ &\leqq (\alpha^2 + \beta^2 - \alpha^2)\|H_\beta[f]\|,\end{aligned}$$

so that

$$\alpha^2\|H_\alpha[f]\| \leqq \beta^2\|H_\beta[f]\|, \qquad \alpha < \beta. \tag{4.64}$$

This inequality states that

$$\alpha^2\|H_\alpha[f]\| \equiv \alpha^2\|f - \Pi_\alpha[f]\|$$

is an increasing function of α. Hence it can tend to zero as $\alpha\to\infty$ if and only if it is identically zero, i.e., if and only if $f(x)$ is invariant under all Picard transformations. We have consequently proved

THEOREM 4.6. *If $f(x)\epsilon M(\Pi)$, and*

$$\lim_{\alpha\to\infty} \alpha^2\|f - \Pi_\alpha[f]\| = 0, \tag{4.65}$$

then $f(x)=Ax+B$.

It follows that there exists a non-negative constant C for every $f(x)\epsilon M(\Pi)$ such that

$$\alpha^2\|f - \Pi_\alpha[f]\| \geqq C \tag{4.66}$$

for infinitely many values of $\alpha \to \infty$, and $C=0$ if and only if $f(x)=Ax+B$. Hence in a space $M(\Pi)$ whose metric is subject to the restrictions stated above, the degree of approximation of $f(x)$ by its Picard transform is of the second order with respect to $1/\alpha$ at the best. This order is actually reached, however, namely by all elements which are themselves transforms of elements of M, i.e., for every $g=\Pi_\beta[f]$. Indeed, formula (4.63) tells us that

$$\alpha^2\|\Pi_\alpha[\Pi_\beta[f]]\| = \beta^2\|H_\beta[\Pi_\alpha[f]]\|. \tag{4.67}$$

The right-hand side does not exceed $2\beta^2\|f\|$ independently of α. Hence the left-hand side remains bounded as $\alpha \to \infty$, i.e.,

$$\limsup_{\alpha\to\infty} \alpha^2\|g - \Pi_\alpha[g]\| \leqq 2\beta^2\|f\|, \qquad g = \Pi_\beta[f]. \tag{4.68}$$

This proves the assertion.

5. The Dirichlet case

5.1. The kernel in the Dirichlet case differs fundamentally in some respects from the kernels in the cases which we have discussed so far. Thus it satisfies neither (K_1) nor (K_2). One is constantly hampered by these defects when trying to extend the preceding theory to the Dirichlet case. The difficulties start right at the beginning, viz., with the determination of $S(D)$. It is by no means sufficient that (S_1) is satisfied in order that (S_2) be also satisfied as well as the functional equation

$$D_\alpha[D_\beta[f]] = D_\gamma[f], \qquad \gamma = \min(\alpha, \beta). \tag{5.11}$$

Both the originators of the zero element and the invariant elements form linear manifolds which are difficult to characterize. Finally if we come to the question of metric sub-spaces $M(D)$, it turns out that (M_3), which was basic in the previous discussion, is no longer valid in the cases of main interest. The only instance to which our methods obviously apply is the space $L_2(-\infty, \infty)$. Here the transforms exist, belong to the same space, and satisfy (5.11). Problems (i) and (ii) can be completely solved. The space is metric and (M_3) holds. It is not possible to extend all of what we are doing to the case $L_p(-\infty, \infty)$, $p\neq 2$, but we shall note below what results are valid in the more general case. In view of this situation the space will be taken to be $L_2(-\infty, \infty)$ unless otherwise stated.

5.2. The solutions of problem (i) can be obtained by the method of Fourier transforms along the lines given in §1.2. We have

$$T[x; K(u)] = \begin{cases} 0 & \text{for } |x| > 1, \\ (2\pi)^{-1/2} & \text{for } |x| < 1. \end{cases} \tag{5.21}$$

Hence we are confronted with case (2) in the notation of §1.2. It follows that a necessary and sufficient condition that

$$D_\alpha[f] = 0, \qquad f \epsilon L_2, \tag{5.22}$$

is that

$$f(x) = (2\pi)^{-1/2} \operatorname*{l.i.m.}_{a\to\infty} \left\{ \int_{-a}^{-\alpha} + \int_{\alpha}^{a} \right\} e^{ixu} F(u) du, \tag{5.23}$$

where $F(u)$ is an arbitrary function in L_2. The set $\mathfrak{M}$ of all such functions $f(x)$ is obviously a linear manifold in L_2.

5.3. The same method applies to problem (ii). Suppose

$$D_\alpha[g] = g. \tag{5.31}$$

We have again case (2) of §1.3. It follows that a necessary and sufficient condition in order that $g(x)$ shall satisfy (5.31) is that

$$g(x) = (2\pi)^{-1/2} \int_{-\alpha}^{\alpha} e^{ixu} G(u) du, \tag{5.32}$$

where $G(u)$ is an arbitrary function of L_2. The set of all such functions forms a linear manifold $\mathfrak{F}$. We note that $\mathfrak{F}$ and $\mathfrak{M}$ are orthogonal complements of each other in L_2, since

$$\int_{-\infty}^{\infty} f(x)\overline{g(x)}\, dx = \int_{-\infty}^{\infty} T[x; f]\overline{T[x; g]} dx = 0.$$

The discussion and results of §§5.2 and 5.3 extend without difficulty to the case in which we replace L_2 by L_p, $1 < p < 2$. The case $p = 1$ is not accessible because $D_\alpha[f]$ need not be in L_1 when $f \epsilon L_1$.* In case $p > 2$ the method breaks down because the method of Fourier transforms fails.

5.4. Supposing $f(x) \epsilon L_2$, let us put $T[x; f] = F(x)$. A simple calculation shows that

$$D_\alpha[f] = (2\pi)^{-1/2} \int_{-\alpha}^{\alpha} e^{ixu} F(u) du. \tag{5.41}$$

We have consequently

* For the properties of $D_\alpha[f]$ in $L_1(-\infty, \infty)$, $1 \leq p \leq \infty$, see E. Hille and J. D. Tamarkin, Bulletin of the American Mathematical Society, vol. 39 (1933), pp. 768–774.

$$\|D_\alpha[f]\|^2 = \int_{-\alpha}^{\alpha} |F(u)|^2 du \leqq \|F\|^2 = \|f\|^2, \tag{5.42}$$

or

$$\|D_\alpha[f]\| \leqq \|f\|, \tag{5.43}$$

so that (M_3) holds. We then get from (5.41) that

$$\|D_\alpha[f]\| \leqq \|D_\beta[f]\|, \qquad \alpha < \beta. \tag{5.44}$$

It is obvious that $\|D_\alpha[f]\|$ is continuous, and

$$\lim_{\alpha\to 0} \|D_\alpha[f]\| = 0, \qquad \lim_{\alpha\to\infty} \|D_\alpha[f]\| = \|f\|. \tag{5.45}$$

Further, $\{D_\alpha[f]\}$ is a continuous family of continuous transformations defined over L_2. Let us put

$$E_\alpha[f] = f - D_\alpha[f]. \tag{5.46}$$

$E_\alpha[f]$ is also in L_2. Its Fourier transform equals $F(x)$ for $|x| > \alpha$, and zero for $|x| < \alpha$. Hence

$$\|E_\alpha[f]\|^2 = \left\{ \int_{-\infty}^{-\alpha} + \int_{\alpha}^{\infty} \right\} |F(u)|^2 du. \tag{5.47}$$

It follows that

$$\|E_\alpha[f]\| \geqq \|E_\beta[f]\|, \qquad \alpha < \beta, \tag{5.48}$$

$$\lim_{\alpha\to\infty} \|E_\alpha[f]\| = 0. \tag{5.49}$$

It is clear that (5.49) does not hold uniformly for all $f(x)$ having norms under a fixed bound, nor is it possible to assign any limits one way or the other to the degree of approximation with respect to $1/\alpha$.

Formula (5.41) remains valid for $f(x)\epsilon L_p$, $1<p<2$, but while it is true that $D_\alpha[f]$ is a bounded transformation in L_p, it does not seem likely that the bound should be equal to unity. It follows that (5.44) is likely to be false for $p\neq 2$.

5.5. Let us define

$$D_\alpha[f] = 0, \qquad \alpha < 0. \tag{5.51}$$

The family of transformations D_α is then defined for $-\infty < \alpha < \infty$, D_α is zero for $\alpha < 0$, tends to the identity as $\alpha \to \infty$, and is continuous for all values of α. These properties together with formula (5.11) express the fact that $\{D_\alpha[f]\}$ is a *family of projection operators forming the resolution of the identity of a self-adjoint transformation* H, in the terminology of J. von Neumann and M. H. Stone. We shall show that

$$H[f] = \tilde{f}'(x) \sim \tilde{f}'(x), \tag{5.52}$$

where

$$\tilde{F}(x) = -\frac{1}{\pi}\,\text{P.V.}\int_{-\infty}^{\infty} F(u+x)\,\frac{du}{u}, \tag{5.53}$$

and P.V. denotes that the Cauchy principal value of the integral is to be taken at $u=0$. We recall that $\tilde{F}(x)$ exists for almost all x and is in L_2 if $F(x)$ is in L_2. In the following $f(x)$ is an absolutely continuous function in L_2 whose derivative, $f'(x)$, is also in L_2. We have

$$\begin{aligned} T[\alpha;\tilde{f}'] &= -\,i \operatorname{sgn} \alpha T[\alpha; f'] = |\alpha|\, T[\alpha; f], \\ T[\alpha;\tilde{f}'] &= i\alpha T[\alpha;\tilde{f}] \qquad\quad = |\alpha|\, T[\alpha; f]. \end{aligned} \tag{5.54}$$

These relations also prove the equivalence of the conjugate of the derivative and the derivative of the conjugate function. With the usual notation for the inner product, and assuming $g(x)\epsilon L$,

$$\begin{aligned} (\tilde{f}', g) &= (|\alpha|\, T[\alpha; f],\ T[\alpha; g]) = \int_{-\infty}^{\infty} |\alpha|\, F(\alpha)\overline{G(\alpha)}d\alpha \\ &= \int_0^{\infty} \alpha d_\alpha \int_{-\alpha}^{\alpha} F(u)\overline{G(u)}du = \int_0^{\infty} \alpha d_\alpha (D_\alpha[f], g) \end{aligned}$$

by formula (5.41). It follows that

$$(\tilde{f}', g) = \int_{-\infty}^{\infty} \alpha d_\alpha (D_\alpha[f], g). \tag{5.55}$$

This relation proves formula (5.52).

YALE UNIVERSITY,
NEW HAVEN, CONN.

Reprinted from the Proceedings of the NATIONAL ACADEMY OF SCIENCES,
Vol. 24, No. 3, pp. 159–161. March, 1938.

ON SEMI-GROUPS OF TRANSFORMATIONS IN HILBERT SPACE

BY EINAR HILLE

DEPARTMENT OF MATHEMATICS, YALE UNIVERSITY

Communicated February 9, 1938

1. Let E be a normed complete linear vector space, in other words a space (B) in the terminology of Banach. Let $x \epsilon E$ and let $T_\alpha(x)$ be a linear transformation on E to E defined for every $\alpha > 0$. If

$$T_\alpha(T_\beta(x)) = T_{\alpha + \beta}(x), \tag{1}$$

we say that $\{T_\alpha(x)\}$ forms a *semi-group*.

Assuming in addition that

$$\| T_\alpha(x) \| \leqq \| x \|, \tag{2}$$

I have investigated some of the properties of these transformations.[1] Continuing this study, I have found the case in which E is a Hilbert space and $T_\alpha(x)$ is a self-adjoint, positive definite transformation of particular interest. In this case there exists a representation of $T_\alpha(x)$ analogous to that found by M. H. Stone for unitary transformations forming a group.[2]

2. The main result is the following theorem.

Let $T_\alpha(x)$, $\alpha > 0$, *be a family of self-adjoint, positive definite transformations on* $\mathfrak{H}$ *to* $\mathfrak{H}$ *satisfying* (1) *and* (2). *Then there exists a self-adjoint transformation* $A(x)$, *positive definite but not necessarily bounded, with its resolution of the identity* $E(\lambda)$, *such that*

$$(T_\alpha(x), x) = \int_0^\infty e^{-\alpha\lambda} d_\lambda(E(\lambda)x, x). \tag{3}$$

It follows in particular that for a fixed x, $(T_\alpha(x), x)$ is an analytic function of α, holomorphic for $\Re(\alpha) > 0$ and continuous for $\Re(\alpha) \geqq 0$. This would seem to be the reason why the proof can be carried through without any assumptions regarding the continuity or measurability of $(T_\alpha(x), x)$.

A detailed proof will be published elsewhere, but the following outline of the argument will probably be found sufficient to enable the interested reader to fill in the omissions. The basic observation is that

$$\Delta_h^n(T_\alpha(x), x) = \sum_{k=0}^{n} (-1)^k \binom{n}{k} (T_{\alpha + kh}(x), x) \geqq 0$$

for every $\alpha \geqq 0$, $h > 0$, $n \geqq 0$. Hence $(T_\alpha(x), x)$ is completely monotone in $0 \leqq \alpha < \infty$ and by the Bernstein-Widder theorem[3]

$$(T_\alpha(x), x) = \int_0^\infty e^{-\alpha\lambda} dV(\lambda; x), \tag{4}$$

where $V(\lambda; x)$ is a never decreasing function of λ and $V(0; x) = 0$. By (2) $0 \leqq (T_\alpha(x), x) \leqq (x, x)$ so that $V(\lambda; x) \leqq (x, x)$.

The real inversion formulas for the Laplace integral show that $V(\lambda; x)$ is a linear functional in $(T_\alpha(x), x)$. The latter being bilinear in x, we conclude the existence of a bilinear functional $V(\lambda; x, y)$ such that

$$(T_\alpha(x), y) = \int_0^\infty e^{-\alpha\lambda} dV(\lambda; x, y), \tag{5}$$

and $V(\lambda; x, x) = V(\lambda; x)$. Further

$$|V(\lambda; x, y)|^2 \leqq (x, x)(y, y),$$

whence it follows that $V(\lambda; x, y) = (E(\lambda)x, y)$. The transformations $E(\lambda)x$ are evidently self-adjoint. Replacing x by $E(\mu)x$ in (5) and putting $\alpha = 0$, we easily conclude that $E(\lambda)x$ is a resolution of the identity. We then define the corresponding self-adjoint transformation $A(x)$ in the usual manner by

$$(A(x), y) = \int_0^\infty \lambda \, d_\lambda(E(\lambda)x, y),$$

the domain of $A(x)$ being that subset of $\mathfrak{H}$ for which

$$\int_0^\infty \lambda^2 d_\lambda(E(\lambda)x, x) < \infty.$$

That $A(x)$ is positive definite follows from the fact that $E(\lambda) = 0$ for $\lambda \leqq 0$. Incidentally we observe that it is permitted to interpret $A(x)$ as a determination of $-\log T_1(x)$.

3. Among the various transformations forming semi-groups we select the Poisson integral for the half-plane, i.e.,

$$P_\alpha(f) = \frac{\alpha}{\pi}\int_{-\infty}^{\infty} \frac{f(t+u)}{u^2+\alpha^2}\,du. \tag{6}$$

If $f(t)\,\epsilon L_2(-\infty, \infty)$ so does $P_\alpha(f)$ and conditions (1) and (2) are satisfied. Moreover, $P_\alpha(f)$ is a self-adjoint, positive definite transformation. A straightforward calculation gives

$$(P_\alpha(f), f) = \int_{-\infty}^{\infty} e^{-\alpha|\lambda|}\,|F(\lambda)|^2\,d\lambda,$$

where $F(\lambda)$ is the Fourier transform of $f(t)$. Putting

$$D_\lambda(f) = \frac{1}{\pi}\int_{-\infty}^{\infty} \frac{\sin \lambda u}{u} f(t+u)\,du \tag{7}$$

for $\lambda > 0$, $D_\lambda(f) = 0$ for $\lambda \leqq 0$, and utilizing the relation

$$(D_\lambda(f), f) = (D_\lambda(f), D_\lambda(f)) = \int_{-\lambda}^{\lambda} |F(u)|^2\,du,\ \lambda > 0,$$

we find

$$E(\lambda)f = D_\lambda(f),\ A(f) = \tilde{f}', \tag{8}$$

where $\tilde{g}(t)$ is the conjugate function of $g(t)$.

[1] E. Hille, "Notes on Linear Transformations. I," *Trans. Amer. Math. Soc.*, **39**, 131–153 (1936).

[2] M. H. Stone, "Linear Transformations in Hilbert Space," these PROCEEDINGS, **16**, 172–175 (1930), and "On One-Parameter Unitary Groups in Hilbert Space," *Ann. Math.* (2) **33**, 643–648 (1932). See also J. von Neumann, "Über einen Satz von M. H. Stone," *Ann. Math., Ibid.*, 567–573, and F. Riesz, "Über Satze von Stone und Bochner," *Acta Szeged*, **6**, 184–198 (1933).

[3] S. Bernstein, "Sur les fonctions absolument monotones," *Acta Math.*, **52**, 1–66 (1929). D. V. Widder, "Necessary and Sufficient Conditions for the Representation of a Function as a Laplace Integral," *Trans. Amer. Math. Soc.*, **33**, 851–892 (1931).

[4] See loc. cit., note[1] for the properties of $P_\alpha(f)$ and $D_\lambda(f)$.

ANNALS OF MATHEMATICS
Vol. 40, No. 1, January, 1939

NOTES ON LINEAR TRANSFORMATIONS. II. ANALYTICITY OF SEMI-GROUPS[1]

BY EINAR HILLE

(Received June 3, 1938)

1. PROBLEM AND METHODS

1.1 Introduction

In the first note of this series (see E. Hille [2]) the author showed that certain well-known functional transformations of the form

$$(1.1.1) \qquad U_\lambda x = \lambda \int_{-\infty}^{\infty} K(\lambda u)x(t + u)\, du$$

satisfy functional equations with respect to the parameter λ.

In particular, for suitably restricted classes of functions $x(t)$, the two transformations[2]

$$(1.1.2) \qquad W_\alpha x = (\pi\alpha)^{-\frac{1}{2}} \int_{-\infty}^{\infty} \exp(-u^2/\alpha)x(t + u)\, du,$$

and

$$(1.1.3) \qquad P_\alpha x = \frac{\alpha}{\pi} \int_{-\infty}^{\infty} \frac{x(t + u)}{u^2 + \alpha^2}\, du,$$

were found to satisfy the same equation

$$(1.1.4) \qquad T_\alpha[T_\beta x] = T_{\alpha+\beta}x, \qquad \alpha, \beta > 0.$$

Here $W_\alpha x$ is the Gauss-Weierstrass singular integral and $P_\alpha x$ is the Poisson integral for the upper half-plane. Poisson's integral for the circle leads to the same functional equation provided we replace the customary parameter r by $e^{-\alpha}$. The equation is also encountered in the applications of Abel's method of summation or its various generalizations (convergence factors $e^{-\alpha\lambda_n}$). As an example of a different type, leading to the same equation, we mention the theory of fractional integration.

It was observed in my paper [2] that $W_\alpha x$ and $P_\alpha x$ are analytic functions of the parameter in the right half-plane and the same property holds in the case

---

[1] Presented to the American Mathematical Society, February 26 and April 16, 1938. A preliminary communication (see the list of References at the end of this paper, E. Hille [3]) was made to the National Academy of Sciences on February 9, 1938. The results of this note are developed more fully in §§4.1 and 4.2 of the present paper.

[2] These transformations can evidently be reduced to the form of (1.1.1) by a suitable change of the parameter.

of the other examples just mentioned. These instances are sufficiently important to warrant an investigation of the analyticity properties of the solutions of (1.1.4). A first attack upon this problem will be made in the present paper.

1.2. Problems and Results

We consider an abstract metric space E of Banach's type and a one-parameter family of bounded transformations T_α on E to E. The parameter α is restricted to a real or complex point set Z. *If T_α satisfies* (1.1.4) *for all α, β, and $\alpha + \beta$ in Z, we say that T_α has the semi-group property in Z. In particular, we call the family a semi-group if its domain of definition Z is invariant under addition and $\alpha = 0$ is a non-isolated point of the domain where $T_0 = I$.*

Our problem is to determine sufficient conditions under which the family has the semi-group property in a domain Z and T_α is a holomorphic function of α in Z in a sense to be made precise later. In the present paper we concentrate the attention on one feature of this problem. We suppose that a semi-group is defined on the positive real axis and ask when it admits of an analytic extension to the complex plane preserving the semi-group property. As is to be expected, the last proviso is superflous; an analytic extension always preserves the semi-group property (Theorem 5 below). We limit our problem still further by being interested only in the case in which an analytic extension exists in a right half-plane. This case seems to be the most important one and is fairly easy to characterize in simple terms.

It is not difficult to give examples of semi-groups which are not measurable, bounded or continuous with respect to the parameter. In a study of the analyticity problem it is necessary to exclude these pathological cases. Even if this is done, analytic extensions do not have to exist. It appears from the results of this paper that a decisive rôle is played in the problem by the properties of the resolvent $R(\lambda, \alpha) = [\lambda I - T_\alpha]^{-1}$ for small positive values of α.

Since T_α is a bounded transformation by assumption, the spectrum $S(\alpha)$ of T_α is restricted to a neighborhood of the origin in the complex λ-plane. If for small values of α the spectrum of T_α avoids completely a fixed sector $\psi < \arg \lambda < 2\pi - \psi$, $0 < \psi < \pi/2$, and as $\lambda \to 0$ in this sector the norm of $R(\lambda, \alpha)$ does not grow faster than a negative power of $|\lambda|$, the exponent of which may vary with α but is at most $O(1/\alpha)$, then there exists an analytic extension of the semi-group in a half-plane. If the exponent is $o(1/\alpha)$ instead, this half-plane is the right half-plane and the extension is also a semi-group.

As a rule we are only interested in analytical semi-groups, but we note that if for small α's the spectrum of T_α can be enclosed in a fixed circle which leaves $\lambda = 0$ on the outside, then the original semi-group can be extended to the whole plane and generates a group.[2a]

[2a] [Added in proof:] Cf. also the author's paper "Analytical semi-groups in the theory of linear transformations" in the forthcoming Proceedings of the Ninth Scandinavian Mathematical Congress.

The paper is divided into four chapters. The first is devoted to the general theory of one-parameter families of transformations. The first two paragraphs contain a discussion of the underlying topologies and the analysis of functional operators. There is probably not much which is strictly new here, but the results are needed for the later discussion and they are included for the convenience of the reader. We are mainly interested in families analytic in the parameter and of exponential order of growth in a right half-plane. Such transformations are representable by various classes of operational binomial series the theory of which is outlined in §2.3. Families with the semi-group property appear in the last two paragraphs of the chapter. In §2.4 I have collected all material known to me concerning the relationships between the semi-group property on one hand and measurability, boundedness, and continuity on the other (see also §5.2). In §2.5 analyticity is brought in.

Chapter 3 is devoted to analytical semi-groups bounded in a half-plane. Here we use the term bounded in the sense of exponential order of growth of the norm. In §3.1 we show that holomorphism and boundedness in a half-plane of a semi-group impose severe restrictions on the resolvent $R(\lambda, \alpha)$ and the spectrum $S(\alpha)$ of T_α for small positive values of α. A further study of the spectral properties of the transformations in a semi-group is given in §§3.2 and 3.4.

The main theorem, already sketched above, is proved in §3.3. We show that if a semi-group $\{T_\alpha\}$, defined for $\alpha > 0$, and satisfying mild restrictions as to measurability of the norm and continuity of the transformations, in addition, has the property that the resolvents and spectra are limited in the manner indicated above for small values of α, then there exists an analytic extension in a right half-plane. For this purpose we study the integral

$$W(\zeta, \alpha) = \frac{1}{2\pi i} \int_C R(\lambda, \alpha)\lambda^{\zeta/\alpha}\, d\lambda,$$

where the contour of integration surrounds the spectrum of T_α, beginning and ending at the origin. The integral defines an operator function holomorphic in a right half-plane which for $\zeta = n\alpha$ equals $T_{n\alpha}$. We can always form this integral which interpolates the sequence of operators formed by the integral powers of T_α. Thus the crux of the problem is to show that $W(\zeta, \alpha)$ is independent of α and agrees with T_ζ for sufficiently large real values of ζ. This succeeds with the aid of an important theorem due to F. Carlson [1], see §5.1. The theory of binomial series plays a conspicuous rôle in this chapter.

The case in which the spectrum of T_α, $\alpha > 0$, is always positive and the resolvent is of finite order for approach of λ towards the spectrum is particularly interesting (§3.5). Here we have analytic extensions representable by various classes of generalized Laplace-Stieltjes integrals. Finally we prove an embedding theorem in §3.6: any bounded transformation whose spectrum is enclosed in a sector of opening less than π and vertex at the origin and whose resolvent is of finite order outside of this sector can be embedded in a family having the semi-group property. This family is formed by the powers of T.

Chapter 4 gives various special instances of the general theory: self-adjoint transformations in Hilbert space, the Gauss-Weierstrass and Poisson transformations, Abel summability, and fractional integration.

During the preparation of this manuscript I have profited much from discussions with Professors N. Dunford and J. D. Tamarkin whose constructive criticism has contributed much to the paper in particular in §2.2.

2. One-Parameter Families of Transformations

2.1. The Basic Topologies

We shall be concerned with transformations defined in an abstract space E. We assume that E is a space (B) in the terminology of Banach, i.e., E is a normed linear vector space, complete with respect to its metric. The elements of E will be denoted by x, y, etc. The norm of x is denoted by $\| x \|$. All complex numbers are admitted as multipliers of the elements of E.

Together with the space E we also consider the conjugate space $\bar{E}$ whose elements are the bounded linear functionals L on E. We recall that $\bar{E}$ can be metrized by taking $\| L \| = \sup_{\|x\|=1} | L(x) |$.

The metric defines the strong topology in E. We have also the weak topology in which the relation $x_n \to x_0$ means that $L(x_n - x_0) \to 0$ for every $L \in \bar{E}$.

The set of all linear bounded transformations U on E to E forms a ring $\mathfrak{E}$, non-commutative and having divisors of zero. We recall that there are three topologies of special interest in $\mathfrak{E}$, viz. the weak, the strong, and the uniform topologies.[3] In the weak topology the relation $U_n \to U_0$ means that $L((U_n - U_0)x) \to 0$ for each $x \in E$, $L \in \bar{E}$. In the strong topology it means $\| (U_n - U_0)x \| \to 0$ for each $x \in E$, and in the uniform topology $\| U_n - U_0 \| \to 0$, i.e., $\sup_{\|x\|=1} \| (U_n - U_0)x \| \to 0$.

2.2. The Analysis of Functional Operators

In our study of semi-groups we shall have to apply various processes of analysis to families of linear transformations in their dependence upon parameters. For the convenience of the reader we collect in this and the next paragraphs what we shall need in the way of notions and results.

Let $U(\zeta)$, $\zeta = \xi + i\eta$, denote a family of transformations in $\mathfrak{E}$, defined for ζ in a set Z of the complex ζ-plane. We say that $U(\zeta)$ is *weakly measurable* in Z, if $L(U(\zeta)x)$ is a measurable function of ζ in Z for each $x \in E$, $L \in \bar{E}$. We say that $U(\zeta)$ is *continuous* at $\zeta = \zeta_0$ in the sense of one of the topologies (weak, strong or uniform) if the relation $U(\zeta_n) \to U(\zeta_0)$ holds in the sense of that topology whenever $\zeta_n \to \zeta_0$. The meaning of continuity in a domain and in the sense of the various topologies is obvious. We note that uniform continuity implies strong continuity which in turn implies the weak kind. If Z is a set

[3] These notions are due to J. von Neumann [1].

on the real axis we can also define continuity on the right or on the left, semi-continuity etc. Thus, in particular, $U(\xi)$ is strongly continuous on the right at $\xi = \xi_0$ if $\| [U(\xi_0 + h) - U(\xi_0)]x \| \to 0$ for each $x \in E$ as $h \to 0$ through positive values.

Again if $U(\xi)$ is defined for $\alpha \leqq \xi \leqq \beta$ we can define the notions of bounded variation. Thus, for example, $U(\xi)$ is of bounded variation in the strong sense in $[\alpha, \beta]$ if for any sub-division of $[\alpha, \beta]$ and any $x \in E$

$$\sum_{k=1}^{n} \| [U(\xi_k) - U(\xi_{k-1})]x \| \leqq M(x).$$

We shall need several notions of integration. The integrals will either be contour integrals in the complex plane or Riemann-Stieltjes integrals on an interval (α, β). The integrand will be either an operator function $U(\zeta)$ in $\mathfrak{E}$ or elements $U(\zeta)x$ in E or bilinear functionals $L(U(\zeta)x)$ in the product space $(E, \bar{E})$.

In the case of an operator integral we consider sums of the usual type

$$U_n = \sum_{k=1}^{n} U(\zeta_{k,n}^{*})(\zeta_{k,n} - \zeta_{k-1,n}).$$

Here the ζ's are points on a rectifiable oriented arc C in the complex plane and $U(\zeta)$ is defined on C. Suppose that as $\max | \zeta_{k,n} - \zeta_{k-1,n} | \to 0$ the sums converge (in the weak, strong or uniform sense) to a limit independent of the mode of sub-division. This means that there exists an operator U^* in $\mathfrak{E}$ such that U_n converges to U^* in the topology in question. We refer to U^* as the *weak, strong or uniform integral*, as the case may be, of $U(\zeta)$ along C and use the notation $\int_C U(\zeta)d\zeta$, leaving the character of the integral to be surmised from the context. The space $\mathfrak{E}$ is complete in the strong as well as in the uniform sense, but may conceivably not be weakly complete. This fact may sometimes render the notion of weak convergence somewhat illusory.

Thus if $U(\zeta)$ is strongly continuous along C, the sums U_n converge strongly, i.e., $U(\zeta)$ is integrable in the strong sense along C. The same is true if we replace the word "strong" throughout by the word "uniform." But if we know that $U(\zeta)$ is weakly continuous along C, we can merely show that $L(U_n x)$ converges to a bilinear functional $F(x, L)$ on the product space $(E, \bar{E})$, and in general we do not know for sure that $F(x, L)$ is of the form $L(U^*x)$ with $U^* \in \mathfrak{E}$. This complicates matters somewhat, but, fortunately, in the case of greatest interest to us, that in which $U(\zeta)$ is holomorphic, the three notions of integration coincide as we shall see below.

The strong convergence of the Riemann sums formed out of the operators $U(\zeta)$ on C in $\mathfrak{E}$ implies and is implied by the strong convergence of the Riemann sums formed out of the corresponding elements $U(\zeta)x$ of E. Thus we have also integrals of the type $\int_C [U(\zeta)x]d\zeta$ and

$$\left[\int_C U(\zeta)d\zeta\right]x = \int_C [U(\zeta)x]d\zeta. \tag{2.2.1}$$

Here both integrals are taken in the strong sense, but if the integral on the left exists in the weak sense, so does the integral on the right. The relation

$$(2.2.2)\qquad L\left\{\left[\int_C U(\zeta)\,d\zeta\right]x\right\} = L\left\{\int_C [U(\zeta)x]\,d\zeta\right\} = \int_C L[U(\zeta)x]\,d\zeta$$

holds for weak integrals. For strong integrals we have further

$$(2.2.3)\qquad \left\| \int_C [U(\zeta)x]\,d\zeta \right\| \leqq \int_C \| [U(\zeta)x] \| \, | d\zeta | ,$$

and for uniform integrals even

$$(2.2.4)\qquad \left\| \int_C U(\zeta)\,d\zeta \right\| \leqq \int_C \| U(\zeta) \| \, | d\zeta | .$$

We leave it to the reader to develop the theory of Riemann-Stieltjes integrals. We shall merely be concerned with the case in which the integrand is uniformly continuous, the integrator is uniformly of bounded variation and the limit of the Riemann-Stieltjes sum is taken in the uniform sense.

We say that $U(\zeta)$ is *differentiable at* $\zeta = \zeta_0$ in the sense of the topology if $\lim_{h\to 0}[U(\zeta_0 + h) - U(\zeta_0)]/h$ exists in the sense of that topology and is independent of the mode of approach to zero of h. If $U(\zeta)$ is uniquely defined, continuous, and differentiable at all points of a domain Z, we say that $U(\zeta)$ is *holomorphic* in Z in the sense of the topology in question. It is clear that uniform holomorphism implies the strong kind which in turn implies the weak one. But here the converse is also true and even still more.

THEOREM 1. *If an operator function $U(\zeta)$ has the property that $L[U(\zeta)x]$ is holomorphic in a fixed domain Z for every $x \in E$, $L \in \bar{E}$, then $U(\zeta)$ is holomorphic in Z in the uniform sense.*[4]

PROOF. Let Z_0 be any bounded domain interior to Z and having a positive distance from the complementary set of Z. We shall base our proof upon the fact that if a numerical function is holomorphic in Z its difference quotient $[f(\zeta + h) - f(\zeta)]/h$ converges uniformly in Z_0. Thus we can find a constant $M = M(x, L)$ such that for ζ, $\zeta + h$, and $\zeta + g$ in Z_0

$$| L\{([U(\zeta + g) - U(\zeta)]/g - [U(\zeta + h) - U(\zeta)]/h)x\} | \leqq | g - h | M(x, L).$$

But a theorem of Banach ([1], p. 80, theorem 6) says that if the least upper bound of $| L(y) |$ for y in a fixed set $Y \in E$ is finite for every L in $\bar{E}$, then Y must be a bounded set. Letting Y be the set of elements

$$([U(\zeta + g) - U(\zeta)]/g - [U(\zeta + h) - U(\zeta)]/h)x/(g - h),$$

where x is fixed and ζ, $\zeta + g$ and $\zeta + h$ range over Z_0, we see that we can find a constant $N = N(x)$ such that

$$\| ([U(\zeta + g) - U(\zeta)]/g - [U(\zeta + h) - U(\zeta)]/h)x \| \leqq | g - h | N(x).$$

[4] This theorem is due to Professor Nelson Dunford. I publish it here with his permission.

This inequality shows that the difference quotient $[U(\zeta + h) - U(\zeta)]/h$ converges strongly in Z_0. Since $\mathfrak{E}$ is complete in the strong sense we conclude that $U(\zeta)$ is strongly differentiable in Z_0 and consequently also holomorphic in the strong sense. Z_0 being arbitrary, this conclusion extends to Z.

But we can go further. Another theorem of Banach ([1] p. 80, theorem 5) says that if the least upper bound of $\| Vx \|$ for V ranging over a fixed set $\mathfrak{E}_1 \subset \mathfrak{E}$ is finite for every fixed x in E, then $\mathfrak{E}_1$ must be bounded in the uniform topology of $\mathfrak{E}$. This proves the existence of a constant C such that

$$\| [U(\zeta + g) - U(\zeta)]/g - [U(\zeta + h) - U(\zeta)]/h \| \leqq C \,|\, g - h \,|$$

for ζ, $\zeta + g$, and $\zeta + h$ ranging over Z_0. From this we conclude uniform differentiability in Z_0 and consequently that $U(\zeta)$ is holomorphic in Z in the uniform sense. This completes the proof of the theorem. Thus the three notions of holomorphism coincide, and we can say simply that $U(\zeta)$ is holomorphic in Z without specifying the topology.

This implies that if C is a simple rectifiable arc in Z and $U(\zeta)$ is holomorphic in Z, then the integrals $\int_C U(\zeta)d\zeta$ and $\int_C [U(\zeta)x]\, d\zeta$ exist as uniform limit in $\mathfrak{E}$ and strong limit in E respectively. But the theorem also gives us a convenient means of extending ordinary complex function theory to operational function theory in the strong topology of E and the uniform topology of $\mathfrak{E}$. In the following we shall need especially the validity of Cauchy's theorem and of Cauchy's formulas and the possibility of expanding a holomorphic function in power series.[4a]

If C is a closed contour satisfying the usual restrictions and if $U(\zeta)$ is holomorphic inside and on C, then $\int_C L[U(\zeta)x]\, d\zeta = 0$ for every x and L, i.e.,

$$L\left\{\left[\int_C U(\zeta)\, d\zeta\right] x\right\} = 0,$$

whence we conclude that

$$\int_C [U(\zeta)x] d\zeta = 0 \text{ and } \int_C U(\zeta)\, d\zeta = 0. \tag{2.2.5}$$

Thus Cauchy's theorem holds for holomorphic operators.

The same type of argument shows that

$$U^{(n)}(\zeta_0) = \frac{n!}{2\pi i} \int_C \frac{U(\zeta)\, d\zeta}{(\zeta - \zeta_0)^{n+1}}, \qquad n = 0, 1, 2, \cdots, \tag{2.2.6}$$

[4a] [Added in proof:] N. Wiener [1] seems to have been the first to consider analytical function theory in the strong sense in a complex space (B). For recent results and literature, consult A. E. Taylor [1].

so that Cauchy's formulas apply. Using (2.2.4) we see that if $U(\zeta)$ is holomorphic in $|\zeta - \zeta_0| \leqq \rho$ then

$$\begin{aligned} \| U^{(n)}(\zeta_0) \| &\leqq n!\rho^{-n} \frac{1}{2\pi} \int_0^{2\pi} \| U(\zeta_0 + \rho e^{i\theta}) \| \, d\theta \\ &\leqq n!\rho^{-n} \sup_{0 \leqq \theta < 2\pi} \| U(\zeta_0 + \rho e^{i\theta}) \| . \end{aligned} \tag{2.2.7}$$

For $n = 0$ this inequality shows that $\| U(\zeta) \|$ is a subharmonic function in the domain of holomorphism of $U(\zeta)$.

For $|\zeta - \zeta_0| < \rho$ we have

$$U(\zeta) = \sum_{n=0}^{\infty} \frac{1}{n!} U^{(n)}(\zeta_0)(\zeta - \zeta_0)^n \tag{2.2.8}$$

in the sense of uniform convergence since the estimates (2.2.7) show that the sum of the norms of the terms converges for such values. A similar argument applies to the series of Laurent.

2.3. Operational Binomial Series

In chapter 3 of this paper steady use will be made of expansions of operator functions in binomial series. The main properties of such series will be listed in the present paragraph.[5]

Consider the series

$$U(\zeta; x, L) \equiv \sum_{n=0}^{\infty} L(U_n x) \binom{\zeta}{n}, \qquad U_n \in \mathfrak{E}, \tag{2.3.1}$$

for fixed x and L. The series has an abscissa of convergence $\sigma_0(x, L)$ given by

$$\sigma_0(x, L) = -1 + \limsup_{n \to \infty} \log | L(V_n x) | / \log (n + 1), \tag{2.3.2}$$

where

$$V_n = U_0 - U_1 + U_2 - \cdots + (-1)^n U_n ,$$

provided the abscissa exceeds -1. This is the only case of interest to us. It is well known that the series defines an analytic function of ζ, holomorphic in the half-plane of convergence $\Re(\zeta) > \sigma_0(x, L)$. Let

$$\sigma_0 = \sup \sigma_0(x, L) \quad \text{when} \quad \| x \| = 1, \| L \| = 1.$$

We suppose that $\sigma_0 < \infty$. It follows that the series

$$\sum_{n=0}^{\infty} U_n \binom{\zeta}{n} \tag{2.3.3}$$

converges weakly for $\Re(\zeta) > \sigma_0$. If it has a weak limit $U(\zeta)$ in this domain, $U(\zeta)$ is holomorphic for $\Re(\zeta) > \sigma_0$. We leave this question open, but show the

[5] For the general theory of binomial series, also known as binomial coefficient series or Newton's interpolation series, the reader is referred to N. E. Nörlund [2]. The basic memoirs in the field are those of F. Carlson [2] and Nörlund [1].

existence of a uniform limit for $\Re(\zeta) > \sigma_0 + 1$, so that the series defines a holomorphic operator function in $\mathfrak{E}$ at least for $\Re(\zeta) > \sigma_0 + 1$. We call this function $U(\zeta)$.

For this purpose we shall estimate the coefficients. It follows from (2.3.2) that to every $\sigma > \sigma_0$ we can find a constant $M = M(\sigma, x, L)$ such that

$$| L(V_n x) | \leqq M(\sigma, x, L) \frac{\Gamma(n + \sigma + 2)}{\Gamma(\sigma + 2)\Gamma(n + 1)}$$

and the same estimate is clearly true for $| L(U_n x) |$. Using Banach's theorems quoted above, we conclude that

$$\| U_n \| \leqq B(\sigma) \frac{\Gamma(n + \sigma + 2)}{\Gamma(\sigma + 2)\Gamma(n + 1)}. \tag{2.3.4}$$

Consequently

$$\| U(\zeta) \| \leqq B(\sigma) \sum_{n=0}^{\infty} \frac{\Gamma(n + \sigma + 2)}{\Gamma(\sigma + 2)\Gamma(n + 1)} \left| \binom{\zeta}{n} \right| \tag{2.3.5}$$

and this series converges for $\Re(\zeta) > \sigma + 1$. Hence the series (2.3.3) converges strongly and uniformly at least for $\Re(\zeta) > \sigma + 1$.

The expansion of a numerical function in binomial series is not unique owing to the existence of null series. If m is an integer $\geqq 0$, the series

$$\psi_m(\zeta) = \sum_{n=m}^{\infty} (-1)^{n+m} \binom{n}{m} \binom{\zeta}{n}$$

converges for $\Re(\zeta) > m$ to the sum zero. It has the same sum for $\zeta = 0, 1, \cdots, m - 1$, and equals 1 for $\zeta = m$. This fact can be used to obtain the so called reduced series for $U(\zeta)$. We shall suppose that $U(\zeta)$ is known in advance to be holomorphic in the right half-plane and to be defined at the origin, $U(0) \in \mathfrak{E}$. We can then find coefficients $c_0, c_1, \cdots, c_p$, where $p = [\sigma_0]$ so that the function

$$U(\zeta) + \sum_{m=0}^{p} c_m \psi_m(\zeta) U(m)$$

has a binomial expansion weakly convergent for $\Re(\zeta) > \sigma_0$ and takes on the value $U(m)$ for $\zeta = m$, $m = 0, 1, \cdots, p$. The resulting binomial series is known as the reduced series. We shall always suppose that this process has been carried out in advance, so that the binomial series with which we are dealing are reduced series.

In the case of a reduced series, the coefficients are uniquely determined by the values of $U(\zeta)$ at the non-negative integers. We find by substitution

$$U_n = \sum_{k=0}^{n} (-1)^{n-k} \binom{n}{k} U(k), \qquad n = 0, 1, 2, \cdots. \tag{2.3.6}$$

Let us return to (2.3.5). Using an estimate due to F. Carlson [2] we find that for every $\beta > \sigma_0 + 1$ and for $-\pi/2 \leqq \theta \leqq \pi/2$

$$\| U(\beta + re^{i\theta}) \| \leqq B(\beta) \exp [rl(\theta)] \, r^{\beta + \frac{1}{2}} (1 + r \cos \theta)^{-\frac{1}{2}}, \tag{2.3.7}$$

where

$$l(\theta) = \theta \sin \theta + \cos \theta \log (2 \cos \theta). \tag{2.3.8}$$

Carlson has shown, conversely, that if an analytic function $f(\zeta)$ is holomorphic for $\Re(\zeta) \geqq \beta$ and

$$|f(\beta + re^{i\theta})| \leqq M \exp [rl(\theta)] (1 + r)^a, \tag{2.3.9}$$

then $f(\zeta)$ can be represented by a binomial series whose abscissa of convergence is less than or equal to the larger of the two numbers a and β. From this result we get immediately the following

THEOREM 2. *Let $U(\zeta)$ be an operator function holomorphic for $\Re(\zeta) > 0$ and having a value $U(0) \in \mathfrak{E}$ at the origin. Further, for every $\beta > 0$ let*

$$\| U(\beta + re^{i\theta}) \| \leqq M(\beta) \exp [rl(\theta)] (1 + r)^a. \tag{2.3.10}$$

Then

$$U(\zeta) = \sum_{n=0}^{\infty} \left\{ \sum_{k=0}^{n} (-1)^{n-k} \binom{n}{k} U(k) \right\} \binom{\zeta}{n}, \tag{2.3.11}$$

where the series is weakly convergent at least for $\Re(\zeta)$ greater than the larger of the numbers 0 and a, and is strongly as well as uniformly convergent at least for $\Re(\zeta)$ greater than the larger of the numbers 0 and $1 + a$.[6]

PROOF. We apply Carlson's theorem to the function $L[U(\zeta)x]$ with an arbitrary $x \in E$, $L \in \bar{E}$, and conclude that the series (2.3.11) is weakly convergent in the half-plane stated in the theorem. We have then merely to apply the argument which led to the estimate (2.3.5) above in order to complete the proof.

A numerical binomial series can be summed by Cesàro's method of summation, but Nörlund[7] has given a much more powerful method which applies in the largest half-plane in which the function represented by the series is of finite exponential order. If $f(\zeta)$ is holomorphic for $\Re(\zeta) \geqq \beta$, and

$$|f(\beta + re^{i\theta})| \leqq A \exp [Br], \qquad -\frac{\pi}{2} \leqq \theta \leqq \frac{\pi}{2},$$

then we can clearly determine ω so small that $f(\omega\zeta)$ satisfies (2.3.9) and is consequently representable by a convergent binomial series in the variable ζ. This gives a convergent binomial series in the variable ζ/ω for the original function $f(\zeta)$. Since $\min l(\theta) = \log 2$ it is sufficient to take $\omega < (\log 2)/B$. In particular, if $f(\zeta)$ is representable by an ordinary binomial series in the variable ζ convergent for $\Re(\zeta) > \sigma_0$, it also admits of a representation in terms of a binomial series in the variable ζ/ω for $0 < \omega < 1$. Let the abscissa of convergence of the latter

[6] The reader should keep in mind that the statement "uniformly convergent" refers to the topology of $\mathfrak{E}$ and not to the ζ-plane. Actually the uniform convergence holds uniformly with respect to ζ in any finite portion of the half-plane $\Re(\zeta) \geqq \max (0, 1 + a) + \epsilon$.

[7] For the following discussion compare Nörlund [1], pp. 37–44.

series be denoted by $\sigma_0(\omega)$. Nörlund has given the following method for the determination of $\sigma_0(\omega)$. Define

$$\gamma(\xi) = \lim \sup_{|\eta| \to \infty} \frac{1}{|\eta|} \log |f(\xi + i\eta)|.$$

There exists a largest value $\beta_0 \leqq \beta$ such that $f(\xi + i\eta)$ is holomorphic and $\gamma(\xi)$ is finite for $\xi > \beta_0$. Then $\gamma(\xi)$ is non-negative, monotone decreasing, continuous, and convex for $\xi > \beta_0$. We can find the largest interval (ω_0, ω_1) such that for $0 \leqq \omega_0 < \omega < \omega_1 \leqq +\infty$ the equation

$$\gamma(\sigma) = \frac{\pi}{2\omega}$$

has a unique solution, $\sigma = \sigma_0(\omega)$. This is the required abscissa of convergence. For $0 < \omega \leqq \omega_0$ we have $\sigma_0(\omega) = \sigma_0(\omega_0)$, for $\omega > \omega_1$, $\sigma_0(\omega) = +\infty$. Further $\sigma_0(\omega)$ is monotone increasing and continuous for $\omega_0 < \omega < \omega_1$.

Applying this method to operator functions we are led to the following theorem.

THEOREM 3. *Let $U(\zeta)$ be an operator function belonging to $\mathfrak{E}$, holomorphic for $\Re(\zeta) > 0$, and $U(0) \in \mathfrak{E}$. Suppose that for every $\delta > 0$, $\| U(\delta + re^{i\theta}) \| \leqq A_\delta$ $\exp [B_\delta r]$, and put*

$$\gamma(\xi) = \lim \sup_{|\eta| \to \infty} \frac{1}{|\eta|} \log \| U(\xi + i\eta)\|, \qquad \xi > 0. \tag{2.3.12}$$

Then $\gamma(\xi)$ is non-negative, monotone decreasing, continuous, and convex for $\xi > 0$. Determine the largest interval (ω_0, ω_1) so that the equation

$$\gamma(\sigma) = \frac{\pi}{2\omega}$$

has a unique solution, $\sigma = \sigma_0(\omega)$, for $\omega_0 < \omega < \omega_1$. $\sigma_0(\omega)$ is monotone increasing and tends to zero when $\omega \to \omega_0$. Further,

$$U(\zeta) = \sum_{n=0}^{\infty} \left\{ \sum_{k=0}^{n} (-1)^{n-k} \binom{n}{k} U(k\omega) \right\} \binom{\zeta/\omega}{n}, \tag{2.3.13}$$

where the series is weakly convergent for $\Re(\zeta) > 0$ when $0 < \omega \leqq \omega_0$, and for $\Re(\zeta) > \sigma_0(\omega)$ when $\omega_0 < \omega < \omega_1$, and fails to converge weakly when $\omega > \omega_1$. The abscissas of strong and of uniform convergence exceed the abscissa of weak convergence by at most ω.

PROOF. We have merely to apply Nörlund's theorem quoted above to the family of functions $L[U(\zeta)x]$, $\| L \| = 1$, $\| x \| = 1$. Each of these functions gives rise to a growth function $\gamma(\xi; x, L)$ and an abscissa of convergence $\sigma_0(\omega; x, L)$. Further

$$\gamma(\xi) = \sup \gamma(\xi; x, L), \qquad \sigma_0(\omega) = \sup \sigma_0(\omega; x, L).$$

From these relations follow both the analytical properties of $\gamma(\xi)$ and $\sigma_\theta(\omega)$ and the properties of weak convergence stated in the theorem. The position of the other abscissas follows from the estimate of the coefficients which is obtained from the knowledge of the abscissa of weak convergence.

2.4. Semi-Groups of Transformations

After these preliminaries we are prepared to take up the study of semi-groups.

Let $\mathfrak{S} = \{T_\alpha\}$ be a one-parameter family of transformations having the following properties:

(T_1) *$T_\alpha \in \mathfrak{E}$ for $\alpha \geqq 0$ and T_0 is the identity.*

(T_2) *For each x in E we have*

$$T_\alpha[T_\beta x] = T_\beta[T_\alpha x] = T_{\alpha+\beta}x, \qquad \alpha \geqq 0, \qquad \beta \geqq 0. \tag{2.4.1}$$

Under these circumstances we call $\mathfrak{S}$ a one-parameter *semi-group*. It is evident that a semi-group has several properties in common with Abelian groups. Thus there exists a unit-element, the identical transformation. Further, the composition, in our case addition, is unrestricted, associative, and commutative. But ordinarily the elements of $\mathfrak{S}$ do not have inverses, that is unless T_α produces a one-to-one map of E upon itself.

We have supposed here that T_α is defined for $0 \leqq \alpha < \infty$. If we suppose instead that T_α is defined merely for $0 \leqq \alpha \leqq a$ and that (2.4.1) holds whenever α, β, and $\alpha + \beta$ belong to the interval $[0, a]$, then we can extend $\{T_\alpha\}$ to become a semi-group. Indeed, it is evidently possible to define transformations T_α for $\alpha > a$ with the aid of (2.4.1) in such a manner that the resulting extended set forms a semi-group.

A different situation presents itself when T_α is defined for $0 < a \leqq \alpha \leqq b < \infty$ and (2.4.1) holds when α, β, and $\alpha + \beta$ belong to $[a, b]$. The latter assumption presupposes that $b > 2a$. Here an extension for $\alpha < a$ or $b < \alpha$ does not appear to be always possible (cf. §2.5). We shall call such a set $\{T_\alpha\}$ an *incomplete semi-group*.

Let $\mathfrak{S}$ be a semi-group. We say that $\mathfrak{S}$ is *weakly measurable* when the operator function T_α is weakly measurable for $\alpha \geqq 0$. The property of being weakly measurable is not implied by the semi-group property. Indeed, let $f(u)$ be a real, non-measurable solution[8] of the equation $f(u + v) = f(u) + f(v)$. Then

$$T_\alpha x = e^{if(\alpha)}x, \qquad \alpha \geqq 0, \tag{2.4.2}$$

defines a semi-group which obviously is not weakly measurable. Since $\| T_\alpha x \| = \| x \|$, $\| T_\alpha \| = 1$, we conclude that measurability, or even continuity and analyticity, of $\| T_\alpha \|$ does not imply the weak measurability of a semi-group.

We say that $\mathfrak{S}$ is *bounded on a finite interval* (a, b), $0 \leqq a < b < \infty$, if

$$\| T_\alpha \| \leqq M, \qquad a \leqq \alpha \leqq b. \tag{2.4.3}$$

---

[8] Such solutions exist according to Hamel [1].

It is *bounded on an infinite interval* (a, ∞) if

$$\| T_\alpha \| \leqq Me^{B\alpha}, \qquad a \leqq \alpha < \infty. \tag{2.4.4}$$

It is clear that (2.4.4) implies (2.4.3), but the converse is also true by virtue of the semi-group property, the resulting value of B being $\leqq (\log M)/b$. In particular, the inequality

$$\limsup_{\alpha \to 0} \| T_\alpha \| = A < \infty \tag{2.4.5}$$

implies the boundedness of $\mathfrak{S}$ on $(0, \epsilon)$ and hence also on $(0, \infty)$.

The property of being bounded is not implied by the semi-group property. Indeed, suppress the factor i in the exponent in formula (2.4.2) and the result is a non-measurable semi-group which is not bounded in any interval. Whereas boundedness does not imply measurability, measurability of the norm implies at least some types of boundedness.

THEOREM 4. *If* $\| T_\alpha \|$ *is measurable for* $\alpha > 0$, *then* $\| T_\alpha \|$ *is bounded on any interval* $[a, b]$, $0 < a < b < \infty$.

PROOF. Put

$$e(\alpha) = \log \| T_\alpha \|, \qquad p(\alpha) = \tfrac{1}{2}[e(\alpha) + | e(\alpha) |]. \tag{2.4.6}$$

By assumption these are measurable functions of α. From the semi-group property follows that

$$e(\alpha + \beta) \leqq e(\alpha) + e(\beta), \tag{2.4.7}$$

and a simple argument shows that the same inequality is satisfied by $p(\alpha)$. We want to show that $p(\alpha)$ is bounded on $[a, b]$, $a > 0$. Suppose contrariwise. Then there exists a $\gamma \geqq a$ and a sequence $\alpha_n \to \gamma$ such that $p(\alpha_n) \geqq 2n$. Now let α and β be arbitrary positive numbers whose sum equals α_n, then $p(\alpha) + p(\beta) \geqq 2n$ by (2.4.7). This implies that the subset on $[0, \alpha_n]$ where $p(\alpha) \geqq n$ has a measure $\geqq \alpha_n/2$ for every n. Consequently $p(\alpha) = +\infty$ on a set of measure $\geqq a/2$. But this is impossible, because $T_\alpha \in \mathfrak{E}$, i.e., $\| T_\alpha \|$ has a finite definite value for every fixed $\alpha > 0$. It follows that $p(\alpha)$ is bounded in $[a, b]$ and consequently also $\| T_\alpha \|$.

The argument does not permit us to show that measurability of $\| T_\alpha \|$ implies (2.4.5), i.e., boundedness of $\mathfrak{S}$ on $(0, \infty)$. For this question see §5.2.

We have already defined various types of continuity for operator functions. Suppose that $\mathfrak{S}$ is a semi-group and that T_α is continuous on the right in the weak sense at $\alpha = \alpha_0 \geqq 0$, i.e., for every $x \in E$, $L \in \bar{E}$,

$$L(T_{\alpha_0+h}x) \to L(T_{\alpha_0}x) \quad \text{as} \quad h \downarrow 0.$$

Let $\alpha > \alpha_0$. Then

$$L(T_{\alpha+h}x) = L[T_{\alpha_0+h}(T_{\alpha-\alpha_0}x)] \to L[T_{\alpha_0}(T_{\alpha-\alpha_0}x)] = L(T_\alpha x),$$

i.e., continuity on the right at $\alpha = \alpha_0$ implies the same property for $\alpha > \alpha_0$.

The semi-group property alone does not imply any kind of continuity of T_α

as is obvious from our previous examples involving non-measurable semi-groups. But N. Dunford [1] has proved the following

THEOREM. *If $\mathfrak{S}$ is bounded on* $(0, \infty)$ *and weakly measurable and if, in addition, for every* $x \in E$ *the set* $\{T_\alpha x\}$, $\alpha > 0$, *is separable, in particular, if* E *is separable, then* T_α *is strongly continuous on the right for* $\alpha > 0$. *If the set* $\{T_\alpha\}$ *of points in* $\mathfrak{E}$ *is separable in the uniform topology of* $\mathfrak{E}$, *then* T_α *is uniformly continuous on the right for* $\alpha > 0$.

Finally we note that continuity of a semi-group does not imply differentiability. This is shown by the following example. Let E be the Lebesgue space $L_2(-\infty, \infty)$ with the customary metric and put

$$T_\alpha x = e^{i\alpha t} x(t), \qquad \alpha \geqq 0.$$

Then T_α generates a semi-group $\mathfrak{S}$ which is uniformly continuous. But if we form the difference quotients

$$\frac{1}{h}[T_{\alpha+h} - T_\alpha]x = \frac{1}{h}\{e^{iht} - 1\}e^{i\alpha t}x(t),$$

we have a sequence of transformations the norms of which tend to infinity as $h \to 0$. It follows that T_α is not differentiable even in the weak sense in spite of the fact that $T_\alpha x$ is an analytic function of α for fixed x and t.

2.5. Analytic Extension and the Semi-Group Property

So far we have only considered semi-groups defined on the positive real axis. Suppose now that we have an operator function $W(\zeta)$ defined in a domain Z_0 of the complex plane and such that

$$W(\zeta_1)[W(\zeta_2)] = W(\zeta_2)[W(\zeta_1)] = W(\zeta_1 + \zeta_2), \tag{2.5.1}$$

whenever ζ_1, ζ_2, and $\zeta_1 + \zeta_2$ belong to Z_0. We then say that $W(\zeta)$ has the *semi-group property in* Z_0. *If*, in particular, Z_0 *is closed under addition*, i.e., $\zeta_1 + \zeta_2$ belongs to Z_0 whenever ζ_1 and ζ_2 do so, *and if*, in addition, $\zeta = 0$ *is a boundary point of* Z_0 *and* $W(0) = I$, *we say that* $W(\zeta)$ *defines a semi-group in* Z_0.

If a semi-group $\mathfrak{S} = \{T_\alpha\}$ be given for $\alpha \geqq 0$, it is always possible to embed it in a semi-group defined for the right half-plane. It is sufficient to take $W(\xi + i\eta) = e^{i\eta}T_\xi$. This extension is obviously not analytic except in the special case in which $T_\xi = e^\xi$

On the other hand, if an incomplete semi-group admits of an analytic extension, then the extension has the semi-group property in the domain of analyticity. Before we prove this result, let us define the *linear extension* Z of Z_0. A point ζ shall belong to Z if and only if there are points $\zeta_1, \zeta_2, \cdots, \zeta_k$ in Z_0 and positive integers $m_1, m_2, \cdots, m_k$ such that

$$\zeta = m_1\zeta_1 + m_2\zeta_2 + \cdots + m_k\zeta_k.$$

We shall prove the following

THEOREM 5. *Let* $\{T_\alpha\}$, $0 < a < \alpha < b < \infty$ $(2a < b)$ *be an incomplete semi-group. Let* $W(\zeta)$ *be an operator function defined in a domain* Z_0 *of the complex* ζ*-plane, containing the segment* (a, b) *of the real axis, let* $W(\zeta)$ *be holomorphic in* Z_0 *and* $W(\xi) = T_\xi$, $a < \xi < b$. *Then* $W(\zeta)$ *has the semi-group property in* Z_0. *Moreover,* $W(\zeta)$ *can be defined as a holomorphic function having the semi-group property in the linear extension* Z *of* Z_0.

PROOF. Let α be a fixed real number, $a < \alpha < b/2$, and form the two operator functions

$$W(\zeta)[W(\alpha)] \quad \text{and} \quad W(\alpha)[W(\zeta)].$$

They exist and are holomorphic functions of ζ in Z_0. For $\zeta = \xi$ real, $a < \xi < b/2$, they coincide, the common value being $W(\xi + \alpha)$. They are consequently identical within their common domain of analyticity and coincide with $W(\zeta + \alpha)$ as long as also $\zeta + \alpha \in Z_0$. Let $Z_0(\alpha)$ be the domain $Z_0 + \alpha$, i.e., $\zeta \in Z_0(\alpha)$ if $\zeta - \alpha \in Z_0$.

In that portion of $Z_0(\alpha)$ which is outside of Z_0 we define

$$W(\zeta) = W(\zeta - \alpha)[W(\alpha)]. \tag{2.5.2}$$

This definition makes sense since $\zeta - \alpha \in Z_0$. It defines $W(\zeta)$ as a holomorphic function of ζ in $Z_0(\alpha)$. Suppose now that $\zeta \in Z_0(\alpha) \cdot Z_0(\beta)$, where $a < \alpha < \beta < b/2$. Since

$$W(\xi - \alpha)[W(\alpha)] = W(\xi - \beta)[W(\beta)]$$

for $a + \beta < \xi < b/2 + \alpha$, we conclude that

$$W(\zeta - \alpha)[W(\alpha)] = W(\zeta - \beta)[W(\beta)], \qquad \zeta \in Z_0(\alpha) \cdot Z_0(\beta).$$

Hence the definition of $W(\zeta)$ is unique in the union of all sets $Z_0(\alpha)$, $a < \alpha < b/2$, and is independent of α.

Now consider the operator function

$$W(\zeta - \alpha)[W(\alpha)].$$

This function is defined for $\alpha \in Z_0$, $\zeta - \alpha \in Z_0$, and is a holomorphic function of α as well as of ζ in the domain of definition. Let $a < \alpha_0 < b/2$ and consider a small circular region $\gamma_1 : |\alpha - \alpha_0| < \rho$. Let ζ_0 be such that $\zeta_0 - \alpha_0 \in Z_0$ and consider the circular region $\gamma_2 : |\zeta - \zeta_0| < \rho$. If ρ is suitably chosen, all points of γ_1 are in Z_0 and all points of γ_2 are in $Z_0(\alpha_0)$. When α is real and in γ_1, and $\zeta \in \gamma_2$, then $W(\zeta - \alpha)[W(\alpha)] = W(\zeta)$, i.e., is independent of α. It follows that this relation holds everywhere in γ_1 and hence throughout the domain of analyticity of $W(\zeta)$. Thus the definition

$$W(\zeta) = W(\zeta - \alpha)[W(\alpha)]$$

holds for $\alpha \in Z_0$, $\zeta - \alpha \in Z_0$. It defines $W(\zeta)$ uniquely as a holomorphic operator function satisfying the semi-group property. In particular, the semi-group property is found to hold in Z_0.

At this stage of the process the function $W(\zeta)$ is defined for ζ in Z_1, where Z_1 is the set of all points ζ of the form $\zeta = m_1\zeta_1 + m_2\zeta_2$ with ζ_1 and ζ_2 in Z_0, m_1 and m_2 being non-negative integers, $1 \leqq m_1 + m_2 \leqq 2$. In particular, the line-segment $a < \xi < 2b$ belongs to Z_1. We can now apply exactly the same argument and the same process of extension to the domain Z_1 as we have just applied to Z_0. This leads to the definition of $W(\zeta)$ as a holomorphic function satisfying the semi-group property in a domain Z_2. Here Z_2 contains all points of the form $\zeta = \sum_1^4 m_k\zeta_k$ where ζ_k's are in Z_0 and the m_k's are non-negative integers having a sum between one and four. The procedure can then be repeated indefinitely. It is clear that every point of Z, the linear extension of Z_0 defined above, is reached after a finite number of extensions and belongs to the domain of holomorphism of $W(\zeta)$. Further, the semi-group property holds in Z by virtue of the process of extension. This completes the proof of the theorem. We have assumed tacitly that Z_0 is not its own linear extension in which case of course the last part of the proof becomes superfluous.

We see in particular that the line-segment (a, ∞) belongs to Z, i.e., an incomplete semi-group defined on $a < \xi < b$ $(2a < b)$ can always be extended for $\xi \geqq b$ in such a manner that the semi-group property is preserved, provided T_ξ is analytic on the line-segment $a < \xi < b$. It would seem likely that in this statement analyticity can be replaced by quasi-analyticity.

Theorem 5 shows that the domain of holomorphism Z of an operator function $W(\zeta)$, which defines a semi-group, is closed under addition. If it contains a point ζ_0 it must contain all points of the form $\zeta_0 + \zeta_1$ where $\zeta_1 \in Z$. In the simplest case, that in which Z contains the positive real axis, we know that all points $\zeta_0 + h$, $h > 0$, are in Z. In any case we can find a sequence $\{\zeta_n\}$ such that $\zeta_n \in Z$, $|\zeta_n| > |\zeta_{n+1}|$, $\zeta_n \to 0$, and $\arg \zeta_n$ tends to a finite limit. If the distance of ζ_0 from the boundary of Z is supposed to be δ, then all points ζ belong to Z which have a distance less than δ from any one of the points $\zeta_0 + k\zeta_n$, where $k = 0, 1, 2, \cdots, n = 1, 2, 3, \cdots$. In particular the rays $\arg(\zeta - \zeta_0) = \arg \zeta_n$ belong to Z for n sufficiently large. We conclude from these facts that *Z is simply connected.*

It is obvious that Z contains sectors of the form $\theta_1 < \arg(\zeta - \zeta_0) < \theta_2$. If $\theta_2 - \theta_1$ can be taken greater than π, it is readily seen that Z must be the whole plane. In this case $W(\zeta)$ defines a group rather than a semi-group. Let us exclude this case. Z is then contained in some half-plane $\theta_1 < \arg \zeta < \theta_1 + \pi$, and it is no restriction to assume $\theta_1 = -\pi/2$, so that Z is some subset of the right half-plane. We see thus that the right half-plane can be regarded as the maximum domain of holomorphism of a semi-group.

The boundary Γ of Z is the natural boundary of $W(\zeta)$ in the following sense. To every point ζ_0 on Γ we can find elements $x \in E$, $L \in \bar{E}$ such that $L[W(\zeta)x]$ admits of ζ_0 as a singular point. Judging from simple examples it should be possible to determine x and L so that $L[W(\zeta)x]$ has Γ as its natural boundary and it may even be possible to determine an L to every x so that this is the case. I am not able to prove these surmises, however.

We shall say that a family of operators $\{W(\zeta)\}$ having the semi-group property in a domain Δ is bounded in Δ if there exist constants A and B such that

$$\| W(\zeta) \| \leqq A \exp [B \mid \zeta \mid], \qquad \zeta \in \Delta. \tag{2.5.3}$$

Let us consider the most important case, that in which $\{W(\zeta)\}$ forms a semi-group in the right half-plane, i.e., the maximal domain of existence. Let S_ϵ be the sector $\epsilon \leqq \mid \zeta \mid, \mid \arg \zeta \mid \leqq \pi/2 - \epsilon, \epsilon > 0$. It is readily seen that $\mathfrak{S}$ is bounded in S_ϵ for every $\epsilon > 0$ when $W(\zeta)$ is holomorphic. Indeed, if this were not the case, then there would exist a monotone increasing function $\omega(r) \to \infty$ with r and a sequence of points $\zeta_n \in S_\epsilon$, where $\mid \zeta_n \mid = r_n \to \infty$ such that

$$\| W(\zeta_n) \| \geqq \exp [r_n \omega(r_n)].$$

But

$$\| W(\zeta/m) \|^m \geqq \| W(\zeta) \| .$$

Hence choosing $\zeta = \zeta_n$, $m = [r_n]$, we get

$$\| W(\zeta_n/[r_n]) \| \geqq \exp (\omega(r_n))$$

and this would imply that $\| W(\zeta) \|$ would not be bounded in the domain $1 < \mid \zeta \mid < 1 + \delta, \mid \arg \zeta \mid < \pi/2 - \epsilon$. This contradicts the assumption that $W(\zeta)$ is holomorphic for $\xi > 0$. Here we conclude that $\mathfrak{S}$ is bounded in S_ϵ for every $\epsilon > 0$.

On the other hand, it does not seem to follow from the semi-group property that $W(\zeta)$ is necessarily bounded in any right half-plane. Nor is the rate of growth of $\| W(\xi + i\eta) \|$ for fixed ξ as $\mid \eta \mid \to \infty$ or for fixed η as $\xi \downarrow 0$ subject to any obvious restrictions.

3. Analytical Semi-Groups Bounded in a Half-Plane

3.1. Necessary Conditions

Let the operator function $W(\zeta)$ define a semi-group bounded and holomorphic in every half-plane $\xi \geqq \delta > 0$. We shall characterize $W(\zeta)$ by means of the properties of its resolvent for small real values of ζ.

Theorem 3 shows that $W(\zeta)$ can be represented by a binomial series in the variable ζ/ω for sufficiently small values of ω. Since

$$\sum_{k=0}^{\infty} (-1)^k \binom{n}{k} W(k\omega) = [I - W(\omega)]^n, \tag{3.1.1}$$

the series takes the form

$$W(\zeta) = \sum_{k=0}^{\infty} (-1)^n [I - W(\omega)]^n \binom{\zeta/\omega}{n} \tag{3.1.2}$$

or symbolically

$$W(\zeta) = \{I - [I - W(\omega)]\}^{\zeta/\omega} = [W(\omega)]^{\zeta/\omega}. \tag{3.3.3}$$

Here, however, we have to define the power with the aid of the preceding series when the latter is convergent.

Let the abscissa of weak convergence be $\sigma_0(\omega)$. It is determined by the method given in Theorem 3. We know that $\sigma_0(\omega) \downarrow 0$ as $\omega \downarrow 0$. If $\omega < \omega_1$ and $\sigma > \sigma_0(\omega)$, formula (2.3.4) shows that

$$\| [I - W(\omega)]^n \| \leqq B(\sigma, \omega) \frac{\Gamma(\sigma/\omega + n + 2)}{\Gamma(\sigma/\omega + 2)\Gamma(n + 1)}. \tag{3.1.4}$$

Let us now consider the series

$$-\sum_{n=1}^{\infty} (1 - \lambda)^{-n-1}[I - W(\omega)]^n \equiv R(\lambda, \omega). \tag{3.1.5}$$

We have

$$\| R(\lambda, \omega) \| \leqq B(\sigma, \omega) \left\{1 - \frac{1}{|\lambda - 1|}\right\}^{-2-\sigma/\omega} \tag{3.1.6}$$

by (3.1.3). It follows that the series converges in the sense of the uniform topology for $|\lambda - 1| > 1$.

Following Hadamard we define the *order of an operator* power series $\sum_0^\infty U_n \zeta^n$ on its circle of convergence, which we take to have the radius one, as

$$1 + \limsup_{n\to\infty} \log \| U_n \| / \log n.$$

Using (3.1.4) we see that $R(\lambda, \omega)$ is of finite order $\rho(\omega)$ on the circle $|\lambda - 1| = 1$ and

$$\rho(\omega) \leqq \frac{1}{\omega} \sigma_0(\omega) + 2. \tag{3.1.7}$$

A simple calculation shows that for $|\lambda - 1| > 1$ we have

$$R(\lambda, \omega)[\lambda I - W(\omega)] = [\lambda I - W(\omega)]R(\lambda, \omega) = I. \tag{3.1.8}$$

In other words, $R(\lambda, \omega)$ is the *resolvent* of the transformation $W(\omega)$ and this resolvent exists as a bounded transformation in $\mathfrak{E}$ outside of the circle $|\lambda - 1| = 1$ for $0 < \omega < \omega_1$ and is of finite order on the circle.[9] Thus we have proved the following theorem.

THEOREM 6. *Let $W(\zeta)$ be holomorphic and define a semi-group in the right half-plane and let $W(\zeta)$ be bounded in the sense of formula* (2.5.3) *in every half-plane $\xi \geqq \delta > 0$. Let*

$$\gamma(\xi) = \limsup_{|\eta|\to\infty} \frac{1}{|\eta|} \log \| W(\xi + i\eta) \|, \qquad \xi > 0,$$

[9] For the definition and properties of the resolvent see M. H. Stone [1], pp. 128–141. The extension from Hilbert space to a general space of type (B) does not cause any difficulties.

and put $\gamma(\infty) = \lim_{\xi\to\infty} \gamma(\xi)$. *Finally let*

$$\omega_1 = \frac{\pi}{2\gamma(\infty)}.$$

Then for $0 < \omega < \omega_1$ *the resolvent* $R(\lambda, \omega)$ *exists and is holomorphic for* $|\lambda - 1| > 1$. *Moreover,* $R(\lambda, \omega)$ *is of finite order* $\rho(\omega) \leqq (1/\omega)\sigma_0(\omega) + 2$ *on the circle* $|\lambda - 1| = 1$ *and satisfies* (3.1.6) *for any* $\sigma > \sigma_0(\omega)$.

3.2. Spectral Properties of Semi-Groups, I

Suppose that $\mathfrak{S} : \{T_\alpha\}$ defines a semi-group on the positive real axis. In the complex λ-plane let $S(\alpha)$ and $R(\alpha)$ denote the *spectrum* and the *resolvent set* respectively of the transformation T_α. We recall that if $\lambda_1 \epsilon R(\alpha)$, then $R(\lambda_1, \alpha) \epsilon \mathfrak{E}$, whereas if $\lambda_2 \epsilon S(\alpha)$, then $R(\lambda_2, \alpha)$ may cease to exist altogether and in any case does not belong to $\mathfrak{E}$.

We have obviously

$$R(\lambda^n, n\alpha) = \prod_{k=1}^{n} R(\epsilon^k\lambda, \alpha), \qquad \epsilon = \exp(2\pi i/n), \tag{3.2.1}$$

$$R(\lambda, \alpha) = R(\lambda^n, n\alpha) \sum_{j=0}^{n-1} \lambda^j T_{j\alpha}, \tag{3.2.2}$$

when the resolvents involved belong to $\mathfrak{E}$. The first formula shows that if all the resolvents on the right hand side exist so does the left hand side; hence if the left side does not exist in $\mathfrak{E}$, at least one factor on the right is not in $\mathfrak{E}$. The second formula shows that the existence of $R(\lambda^n, n\alpha)$ implies that of $R(\lambda, \alpha)$, but not vice versa. In other words, if $\lambda_1 \epsilon S(\alpha)$ then $\lambda_1^n \epsilon S(n\alpha)$ and at least one of the nth roots of λ_1 belongs to $S(\alpha/n)$, whereas if $\lambda_2 \epsilon R(\alpha)$ then all the nth roots of λ_2 belong to $R(\alpha/n)$ but λ_2^n need not belong to $R(n\alpha)$. We can express this symbolically by

$$[R(\alpha/n)]^n \supset R(\alpha), \tag{3.2.3}$$

$$[S(\alpha/n)]^n \subset S(\alpha), \tag{3.2.4}$$

where we use the symbol D^n to denote the domain of the nth powers of the quantities in a given domain D.

Some applications of these relations will be useful later. Since T_α is a bounded transformation, the series

$$\sum_{n=0}^{\infty} \lambda^{-n-1} T_{n\alpha} = R(\lambda. \alpha) \tag{3.2.5}$$

converges uniformly for $|\lambda| > \|T_\alpha\|$. That its sum is $R(\lambda, \alpha)$ is verified by multiplying on the left or on the right by $\lambda I - T_\alpha$ which gives the result I in either case. Thus $S(\alpha)$ is located in the circle $|\lambda| \leqq \|T_\alpha\|$. From (3.2.4) we conclude that $S(\alpha/n)$ is located in the circle $|\lambda| \leqq \|T_\alpha\|^{1/n}$. This is better than the obvious estimate $|\lambda| \leqq \| T_{\alpha/n} \|$ Since $\| T_\alpha \|$ is bounded on an interval of the form $0 < a/2 \leqq \alpha \leqq a$ when $\| T_\alpha \|$ is measurable for $\alpha > 0$ by Theorem 4, we see that this hypothesis implies that $S(\alpha)$ is uniformly bounded for $0 < \alpha \leqq a$.

Theorem 6 shows that the resolvent sets corresponding to holomorphic semi-groups are subject to special restrictions. Let us suppose that $S(\alpha)$ is located in the right half-plane for $0 < \alpha \leqq a$. Formula (3.2.4) shows that this result is self-improving. In particular $S(\alpha 2^{-n})$ is located in the sector $|\arg \lambda| \leqq \pi 2^{-n-1}$, $n = 0, 1, 2, \cdots$. Indeed, for $n = 0$ this is merely our assumption. For $n = 1$ we know that $S(\alpha/2)$ can consist only of square-roots of quantities in $S(\alpha)$ and here only the principal determination of the square-root can be used. Hence $S(\alpha/2)$ is located in the sector $|\arg \lambda| \leqq \pi/4$. The proof is then accomplished by complete induction. We can also formulate the result as follows. The assumption that $S(\alpha)$ belongs to the right half-plane for $0 < \alpha \leqq a$ implies that $S(\alpha)$ belongs to the sector $|\arg \lambda| \leqq \pi 2^{-n-1}$ for $2^{-n-1}a < \alpha \leqq 2^{-n}a$.

Let us now combine this result with the previous one. Suppose that $\| T_\alpha \|$ is measurable for $\alpha > 0$ and $S(\alpha)$ is located in the right half-plane for $0 < \alpha \leqq a$. Then we can find an a_0 such that $S(\alpha)$ is located in the circle $|\lambda - 1| \leqq 1$ for $0 < \alpha \leqq a_0$. This result is also self-improving. Using formula (3.2.4) once more we obtain the following theorem.

THEOREM 7. *If $\| T_\alpha \|$ is measurable for $\alpha > 0$ and the spectrum $S(\alpha)$ of T_α is located in the right half-plane for $0 < \alpha \leqq a$ then there exists an a_0 with the following property. For $2^{-n-1}a_0 < \alpha \leqq 2^{-n}a_0$, $S(\alpha)$ is located in the domain,* $\lambda = re^{i\theta}$,

$$r^{2^n} \leqq 2 \cos (2^n\theta),\ |\theta| \leqq \pi 2^{-n-1}, \qquad n = 0, 1, 2, \cdots. \tag{3.2.6}$$

We are also interested in the order of the resolvent on the circle $|\lambda - 1| = 1$. Theorem 7 shows that the only point that really matters is the origin. Suppose now that for a particular value of α the resolvent is of finite order at the origin. Can anything be said concerning the order of the resolvent at $\lambda = 0$ for other values of α? Theorem 8 below gives an answer to this question.

Consider a sector $V_\psi : 0 < |\lambda| \leqq 1, \psi \leqq \arg \lambda \leqq 2\pi - \psi$, where $0 < \psi < \pi/2$ so the opening of the sector is greater than π. Consider an operator function $U(\lambda)$ holomorphic in V_ψ. Put

$$M(r) = \sup \| U(re^{i\theta}) \|, \qquad \psi \leqq \theta \leqq 2\pi - \psi. \tag{3.2.7}$$

We say that $U(\lambda)$ is of *sub-exponential order* in V_ψ if

$$\liminf_{r\to 0} r^\gamma \log M(r) = 0, \qquad \pi\gamma = 2(\pi - \psi), \tag{3.2.8}$$

whereas $U(\lambda)$ is of *finite order* τ in V_ψ if

$$\limsup_{r\to 0} \log M(r)/\log \frac{1}{r} = \tau. \tag{3.2.9}$$

With this terminology agreed upon, we have the following result.

THEOREM 8. *Let $R(\lambda, \alpha)$ be holomorphic and of subexponential order in a fixed sector V_ψ for $0 < \alpha \leqq a$, and of finite order $\tau(a)$ for $\alpha = a$. Then $R(\lambda, \alpha)$ is of finite order $\tau(\alpha_n)$ for $\alpha = \alpha_n = a2^{-n}$, $n = 0, 1, 2, \cdots$, and $\alpha_n \tau(\alpha_n)$ does not increase with n.*

PROOF. Consider the relation

$$R(\sqrt{\lambda}, \alpha/2) = (\sqrt{\lambda} I + T_{\alpha/2}) R(\lambda, \alpha).$$

We know that if $\lambda \in R(\alpha)$ then $\pm\sqrt{\lambda} \in R(\alpha/2)$. Now choose $\lambda = re^{2i\psi}$, $\sqrt{\lambda} = \sqrt{r}e^{i\psi}$. A simple calculation shows that

$$\limsup_{r\to 0} \log \| R(\sqrt{r}e^{i\psi}, \alpha/2) \| / \log \frac{1}{r} \leqq \limsup_{r\to 0} \log \| R(re^{2i\psi}, \alpha) \| / \log \frac{1}{r} \leqq \tau,$$

if $R(\lambda, \alpha)$ is of finite order $\tau = \tau(\alpha)$ in V_ψ. Thus to every $\epsilon > 0$ and for $\theta = \psi$ we have

$$\| R(re^{i\theta}, \alpha/2) \| \leqq A(\alpha, \epsilon) r^{-2\tau-\epsilon},$$

and the same inequality is evidently valid for $\theta = 2\pi - \psi$.

Now form the operator function

$$\lambda^{2\tau+\epsilon} R(\lambda, \alpha/2).$$

It is holomorphic in the sector V_ψ and we have just seen that its norm is bounded on the border of the sector. Since it is of subexponential order in V_ψ, the usual Phragmén-Lindelöf argument shows that its norm is bounded in V_ψ for every $\epsilon > 0$. It follows that $R(\lambda, \alpha/2)$ is also of finite order in V_ψ and, if its order be denoted by $\tau(\alpha/2)$, we have $\tau(\alpha/2) \leqq 2\tau(\alpha)$. The proof of the theorem then follows by complete induction.

3.3. Sufficient Conditions

We are interested in the problem of finding sufficient conditions in order that a semi-group, given on the positive real axis, will admit of an analytic extension in some right half-plane. Theorem 6 gives necessary conditions for the existence of such an extension bounded in every half-plane $\xi \geqq \delta > 0$. In addition, the operator function defining the semi-group must satisfy obvious conditions of measurability, continuity, and differentiability on the real axis. We shall prove, however, that a much weaker set of conditions ensures the existence of the desired analytic extension. We shall be concerned with operator functions T_α satisfying the following six postulates:

(T_1) *and* (T_2) *as in* §2.4.

(T_3) *$\| T_\alpha \|$ is measurable for $\alpha > 0$.*

(T_4) *T_α is weakly continuous on the right at $\alpha = \alpha_0$, where α_0 is fixed, $\geqq 0$, but arbitrary.*

(T_5) *$S(\alpha)$, the spectrum of T_α, is located in the right half-plane for $0 < \alpha \leqq a_1$.*

(T_6) *$R(\lambda, \alpha)$, the resolvent of T_α, is of sub-exponential order in a fixed sector V_ψ for $0 < \alpha \leqq a_2$ and is of finite order for $\alpha = a_2$.*

These conditions will turn out to be sufficient. While they compare favorably with the necessary conditions, they are not likely to be independent or reduced to minimum terms. Various modifications are possible. Thus if the space E is separable we can replace (T_3) and (T_4) by

(T_3^*) *$\| T_\alpha \|$ is bounded on $0 < \alpha \leqq a$.*

(T_4^*) *T_α is weakly measurable for $\alpha > 0$.*

By virtue of Dunford's theorem these conditions imply (T_3) and (T_4) when E is separable. We shall now prove

THEOREM 9. *If $\mathfrak{S}: \{T_\alpha\}$ satisfies postulates (T_1) to (T_6), then there exists an operator function $W(\zeta)$ and a constant $c \geqq 0$ such that $W(\zeta)$ is holomorphic for $\xi > c$ and bounded for $\xi \geqq c + \delta$, $\delta > 0$, and $W(\xi) = T_\xi$ for $\xi \geqq c$. $W(\zeta)$ has the semi-group property in its domain of existence.*

PROOF. It is enough to prove the existence of $W(\zeta)$, the semi-group property follows from Theorem 5. We note first that (T_3) to (T_6) imply the validity of Theorems 7 and 8. Put $\alpha_n = a_2 2^{-n}$. Here we can choose n so large that $S(\alpha_n)$ is located entirely within the circle $|\lambda - 1| = 1$, except for the origin, and of course entirely outside of the sector V_ψ. For such an n we write $\alpha_n = h$ and form the integral operator

$$W(\zeta, h) = \frac{1}{2\pi i}\int_C R(\lambda, h)\lambda^{\zeta/h}\, d\lambda, \tag{3.3.1}$$

where the integral starts and ends at the origin and surrounds $S(h)$ once in the positive sense. We can integrate, for instance, along the circle $|\lambda - 1| = 1$. The power is given its principal determination. By Theorem 8 the resolvent $R(\lambda, h)$ is of finite order $\tau(h)$ at the origin. Hence the integral will exist for

$$\xi > h[\tau(h) - 1] \tag{3.3.2}$$

and defines an analytic function, holomorphic in this half-plane.[10] Since $R(\lambda, h)$ is holomorphic outside of $|\lambda - 1| = 1$ we have

$$R(\lambda, h) = -\sum_{n=0}^{\infty} (1 - \lambda)^{-n-1}(I - T_h)^n, \quad |\lambda - 1| > 1. \tag{3.3.3}$$

(Compare formula (3.1.5).) Cauchy's formula gives

$$(I - T_h)^n = \frac{1}{2\pi i}\int_C R(\lambda, h)(1 - \lambda)^n\, d\lambda,$$

where we integrate along any circle $|\lambda - 1| = \rho > 1$. Choosing $\rho = 1 + \frac{1}{n}$ and using the estimate

[10] The integral (3.3.1) is analogous to that of S. Pincherle [1] the existence of which is necessary and sufficient for the representability of an analytic function by means of a binomial series. The procedure in the present paragraph was suggested by pp. 43–46 of F. Carlson [2].

$$\| R(\lambda, h) \| \leqq A(h, \epsilon) \mid \lambda \mid^{-\tau(h)-\epsilon},$$

which is valid along the contour of integration, we get

$$\| (I - T_h)^n \| \leqq B(h, \epsilon) n^{\tau(h)-1+\epsilon} \tag{3.3.4}$$

Substituting (3.3.3) into (3.3.1) and integrating termwise, which is certainly justified for ζ's with sufficiently large real parts, we get

$$W(\zeta, h) = \sum_{n=0}^{\infty} (-1)^n (I - T_h)^n \binom{\zeta/h}{n}. \tag{3.3.5}$$

Formula (3.3.4) shows that this series converges in the uniform sense at least for

$$\xi > h\,[\tau(h) - 1] \equiv \sigma_h\,. \tag{3.3.6}$$

Carlson's estimate (2.3.7) shows that for $\beta > \sigma_h$ and $-\pi/2 \leqq \theta \leqq \pi/2$

$$\| W(\beta + re^{i\theta}, h) \| \leqq M(\beta, h) \exp\left[\frac{r}{h}\, l(\theta)\right] r^{\sigma_h + \frac{1}{2} + \epsilon} \tag{3.3.7}$$

Finally we note that

$$W(kh, h) = T_{kh}\,, \tag{3.3.8}$$

where k is any integer greater than $\tau(h) - 1$.

The preceding analysis applies to any h of the form $h = \alpha_n = a_2 2^{-n}$ with sufficiently large n. Let us put

$$W_n(\zeta) = W(\zeta, \alpha_n) \tag{3.3.9}$$

and write σ_n and τ_n for $\sigma(h)$ and $\tau(h)$ when $h = \alpha_n$.

We shall prove that $W_n(\zeta)$ and $W_{n+1}(\zeta)$ are identical in their common domain of analyticity. Let $\sigma > \max(0, \sigma_n, \sigma_{n+1})$ and form the function

$$\Delta_n(\zeta) = (1 + \zeta)^{-\sigma-1}[W_n(\zeta) - W_{n+1}(\zeta)]. \tag{3.3.10}$$

The preceding discussion shows that $\Delta_n(\zeta)$ has the following properties:

(i) $\Delta_n(\zeta)$ is holomorphic for $\xi \geqq \sigma$.

(ii) $\| \Delta_n(\sigma + re^{i\theta}) \| \leqq M_n \exp\left[\frac{2r}{\alpha_n}\, l(\theta)\right]$.

(iii) $\Delta_n(k\alpha_n) = 0$ for all integers $k > \sigma/\alpha_n$.

(iv) $\frac{1}{\delta}\left[\frac{\pi}{2} - l\left(\frac{\pi}{2} - \delta\right)\right] \to +\infty$ as $\delta \to 0$.

These properties together imply that $\Delta_n(\zeta)$ is identically zero.[11] Hence

$$W_n(\zeta) \equiv W_{n+1}(\zeta), \qquad \xi > \max(\sigma_n, \sigma_{n+1}).$$

[11] See F. Carlson [1], theorem E on p. 60. For the reader's convenience the special case of this theorem which is needed here is reproduced in the Appendix of the present paper. Carlson's theorem applies directly to the functions $L[\Delta_n(\zeta)x]$, $x \in E$, $L \in \bar{E}$, and shows that they are all identically zero, whence it follows that $\Delta_n(\zeta) \equiv 0$.

It follows that all functions $W_n(\zeta)$ are identical, $n \geqq n_0$, and define the same analytic function $W(\zeta)$ which is represented by the convergent binomial series (3.3.5) for $h = \alpha_n$, $\xi > \sigma_n$.

Theorem 8 shows that $\alpha_n \tau(\alpha_n)$ tends to a finite non-negative limit, c say, as $n \to \infty$. Since $\sigma_n = \alpha_n[\tau(\alpha_n) - 1]$ we conclude that $W(\zeta)$ is holomorphic for $\xi > c$. It is bounded for $\xi \geqq c + \delta$, $\delta > 0$, on account of (3.3.7), since we can choose n so large that $\sigma_n < c + \delta$.

Formula (3.3.8) says that $W_n(\xi) = T_\xi$ for $\xi = k\alpha_n = k2^{-n}a_2$ provided these values exceed σ_n. It follows that $W(\xi) = T_\xi$ for every $\xi > c$ which is of the form $k2^{-n}a_e$, where k and n are arbitrary positive integers. These points are everywhere dense on (c, ∞). But T_ξ is by assumption (T_4) weakly continuous on the right for $\xi = \alpha_0$, hence also for every $\xi > \alpha_0$. From this we conclude that $L[W(\xi)x] \equiv L[T_\xi x]$ for $\xi > c$ and for every $x \in E$, $L \in \bar{E}$, whence it follows that $W(\xi) = T_\xi$, $\xi > c$. This completes the proof of the theorem.

Theorems 6 and 9 are not quite comparable unless $c = 0$. This will certainly be the case if

$$\alpha_n \tau(\alpha_n) \to 0 \quad \text{as} \quad n \to \infty. \tag{3.3.11}$$

It does not seem obvious that the domain of existence of $W(\zeta)$ as a holomorphic operator function in $\mathfrak{E}$ has to extend all the way to the imaginary axis, i.e., that c must be zero.

3.4. Spectral Properties of Semi-Groups, II

In the present paragraph we continue the discussion of the spectral properties of semi-groups which was started in §3.2. We assume now that (T_1) to (T_6) are satisfied, so that T_ξ has an analytic extension $W(\xi + i\eta)$.

It follows that all the results of §3.1 apply provided we replace the half-plane $\xi > 0$ by $\xi > c$. We see consequently that there exists an ω_1 such that for $0 < \omega < \omega_1$, the resolvent $R(\lambda, \omega)$ is holomorphic for $|\lambda - 1| > 1$ and of finite order $\rho(\omega)$ on the circle. This implies that $R(\lambda, \omega)$ is of finite order in the sector $V_{\pi/2} : 0 < |\lambda| \leqq 1, \pi/2 \leqq \arg \lambda \leqq 3\pi/2$ and this order cannot exceed $2\rho(\omega)$ as is seen by comparing the distances $|\lambda|$ and $|\lambda - 1| - 1$ from λ to the origin and to the circle respectively. Using the argument employed in the proof of Theorem 8, we easily verify that $R(\lambda, \omega)$ is of finite order in the sector $V_{\pi/4} : 0 < |\lambda| \leqq 1, \pi/4 \leqq \arg \lambda \leqq 7\pi/4$ for $\omega < \frac{1}{2}\omega_1$ provided $R(\lambda, \omega)$ is of sub-exponential order in this sector. The same type of argument applies to any other sector V_ψ with $0 < \psi < \pi/2$.

If $\tau(\omega)$ is the order of $R(\lambda, \omega)$ in some sector V_ψ it seems likely that $\omega\tau(\omega)$ is an increasing function of ω, but I can not prove this surmise.

Using Theorem 3 we see that

$$W(\zeta) = \sum_{n=0}^{\infty} (-1)^n [I - W(\omega)]^n \binom{\zeta/\omega}{n} \tag{3.4.1}$$

for $0 < \omega < \omega_1$, $\xi > \sigma_0(\omega)$. Here $\sigma_0(\omega) \leqq \omega[\tau(\omega) - 1]$ and decreases to c as $\omega \to 0$. We have evidently also

$$(3.4.2) \qquad W(\zeta) = \frac{1}{2\pi i} \int_C R(\lambda, \omega) \lambda^{\zeta/\omega} \, d\lambda$$

for the same values of the variables. Here we can let C be any contour surrounding $S(\omega)$ once in the positive sense which starts and ends at the origin. Taking C to be the circle $|\lambda - r_\omega| = r_\omega$ we get the following result.

THEOREM 10. *There exists an $\Omega \geqq \omega_1$ such that for $0 < \omega < \Omega$ we can find circles $|\lambda - r_\omega| = r_\omega$ containing $S(\omega)$. For any such r_ω*

$$(3.4.3) \qquad W(\zeta) = [r_\omega]^{\zeta/\omega} \sum_{n=0}^{\infty} (-1)^n [I - W(\omega)/r_\omega]^n \binom{\zeta/\omega}{n},$$

where the series converges in the uniform sense at least for $\xi > \omega[\tau(\omega) - 1]$. If $r(\omega) = \inf r_\omega$ for fixed ω, then $(1/\omega) \log r(\omega)$ is an increasing function of ω.

PROOF. If r_ω is chosen so that the circle in question contains $S(\omega)$ in its interior except for the point $\lambda = 0$ we have

$$(3.4.4) \qquad R(\lambda, \omega) = -\sum_{n=0}^{\infty} (r_\omega - \lambda)^{-n-1} [r_\omega I - W(\omega)]^n$$

for $|\lambda - r_\omega| > r_\omega$. If we suppose in addition that there exist a sector V_ψ with $\psi < \pi/2$ in which $R(\lambda, \omega)$ is of finite order $\tau(\omega)$, then we can substitute the series (3.4.4) in formula (3.4.2) and integrate term-wise. The resulting expansion is (3.4.3). The convergence is settled as in the proof of Theorem 9.

Now take a fixed value of ω, $\omega = \alpha$ say, and the corresponding series (3.4.4) with $r_\alpha > r(\alpha)$. Thus

$$(3.4.5) \qquad W(\zeta) [r_\alpha]^{-\zeta/\alpha} = \sum_{n=0}^{\infty} (-1)^n [I - W(\alpha)/r_\alpha]^n \binom{\zeta/\alpha}{n}.$$

By Theorem 3 the same function also admits of an expansion in a binomial series in terms of the variable ζ/ω when $\omega < \alpha$, the abscissa of convergence of which is not greater than that of (3.4.5). This new series is

$$W(\zeta) [r_\alpha]^{-\zeta/\alpha} = \sum_{n=0}^{\infty} (-1)^n [I - W(\omega) r_\alpha^{-\omega/\alpha}]^n \binom{\zeta/\omega}{n}.$$

This is evidently an expansion of type (3.4.3), i.e. $r(\omega) \leqq r_\alpha^{\omega/\alpha}$. This holds for every $r_\alpha > r(\alpha)$, hence $r(\omega) \leqq [r(\alpha)]^{\omega/\alpha}$ or $(1/\omega) \log r(\omega)$ is an increasing function of ω. This completes the proof of the theorem.

The quantity Ω introduced in Theorem 10 may be finite or infinite. In the latter case we have the following result.

THEOREM 11. *If $\Omega = \infty$, i.e., if $S(\omega)$ is located in the right half-plane for every $\omega > 0$, then $S(\omega)$ is a real non-negative point set for every $\omega > 0$.*

PROOF. In order to see this, it is enough to refer to formula (3.2.4) according to which $[S(\omega)]^n \subset S(n\omega)$. If a non-real value of λ should belong to some $S(\omega)$, then we can choose n so large that the nth power of λ lies in the left half-plane. Then $\lambda^n \epsilon S(n\omega)$ or $S(n\omega)$ has a point in the left half-plane against our assumption.

Suppose now that Ω is finite. We understand here by Ω the least upper bound of the values of ω for which $S(\omega)$ can be enclosed in a circle $|\lambda - r_\omega| = r_\omega$ and $R(\lambda, \omega)$ is of finite order in some sector V_ψ where $\psi = \psi(\omega) < \pi/2$. Then there cannot exist an a such that $\Omega < a < 2\Omega$ for which $S(a)$ can be enclosed in a circle $|\lambda - r_a| = r_a$ and $R(\lambda, \omega)$ is of finite order in some sector V_ψ for $\omega = a$ and of sub-exponential order for $\omega < a$. Indeed, Theorem 8 applies for such an a, and the argument used in proving Theorems 9 and 10 can be employed again to show that $R(\lambda, \omega)$ is holomorphic outside of the circle $|\lambda - r_a| = r_a$ and of finite order in V_ψ for $0 < \omega < a$. This contradicts the definition of Ω.

This does not exclude the possibility that $S(\omega)$ can be enclosed in a circle $|\lambda - r_\omega| = r_\omega$ and that $R(\lambda, \omega)$ is of finite order in V_ψ for some $\omega > 2\Omega$. The following example will make the situation clear. Let $\mathfrak{S} : \{W(\zeta)\}$ be a semi-group, holomorphic in $\xi > 0$, such that for each $\xi > 0$ the spectrum of $W(\xi)$ consists of the line-segment $(0, 1)$. We shall encounter such semi-groups in the next chapter. Then form $\mathfrak{S}_1 : \{U(\zeta)\}$ where $U(\zeta) = e^{i\zeta}W(\zeta)$. This is evidently also a semi-group holomorphic in $\xi > 0$. The spectrum of $U(\xi)$ consists of the line-segment joining the point $e^{i\xi}$ with the origin in the λ-plane. We can choose $W(\zeta)$ so that the order of $W(\xi)$ is one in any fixed sector V_ψ for every $\xi > 0$. For $\mathfrak{S}_1$ the constant Ω equals $\pi/2$, but for $(2k + 1)\pi/2 < \omega < (2k + 3)\pi/2$, $k = 1, 2, 3, \cdots$, $S(\omega)$ is located in the right half-plane and can be enclosed in a circle $|\lambda - r_\omega| = r_\omega$.

The last result of the present section is an extension of Theorem 7.

THEOREM 12. *Let $a \leqq \Omega$ and let $r(a)$ be determined as in Theorem 10. For $\omega < a$ the spectrum $S(\omega)$ of T_ω is enclosed in the region $S(\omega, a)$:*

$$|\lambda| \leqq \left\{2\, r(a) \cos \frac{a}{\omega}\, \theta\right\}^{\omega/a}, \qquad |\theta| \leqq \frac{\omega\pi}{2a}. \tag{3.4.6}$$

PROOF. We shall show that $R(\lambda, \omega)$ is holomorphic outside of $S(\omega, a)$. Let m be the least integer greater than $c/\omega - \frac{1}{2}$. Then for values of λ to be specified later[12]

$$\begin{aligned}\sum_{m+1}^{\infty} T_{n\omega}\lambda^{-n} &= \int_{\mu}^{\infty} W(\omega t)\lambda^{-t}\, dt \\ &\quad - i\lambda^{-\mu} \int_0^{\infty} \{W(\omega\mu + i\omega u)\lambda^{-iu} - W(\omega\mu - i\omega u)\lambda^{iu}\}\, [1 + e^{2\pi u}]^{-1}\, du \\ &\equiv I_1 + I_2.\end{aligned}$$

Here $\mu = m + \frac{1}{2}$ and the first integral is to be taken along a ray $\arg(t - \mu) = \psi$ to be chosen later. Substituting the estimate of $\| W(\zeta) \|$ which is obtained

[12] See E. Lindelöf [1], p. 129 et seq.

from formula (3.4.5) by the usual evaluation of the binomial series, we get after simple calculations

$$\| I_2 \| \leqq M \, | \lambda |^{-\mu}, \, 0 < | \lambda | < \infty, \, | \arg \lambda | < \frac{3\pi}{2},$$

and I_2 has no other singularities than 0 and ∞ as long as the argument of λ is restricted in the manner indicated.

Thus the only singularities which are of interest to us arise from I_1. The norm of the integrand in I_1 is dominated by

$$\rho^{\sigma} \exp \left\{ \rho \left[\frac{\omega}{a} \, l(\psi) + \cos \psi \cdot \log Q + \theta \sin \psi \right] \right\},$$

where $Q = r_a^{\omega/a} r^{-1}$, $r = | \lambda |$, $\theta = \arg \lambda$, and $\sigma > \sigma_a + \frac{1}{2}$. If $\frac{\omega \pi}{2a} \leqq | \theta | \leqq \pi$, and $2Q^{a/\omega} \leqq 1$, any choice of ψ such that $\operatorname{sgn} \psi = - \operatorname{sgn} \theta$ will give a convergent integral. If $2Q^{a/\omega} > 1$, we take $\cos \psi = \frac{1}{2} Q^{-a/\omega}$, $\operatorname{sgn} \psi = - \operatorname{sgn} \theta$, and get the same result. Thus for $\frac{\omega \pi}{2a} \leqq | \theta | \leqq \pi$ we find that $\| I_1 \|$ is finite and bounded in the interior of the sector. In particular, I_1 is holomorphic in this sector.

Suppose now that $| \theta | < \frac{\omega \pi}{2a}$. The multiplier of ρ in the exponent is

$$\left(\frac{\omega}{a} \psi + \theta \right) \sin \psi + \log \left\{ \left[2 r_a \cos \frac{a}{\omega} \theta \right]^{\omega/a} r^{-1} \right\} \cos \psi.$$

Let us choose $\psi = - \frac{a}{\omega} \theta$. The exponent then becomes

$$\log (R/r) \cos \frac{a}{\omega} \theta,$$

where

$$R = \left[2 \, r_a \cos \frac{a}{\omega} \theta \right]^{\omega/a}$$

The exponent is negative for $r > R$. If this condition is satisfied

$$\| I_1 \| \leqq \Gamma(\sigma + 1) \left[\log (R/r) \cos \frac{a}{\omega} \theta \right]^{-\sigma - 1}$$

It follows that $R(\lambda, \omega)$ is holomorphic outside of the curve $r = R$ for every permissible choice of r_a. Hence we can replace r_a by its greatest lower bound $r(a)$ and the theorem is proved.[13]

---

[13] This proof is based upon an argument given by F. Carlson [2], pp. 49–52.

3.5. Semi-Groups with Real Spectra

We shall now consider semi-groups such that each transformation T_α has a real spectrum. We suppose that conditions (T_1) to (T_4) are satisfied and replace (T_5) and (T_6) by the following assumptions.

(T_5') *The spectrum of* T_α *is real positive for every* $\alpha > 0$.

(T_6') *If* δ *denotes the distance of* λ *from* $S(\alpha)$, *then there exist constants* $K(\alpha)$ *and* $\pi(\alpha)$ *such that for every* $\alpha > 0$

$$\| R(\lambda, \alpha) \| \leqq K(\alpha)\delta^{-\pi(\alpha)}$$

It is clear that these assumptions imply (T_5) and (T_6) respectively so that Theorems 9, 10, and 12 hold. In particular, there exists an analytic extension $W(\zeta)$ of T_α defined for $\xi > c$ where

$$0 \leqq c \leqq \liminf_{\alpha \to 0} \alpha\pi(\alpha). \tag{3.5.1}$$

This function $W(\zeta)$ is representable by the various binomial series which we have encountered in §§3.3 and 3.4, but in the present case other representations are available, namely in terms of Laplace-Stieltjes integrals.

The spectrum of T_α is supposed to be real positive. We can always reduce the discussion to the case in which the spectrum is restricted to the interval [0, 1]. In order to do so, let us recall the quantity $r(\alpha) = \inf r_\alpha$ introduced in Theorem 10. Put

$$2r(\alpha) = M(\alpha), \qquad M(1) = M. \tag{3.5.2}$$

Then $M(\alpha)$ is the least upper bound of the spectral values of T_α. We shall show that

$$M(\alpha) \leqq M^\alpha.$$

Theorem 12 shows that this is true for $0 < \alpha \leqq 1$. Suppose that it has been shown to be true for $0 < \alpha \leqq a$. In §3.2 it was proved that if $\lambda_0 \in S(\alpha)$, then either $+\sqrt{\lambda_0}$ or $-\sqrt{\lambda_0}$ belongs to $S(\alpha/2)$. In the case of real positive spectra we must take the positive square root. Suppose $a < \alpha \leqq 2a$ and that $\lambda_0 > M^\alpha$. Then $\sqrt{\lambda_0} > M^{\alpha/2}$ and a fortiori, $M(\alpha/2) > M^{\alpha/2}$. This contradicts our assumption. Hence $M(\alpha) \leqq M^\alpha$ for all $\alpha > 0$.

Let us now define

$$U(\zeta) = M^{-\zeta} W(\zeta). \tag{3.5.3}$$

This is also an operator function having the semi-group property in the half-plane $\xi > c$. If $\lambda_0 \in S_W(\alpha)$ then $M^{-\alpha}\lambda_0 \in S_U(\alpha)$ and vice versa. It follows that $S_U(\alpha)$ is enclosed in the interval $[0, M(\alpha)M^{-\alpha}] \subset [0, 1]$. Hence the spectrum of $U(\alpha)$ always belongs to the interval [0, 1].

We have now

$$U(\zeta) = \frac{1}{2\pi i} \int_C R_U(\lambda, \alpha)\lambda^{\zeta/\alpha}\, d\lambda, \tag{3.5.4}$$

where C is any rectifiable contour surrounding the interval [0, 1] once in the positive sense, beginning and ending at the origin, and $\alpha > 0$ is arbitrary. The integral has a sense at least for $\xi > \alpha[\pi(\alpha) - 1]$.

We take a particular contour C in order to estimate the norm of the integral operator. Let ρ be a small number to be disposed of later and let C consist of the two straight line segments joining the origin with the points $1 + \rho i$ and $1 - \rho i$ and the semi-circle of radius ρ joining the points and passing to the right of $+1$. The rectilinear paths give contributions to the norm of the form

$$K_1(\alpha)\,[1 + \rho^2]^{\xi/2\alpha} \exp\left[\frac{|\eta|}{\alpha} \arctan \rho\right] \{\xi - \alpha\,[\pi(\alpha) - 1]\}^{-1} \rho^{-\pi(\alpha)},$$

whereas the circular arc gives

$$K_2(\alpha)(1 + \rho)^{\xi/\alpha} \exp\left[\frac{|\eta|}{\alpha} \arctan \rho\right] \rho^{-\pi(\alpha)+1}.$$

If ξ is fixed and $> \alpha\,[\pi(\alpha) - 1]$ we can choose $\rho = \alpha/|\eta|$ and obtain

$$\| U(\xi + i\eta) \| \leqq K(\alpha, \xi)\, |\eta|^{\pi(\alpha)}, \tag{3.5.5}$$

whereas for $\zeta = c(\alpha) + re^{i\theta}$, $|\theta| < \pi/2$, we can take $\rho = \alpha/r$ and get

$$\| U(c(\alpha) + re^{i\theta}) \| \leqq K(\alpha, \theta) r^{\pi(\alpha)-1}, \tag{3.5.6}$$

where $c(\alpha)$ is any quantity greater than $\alpha[\pi(\alpha) - 1]$. These are the required estimates.

Let us now introduce the Lindelöf mu-function.

$$\mu(\xi) = \limsup_{|\eta|\to\infty} \frac{\log \| U(\xi + i\eta) \|}{\log |\eta|}, \tag{3.5.7}$$

which is defined for $\xi > c$. It follows from (3.5.5) that $\mu(\xi) \leqq \pi(\alpha)$ if $\xi > \alpha\,[\pi(\alpha) - 1]$. It is known that $\mu(\xi)$ is a convex, monotone decreasing function, continuous for $\xi > c$. $W(\zeta)$ and $U(\zeta)$ have the same mu-functions.

Let $\xi_0 > c$, $\beta > \mu(\xi_0) + 1$, and form

$$A_\beta(t) = \frac{1}{2\pi i} \int_{\xi_0 - i\infty}^{\xi_0 + i\infty} e^{t\zeta} U(\zeta) \zeta^{-1-\beta}\, d\zeta. \tag{3.5.8}$$

This defines an operator function depending upon two parameters t and β. It is easy to see that $A_\beta(t) = 0$ for $t \leqq 0$ and

$$\| A_\beta(t) \| \leqq K_1 e^{t\xi_0}, \quad \int_0^t \left\| \frac{d}{du} A_\beta(u) \right\| du \leqq K_2 e^{t\xi_0}. \tag{3.5.9}$$

Thus $A_\beta(t)$ is absolutely continuous even in the uniform sense.

Next form the integral

$$\int_0^\infty e^{-\zeta t}\, dA_\beta(t) \equiv \int_0^\infty e^{-\zeta t} \frac{d}{dt} A_\beta(t)\, dt. \tag{3.5.10}$$

which is actually of the Laplace type though we use the Laplace-Stieltjes notation to indicate the possibilities of generalizations. The estimates (3.5.9) suffice to show the absolute convergence of this integral for $\xi > \xi_0$ in the sense that

$$\int_0^\infty e^{-\xi t} \left\| \frac{d}{dt} A_\beta(t) \right\| dt < \infty. \tag{3.5.11}$$

The argument which is familiar in the theory of ordinary Laplace-Stieltjes integrals can be used to show that the value of the integral in (3.5.10) is $\zeta^{-\beta} U(\zeta)$. Thus

$$U(\zeta) = \zeta^\beta \int_0^\infty e^{-\zeta t}\, dA_\beta(t) \tag{3.5.12}$$

for $\xi > \xi_0$, $\beta > \mu(\xi_0) + 1$.

The last condition is unnecessarily restrictive. $A_\beta(t)$ has a sense at least for $\beta > \mu(\xi_0)$, but may cease to be absolutely continuous for $\beta < \mu(\xi_0) + \frac{1}{2}$, and may even lose the property of being of bounded variation.[14] But formula (3.5.12) is still valid if the integral is suitably interpreted. It suffices to take

$$\int_0^\infty e^{-\zeta t}\, dA_\beta(t) = \zeta \int_0^\infty e^{-\zeta t} A_\beta(t)\, dt. \tag{3.5.13}$$

The integral on the right converges absolutely for $\xi > \xi_0$ in the sense that it converges if we replace $e^{-\zeta t}$ by its absolute value and $A_\beta(t)$ by its norm.

The dependence of $A_\beta(t)$ upon the parameter β is made evident by the formula

$$A_{\alpha+h}(t) = \frac{1}{\Gamma(h)} \int_0^t (t-s)^{h-1} A_\alpha(s)\, ds.$$

We have thus proved

THEOREM 13. *Let $\mathfrak{S}: \{T_\alpha\}$ satisfy postulates* (T_1) *to* (T_4), (T_5') *and* (T_6'). *Then there exists an analytic extension $W(\zeta)$ of T_α, holomorphic and having the semi-group property for $\xi > c$, where c satisfies* (3.5.1). *The Lindelöf mu-function $\mu(\xi)$ of $W(\zeta)$ exists and is bounded for $\xi \geqq c + \delta$. For $\xi > \xi_0 > c$ there exist representations of the form*

$$W(\zeta) = M^\zeta \zeta^{1+\beta} \int_0^\infty e^{-\zeta t} A_\beta(t)\, dt, \tag{3.5.14}$$

where $A_\beta(t)$ is given by (3.5.8) *at least for $\beta > \mu(\xi_0)$. If the integrand is replaced by its norm, the resulting integral converges for $\xi > \xi_0$.*

The most interesting special case is that in which $\pi(\alpha)$ is bounded $\leqq \pi_0$ for $\alpha > 0$. We have then $c = 0$, $\mu(\xi) \leqq \pi_0$ for $\xi > 0$, and (3.5.14) is valid at least for $\beta > \pi_0$, the integral being convergent for $\xi > 0$.

[14] Cf. E. Hille & J. D. Tamarkin [2] for the corresponding phenomenon in connection with ordinary Laplace-Stieltjes integrals.

3.6. The Embedding Theorem

We shall take up the question of constructing one-parameter families of transformations having the semi-group property in a half-plane. We shall show that under suitable restrictions on the resolvent of a transformation T it is possible to embed T in such a family of transformations by a suitable process of interpolation performed on the positive integral powers of T.

THEOREM 14. *Let $T \in \mathfrak{S}$ and let $R(\lambda)$, the resolvent of T, be holomorphic outside of the sector $V_s : 0 \leqq |\lambda| \leqq K, \psi_1 \leqq \arg \lambda \leqq \psi_2, \psi_2 - \psi_1 < \pi$, and let $\| R(\lambda) \| \leqq M |\lambda|^{-\tau}$ for $|\lambda| < K$ in the complementary sector. Let*

$$W(\zeta) = \frac{1}{2\pi i} \int_C R(\lambda)\lambda^\zeta \, d\lambda, \tag{3.6.1}$$

where C surrounds V_s once in the positive sense, beginning and ending at the origin. Then $W(\zeta)$ is holomorphic and has the semi-group property at least for $\xi > \tau - 1$ and $W(n) = T^n$ for $n > \tau - 1$. $W(\zeta)$ generates a semi-group if $\tau = 1$ and a group if $\tau < 1$.

PROOF. Let $\psi_0 = \frac{1}{2}(\psi_1 + \psi_2)$ and suppose $-\pi \leqq \psi_0 < +\pi$. On C we suppose $\psi_0 - \pi/2 \leqq \arg \lambda \leqq \psi_0 + \pi/2$ and take $\lambda^\zeta = \exp [\zeta (\log |\lambda| + i \arg \lambda)]$. The integral in (3.6.1) is easily seen to exist as a holomorphic function for $\xi > \tau - 1$ and $W(n) = T^n$ in the same half-plane by Cauchy's theorem. In order to see the semi-group property we had better expand $W(\zeta)$ in a binomial series. Let $a = \rho_0 e^{i\psi_0}$ where $\rho_0 > K/2$. A simple calculation shows that

$$W(\zeta) = a^\zeta \sum_{n=0}^{\infty} (-1)^n (I - T/a)^n \binom{\zeta}{n}, \tag{3.6.2}$$

where the sum of the norms converges for $\xi > \tau - 1$. Now form $W(\zeta_1)[W(\zeta_2)]$ with $\Re(\zeta_\nu) > \tau - 1$ and substitute the series. Since the powers of $(I - T/a)$ commute with each other and with the scalar factors, we obtain the double series

$$a^{\zeta_1 + \zeta_2} \sum_{j=0}^{\infty} \sum_{k=0}^{\infty} (-1)^{j+k} (I - T/a)^{j+k} \binom{\zeta_1}{j} \binom{\zeta_2}{k}$$

which converges if every term is replaced by its norm. Hence we can collect terms freely and obtain the required result $W(\zeta_1 + \zeta_2)$.

$W(\zeta)$ is holomorphic in the right half-plane if $\tau = 1$ and hence generates a semi-group in this case. If $\tau < 1$, we find that $W(\zeta)$ generates a group instead. Indeed, $W(\zeta)$ is then holomorphic in a strip in the left half-plane and can consequently be continued analytically in the whole plane by Theorem 5. This case arises in particular if $\lambda = 0$ does not belong to the spectrum of T.

A similar remark applies to Theorems 9 and 13. If for any choice of $\alpha > 0$ the quantity $\tau(\alpha)$ or $\pi(\alpha)$ is less than one the corresponding function $W(\zeta)$ generates a group.

The function $W(\zeta)$ defined by (3.6.1) is not the only solution of the interpolation problem. If k is any integer, we can clearly multiply $W(\zeta)$ by $e^{2k\pi i \zeta}$ and still get a solution. It is not unlikely that these are the only measurable solutions, but I cannot prove that this is the case. At any rate the solution $W(\zeta)$ has an extremal property. If $U(\zeta)$ is any other solution of the interpolation problem which is holomorphic in a right half-plane, then $U(\zeta) - W(\zeta)$ vanishes at all the integers in their common half-plane of existence. If this difference $D(\zeta)$ is not to vanish identically, its norm must satisfy the following inequality[15]

$$\limsup_{|\eta|\to\infty} \frac{1}{|\eta|} \log \| D(\xi + i\eta) \| \geqq \pi.$$

On the other hand, direct estimates of the integral (3.6.1) shows that

$$\limsup_{|\eta|\to\infty} \frac{1}{|\eta|} \log \| W(\xi + i\eta) \| \leqq \max (|\psi_1|, |\psi_2|). \tag{3.6.3}$$

Hence if $-\pi < \psi_1 < \psi_2 < \pi$, we can claim that $W(\zeta)$ *gives the solution of the interpolation problem with the minimum rate of growth of the norm on vertical lines.* These limits are precise. For a suitable choice of a transformation T with a real negative spectrum, say on the interval $[-1, 0]$, so that $\psi_1 = \psi_2 = \pi$, the two transformations

$$e^{\pm\pi i\zeta} \sum_{n=0}^{\infty} (-1)^n (I + T)^n \binom{\zeta}{n}$$

may have the same rate of growth.

4. Special Instances of the General Theory

4.1. Self-Adjoint Semi-Groups in Hilbert Space

Suppose that $E = \mathfrak{H}$, the Hilbert space, and let $\mathfrak{S} : \{T_\alpha\}$ be a semi-group of transformations on $\mathfrak{H}$ to $\mathfrak{H}$ defined for $\alpha \geqq 0$. In addition we suppose that

(T_7) *T_α is self-adjoint for every $\alpha > 0$.*

This means that

$$(T_\alpha x, y) = (x, T_\alpha y), \qquad x, y \in \mathfrak{H}, \tag{4.1.1}$$

where (u, v) is the fundamental bilinear functional which defines the metric of the space. As the first consequence of this relation we get

$$\begin{aligned}(T_\alpha x, x) &= (T_{\alpha/2} T_{\alpha/2} x, x) \\ &= (T_{\alpha/2} x, T_{\alpha/2} x) = \| T_{\alpha/2} x \|^2 \geqq 0,\end{aligned} \tag{4.1.2}$$

i.e., T_α is *positive definite* for every $\alpha > 0$.

Since $| (T_\alpha x, T_\beta x) | \leqq \| T_\alpha x \| \, \| T_\beta x \|$, we get

$$(T_{\alpha+\beta} x, x)^2 \leqq (T_{2\alpha} x, x)(T_{2\beta} x, x). \tag{4.1.3}$$

Hence putting

$$\log (T_\alpha x, x) = \varphi(\alpha; x), \tag{4.1.4}$$

[15] Cf. F. Carlson [1], Theorem C, p. 58.

we find that

$$\varphi(\tfrac{1}{2}(\alpha + \beta);x) \leqq \tfrac{1}{2}[\varphi(\alpha; x) + \varphi(\beta; x)]. \tag{4.1.5}$$

In other words, $\varphi(\alpha; x)$ is convex with respect to α.[16]

In postulate (T_3) we assume that $\| T_\alpha \|$ is measurable for $\alpha > 0$. In the present case it is more convenient to replace this postulate by either of the following assumptions.

(T_3') *$\| T_\alpha x \|$ is measurable for every $x \in \mathfrak{H}$.*

(T_3'') *The limits of indetermination of $\| T_\alpha x \|$ as $\alpha \to 0$ are finite for every $x \in \mathfrak{H}$.*

Assumption (T_3') implies that $\varphi(\alpha; x)$ is measurable for $\alpha > 0$. A convex function is continuous in the interior of its interval of definition if it is measurable. Hence $\varphi(\alpha; x)$ is continuous for $\alpha > 0$, x fixed, and the same is true of $(T_\alpha x, x)$ and $\| T_\alpha x \|$. From this we conclude that T_α is weakly continuous, i.e., (T_3') implies (T_4). That (T_3') implies (T_3) follows from the separability of the space.

If we use (T_3'') instead of (T_3') we notice that $\varphi(\alpha; x)$ is a convex function bounded above at the origin and hence bounded above in every finite interval by the semi-group property. Such a function is continuous and we can proceed as above.

From the fact that T_α is a positive definite transformation follows that its spectrum is real and non-negative. If M is defined by formula (3.5.2), we know that $S(\alpha)$ is located in the interval $[0, M^\alpha]$. Thus postulates (T_5) and (T_5') are satisfied.

Put $\lambda = \mu + i\nu$. Then

$$\| (T_\alpha - \lambda I)x \|^2 = (T_{2\alpha}x, x) - 2\mu(T_\alpha x, x) + (\mu^2 + \nu^2)(x, x). \tag{4.1.6}$$

Disregarding for a moment the term in ν^2, what remains is a quadratic polynomial in μ which is never negative by virtue of (4.1.3). We have obviously

$$\| T_\alpha - \lambda I \| \geqq | \lambda |, \qquad \mu \leqq 0, \tag{4.1.7}$$

$$\| T_\alpha - \lambda I \| \geqq | \nu |, \qquad \mu > 0, \tag{4.1.8}$$

---

[16] The convexity of log $(T_\alpha x, x)$ was proved by B. v. Sz. Nagy [1] in the case of groups of self-adjoint transformations in Hilbert space and he based one of his proofs of the representation theorem for such groups upon this property. Dr. B. v. Sz. Nagy kindly called my attention to his paper in a letter of April 7, 1938, and indicated how his methods could be used to prove a more general representation theorem for semi-groups than the one stated and proved in outline in my note [3]. See Theorem 15 below. My original method had no points of contact with those of B. v. Sz. Nagy. In the present paragraph I develop both methods and a mixed one having elements in common with both of them. Postulate (T_3'') below was used by B. v. Sz. Nagy, but that Theorem 15 is valid under the assumptions (T_1), (T_2), (T_3''), and (T_7) was pointed out to the author by J. von Neumann in a letter of February 6, 1938, i.e., before the author was aware of B. v. Sz. Nagy's investigation.

but if $\mu > M^\alpha$ we can replace the second inequality by

$$(4.1.9) \qquad \| T_\alpha - \lambda I \| \geqq | \lambda - M^\alpha |,$$

since

$$(T_\alpha x, x) = \| T_{\alpha/2} x \|^2 \leqq M^\alpha(x, x).$$

These inequalities taken together show that postulate (T_6') is satisfied and that

$$(4.1.10) \qquad \| R(\lambda, \alpha) \| \leqq 1/\delta,$$

where δ is the distance from λ to $S(\alpha)$.

We have consequently shown that postulates (T_1), (T_2), (T_3') or (T_3''), and (T_7) are sufficient for the validity of Theorem 13. Thus T_α admits of an analytic extension $W(\zeta)$ in the right half-plane and

$$(4.1.11) \qquad W(\zeta) = M^\zeta \zeta^{1+\beta} \int_0^\infty e^{-\zeta t} A_\beta(t)\, dt$$

for any $\beta > 1$.

Actually the representation is valid for $\beta \geqq 0$. It does not seem possible to prove this by a further refinement of the estimates of the norm of $W(\zeta)$. There are, however, various possibilities of proving (4.1.11) directly with little or no reference to the general theory developed in this paper. Thanks to the special properties of self-adjoint transformations in Hilbert space, special methods lead to a sharper result.

One possibility is offered by the canonical representation of self-adjoint transformations in Hilbert space, i.e.,

$$(Tx, y) = \int_{-\infty}^{\infty} s\, d(F(s)x, y),$$

where $F(s)$ is the resolution of the identity corresponding to T. Since T is positive definite and bounded and its spectrum belongs to the interval $[0, M]$ this formula reduces to

$$(4.1.12) \qquad (Tx, y) = \int_0^M s\, d(F(s)x, y).$$

Let us now form

$$(4.1.13) \qquad U(\zeta) = \int_0^M s^\zeta\, dF(s), \qquad \Re(\zeta) \geqq 0.$$

Using the properties of $F(s)$ we see that $U(\zeta)$ defines a semi-group in the right half-plane where it is holomorphic and satisfies the inequality

$$\| U(\xi + i\eta) \| \leqq M^\xi.$$

Further $U(n) = T^n$ for $n = 0, 1, 2, \cdots$. On the other hand, the function $W(\zeta)$, which we have constructed, also interpolates the sequence $\{T^n\}$ and has the smallest rate of growth for its norm on vertical lines (cf. the remarks at the end of §3.6). It follows that $U(\zeta) = W(\zeta)$.

Hence

$$W(\zeta) = M^{\zeta} \int_0^{\infty} e^{-\zeta u}\, dE(u), \tag{4.1.14}$$

where $E(u) = F(Me^{-u})$. It is clear that $E(u)$ is also a resolution of the identity; the corresponding transformation can be interpreted as a determination of $-\log T$. As a rule this is not a bounded transformation. Its domain is that sub-set of $\mathfrak{H}$ for which

$$\int_0^{\infty} u^2\, d_u(E(u)x, x) < \infty .$$

We note that $-\log T$ is positive definite since $E(u) = 0$ for $u \leqq 0$.

We can arrive at the same result without any reference to the preceding general theory by observing that $U(\xi) = T_{\xi}$ for dyadic-rational values of ξ on account of the semi-group property and the positive definiteness of the transformations. Since $U(\xi)$ is weakly continuous by definition for $\xi > 0$ and T_{ξ} has been shown to have this property if we assume (T_3') or (T_3''), we see that $T_{\xi} = U(\xi)$ for all ξ. This is the argument of B. v. Sz. Nagy referred to in footnote 16.

Another possibility is offered by the properties of monotony which characterize $(T_{\alpha}x, x)$ as function of α for fixed x. Let us put

$$T_{\alpha} = M^{\alpha} V_{\alpha} . \tag{4.1.15}$$

Then $\mathfrak{S}_1 : \{V_{\alpha}\}$ is a semi-group of self-adjoint transformations having their spectra in the interval [0, 1]. Let $h > 0$, $\alpha \geqq 0$, and form

$$\Delta_h^n V_{\alpha} \equiv \sum_{k=0}^{n} (-1)^k \binom{n}{k} T_{\alpha+kh} = (I - V_h)^n V_{\alpha}. \tag{4.1.16}$$

This is a self-adjoint, positive definite transformation. In order to prove this, we notice that $(1 - t)^n$ is positive for $0 < t < 1$, whence it follows that $(I - V_h)^n$, which is obviously self-adjoint, is also positive definite.[17] Since V_{α} has the same properties and commutes with the powers of $I - V_h$ we conclude that (4.1.16) defines a self-adjoint, positive definite transformation. Hence

$$\left(\Delta_h^n V_{\alpha}x, x\right) \geqq 0$$

for $n = 0, 1, 2, \cdots$, and every $h > 0$, $\alpha \geqq 0$. In other words, $(V_{\alpha}x, x)$ is a completely monotone function of α for fixed x. It follows from the investigations of F. Hausdorff [1], S. Bernstein [1], and D. V. Widder [1], that

$$(V_{\alpha}x, x) = \int_0^{\infty} e^{-\alpha u}\, d_u E(u; x), \tag{4.1.17}$$

[17] See F. Riesz [11, pp. 31–33.

where $E(u; x)$ is a never decreasing function of u, $E(0; x) = 0$ and $E(u; x) \to (x, x)$ as $u \to \infty$.

The simplest way of completing the proof is to use formula (4.1.2) which tells us that

$$(V_n x, x) = \int_0^\infty e^{-nu} d_u(E(u)x, x)$$

for every positive integer n. Comparing with (4.1.17) and using the theorem of Lerch for Laplace-Stieltjes integrals, we see that $E(u; x) = (E(u)x, x)$ at all points of continuity. Thus we are again led to formula (4.1.14).

It is also possible to prove that $E(u; x) = (E(u)x, x)$, where $E(u)$ is a resolution of the identity, without using the canonical representation of T. The proof is too long to give in detail, but some indications of the various steps are not quite devoid of interest. One obtains from (4.1.17) the fact that

$$(V_\alpha x, y) = \int_0^\infty e^{-\alpha u} d_u E(u; x, y).$$

Here $E(u; x, y)$ is a bilinear functional in x and y which is of uniformly bounded variation with respect to u on $[0, \infty]$ if $\| x \| = \| y \| = 1$. Further $E(u; x, x) = E(u; x)$. With the aid of these facts one proves that $E(u; x, y) = (R(u)x, y)$, where $R(u)$ is a self-adjoint, positive definite transformation such that $0 \leqq (R(u)x, x) \leqq (x, x)$. Next one proves that $R(u)$ commutes with V_α, for instance, with the aid of the inversion formula of Laplace integrals and the semi-group property of V_α. This fact and the relation

$$(V_{\alpha+\beta} x, y) = \int_0^\infty e^{-\beta u} d_u(R(u) V_\alpha x, y)$$

give the identity

$$\int_0^u e^{-\alpha t} d_t(R(t)x, y) = \int_0^\infty e^{-\alpha t} d_t(R(u)R(t)x, y),$$

from which we conclude that $R(u)R(t) = R(s)$, where s is the smaller of the numbers t and u. These relations show that $R(u)$ is a resolution of the identity, provided we normalize in the conventional manner at the points of discontinuity.

We have consequently proved in three different ways the following result.

THEOREM 15. *Let* $\mathfrak{S} : \{T_\alpha\}$ *be a semi-group of transformations on* $\mathfrak{H}$ *to* $\mathfrak{H}$ *satisfying postulates* (T_1), (T_2), (T_3') *or* (T_3''), *and* (T_7). *Then* T_α *admits of an analytic extension* $W(\zeta)$ *in the right half-plane, and*

$$W(\zeta) = M^\zeta \int_0^\infty e^{-\zeta u} dE(u),$$

where $E(u)$ is the resolution of the identity of a positive definite self-adjoint transformation which may be regarded as a determination of $-\log T$.[18]

4.2. Special Examples

To illustrate Theorem 15 we shall discuss the transformations of Gauss-Weierstrass and Poisson defined by formulas (1.1.2) and (1.1.3) of the Introduction.

Let $x(t) \in L_2(-\infty, \infty)$ and form

$$(4.2.1) \qquad W_\zeta x(t) = (\pi\zeta)^{-\frac{1}{2}} \int_{-\infty}^{\infty} \exp[-u^2/\zeta] x(t+u)\, du, \qquad \Re(\zeta) > 0.$$

It is a simple matter to show that $W_\zeta x(t) \in L_2(-\infty, \infty)$ and that $\mathfrak{S} = \{W_\zeta\}$ is a semi-group in the right half-plane. If $\zeta = \xi$ is real, it is easily seen that the transformation is self-adjoint and hence positive definite.

In order to study the spectrum of W_ζ and to determine its canonical representation as a Laplace-Stieltjes integral, we shall need a couple of other functional transformations of $x(t)$. We recall that the Fourier transform $X(t)$ of $x(t)$ is defined as

$$X(t) = \text{l.i.m.}\, (2\pi)^{-\frac{1}{2}} \int_{-a}^{a} e^{-itu} x(u)\, du,$$

and that $X(t) \in L_2$ and $\| X(t) \| = \| x(t) \|$. Further we need the Dirichlet transform $D_\alpha x(t)$ defined by

$$(4.2.2) \qquad D_\alpha x(t) = \frac{1}{\pi} \int_{-\infty}^{\infty} \frac{\sin \alpha u}{u} x(t+u)\, du, \qquad \alpha > 0.$$

This function has the properties[19] $D_\alpha x(t) \in L_2$, $\| D_\alpha x(t) \| \leqq \| x(t) \|$, and

$$(4.2.3) \qquad (D_\alpha x(t), x(t)) = \int_{-\alpha}^{\alpha} | X(t) |^2\, dt.$$

We have then

$$(W_\zeta x, x) = (\pi\zeta)^{-\frac{1}{2}} \int_{-\infty}^{\infty} \exp[-u^2/\zeta]\, du \int_{-\infty}^{\infty} x(t+u)\overline{x(t)}\, dt$$

$$= \int_{-\infty}^{\infty} \exp[-\zeta s^2/4] \, | X(s) |^2\, ds,$$

where we have used a classical relation in the theory of Fourier transforms. This formula shows that for ξ real we have

$$0 \leqq (W_\xi x, x) \leqq (x, x).$$

[18] In my note [3] a somewhat weaker theorem was stated and proved with the aid of the last of the three methods presented above. The theorem involved a boundedness assumption, viz. $\| I - T_\alpha \| \leqq 1$ for all α, which implies $M = 1$.

[19] For the properties of the Dirichlet transform used here, see E. Hille [2], pp. 150–153.

Moreover, if we give any λ, $0 < \lambda < 1$, and any $\xi > 0$, we can find an $x(t)$ with $\| x(t) \| = 1$ such that $(W_\xi x(t), x(t)) = \lambda$. For this purpose it is enough to choose

$$x(t) = \frac{\sin bt}{\sqrt{(b\pi)}t}, \qquad (W_\xi x, x) = A^{-1} \int_0^A e^{-u^2} du, \quad A = \tfrac{1}{2} b \sqrt{\xi},$$

and to determine A so that the integral equals λA which is possible when $0 < \lambda < 1$. Every point of the interval [0, 1] belongs to the spectrum of W_ξ.

Using formula (4.2.3) we see that

$$(W_\zeta x, x) = \int_0^\infty e^{-\zeta s^2/4} d_s(D_s x, x)$$

or

$$(W_\zeta x, x) = \int_0^\infty e^{-\zeta u} d_u(D_{2\sqrt{u}} x, x). \tag{4.2.4}$$

This is the desired representation, the existence of which is ensured a priori by Theorem 15. From it also follows the structure of the spectrum.

We have $E_W(u) = D_{2\sqrt{u}}$. To this resolution of the identity corresponds the transformation

$$A_W = \int_0^\infty u \, d_u D_{2\sqrt{u}}, \tag{4.2.5}$$

whose domain of definition is that set of functions $x(t) \in L_2$ for which

$$\int_0^\infty u^2 d_u(D_{2\sqrt{u}} x, x) < \infty.$$

Using (4.2.3) once more, we see that the integral equals

$$\frac{1}{16} \int_{-\infty}^\infty t^4 | X(t) |^2 dt = \frac{1}{16} \int_{-\infty}^\infty | x''(t) |^2 dt.$$

Hence the domain of existence of A_W is the sub-set of L_2 which consists of the functions which are twice differentiable and whose first and second order derivatives also belong to L_2. Calculating the value of $(A_W x, y)$ we find that

$$A_W = -\frac{1}{4} \frac{d^2}{dt^2}. \tag{4.2.6}$$

Since $A_W = -\log W$, we have formally

$$W = \exp\left[\frac{1}{4} \frac{d^2}{dt^2}\right], \tag{4.2.7}$$

which is, of course, valid only for a severely restricted class of functions $x(t)$. But the corresponding inversion formula for the Gauss-Weierstrass integral, viz.

$$x(t) = \exp\left[-\frac{1}{4}\frac{d^2}{dt^2}\right] Wx(t),$$

figures in the literature and is not quite devoid of interest.[20]

In the case of Poisson's integral for the half-plane

$$P_\zeta x(t) = \frac{\zeta}{\pi}\int_{-\infty}^{\infty} \frac{x(t+u)}{\zeta^2+u^2}\,du \tag{4.2.8}$$

we arrive at similar formulas. We find that

$$(P_\zeta x, x) = \int_{-\infty}^{\infty} e^{-\zeta|s|}\,|X(s)|^2\,ds.$$

This formula shows that P_ξ is a positive definite, self-adjoint transformation whose spectrum fills out the interval [0, 1]. It also gives the canonical representation

$$(P_\zeta x, x) = \int_0^{\infty} e^{-\zeta u}\,d_u(D_u x, x). \tag{4.2.9}$$

Thus $E_P(u) = D_u$ and corresponds to the transformation

$$A_P = \int_0^{\infty} u\,d_u D_u. \tag{4.2.10}$$

A simple calculation shows that the domain of existence of A_P is that sub-set of L_2 which consists of the absolutely continuous functions whose first derivatives also belong to L_2. Using the fact that the inverse Fourier transform of $|x|X(t)$ is the derivative of the conjugate function of $x(t)$ which is equivalent to the conjugate of the derivative, we find that

$$A_P x = \tilde{x}'. \tag{4.2.11}$$

We note that $A_P^2 = 4A_W$ almost everywhere in the common domain of definition of the two transformations A_P^2 and A_W.

So far we have taken $E = L_2$, but W_ζ and P_ζ define semi-groups for the right half-plane in any Lebesgue space $L_p(-\infty, \infty)$, $1 \leqq p \leqq \infty$. The Dirichlet transformation is a bounded transformation on L_p to L_p for $1 < p < \infty$, but neither for $p = 1$ nor for $p = \infty$. As a consequence of this fact we find that the representations

$$W_\zeta = \int_0^{\infty} e^{-\zeta u}\,dD_{2\sqrt{u}}, \qquad P_\zeta = \int_0^{\infty} e^{-\zeta u}\,dD_u \tag{4.2.12}$$

[20] See some remarks in E. Hille [1], p. 428, concerning the validity of this formula and references to the papers of Eddington and C. W. Oseen.

remain valid for $1 < p < \infty$, but for $p = 1$ or ∞ we have to integrate by parts which brings in kernels of the Fejér type.

4.3. Semi-Groups and Abel Summability

We encounter semi-groups in the various applications of Abel's method of summability and its generalizations. We shall consider some instances.

We start with applications to sequences. The elements of the space E are then vectors $X = (x_0, x_1, \cdots, x_n, \cdots)$. We suppose that the space and its metric are such that if $\{a_n\}$ is any sequence of complex numbers such that $\sup |a_n| < \infty$, then $X \in E$ implies that $(a_n x_n) \in E$ and

$$\| (a_n x_n) \| \leqq \sup | a_n | \, \| (x_n) \| . \tag{4.3.1}$$

Let $\{\lambda_n\}$ be an arbitrary sequence of non-negative real numbers and define

$$T_\zeta X = (e^{-\zeta \lambda_n} x_n). \tag{4.3.2}$$

Then $\{T_\zeta\}$ is a semi-group in the right half-plane. In the classical case of Abel, $\lambda_n = n$, and in the various generalizations of Abel summability it is customary to assume that $\lambda_n < \lambda_{n+1}$ and $\lambda_n \to \infty$.

It is easily seen that the spectrum $S(\alpha)$ of T_α consists of the points $e^{-\alpha\lambda_n}$, $n = 0, 1, 2, \cdots$, and their limit points. Thus the spectrum is real positive and restricted to the interval $[0, 1]$. It is easy to verify that T_α is uniformly continuous for $\alpha > 0$, whence it follows that (T_3) and (T_4) are satisfied. A simple calculation shows that (T_6') also holds with $K(\alpha) = 1$ and $\pi(\alpha) = 1$. Thus all the assumptions of Theorem 13 hold and formula (3.5.14) holds with $\beta > 1$, $M = 1$. Actually we can take $\beta \geqq 0$. Indeed we have

$$T_\zeta = \int_0^\infty e^{-\zeta u} \, dA_0(u),$$

where $A_0(u)$ is the evidently bounded operation which consists in replacing the vector $(x_0, x_1, \cdots, x_n, \cdots)$ by the vector $(x_0, x_1, \cdots, x_{[u]}, 0, 0, \cdots)$. Here $[u]$ denotes the integral part of u, except when $u = 0$, when all components are to be replaced by 0.

We get somewhat more interesting formulas in the case of function spaces. Let $E = L_p(-\pi, \pi)$, $1 \leqq p < \infty$, with the customary metric. Let $x(t) \in L_p$ and put

$$x(t) \sim \sum_{-\infty}^{\infty} x_n e^{nit}. \tag{4.3.3}$$

Let us form

$$T_\zeta x(t) \sim \sum_{-\infty}^{\infty} x_n e^{-\zeta\lambda_{|n|}} e^{nit}, \tag{4.3.4}$$

where $\Re(\zeta) > 0$. The notation suggests that the series on the right is a Fourier series, but this is by no means the case unless the λ_n-sequence is subject to special conditions. Necessary and sufficient conditions in order that T_ζ shall define a bounded transformation on L_p to L_p for every ζ in the right half-plane are given by the theory of factor sequences of Fourier series (see A. Zygmund [1], pp. 100–105). In order to simplify matters we suppose that λ_n is monotone increasing and

$$\lambda_n/\log n \to +\infty \quad \text{as} \quad n \to \infty.$$

These conditions are sufficient in order to ensure that T_ζ shall have the desired property.

Under these circumstances T_ζ defines a semi-group in the right half-plane. The canonical form of T_ζ is easily obtained from (4.3.4). We define an operator $A_0(u)$ which carries the Fourier series (4.3.3) into the Nth partial sum of the series where N is the largest value of n for which $\lambda_n \leqq u$. Then

$$T_\zeta = \int_0^\infty e^{-\zeta u}\, dA_0(u).$$

Take the simplest case of all, $p = 2$, $\lambda_n = n$, i.e.,

$$\text{(4.3.5)} \qquad \Pi_\zeta x(t) = \sum_{-\infty}^{\infty} x_n e^{-|n|\zeta} e^{nit},$$

which leads to Poisson's integral for the circle if we put $e^{-\zeta} = r$. $A_0(u)$ is then simply $E_\Pi(u)$, the resolution of the identity of Theorem 15, and a simple calculation shows that the corresponding operator A_Π is defined by

$$(A_\Pi x, y) = \sum_{n=-\infty}^{\infty} |\, n \,|\, x_n \bar{y}_n.$$

It follows that

$$\text{(4.3.6)} \qquad A_\Pi x = \tilde{x}',$$

i.e., we arrive at formally the same operation as in the case of Poisson's integral for the half-plane, see formula (4.2.11). Actually the operators, though closely related, are completely distinct.

We can also describe A_Π as the operator which multiplies the nth Fourier coefficient of $x(t)$ by $|\, n \,|$. This property is characteristic for the Poisson-Abel transformation and is independent of the coördinate system, i.e., of the orthogonal system. Let (a, b) be an arbitrary finite or infinite interval, $\{\omega_n(t)\}$ an ortho-normal system, complete in $L_2(a, b)$. Let $x(t) \in L_2(a, b)$ and

$$\text{(4.3.7)} \qquad x(t) \sim \sum_{n=1}^{\infty} x_n \omega_n(t).$$

Define

$$\text{(4.3.8)} \qquad Q_\zeta x(t) \sim \sum_{n=1}^{\infty} x_n e^{-n\zeta} \omega_n(t).$$

This is again a semi-group defined in the right half-plane and admits of a representation of the form

$$Q_\zeta = \int_0^\infty e^{-\zeta u}\, dE_Q(u)$$

Here $E_Q(u)$ is the operator which replaces the orthogonal series (4.3.7) by its Nth partial sum where $N = [u]$. The corresponding operator A_Q carries the series (4.3.7) into the series

$$\sum_{n=1}^{\infty} n x_n \omega_n(t),$$

i.e., multiplies the nth Fourier coefficient of $x(t)$ by n.

The case of Hermitian series is particularly interesting. We have $a = -\infty$, $b = +\infty$, and $\omega_n(t)$ equals

$$\eta_n(t) = \{\pi^{\frac{1}{2}} 2^n n!\}^{-\frac{1}{2}} H_n(t) e^{-t^2/2}. \tag{4.3.9}$$

Denote the corresponding Abel-Poisson operator by H_ζ, and let A_H denote $-\log H_1$ as usual. It turns out that A_H is the differential operator

$$A_H x(t) = -\tfrac{1}{2}[x''(t) + (1 - t^2)x(t)] \tag{4.3.10}$$

which enters in so many other problems involving Hermitian orthogonal functions. It is enough to recall that

$$A_H \eta_n(t) = n\eta_n(t). \tag{4.3.11}$$

This relation also reveals how (4.3.10) can be proved.

These examples could be multiplied ad infinitum, but the instances already given are enough to show the possibilities of obtaining formulas of varying degree of interest.

4.4. Fractional Integration

The best known functional operation having the semi-group property is probably the Riemann-Liouville fractional integral.

Let E be a Lebesgue space $L_p(0, a)$, $1 \leqq p \leqq \infty$, $a < \infty$, and form

$$I^\zeta f = \frac{1}{\Gamma(\zeta)} \int_0^t (t - u)^{\zeta - 1} f(u)\, du \equiv f_\zeta(t), \tag{4.4.1}$$

where $\Re(\zeta) > 0$. It is well known that $f_\zeta(t) \in L_p(0, a)$ and that $I^\zeta f$ has the semi-group property in the right half-plane. Stirling's formula for the gamma function shows that I^ζ, which is a holomorphic operator function in the right half-plane, is bounded in the sense of formula (2.5.3) in the half-plane. Indeed, we find $\gamma(\xi) = \pi/2$ [see formula (2.3.12)] whence it follows that $\omega_0 = \omega_1 = 1$. Consequently there exist representations by means of binomial series in the variable ζ/ω for $0 < \omega < 1$ which are weakly convergent for $\Re(\zeta) > 0$ and strongly as well as uniformly convergent at least for $\Re(\zeta) > \omega$.

But in the present case we have convergent representations by binomial series in terms of the variable ζ itself. We get the most elegant representation in terms of the variable $\zeta - 1$, however. For this purpose we use the known series[21]

$$\frac{v^{\zeta-1}}{\Gamma(\zeta)} = \sum_{n=0}^{\infty} (-1)^n L_n(v) \binom{\zeta - 1}{n}.$$

Here $L_n(v)$ denotes the nth polynomial of Laguerre. When $v > 0$ the series converges for $\xi > \frac{1}{4}$ and absolutely for $\xi > \frac{3}{4}$. When $v = 0$ we must have $\xi > 1$ and the series is absolutely convergent. With the aid of this series we obtain the expansion

$$I^\zeta f = \sum_{n=0}^{\infty} (-1)^n \binom{\zeta - 1}{n} \int_0^t L^n(t - u) f(u)\, du. \tag{4.4.2}$$

Using the well known estimate

$$| L_n(t) | \leqq e^{t/2},$$

we see that the series converges strongly as well as uniformly at least for $\Re(\zeta) > 1$.

The resolvent of I^α is known. We have[22]

$$R(\lambda, \alpha) f = \frac{1}{\lambda} \frac{d}{dt} \int_0^t E_\alpha[(t - u)^\alpha / \lambda] f(u)\, du, \tag{4.4.3}$$

where

$$E_\alpha(z) = \sum_{n=0}^{\infty} \frac{z^n}{\Gamma(n\alpha + 1)}$$

is the Mittag-Leffler function.

This formula shows that $R(\lambda, \alpha)$ is an entire function of $1/\lambda$ for every $\alpha > 0$, so that the spectrum reduces to just one point, namely $\lambda = 0$. Supposing $0 < \alpha < 2$, let V_α denote the sector $\frac{\alpha}{2}\pi < \arg z < \left(2 - \frac{\alpha}{2}\right)\pi$ and use the same notation in the λ-plane. From the classical estimate

$$E_\alpha(z) = \frac{1}{\Gamma(-\alpha) z} + O\left(\frac{1}{|z|^2}\right),$$

valid for z in V_α, we conclude that

$$\| R(\lambda, \alpha) \| \leqq M(\alpha) \, | \lambda |^{-1}, \qquad \lambda \in V_\alpha. \tag{4.4.4}$$

In the complementary sector we have

$$\| R(\lambda, \alpha) \| \leqq M(\alpha) \, | \lambda |^{-1-1/\alpha} \exp [a \, | \lambda |^{-1/\alpha}]$$

and this estimate is not capable of essential improvement.

---

21 Due to S. Wigert [1], p. 18. Cf. N. E. Nörlund [2], pp. 161–165.

22 See E. Hille and J. D. Tamarkin [1], pp. 524–525.

It follows from these estimates that the spectrum of I^{α} satisfies the assumptions of Theorem 9 for $0 < \alpha < 1$.

Finally let us consider an example which does not fit into the general theory of semi-groups developed in this paper. In a recent important memoir M. Riesz [1] has considered a functional transformation of the Riemann-Liouville type. Let Ω_m be a Euclidean space of $m \geqq 1$ dimensions. Let r_{PQ} denote the distance between the two points P and Q and form

$$J^{\alpha}f(P) = \frac{\Gamma\left(\frac{m-\alpha}{2}\right)}{\pi^{m/2} 2^{\alpha} \Gamma\left(\frac{\alpha}{2}\right)} \int_{\Omega_m} f(Q) r_{PQ}^{\alpha - m} \, dQ, \tag{4.4.5}$$

where the only assumption on f is that the integral shall be absolutely convergent for all finite P in Ω_m. Riesz takes α to be positive and shows that

$$J^{\alpha}[J^{\beta}f(P)] = J^{\alpha+\beta}f(P), \qquad \alpha > 0, \qquad \beta > 0, \qquad \alpha + \beta < m. \tag{4.4.6}$$

Formula (4.4.5) also has a sense for complex values of α. Indeed, if $J^{\zeta}f(P)$ has a sense for $\zeta = \xi$ then it also exists for $\zeta = \xi + i\eta$, where η is arbitrary. Formula (4.4.6) also holds if we suppose the real parts of the exponents restricted to the interval $(0, m)$. Thus we are dealing with an operator family having the semi-group property in a strip of the complex plane. But it is clear that the operator J^{α} as a rule does not define a bounded transformation on and to a Banach space, so that our theory does not apply.

Appendix

5.1. Carlson's Theorem

We shall state and prove that part of Theorem E, due to F. Carlson, which was used in §3.3.

THEOREM E. *Let $f(z)$ be holomorphic in $x \geqq 0$. In addition suppose that* (i) *on the imaginary axis*

$$|f(\pm ir)| \leqq Ce^{\pi r}$$

(ii) *for* $-\pi/2 \leqq \theta \leqq \pi/2$

$$|f(re^{i\theta})| \leqq Ce^{\lambda(\theta) r}, \qquad \lambda(\theta) \leqq M,$$

where $\lambda(-\theta) = \lambda(\theta)$ *and*

$$\limsup_{\delta \to 0} \frac{1}{\delta}\left[\pi - \lambda\left(\frac{\pi}{2} - \delta\right)\right] = +\infty,$$

and (iii) $f(n) = 0$ *for* $n = 1, 2, 3, \cdots$. *Then* $f(z) \equiv 0$.

PROOF. Let $\omega > 0$ and form

$$H(z, \omega) = zf(z)e^{\omega z}/\sin \pi z.$$

This function is holomorphic for $x \geqq 0$. Take a ray $\arg z = \pi/2 - \delta$. Then

$$| H(z, \omega) | \leqq B \exp\left\{r\left[\lambda\left(\frac{\pi}{2} - \delta\right) + \omega \sin \delta - \pi \cos \delta\right]\right\}.$$

$$= B \exp\left\{-r \sin \delta\left[\frac{\pi - \lambda\left(\frac{\pi}{2} - \delta\right)}{\sin \delta} - \omega - \pi \frac{1 - \cos \delta}{\sin \delta}\right]\right\}$$

Assumption (ii) shows that for fixed ω and suitably chosen small positive δ the expression inside the second square brackets can be made as large positive as we please. Hence we can choose a sequence $\delta_n \to 0$ and find numbers $\omega_n \to \infty$ as $n \to \infty$ such that

$$| H(re^{\pm i\theta_n}, \omega) | \leqq B$$

for $\theta_n = \pi/2 - \delta_n$, $0 \leqq \omega \leqq \omega_n$.

In the sector $V_n : | \arg z | \leqq \theta_n$, we can find a constant A_n such that $| H(z, \omega) | \leqq M \exp (A_n r)$, $z \in V_n$, $0 \leqq \omega \leqq \omega_n$. An application of the classical Phragmén-Lindelöf theorem to the function $H(z, \omega)$ in the sector V_n shows that

$$| H(z, \omega) | \leqq B, \qquad z \in V_n, \qquad 0 \leqq \omega \leqq \omega_n .$$

But this implies that $| H(z, \omega) | \leqq B$ for $x \geqq 0$ and all values of ω. Hence

$$| zf(z) | \leqq B | \sin \pi z | e^{-\omega x}$$

for all $\omega > 0$. But this means that $|f(z)|$ is arbitrarily small for $x \geqq \epsilon > 0$. Consequently $f(z) \equiv 0$.

5.2 Extensions of Theorem 4[23]

In the present paragraph we shall extend Theorem 4 of §2.4 in three different directions. We shall show that a semi-group whose norm is measurable need not have a finite least upper bound on the interval $(0, a)$. On the other hand, it does have a positive greatest lower bound which tends to a limit $\geqq 1$ as $a \to 0$. Finally, a group whose norm is measurable has a finite least upper bound on any finite interval $(-a, a)$.

Theorem 4_1. *There exists a space E of type (B) and an operator function $W(\zeta)$ on E to E which is holomorphic and has the semi-group property for $\Re(\zeta) > 0$, such that $|| W(\zeta) || \to \infty$ as $\zeta \to 0$ along the positive real axis.*

Proof. We can take for E the space $L(-\pi, \pi)$. We shall use a modification of the transformations defined by formula (4.3.4) obtained by introducing gaps in the factor sequence. Let $\{p_k\}$ be a monotone increasing sequence of positive integers such that $\sum 1/p_k$ is divergent, and form the function

$$F(t; \zeta) = \sum_{k=1}^{\infty} e^{-p_k \zeta} \cos p_k t, \qquad \Re(\zeta) > 0. \tag{5.2.1}$$

[23] Added in proof October 22, 1938.

If $x(t) \in L(-\pi, \pi)$ we define

$$(5.2.2) \qquad W(\zeta)[x(t)] = (2\pi)^{-1} \int_{-\pi}^{\pi} F(t - u; \zeta) x(u)\, du.$$

It is a simple matter to show that this is a bounded transformation on L to L whose norm equals

$$(5.2.3) \qquad \| W(\zeta) \| = \pi^{-1} \int_0^{\pi} | F(t; \zeta) | \, dt.$$

The transformation is evidently holomorphic and has the semi-group property in the right half plane.

We shall now impose the additional restriction that $p_k \equiv 1 \pmod{4}$ for every k. Then

$$\begin{aligned} \pi \| W(\xi) \| &> \int_0^{\pi/2} | F(t; \xi) | \, dt \\ &= \int_0^{\pi/2} \left| \sum_1^{\infty} e^{-p_k \xi} - \sum_1^{\infty} [1 - \cos p_k t] e^{-p_k \xi} \right| dt \\ &\geqq (\pi/2) \sum_1^{\infty} e^{-p_k \xi} - \sum_1^{\infty} [(\pi/2) - (1/p_k)] e^{-p_k \xi} \\ &= \sum_1^{\infty} e^{-p_k \xi}/p_k > e^{-1} \sum_1^{[1/\xi]} 1/p_k, \end{aligned}$$

and this expression tends to $+\infty$ as $\xi \to 0$.

The existence of such functions shows that the boundedness of $\| T_\alpha \|$ on an interval $(0, a)$ is not a consequence of the semi-group property and the measurability of $\| T_\alpha \|$. Thus Theorem 4 is the best of its kind.

THEOREM 4_2. *If T_α has the semi-group property and $\| T_\alpha \|$ is measurable, both for $\alpha > 0$, and if $\lambda(a) = \inf \| T_\alpha \|$ when $0 < \alpha \leqq a$, then $\lambda(a)$ is positive for every a and increases to a limit $\geqq 1$ as a decreases to zero, unless T_α is identically zero for $\alpha > 0$.*

PROOF. It is clear that $\lambda(a) \geqq 0$ is a never-increasing function of a. Suppose that $\lambda(a_0) = 0$. There is then a sequence $\{\alpha_n\}$ tending to a_0 such that, in the notation of §2.4, $e(\alpha_n) \leqq -n$. By Theorem 4 we can find a finite M such that $e(\beta) < M$ for $1 \leqq \beta \leqq 2$. Formula (2.4.7) shows that for such β's we have $e(\alpha_n + \beta) \leqq -n + M$. We conclude from this the existence of a fixed measurable set of measure $\geqq 1$ in which the inequality $e(\alpha) < -n + M$ holds for every n. This implies $e(\alpha) = -\infty$ in the set in question. This is impossible unless T_α is identically zero for every $\alpha > 0$. We have consequently $\lambda(a) > 0$ for every a. Using (2.4.7) once more, and putting $\liminf_{\alpha \to 0} e(\alpha) = e_0$, we see that $e_0 \leqq 2 e_0$ or $e_0 \geqq 0$. This completes the proof of the theorem.

THEOREM 4_3. *If T_α defines a group on $(-\infty, \infty)$ and $\| T_\alpha \|$ is measurable, then $\| T_\alpha \|$ is bounded on every finite interval $(-a, a)$.*

PROOF. We proceed as in the proof of Theorem 4. If the theorem is false we can find a sequence $\alpha_n \to \gamma$ such that $p(\alpha_n) \geqq 2n$. We let α and β be arbitrary numbers from the interval $(-a, a)$ such that $\alpha + \beta = \alpha_n$. We can then conclude that $p(\alpha) > n$ for every n on a fixed measurable α-set of positive measure. Hence $p(\alpha) = +\infty$ on this set. This being impossible, $p(\alpha)$ must be bounded on $(-a, a)$ and consequently also $\| T_\alpha \|$.

REFERENCES

BANACH, S. [1] Théorie des opérations linéaires. Monografje mat. I. Warsaw, 1932.

BERNSTEIN, S. [1] Sur les fonctions absolument monotones. Acta Mathematica, 52 (1929) 1–66.

CARLSON, F. [1] Sur une classe de série de Taylor. Thèse, Upsal, 1914, 76 pp.

[2] Sur les séries de coefficients binomiaux. Nova Acta R. Soc. Sci. Upsaliensis, (4) 4 (1915), No. 3, 61 pp.

DUNFORD, N. [1] On one-parameter groups of linear transformations. Annals of Mathematics, (2) 39 (1938) 569–573.

HAMEL, G. [1] Eine Basis aller Zahlen und unstetiger Lösungen der Funktionalgleichung $f(x + y) = f(x) + f(y)$. Math. Annalen, 60 (1905) 459–462.

HAUSDORFF, F. [1] Momentprobleme für ein endliches Intervall. Math. Zeitschrift, 16 (1923) 220–248.

HILLE, E. [1] A class of reciprocal functions. Annals of Mathematics, (2) 27 (1926) 427–464.

[2] Notes on linear transformations. I. Trans. Amer. Math. Soc., 39 (1936), 131–153.

[3] On semi-groups of transformations in Hilbert space. Proc. Nat. Acad. Sci., 24 (1938) 159–161.

HILLE, E. & TAMARKIN, J. D. [1] On the theory of linear integral equations. I. Annals of Mathematics, (2) 31 (1930) 479–528.

[2] On the theory of Laplace integrals. Proc. Nat. Acad. Sci., 19 (1933) 908–912; 20 (1934) 140–144.

LINDELÖF, E. [1] Le calcul des résidus et ses applications à la théorie des fonctions. Paris, Gauthier-Villars, 1905.

NEUMANN, J. VON [1] Zur Algebra der Funktionaloperationen und Theorie der normalen Operatoren. Math. Annalen, 102 (1929) 370–427.

NÖRLUND, N. E. [1] Sur les formules d'interpolation de Stirling et de Newton. Ann. Ecole Norm. Sup., (3) 39 (1922) 343–403; 40 (1923) 35–54.

[2] Leçons sur les séries d'interpolation. Paris, Gauthier-Villars, 1926.

PINCHERLE, S. [1] Sur les fonctions déterminantes. Ann. Ecole Norm. Sup., (3) 22 (1905) 9–68.

RIESZ, F. [1] Über die linearen Transformationen des komplexen Hilbertschen Raumes. Acta Litt. ac Sci., Szeged, 5 (1930) 23–54.

RIESZ, M. [1] Intégrales de Riemann-Liouville et potentiels. Acta. Litt. ac Sci., Szeged, 9 (1938) 1–42.

STONE, M. H. [1] Linear transformations in Hilbert space and their applications to analysis. Amer. Math. Soc. Coll. Publ. XV. New York, 1932.

SZ. NAGY, B. VON [1] Über messbare Darstellungen Liescher Gruppen. Math. Annalen, 112 (1936) 268–296.

TAYLOR, A. E. [1] Linear operations which depend analytically upon a parameter. Annals of Mathematics, (2) 39 (1938) 574–593.

WIDDER, D. V. [1] Necessary and sufficient conditions for the representation of a function as a Laplace integral. Trans. Amer. Math. Soc., 33 (1931) 851–892.

WIENER, N. [1] Note on a paper of M. Banach. Fundamenta Mathematicae, 4 (1923) 136–143.

WIGERT, S. [1] Contributions à la théorie des polynomes d'Abel-Laguerre. Arkiv f. Mat., 15 (1921) no. 25, 22 pp.

ZYGMUND, A. [1] Trigonometrical series. Monografje Mat., V. Warsaw, 1935.

Reprinted from the Proceedings of the National Academy of Sciences,
Vol. 28, No. 5, pp. 175–178. May, 1942

REPRESENTATION OF ONE-PARAMETER SEMI-GROUPS OF LINEAR TRANSFORMATIONS

By Einar Hille

Department of Mathematics, Yale University

Communicated April 7, 1942

1. Let E be a separable normed complete linear vector space. Let T_s be defined for $s > 0$ as a linear bounded transformation on E to E such that

$$T_s T_t = T_t T_s = T_{s+t},\ s > 0,\ t > 0. \tag{1.1}$$

We suppose further:

(1) T_s is weakly measurable for $s > 0$,
(2) $||T_s|| \leq 1,\ s > 0$,
(3) $T_s(E)$ is dense in E.

Here (2) can be replaced by $||T_s|| \leq M$ for $0 < s < 1$ without inconvenience but with some unessential modifications of the results. It is enough that (3) holds for a single s, it will then hold for all. If (3) is not satisfied in the original space, we can often find a subspace where it holds and the discussion then applies to that subspace.[1]

On the basis of these assumptions we have proved strong continuity,[2] differentiability and representation theorems for semi-groups. The prototype of such results is the theorem of M. H. Stone on unitary groups in Hilbert space.[3] Our method is essentially that of Stone. Corresponding results for groups on spaces of type (B) have been announced by I. Gelfand and M. Fukamiya.[4]

We form

$$R(\lambda)x = -\int_0^\infty e^{-\lambda s}T_s x ds, \qquad x \in E, \mathfrak{R}(\lambda) > 0 \tag{1.2}$$

where the integral exists for instance in the sense of Bochner and Dunford.[5] This formula defines a linear transformation $R(\lambda)$ on E to E for $\mathfrak{R}(\lambda) > 0$ and $||R(\lambda)|| \leq [\mathfrak{R}(\lambda)]^{-1}$. The range $R(\lambda)(E)$ of $R(\lambda)$ is dense in E for every λ. Further $(\lambda - \mu)R(\lambda)R(\mu) = R(\lambda) - R(\mu)$ for $\mathfrak{R}(\lambda) > 0, \mathfrak{R}(\mu) > 0$ and $R(\lambda_0)x = 0$ implies $x = 0$.

$R(\lambda)$ is the resolvent of a closed linear transformation A whose domain of definition, $D(A)$, say, includes $R(\lambda)(E)$ for every λ and is, therefore, dense in E. We have

$$R(\lambda)(A - \lambda I) = I \text{ in } D(A), \ (A - \lambda I)R(\lambda) = I \text{ in } E, \tag{1.3}$$

and for every $x \in D(A)$ we have in the sense of strong convergence

$$Ax = \lim_{h \to 0} A_h x, \ A_h = \frac{1}{h}[T_h - I]. \tag{1.4}$$

The point spectrum of A may be dense in $\mathfrak{R}(\lambda) \leq 0$. From the fact that $D(A)$ is dense in E we conclude that for $h \to 0$

$$T_h x \to x, \ x \in E, \tag{1.5}$$

again in the sense of strong convergence.

2. With the aid of (1.2) we can derive a number of representations of the semi-group. The classical inversion formula of Laplace-Stieltjes integrals can be made to yield the result that for $c > 0$, $s \geq 0$ and $x \in E$

$$\int_0^s T_u x du = -\lim_{\omega \to \infty} (2\pi i)^{-1} \int_{c - i\omega}^{c + i\omega} e^{s\lambda} R(\lambda) x \frac{d\lambda}{\lambda}, \tag{2.1}$$

where the limit exists in the strong sense. If $x \in D(A)$, the integral obtained by letting $\omega \to \infty$ is convergent. Further, for $s > 0$, $x \in D(A)$

$$T_s x = -\lim_{\omega \to \infty} (2\pi i)^{-1} \int_{c - i\omega}^{c + i\omega} e^{s\lambda} R(\lambda) x d\lambda \tag{2.2}$$

in the strong sense. For $s = 0$, the formula gives $^1/_2 x$ instead of x. If the limit in (2.2) be replaced by the $(C, 1)$-limit, we can omit the restriction $x \in D(A)$.

The right-hand side of formula (2.2) can be interpreted as a definition of $\exp(sA)x$ in the sense of operational calculus,[6] so that

$$T_s x = \exp(sA)x,\ x \in D(A),\ s > 0. \tag{2.3}$$

Other interpretations of $\exp(sA)$ are obtainable by our methods. Thus

$$\int_0^s T_u x du = \lim_{h \to 0} \int_0^s \exp(uA_h) x du \tag{2.4}$$

for $s \geq 0$ and every $x \in E$, while

$$T_s x = \lim_{h \to 0} \exp(sA_h)x \tag{2.5}$$

for $x \in D(A)$. Here

$$\exp(sA_h)x = \sum_0^\infty \frac{s^n}{n!} A_h^n x. \tag{2.6}$$

The proof of these relations is obtained by observing that $\exp(sA_h)$ defines a semi-group for $s > 0$, the corresponding differential operator being A_h with a resolvent $R_h(\lambda)$, and the latter can be shown to converge uniformly to $R(\lambda)$, sufficiently regularly with respect to λ, so that if $R(\lambda)$ is replaced by $R_h(\lambda)$ in (2.1), we can pass to the limit with h under the sign of integration.[7]

For the case of a group I. Gelfand has proposed the interpretation

$$\exp(sA)x = \sum_0^\infty \frac{s^n}{n!} A^n x. \tag{2.7}$$

Gelfand's method appears to break down for semi-groups and it is not clear, in general, that there will exist elements belonging to all sets $D(A^n)$ for which the series is convergent in any sense whatsoever. For special semi-groups the situation may be different. Thus if T_s is defined on a Lebesgue space and T_s commutes with real translations on the point variable, then it is an easy matter to show that there exists a subspace E_0 dense on E such that the series in (2.7) converges for every $x \in E_0$ and every finite real or complex s in the sense that the sum of the norms of the terms is convergent.

[1] Condition (3) is used in proving that $R(\lambda)(E)$ and hence also $D(A)$ are dense in E. We also use the separability of the space at this point.

[2] Strong continuity for $s > 0$ but not for $s = 0$ has been proved by Dunford, N., "On One-Parameter Groups of Linear Transformations," *Ann. Math.*, ser. 2, **39**, 569–573 (1938), under more general assumptions.

[3] These PROCEEDINGS, **16**, 173–174 (1930), and *Ann. Math.*, ser. 2, **33**, 643–648 (1932).

[4] Gelfand, I., "On One-Parametrical Groups of Operators in a Normed Space," *Compt. Rend. Acad. Sci. U. R. S. S.*, **25**, 713–718 (1939); Fukamiya, M., "On One-Parameter Groups of Operators," *Proc. Imp. Acad. Tokyo*, **16**, 262–265 (1940). Fukamiya assumes separability; his proof, partly somewhat fragmentary, uses the method of

Stone, but does not go beyond the existence of A and $R(\lambda)$ and stops short of representation theorems. Gelfand in part uses an unpublished notion of integration of his own; he claims to be able to dispense with separability, and arrives at the representation (2.7) below, valid in a subspace dense in E.

[5] Bochner, S., "Integration von Funktionen, deren Werte die Elemente eines Vektorraumes sind," *Fund. Math.*, **20,** 262–276 (1933); Dunford, N., "Integration in General Analysis," *Trans. Amer. Math. Soc.*, **37,** 441–453 (1935). Separability plus weak measurability implies that the integrand is measurable in the sense of Bochner (see Pettis, B. J., "Integration in Vector Spaces," *Trans. Amer. Math. Soc.*, **44,** 277–304 (1938), (Theorem 1.1 and Corollary 1.11) and the norm is measurable and dominated by an integrable function.

[6] Cf. Stone, M. H., *Linear Transformations in Hilbert Space*, New York, 1932, Chapter VI.

[7] The author has found a simpler proof of (2.5) which is valid for all x and E—Condition (3) can be replaced by the weaker condition (3*): The least linear hull of the range spaces is dense in E. This condition is necessary for the validity of (1.5). [Added in proof.]

Reprinted from the Proceedings of the NATIONAL ACADEMY OF SCIENCES,
Vol. 28, No. 10, pp. 421–424. October, 1942

ON THE ANALYTICAL THEORY OF SEMI-GROUPS

BY EINAR HILLE

DEPARTMENT OF MATHEMATICS, YALE UNIVERSITY

Communicated September 2, 1942

1. Let E be a complex Banach space, $\{T_s\}$ a one-parameter family of bounded linear transformations on E to E defined for $s > 0$ and having the semi-group property

$$T_sT_t = T_tT_s = T_{s+t},\ s > 0,\ t > 0. \qquad (1)$$

In an earlier note[1] the author announced representation theorems for such semi-groups. The present note elaborates one of these results and is further concerned with the behavior of T_h for small positive h.

2. We start with a result on the degree of approximation of I by T_h. We write $A_h = (1/h)[T_h - I]$ and put $Ax = \lim_{h \to 0} A_hx$ whenever the limit exists in the strong sense. The set of elements x for which the limit exists is a linear subspace $D(A)$ which may reduce to the zero-element.

THEOREM 1. *Let $\|T_s\|$ be bounded in every finite interval and put* $\sup_{0 < s < \omega} \|T_s\| = M(\omega)$. *If for a particular x, $\|(T_h - I)x\| \to 0$ in such a manner that* $\liminf_{h \to 0}\|A_hx\| = 0$ *then $A_sx \equiv 0$, i.e., $T_sx \equiv x$ for all $s > 0$. If $\|(T_h - I)x\| \to 0$ with h for all $x \in E$ and $y \in D(A)$, then $\|A_hy\| \leq M(h)\|Ay\|$ or $\|(T_h - I)y\| \leq hM(h)\|Ay\|$.*

Roughly speaking the theorem asserts that only invariant elements x admit of an approximation by T_hx which is of a higher degree than the first in h and that first degree approximation is reached by all elements of $D(A)$. The boundedness assumption on the norm is satisfied if, for instance, $\|T_s\|$ is measurable and $\limsup_{h \to 0}\|T_h\| < \infty$. If $s = nh + \delta, 0 \leq \delta < h$, a simple calculation gives

$$sA_sx = \delta A_\delta x + (T_\delta + T_{\delta + h} + \ldots + T_{\delta + (n-1)h})hA_hx,$$

whence

$$\|A_sx\| \leq M(s) \liminf_{h \to 0}\|A_hx\| \qquad (2)$$

from which the first assertion follows. A simple case of this part of the theorem was proved in an earlier paper.[2] For the second part we note that if $y \epsilon D(A)$ so does $T_s y$ and $AT_s y = T_s Ay = \frac{d}{ds} T_s y$ is continuous on the right in s. Hence

$$\| (T_h - I)y \| = \| \int_0^h T_s Ay ds \| \leq hM(h) \| Ay \|. \quad (3)$$

This theorem has a number of consequences in various branches of analysis, in particular in the theory of singular integrals and the summation of orthogonal series. Abel-Poisson summability of Fourier series provides an interesting example.

THEOREM 2. *If* $f(u) \epsilon L_p(-\pi, \pi)$, $1 \leq p \leq \infty$, *or* $C[-\pi, \pi]$ *and its Poisson transform is*

$$f(u; r) = \frac{1}{2\pi} \int_{-\pi}^{\pi} \frac{(1 - r^2)f(t + u)dt}{1 - 2r \cos t + r^2}, \, 0 < r < 1, \quad (4)$$

then $\lim \inf_{r \to 1} (1 - r)^{-1} \| f(u) - f(u; r) \| = 0$ *implies that* $\tilde{f}(u)$ *is a constant. Further* $\| f(u) - f(u; r) \| \leq \log (1/r) \| \tilde{f}'(u) \|$ *whenever* $\tilde{f}'(u)$ *belongs to the space.*

If E is a separable space or, more generally, if the set $\{T_s x\}$, $0 < s < \infty$, is separable for every fixed $x \epsilon E$, and if $\omega(h)$ is a given modulus of continuity, then it is possible to find elements $y \epsilon E$ such that $\| (T_h - I)y \| \leq C\omega(h)$. For this purpose it is enough to choose a numerically valued function $K(s)$ such that (i) $K(s) \epsilon L(0, \infty)$, (ii) $K(0) = 0$ and (iii) $| K(s + h) - K(s) | \leq \omega(h)$ for $0 \leq s < \infty$, and to take

$$y = \int_0^\infty K(s) T_s x ds \quad (5)$$

where the integral exists in the sense of Bochner.[3]

3. In the preceding note[1] a representation theorem was announced [formula (2.5) compared with footnote 7]. The assumptions of this theorem are: (i) T_s is weakly measurable, (ii) $\| T_s \| \leq 1$, (iii) E is separable and (iv) $\Sigma_s T_s(E)$ is dense in E. These assumptions imply, as was stated in our note, that (1) $\| (T_h - I)x \| \to 0$ with h for all x, and (2) $D(A)$ is dense in E. Now, conversely, if (1) is satisfied, then T_s is strongly continuous on the right for $s \geq 0$ if $T_0 = I$. This implies (i) and (iv) and also that the set $\{T_s x\}$, $0 \leq s$, is separable for every fixed s. It therefore turns out that (1) + (ii) are sufficient hypotheses for the representation theorem which can be formulated as follows:[4]

THEOREM 3. *If* $\| T_s \| \leq 1$ *for* $s > 0$ *and* $\| (T_h - I)x \| \to 0$ *with* h *for all* x, *then* $\exp (sA_h)$ *is a bounded linear transformation such that* (1) $\| \exp (sA_h) \| \leq 1$, $s \geq 0$, $h > 0$, *and* (2)

$$\lim_{h \to 0} \| \exp (sA_h)x - T_s x \| = 0, \tag{6}$$

uniformly in s, $0 \leq s \leq \omega < \infty$.

The proof of (1) follows from

$$\exp (sA_h) = \exp (-s/h) \exp ((s/h)T_h).$$

The proof of (2) may be based upon the well-known fact that

$$\lim_{t \to \infty} e^{-t} \sum \frac{t^n}{n!} = 1 \tag{7}$$

if the summation is extended merely over the terms for which $(1 - \delta)t \leq n \leq (1 + \delta)t$, δ fixed positive, while the summation over the complementary values of n gives the limit zero. We have

$$\exp (sA_h)x - T_s x = \exp (-s/h) \sum_0^\infty \frac{1}{n!}\left(\frac{s}{h}\right)^n [T_{nh}x - T_s x]. \tag{8}$$

If $x \in D(A)$, $T_s x$ is absolutely continuous in s. We can then choose a δ such that $\| T_{nh}x - T_s x\| \leq \epsilon \| x \|$ for $(1 - \delta)s \leq nh \leq (1 + \delta)s$, where δ is independent of s in $0 \leq s \leq \omega$. Outside of this n-range we have $\| T_{nh}x - T_s x\| \leq 2 \|x\|$. Hence the right-hand side tends strongly to zero with h. This proves (2) for $x \in D(A)$. But $D(A)$ is dense in E and $\exp (sA_h)$ is uniformly bounded with respect to h. Hence (2) holds for all $x \in E$.

A special case is the following generalization of Taylor's theorem.[5]

Theorem 4. *If $f(u)$ is uniformly continuous in $0 \leq u < \infty$, then*

$$f(u + s) = \lim_{h \to 0} \sum_{n=0}^{\infty} \frac{1}{n!}\left(\frac{s}{h}\right)^n \sum_{k=0}^{n} (-1)^{n-k} \binom{n}{k} f(u + kh), \tag{9}$$

where the limit exists uniformly with respect to u for all u and uniformly with respect to s for $0 \leq s \leq \omega$. If instead $f(u) \in L_p(0, \infty)$, p fixed, $1 \leq p < \infty$, then the limit exists in the sense of convergence in the mean of order p.

4. The relations between the spectral properties of A and of T_s are of importance. We use the symbols $S(U)$, $PS(U)$, $CS(U)$ and $RS(U)$ to denote the spectrum of U and its point, continuous and residual components. The resolvent of U is denoted by $R(\lambda, U)$.

Theorem 5. *Under the assumptions of Theorem 3, $\alpha \in S(A)$ implies $e^{\alpha s} \in S(T_s)$ for all $s > 0$. In particular, $e^{\alpha s} \in PS(T_s)$ if $\alpha \in PS(A)$.*

The proof follows, for instance, from the identity

$$(T_s - e^{\alpha s}I)x = \int_0^s e^{\alpha(s-t)}(A - \alpha I)T_t x dt, \quad x \in D(A). \tag{10}$$

This formula shows that $(A - \alpha I)x = 0$ implies $(T_s - e^{\alpha s}I)x = 0$ which takes care of the point spectrum. If $\alpha \in CS(A)$, $\| (A - \alpha I)x\|$ is not bounded away from zero on the unit sphere and the same is obviously true for $\| (T_s - e^{\alpha s}I)x \|$. Thus $R(e^{\alpha s}, T_s)$ either does not exist or is unbounded.

Finally, if $(A - \alpha I)$ maps $D(A)$ upon a space E_α non-dense in E, then $(T_s - e^{\alpha s}I)$ maps E upon a subspace of the closure of E_α. Thus if $R(e^{\alpha s}, T_s)$ exists, its domain of definition cannot be dense in E and $e^{\alpha s}$ belongs either to $PS(T_s)$ or $RS(T_s)$.

Theorem 5 admits of a limited converse. If it is known that $e^{\alpha s} \epsilon S(T_s)$ for $s = s_1$ and s_2 where s_1/s_2 is irrational, and if, in addition, $e^{\alpha s_1}$ and $e^{\alpha s_2}$ are spectral values of the same type in a certain narrow sense (for instance, if both belong to the point-spectrum and have a common characteristic element), then $e^{\alpha s} \epsilon S(T_s)$ for all $s > 0$ and $\alpha \epsilon S(A)$.

The spectral behavior of A and the approximation of I by T_h are closely related. The following theorem indicates such a relation.

THEOREM 6. *If* $\| T_h - I\| \to 0$ *with* h *and* $\| T_h - I\| < 1 - 1/e$ *for* $h < \rho$, *and if* $|\alpha| > 1/\rho$, *then* $\|(A - \alpha I)x\|$ *is bounded away from zero on the intersection of* $D(A)$ *with the unit-sphere. In particular,* $PS(A)$ *and* $CS(A)$ *are located inside the circle* $|\alpha| = 1/\rho$.[6]

The assumption $\| T_h - I\| \to 0$ is more than sufficient to ensure the validity of (10). Denoting $\int_0^s |\exp[\alpha(s - t)]| dt$ by $M(\alpha, s)$, we then get

$$M(\alpha, s) \|(A - \alpha I)x\| \geq |1 - e^{\alpha s}| \|x\| - \|(T_s - I)x\|. \quad (11)$$

For $s = 1/|\alpha| < \rho$, the right-hand side is positive and the theorem follows.

[1] "Representation of One-Parameter Semi-Groups of Linear Transformations," these PROCEEDINGS, **28**, 175–178 (1942).

[2] "Notes on Linear Transformations," *Trans. Amer. Math. Soc.*, **39**, 131–153 (1936).

[3] Compare I. Gelfand's construction of elements in $D(A)$ in "On One-Parametrical Groups of Operators in a Normed Space," *Compt. Rend. Acad. Sci. U. R. S. S.*, **25**, 713–718 (1939).

[4] For Theorems 3–6 the author had originally found proofs based upon the resolvent theory sketched in the earlier note. This method has the advantage of great general applicability, but it is fairly complicated and in special cases stronger results are obtained more easily by direct methods such as those indicated in the present note. The proof of Theorem 3 given here is an adaptation to the abstract case of an argument suggested by Professor G. Szegö for the proof of Theorem 4. A still simpler and more direct proof has just been found by Professor Nelson Dunford.

[5] Professor Dunford has called my attention to the fact that Theorem 4 gives a new and fairly direct proof of Weierstrass' approximation theorem.

[6] This theorem suggests strongly that $RS(A)$ is also bounded and, as a consequence, that A is a bounded operator. My methods do yot yield this result, but it has been proved very elegantly by Dunford whose proofs will appear elsewhere.

Reprinted from the Proceedings of the NATIONAL ACADEMY OF SCIENCES,
Vol. 30, No. 3 pp. 58–60. March, 1944

ON THE THEORY OF CHARACTERS OF GROUPS AND SEMI-GROUPS IN NORMED VECTOR RINGS

BY EINAR HILLE

DEPARTMENT OF MATHEMATICS, YALE UNIVERSITY

Communicated February 4, 1944

1. Recently I. Gelfand[1] has defined a class of characters of Abelian groups embedded in a normed commutative vector ring. The present note contains an elaboration of this theory with a strengthening of the main theorem and extensions to semi-groups.

Let $\mathfrak{R}$ be a normed commutative vector ring with unit element e, $\mathbf{\mathfrak{M}}$ the set of maximal ideals $\mathfrak{M}$ in $\mathfrak{R}$, $\mathfrak{R}/\mathfrak{M}$ the corresponding residue class rings each of which is a field. Defining the norm $||X||$ of a residue class X, modulo $\mathfrak{M}$, as min $||x||$ for $x \in X$, Gelfand shows that $\mathfrak{R}/\mathfrak{M}$ is a normed field and hence isomorphic with the complex field. Let X_α be the class of elements congruent to αe (mod. $\mathfrak{M}$), α arbitrary complex number. Then $\bigcup X_\alpha = \mathfrak{R}$ and $X_\alpha + X_\beta = X_{\alpha+\beta}, X_\alpha X_\beta = X_{\alpha\beta}$.

Define[2]

$$\mu(x;\ \mathfrak{M}) = \alpha, \qquad x \equiv \alpha e \text{ (mod. } \mathfrak{M}). \tag{1}$$

Since $x - \alpha e$ is singular for every $x \in X_\alpha$, the spectrum $\sigma(x)$ of x contains the point $\lambda = \alpha$. Conversely, if $\alpha \in \sigma(x)$, there exists an $\mathfrak{M}$ such that $x \equiv \alpha e$ (mod. $\mathfrak{M}$). Since $|\mu(x;\ \mathfrak{M})| \leqq \max |\sigma(x)| \leqq ||x||$, $\mu(x;\ \mathfrak{M})$ is a bounded functional which is additive as well as multiplicative

$$\mu(x + y;\ \mathfrak{M}) = \mu(x;\ \mathfrak{M}) + \mu(y;\ \mathfrak{M}), \mu(xy;\ \mathfrak{M}) = \mu(x;\ \mathfrak{M})\mu(y;\ \mathfrak{M}). \tag{2}$$

In particular, $\mu(x;\ \mathfrak{M})$ is continuous.

Finally the notion of quasi-nilpotency will be needed. An element q of $\mathfrak{R}$ is quasi-nilpotent if $e - \alpha q$ has an inverse for every complex α or, which is equivalent in a normed vector ring, if $||q^n||^{1/n} \to 0$ when $n \to \infty$. The quasi-nilpotent elements form an ideal, the radical of $\mathfrak{R}$, which equals the intersection of all maximal ideals.

After these introductory remarks we proceed to the subject matter proper.

2. Let $\mathfrak{G}$ be a set of elements in the ring $\mathfrak{R}$ which form a group under the operation of ring multiplication. For every $\mathfrak{M} \in \mathbf{\mathfrak{M}}$ the bounded linear multiplicative functional $\mu(x;\ \mathfrak{M})$ for x in $\mathfrak{G}$ defines a continuous one-dimensional representation $\Gamma(\mathfrak{M})$ of $\mathfrak{G}$, $\mu(x;\ \mathfrak{M})$ being the characters of the representation.

THEOREM 1. *$\Gamma(\mathfrak{M})$ is unitary in the sense that $|\mu(x;\ \mathfrak{M})| = 1$ on $\mathfrak{G}$ for every $\mathfrak{M} \in \mathbf{\mathfrak{M}}$ if and only if for all $x \in \mathfrak{G}$*

$$\lim_{n\to\infty} ||x^n||^{1/n} = 1, \qquad \lim_{n\to\infty} ||x^{-n}||^{1/n} = 1. \tag{3}$$

Condition (3) expresses that the spectrum of x lies on the unit-circle. Since $\sigma(x)$ is the range of $\mu(x; \mathfrak{M})$ for $\mathfrak{M} \in \mathbf{M}$ and fixed x, this is all that should be proved.

The main question in the representation theory is whether or not there are enough representations available so that any given pair of distinct elements of the group have distinct characters in at least one representation. The limitations of the present theory are clearly brought out by

THEOREM 2. *If $x \neq y$ are two elements of $\mathfrak{G}$, then $\mu(x; \mathfrak{M}) = \mu(y; \mathfrak{M})$ for all $\mathfrak{M} \in \mathbf{M}$ if and only if $x - y = q$, where q is quasi-nilpotent. This case can arise if and only if $\mathfrak{G}$ contains an element of the form $e + q$ and is excluded if $\mathfrak{R}$ is without radical.*

We omit the proof which is an immediate consequence of the properties of the functional $\mu(x; \mathfrak{M})$ listed above.

Gelfand has shown that even if $\mathfrak{R}$ has a radical it is possible to exclude the exceptional elements of the form $e + q$ from the group $\mathfrak{G}$ by requiring that all cyclical subgroups form bounded pointsets in the vector space $\mathfrak{R}$. The following condition is less restrictive.

THEOREM 3. *If $||x^{\pm n}|| = o(n)$ when $n \to \infty$, then x cannot be of the form $e + q$ where q is quasi-nilpotent and $q \neq 0$. The conclusion becomes false if $o(n)$ be replaced by $O(n)$.*

If $x = e + q$ then the resolvent $R(\lambda; e + q) = ((\lambda - 1)e - q)^{-1}$ admits of the three expansions

$$\begin{aligned} -(e + q)^{-1} - \textstyle\sum_1^\infty (e + q)^{-n-1}\lambda^n, \qquad & |\lambda| < 1; \\ e\lambda^{-1} + \textstyle\sum_1^\infty (e + q)^n \lambda^{-n-1}, \qquad & |\lambda| > 1; \\ e(\lambda - 1)^{-1} + \textstyle\sum_1^\infty q^n (\lambda - 1)^{-n-1}, \qquad & \lambda \neq 1. \end{aligned}$$

In the first two series the coefficients are $o(n)$ in norm. A theorem in classical function theory, found by G. Pólya,[3] asserts that if $f(z)$ is an entire function of $w = 1/(z - 1)$ and if in the power series expansions of $f(z)$ about 0 and about ∞ the coefficients are $O(n^\nu)$, then $f(z)$ is a polynomial in w of degree $\leq \nu + 1$. If the coefficients are $o(n^\nu)$ instead, the degree will be $< \nu + 1$. By an obvious extension of this theorem to vector valued functions, one concludes that $R(\lambda; e + q)$ has a simple pole at $\lambda = 1$ which requires $q = 0$. On the other hand, if q is a nilpotent element such that $q \neq 0$, $q^2 = 0$, then $(e + q)^n = e + nq$ which is $O(n)$ in norm. Thus the stated condition is the best of its kind.

It should be observed that a condition of the form $||x^{\pm n}|| = o(n^m)$, if satisfied by an element of the form $e + q$, requires $q^m = 0$ and this condition is again the best possible.[4]

Combining the preceding theorems we obtain

THEOREM 4. *If the elements of* $\mathfrak{G}$ *satisfy the condition of Theorem 3, then all representations* $\Gamma(\mathfrak{M})$ *are unitary and if* $x \neq y$, *there is a representation in which* $\mu(x;\ \mathfrak{M}) \neq \mu(y;\ \mathfrak{M})$.

3. These considerations may be extended in part to semi-groups. Let $\mathfrak{S}$ be a set of elements in $\mathfrak{R}$ which form a semi-group under ring multiplication so that $x \in \mathfrak{S}$, $y \in \mathfrak{S}$ implies $xy \in \mathfrak{S}$. We shall require that $\mathfrak{S}$ be closed under inversion so that if x is regular and is in $\mathfrak{S}$ then x^{-1} is also in $\mathfrak{S}$.

The functionals $\mu(x;\ \mathfrak{M})$ also provide one-dimensional representations $\Sigma(\mathfrak{M})$ of $\mathfrak{S}$. It is obviously meaningless to require that $\Sigma(\mathfrak{M})$ be unitary. If $x \neq y$ are two elements of $\mathfrak{S}$ we still have $\mu(x;\ \mathfrak{M}) = \mu(y;\ \mathfrak{M})$ for all $\mathfrak{M}$ if and only if $x - y = q$. Thus this case can arise only if $\mathfrak{R}$ has a radical, but when present the situation is much more complicated than in the case of groups. Thus as a rule there is no longer a single exceptional class; every singular element $x \in \mathfrak{S}$ may give rise to a class $\{x + q\}$ the elements of which cannot be separated by the representations. Since the condition of Theorem 3 can be applied to regular elements only, the problem of excluding such singular exceptional classes is a much more difficult one which apparently cannot be handled by the methods of the present note.

[1] Gelfand, I., "Zur Theorie der Charaktere der Abelschen topologischen Gruppen," *Rec. Math.*, N. S., **9** (**51**), 49–50 (1941). For the theory of normed vector rings see Gelfand, "Normierte Ringe," *Ibid.*, 3–24, and Lorch, E. R., "The Theory of Analytic Functions in Normed Abelian Vector Rings," *Trans. Amer. Math. Soc.*, **54**, 414–425 (1943).

[2] This is $x(\mathfrak{M})$ in Gelfand's notation. His $x(\mathfrak{M})$, however, is a function on maximal ideals to complex numbers corresponding to a fixed element x of the ring while $\mu(x;\ \mathfrak{M})$ is a function on the ring to complex numbers corresponding to a fixed maximal ideal.

[3] Pólya, G., "Aufgabe 106," *Jahresbericht D.M.V.*, **40**, *81* (1931). Several solutions are published, *Ibid.*, **42** (1933); that by Szegö, G., pp. *15–16*, is short and self-contained. See also the closely related problem 105: an entire function of order one and minimal type which is $O(n^m)$ on the integers is a polynomial of degree $\leq m$.

[4] Cf. Gelfand, I., "Ideale und primäre Ideale in normierten Ringen," *Rec. Math.*, N. S., **9** (**51**), 41–48 (1941), where similar conclusions are drawn less directly from a deeper theorem due to F. and R. Nevanlinna.

THE DIFFERENTIABILITY AND UNIQUENESS OF CONTINUOUS SOLUTIONS OF ADDITION FORMULAS

NELSON DUNFORD AND EINAR HILLE

The problem of representing a one-parameter group of operators (that is, a family T_ξ, $-\infty<\xi<\infty$, of bounded linear operators on a Banach space which satisfies $T_{\xi+\zeta}=T_\xi T_\zeta$) reduces according to several well known methods of attack to establishing differentiability of the function T_ξ at $\xi=0$. The derivative $Ax=\lim_{\xi\to 0}\xi^{-1}(T_\xi-I)x$ exists as a closed operator with domain $D(A)$ dense, providing T_ξ is continuous in the *strong* operator topology (that is, $\lim_{\xi\to\xi_0}T_\xi x=T_{\xi_0}x$, $x\in\mathfrak{X}$). It is then possible to assign a meaning to exp (ξA) in a natural way and so that $T_\xi=\exp(\xi A)$, $-\infty<\xi<\infty$. The operator A is bounded if and only if T_ξ is continuous in ξ in the *uniform* operator topology (that is, $\lim_{\xi\to\xi_0}|T_\xi-T_{\xi_0}|=0$) in which case $A=\lim_{\xi\to 0}\xi^{-1}(T_\xi-I)$ exists in the uniform topology. This implies that T_ξ is an entire function of ξ; conversely, if T_ξ is analytic anywhere, then A is bounded. These considerations extend to the semi-group case in which $T_{\xi+\zeta}=T_\xi T_\zeta$ is known to hold only for positive values of the parameters, although

---

Presented to the Society, August 14, 1944; received by the editors February 17, 1947.

the number of distinct cases is much larger and, in particular, analyticity does not imply that A is bounded.

It is a matter of natural curiosity to ask whether or not similar results hold if the semi-group law $f(\xi+\zeta)=f(\xi)f(\zeta)$ is replaced by an arbitrary addition formula. The results presented in the following correspond to the case of an analytical group (that is, continuity in the uniform topology). The strong operator topology leads to particular difficulties which have been overcome only in part, but we hope to return to this case on a future occasion.

In this note we consider the differentiability and uniqueness of continuous solutions $f(\xi)$ of the equation

$$f(\xi+\zeta)=G[f(\xi),f(\zeta)], \qquad 0\leqq \xi,\zeta,\xi+\zeta\leqq\omega, \tag{1}$$

where $G[\alpha,\beta]$ is a symmetric complex function analytic for α,β in the closure of a domain Δ bounded by a rectifiable Jordan curve. The solutions considered are functions $f(\xi)$ on $0\leqq\xi\leqq\omega$ to a commutative complex Banach algebra B with unit e. We define $G[u,v]$ only for those $u,v\in B(\Delta)$, the subset of B consisting of elements x whose spectrum $\sigma x\subset\Delta$. For such u,v we define $G[u,v]$ by the double resolvent integral

$$G[u,v]=\frac{1}{(2\pi i)^2}\int_{\Gamma_u}\int_{\Gamma_v}G[\alpha,\beta]R(\alpha,u)R(\beta,v)d\alpha d\beta, \tag{2}$$

where $R(\alpha,u)=(\alpha e-u)^{-1}$ and Γ_u, Γ_v are oriented envelopes in Δ of σu, σv respectively. Thus by a solution of (1) is meant a function $f(\xi)$ on $0\leqq\xi\leqq\omega$ to $B(\Delta)$ which satisfies (1).

THEOREM. *If $f(\xi)$ is a continuous B-valued solution of* (1) *and if $G_1[f(0),f(0)]$ has an inverse ($G_1[\alpha,\beta]=(\partial/\partial\alpha)G[\alpha,\beta]$), then $f(\xi)$ has derivatives of all orders and $f'(\xi)=G_1[f(0),f(\xi)]f'(0)$. If $g(\xi)$ is any other continuous solution of* (1) *with $g(0)=f(0)$ and $g'(0)=f'(0)$, then $g(\xi)\equiv f(\xi)$. If $\phi(\xi)$ is a nonconstant scalar analytic solution of* (1) *then the only continuous B-valued solution of* (1) *with $f(0)=\phi(0)e$ is given by $f(\xi)=\phi(f'(0)\xi)$ in a neighborhood $0\leqq\xi\leqq\rho$.*

LEMMA. *Let $Q[\alpha,\beta,\gamma]=(G[\alpha,\gamma]-G[\beta,\gamma])(\alpha-\beta)^{-1}$, $\alpha,\beta,\gamma\in\Delta$. If $f(\xi)$ is continuous on $0\leqq\xi\leqq\omega$ to $B(\Delta)$ and $0\leqq\eta,\zeta\leqq\omega$ then uniformly with respect to ζ we have*

$$\lim_{\xi\to\eta}Q[f(\xi),f(\eta),f(\zeta)]=G_1[f(\eta),f(\zeta)], \tag{3}$$

$$Q(u,v,w)(u-v)=G(u,w)-G(v,w), \qquad u,v,w\in B(\Delta). \tag{4}$$

First note that since $f(\xi)$ is continuous in $[0, \omega]$ the range of $f(\xi)$ is a closed connected compact set $R \subset B(\Delta)$. If $\Phi = \bigcup \sigma x$, $x \in R$, then Φ is a closed subset of Δ and if Γ is an oriented envelope of Φ in Δ, then $R[\alpha, f(\xi)]$ is a continuous function of (α, ξ) for $\alpha \in \Gamma$, $0 \leqq \xi \leqq \omega$. There is consequently a finite positive constant $M = M(\Gamma)$ such that

$$|R[\alpha, f(\xi)]| \leqq M(\Gamma), \qquad \alpha \in \Gamma, 0 \leqq \xi \leqq \omega. \tag{5}$$

Further, $\lim_{\xi \to \eta} R[\alpha, f(\xi)] = R[\alpha, f(\eta)]$ uniformly with respect to α on Γ. It follows that uniformly in ζ, $0 \leqq \zeta \leqq \omega$,

$$\begin{aligned}
&\lim_{\xi \to \eta} Q[f(\xi), f(\eta), f(\zeta)] \\
&= -\frac{1}{8\pi^3 i}\int_\Gamma\int_\Gamma\int_\Gamma Q[\alpha, \beta, \gamma]R[\alpha, f(\eta)]R[\beta, f(\eta)]R[\gamma, f(\zeta)]d\alpha d\beta d\gamma \\
&= \frac{1}{8\pi^3 i}\int_\Gamma\int_\Gamma\int_\Gamma Q(\alpha, \beta, \gamma)\frac{R[\alpha, f(\eta)] - R[\beta, f(\eta)]}{\alpha - \beta}R[\gamma, f(\zeta)]d\alpha d\beta d\gamma \\
&= \frac{1}{8\pi^3 i}\int_{\Gamma_1}\int_{\Gamma_1} d\beta d\gamma R[\gamma, f(\zeta)]\left\{\int_\Gamma \frac{Q(\alpha, \beta, \gamma)R[\alpha, f(\eta)]}{\alpha - \beta}d\alpha \right. \\
&\qquad\qquad \left. - R[\beta, f(\eta)]\int_\Gamma \frac{Q(\alpha, \beta, \gamma)}{\alpha - \beta}d\alpha\right\}.
\end{aligned}$$

Here Γ, Γ_1, are oriented envelopes of Φ in Δ and Γ_1 is interior to Γ. Now

$$\int_\Gamma \frac{Q(\alpha, \beta, \gamma)}{\alpha - \beta}d\alpha = 2\pi i Q(\beta, \beta, \gamma) = 2\pi i G_1(\beta, \gamma)$$

so

$$\begin{aligned}
&\lim_{\xi \to \eta} Q[f(\xi), f(\eta), f(\zeta)] \\
&\qquad = \frac{1}{(2\pi i)^2}\int_{\Gamma_1}\int_{\Gamma_1} G_1(\beta, \gamma)R[\beta, f(\eta)]R[\gamma, f(\zeta)]d\beta d\gamma + U \\
&\qquad = G_1[f(\eta), f(\zeta)] + U.
\end{aligned}$$

Here

$$\begin{aligned}
U &= \frac{1}{8\pi^3 i}\int_{\Gamma_1}\int_{\Gamma_1} d\beta d\gamma R[\gamma, f(\zeta)]\int_\Gamma \frac{Q(\alpha, \beta, \gamma)}{\alpha - \beta}R[\alpha, f(\eta)]d\alpha \\
&= \frac{1}{8\pi^3 i}\int_{\Gamma_1} d\gamma R[\gamma, f(\zeta)]\int_\Gamma d\alpha R[\alpha, f(\eta)]\int_{\Gamma_1}\frac{Q(\alpha, \beta, \gamma)}{\alpha - \beta}d\beta.
\end{aligned}$$

But for $\alpha \in \Gamma$ the last integral is zero so $U=0$. This completes the proof of (3). Equation (4) may be proved in a fashion analogous to the method used for establishing the multiplicative law in the operational calculus involving functions of one variable.

We now proceed to the proof of the theorem. From the contour integral definition of $G_1[f(0), f(\zeta)]$ we see that it is continuous in ζ and hence

$$\lim_{\alpha\to 0} \frac{1}{\alpha}\int_0^\alpha G_1[f(0), f(\zeta)]d\zeta = G_1[f(0), f(0)].$$

Thus we may fix $\alpha<\omega$ so that the integral on the left has an inverse in B. From the lemma we have

$$\lim_{\xi\to 0} \frac{1}{\alpha}\int_0^\alpha Q[f(\xi), f(0), f(\zeta)]d\zeta = \frac{1}{\alpha}\int_0^\alpha G_1[f(0), f(\zeta)]d\zeta.$$

Hence

$$\lim_{\xi\to 0} \left\{\frac{1}{\alpha}\int_0^\alpha Q[f(\xi), f(0), f(\zeta)]d\zeta\right\}^{-1} = \left\{\frac{1}{\alpha}\int_0^\alpha G_1[f(0), f(\zeta)]d\zeta\right\}^{-1}. \tag{6}$$

From (4) we have

$$[f(\xi) - f(\eta)]Q[f(\xi), f(\eta), f(\zeta)] = G[f(\xi), f(\zeta)] - G[f(\eta), f(\zeta)] \tag{7}$$

whence

$$\begin{aligned} \frac{1}{\xi}[f(\xi) - f(0)]\frac{1}{\alpha}\int_0^\alpha Q[f(\xi), f(0), f(\zeta)]d\zeta &= \frac{1}{\alpha}\int_0^\alpha \frac{1}{\xi}\{G[f(\xi), f(\zeta)] - G[f(0), f(\zeta)]\}d\zeta \\ &= \frac{1}{\alpha\xi}\int_0^\alpha [f(\xi+\zeta) - f(\zeta)]d\zeta \\ &= \frac{1}{\alpha\xi}\left\{\int_\alpha^{\alpha+\xi} f(\tau)d\tau - \int_0^\xi f(\tau)d\tau\right\} \\ &\to \frac{1}{\alpha}[f(\alpha) - f(0)] \qquad \text{as } \xi\to 0. \end{aligned} \tag{8}$$

Thus (6) and (8) give the existence of

$$\lim_{\xi\to 0}\frac{1}{\xi}[f(\xi)-f(0)] = \left\{\frac{1}{\alpha}\int_0^\alpha G_1[f(0), f(\zeta)]d\zeta\right\}^{-1}\frac{1}{\alpha}[f(\alpha)-f(0)].$$

Thus $f(\xi)$ is differentiable at $\xi=0$. Applying the lemma once more we have

$$\begin{aligned}\frac{1}{\eta}[f(\xi+\eta)-f(\xi)] &= \frac{1}{\eta}\{G[f(\eta), f(\xi)] - G[f(0), f(\xi)]\}\\ &= \frac{1}{\eta}[f(\eta)-f(0)]Q[f(\eta), f(0), f(\xi)]\\ &\to f'(0)G_1[f(0), f(\xi)]\end{aligned}$$

uniformly for $0\leqq\xi\leqq\omega$. The existence of the higher derivatives is readily established. We shall indicate the argument for the case of the second derivative. We have

$$f'(\xi) = f'(0)\frac{1}{(2\pi i)^2}\int_\Gamma\int_\Gamma G_1[\alpha, \beta]R[\alpha, f(0)]R[\beta, f(\xi)]d\alpha d\beta.$$

It is readily shown that uniformly for β on Γ we have as $\eta\to 0$

$$\frac{1}{\eta}\{R[\beta, f(\xi+\eta)] - R[\beta, f(\xi)]\} \to -f'(\xi)\{R[\beta, f(\xi)]\}^2.$$

Hence

$$f''(\xi) = -f'(0)f'(\xi)\frac{1}{(2\pi i)^2}\int_\Gamma\int_\Gamma G_1(\alpha, \beta)R[\alpha, f(0)]R[\beta, f(\xi)]^2 d\alpha d\beta$$

and a contour integral argument of familiar type shows that this expression equals $f'(0)f'(\xi)G_{11}[f(0), f(\xi)]$ where $G_{11}[\alpha, \beta] = (\partial^2/\partial\alpha\partial\beta)G[\alpha, \beta]$ and $G_{11}(u, v)$ is defined by the usual contour integral for $u, v\in B(\Delta)$. We see in particular that $f''(0) = [f'(0)]^2G_{11}[f(0), f(0)]$ and hence is uniquely determined by $f(0)$ and $f'(0)$. Similarly it may be shown that all higher derivatives exist and are uniquely determined by $f(0)$ and $f'(0)$.

Now suppose that $g(\xi)$ is another continuous solution of equation (1) with $g(0)=f(0)$ and $g'(0)=f'(0)$. From the preceding remarks we see that $g^{(n)}(0)=f^{(n)}(0)$, $n=1, 2, \cdots$. From the contour integral representation of the function $Q[f(\xi), g(\xi), f(\eta)]$ combined with (5) we see that there is a finite $K=K(f, g)$ such that for $0\leqq\xi, \eta\leqq\omega$

$$|Q[f(\xi), g(\xi), f(\eta)]| \leqq K, \qquad |Q[g(\eta), f(\eta), g(\xi)]| \leqq K.$$

Since

$$G[f(\xi), f(\eta)] - G[g(\xi), f(\eta)] = Q[f(\xi), g(\xi), f(\eta)][f(\xi) - g(\xi)]$$

we have

$$(9) \qquad |G[f(\xi), f(\eta)] - G[g(\xi), f(\eta)]| \leqq K | [f(\xi) - g(\xi)] |$$

and a similar inequality in which f, g and ξ, η are interchanged. Placing $h(\xi) = f(\xi) - g(\xi)$ we have

$$\begin{aligned} h(\xi + \eta) &= f(\xi + \eta) - g(\xi + \eta) \\ &= G[f(\xi), f(\eta)] - G[g(\xi), g(\eta)] \\ &= G[f(\xi), f(\eta)] - G[g(\xi), f(\eta)] \\ &\quad + G[g(\xi), f(\eta)] - G[g(\xi), g(\eta)]. \end{aligned}$$

Hence from (9) we have

$$| h(\xi + \eta) | \leqq K\{ | f(\xi) - g(\xi) | + | f(\eta) - g(\eta) | = K\{ | h(\xi) | + | h(\eta) | \}$$

whence

$$| h(2\xi) | \leqq 2K | h(\xi) |.$$

By repeated use of this inequality we get

$$(10) \qquad h(\xi) \leqq (2K)^m | h(\xi 2^{-m}) |, \qquad m = 1, 2, 3, \cdots.$$

Now consider

$$g_n(\xi) = g(\xi) - \sum_{\nu=0}^{n} g^{(\nu)}(0) \frac{\xi^\nu}{\nu!}$$

and let

$$M_n[g] = \max | g_n^{(n)}(\xi) | = \max | g^{(n)}(\xi) - g^{(n)}(0) |$$

in the interval $[0, \omega]$. Since

$$g_n(\xi) = \int_0^\xi \cdots \int_0^{\xi_{n-2}} \int_0^{\xi_{n-1}} g_n^{(n)}(\xi_n) d\xi_n d\xi_{n-1} \cdots d\xi_1,$$

we have

$$| g_n(\xi) | \leqq \frac{\xi^n}{n!} M_n[g], \qquad 0 \leqq \xi \leqq \omega.$$

Since

$$| h(\xi) | = | f(\xi) - g(\xi) | = | f_n(\xi) - g_n(\xi) | \leqq | f_n(\xi) | + | g_n(\xi) |,$$

we see that $|h(\xi)| \leqq 2\xi^n M_n/n!$, where M_n is the larger of $M_n[f]$ and

$M_n[g]$. A combination of this inequality with (10) yields

$$|h(\xi)| \leqq \frac{2}{n!} M_n \omega^n (2^{1-n}K)^m, \qquad 0 \leqq \xi \leqq \omega,\ m, n = 1, 2, \cdots.$$

Here we fix n so that $2^{1-n}K<1$. Since m is independent of n and may be taken arbitrarily large we see that $h(\xi)\equiv 0$ and thus that $g(\xi)=f(\xi)$, $0\leqq\xi\leqq\omega$.

Suppose now that $\phi(\xi)$ is a nonconstant analytic scalar solution of (1). Since ϕ is analytic at $\zeta=0$ we have $\phi(\zeta)=\sum\alpha_n\zeta^n|\zeta|<\rho$ and

$$\phi(\zeta_1+\zeta_2) = G[\phi(\zeta_1), \phi(\zeta_2)], \qquad |\zeta_1|, |\zeta_2|, |\zeta_1+\zeta_2| < \rho. \tag{11}$$

Differentiating $\phi(\zeta)=G[\phi(\zeta), \phi(0)]$ we get $\phi'(\zeta)=\phi'(\zeta)G_1[\phi(\zeta), \phi(0)]$ and since $\phi'(\zeta)\not\equiv 0$ we have $G_1[\phi(\zeta), \phi(0)]\equiv 1$ and, in particular,

$$G_1[\phi(0), \phi(0)] = 1.$$

Differentiating (11) by parts with respect to ζ_1 and putting $\zeta_1=0$ in the result one gets

$$\phi'(\xi) = \phi'(0)G_1[\phi(0), \phi(\xi)]$$

whence $\phi'(0)\neq 0$. But if $\phi(\zeta)$ satisfies the conditions stated so does $\phi(\alpha\zeta)$ for any $\alpha\neq 0$. We can consequently normalize $\phi(\zeta)$ by assuming that $\phi'(0)=1$. Suppose now that $f(\xi)$ is a continuous, and hence differentiable, solution of (1) such that $f(0)=\phi(0)e$, $f'(0)=a$. Then $G_1[f(0), f(0)]=G_1[\phi(0), \phi(0)]e=e$ and hence it has an inverse. On the other hand the function $\phi(a\zeta)$ is given by the series $\sum\alpha_n a^n\zeta^n$ at least for $|\zeta|<\rho/|a|$ and for such values of ζ we have also $\sigma\phi(a\zeta)=\phi(\zeta\sigma a)$ $\subset\Delta$ since $|\sigma a|\leqq|a|$. From the construction of the series and the properties mentioned it follows that it satisfies (1) for $|\zeta_1|$, $|\zeta_2|$, $|\zeta_1+\zeta_2|<\rho/|a|$. Further, $\phi(0)=\phi(0)e=f(0)$, $\phi'(0)=a=f'(0)$ and thus $f(\xi)=\phi(a\xi)$.

THE INSTITUTE FOR ADVANCED STUDY AND
YALE UNIVERSITY

LIE THEORY OF SEMI-GROUPS OF LINEAR TRANSFORMATIONS

EINAR HILLE

1. **Introduction.** In the Colloquium Lectures which I had the honor of delivering to the Society at the Wellesley meeting in August 1944, an outline was given of a theory of *one-parameter semi-groups of linear bounded operators on a complex* (B)*-space* $\mathfrak{X}$ *to itself.* The problem here is the study of a family of linear bounded transformations $\mathfrak{S}=\{T(\alpha)\}$, defined for $\alpha>0$, with the product law

$$T(\alpha)T(\beta) = T(\alpha+\beta). \tag{1.1}$$

Such families arise in the most varied branches of classical and of modern analysis and are interesting for their own sake as well as for the many applications.

An extension to the n-parameter case was presented to the Society in October 1944 (abstract 51-1-15). Here the parameter $a=(\alpha_1, \alpha_2, \cdots, \alpha_n)$ is a vector in n-dimensional real euclidean space E_n, the operators $T(a)$ are defined for non-negative values of the components of a, and the *product law* reads

$$T(a)T(b) = T(a+b) \tag{1.2}$$

with $a+b=(\alpha_1+\beta_1, \cdots, \alpha_n+\beta_n)$. These operators commute. If $\|T(a)\|$ is bounded for small a, if certain unions of range spaces $T(a)[\mathfrak{X}]$ are dense in the space $\mathfrak{X}$, and if $T(a)$ is a strongly measurable function of a, then $T(a)$ is actually strongly continuous for all a and $T(h)x\to x$ for each x when $h\to 0$. Further $T(a)$ is the direct product of n commuting one-parameter semi-groups

$$T(a) = T_1(\alpha_1)T_2(\alpha_2)\cdots T_n(\alpha_n). \tag{1.3}$$

It turned out later that the analysis could be extended, at least in part, to the case in which the parameter set is an *open positive cone* $\mathfrak{C}$ in a (B)-space $\mathfrak{P}$. Here $\mathfrak{C}$ is an open set, if a and b are in $\mathfrak{C}$ so are $\alpha a+\beta b$ for $0\leqq\alpha, 0\leqq\beta, 0<\alpha+\beta$. The product law is still given by (1.2).

These investigations with many extensions and numerous applications have now appeared in book form ([**6**] in the References at the end of this address). The earliest results on continuity in the one-parameter case are due to N. Dunford [**2**] and extensions to the

---

The Retiring Presidential address delivered before the Annual Meeting of the Society in Columbus, Ohio, on December 29, 1948; received by the editors December 7, 1948.

n-parameter case have also been found by N. Dunford and I. E. Segal [**3**]. A number of results referring to the one-parameter case with applications to stochastic processes and Brownian motion have been found independently by K. Yosida whose work is now in course of publication [**10, 11, 12**].

So far only commutative operators have been considered and the product law (1.2) is the simplest possible. The non-commutative case has resisted numerous attacks in the past and it is only a few months ago that any headway was made with this problem. I shall have the pleasure of outlining the new theory here; it is a blend of the classical theory of Lie groups with the recent theory of one-parameter semi-groups.

2. **Assumptions.** We shall be concerned with an n-parameter family $\mathfrak{S}$ of linear bounded operators $T(a)$ on a (B)-space $\mathfrak{X}$ to itself. These operators shall form a semi-group, that is, if $T(a)$ and $T(b)$ are in $\mathfrak{S}$ so are their products $T(a)T(b)$ and $T(b)T(a)$ which ordinarily are distinct.

In a first study of the problem, we are entitled to restrict ourselves to comparatively simple situations. Our assumptions will be chosen accordingly, but they will be introduced as needed so as to bring out what parts of the theory require heavy machinery and what may be proved under less restrictive assumptions. No claim is made that the assumptions have their definitive form.

We are dealing with three different spaces, the parameter, the operand, and the operator spaces, and a more or less complicated product law. Consequently we shall need assumptions of diverse nature which may be classified under the following four headings:

(1) The parameter set, that is, the values of a for which $T(a)$ is defined.

(2) The properties of $T(a)$ for fixed a.

(3) The properties of $T(a)$ as function of a.

(4) The product law, that is, the function

$$c = F(a, b) \tag{2.1}$$

which is defined by the relation

$$T(c) = T(a)T(b). \tag{2.2}$$

The assumptions relating to (1) and (2) will be kept fixed throughout the following discussion. They are:

A[1]. *$T(a)$ is defined for $a = (\alpha_1, \cdots, \alpha_n)$ in $\overline{E}_n^+$, that part of E_n in which $\alpha_j \geqq 0$, $j = 1, \cdots, n$.*

A^2. *$T(a)$ is a linear bounded operator on $\mathfrak{X}$ to itself, $T(0)=I$, the identity operator, and $\|T(a)\| \leq 1$ for a in $\overline{E}_n^+$.*

The initial assumption under (3) is:

A_1^3. *$T(a)$ is a strongly measurable function of a in $\overline{E}_n^+$.*

The product law will require a number of different assumptions; we start with the following which will be used in the next section. The Euclidean length of the vector b is denoted by $|b|$.

A_1^4. *$F(a, b)$ is a continuous function on $\overline{E}_n^+ \times \overline{E}_n^+$ to $\overline{E}_n^+$ and*

$$F(a, 0) = F(0, a) = a, \tag{2.3}$$

$$F(a, F(b, c)) = F(F(a, b), c). \tag{2.4}$$

A_2^4. *To every $R>0$ there is a positive $\delta=\delta(R)$ such that $F(a, h_1) \neq F(a, h_2)$ when $h_1 \neq h_2$ provided $|a|<R$, $|h_1|<\delta$, $|h_2|<\delta$.*

A_3^4. *To every bounded set K whose closure is in $E_n^+ (\alpha_j>0, j=1, \cdots, n)$ there is a positive $\delta=\delta(K)$ such that for $c \in K$, $|h|<\delta$, the equation $F(h, b)=c$ has a unique solution $b=\psi(c, h)$ in E_n^+ which is a continuous function of (c, h) such that for fixed values of c measurable sets correspond to measurable sets.*

Some comments are in order at this juncture. We are concerned with a full semi-group and not with a semi-group germ. Assumption A^1 is then a natural generalization from one to n dimensions, but it should be realized that *a parameter set is admissible if and only if it is closed under the product operation.* Thus A^1 implies a restriction on the product law and for other choices of $F(a, b)$ we may have to consider other configurations in E_n besides $\overline{E}_n^+$. The complex euclidean space should also be considered. Actually some portions of the theory extend without material change to the case in which the parameter set is the closure of an open positive cone in an arbitrary complex (B)-space $\mathfrak{P}$.

The boundedness assumption in A^2 is very convenient for a first study of the problem. It is a relict, however, of the days when unitary operators in a Hilbert space stood in the foreground and it has the disadvantage of obscuring the fact that the norm of a semi-group operator, while bounded on bounded sets having a positive distance from the boundary of the parameter set, may very well become unbounded when the parameter approaches the boundary. Such questions will have to be relegated to a later study, however.

Assumption A_1^3 will be discussed in the next section. Condition (2.3) expresses that $T(a)I=IT(a)=T(a)$ and (2.4) implies the *associative law* $T(a)[T(b)T(c)]=[T(a)T(b)]T(c)$. Assumption A_1^4 alone does not take us anywhere and has to be supplemented by other conditions; A_2^4 and A_3^4 suffice for questions of continuity but for the existence of one-parameter sub-semi-groups we shall need Lipschitz conditions and so on.

3. **Continuity.** In the case of a one-parameter semi-group boundedness of the norm together with strong measurability of the operator function $T(\alpha)$ implies strong continuity of $T(\alpha)$ for $\alpha>0$.[1] It does not imply strong continuity at the origin or continuity in the uniform operator topology for $\alpha>0$. This result extends to the present situation.

THEOREM 3.1. *Under the assumptions of* §2, *omitting* A_2^4, *the operator* $T(a)$ *is strongly continuous in* E_n^+, *that is, for* $\alpha_j>0$, $j=1, \cdots, n$.

The proof uses the same principles as in the one-dimensional case, that is, the product law plus continuity of a definite integral with respect to translations. Let D be a bounded domain the closure of which is in E_n^+. By virtue of A_3^4 there exists a positive $\eta=\eta(D)$ such that for each c in D and each $a=(\alpha_1, \cdots, \alpha_n)$ with $0\leqq\alpha_j\leqq\eta$, $j=1, \cdots, n$, the equation

$$c = F(a, b) \tag{3.1}$$

has a unique solution

$$b = \psi(c, a) \tag{3.2}$$

which is a continuous function of (c, a) in the product set in question. In particular, $|\psi(c_1, a)-\psi(c_2, a)|$ is small when $|c_1-c_2|$ is small and this holds uniformly in a.

For c_1 and c_2 in D we have then

$$T(c_1)x - T(c_2)x = T(a)[T(\psi(c_1, a))x - T(\psi(c_2, a))x].$$

The right side being independent of a, we may integrate the identity with respect to a over the cube $C(\eta)$, $\frac{1}{2}\eta\leqq\alpha_j\leqq\eta$, $j=1, \cdots, n$, obtaining

$$(\tfrac{1}{2}\eta)^n[T(c_1)x - T(c_2)x] = \int_{C(\eta)} T(a)[T(\psi(c_1, a))x - T(\psi(c_2, a))x]da$$

whence

$$(\tfrac{1}{2}\eta)^n\|T(c_1)x - T(c_2)x\| \leqq \int_{C(\eta)} \|T(\psi(c_1, a))x - T(\psi(c_2, a))x\|da.$$

[1] R. S. Phillips has recently shown that the boundedness assumption is superfluous and strong measurability alone is necessary and sufficient for strong continuity. [Added in proof March, 1950.]

Since $T(b)$ is a strongly measurable function of b and in the correspondence $b=\psi(c, a)$ measurable b-sets are the images of measurable a-sets, the operator $T(\psi(c, a))$ is strongly measurable in a, so that the integrand is a bounded Lebesgue measurable function of a. From the fact that $\psi(c_2, a)\to\psi(c_1, a)$ when $c_2\to c_1$, the convergence being uniform with respect to a in $C(\eta)$, one may infer that the integral tends to zero when $c_2\to c_1$. It follows that $T(c)$ is strongly continuous at $c=c_1$ and hence everywhere in E_n^+.

Ordinarily we cannot prove continuity in the uniform topology, but if we assume uniform measurability at the outset, then x may be suppressed everywhere in the proof and uniform continuity results.

Assumption A^2 is evidently used twice in the proof, but in both places it could be replaced by the weaker assumption that $\|T(a)\|$ is bounded in every bounded domain whose closure lies in E_n^+. In the case of a separable space $\mathfrak{X}$, weak measurability implies the strong kind so that "strongly" could be replaced by "weakly" in A_1^3. In this case it is also likely that we may dispense with the boundedness condition on the norm altogether, merely assuming $\|T(a)\|$ to be finite in E_n^+. This is suggested by the following considerations.

If $\mathfrak{X}$ is separable and $T(a)$ is weakly measurable in E_n^+, then $\log \|T(a)\|$ is measurable Lebesgue and different from $+\infty$. Further it satisfies the inequality

$$f(c) \leqq f(a) + f(b), \qquad c = F(a, b), \tag{3.3}$$

which is the proper generalization of the subadditive inequality

$$f(a + b) \leqq f(a) + f(b) \tag{3.4}$$

corresponding to the case $F(a, b)=a+b$. For the latter it is known that a solution, defined in E_n^+, which is measurable and different from $+\infty$, is bounded above in every bounded interior domain D whose closure lies in E_n^+ (see [6, p. 135] for the case $n=1$, the extension to arbitrary n has been given by R. A. Rosenbaum). In principle the method of the proof also extends to (3.3) but the analytical and topological difficulties are considerable so the discussion of this question has to be postponed to another occasion.

One can get fairly trivial examples indicating that $T(a)$ need not be continuous on the boundary of E_n^+ under the assumptions of Theorem 3.1, not even in the weak sense.[2] In particular, continuity is apt to fail at the origin. A common cause of such failure is that the union of the range spaces, $\bigcup_a T(a)[\mathfrak{X}]$, is not dense in $\mathfrak{X}$. In the one-

[2] In the one-parameter case it may be shown that weak continuity to the right at the origin implies strong continuity. [Added in proof.]

parameter case it would be sufficient to add the assumption that $\bigcup_\alpha T(\alpha)[\mathfrak{X}]$ is dense in $\mathfrak{X}$ in order to obtain right-hand continuity at $\alpha=0$. It is not clear at the time of writing this if such a condition would suffice as additional assumption to ensure continuity of $T(a)$ on the boundary of E_n^+. On the other hand, if $n=1$ the assumption that $T(\alpha)x \to x$ for each x when $\alpha \to 0$ suffices for strong continuity in $\overline{E}_1^+$ and this result extends to the general case. We assume

A_2^3. *$T(a)$ is strongly continuous at $a=0$.*

THEOREM 3.2. *If $T(a)$ satisfies A_2^3 and the conditions of §2, except A_1^3 and A_3^4, then $T(a)$ is strongly continuous in $\overline{E}_n^+$.*

If x and $\epsilon>0$ are given, we can find an η so small that $\|T(h)x-x\| < \epsilon$ for $h \in S(0)=E(|h|<\eta,\ h\in\overline{E}_n^+)$. The mapping

$$T_a:\quad h \to F(a,h), \qquad a \in \overline{E}_n^+,$$

takes $S(0)$ into a set $S(a)$. If $|a_0|<R$ and $\eta<\delta(R)$, as we may assume, condition A_2^4 asserts that the mapping T_{a_0} is one-to-one so that $S(a_0)$ is an n-cell and Int $[S(a_0)]\neq\emptyset$. Thus $S(a_0)$ contains a sphere of center c_0 and radius ρ, say. But $F(a,h)$ is continuous by A_1^4 so there exists a δ_0 such that $|F(a,h)-F(a_0,h_0)|<\rho$ if $|a-a_0|<\delta_0$, $|h-h_0| < \delta_0$ where $h, h_0 \in S(0)$. Here we choose h_0 so that $c_0=F(a_0,h_0)$. We can then be sure that $S(a)\cap S(a_0)\neq\emptyset$ if $|a-a_0|<\delta_0$. Hence if $c=F(a,h_1)=F(a_0,h_2)$ is a common point of $S(a)$ and $S(a_0)$ we have

$$\begin{aligned}\|T(a)x - T(a_0)x\| &\leqq \|T(c)x - T(a)x\| + \|T(c)x - T(a_0)x\| \\ &= \|T(a)[T(h_1)x - x]\| + \|T(a_0)[T(h_2)x - x]\| \\ &\leqq \|T(h_1)x - x\| + \|T(h_2)x - x\| \leqq 2\epsilon\end{aligned}$$

where $|a-a_0|<\delta_0$. This proves that $T(a)x$ is continuous at $a=a_0$. Hence $T(a)$ is strongly continuous in $\overline{E}_n^+$.

4. **A functional equation.** In the study of one-parameter sub-semi-groups we encounter the functional equation

$$g(\rho+\sigma) = F[g(\rho), g(\sigma)] \tag{4.1}$$

where $g(\rho)$ is a function on positive numbers to $\overline{E}_n^+$. For this problem we need further information concerning $F(a,b)$.

A_4^4. *There exists a fixed positive constant B such that for all points in $\overline{E}_n^+$ we have*

$$\begin{aligned}|F(a_1,b) - F(a_2,b)| &\leqq [1+B|b|]\,|a_1-a_2|, \\ |F(a,b_1) - F(a,b_2)| &\leqq [1+B|a|]\,|b_1-b_2|.\end{aligned} \tag{4.2}$$

A_5^4. *There exists a positive monotone increasing continuous function* $\omega(\xi)$, $0<\xi<\infty$, *which tends to zero with* ξ, *such that*

$$(4.3)\qquad \begin{aligned} F(a, b) &= a + b + G(a, b), \\ |G(a, b)| &\leq r\omega[|a + b|], \qquad r = \min(|a|, |b|). \end{aligned}$$

It would be possible to combine these inequalities in such a manner that they refer to the behavior of $|F(a, b_1)-F(a, b_2)-b_1+b_2|$ and $|F(a_1, b)-F(a_2, b)-a_1+a_2|$. Such inequalities are basic in the study of analytical group germs in a (B)-space due to G. Birkhoff [1]. Our conditions seem to be slightly better adapted to the needs of the methods used below. They are also closely related to the assumptions of P. A. Smith [8, 9]. See further I. E. Segal [7] whose work suggests that the Lipschitz condition might be inessential but it is not clear to me at the moment how his methods could be brought to bear on the present problem. In making comparisons the reader should keep in mind that we are dealing with a fixed coordinate system, a semi-group rather than a group, and a situation in the large. All the results of the present section hold for the case in which E_n is replaced by an arbitrary (B)-space and $\overline{E}_n^+$ by the closure of an open positive cone.

With an arbitrary element $b\in\overline{E}_n^+$ we also consider its successive "powers" defined by

$$(4.4)\qquad b^{(1)} = b, \quad b^{(m)} = F(b, b^{(m-1)}).$$

When b and m are bulky expressions we shall write $(b; m)$ for $b^{(m)}$. A basic property of the powers is given in the following

LEMMA. *Let* τ *be defined by* $\omega(\tau)=\frac{1}{2}$. *If* $b\in\overline{E}_n^+$ *and* m *is a positive integer such that* $m|b|\leq\frac{1}{2}\tau$, *then*

$$(4.5)\qquad b^{(m)} = mb + R_m(b), \qquad |R_m(b)| \leq m|b|\omega(2m|b|).$$

The lemma holds for $m=2$ by A_5^4 and is proved by induction using the inequality

$$|R_{m+1}(b)| \leq |R_m(b)| + |b|\omega[(k+1)|b| + |R_m(b)|].$$

We can now state and prove

THEOREM 4.1. *Under the assumptions* A_1^4, A_4^4, *and* A_5^4 *the equation* (4.1) *with the initial condition*

$$(4.6)\qquad \lim_{\rho\to 0} \rho^{-1}g(\rho) = a \in \overline{E}_n^+$$

has a unique solution $g(\rho)=f(\rho a)$ *in* $\overline{E}_n^+$. *The solution is an absolutely continuous function of* ρ *with bounded derived numbers and for* $0 \leqq \rho \leqq \sigma$

$$| g(\sigma) - g(\rho) | \leqq B^{-1}[e^{B\sigma|a|} - e^{B\rho|a|}]. \tag{4.7}$$

Starting with an arbitrary element b of $\overline{E}_n^+$ we form a sequence of powers $(b/\nu)^{(\mu)} = (b/\nu;\ \mu)$ where μ and ν are positive integers. Using A_4^4 and induction on μ one shows that

$$| (b/\nu;\ \mu + 1) - (b/\nu;\ \mu) | \leqq \frac{|b|}{\nu}\left\{1 + \frac{B|b|}{\nu}\right\}^{\mu},$$

whence we obtain

$$| (b/\nu;\ \mu) | \leqq B^{-1}[e^{(\mu/\nu)B|b|} - 1] \tag{4.8}$$

for all μ, ν.

In the following $\mu = \nu$ will be a power of 2 and we shall investigate the convergence of the sequence $\{(2^{-j}b;\ 2^j)\}$. Here $(2^{-j}b;\ 2^j)$ is obviously the square of $(2^{-j}b;\ 2^{j-1})$. Using this fact, (4.8), and A_4^4 repeatedly, we obtain for $j < k$

$$\begin{aligned} \Delta_{j,k}(b) &= | (2^{-k}b;\ 2^k) - (2^{-j}b;\ 2^j) | \\ &\leqq 2^j e^{B|b|} | (2^{-k}b;\ 2^{k-j}) - 2^{-j}b |. \end{aligned}$$

Let R be a fixed positive number, arbitrarily large, and restrict b to the sphere $|b| \leqq R$. Suppose that j is so large that $2^{1-j}R \leqq \tau$. We can then apply the lemma with b replaced by $2^{-k}b$ and m by 2^{k-j}. After some simplification we obtain

$$\Delta_{j,k}(b) \leqq |b| e^{B|b|}\omega(2^{1-j}|b|). \tag{4.9}$$

It follows that the sequence $\{(2^{-j}b;\ 2^j)\}$ converges to a limit and we set

$$f(b) = \lim_{j\to\infty} (2^{-j}b;\ 2^j). \tag{4.10}$$

The convergence being uniform with respect to b for $|b| \leqq R$, we conclude that $f(b)$ is a continuous function of b.

In particular the limit exists for $b = \rho a$, uniformly with respect to ρ for $0 \leqq \rho \leqq R < \infty$ so that $f(\rho a)$ is a continuous function of ρ.

From (4.10) we conclude also that

$$f(b;\ m) \equiv \lim_{j\to\infty} (2^{-j}b;\ m2^j) \tag{4.11}$$

exists and equals the mth power of $f(b)$. But this implies that

$$\lim_{j\to\infty} (2^{-j}m^{-1}b;\ m2^j) = f(bm^{-1};\ m)$$

exists. We want to show that the latter limit is actually independent of m and consequently equals $f(b;\ 1) = f(b)$. For this purpose we consider

$$\Delta_{k,m}(b, j) = \left| (2^{-j}k^{-1}b;\ k2^j) - (2^{-j}m^{-1}b;\ m2^j) \right|.$$

Here we use the fact that the terms on the right are squares together with A_4^4 and (4.8) to obtain the estimate

$$\Delta_{k,m}(b, j) \leqq 2e^{\frac{1}{2}B|b|}\Delta_{k,m}(2^{-1}b, j-1) \leqq \cdots$$
$$\leqq 2^j e^{B|b|}\Delta_{k,m}(2^{-j}b, 0).$$

If $2^{1-j}|b| \leqq \tau$, we can apply the lemma once more and see that

$$2^j\Delta_{k,m}(2^{-j}b, 0) < 2\omega(2^{1-j}|b|) \to 0$$

when $j \to \infty$. It follows that

$$f(bk^{-1};\ k) = f(bm^{-1};\ m) = f(b;\ 1) = f(b)$$

as asserted. Hence with $b = ka$ we have $f(a;\ k) = f(ka)$ for every positive integer k.

From this it follows that $f(\rho a)$ satisfies (4.1), to start with for positive integral values of ρ and σ which implies that it also holds for rational values and finally, $f(\rho a)$ being continuous, for all positive real values.

The argument used for the convergence proof also gives the inequality

$$\left| (2^{-k}b;\ 2^k) - (2^{-k}c;\ 2^k) \right| \leqq 2^k e^{B|c|} \left| 2^{-k}b - 2^{-k}c \right|$$
$$= e^{B|c|} \left| b - c \right|$$

if $|b| \leqq |c|$. Passing to the limit with k we obtain the Lipschitz condition

$$(4.12) \qquad \left| f(b) - f(c) \right| \leqq e^{B|c|} \left| b - c \right|, \qquad |b| \leqq |c|.$$

It follows that $f(\rho a)$ satisfies a Lipschitz condition with respect to ρ. Formula (4.7) is an immediate consequence of (2.3), (4.1), A_4^4, and the estimate

$$(4.13) \qquad \left| f(b) \right| \leqq B^{-1}\left[e^{B|b|} - 1\right]$$

which follows from (4.8).

In order to verify the initial condition we revert to the lemma once more. We have

$$(2^{-j}\rho a;\ 2^j) = \rho a + R_{2^j}(2^{-j}\rho a)$$

and if $2\rho|a| \leqq \tau$ the norm of the remainder does not exceed $\rho|a|\omega(2\rho|a|)$. When j tends to infinity, the left member tends to $f(\rho a)$. It follows that

$$|f(\rho a) - \rho a| \leqq \rho|a|\omega(2\rho|a|) \tag{4.14}$$

for $2\rho|a| \leqq \tau$ so the initial condition is satisfied.

For the uniqueness proof we shall need only A_4^4. Suppose that $g(\rho)$ and $h(\rho)$ are two solutions of (4.1) and (4.6) in $\overline{E}_n^+$. We may assume that they both satisfy (4.13) with b replaced by ρa; if necessary we restrict ρ to a fixed finite interval and replace B by a larger constant. We have then by the usual square root argument

$$\begin{aligned} |g(\rho) - h(\rho)| &= |F[g(\tfrac{1}{2}\rho), g(\tfrac{1}{2}\rho)] - F[h(\tfrac{1}{2}\rho), h(\tfrac{1}{2}\rho)]| \\ &\leqq 2e^{\frac{1}{2}B\rho|a|}|g(\tfrac{1}{2}\rho) - h(\tfrac{1}{2}\rho)|, \end{aligned}$$

whence by iteration

$$|g(\rho) - h(\rho)| \leqq 2^k e^{B\rho|a|}|g(2^{-k}\rho) - h(2^{-k}\rho)|$$

and this tends to zero when $k \to \infty$ by virtue of the common initial condition. Hence $g(\rho) \equiv h(\rho)$ and our sketch of the proof of Theorem 4.1 is complete.

This theorem may be strengthened in a direction which will be useful below.

THEOREM 4.2. *If $\omega(\xi)$ satisfies the integrability condition*

$$\int_0^1 \xi^{-1}\omega(\xi)d\xi < \infty, \tag{4.15}$$

and if $g(\rho)$ is a solution of (4.1) in $\overline{E}_n^+$ which tends to zero with ρ, then there exists an a in $\overline{E}_n^+$ such that (4.6) holds.

Suppose that ρ_0 is so small that $2|g(\rho)| \leqq \tau$ for $\rho \leqq \rho_0$. Repeated use of A_5^4 gives

$$\begin{aligned} g(\rho) &= 2^k g(2^{-k}\rho) + R_k(\rho), \\ |R_k(\rho)| &\leqq \sum_{\nu=1}^{k} 2^{\nu-1}|g(2^{-\nu}\rho)|\omega[2|g(2^{-\nu}\rho)|]. \end{aligned}$$

On the other hand

$$\begin{aligned} |g(\rho)| &\geqq \{1 - \tfrac{1}{2}\omega[2|g(2^{-1}\rho)|]\}2|g(2^{-1}\rho)| \geqq \cdots \\ &\geqq \prod_{j=1}^{\nu}\{1 - \tfrac{1}{2}\omega[2|g(2^{-j}\rho)|]\}2^\nu|g(2^{-\nu}\rho)|. \end{aligned} \tag{4.16}$$

It follows that

$$| R_k(\rho) | \leqq \tfrac{1}{2} | g(\rho) | \sum_{\nu=1}^{k} \prod_{j=1}^{\nu} \{1 - \tfrac{1}{2}\omega[2 | g(2^{-j}\rho) |]\}^{-1}\omega[2 | g(2^{-\nu}\rho) |].$$

Here the right-hand member tends to a finite limit when $k \to \infty$ if and only if

$$\sum_{\nu=1}^{\infty} \omega[2 | g(2^{-\nu}\rho) |] < \sum_{\nu=1}^{\infty} \omega[(\tfrac{2}{3})^{\nu}\tau]$$

is convergent where the inequality is a trivial consequence of (4.16). But the second series converges by virtue of (4.15). It follows that $R_k(\rho)$ converges uniformly to a continuous limit for $0 \leqq \rho \leqq \rho_0$ when $k \to \infty$ and this implies that

$$\lim_{k \to \infty} 2^k g(2^{-k}\rho) \equiv l(\rho) \tag{4.17}$$

exists as a continuous function of ρ. Furthermore, one sees that if $l(\rho) = \rho m(\rho)$ then $m(2\rho) = m(\rho)$.

In exactly the same manner one proves that

$$\lim_{k \to \infty} 3^k g(3^{-k}\rho)$$

exists and a more detailed analysis shows that it also equals $l(\rho)$. This forces $m(\rho)$ to have the additional property $m(3\rho) = m(\rho)$, and log 2 and log 3 being incommensurable, this makes the continuous function $m(\rho)$ equal to a constant, a say. Since $g(\rho) \in \overline{E}_n^+$ by assumption, a will have the same property. From $l(\rho) = \rho a$, we conclude that (4.6) holds.

Condition (4.15) is not particularly restrictive and it is fully utilized in the proof of the theorem, but we have of course no evidence whatsoever that it is a necessary condition.

5. **The one-parameter sub-semi-groups.** In view of the discussion in §3, it is natural to replace A_1^3 by

A_3^3. *$T(a)$ is a strongly continuous function of a in $\overline{E}_n^+$.*

The basic theorem on sub-semi-groups reads

THEOREM 5.1. *Let the semi-group $\mathfrak{S} = \{T(a)\}$ satisfy assumptions A^1, A^2, A_3^3, A_1^4, A_4^4, and A_5^4. If a is any element of $\overline{E}_n^+$, then the sequence of operators $\{[T(2^{-j}a)]^{2^j}\}$ converges strongly to a limit $S(a) = T[f(a)]$ where $f(a)$ is defined by* (4.10). *The operators $\{S(\rho a)\}$, $0 < \rho < \infty$, form a sub-semi-group $\mathfrak{S}_a$ of $\mathfrak{S}$ and $\mathfrak{S}_{\alpha a} = \mathfrak{S}_a$ if $\alpha > 0$. If* (4.15) *also holds, then conversely if $\mathfrak{T} = \{S(\rho)\}$, $0 < \rho < \infty$, is a sub-semi-group of $\mathfrak{S}$, if*

$S(\rho)=T[g(\rho)]$ and $\lim_{\rho\to 0} g(\rho)=0$, then there exists an $a\in\overline{E}_n^+$ such that $\mathfrak{T}=\mathfrak{S}_a$.

The proof follows directly from Theorem 4.1 and 4.2. Referring back to A_1^4 and the definition of the power $(2^{-j}a;\ 2^j)$ one sees that

$$[T(2^{-j}a)]^{2^j} = T[(2^{-j}a;\ 2^j)]. \tag{5.1}$$

By (4.10) we know that $(2^{-j}a;\ 2^j)\to f(a)$ when $j\to\infty$ and by A_3^3

$$T[(2^{-j}a;\ 2^j)]x \to T[f(a)]x \equiv S(a)x \tag{5.2}$$

for each x. That the operators $S(\rho a)$, $0<\rho<\infty$, form a semi-group with the product law $S(\rho a)S(\sigma a)=S((\rho+\sigma)a)$ follows from the fact that $f(\rho a)$ satisfies equation (4.1).

If (4.15) holds, then we can apply Theorem 4.2. If then $\mathfrak{T}=\{S(\rho)\}$ is a one-parameter sub-semi-group of $\mathfrak{S}$ with the canonical product law, we can find a function $g(\rho)$ satisfying (4.1) such that $S(\rho) =T[g(\rho)]$. By assumption $g(\rho)\to 0$ with ρ so there is an a in $\overline{E}_n^+$ such that $g(\rho) =f(\rho a)$ and, hence, $\mathfrak{T}=\mathfrak{S}_a$.

The equation

$$p = f(\rho a), \qquad 0 < \rho < \infty' \tag{5.3}$$

defines a path Γ_a in $\overline{E}_n^+$, starting at the origin where it is tangent to the vector $p=\rho a$. The properties of $f(\rho a)$ listed in Theorem 4.1 show that Γ_a has a tangent almost everywhere and every arc of Γ_a corresponding to $0<\rho<R$ is rectifiable. If $F(p,\ q)$ has continuous partial derivatives with respect to the components of q, we may show that $f(\rho a)=[\phi_1(\rho),\ \cdots,\ \phi_n(\rho)]$ is a solution of the system of first order differential equations

$$\begin{cases} \dfrac{d\phi_j}{d\rho} = \displaystyle\sum_{k=1}^{n} \alpha_k F_{j,k}(\phi_1,\ \cdots,\ \phi_n;\ 0,\ \cdots,\ 0), \\ \phi_j(0) = 0. \end{cases} \qquad i = 1,\ \cdots,\ n, \tag{5.4}$$

Here $F_{j,k}$ is the derivative of the jth component of $F(p,\ q)$ with respect to the kth component of q and $a=(\alpha_1,\ \cdots,\ \alpha_n)$. This system, when available, usually offers more convenient determination of Γ_a than the functional equation (4.1) with the initial condition (4.6). It should be observed, however, that the latter define Γ_a uniquely in situations where the uniqueness theorems for differential equations do not apply and even the differential equations themselves may fail to exist.

Every path Γ_a is confined to $\overline{E}_n^+$ and condition (4.15) may be used to show that Γ_a cannot return to the origin when ρ tends to a finite limit or to infinity. Under the same assumption, Γ_a is a simple arc. The general question of what happens to $f(\rho a)$ when $\rho\to\infty$ is very important. A particularly simple case is that in which $f(\rho a)$ tends to a finite limit p_0 for then $p_0=F(p_0, p_0)$ and $T(p_0)$ is a *projection operator.*

A transformation semi-group, in contrast to a group, may contain projections and their parameters are determined by the equation

$$F(p, p) = p \tag{5.5}$$

which defines a locus P in $\overline{E}_n^+$. P always contains $p=0$ and may reduce to this point. The origin is an isolated point of P since $|F(p, p)-p|>0$ as long as $p\neq 0$ and $\omega(2|p|)<1$ as we see from A_5^4. If a path Γ_a has an interior point $p_0=f(\omega a)$ in common with P, then the operator $T[f(\rho a)]$ is periodic with period ω for $\rho>\omega$. This cannot happen, however, if condition (4.15) holds and if Int $\Gamma_a \epsilon E_n$. A path Γ_a may very well have its end point on P and P may be made up of such terminal points. This happens in the case of the projective semi-group on positive numbers for which all paths Γ_a are straight line segments joining the origin with the surface P which is a portion of a hyperbolic paraboloid in E_3.

The determination of all points p such that $p=f(a)$ for some a appears to be very difficult. It is clear that if $p=f(a)$, then we must be able to determine a sequence of points $\{p_k\}$ in $\overline{E}_n^+$ such that $p_0=p$ and

$$F(p_k, p_k) = p_{k-1}, \qquad k = 1, 2, 3, \cdots. \tag{5.6}$$

Since $p_k=f(2^{-k}a)$ we must also have

$$\lim_{k\to\infty} 2^k p_k = a. \tag{5.7}$$

It is possible to determine conditions under which this process may be carried through, but so far the results have been rather disappointing. In our theory it is much easier to determine the paths from the origin than to find the path, if any, which joins a given point p with the origin.

Let us observe that if $p=f(a)$ then

$$T(p)x = \lim_{\delta\to 0} \exp\left\{\frac{1}{\delta}\left[T(f(\delta a)) - I\right]\right\} x, \qquad x \in \mathfrak{X}, \tag{5.8}$$

in the sense of strong convergence. Cf. [6, p. 189].

6. **The infinitesimal generators.** With each one-parameter sub-semi-group $\mathfrak{S}_a$ defined above there is associated an *infinitesimal gen-*

erator $A(a)$ of $\mathfrak{S}$ which is defined by

$$\lim_{\delta\to 0}\frac{1}{\delta}\,[T(f(\delta a)) - I]x = A(a)x \tag{6.1}$$

whenever the limit exists. The domain of $A(a)$ will be denoted by $\mathfrak{D}(a)$; it is clearly a linear subspace of $\mathfrak{X}$.

THEOREM 6.1. *Under the assumptions of Theorem* 5.1 *the set* $\mathfrak{D}(a)$ *is dense in* $\mathfrak{X}$ *for each* a *in* $\overline{E}_n^+$. *In particular,* $\mathfrak{D}(a)$ *contains all elements of the form*

$$\int_\alpha^\beta T[f(\rho a)]y d\rho, \qquad y\in\mathfrak{X},\ 0\leq\alpha<\beta<\infty. \tag{6.2}$$

This is a well known result in the theory of one-parameter semi-groups (N. Dunford [2], cf. [6, p. 185]). We also observe that for x in $\mathfrak{D}(a)$ we have

$$\frac{d}{d\rho}T[f(\rho a)]x = T[f(\rho a)]A(a)x = A(a)T[f(\rho a)]x. \tag{6.3}$$

The operator $A(a)$ which is closed is ordinarily unbounded on $\mathfrak{D}(a)$. Its resolvent is given by the Laplace transform

$$R[\lambda; A(a)]x = \int_0^\infty e^{-\lambda\rho}T[f(\rho a)]x d\rho, \qquad \Re(\lambda)>0. \tag{6.4}$$

The spectrum of $A(a)$ may very well fill the complementary half-plane $\Re(\lambda)\leq 0$. $\mathfrak{D}(a)$ is the range of $R[\lambda; A(a)]$ for any fixed λ with $\Re(\lambda)>0$. If $R[\lambda; A(a)]$ is known, it determines $T[f(\rho a)]$ uniquely with the aid of the inversion formulas for the Laplace transform.

The mapping $a\to A(a)$ defines a correspondence between the vectors a of $\overline{E}_n^+$ and the infinitesimal generators $A(a)$ of $\mathfrak{S}$. Under suitable assumptions this correspondence is actually an isomorphism under the operations of addition and multiplication by positive numbers. Our previous postulates suffice for the scalar multiplication, however, since (6.1) shows that we have

THEOREM 6.2. *Under the assumptions of Theorem* 5.1 *we have* $\mathfrak{D}(\alpha a)=\mathfrak{D}(a)$ *for* $\alpha>0$ *and* $A(\alpha a)=\alpha A(a)$.

For the addition and for the fundamental theorems we need further restrictions and the following are convenient assumptions.

A_4^3. *There exists a positive* ρ_0 *such that* $|a|<\rho_0$, $|b|<\rho_0$, $a\neq b$ *implies that* $T(a)\neq T(b)$.

A_6^4. *$F(a, b)$ has continuous partial derivatives with respect to the components of a and b up to and including the third order.*

At every point of $\overline{E}_n^+$ the function $F(a, b)$ may then be expanded in a Taylor series up to terms of the third order. Because of A_1^4 these expansions have a special form: at the origin we have in particular

$$
(6.5)\qquad \begin{gathered} \gamma_i = \alpha_i + \beta_i + \sum_j \sum_k \alpha_{jk}^i \alpha_j \beta_k + R_{3,i}, \qquad i = 1, 2, \cdots, n, \\ a = (\alpha_i), \quad b = (\beta_i), \quad c = (\gamma_i), \quad c = F(a, b). \end{gathered}
$$

The crux of the problem before us is to construct elements of $\mathfrak{X}$ dense in $\mathfrak{X}$ belonging to the domains of definitions of $A(a)$ and of $A(a)A(b)$ for every choice of a and b in $\overline{E}_n^+$. For this purpose we shall use a modification of artifices due to N. Dunford [2], I. Gelfand [5], and L. Gårding [4]. The same device gives elements belonging to the domain of existence of the product of three or more infinitesimal generators, provided we assume the existence of enough derivatives.

Let $K(c)$ be a numerically-valued function of class $C^{(m)}$, $m \geqq 3$, defined in $\overline{E}_n^+$. Let D be a bounded domain whose closure lies in E_n^+ and let $K(c)$ be integrable over D and vanish outside of D. If D_η is a homeomorphic image of D in E_n^+ such that no two corresponding points are at a distance of more than η apart, then we suppose that

$$
(6.6)\qquad \int | K(c) | \, dc = o(\eta) \qquad \text{when } \eta \to 0,
$$

where the integration is extended over that part of D which is not in D_η. We require that the partials of order not greater than m have the same integrability properties. The class of all such kernels $K(c)$ will be denoted by $\mathfrak{K}$. Specifically we may choose D as the cube $0<\sigma<\gamma_j<\tau<\infty, j=1, \cdots, n$, and define $K(c)$ in the cube as

$$
(6.7)\qquad \begin{aligned} K(c) &= K(c; \sigma, \tau) \\ &= [C(\tau - \sigma)]^{-n} \exp \left\{ -(\tau - \sigma)^2 \sum_{j=1}^{n} [(\tau - \gamma_j)(\gamma_j - \sigma)]^{-1} \right\} \end{aligned}
$$

where $C = \int_0^1 \exp \{ -[\gamma(1 - \gamma)]^{-1} \} d\gamma$.

For any choice of $K(c)$ in $\mathfrak{K}$ we define

$$
(6.8)\qquad y = K[x] = \int_D K(c) T(c) x \, dc, \qquad x \in \mathfrak{X}.
$$

For fixed $K(c)$ this is a linear bounded transformation of $\mathfrak{X}$ to $\mathfrak{X}$ the norm of which does not exceed the integral of $|K(c)|$ over D. If $K(c)=K(c;\sigma,\tau)$ the norm is at most one.

THEOREM 6.3. *Under the assumptions* A^1, A^2, A_3^3, *and* A_1^4 *to* A_6^4 *the set* $[\bigcap_a \mathfrak{D}(a)] \cap [\bigcap_{a,b} \mathfrak{D}[A(a)A(b)]]$ *is dense in* $\mathfrak{X}$ *and contains the set* $\mathfrak{K}[\mathfrak{X}]$ *of elements of the form* (6.8).

That $\mathfrak{K}[\mathfrak{X}]$ is dense in $\mathfrak{X}$ follows from

$$\lim_{\sigma\to 0}\lim_{\tau\to\sigma} K(c;\sigma,\tau)[x] = x \tag{6.9}$$

in the sense of strong convergence for every x.

We proceed to indicate briefly how one shows that $K[x]\in\mathfrak{D}(a)$ where, without restricting the generality, we may take $|a|=1$. We form

$$\begin{aligned}\frac{1}{\delta}&\{T[f(\delta a)] - I\}y \\ &= \frac{1}{\delta}\int_D K(c)\{T[F(f(\delta a),c)] - T(c)\}xdc.\end{aligned} \tag{6.10}$$

In order to find the limit of the right member, we have to study the mapping U_δ defined by $b=F[f(\delta a),c]$ which takes D into a set $D(\delta)$. Condition A_3^4 asserts that the correspondence is one-to-one for sufficiently small values of δ since $|f(\delta a)|\leqq B^{-1}[e^{B\delta}-1]<2\delta$ if $B\delta<1$. Further

$$|b-c| < (1+B|c|)|f(\delta a)| < M\delta$$

where M depends only upon B and D. The mapping is consequently a homeomorphism involving only a small distortion of D when δ is small. It follows that (6.6) holds with η replaced by δ and D_η by $D(\delta)$.

By A_3^4 we can solve the equation $b=F[f(\delta a),c]$ for c when δ is small obtaining the unique solution $c=\psi(b,f(\delta a))$. This solution is continuous by A_3^4; the added condition A_6^4 also makes it differentiable. It follows from the theorem on implicit functions that the solution $c=\psi(b,s)$ of the equation $F(s,c)=b$ has continuous partial derivatives with respect to the components of b and s of order not greater than 3 which is the limit for the existence of partials of $F(p,q)$ postulated in A_6^4. In this case $f(\delta a)$ has continuous derivatives with respect to δ of order not greater than 3 as is seen from equation (5.4). It follows that $\psi(b,f(\delta a))$ has continuous partials with respect to the components of b and with respect to δ of order not greater than 3. Further, the Jacobian

$$J(c; b) = J[\psi(b, f(\delta a)); b] \tag{6.11}$$

of the inverse transformation U_δ^{-1} from $D(\delta)$ to D is near to one uniformly in D when δ is small and has partial derivatives with respect to the components of b and with respect to δ of order not greater than 2.

From this we conclude that the right member of (6.10) may be written

$$\begin{aligned}
\delta^{-1}\int_{D(\delta)} K(c)T(b)xJ(c; b)db &- \delta^{-1}\int_D K(b)T(b)xdb \\
&= \int_{D_1} \delta^{-1}[K(c)J(c; b) - K(b)]T(b)xdb \\
&\quad + \delta^{-1}\int_{D_2} K(c)J(c; b)T(b)xdb - \delta^{-1}\int_{D_3} K(b)T(b)xdb \\
&= J_1 + J_2 + J_3,
\end{aligned}$$

where $D_1 = D \cap D(\delta)$, $D_2 = D(\delta) - D_1$, $D_3 = D - D_1$. Here the norms of J_2 and J_3 do not exceed $\|x\|$ times

$$\delta^{-1}\int_{D_2} |K(c)|\, J(c; b)db \quad \text{and} \quad \delta^{-1}\int_{D_3} |K(b)|\, db$$

respectively. Both of these expressions tend to zero with δ by (6.6). It follows that (6.10) tends to a limit so that $y \in \mathfrak{D}(a)$ and

$$A(a)y = \int_D K_1(b; a)T(b)xdb, \tag{6.12}$$

$$K_1(b; a) = \frac{\partial}{\partial\delta}\{K[\psi(b, f(\delta a))]J[\psi(b, f(\delta a); b]\}_{\delta=0}.$$

For D near to the origin this becomes

$$\begin{aligned}
K_1(b; a) &= -K(b)\Big[\sum_i\sum_j \alpha^i_{ji}\alpha_j + O(|b|)\Big] \\
&\quad - \sum_i K_i(b)\Big[\alpha_i + \sum_j\sum_k \alpha^i_{jk}\alpha_j\beta_k + O(|b|^2)\Big],
\end{aligned} \tag{6.13}$$

where $K_i(b)$ is the partial of $K(b)$ with respect to β_i. It should be observed that the remainder terms are independent of the kernel. We conclude that $A(a)K$ is a linear bounded transformation on $\mathfrak{X}$ to $\mathfrak{X}$ and

$$\|A(a)K\| \leqq \int_D |K_1(b; a)| \, db.$$

Moreover, it is not difficult to see that this bound is a bounded function of a for $|a| = 1$.

We have now to consider the existence of $A(a_1)A(a_2)y$. Here $A(a_2)y$ is given by (6.12) with a replaced by a_2. This integral is of the same type as (6.8) with a kernel $K_1(c; a_2)$ instead of $K(c)$. Here $K_1(c; a_2)$ has the same properties as $K(c)$ except for differentiability; however, $K_1(c; a_2) \in C^{(2)}$ at least and this more than suffices for our needs. The argument given above may consequently be used also to prove the existence of $A(a_1)A(a_2)y$. Further we see that $A(a_1)A(a_2)K$ is a bounded linear operator and the bound is a bounded function of a_1 and a_2 on the unit sphere in $\overline{E}_n^+$. This completes the proof.

We have seen that the set $\mathfrak{R}[\mathfrak{X}]$ is dense in $\mathfrak{D}(a)$. Actually a stronger statement can be made and we can make assertions about the graphs of the operators in the relevant product spaces.

THEOREM 6.4. *The graph* $[y, A(a)y]$, $y \in \mathfrak{R}[\mathfrak{X}]$, *is dense in the graph* $[x, A(a)x]$, $x \in \mathfrak{D}(a)$, *in* $\mathfrak{X} \times \mathfrak{X}$. *More generally, the graph* $[y, A(a_1)y, \cdots, A(a_k)y]$, $y \in \mathfrak{R}[\mathfrak{X}]$, *is dense in the graph* $[x, A(a_1)x, \cdots, A(a_k)x]$, $x \in \bigcap_1^k \mathfrak{D}(a_j)$, *in* $\mathfrak{X} \times \cdots \times \mathfrak{X}$ *($k+1$ factors).*

The proof is long and laborious so we shall merely sketch the argument for $k=1$ and indicate briefly the extension to more dimensions. Since $A(a)$ is a closed linear operator the graph $\mathfrak{G}_1 = [x, A(a)x]$, $x \in \mathfrak{D}(a)$, with points g_1 may be made into a (B)-space under the norm $\|g_1\| = \|x\| + \|A(a)x\|$ with obvious definition of the algebraic operations. If the subset $\mathfrak{G}_{10} = [y, A(a)y]$, $y \in \mathfrak{R}[\mathfrak{X}]$, is non-dense in $\mathfrak{G}_1$, then there exists a linear bounded functional on $\mathfrak{G}_1$ which vanishes on $\mathfrak{G}_{10}$ without vanishing identically. Any bounded linear functional on $\mathfrak{G}_1$ is of the form

$$g_1^*(g_1) = x_1^*(x) + x_2^*[A(a)x]$$

where x_1^* and x_2^* are arbitrary linear bounded functionals on $\mathfrak{X}$. We have then for some special choice of x_1^* and x_2^* that

$$x_1^*(y) + x_2^*[A(a)y] = 0, \qquad y \in \mathfrak{R}[\mathfrak{X}], \tag{6.14}$$

and by assumption this does not hold for all x in $\mathfrak{D}(a)$ if we replace y by x. We now make a special choice of y in $\mathfrak{R}[\mathfrak{X}]$. We take as domain of integration the cube $C(\delta, \epsilon)$: $\delta < \gamma_i < \epsilon$, $i = 1, 2, \cdots, n$, and set $K(c) = 0$ outside the cube and equal to $\prod_1^n K_0(\gamma_i)$ in the cube where

$$K_0(\gamma) = \exp\{-\eta(\epsilon - \delta)[(\epsilon - \gamma)(\gamma - \delta)]^{-1}\}, \qquad \eta > 0.$$

The corresponding elements $y = y(x;\ \delta,\ \epsilon,\ \eta)$, $x \in \mathfrak{D}(a)$, are in $\mathfrak{R}[\mathfrak{X}]$. Substituting this value of y in (6.14) gives an identity in δ, ϵ, η, and x. By continuity we may let $\delta \to 0$ and afterwards $\eta \to 0$. Under these two operations $x_1^*(y)$ is carried into

$$\int_{C(0,\epsilon)} x_1^*[T(c)x]dc \tag{6.15}$$

since $K_0(\gamma) \to 1$ boundedly in $C(0, \epsilon)$. For the discussion of $x_2^*[A(a)y]$ we use (6.13) which gives the corresponding kernel. This expression involves two terms of which the first one leads to

$$\int_{C(0,\epsilon)} \left\{ \sum_i \sum_j \overset{i}{\alpha}_{ji}\alpha_j + O(|c|) \right\} x_2^*[T(c)x]dc. \tag{6.16}$$

We recall that the remainder term is independent of the kernel. The second term of (6.13) gives rise to n singular integrals of which the first one, $i = 1$, involves the integral

$$\eta \int_0^\epsilon K_0(\gamma_1)[(\gamma_1 - \epsilon)^{-2} - \gamma_1^{-2}] f_1(c) d\gamma_1$$

which is multiplied by $\prod_2^n K_0(\gamma_j)$ integrated with respect to the remaining $n-1$ variables from 0 to ϵ. Here

$$f_i(c) = \left[\alpha_i + \sum_j \sum_k \overset{i}{\alpha}_{jk}\alpha_j\gamma_k + O(|c|^2) \right] x_2^*[T(c)x].$$

Passing to the limit with η gives the result

$$f_1(\epsilon, \gamma_2, \cdots, \gamma_n) - f_1(0, \gamma_2, \cdots, \gamma_n)$$

as is easily seen. At the same time $\prod_2^n K_0(\gamma_j) \to 1$. Thus we see that the limit of $x_2^*[A(a)y]$ under the limit processes $\delta \to 0$, $\eta \to 0$ becomes the integral over the $(n-1)$-dimensional boundary of the cube $C(0, \epsilon)$ of the function which on the face $\gamma_i = \epsilon$ equals $f_i(c)$ and on the face $\gamma_i = 0$ equals $-f_i(c)$.

After performing these operations on $x_1^*(y) + x_2^*[A(a)y]$ we multiply by ϵ^{-n} and let $\epsilon \to 0$. The contributions from (6.15) and (6.16) add up to

$$x_1^*(x) - \left\{ \sum_i \sum_j \overset{i}{\alpha}_{ji}\alpha_j \right\} x_2^*(x). \tag{6.17}$$

The surface integral leads to two terms of which one arises from the partials of the factors $\alpha_i + \sum_j \sum_k \alpha_{jk}^i \alpha_j \gamma_k + O(|c|^2)$ and gives a limit which cancels the second term of (6.17). The second term involves combinations of difference quotients of $x_2^*[T(c)x]$. Since everything else tends to a limit when $\epsilon \to 0$, this term must also tend to a limit which we denote by $X^*(x)$. We have then

$$x_1^*(x) + X^*(x) = 0, \qquad x \in \mathfrak{D}(a). \tag{6.18}$$

Since $\mathfrak{D}(a)$ is dense in $\mathfrak{X}$, this must hold for all x so that $X^* = -x_1^*$. But for $x = y$ in $\mathfrak{R}[\mathfrak{X}]$ we have $X^*(y) = x_2^*[A(a)y]$ and $\mathfrak{R}[\mathfrak{X}]$ is dense in $\mathfrak{D}(a)$ so we must have $X^*(x) = x_2^*[A(a)x]$ for all x in $\mathfrak{D}(a)$. But this shows that (6.14) holds with y replaced by any x of $\mathfrak{D}(a)$ so that $g_1^*(g_1) = 0$ for all g_1. Thus $\mathfrak{G}_{10}$ is dense in $\mathfrak{G}_1$.

The extension from 1 to k does not offer any new difficulties. The graph $\mathfrak{G}_k = [x, A(a_1)x, \cdots, A(a_k)x]$, $x \in \bigcap_1^k \mathfrak{D}(a_j)$, becomes a (B)-space under the norm $\|g_k\| = \|x\| + \sum_1^k \|A(a_j)x\|$ and the linear functionals on $\mathfrak{G}_k$ are of the form

$$g_k^*(g_k) = x_1^*(x) + x_2^*[A(a_1)x] + \cdots + x_{k+1}^*[A(a_k)x].$$

Proceeding as above with the same choice of $y = y(x;\ \delta,\ \epsilon,\ \eta)$, $x \in \bigcap_1^k \mathfrak{D}(a_j)$, and passing to the limit with the parameters, one obtains a relation of type (6.18) holding for all x under consideration. But for $x = y \in \mathfrak{R}[\mathfrak{X}]$ we have

$$X^*(y) = x_2^*[A(a_1)y] + \cdots + x_{k+1}^*[A(a_k)y]. \tag{6.19}$$

Here X^* has a unique extension when we pass from the dense set $\mathfrak{R}[\mathfrak{X}]$ to the set $\bigcap_1^k \mathfrak{D}(a_j)$ and the right-hand side of (6.19) has the obvious extension obtained by replacing y by x. It follows that (6.19) holds for all points of $\bigcap_1^k \mathfrak{D}(a_j)$ so that g_k^* is the zero functional and the subgraph is dense in $\mathfrak{G}_k$.

We now define an operator $U(a)$ by

$$U(a)x = \lim_{\delta \to 0} \delta^{-1}[T(\delta a) - I]x \tag{6.20}$$

whenever the limit exists.

THEOREM 6.5. *$U(a)y$ exists and equals $A(a)y$ if $y \in \mathfrak{R}[\mathfrak{X}]$.*

The existence of $U(a)y$ is proved by the argument used in the proof of Theorem 6.3. We have merely to replace $f(\delta a)$ by δa throughout. But $f(\delta a) - \delta a = O(\delta^2)$ whence it follows that

$$\psi(b, f(\delta a)) - \psi(b, \delta a) = O(\delta^2),$$

$$J[\psi(b, f(\delta a)); b] - J[\psi(b, \delta a); b] = O(\delta^2),$$

so the right member of (6.13) is unchanged when we replace $f(\delta a)$ by δa. Hence $U(a)y = A(a)y$ for $y \in \mathfrak{K}[\mathfrak{X}]$. If it could be shown that $U(a)$ is a closed operator, then it would follow from Theorem 6.4 that the domain of $U(a)$ contains that of $A(a)$ and that equality between the operators holds in $\mathfrak{D}(a)$. At present we cannot decide this question and it is not vital for the following discussion.

THEOREM 6.6. *If* $y = K(c)[x]$ *and if* a_1, a_2, b *are points of* $\overline{E}_n^+$ *with* $|a_1| < \delta(D)$, $|a_2| < \delta(D)$, $|b| \leqq 1$, *then there exists a constant C depending only upon* $K(c)$ *such that*

$$(6.21) \qquad |[T(a_1) - T(a_2)]y| \leqq C|a_1 - a_2| \|x\|$$

and the same relation holds with y replaced by $A(b)y$.

This is proved by the method of Theorem 6.3. We omit the details.

We can now prove that the correspondence $a \to A(a)$ is a homomorphism under addition in the following sense.

THEOREM 6.7. $\mathfrak{D}(a_1 + a_2)$ *contains* $\mathfrak{D}(a_1) \cap \mathfrak{D}(a_2)$ *and in the latter set*

$$(6.22) \qquad A(a_1 + a_2)x = A(a_1)x + A(a_2)x.$$

We start by proving (6.22) for $x = y \in \mathfrak{K}[\mathfrak{X}]$. In this set it is sufficient to prove the corresponding relation with A replaced by U and this is accomplished if we can show that

$$[T(\delta(a_1 + a_2)) - T(\delta a_1) - T(\delta a_2) + I]y = o(\delta)$$

when $\delta \to 0$. The left member equals

$$\{T(\delta(a_1 + a_2)) - T[F(\delta a_1, \delta a_2)]\}y + [T(\delta a_1) - I][T(\delta a_2) - I]y.$$

By Theorem 6.6 the norm of the first term does not exceed

$$C|\delta(a_1 + a_2) - F(\delta a_1, \delta a_2)| \|x\|$$

which is $O(\delta^2)$ by A_5^4 since $\omega(\xi) = O(\xi)$. Since $y \in \mathfrak{D}(a_2)$, we have $[T(\delta a_2) - I]y = \delta A(a_2)y + o(\delta)$, so that

$$\begin{aligned}\|[T(\delta a_1) - I][T(\delta a_2) - I]y\| &\leqq \delta \|[T(\delta a_1) - I]A(a_2)y\| \\ &+ o(\delta)\|T(\delta a_1) - I\| \leqq C\delta^2|a_1| \|x\| + o(\delta)\end{aligned}$$

from which the assertion follows. In order to extend the validity of (6.22) from $\mathfrak{K}[\mathfrak{X}]$ to $\mathfrak{D}(a_1) \cap \mathfrak{D}(a_2)$ we argue as follows. By Theorem 6.4 the subgraph $[y, A(a_1)y, A(a_2)y]$, $y \in \mathfrak{K}[\mathfrak{X}]$, is dense in the graph $[x, A(a_1)x, A(a_2)x]$, $x \in \mathfrak{D}(a_1) \cap \mathfrak{D}(a_2)$. Hence for any x in the latter set we may find a sequence $y_n \in \mathfrak{K}[\mathfrak{X}]$ such that $y_n \to x$, $A(a_1)y_n \to A(a_1)x$, $A(a_2)y_n \to A(a_1)x$. It follows that

$$A(a_1 + a_2)y_n = A(a_1)y_n + A(a_2)y_n \to A(a_1)x + A(a_2)x.$$

Since $A(a_1+a_2)$ is also a closed operator, it follows that $A(a_1+a_2)x$ exists and that (6.22) holds.

We come now to the main theorem of this section.

THEOREM 6.8. *Let*

$$a = (\alpha_1, \alpha_2, \cdots, \alpha_n) = \alpha_1 e_1 + \alpha_2 e_2 + \cdots + \alpha_n e_n$$

and set $A(e_k) = A_k$. *Then* $\mathfrak{D}(a)$ *contains the set* $\bigcap_1^n \mathfrak{D}(e_k) \equiv \mathfrak{D}_n$ *and for* x *in the latter set*

$$A(a)x = \alpha_1 A_1 x + \alpha_2 A_2 x + \cdots + \alpha_n A_n x. \tag{6.23}$$

The basic infinitesimal generators $A_1, A_2, \cdots, A_n$ *are linearly independent in* $\mathfrak{D}_n$.

The validity of (6.23) is an immediate consequence of the preceding theorem. A linear relation with constant coefficients between $A_1x, A_2x, \cdots, A_nx$ valid for all x in $\mathfrak{D}_n$ implies the existence of two distinct vectors a and b in $\overline{E}_n^+$ such that $A(a)x = A(b)x$ in $\mathfrak{D}_n$. We know that the subgraph $[y, A(a)y]$, $y \in \mathfrak{R}[\mathfrak{X}]$, is dense in the graph of $A(a)$. A fortiori this is true for the graph $[x, A(a)x]$, $x \in \mathfrak{D}_n$. Now for $\Re(\lambda) > 0$ the set $[\lambda I - A(a)][\mathfrak{D}(a)]$ is dense in $\mathfrak{X}$ so the same must be true for $[\lambda I - A(a)][\mathfrak{D}_n] = [\lambda I - A(b)][\mathfrak{D}_n]$. From this we conclude that the linear bounded operators $R[\lambda; A(a)]$ and $R[\lambda; A(b)]$ coincide in a dense set and hence everywhere. By the inversion formulas for the Laplace transform, applied to formula (6.4), we conclude that the corresponding one-parameter semi-group operators $T[f(\rho a)]$ and $T[f(\rho b)]$ are identical for all values of ρ. But for all small values of ρ we have $f(\rho a) \neq f(\rho b)$ and $|f(\rho a)| < \rho_0$, $|f(\rho b)| < \rho_0$. Thus the result contradicts A_4^3 so we conclude that $A(a)$ and $A(b)$ are distinct operators on $\mathfrak{D}_n$ when $a \neq b$.

The infinitesimal generators $A(a)$ of $\mathfrak{S}$ form an n-dimensional system which is closed under addition and multiplication by positive numbers. On the other hand, multiplication by negative numbers is not allowed since a and $-a$ are not simultaneously in $\overline{E}_n$ if $a \neq 0$. In this respect there is a striking difference between the semi-module of infinitesimal generators of a semi-group and the Lie ring of generators of a group.

7. **The fundamental theorems.** We come now to the analogues of the three fundamental theorems of Lie.

THEOREM 7.1. *For* $y \in \mathfrak{R}[\mathfrak{X}]$ *and small values of* a *in* E_n^+

$$(7.1)\qquad \frac{\partial}{\partial \alpha_j} T(a)y = \sum_{k=1}^{n} \Gamma_{jk}(a)T(a)A_k y, \qquad j = 1, 2, \cdots, n,$$

where the matrix $(\Gamma_{jk}(a))$ *tends to the unit matrix when* $a \to 0$. *If the operator in the left member is closed, then its domain contains* $\mathfrak{D}_n$ *and* (7.1) *holds in* $\mathfrak{D}_n$.

The left side of (7.1) is the limit when $\delta \to 0$ of

$$(7.2)\qquad \delta^{-1}[T(a + \delta e_j) - T(a)]y.$$

Using A_3^4 we see that the equation $F(a, h) = a + \delta e_j$ has a unique solution in $\overline{E}_n^+$, viz.

$$h = \psi(a + \delta e_j, a) = \chi_j(a, \delta) = c_j(a)\delta + O(\delta^2),$$

where the components $\Gamma_{jk}(a)$ of the vector $c_j(a)$ are determined by the linear system of equations

$$(7.3)\qquad \sum_{k=1}^{n} F_{ik}(a, 0)\Gamma_{jk}(a) = \delta_{ij}, \qquad i = 1, 2, \cdots, n.$$

Here F_{ik} as in formula (5.4) is the partial derivative of the ith component of $F(a, b)$ with respect to the kth component of b in which we set $b = 0$. Since $F_{ik}(a, 0) \to \delta_{ik}$ when $|a| \to 0$, the determinant of the system (7.3) is different from zero for small values of $|a|$ so that the $\Gamma_{jk}(a)$ may be determined. Further $\Gamma_{jk}(a) \to \delta_{jk}$ when $|a| \to 0$. Thus

$$(7.4)\qquad \begin{aligned} &\delta^{-1}[T(a + \delta e_j) - T(a)]y \\ &\qquad = T(a)\delta^{-1}\{T[\chi_j(a, \delta)] - I\}y \\ &\qquad = T(a)\delta^{-1}\{T[\delta c_j(a)] - I\}y \\ &\qquad\quad + T(a)\delta^{-1}\{T[\chi_j(a, \delta)] - T[\delta c_j(a)]\}y. \end{aligned}$$

The first term in the last member tends to $T(a)A[c_j(a)]y$ when $\delta \to 0$ by Theorem 6.5 while the second term tends to θ by Theorem 6.6. Hence the limit exists and by Theorem 6.8 it equals

$$T(a)A[c_j(a)]y = \sum_{k=1}^{n} \Gamma_{jk}(a)A_k y.$$

The left side of this equation involves a closed operator, but it is not a priori obvious that the partials of $T(a)$ are closed operators; if they are, then Theorem 6.4 shows immediately that (7.1) holds in $\mathfrak{D}_n$.

Theorem 7.1 also gives us

$$(7.5)\qquad T(a)A_i y = \sum_{k=1}^{n} \Delta_{ik}(a) \frac{\partial}{\partial \alpha_k} T(a)y$$

where $(\Delta_{ik}(a))$ is the inverse of the matrix $(\Gamma_{jk}(a))$. It follows from (7.3) that

$$\Delta_{ik}(a) = F_{ki}(a, 0). \tag{7.6}$$

The second and third fundamental theorems involve the *structural constants* of the semi-group defined by

$$\overset{i}{\gamma}_{jk} = \overset{i}{\alpha}_{jk} - \overset{i}{\alpha}_{kj}, \qquad \overset{i}{\alpha}_{jk} = \left(\frac{\partial^2 F_i}{\partial \alpha_j \partial \beta_k}\right)_{0,0} \tag{7.7}$$

in the notation of (6.5). We note that

$$\frac{\partial}{\partial \alpha_j} \Delta_{ik}(a) \to \overset{k}{\alpha}_{ji}, \qquad |a| \to 0. \tag{7.8}$$

THEOREM 7.2. *If* $y \in \mathfrak{R}[\mathfrak{X}]$

$$[A_i, A_j]y \equiv (A_iA_j - A_jA_i)y = \sum_{m=1}^{n} \overset{m}{\gamma}_{ij} A_m y. \tag{7.9}$$

If the operator in the first member is closed, then the relation holds in $\mathfrak{D}_n$.

For the proof we use (7.5) twice forming

$$\begin{aligned} T(a)A_iA_jy &= \sum_{k=1}^{n} \Delta_{ik}(a) \frac{\partial}{\partial \alpha_k} [T(a)A_jy] \\ &= \sum_{k=1}^{n} \Delta_{ik}(a) \frac{\partial}{\partial \alpha_k} \left\{ \sum_{m=1}^{n} \Delta_{jm}(a) \frac{\partial}{\partial \alpha_m} T(a)y \right\}. \end{aligned}$$

Interchanging i and j, subtracting and simplifying we get

$$T(a)[A_i, A_j]y = \sum_{m=1}^{n} \overset{m}{\Delta}_{ij}(a) \frac{\partial}{\partial \alpha_m} T(a)y$$

where

$$\overset{m}{\Delta}_{ij}(a) = \sum_{k=1}^{n} \left\{ \Delta_{ik}(a) \frac{\partial}{\partial \alpha_k} \Delta_{jm}(a) - \Delta_{jk}(a) \frac{\partial}{\partial \alpha_k} \Delta_{im}(a) \right\},$$

which tends to $\overset{m}{\gamma}_{ij}$ when $|a| \to 0$ by (7.8) since $\Delta_{ik}(a) \to \delta_{ik}$. Formula (7.1) shows that the partial of $T(a)y$ with respect to α_m tends to $A_m y$. It should be observed that the second order partials of $T(a)y$ which arise in the process, but cancel in the subtraction, actually exist when $y \in \mathfrak{R}[\mathfrak{X}]$. This follows from the fact that in (7.4) we may replace y by $A_i y$ and still carry through the limit process. The second partials may consequently be found by formal differentiation of formula (7.1). If $[A_i, A_j]$ is a closed operator, then Theorem 6.4 shows that (7.9) holds in $\mathfrak{D}_n$.

The second fundamental theorem asserts that each *commutator* $[A_i, A_j]$ is a linear combination of basic infinitesimal generators, but in the semi-group case it does not follow that $[A_i, A_j]$ is also an infinitesimal generator, that is, we cannot always find a vector c in $\bar{E}_n^+$ such that $[A_i, A_j]y = A(c)y$. In particular, we note that if $[A_i, A_j]$ is an infinitesimal generator, $[A_j, A_i]$ cannot be one.

THEOREM 7.3. *The structural constants satisfy*

$$\gamma^{i}_{jk} = -\gamma^{i}_{kj}, \tag{7.10}$$

$$\sum_{m=1}^{n} [\gamma^{p}_{im}\gamma^{m}_{jk} + \gamma^{p}_{jm}\gamma^{m}_{ki} + \gamma^{p}_{km}\gamma^{m}_{ij}] = 0. \tag{7.11}$$

Here (7.10) follows from (7.7) while (7.11) follows from the relation

$$[A_i, [A_j, A_k]] + [A_j, [A_k, A_i]] + [A_k, [A_i, A_j]] = \Theta \tag{7.12}$$

which holds when the operator on the left acts on the subspace $\mathfrak{R}[\mathfrak{X}]$.

8. **Conclusions.** The preceding theory raises perhaps more questions than it answers. Let us list some directions in which further research is desirable.

(1) Determine the set of points c such that $c = f(a)$ for some a. In particular, are all finite boundary points of this set "accessible" and found by solving the equation $F(p, p) = p$?

(2) Extend the investigation to other parameter sets.

(3) Prove that the partials of $T(a)$ and the commutators are closed operators.

(4) Formulate and prove converses of the fundamental theorems.

(5) Is it possible to embed the given semi-group $\mathfrak{S}$ in a group of, in general, unbounded operators, the group being generated by the set $\sum \alpha_k A_k$ with real α's not necessarily positive?

REFERENCES

1. G. Birkhoff, *Analytical groups*, Trans. Amer. Math. Soc. vol. 43 (1938) pp. 61–101.

2. N. Dunford, *On one parameter groups of linear transformations*, Ann. of Math. (2) vol. 39 (1938) pp. 569–573.

3. N. Dunford and I. E. Segal, *Semi-groups cf operators and the Weierstrass theorem*, Bull. Amer. Math. Soc. vol. 52 (1946) pp. 911–914.

4. L. Gårding, *Note on continuous representations of Lie groups*, Proc. Nat. Acad. Sci. U. S. A. vol. 33 (1947) pp. 331–332.

5. I. Gelfand, *On one-parametrical groups of operators in a normed space*, C. R. (Doklady) Acad. Sci. URSS. N.S. vol. 25 (1939) pp. 713–718.

6. E. Hille, *Functional analysis and semi-groups*, Amer. Math. Soc. Colloquium Publications, vol. 30, New York, 1948, xii+528 pp.

7. I. E. Segal, *Topological groups in which multiplication on one side is differentiable*, Bull. Amer. Math. Soc. vol. 52 (1946) pp. 481–487.

8. P. A. Smith, *Foundations of the theory of Lie groups with real parameters*, Ann. of Math. (2) vol. 44 (1943) pp. 481–513.

9. ———, *Foundations of Lie groups*, Ann. of Math. (2) vol. 48 (1947) pp. 29–42.

10. K. Yosida, *On the differentiability and the representation of one-parameter semi-group of linear operators*, Journal of the Mathematical Society of Japan vol. 1 (1948) pp. 15–21.

11. ———, *An operator-theoretical treatment of temporally homogeneous Markoff process*, Journal of the Mathematical Society of Japan.

12. ———, *Brownian motion on the surface of the 3-sphere*, Ann. Math. Statist. vol. 20 (1949) pp. 292–296.

YALE UNIVERSITY

Chapter 13
Ordinary Differential Equations

The papers included in this chapter are

[92] Non-oscillation theorems

[102] Behavior of solutions of linear second order differential equations

[122] Remarques sur les systèmes des équations différentielles linéaires à une infinité d'inconnues

[133] Pathology of infinite systems of linear first order differential equations with constant coefficients

[136] Green's transforms and singular boundary value problems

[143] On the Thomas-Fermi equation

An introduction to Hille's work on ordinary differential equations can be found starting on page 691.

NON-OSCILLATION THEOREMS

BY

EINAR HILLE

1. Introduction. In the following we shall be concerned with the differential equation

$$(1.1) \qquad y'' - F(x)y = 0,$$

where $F(x)$ is a real-valued function defined for $x>0$ and belonging to $L(\epsilon, 1/\epsilon)$ for each $\epsilon>0$. A solution of (1.1) is a real-valued function $y(x)$, absolutely continuous together with its first derivative, which satisfies the equation for almost all x, in particular at all points of continuity of $F(x)$.

We shall say that the equation is *non-oscillatory* in (a, ∞), $a\geqq 0$, if no solution can change its sign more than once in the interval. Since the zeros of linearly independent solutions separate each other, it is sufficient that there exists a solution without zeros in the interval in order that the equation be non-oscillatory there. It is well known that (1.1) is non-oscillatory in $(0, \infty)$ if $F(x)\geqq 0$, but it may also have this property when $F(x)\leqq 0$ as is shown by the example

$$(1.2) \qquad y'' + \gamma x^{-2}y = 0$$

with

$$(1.3) \qquad y(x) = C_1x^{\rho} + C_2x^{1-\rho}, \qquad \rho^2 - \rho + \gamma = 0.$$

This equation is oscillatory for $\gamma>\frac{1}{4}$, but non-oscillatory for $\gamma\leqq\frac{1}{4}$. This example will play an important role in the following.

In this note we shall study two distinct but related problems.

I. *When does equation* (1.1) *have a solution which tends to a limit* $\neq 0, \infty$ *when* $x\to\infty$?

II. *If* $F(x) = -f(x)$ *where* $f(x)$ *is non-negative, when is equation* (1.1) *non-oscillatory in* (a, ∞) *for a sufficiently large*?

It should be noted that if $F(x) = -f(x)\leqq 0$, then a solution of Problem I is also a solution of Problem II but not vice versa. The impetus to the present investigation was given by a recent paper by Aurel Wintner ([10] in the appended bibliography) in which these problems were mentioned. Wintner proved (loc. cit. pp. 96–97) that $f(x)\in L(1, \infty)$ is a necessary condition for both I and II; in studying problem I he restricted himself to the case in which $F(x) = f(x)\geqq 0$. Example (1.2) shows, however, that the condition is not sufficient in either case.

Problem I, though not trivial lies fairly close to the surface. Assuming

Presented to the Society, September 5, 1947; received by the editors June 3, 1947.

$F(x)$ to be of constant sign at least for large values of x, it is not difficult to strengthen Wintner's necessary condition to $xF(x) \in L(1, \infty)$ and it is an easy matter to show that this condition is also sufficient. Actually a special case of Problem I was attacked by the present writer in 1924 [4, p. 491] in connection with the equation

$$w'' + [a^2 - F(z)]w = 0 \tag{1.4}$$

in which the perturbation term $F(z)$ is supposed to be holomorphic in a right half-plane and $|F(x+iy)| \leqq M(x)$ where $M(x)$ is monotone decreasing and satisfies a suitable condition of integrability[1]. The case which has a bearing on Problem I is that in which $a=0$; assuming $xM(x) \in L(0, \infty)$, I could prove the existence of a solution $w_1(z)$ of (1.4) such that $w_1(z)=1+o(1)$ for large $|z|$. This is the *exceptional* or *sub-dominant solution*; assuming the stronger condition $x^2M(x) \in L(0, \infty)$, I could isolate a dominant solution $w_2(z)=z+o(1)$ for large $|z|$. The proof is based upon the method of successive approximations and applies just as well if $F(\cdot)$ is defined only for real values of the variable. The assumption that $M(x)$ is monotone turns out to be superflous and it is an easy matter to show that $xF(x) \in L(1, \infty)$ suffices for the existence of the exceptional solution. The details of the discussion are given in §2 below.

Problem II on the other hand is much more refractory. Wintner's necessary condition, $f(x) \in L(1, \infty)$, can be strengthened to, for instance, $x^\sigma f(x) \in L(1, \infty)$ for each $\sigma<1$, but not for $\sigma=1$. Example (1.2) shows that even this stronger condition is not sufficient. More significant is the observation that the function

$$g(x) = x\int_x^\infty f(t)dt \tag{1.5}$$

stays bounded for large x if (1.1) is non-oscillatory. More precisely, if the inferior and superior limits of $g(x)$ for $x \to \infty$ are denoted by g_* and g^* respectively, then $g_* \leqq \frac{1}{4}$, $g^* \leqq 1$ are necessary and $g^* < \frac{1}{4}$ sufficient conditions in order that (1.1) be non-oscillatory for large x. Here all inequalities are sharp. The proof of this result is given in §3, the necessary counter examples are constructed in §4.

Example (1.2) was taken as the point of departure for a study of Problem II by A. Kneser [5, pp. 414–418] in 1892[2]. He proved that if

[1] There is a considerable literature on this and related problems. Of recent papers, reference should be made to R. Bellman [1] and N. Levinson [6]. The idea of reducing the study of (1.4) to a singular integral equation which is solved by the method of successive approximations goes back to É. Cotton [2]. The case $a=0$, in which at most one solution can remain bounded, seems to have been disregarded in the literature.

[2] I am indebted to a referee for calling my attention to the fact that Problem II goes back to Kneser. A further search of the literature led to Riemann-Weber [7] where the discus-

$\lim_{x\to\infty}x^2f(x)=\gamma$, then (1.1) is oscillatory for $\gamma>\frac{1}{4}$ and non-oscillatory for $\gamma<\frac{1}{4}$. If $\gamma=\frac{1}{4}$ no conclusion can be drawn unless $x^2f(x)<\frac{1}{4}$ for all large x. In §5 we shall exhibit an infinite sequence of differential equations which lead to successive refinements of Kneser's criterion of a nature similar to the logarithmic scale in the theory of convergence of infinite series. In this discussion one may replace limits by inferior and superior limits, but no conclusion can be drawn if $\frac{1}{4}$ lies in the interval of indetermination of the critical ratio in question.

In the study of Problem II the constant $\frac{1}{4}$ plays a peculiar role. It is the value of γ for which the quadratic equation $\rho^2-\rho+\gamma=0$ has equal roots. This equation was noted above in (1.3) where it occurred as the indicial equation in the sense of Fuchs of equation (1.2). It plays a similar role for the other equations of the logarithmic scale, but it also enters the argument in a different way. The main tool in our study of Problem II is the associated Riccati equation

$$v' + v^2 + f(x) = 0, \qquad v = \frac{y'}{y}, \tag{1.6}$$

and the corresponding singular integral equation which in the non-oscillatory case can be written in the form

$$u(x) = x\int_x^\infty u^2(t)\frac{dt}{t^2} + g(x), \qquad u(x) = xv(x), \tag{1.7}$$

with $g(x)$ defined by (1.5). This equation shows that the behavior of $g(x)$ for large x is decisive for that of $u(x)$; in particular, $\lim u(x)=\rho$ can exist if and only if $\lim g(x)=\gamma$ exists in which case $\rho^2-\rho+\gamma=0$ and $\gamma\leqq\frac{1}{4}$. It is perhaps of some interest to observe that here is a linear problem which, apparently, is best studied by nonlinear methods.

Finally, in §6 we give a partial extension of Problem II to the complex plane which also is based upon equation (1.7).

2. **Problem I.** Let $F(x)$ be a real-valued function, defined for $x>0$ and belonging to $L(\epsilon, 1/\epsilon)$ for each $\epsilon>0$. We shall say that *the differential equation*

sion of oscillation theorems (pp. 53–72, 5th ed., omitted in the 7th Frank-v. Mises ed.) is based on Kneser. The case in which $f(x)\to 0$ is considered on pp. 60–62, but here the discussion goes beyond Kneser. If $\gamma=\frac{1}{4}$, Weber puts $y=x^{1/2}\eta$, $\xi=\log x$; this gives a new differential equation to which Kneser's criterion may be applied. The resulting theorem is equivalent to our Theorem 10 below. Weber also indicates how the iteration of this process leads to an infinite chain of conditions. The existence of this passage in Riemann-Weber had escaped my memory at the time of deriving Theorems 10 and 11. Though these theorems are not essentially new, I have not suppressed them from the text since they provide necessary background for the rest of the discussion. For further literature relating to Problem II, see also W. B. Fite [3, p. 343] and A. Wiman [9, pp. 4–5]. Wiman has added much to our knowledge of the fine structure of the solutions, but for the case of Problem II he does not go beyond Kneser, nor does he claim to do so. [Revised September 9, 1947.]

$$y'' - F(x)y = 0 \tag{2.1}$$

has property I *if it admits of a solution* $y_1(x)$ *such that*

$$\lim_{x\to\infty} y_1(x) = 1. \tag{2.2}$$

LEMMA 1. *If* (2.1) *has property* I, *then its general solution is of the form*

$$y(x) = C_1[1 + \epsilon_1(x)] + C_2x[1 + \epsilon_2(x)], \tag{2.3}$$

where C_1 *and* C_2 *are arbitrary constants and* $\lim_{x\to\infty}\epsilon_k(x) = 0$.

Proof. By assumption there exists a solution $y_1(x) = 1 + \epsilon_1(x)$. Let a be so large that $|\epsilon_1(x)| < 1$ for $x \geqq a$. Then

$$y_2(x) = y_1(x)\int_a^x \frac{dt}{[y_1(t)]^2} = x[1 + \epsilon_2(x)]$$

exists and is also a solution of (2.1) as is well known and easily verified.

LEMMA 2. *If* (2.1) *has property* I, *then* (2.1) *is non-oscillatory for large* x. *Moreover, if* $F(x)$ *keeps a constant sign in* (c, ∞), $c > 0$, *then every solution* $y(x)$ *is ultimately monotone and* $y'(x)$ *tends to a finite limit when* $x \to \infty$. *Further,* $y(x)$ *is ultimately convex or concave towards the* x*-axis according as* $F(x) \geqq 0$ *or* $F(x) \leqq 0$.

Proof. If a is chosen as above, then $y_1(x) \neq 0$ for $x \geqq a$ and no solution can have more than one zero in (a, ∞). Suppose that $F(x)$ keeps a constant sign for $x \geqq c$, say $F(x) \geqq 0$, and that a solution $y(x)$ is positive for $c \leqq b \leqq x$. If $b \leqq x_1 < x_2 < \infty$, then

$$y'(x_2) - y'(x_1) = \int_{x_1}^{x_2} F(t)y(t)dt \tag{2.4}$$

is never negative. Hence $y'(x)$ is never decreasing for $x \geqq b$ and tends to a limit when $x \to \infty$. This limit must be finite because $y(x) < Kx$ for large x by (2.3). The convexity properties are obvious.

THEOREM 1. *If* $F(x)$ *keeps a constant sign for large* x *and if* (2.1) *has property* I, *then* $xF(x) \in L(1, \infty)$.

Proof. We take $C_1 = 0$, $C_2 = 1$ in (2.3) and substitute in (2.4). Since the left side tends to a finite limit when $x_2 \to \infty$, so does the right, that is $F(x)y(x) \in L(a, \infty)$ because $F(x)y(x)$ ultimately keeps a constant sign. But $y(x) > \frac{1}{2}x$ for large x and $F(x) \in L(1, a)$. Hence $xF(x) \in L(1, \infty)$.

THEOREM 2. *If* $xF(x) \in L(1, \infty)$, *then* (2.1) *has property* I. *Moreover*

$$|y_1(x) - 1| \leqq \exp[G(x)] - 1, \qquad G(x) = \int_x^\infty t|F(t)|dt. \tag{2.5}$$

Proof. Consider the integral equation

$$Y(x) = 1 + \int_x^\infty (t - x)F(t)Y(t)dt \tag{2.6}$$

and define

$$Y_0(x) = 1, \qquad Y_k(x) = 1 + \int_x^\infty (t - x)F(t)Y_{k-1}(t)dt.$$

We have obviously

$$|Y_1(x) - Y_0(x)| \leqq \int_x^\infty (t - x)|F(t)|dt \leqq G(x).$$

Using the fact that $G'(x) = -x|F(x)|$ for almost all x, it is a simple matter to verify the estimate

$$|Y_k(x) - Y_{k-1}(x)| \leqq \frac{1}{k!}[G(x)]^k, \qquad k = 1, 2, 3, \cdots.$$

From this it follows that $Y_k(x)$ converges to a limit $Y(x)$ uniformly for $x \geqq \epsilon > 0$, and $Y(x)$ satisfies (2.5) and (2.6). Differentiating (2.6) twice with respect to x, we see that $Y(x)$ satisfies (2.1) for almost all positive x. Thus $Y(x)$ is the desired solution $y_1(x)$ and (2.1) has property I(3). This completes the proof.

COROLLARY. *If $F(x)$ keeps a constant sign for large x, then $xF(x) \in L(1, \infty)$ is necessary and sufficient in order that equation* (2.1) *have property* I.

This is a solution of Problem I. The condition remains sufficient if $F(x)$ is allowed to be complex-valued, but the necessity is lost already when we drop the assumption that $F(x)$ keeps a constant sign for large x. This is shown by the following simple example. The function

$$y(x) = 1 - \frac{\sin x}{x}$$

satisfies a differential equation of type (2.1) with

$$F(x) = \frac{(x^2 - 2)\sin x + 2x\cos x}{x^2(x - \sin x)}$$

as is shown by computing y''/y. The corresponding equation has property I

(3) Differentiation in (2.6) shows that $y_1'(x) \to 0$ when $x \to \infty$ and more precise information is readily obtained from the equation if desired. Similarly, in the case of the dominant solution $y_2(x)$ of Theorem 3 we have $y_2'(x) \to 1$. It should be added that the estimates (2.5) and (2.11) differ from the corresponding estimates on p. 491 of [4]. The difference is not essential, but may conceivably be due to an error of calculation in the older estimates.

but neither $F(x)$ nor $xF(x)$ belongs to $L(1, \infty)$.

For large values of x the inequality (2.5) may be replaced by the more favorable estimate

$$(2.7)\qquad \begin{aligned} | y_1(x) - 1 | &< F_2(x)[1 - F_2(x)]^{-1}, \\ F_2(x) &= \int_x^\infty (t - x) | F(t) | dt, \end{aligned}$$

which is valid for $F_2(x) < 1$. This follows from the estimate $| Y_k(x) - Y_{k-1}(x) | < [F_2(x)]^k$ which is easily verified.

We shall say that *the differential equation* (2.1) *has property* I* *if there exists a solution* $y_2(x)$ *such that*

$$(2.8)\qquad \lim_{x\to\infty} [y_2(x) - x] = 0.$$

THEOREM 3. *Property* I* *implies property* I *but not vice versa. The equation has property* I* *if* $x^2F(x) \in L(1, \infty)$ *and, if* $F(x)$ *keeps a constant sign for large* x, *then the condition is necessary as well as sufficient.*

Proof. If (2.1) has property I* then (as in the proof of Lemma 1)

$$(2.9)\qquad y_1(x) = y_2(x) \int_x^\infty [y_2(t)]^{-2} dt = 1 + o\left(\frac{1}{x}\right)$$

is a solution of (2.1), so that (2.1) has property I. In particular, if $F(x)$ keeps a constant sign for large x, we see that $xF(x) \in L(1, \infty)$. This condition also implies property I, but, as we shall see below, the stronger condition $x^2F(x) \in L(1, \infty)$ is required for property I* when $F(x)$ keeps a constant sign.

In order to prove that $x^2F(x) \in L(1, \infty)$ is a sufficient condition for property I*, we proceed as in the proof of Theorem 2, replacing (2.6) by the new singular integral equation

$$(2.10)\qquad Z(x) = x + \int_x^\infty (t - x)F(t)Z(t)dt.$$

The details can be left to the reader; we note, however, the resulting estimate

$$(2.11)\qquad | y_2(x) - x | < \exp [H(x)] - 1, \qquad H(x) = \int_x^\infty t^2 | F(t) | dt.$$

The proof of the necessity is more interesting. Suppose that $F(x)$ keeps a constant sign for large x and that (2.1) has property I*. As observed above, it has also property I and if $y_1(x)$ denotes the solution defined by (2.9) we have conversely

$$y_2(x) = y_1(x) \int_a^x [y_1(t)]^{-2} dt + Cy_1(x).$$

Here we choose a so large that (i) sgn $F(x)$ is constant and (ii) $|y_1(x)-1|<\frac{1}{5}$ for $x>a$. After a has been chosen, C is uniquely determined but its actual value is of no importance for the following. We set

$$y_1(x) = 1 + \gamma(x)$$

and expand $[1+\gamma(t)]^{-2}$ in powers of $\gamma(t)$, using two terms and the exact form of the remainder. The result may be written

$$\begin{aligned} y_2(x) - (x-a) - C[1+\gamma(x)] &= \left[\gamma(x)(x-a) - \int_a^x \gamma(t)dt\right] \\ &\quad - [1+2\gamma(x)]\int_a^x \gamma(t)dt + [1+\gamma(x)]\int_a^x \frac{3\gamma^2(t)+2\gamma^3(t)}{[1+\gamma(t)]^2}dt. \end{aligned} \tag{2.12}$$

By assumption, the left side tends to the limit $a-C$ when $x\to\infty$ so the right member must also tend to the same limit. We shall show that this implies that $\int_a^x\gamma(t)dt$ tends to a finite limit.

By (2.6), which admits of $y_1(x)$ as its unique solution, we have

$$\gamma(x) = \int_x^\infty (t-x)F(t)y_1(t)dt, \qquad \gamma'(x) = -\int_x^\infty F(t)y_1(t)dt \tag{2.13}$$

so that sgn $\gamma(x)=$sgn $F(x)$, sgn $\gamma'(x)=-$sgn $F(x)$ for $x>a$ and $|\gamma(x)|$ is monotone decreasing to zero.

There are two cases to consider.

(i) $F(x)\geqq 0$ for $x>a$. In this case the first term in the right member of (2.12) is negative while the third term does not exceed 0.567 times $\int_a^x\gamma(t)dt$. It follows that the latter integral must tend to a finite limit when $x\to\infty$ and this implies that $\gamma(x)\in L(1,\infty)$.

(ii) $F(x)\leqq 0$ for $x>a$. In this case all three terms in the right member of (2.12) are positive and the same conclusion holds.

But

$$\int_a^\infty |\gamma(t)|\,dt = \int_a^\infty\int_t^\infty (s-t)\,|F(s)|\,y_1(s)dsdt$$

exists if and only if

$$\int_a^\infty\int_t^\infty (s-t)\,|F(s)|\,dsdt = \frac{1}{2}\int_a^\infty (s-a)^2\,|F(s)|\,ds$$

exists, that is, if and only if $x^2F(x)\in L(1,\infty)$. This completes the proof.

If (2.1) has property I* then

$$\lim_{x\to\infty}[C_1y_1(x) + C_2y_2(x) - C_1 - C_2x] = 0. \tag{2.14}$$

Expressed in geometrical language, this leads to the following.

COROLLARY. *There is a one-to-one correspondence between the integral curves of* (2.1) *and the non-vertical straight lines. Every integral curve has a unique slanting asymptote, distinct integral curves having distinct asymptotes, and every slanting straight line is the asymptote of a unique integral curve.*

The theory of Bessel functions provides illustrations of the results of the present section. If $F(x)=x^{-2-\alpha}$, then the corresponding equation has property I for $\alpha>0$ and property I* for $\alpha>1$. The corresponding subdominant solution $y_1(x)$ is a constant multiple of

$$x^{1/2}J_{1/\alpha}\left(\frac{2i}{\alpha}\, x^{-\alpha/2}\right),$$

a result which goes back to Euler. Similarly if $F(x)=-e^{-2x}$ we have $y_1(x)=J_0(e^{-x})$. Dominant solutions involve functions of the second kind. See G. N. Watson [8, pp. 96 and 99].

3. **Problem II.** Let $f(x)$ be a non-negative function defined for $x>0$ and belonging to $L(\epsilon, 1/\epsilon)$ for each $\epsilon>0$. We shall say that *the differential equation*

$$y'' + f(x)y = 0 \tag{3.1}$$

has the property II *if it is non-oscillatory for large* x.

LEMMA 3. *If* (3.1) *has property* II *and if* $y(x)$ *is a solution which is positive for* $x\geqq a$, *then* $y(x)$ *is monotone increasing and concave downwards for* $x>a$. *Further* $y'(x)$ *is positive and monotone decreasing towards a limit* $\geqq 0$.

Proof. From (2.4) we get

$$y'(x_2) - y'(x_1) = -\int_{x_1}^{x_2} f(t)y(t)dt \tag{3.2}$$

which is non-positive for $a<x_1<x_2$. It follows that $y'(x)$ is never increasing for $a<x$ and $y(x)$ is concave downwards. Since the graph of $y=y(x)$ lies below the curve tangent and does not intersect the x-axis for $x>a$, we must have $y'(x)>0$.

Formula (3.2) shows that $y(x)f(x)\in L(1, \infty)$, but all that can be concluded from Lemma 3 concerning the growth of $y(x)$ is that $|y(x)|\leqq Kx$ for large x. If the converse inequality should hold for a particular $y(x)$, then $xf(x)\in L(1, \infty)$ as we know from the preceding discussion, but this property is not necessary for the equation to have property II. For a further study of this matter we resort to the associated Riccati equation

$$v' + v^2 + f(x) = 0, \qquad v = y'/y. \tag{3.3}$$

LEMMA 4. *If* (3.1) *has property* II *and if* $y(x)\neq 0$ *for* $x\geqq a$, *then*

$$(3.4)\qquad 0 < (x+c)v(x) \leqq 1, \quad c = -a + 1/v(a), \quad v(x) = y'(x)/y(x).$$

Proof. For $x \geqq a$ the function $v(x)$ is absolutely continuous and satisfies (3.3) for almost all x. By Lemma 3, $v(x) > 0$ for $x \geqq a$ and (3.3) shows that it is a never increasing function of x. Moreover, $v' + v^2 \leqq 0$ whence it follows that

$$\frac{d}{dx}\left[-\frac{1}{v} + x\right] \leqq 0$$

and

$$-1/v(x) + x \leqq -1/v(a) + a, \qquad a \leqq x,$$

which is (3.4).

COROLLARY. *We have*

$$(3.5)\qquad 0 \leqq \limsup_{x\to\infty} xv(x) \leqq 1.$$

Next we proceed to strengthen Wintner's necessary condition.

LEMMA 5. *If $\mu(x)$ is a positive never decreasing function, if $\mu(x)x^{-2} \in L(1, \infty)$, and if* (3.1) *has property* II, *then $\mu(x)f(x) \in L(1, \infty)$.*

Proof. We choose a as in Lemma 4, $y(x)$ being a given solution of (3.1). We then multiply (3.3) by $\mu(x)$ and integrate between the limits a and b, obtaining after an integration by parts,

$$(3.6)\qquad \begin{aligned} &v(b)\mu(b) - v(a)\mu(a) - \int_a^b v(t)d\mu(t) \\ &\qquad + \int_a^b \mu(t)v^2(t)dt + \int_a^b \mu(t)f(t)dt = 0 \end{aligned}$$

The assumptions on $\mu(x)$ together with (3.4) show that the second integral in this formula tends to a finite limit when $b \to \infty$ and for the same reason $\mu(b)v(b) \to 0$. Further

$$\begin{aligned} \int_a^b v(t)d\mu(t) &\leqq \int_a^b \frac{d\mu(t)}{t+c} = \frac{\mu(b)}{b+c} - \frac{\mu(a)}{a+c} + \int_a^b \frac{\mu(t)}{(t+c)^2}dt \\ &\to -\frac{\mu(a)}{a+c} + \int_a^\infty \frac{\mu(t)}{(t+c)^2}dt. \end{aligned}$$

It follows that the last integral in (3.6) also tends to a finite limit and the theorem is proved.

Admissible choices of $\mu(x)$ are given by the functions

$$x^\sigma, \ \sigma < 1; \qquad x(\log x)^{-1-\alpha}; \qquad x(\log x)^{-1}(\log\log x)^{-1-\alpha}, \qquad \alpha > 0.$$

In particular we may take $\mu(x)=1$; this leads to the integral equation

$$v(x) = \int_x^\infty v^2(t)dt + \int_x^\infty f(t)dt, \tag{3.7}$$

which is basic for the following discussion.

THEOREM 4. *Equation* (3.1) *has property* II *if and only if the integral equation* (3.7) *has a solution for sufficiently large values of* x.

Proof. The necessity has already been proved. Suppose that there is a finite a such that (3.7) has a solution for $x \geqq a$. From the form of the equation it follows that $v^2(x) \in L(a, \infty)$ and $v(x)$ is a positive, monotone decreasing, absolutely continuous function. Differentiation with respect to x shows that $v(x)$ satisfies (3.3) for almost all x. Hence if we put

$$y(x) = \exp\left[\int_a^x v(t)dt\right],$$

then $y(x)$ satisfies (3.1) for almost all x and for $x \geqq a$ we have $y(x) \geqq 1$ so that (3.1) has property II.

For the discussion of (3.7) let us introduce the following notation

$$u(x) = xv(x), \qquad g(x) = x\int_x^\infty f(t)dt \tag{3.8}$$

in terms of which (3.7) becomes

$$u(x) = x\int_x^\infty u^2(t)\frac{dt}{t^2} + g(x). \tag{3.9}$$

We also write

$$\lim \begin{matrix}\sup \\ \inf\end{matrix} u(x) = \begin{cases} u^* \\ u_* \end{cases}, \qquad \lim \begin{matrix}\sup \\ \inf\end{matrix} g(x) = \begin{cases} g^* \\ g_* \end{cases}. \tag{3.10}$$

THEOREM 5. *If* (3.1) *has property* II, *then* $g_* \leqq \frac{1}{4}$ *and* $g^* \leqq 1$. *Both estimates are the best possible of their kind.*

Proof. Since (3.1) is assumed to have properly II, equation (3.9) has solutions for large values of x and by the corollary of Lemma 4 we have $0 \leqq u_* \leqq u^* \leqq 1$. The first term on the right of (3.9) is positive so that

$$g_* \leqq u_*, \qquad g^* \leqq u^*, \tag{3.11}$$

and the second inequality gives $g^* \leqq 1$ as asserted. Since

$$(u_*)^2 - \epsilon \leqq x\int_x^\infty u^2(t)\,\frac{dt}{t^2} \leqq (u^*)^2 + \epsilon$$

for $x \geqq x_\epsilon$, (3.9) also shows that

$$u_* \geqq (u_*)^2 + g_*, \qquad u^* \leqq (u^*)^2 + g^*. \tag{3.12}$$

The first of these inequalities requires that $g_* \leqq \frac{1}{4}$; it then gives

$$\tfrac{1}{2} - (\tfrac{1}{4} - g_*)^{1/2} \leqq u_* \leqq \tfrac{1}{2} + (\tfrac{1}{4} - g_*)^{1/2}. \tag{3.13}$$

The second inequality under (3.12) is true for all u^* if $g^* \geqq \frac{1}{4}$. It imposes a restriction on u^* if $g^* < \frac{1}{4}$ in which case it shows that either

$$u^* \leqq \tfrac{1}{2} - (\tfrac{1}{4} - g^*)^{1/2} \quad \text{or} \quad \tfrac{1}{2} + (\tfrac{1}{4} - g^*)^{1/2} \leqq u^*. \tag{3.14}$$

This completes the proof of the inequalities. Example (1.2) with $\gamma = \frac{1}{4}$ shows that g_* may equal $\frac{1}{4}$ when the equation has property II. An example with $g^* = 1$ will be constructed in §4.

THEOREM 6. *If* (3.1) *has property* II *and if for a particular solution* $y_1(x)$ *of* (3.1) *the corresponding function* $u_1(x)$ *tends to a limit* ρ *when* $x \to \infty$, *then* $\lim_{x\to\infty} g(x) = \gamma$ *exists and* $\rho^2 - \rho + \gamma = 0$. *Further* $\lim_{x\to\infty} u(x)$ *exists for every solution of* (3.1) *and is either* ρ *or* $1-\rho$. *There exists a solution for which the limit equals* $1-\rho$.

Proof. The first assertion is an immediate consequence of equation (3.9) and the latter also shows that if $\lim_{x\to\infty} u(x)$ exists for any other solution of (3.1), then the limit is either ρ or $1-\rho$. But to prove the existence of the limit requires a more elaborate argument than one would expect at first sight. The crux of the proof lies in showing the existence of a solution $y_2(x)$ with $\lim u_2(x) = 1-\rho$. There are two distinct cases according as $[y_1(x)]^{-2} \in L(a, \infty)$ or not. The first case is present, for a suitable choice of a, if $\frac{1}{2} < \rho \leqq 1$, the second if $0 \leqq \rho < \frac{1}{2}$, while $\rho = \frac{1}{2}$ may belong to either case. We shall carry through the argument in the first case. We form

$$y_2(x) = y_1(x) \int_x^\infty [y_1(t)]^{-2} dt,$$

which exists and is a solution of (3.1) under the present assumptions provided $x > a$. Logarithmic differentiation or use of the Wronskian yields

$$u_2(x) = u_1(x) - x/P(x), \qquad P(x) = y_1(x) y_2(x).$$

But an integration by parts gives

$$\begin{aligned} P(x) &= [y_1(x)]^2 \int_x^\infty [y_1(t)]^{-2} dt \\ &= -x + 2[y_1(x)]^2 \int_x^\infty u_1(t)[y_1(t)]^{-2} dt \\ &= -x + 2\rho P(x) + o[P(x)] \end{aligned}$$

since $u_1(t) \to \rho$ when $t \to \infty$. Hence

$$x/P(x) = 2\rho - 1 + o(1), \qquad \rho \geqq 1/2,$$

and $\lim u_2(x) = 1 - \rho$. If we set $y_2(x)/y_1(x) = R(x)$, then $\lim_{x\to\infty} R(x) = 0$. Hence, if $y(x) = C_1 y_1(x) + C_2 y_2(x)$ where $C_1 \neq 0$, then

$$u(x) = \frac{C_1 u_1(x) + C_2 u_2(x) R(x)}{C_1 + C_2 R(x)} \to \rho.$$

Thus the limit of $u(x)$ always exists and equals ρ unless $C_1 = 0$ in which case it becomes $1 - \rho$. We note that if $\rho = \frac{1}{2}$ so that the two limits coincide, then $y_1(x)$ and $y_2(x)$ are still linearly independent solutions of (3.1).

If $[y_1(x)]^{-2}$ is not in $L(a, \infty)$ for any a, we modify the definition of $y_2(x)$, replacing the integral from x to ∞ by one from a to x having the same integrand. The proof then goes through as in the first case.

The theorem shows that a necessary condition for the existence of $\lim u(x)$ for any solution $y(x)$ of (3.1) is the existence of $\lim g(x)$ which is then necessarily $\leqq \frac{1}{4}$. We shall prove in Theorem 8 below that, conversely, the existence of $\lim g(x)$ implies that of $\lim u(x)$ provided the first limit is $< \frac{1}{4}$.

We shall now prove a comparison theorem which leads to sufficient conditions for property II.

THEOREM 7. *Given the differential equations*

$$(3.15) \qquad Y'' + F(x)Y = 0, \qquad G(x) = x \int_x^\infty F(t)\,dt,$$

$$(3.16) \qquad y'' + f(x)y = 0, \qquad g(x) = x \int_x^\infty f(t)\,dt.$$

If the first equation has property II *and if* $G(x) \geqq g(x)$ *for* $x \geqq a$, *then the second equation also has property* II.

Proof. By Theorem 4 the integral equation

$$(3.17) \qquad U(x) = x \int_x^\infty U^2(t) \frac{dt}{t^2} + G(x)$$

has a solution $U(x)$, defined for $x \geqq b$ say. We now consider equation (3.9) for $x \geqq c = \max(a, b)$ and define successive approximations by

$$u_0(x) = U(x), \qquad u_n(x) = x \int_x^\infty u_{n-1}^2(t) \frac{dt}{t^2} + g(x).$$

Here

$$u_1(x) = x \int_x^\infty U^2(t) \frac{dt}{t^2} + g(x) \leqq x \int_x^\infty U^2(t) \frac{dt}{t^2} + G(x) = U(x) = u_0(x).$$

Since

$$u_n(x) - u_{n-1}(x) = x\int_x^\infty [u_{n-1}^2(t) - u_{n-2}^2(t)]\frac{dt}{t^2},$$

we see that $u_{n-1}(x) \geqq u_n(x) \geqq g(x)$ for all x and all n. Hence $\lim u_n(x) = u(x)$ exists and satisfies (3.9). Using Theorem 4 once more we see that equation (3.16) = (3.1) has property II as asserted.

COROLLARY 1. *Equation* (3.1) *has property* II *if* $g(x) \leqq \frac{1}{4}$ *for* $x \geqq a$. *This holds, in particular, if* $g^* < \frac{1}{4}$.

For $G(x) \equiv \frac{1}{4}$ corresponds to $F(x) = \frac{1}{4}x^{-2}$ and $Y(x) = x^{1/2}(C_1 + C_2 \log x)$ so that the corresponding equation (3.15) has property II. Another sufficient condition will be proved in §5 (Theorem 12).

COROLLARY 2. *If* $U(x)$ *is any solution of* (3.17) *defined for* $x \geqq c$, *and if* $G(x) \geqq g(x)$ *for* $x \geqq c$, *then there exists a solution* $u(x)$ *of* (3.9) *with* $u(x) \leqq U(x)$ *for* $x \geqq (c)$.

This was proved above.

THEOREM 8. *If* $\lim_{x\to\infty} g(x) = \gamma$ *exists and if* $\gamma < \frac{1}{4}$, *then* $\lim_{x\to\infty} u(x) = \lim_{x\to\infty} xy'(x)/y(x)$ *exists for every solution* $y(x)$ *of* (3.1), $y(x) \not\equiv 0$.

Proof. By Theorem 6 it is sufficient to prove the existence of a single solution $u_1(x)$ of (3.9) such that $\lim u_1(x)$ exists. We shall base the proof on Corollary 2. Given an $\epsilon > 0$, we can find an a such that $\gamma - \epsilon \leqq g(x) \leqq \gamma + \epsilon$ for $x \geqq a$. If $\gamma = 0$ we may replace $\gamma - \epsilon$ by 0; we may also suppose that $\gamma + \epsilon \leqq \frac{1}{4}$. Consider the quadratic equation $u^2 - u + g = 0$ with $g = \gamma - \epsilon$ or $\gamma + \epsilon$ and denote the smaller of its roots by ρ and σ respectively so that $\rho < \sigma \leqq \frac{1}{2}$. Together with (3.9) we consider the two auxiliary equations obtained by replacing $g(x)$ by $\gamma - \epsilon$ and $\gamma + \epsilon$ respectively, that is,

$$L(x) = x\int_x^\infty L^2(t)\frac{dt}{t^2} + \gamma - \epsilon, \tag{3.18}$$

$$U(x) = x\int_x^\infty U^2(t)\frac{dt}{t^2} + \gamma + \epsilon. \tag{3.19}$$

The first equation has two constant solutions ρ and $1 - \rho$; every solution tends to a limit when $x \to \infty$ and the limit is $1 - \rho$ unless $L(x) \equiv \rho$. The same description holds for the second equation if we replace ρ by σ.

By Corollary 2 there is a solution of (3.9), $u(x)$ say, such that $u(x) \leqq U(x) \equiv \sigma$, and there is a solution $L(x)$ of (3.18) with $L(x) \leqq u(x)$. But $L(x)$ tends to a limit when $x \to \infty$; the limit being $< \frac{1}{2}$, it must be ρ. Hence $L(x) \equiv \rho$ and $\rho \leqq u(x) \leqq \sigma$ when $x \geqq a$. Since $\sigma - \rho \leqq 2\epsilon^{1/2}$ we see that $\lim_{x\to\infty} u(x)$ must exist and the theorem is proved.

Another application of the ideas underlying Theorem 7 will be given in §6.

4. Counter examples. In order to show that the results proved in §3 are not capable of essential improvement we shall exhibit some counter examples.

A. *There exists a differential equation having property* II *for which* $g_* = 0$, $g^* = 1$.

Construction. If $y = y(x)$ is a positive, monotone increasing function whose graph is concave downwards, and if $y'(x)$ is continuous, then $y(x)$ satisfies a differential equation of type (3.1) having property II. The corresponding function $f(x)$ is easily found. In the following example the graph Γ of $y(x)$ is made up of arcs of parabolas of higher order and straight line segments fitted together so that $y(x)$ and $y'(x)$ are continuous. The function $f(x)$ can then be read off from formula (1.2). Varying the orders of the parabolas and the relative lengths of the arcs, we can modify the properties of $g(x)$ as desired. This is the general idea, the details follow.

We observe first that if the parabola $y = C(x-s)^{1/n}$ goes through the point (x_0, y_0) with the slope p_0, then $C^n = ny_0^{n-1}p_0$, $s = x_0 - y_0/(np_0)$. The curve Γ consists of arcs Γ_n and line segments λ_n following each other in the order $\Gamma_0, \lambda_1, \Gamma_1, \cdots, \Gamma_{n-1}, \lambda_n, \Gamma_n, \cdots$ and λ_n is tangent to Γ_{n-1} and Γ_n. Here Γ_0 is an arc of $y = x^{1/2}$ starting at $x = 0$ and ending at $x = a_1$ while Γ_n is an arc of $y = C_n(x - s_n)^{1/(n+2)}$ starting at $x = a_n + b_n$ and ending at $x = a_{n+1}$, where $a_n < a_n + b_n < a_{n+1}$ and the values of a_n and b_n will be chosen later. The line segment λ_n belongs to the straight line

$$y = C_{n-1}(a_n - s_{n-1})^{-(n+1)/(n+2)}\left[a_n - s_{n-1} + \frac{1}{n+2}(x - a_n)\right].$$

The values of C_n and s_n are uniquely determined by those of a_n and b_n; the value of C_n is immaterial but we need to know that

$$s_n = \sum_{k=1}^{n} \frac{k+1}{k+2} b_k.$$

We have

$$f(x) = \frac{n+1}{(n+2)^2}(x - s_n)^{-2}, \qquad a_n + b_n < x < a_{n+1},$$

and zero elsewhere.

We shall now specialize a_n and b_n. We choose

$$a_n = 2^{n^2}, \qquad b_n = 2^{n^2+n}$$

and find that

$$s_n = \frac{n+1}{n+2} 2^{n^2+n}[1 + O(2^{-2n})].$$

If $2^{n^2} \leqq x \leqq 2^{n^2} + 2^{n^2+n}$, then

$$\int_x^\infty f(t)dt = \frac{n+1}{n+2} 2^{-n^2-n}[1 + O(2^{-n})].$$

It follows that

$$g(2^{n^2}) = O(2^{-n}), \qquad g(2^{n^2+n}) = \frac{n+1}{n+2}[1 + O(2^{-n})].$$

Hence $g_* = 0$ and $g^* = 1$ as asserted.

Among the further properties of this particular function the following should be noted. For $2^{n^2}(1+2^n) \leqq x \leqq 2^{(n+1)^2}$ we have

$$u(x) = \frac{x}{(n+2)(x - s_n)}.$$

A simple computation shows that $u(x)$ decreases from

$$1 + O(2^{-n}) \quad \text{to} \quad \frac{1}{n+2}[1 + O(2^{-n})]$$

in this interval. Hence $u_* = 0$ and $u^* = 1$ which are the extreme limits of indetermination of a function $u(x)$ belonging to a differential equation of property II. In view of the discussion in §5 it is also worth noting that

$$\liminf_{x\to\infty} x^2 f(x) = 0, \qquad \limsup_{x\to\infty} x^2 f(x) = \infty. \tag{4.1}$$

This concludes the discussion of example A.

B. $g_* = 0$ *does not imply property* II.

Construction. We define $y(x)$ for $x \geqq 0$ by its graph which is made up of cosine arcs and horizontal line segments chosen in the following manner.

$$y(x) = 1, \qquad 0 \leqq x \leqq 1,$$

$$y(x) = (-1)^{n+1}, \qquad 2^{n^2+1} \leqq x \leqq 2^{(n+1)^2}, \qquad n = 0, 1, 2, \cdots,$$

$$y(x) = (-1)^n \cos(2^{-n^2}\pi x), \qquad 2^{n^2} \leqq x \leqq 2^{n^2+1}.$$

Here

$$f(x) = 2^{-2n^2}\pi^2, \qquad 2^{n^2} \leqq x \leqq 2^{n^2+1},$$

and zero elsewhere. Further

$$g(2^{n^2+1}) = 2^{-2n}\pi^2[1 + O(2^{-n})]$$

so that $g_* = 0$. Since $y(x)$ has infinitely many zeros, the equation cannot have property II.

C. $g^* = \frac{1}{4}$ *does not imply property* II.

Construction. We take

$$y(x) = x^{1/2}[C_1(\log x)^\rho + C_2(\log x)^{1-\rho}] \tag{4.2}$$

where ρ and $1-\rho$ are the two roots of the quadratic $u^2-u+\gamma_1=0$ and $\gamma_1>\frac{1}{4}$. Here

$$f(x) = \frac{1}{4x^2} + \frac{\gamma_1}{(x \log x)^2}, \tag{4.3}$$

so that $g_* = g^* = \frac{1}{4}$. Since ρ is complex, the solutions are oscillatory and the equation does not have property II.

5. **Extensions of the theorem of Kneser.** As observed in the Introduction, A. Kneser used the methods of Sturm and example (1.2) to derive conditions under which equation (3.1) has property II. With a slight extension, we can formulate his result as follows.

THEOREM 9. *Let*

$$\lim \begin{matrix}\sup\\ \inf\end{matrix}\; x^2 f(x) = \begin{cases}\gamma^* \\ \gamma_*\end{cases}. \tag{5.1}$$

The solutions of (3.1) *are non-oscillatory for large x if $\gamma^*<\frac{1}{4}$, oscillatory if $\gamma_*>\frac{1}{4}$ and no conclusion can be drawn if either γ^* or γ_* equals $\frac{1}{4}$.*

The proof of the first two assertions follows familiar lines and may be omitted here. Example C at the end of §4 is one in which $\gamma_* = \gamma^* = \frac{1}{4}$; the solutions are non-oscillatory if $\gamma_1 \leq \frac{1}{4}$ and oscillatory if $\gamma_1 > \frac{1}{4}$.

Example C suggests a further extension of Kneser's theorem. Cf. Rieman-Weber [7, p. 61] for Theorems 10 and 11.

THEOREM 10. *Let*

$$\lim \begin{matrix}\sup\\ \inf\end{matrix}\; (x \log x)^2 \left[f(x) - \frac{1}{4x^2}\right] = \begin{cases}\gamma_1^*, \\ \gamma_{1*}.\end{cases} \tag{5.2}$$

The solutions of (3.1) *are non-oscillatory for large x if $\gamma_1^* < \frac{1}{4}$, oscillatory if $\gamma_{1*} > \frac{1}{4}$, and no conclusion can be drawn if either γ_1^* or γ_{1*} equals $\frac{1}{4}$.*

The theorem is proved by the usual methods of Sturm using example C as comparison. The limiting cases require further counter examples.

As a matter of fact, examples (1.2) and C are merely the first instances of an infinite sequence of critical comparison equations which form a kind of logarithmic scale. To simplify the formulas, let us introduce some condensed notation. We write

$$L_0(x) = x, \qquad L_p(x) = L_{p-1}(x) \log_p x, \qquad p = 1, 2, 3, \cdots, \tag{5.3}$$

where

$$\log_2 x = \log\log x, \qquad \log_p x = \log\log_{p-1} x.$$

Further we set

$$S_p(x) = \sum_{k=0}^{p} [L_k(x)]^{-2}, \tag{5.4}$$

and define $e_1 = e$, $e_k = \exp(e_{k-1})$. Then the functions

$$y(x) = [L_{p-1}(x)]^{1/2}[C_1(\log_p x)^\rho + C_2(\log_p x)^{1-\rho}], \tag{5.5}$$

where ρ and $1-\rho$ are the two roots of the quadratic equation $u^2 - u + \gamma_p = 0$, satisfy a differential equation of type (3.1) with

$$f(x) = \tfrac{1}{4}S_{p-1}(x) + \gamma_p[L_p(x)]^{-2}, \qquad x > e_{p-1}. \tag{5.6}$$

This observation leads to

THEOREM 11. *Let*

$$\lim \begin{matrix}\sup\\ \inf\end{matrix} [L_p(x)]^2\{f(x) - \tfrac{1}{4}S_{p-1}(x)\} = \begin{cases}\gamma_p^*,\\ \gamma_{p*}.\end{cases} \tag{5.7}$$

The solutions of (3.1) *are non-oscillatory for large* x *if* $\gamma_p^* < \frac{1}{4}$, *oscillatory if* $\gamma_{p*} > \frac{1}{4}$, *and no conclusion can be drawn if either* γ_p^* *or* γ_{p*} *equals* $\frac{1}{4}$.

The proof follows the same lines as the preceding theorems. None of these theorems is particularly good because the limits involved are too much affected by irregularities in $f(x)$. Cf. formula (3.1). These irregularities are smoothed out, to some extent at least, by an integration process which leads to more powerful criteria. Thus, combining the ideas of Theorems 7 and 11, we get

THEOREM 12. *Equation* (3.1) *has property* II *if*

$$g(x) \leqq \tfrac{1}{4}x\int_x^\infty S_p(t)dt \tag{5.8}$$

for x *sufficiently large.*

Corollary 1 of Theorem 7 is the special case $p = 0$ of this theorem.

6. An extension to the complex domain. The use of the singular Riccati integral equation (3.9) is not restricted to real variables. It can also be used to prove non-oscillation theorems in the complex domain. The results obtainable in this manner, though of a somewhat special nature, appear to be basically different from those derived by the present writer in the early nineteen twenties. The following is a sample of what can be done.

THEOREM 13. *Let* $f(z)$ *be holomorphic in a sector* S: $-\theta_1 < \arg z < \theta_2$ *of the complex plane and suppose that*

$$g(z) = z\int_z^\infty f(t)dt \tag{6.1}$$

is well defined in S when the integral is taken along a line parallel to the real axis. Finally, suppose that $|g(z)| \leqq \gamma < \frac{1}{4}$ for $z \in S$. Then there exists a solution $w(z)$ of the differential equation

$$w'' + f(z)w = 0 \tag{6.2}$$

which has no zeros in S.

Proof. It is understood that $0 < \theta_1, \theta_2 < \pi$. We choose α so that $\gamma\alpha = \frac{1}{4}\sin\alpha$ and denote by S_α the intersection of S with the sector $0 < |z| < \infty$, $|\arg z| < \alpha$. Using the method of successive approximations we then construct a solution of

$$u(z) = z\int_z^\infty u^2(t)\frac{dt}{t^2} + g(z) \tag{6.3}$$

for $z \in S_\alpha$. The path of integration is taken parallel to the real axis. We set

$$u_0(z) = 2\gamma, \qquad u_n(z) = z\int_z^\infty u_{n-1}^2(t)\frac{dt}{t^2} + g(z).$$

Suppose that $\max |u_n(z)| = B_n$ for $z \in S_\alpha$. Then

$$B_n \leqq \frac{\alpha}{\sin\alpha} B_{n-1}^2 + \gamma, \tag{6.4}$$

where we have used the formula

$$r\int_0^\infty |re^{i\theta} + s|^{-2}ds = \frac{\theta}{\sin\theta}\cdot$$

In view of the value of B_0 and the definition of α, we conclude from (6.4) that $B_n \leqq 2\gamma$ for all n. Thus the functions $u_n(z)$ are holomorphic in S_α and uniformly bounded. If $z = x$ is real, a simple computation shows that

$$|u_n(x) - u_{n-1}(x)| \leqq 4\gamma |u_{n-1}(x) - u_{n-2}(x)|.$$

Hence the sequence $\{u_n(z)\}$ converges on the positive real axis. By the theorem of Vitali, $\lim_{n\to\infty} u_n(z) = u(z)$ exists and is holomorphic everywhere in S_α. Further $u(z)$ satisfies (6.3).

If S_α exhausts S, we are through. If not, the fact that $|u(z)|$ is bounded in S_α enables us to modify the path of integration in (6.3). Let θ_0 be real, $|\theta_0| < \max(\alpha, \theta_1, \theta_2)$, and replace the path of integration $\arg(t-z) = 0$ by $\arg(t-z) = \theta_0$. Consider the sector $S_\alpha(\theta_0)$ which is the intersection of S with $|\arg z - \theta_0| < \alpha$, $0 < |z| < \infty$. In $S \cap S_\alpha(\theta_0)$ the function $u(z)$ satisfies both the original integral equation and the new one with the modified path. The convergence proof then shows that the same sequence of approximations converges to a holomorphic function in $S_\alpha(\theta_0)$. By a suitable repetition of this process, we can extend $u(z)$ analytically throughout S.

Once we have a solution $u(z)$ of (6.3) which is holomorphic in S, we see that $v(z)=u(z)/z$ satisfies the Riccati equation $v'+v^2+f(z)=0$ everywhere in S, and placing $w(z)=\exp\left[\int_1^z v(t)dt\right]$ we have a solution of (6.2) which is different from zero everywhere in S. This completes the proof.

It is natural to ask if the value zero is of low frequency in S for every solution of (6.2). This is certainly true in simple cases; whether or not it is generally true requires further investigation.

Bibliography

1. R. Bellman, *The boundedness of solutions of linear differential equations*, Duke Math. J. vol. 14 (1947) 83–97.

2. E. Cotton, *Sur les solutions asymptotiques des équations différentielles*, Ann. École Norm. (3) vol. 28 (1911) pp. 473–521.

3. W. B. Fite, *Concerning the zeros of the solutions of certain differential equations*, Trans. Amer. Math. Soc. vol. 19 (1917) pp. 341–352.

4. E. Hille, *A general type of singular point*, Proc. Nat. Acad. Sci. U.S.A. vol. 10 (1924) pp. 488–493.

5. A. Kneser, *Untersuchungen über die reellen Nullstellen der Integrale linearer Differentialgleichungen*, Math. Ann vol. 42 (1893) pp. 409–435.

6. N. Levinson, *The growth of the solutions of a differential equation*, Duke Math. J. vol. 8 (1941) pp. 1–10.

7. B. Riemann-H. Weber, *Die partiellen Differentialgleichungen der Mathematischen Physik*, vol. II, 5th ed., Braunschweig, 1912, xiv+575 pp.

8. G. N. Watson, *A treatise on the theory of Bessel functions*, Camridge, 1944, viii+804 pp.

9. A. Wiman, *Über die reellen Lösungen der linearen Differentialgleichungen zweiter Ordnung*, Arkiv för Matematik, Astronomi och Fysik vol. 12, no. 14 (1917) 22 pp.

10. A. Wintner, *On the Laplace-Fourier transcendents occurring in mathematical physics*, Amer. J. Math. vol. 69 (1947) pp. 87–98.

Yale University,
New Haven, Conn.

ARKIV FÖR MATEMATIK Band 2 nr 2

Read 6 June, 1951

Behavior of solutions of linear second order differential equations

By Einar Hille

1. Introduction. The present note is concerned with the differential equation

$$w'' = \lambda F(x) w, \tag{1.1}$$

where $F(x)$ is defined, positive and continuous for $0 \leqq x < \infty$, while λ is a complex parameter which, except in section 2, is not allowed to take on real values $\leqq 0$. We are mainly interested in qualitative properties of the solutions for large positive values of x including integrability properties on the interval $(0, \infty)$. In section 6 we shall discuss certain extremal problems for this class of differentiel equations.

The results are of some importance for the theory of the partial differential equations of the Fokker-Planck-Kolmogoroff type corresponding to temporally homogeneous stochastic processes. These applications will be published elsewhere. The results also admit of a dynamical formulation and interpretation. This will be used frequently in the following for purposes of exposition. With $x = t$, the equation

$$w'' = \lambda F(t) w \tag{1.2}$$

is the equation of motion in complex vector form of a particle

$$w = u + iv = r e^{i\theta} \tag{1.3}$$

under the influence of a force of magnitude

$$|P| = \varrho F(t) r, \quad \lambda = \varrho e^{i\varphi} = \mu + i\nu, \tag{1.4}$$

making the constant angle φ with the radius vector. We can also write the equations of motion in the form

$$r'' - r(\theta')^2 = \mu F(t) r, \tag{1.5}$$

$$\frac{d}{dt}[r^2 \theta'] = \nu F(t) r^2, \tag{1.6}$$

where the left sides are the radial and the transverse accelerations respectively.

2. Almost uniform motion. As a preliminary step in the discussion we eliminate the fairly trivial case in which $x F(x) \in L(0, \infty)$. This case is basic for the applications referred to above, however.

Theorem 1. *A necessary and sufficient condition that* (1.1) *have a fundamental system of the form*

$$w_1(x) = x[1 + o(1)], \quad w_2(x) = 1 + o(1), \quad x \to \infty, \tag{2.1}$$

for some fixed $\lambda \neq 0$ *is that* $x F(x) \in L(0, \infty)$. *If this condition is satisfied, then* (2.1) *holds for all* λ *and we have also*

$$w_1'(x) = 1 + o(1), \quad w_2'(x) = o(1), \quad x \to \infty. \tag{2.2}$$

In the dynamical interpretation we could refer to this case as *almost uniform motion* (uniform motion corresponds to $F(t) \equiv 0$).

The sufficiency of the condition has been traced back to M. BÔCHER [1], p. 47, the necessity for $\lambda > 0$ to H. WEYL [5], p. 42, but it does not occur explicity in either place. A direct proof for $\lambda = 1$, $F(x)$ real was given by the author [3], pp. 237—238, in 1948; the sufficiency argument is valid also for complex-valued $F(x)$ and the necessity was proved when $F(x)$ is real and keeps a constant sign for large x. A considerably less restrictive sufficient condition for complex-valued $F(x)$ was given by A. WINTNER [7] in 1949 who also gave an alternate proof for the necessity of the condition $x F(x) \in L(0, \infty)$ when $F(x)$ is real and ultimately of constant sign. Wintner referred to the case he studied as *almost free linear motion*. Theorem 1 is a special case of the following more general result:

Theorem 2. *If* $G(x)$ *is continuous for* $0 \leqq x < \infty$ *and there exist a real* β *and a positive* δ *such that* $|\arg[e^{-i\beta} G(x)]| \leqq \frac{1}{2}\pi - \delta$ *for all large* x, *then the differential equation*

$$w'' = G(x)\, w \tag{2.3}$$

has a fundamental system of the form (2.1) *if and only if* $x G(x) \in L(0, \infty)$. *This system also satisfies* (2.2).

Proof. It is only the necessity that calls for a proof. Set

$$e^{-i\beta} G(x) = G_1(x) + i\, G_2(x), \quad w_2(x) = x[1 + \eta_1(x) + i\, \eta_2(x)]$$

and suppose that a is so large that for $x \geqq a$ we have $G_1(x) \geqq 0$,

$$|G_2(x)/G_1(x)| \leqq \cot\delta, \quad |\eta_\nu(x)| \leqq \tfrac{1}{3}\min(1, \tan\delta), \quad \nu = 1, 2.$$

From

$$\Re\{e^{-i\beta}[w_2'(x) - w_2'(a)]\} = \int_a^x s\, G_1(s) \left\{1 + \eta_1(s) - \frac{G_2(s)}{G_1(s)} \eta_2(s)\right\} ds,$$

it follows that

$$\tfrac{5}{3} \int_a^x s\, G_1(s)\, ds \geqq \Re\{e^{-i\beta}[w_2'(x) - w_2'(a)]\} \geqq \tfrac{1}{3} \int_a^x s\, G_1(s)\, ds.$$

Integration of this inequality shows that $w_2(x)$ cannot be $O(x)$ unless $x\,G_1(x) \in L(0, \infty)$ and this implies and is implied by $x\,G(x) \in L(0, \infty)$. This completes the proof of Theorems 1 and 2.

In Theorem 1 we may take as our hypothesis the existence of a solution satisfying any one of the three conditions

$$\text{(i)}\ \ w_1(x) = x\,[1 + o(1)], \quad \text{(ii)}\ \ w_2(x) = 1 + o(1), \quad \text{(iii)}\ \ w_1'(x) = 1 + o(1).$$

They are equivalent and imply

$$\text{(iv)}\ \ w_2'(x) = o(1)$$

as well as $x\,F(x) \in L(0, \infty)$. On the other hand, (iv) does not imply (i)—(iii) or the integrability condition.

3. Direct motion. In the remainder of this paper it will be assumed that $x\,F(x)$ is not in $L(0, \infty)$. Further λ will be restricted to the domain Λ obtained by deleting the origin and the negative real axis from the λ-plane. If λ is real and negative, the solutions of (1.1) are normally oscillatory and their behavior is entirely different from that holding for $\lambda \in \Lambda$. This fairly well known case will not be considered in the following.

The behavior for real positive values of λ is also well known (see, for instance H. WEYL [5], § 1), but it sets the pattern for the rest of Λ so we shall summarize the results. Let $w_0(x) = w_0(x, \lambda)$, $w_1(x) = w_1(x, \lambda)$ be the fundamental system determined by the initial conditions

$$w_0(0) = 0, \ w_0'(0) = 1; \quad w_1(0) = 1, \ w_1'(0) = 0. \tag{3.1}$$

For $x > 0$, $\lambda > 0$, these solutions are positive, monotone increasing and convex downwards so that $w_k(x, \lambda)/x \to \infty$ with x by Theorem 1. Simple counterexamples show that no stronger assertion can be made concerning the rate of growth (cf. section 6 below). In passing we note the Liapounoff-Birkhoff inequalities which in the present case may be given the form

$$\begin{aligned} C_k \exp\left\{-\sqrt{\varrho}\int_0^x [1 + F(s)]\,ds\right\} &\le [w_k(x, \varrho)]^2 + \frac{1}{\varrho}[w_k'(x, \varrho)]^2 \\ &\le C_k \exp\left\{\sqrt{\varrho}\int_0^x [1 + F(s)]\,ds\right\}, \end{aligned} \tag{3.2}$$

where C_k is the initial value of the second member for $x = 0$.

For fixed x the solutions $w_k(x, \lambda)$ are entire functions of λ given by the power series

$$w_k(x, \lambda) = \sum_{n=0}^{\infty} u_{k,n}(x)\,\lambda^n, \quad u_{k,n}(x) = \int_0^x (x - s)\,F(s)\,u_{k,n-1}(s)\,ds \tag{3.3}$$

with $u_{0,0}(x) = x$, $u_{1,0}(x) = 1$. The coefficients being never negative, one sees that

$$|w_k(x, \lambda)| \leqq w_k(x, \varrho), \quad |w_k'(x, \lambda)| \leqq w_k'(x, \varrho). \tag{3.4}$$

This in conjunction with (3.2) shows that $w_k(x, \lambda)$ is an entire function of λ of order $\leqq \frac{1}{2}$. This could also be inferred from (3.3).

For $\lambda > 0$ the formula

$$w_+(x, \lambda) = w_1(x, \lambda) \int_x^\infty \frac{ds}{[w_1(s, \lambda)]^2} \tag{3.5}$$

is obviously meaningful and defines a *subdominant solution* of (1.1). It is positive, monotone decreasing, convex downwards and tends to zero when $x \to \infty$. The first property is obvious; the second follows from

$$w_+'(x, \lambda) = w_1'(x, \lambda) \int_x^\infty \frac{ds}{[w_1(s, \lambda)]^2} - \frac{1}{w_1(x, \lambda)}$$

$$< \int_x^\infty \frac{w_1'(s, \lambda)\, ds}{[w_1(s, \lambda)]^2} - \frac{1}{w_1(x, \lambda)} = 0,$$

and the convexity is implied by (1.1). Finally

$$w_+(x, \lambda) < \frac{w_1(x, \lambda)}{w_1'(x, \lambda)} \int_x^\infty \frac{w_1'(s, \lambda)\, ds}{[w_1(s, \lambda)]^2} = \frac{1}{w_1'(x, \lambda)} \to 0.$$

as $x \to \infty$.

The descriptive properties of $w_k(x, \lambda)$ for complex λ are more complicated than for $\lambda > 0$, but the following results hold.

Theorem 3. *If $x F(x) \notin L(0, \infty)$, if $\lambda = \mu + i\nu \in \Lambda$ and $\nu \neq 0$, then $w_k(x, \lambda)$ describes a spiral $S_k(\lambda)$ from k to ∞ in the complex w-plane as x goes from 0 to $+\infty$, $k = 0, 1$, and $\frac{1}{\nu} \arg w_k(x, \lambda)$ increases steadily from 0 to $+\infty$ with x. $S_k(\lambda)$ has a positive radius of curvature everywhere and is concave towards the origin. If $\mu \geqq 0$, $|w_k(x, \lambda)|$ is monotone increasing and $|w_k(x, \lambda)|/x \to \infty$ with x when $\mu > 0$. For all $\lambda \in \Lambda$, $|w_k(x, \lambda)|^{-1} \in L_2(1, \infty)$.*

Proof. We set $|w_k(x, \lambda)| = r_k(x, \lambda) = r_k(x) = r_k$ and $\arg w_k(x, \lambda) = \theta_k(x, \lambda) = \theta_k(x) = \theta_k$ with $\theta_k(0, \lambda) \equiv 0$. The functions $r_k(t)$ and $\theta_k(t)$ satisfy equations (1.5) and (1.6). From the former we see that for $\mu \geqq 0$ we have $r_k'' > 0$, r_k' increasing, from 1 if $k = 0$, from 0 if $k = 1$. Thus $r_k' > 0$ and r_k is increasing. Further r_k' tends to a limit, γ_k say, as $t \to \infty$. If γ_k is finite, $r_k(t) \sim \gamma_k t$. But $t F(t) \notin L(1, \infty)$ and we have

$$r_k'(t) - r_k'(0) > \mu \int_0^t F(s)\, r_k(s)\, ds,$$

which tends to infinity with t if $\mu > 0$. Hence $r_k'(t) \to \infty$ if $\mu > 0$ and $r_k(t)/t \to \infty$ with t. If $k=0$ the ratio $r_k(t)/t$ is increasing for $t>0$ and if $k=1$ it is increasing for all large t for

$$t\, r_k'(t) - r_k(t) > \mu \int_0^t s\, F(s)\, r_k(s)\, ds - r_k(0).$$

If $\mu = 0$, we still have $r_k'' > 0$, $r_k' > 0$ and increasing so that $r_k(t) > Ct$. An example in section 6 below shows that $r_k(t)$ may actually be $O(t)$ when $\mu = 0$.

A Sturmian comparison argument applied to the equations

$$R'' = \mu F(x) R, \quad r'' = \{\mu F(x) + [\theta'(x)]^2\} r$$

gives

$$w_k(x, |\mu + i\nu|) > |w_k(x, \mu + i\nu)| > w_k(x, \mu), \quad x > 0, \tag{3.6}$$

provided $\mu > 0$, $\nu \neq 0$.

The expression for the transverse acceleration in (1.6) gives

$$[r_k(t)]^2\, \theta_k'(t) = \nu \int_0 F(s)\, [r_k(s)]^2\, ds \tag{3.7}$$

so that $(1/\nu)\, \theta_k'(t)$ is positive. This equation gives the angular momentum of a particle of unit mass at the time t; if $\mu > 0$ it clearly becomes infinite with t and, as we shall see, the same is true everywhere in Λ for $\nu \neq 0$. We note in passing that the left side of (3.7) is never zero for $t > 0$ and this implies that neither $w_k(t, \lambda)$ nor $w_k'(t, \lambda)$ can be zero for $t > 0$, $\lambda \in \Lambda$.

We shall now prove that $(1/\nu) \arg w_k(x, \lambda)$ tends to infinity with x. Suppose, contrariwise, that it tends to a finite limit ω_k instead and set

$$W_k(x) = e^{-i\nu\omega_k}\, w_k(x) = U_k(x) + i\, V_k(x).$$

For a given $\varepsilon > 0$, we can find an $a = a_\varepsilon$ such that $U_k(x) > 0$ and

$$0 \leqq |V_k(x)| \leqq \varepsilon\, U_k(x)$$

for $x \geqq a$. We have

$$W_k'(x) - W_k'(a) = \lambda \int_a^x F(s)\, W_k(s)\, ds. \tag{3.8}$$

Here there are two possibilities. First, the integral may tend to a finite limit as $x \to \infty$ so that $W_k'(x)$ tends to a finite limit. This limit cannot be different from zero since otherwise the case of almost uniform motion would be present. If $W_k'(\infty) = 0$, a second integration gives

$$W_k(x) = W_k(a) - \lambda \int_a^x ds \int_s^\infty F(t)\, W_k(t)\, dt.$$

Here the repeated integral must become infinite with x since otherwise the almost uniform case would turn up again. This implies that $W_k(x)$ becomes infinite in such a manner that its argument tends to zero while in the right member of the equation the term that becomes infinite has an argument differing from that of $-\lambda$ by at most ε. Thus the first possibility leads to a contradiction. Secondly, the integral in (3.8) may become infinite with x. Integration of (3.8) then shows that $W_k(x)$ differs from

$$\lambda \int_a^x (x-t) F(t) W_k(t)\, dt$$

by (linear) terms of lower order. This alternative gives rise to the same type of contradiction: $W_k(x)$ has an argument close to zero, while that of the dominating term in the second member is close to arg λ. Thus the assumption that arg $w_k(x)$ stays bounded must be rejected.

Let us now introduce some notation. We set

$$M_k(x, \lambda) = \int_0^x |w_k'(s, \lambda)|^2\, ds, \tag{3.9}$$

$$N_k(x, \lambda) = \int_0^x F(s)\, |w_k(s, \lambda)|^2\, ds, \tag{3.10}$$

$$L_k(x, \lambda) = M_k(x, \lambda) + \varrho N_k(x, \lambda), \quad \varrho = |\lambda|. \tag{3.11}$$

These are obviously positive increasing functions of x and it will be shown later that they all become infinite with x. We also have

$$\overline{w_k(x, \lambda)}\, w_k'(x, \lambda) = M_k(x, \lambda) + \lambda N_k(x, \lambda). \tag{3.12}$$

This is the Green's transform of the equation (1.1) corresponding to the interval $(0, x)$. See E. HILLE [3], p. 3, and E. L. INCE [4], p. 508. This formula also shows that the product occurring on the left cannot vanish for $x \neq 0$ and $\lambda \in \Lambda$.

As a first application of (3.12) let us verify the assertion concerning the radius of curvature For a complex curve $w = w(x)$ the radius of curvature is given by

$$R = \frac{|w'|^3}{\Im[\bar{w}' w'']}.$$

Using (1.1) and (3.12) this reduces to

$$R_k = \frac{|w_k'(x, \lambda)|^3}{\nu F(x) M_k(x, \lambda)} \tag{3.13}$$

if $\nu > 0$. For $\nu < 0$, the sign should be reversed since the spirals $S_k(\lambda)$ are then described in the negative sense. Note that there are no points of inflection.

Another application of (3.12) is the observation that if

$$\arg \lambda = \varphi = 2\gamma, \quad |\gamma| < \tfrac{1}{2}\pi,$$

then

$$\cos \gamma \, L_k(x, \lambda) = \Re\, [e^{-i\gamma} \overline{w_k(x, \lambda)}\, w_k'(x, \lambda)]. \tag{3.14}$$

Recombined with (3.12) this leads to the basic double inequality

$$\cos \gamma \, L_k(x, \lambda) \leqq |\, w_k(x, \lambda)\, w_k'(x, \lambda)\,| \leqq L_k(x, \lambda). \tag{3.15}$$

As a first consequence of (3.15) we note that

$$\frac{\cos^2 \gamma}{|\, w_k(x)\,|^2} \leqq \frac{|\, w_k'(x)\,|^2}{[L_k(x)]^2} \leqq \frac{L_k'(x)}{[L_k(x)]^2}.$$

Hence

$$\cos^2 \gamma \int_x^\infty \frac{d\,s}{|\, w_k(s, \lambda)\,|^2} < \frac{1}{L_k(x, \lambda)} \tag{3.16}$$

so that $|\, w_k(x, \lambda)\,|^{-1} \in L_2(1, \infty)$ as asserted. From this fact, together with the observation that $\arg w_k(x, \lambda)$ becomes infinite with x, one concludes from (3.7) that $N_k(x, \lambda)$ and hence also $L_k(x, \lambda)$ become infinite with x for every $\lambda \in \Lambda$. The same is true for $M_k(x, \lambda)$ by virtue of (3.16) and the estimate

$$|\, w_k(x, \lambda) - k\,|^2 \leqq x\, M_k(x, \lambda) \tag{3.17}$$

which follows from Schwarz's inequality applied to

$$w_k(x, \lambda) - k = \int_0^x w_k'(s, \lambda)\, d\,s.$$

We also note the inequality

$$|\, w_k(x, \lambda)\,|^2 \leqq k + 2 \int_0^x L_k(s, \lambda)\, d\,s \tag{3.18}$$

which may be read off from (3.15). This completes the proof of Theorem 3.
Some further consequences of (3.15) are listed in

Theorem 4. *Let $\Phi(u)$ be any non-negative function integrable over every finite interval $(0, \omega)$. Let $0 < a < b < \infty$, $A = L_k(a, \lambda)$, $B = L_k(b, \lambda)$. Let ξ, η, ζ be arbitrary non-negative constants. Then*

$$\begin{aligned} \int_a^b \left\{ \xi \frac{\cos^2 \gamma}{|\, w_k(x, \lambda)\,|^2} + \eta\, \varrho \frac{\cos^2 \gamma\, F(x)}{|\, w_k'(x, \lambda)\,|^2} + \zeta \frac{2\sqrt{\varrho}\, \cos \gamma\, \sqrt{F(x)}}{L_k(x, \lambda)} \right\} \Phi[L_k(x, \lambda)]\, d\,x \\ \leqq (\xi + \eta + \zeta) \int_A^B \Phi(u) \frac{d\,u}{u^2}. \end{aligned} \tag{3.19}$$

Proof. The inequality is obviously the sum of three inequalities obtained by setting two of the parameters equal to zero. Here the inequality for $\eta=\zeta=0$ is proved exactly as formula (3.16) which is the special case $\Phi(u)\equiv 1$ and the case $\xi=\zeta=0$ is handled in a similar manner using the inequality

$$\varrho F(x)\,|\,w_k(x,\lambda)\,|^2 < L_k'(x,\lambda).$$

The case $\xi=\eta=0$ follows from the inequality between the geometric and the arithmetic means which gives

$$2\sqrt{\varrho F(x)}\,|\,w_k(x,\lambda)\,w_k'(x,\lambda)\,| \leqq L_k'(x,\lambda)$$

whence

$$2\sqrt{\varrho}\cos\gamma\sqrt{F(x)} < \frac{L_k'(x,\lambda)}{L_k(x.\lambda)}. \tag{3.20}$$

Incidentally, the last inequality also shows that

$$L_k(x,\lambda) > L_k(1,\lambda)\exp\left\{2\sqrt{\varrho}\cos\gamma\int_1^x\sqrt{F(s)}\,ds\right\}. \tag{3.21}$$

Since the three fractions within the braces in (3.19) appear to have similar integrability properties, one might expect them to have similar behavior for large values of x. By analogy with the case of $\lambda>0$ (cf. A. WIMAN [6] p. 17) one might expect that

$$\lim_{x\to\infty}\frac{\sqrt{F(x)}\,w_k(x,\lambda)}{w_k'(x,\lambda)}=\frac{1}{\sqrt{\lambda}}=\varrho^{-\frac{1}{2}}e^{-i\gamma}, \tag{3.22}$$

at least if $F(x)$ is suitably limited. We are not able to prove any such relation, but we can show that

$$\limsup_{x\to\infty}\frac{1}{x}\int_1^x\sqrt{F(s)}\left|\frac{w_k(s,\lambda)}{w_k'(s,\lambda)}\right|ds \leqq \varrho^{-\frac{1}{2}}\sec^2\gamma. \tag{3.23}$$

Indeed, using (3.18) and (3.20) one sees that $\sqrt{\varrho}\cos^2\gamma$ times the integrand does not exceed

$$\frac{1}{2}\frac{L_k'(s,\lambda)}{[L_k(s,\lambda)]^2}\left[k+2\int_0^s L_k(t,\lambda)\,dt\right]$$

and an integration by parts gives the desired inequality.

4. Retrograde motion. The results of the preceding section hold, at least for sufficiently large values of x, for all solutions with one striking exception. Formula (3.5) makes sense for all $\lambda\in\Lambda$ and defines a solution of (1.1). This will be referred to as the *exceptional* or the *subdominant solution.*

Theorem 5. *If* $x F(x) \notin L(0, \infty)$ *and if* $\lambda \in \Lambda$, $\nu \neq 0$, *then the subdominant solution* $w_+(x, \lambda)$ *describes a spiral* $S_+(\lambda)$ *in the complex w-plane as x goes from* 0 *to infinity. The sense of rotation is opposite to that of* $S_1(\lambda)$ *and* $-(1/\nu)\arg w_+(x, \lambda)$ *increases to* $+\infty$ *with x.* $S_+(\lambda)$ *has a positive radius of curvature and is concave towards the origin.* $|w_+(x, \lambda)|$ *is monotone decreasing if* $\mu \geqq 0$ *and tends to zero if* $\mu > 0$. *For each* $\lambda \in \Lambda$ *we have that* $w_+(x, \lambda) w'_+(x, \lambda) \to 0$ *as* $x \to \infty$ *and* $F(x)|w_+(x, \lambda)|^2$ *and* $|w'_+(x, \lambda)|^2$ *belong to* $L(0, \infty)$.

Proof. That $F(x)|w_+(x, \lambda)|^2 \in L(0, \infty)$ follows from (3.5) and (3.16) together with $\varrho F(x)|w_1(x, \lambda)|^2 < L'_1(x, \lambda)$. Further

$$
(4.1) \quad \begin{aligned} |w'_+(x, \lambda)| &= \left| w'_1(x, \lambda) \int_x^\infty \frac{ds}{[w_1(s, \lambda)]^2} - \frac{1}{w_1(x, \lambda)} \right| \\ &< \sec^2 \gamma \frac{|w'_1(x, \lambda)|}{L_1(x, \lambda)} + \frac{1}{|w_1(x, \lambda)|} \end{aligned}
$$

and both terms in the third member are in $L_2(1, \infty)$ so the same is true for the first member. We have also

$$
(4.2) \quad w_+(x, \lambda) w'_+(x, \lambda) = w_1(x, \lambda) w'_1(x, \lambda) \left\{ \int_x^\infty \frac{ds}{[w_1(s, \lambda)]^2} \right\}^2 - \int_x^\infty \frac{ds}{[w_1(s, \lambda)]^2}
$$

and by (3.15) and (3.16) both terms on the right are $O\{[L_1(x, \lambda)]^{-1}\}$ and thus tend to zero when $x \to \infty$.

The Green's transform for the interval (x, ∞) gives

$$
(4.3) \quad \begin{aligned} \overline{w_+(x, \lambda)} w'_+(x, \lambda) &= -\int_x^\infty |w'_+(s, \lambda)|^2 ds - \lambda \int_x^\infty F(s) |w_+(s, \lambda)|^2 ds \\ &\equiv -M_+(x, \lambda) - \lambda N_+(x, \lambda), \end{aligned}
$$

so the first member is different from zero for $0 \leqq x < \infty$, $\lambda \in \Lambda$. In particular, the integral occurring in formula (3.5) is never zero. Further

$$
(4.4) \quad \frac{d}{dx} |w_+(x, \lambda)|^2 = -2[M_+(x, \lambda) + \mu N_+(x, \lambda)]
$$

so that $|w_+(x, \lambda)|$ is monotone decreasing for $\mu \geqq 0$ and tends to a positive limit or zero. If $F(x) \notin L(0, \infty)$ and $\mu > 0$, the first possibility is obviously excluded. It is easily verified, however, that

$$
(4.5) \quad w_0(x, \lambda) = w_+(0, \lambda) w_+(x, \lambda) \int_0^x \frac{ds}{[w_+(s, \lambda)]^2}
$$

so that if $|w_+(x, \lambda)| \to C > 0$ it would follow that $w_0(x, \lambda) = O(x)$ and this is false for $\mu > 0$, but it may be true for $\mu = 0$ as shown by an example in section 6. Next we observe that

$$\arg w_+(x, \lambda) = \arg w_+(0, \lambda) - \nu \int_0^x \frac{N_+(s, \lambda)}{|w_+(s, \lambda)|^2} ds,$$

whence the monotony properties of the argument follow. If now the left member should tend to a finite limit as $x \to \infty$, formula (4.5) may be used to show that $\arg w_0(x, \lambda)$ must also tend to a finite limit and this we know is not true. Finally the radius of curvature R_+ of $S_+(\lambda)$ is obtained from formula (3.13), replacing the subscript k by $+$. This completes the proof.

In passing we note that

$$w_0(x, \lambda) + w_+(x, \lambda) = w_1(x, \lambda) \int_0^\infty \frac{ds}{[w_1(s, \lambda)]^2}, \tag{4.6}$$

where the integral is finite and different from zero for $\lambda \in \Lambda$.

The solutions $w_1(x, \lambda)$ and $w_+(x, \lambda)$ are evidently linearly independent when $\lambda \in \Lambda$. From this we conclude that the subdominant is characterized uniquely up to a multiplicative constant by anyone of the properties listed in Theorem 5. In particular, $C w_+(x, \lambda)$ *is the only solution describing a retrograde spiral if* $S_1(\lambda)$ *is considered as defining the direct motion.* Any other solution will describe a spiral which is ultimately direct.

We note that formula (3.6) has an analogue for $w_+(x, \lambda)$. This is proved in essentially the same manner, but requires the use of part one of Theorem 8 below. The resulting inequality is

$$\left| \frac{w_+(x, \mu + i\nu)}{w_+(0, \mu + i\nu)} \right| < \frac{w_+(x, \mu)}{w_+(0, \mu)}, \quad \mu > 0, \tag{4.7}$$

The other formulas of section 3 may also be extended. Introducing

$$L_+(x, \lambda) = M_+(x, \lambda) + \varrho N_+(x, \lambda), \tag{4.8}$$

we obtain inequalities like

$$\cos \gamma \, L_+(x, \lambda) \leqq |w_+(x, \lambda) w'_+(x, \lambda)| \leqq L_+(x, \lambda), \tag{4.9}$$

$$L_+(x, \lambda) \leqq L_+(0, \lambda) \exp \left\{ -2 \sqrt{\varrho} \cos \gamma \int_0^x \sqrt{F(s)} \, ds \right\}, \tag{4.10}$$

as well as analogues of formulas (3.19) and (3.20).

The three functions L_0, L_1, and L_+ are not unrelated. We conclude from (4.6) that $w_0(x, \lambda)/w_1(x, \lambda)$ and $w'_0(x, \lambda)/w'_1(x, \lambda)$ tend to the same limit as $x \to \infty$, namely the integral in the right member. It follows that $L_0(x, \lambda)/L_1(x, \lambda)$ is bounded away from zero and infinity with bounds depending upon γ. Similarly,

(4.2) shows that $L_1(x, \lambda)\, L_+(x, \lambda)$ is bounded above, the bound depending upon γ. Thus there is essentially only one L-function governing the rate of growth of the solutions of (1.1) or, more precisely, of the products

$$w(x, \lambda)\, w'(x, \lambda) = \tfrac{1}{2} \frac{d}{dx} [w(x, \lambda)]^2$$

which apparently behave in a more regular manner than the solutions themselves when $\mu < 0$.

5. Further study of the subdominant. In the study of Cauchy's problem for the generalized heat equation

$$\frac{\partial^2 U}{dx^2} = F(x) \frac{\partial U}{\partial t}$$

and the adjoint (Fokker-Planck) equation one needs solutions of (1.1) having special properties. These properties are satisfied by the subdominant solution for $\mu > 0$ and cannot possibly be satisfied by any other solution. For $\mu \leqq 0$ we need more information than what is given by Theorem 5. In particular, we want to know if

$$\lim_{x \to \infty} w_+(x, \lambda) = 0, \tag{5.1}$$

$$F(x)\, w_+(x, \lambda) \in L(0, \infty) \tag{5.2}$$

for values of λ in the left half-plane at some distance from the negative real axis. The present section is concerned with these and related questions. We start by introducing some notation and definitions.

We set

$$F_1(x) = \int_0^x F(s)\, ds, \quad F_{\frac{1}{2}}(x) = \int_0^x \sqrt{F(s)}\, ds. \tag{5.3}$$

A positive function $G(x)$ will be said to be of *upper order* ω_2 and *lower order* ω_1 (at infinity) if

$$\lim_{x \to \infty} \begin{matrix} \sup \\ \inf \end{matrix} \frac{\log G(x)}{\log x} = \begin{cases} \omega_2 \\ \omega_1 \end{cases}. \tag{5.4}$$

$F(x)$ satisfies the condition $M(a)$ if

$$\limsup_{x \to \infty} \frac{\log F_1(x)}{F_{\frac{1}{2}}(x)} = a < \infty. \tag{5.5}$$

Finally $P(\eta)$ shall denote the part of the complex λ-plane to the right of the parabola

$$4 \varrho \cos^2 \gamma = \eta^2, \quad \lambda = \varrho\, e^{2 i \gamma}, \tag{5.6}$$

having its focus at the origin and its axis directed along the negatixe real axis.

Theorem 6. *If $L_+(x, \lambda) \in L(0, \infty)$, then the area swept by the radius vector of $w_+(x, \lambda)$ is finite and* (5.1) *holds.*

Proof. From the expression for the transverse acceleration in (1.6) we get

$$|w_+(x, \lambda)|^2 \theta'_+(x, \lambda) = -\nu N_+(x, \lambda)$$

so that a necessary and sufficient condition that the radius vector sweep a finite area is that $N_+(x, \lambda) \in L(0, \infty)$ and this is certainly satisfied if $L_+(x, \lambda) \in L(0, \infty)$ or, what is equivalent, that $[L_1(x, \lambda)]^{-1} \in L(1, \infty)$. Formula (4.9) shows then that $w_+(x, \lambda)$ tends to a limit which must be zero since $x F(x) \notin L(0, \infty)$. Formula (4.10) gives a

Corollary. *The area swept by the radius vector of $w_+(x, \lambda)$ is finite and* (5.1) *holds, both for λ in $P(a)$, if $\exp[F_{\frac{1}{2}}(x)]$ is of lower order $1/a$.*

Theorem 7. *A sufficient condition that $S_+(\lambda)$ be of finite length is that $|w_1(x, \lambda)|^{-1} \in L(1, \infty)$. In this case* (5.1) *also holds.*

This follows from (4.1) combined with (3.15). The condition is not necessary even for real positive λ.

Let us now turn to the validity of (5.2) the dynamical interpretation of which is that *the length of the hodograph of $S_+(\lambda)$ is finite.*

Theorem 8. (5.3) *is valid if anyone of the following conditions is satisfied:*

(i) $\mu > 0$,
(ii) $F(x) \in L(0, \infty)$, $\lambda \in \Lambda$,
(iii) $F(x)$ *satisfies condition* $M(a)$ *and* $\lambda \in P(a)$.

Proof. Case (i) follows from (1.5) with $r = r_+ = |w_+(t, \lambda)|$. If $\mu > 0$, we know that $r''_+ > 0$, $r'_+ < 0$, so that r'_+ tends to a finite limit when $t \to \infty$. Hence $r''_+ \in L(0, \infty)$ and the conclusion is immediate. The fact that also $(\theta'_+)^2 r_+ \in L(0, \infty)$ is used in the proof of (4.7). Case (ii) follows from Theorem 5 and Schwarz's inequality.

Case (iii) is based upon the implications of formula (3.15). We have

$$\int_a^\beta F(x) |w_+(x, \lambda)| \, dx < \sec^2 \gamma \int_a^\beta \frac{F(x) |w_1(x, \lambda)| \, dx}{L_1(x, \lambda)} < \frac{1}{\varrho} \sec^2 \gamma \int_a^\beta \frac{L'_1(x, \lambda) \, dx}{|w_1(x, \lambda)| \, L_1(x, \lambda)}.$$

This means that if $|w_1(x, \lambda)| \geqq [L_1(x, \lambda)]^\varepsilon$ with a fixed $\varepsilon > 0$ in the interval (a, β), then the first member is dominated by $(\varrho \varepsilon)^{-1} \sec^2 \gamma \, [L_1(a, \lambda)]^{-\varepsilon}$. For a given $\varepsilon > 0$, let

$$S_\varepsilon = [x \,\|\, |w_1(x, \lambda)| \geqq [L_1(x, \lambda)]^\varepsilon, \; x \geqq 0]$$

and let E_ε be the complementary set in $(0, \infty)$. Then the integral of $F(x) |w_+(x, \lambda)|$ over the set S_ε is finite. The assertion is consequently proved if we can show that condition $M(a)$ implies the existence for every $\lambda \in P(a)$ of an $\varepsilon_0 = \varepsilon_0(\lambda)$ such that the exceptional set E_ε is bounded for $\varepsilon < \varepsilon_0$. This is proved by showing that the assumptions give two estimates of $|w'_1(x, \lambda)|$ in E_ε and these become inconsistent if ε is small and E_ε is unbounded.

For x in E_ε we have by (3.15)

$$|w_1'(x, \lambda)| > \cos\gamma\, [L_1(x, \lambda)]^{1-\varepsilon}.$$

But

$$w_1'(x, \lambda) = \lambda \int_0^x F(s)\, w_1(s, \lambda)\, ds,$$

so

$$|w_1'(x, \lambda)|^2 \leqq \varrho^2 F_1(x) N_1(x, \lambda) < \varrho F_1(x) L_1(x, \lambda).$$

On the other hand, if $\delta > 0$ is given, condition $M(a)$ combined with (3.2) show that for large x, $x \geqq x_\delta$,

$$F_1(x) \leqq e^{(a+\delta) F_{\frac{1}{2}}(x)} < C_0(\lambda) [L_1(x, \lambda)]^{\frac{1}{2}(a+\delta)\varrho^{-\frac{1}{2}} \sec\gamma}.$$

The resulting double inequality for $|w_1'(x, \lambda)|$ implies that a certain power of $L_1(x, \lambda)$ is bounded away from zero on the set E_ε. If E_ε is unbounded, this requires that the exponent of $L_1(x, \lambda)$ be non-negative. Since δ is arbitrary, this gives

$$4\varrho \cos^2\gamma \leqq \left(\frac{a}{1-2\varepsilon}\right)^2,$$

provided $\varepsilon < \frac{1}{2}$, as we may assume. But if $\lambda = \varrho e^{2i\gamma}$ is a point in $P(a)$, this inequality cannot hold for arbitrarily small values of ε. This shows the existence of an $\varepsilon_0(\lambda)$ such that E_ε is actually bounded for $\varepsilon < \varepsilon_0$ and that

$$|w_1(x, \lambda)| \geqq [L_1(x, \lambda)]^\varepsilon \tag{5.7}$$

holds for all large x when $\varepsilon < \varepsilon_0$. This completes the proof.

It shold be observed that (5.2) also implies

$$w_+(x, \lambda) \left[\frac{d}{dx} \arg w_+(x, \lambda)\right]^2 \in L(0, \infty), \tag{5.8}$$

$$\lim w_+'(x, \lambda) = 0. \tag{5.9}$$

The first relation follows readily from (1.5), the second is implied by

$$w_+'(x, \lambda) - w_+'(0, \lambda) = \lambda \int_0^x F(s)\, w_+(s, \lambda)\, ds$$

and $x F(x) \notin L(0, \infty)$.

The peculiar condition $M(a)$ serves to exclude functions $F(x)$ of highly irregular behavior. It is not necessarily satisfied by increasing functions $F(x)$; for such a function the inferior limit of the quotient in (5.5) is always zero, but the superior limit may very well be infinite. This happens, for instance, if

$$\sqrt{F(x)} = \begin{cases} G_{n-1}, & n-1 \leqq x \leqq n - g_n^{-1}, \\ G_{n-1} + g_n^2 (x - n + g_n^{-1}), & n - g_n^{-1} < x < n, \end{cases} \qquad n = 1, 2, 3, \ldots$$

where

$$G_n = \sum_{k=0}^{n} g_k, \quad g_k = e^{k\, g_{k-1}}, \quad g_0 = 1.$$

While the sudden spurts of $F(x)$ have very little local effect on $F_{\frac{1}{2}}(x)$, they do affect $F_1(x)$ so strongly that $\log F_1(n) > C\, n\, F_{\frac{1}{2}}(n)$.

Condition $M(a)$ is satisfied if there exists a monotone increasing function $Q(x)$ such that

$$[Q'(x)]^2 \leqq F(x) \leqq Q'(x)\, e^{a\, Q(x)} \tag{5.10}$$

as is easily seen. Another sufficient condition is the existence of a positive integrable function $G(u)$ such that

$$F(x) \leqq G[F_{\frac{1}{2}}(x)], \quad \int_0^u G(s)\, ds \leqq e^{a\,u}. \tag{5.11}$$

In particular, taking $G(u)$ as a constant, we see that *if $F(x)$ is bounded then (5.2) holds in some parabolic domain $P(a)$ if $F_{\frac{1}{2}}(\infty) < \infty$ and everywhere in Λ if $F_{\frac{1}{2}}(\infty) = \infty$.*

Theorem 9. *There exists a parabolic domain $P(c)$ in which* (5.1) *and* (5.2) *are both satisfied if either*

(1) *$F(x)$ satisfies condition $M(a)$, $F_1(x)$ is of lower order $1/\sigma$, and $c \geqq \max(a, \sigma a)$, or*

(2) *$\exp[F_{\frac{1}{2}}(x)]$ is of lower order $1/\tau$, $F(x)$ is of upper order ω, and $c \geqq \tau(\omega + 1)$.*

Proof. In the first case $[F_1(x)]^{-\sigma-\varepsilon} \in L(1, \infty)$ for every $\varepsilon > 0$ and condition $M(a)$ gives

$$[F_1(x)]^{-\sigma-\varepsilon} \geqq \exp[-(\sigma + \varepsilon)(a + \delta)\, F_{\frac{1}{2}}(x)]$$

for $x \geqq x_\delta$. But if $\lambda \in P(\sigma a)$, then $2\varrho^{\frac{1}{2}} \cos\gamma \geqq (\sigma + \varepsilon)(a + \delta)$, provided δ and ε are sufficiently small. It then follows from (4.10) that $L_+(x, \lambda) \in L(0, \infty)$ so the conclusion of Theorem 6 applies in $P(\sigma a)$ while that of Theorem 8 holds in $P(a)$.

In the second case the Corollary of Theorem 6 shows that (5.1) holds for $\lambda \in P(\tau)$. For (5.2) we have to go back to the proof of Theorem 8. If E_ε is the exceptional set in which (5.7) fails to hold for a particular fixed ε, then

$$\int_{E_\varepsilon} F(x)\, |w_+(x, \lambda)|\, dx < \sec^2\gamma \int_{E_\varepsilon} F(x)\, [L_1(x, \lambda)]^{\varepsilon - 1}\, dx.$$

For large values of x we have $F(x) < x^{\omega+\delta}$, $\delta > 0$, while

$$\log L_1(x, \lambda) > \log L_1(1, \lambda) + \left(\frac{1}{\tau} - \eta\right) 2\varrho^{\frac{1}{2}} \cos\gamma \log x$$

by (3.21) and the hypothesis. Since $\varepsilon > 0$ is at our disposal, a simple calculation shows that the integral over the exceptional set is finite as soon as $\lambda \in P[\tau(\omega + 1)]$. This completes the proof.

We shall see in the next section that $w_+(x, \lambda)$ need not tend to zero when $x \to \infty$ and λ is purely imaginary, but we have no example of an equation for which $F(x) w_+(x, \lambda)$ fails to be in $L(0, \infty)$ for any $\lambda \in \Lambda$. We have spoken above of the exceptional set E_ε in which (5.7) fails to hold for a particular ε, $0 < \varepsilon < \frac{1}{2}$. We do not know if E_ε can be unbounded no matter how small ε is; at any rate the exceptional intervals must be quite short and far apart because for every $\delta > 0$ the integral of $[L_1(x, \lambda)]^{1-2\varepsilon-\delta}$ over E_ε exists as may be shown with the aid of Theorem 4 taking $\Phi(u) = u^{1-\delta}$.

6. Extremal problems. We shall discuss briefly the question of the extremals for the rate of growth of the solutions. To measure the rate of growth of a solution $w(x, \lambda)$ we use its *upper order*

$$\sigma(\lambda) = \sigma(\lambda, w) = \limsup_{x \to \infty} \frac{\log |w(x, \lambda)|}{\log x}, \tag{6.1}$$

which may be finite or infinite. For a given $F(x)$, the upper order of a solution has only two possible values, $\sigma_n(\lambda)$ and $\sigma_s(\lambda)$, the former corresponding to $w_1(x, \lambda)$, the latter to $w_+(x, \lambda)$. These orders satisfy the following inequalities:

$$\sigma_s(\lambda) \leqq \tfrac{1}{2} \leqq \sigma_n(\lambda) \leqq \sigma_n(\varrho), \qquad \lambda \in \Lambda. \tag{6.2}$$

$$\sigma_s(\lambda) \leqq 0, \quad 1 \leqq \sigma_n(\lambda), \qquad \mu \geqq 0, \tag{6.3}$$

$$\sigma_s(\lambda) \leqq \sigma_s(\mu), \quad \sigma_n(\mu) \leqq \sigma_n(\lambda), \qquad \mu \geqq 0, \tag{6.4}$$

where as usual $\lambda = \mu + i\nu$, $\varrho = |\lambda|$.

The first part of (6.2) follows from

$$|w_+(x, \lambda) - w_+(0, \lambda)|^2 \leqq x \int_0^x |w'_+(s, \lambda)|^2 \, ds \leqq C x,$$

the second from (3.16), and the third from (3.4); (6.3) follows from Theorems 3 and 5, while (6.4) is derived from (3.6) and (4.7).

Here (6.2) is a best possible inequality, for if μ is given, $\mu < 0$, we can choose $a > 0$ and ν such that when $F(x) = (1 + a x)^{-2}$, the orders $\sigma_s(\mu + i\nu)$ and $\sigma_n(\mu + i\nu)$ both differ from $\frac{1}{2}$ by less than any preassigned number (see also below). This of course also implies that $|w_+(x, \lambda)|$ may be monotone increasing to infinity when $\mu < 0$, a behavior totally different from what takes place for $\mu \geqq 0$. Formula (6.3) is also a best possible estimate and here we can show that the bounds are reached everywhere in the right half-plane for a suitable choice of $F(x)$. Thus if

$$F(x) = (x + 2)^{-2} [\log (x + 2)]^{-1} \{1 + (\alpha - 1) [\log (x + 2)]^{-1}\}$$

with $\alpha \geqq 1$ and $|\lambda| \leqq \alpha$, we have

$$|w_1(x, \lambda)| \leqq w_1(x, |\lambda|) \leqq w_1(x, \alpha).$$

But for $\lambda=a$ the equation has the solution $(x+2)[\log(x+2)]^a$ and this is manifestly not the subdominant solution. Its order being unity, it follows that $\sigma_n(\lambda)=1$ for $\mu\geqq 0$. Further

$$w_+(x, a)=[C_a+o(1)][\log(x+2)]^{-a},$$

so that $\sigma_s(a)=0$ and a simple argument shows that $\sigma_s(\lambda)=0$ holds for $\mu\geqq 0$.

Theorems 3 and 5 suggest the possibility of $w_1(x, \nu i)=O(x)$ and $w_+(x, \nu i)= =O(1)$, not $o(1)$, for suitable choices of $F(x)$. This is indeed the case, but examples are harder to construct. We may take, however, as argument the function

$$\theta_+(x)=[\log c]^{\frac{1}{3}}-[\log(x+c)]^{\frac{1}{3}}, \quad c>1,$$

and determine the corresponding absolute value $r_+(x)=r_+(x, 1)$ as the subdominant solution of the differential equation

$$r''=\tfrac{1}{9}(x+c)^{-2}[\log(x+c)]^{-\frac{4}{3}} r.$$

By Theorem 1, $r_+(x)=1+o(1)$ (with a suitable normalization of the solution) and the function $r_+(x)e^{i\theta_+(x)}$ satisfies a differential equation of type (1.1) with $\lambda=i$ and $F(x)>0$ for $x\geqq 0$, provided c is sufficiently large to start with. For large x we have

$$F(x)=\tfrac{1}{3}(x+c)^{-2}[\log(x+c)]^{-\frac{2}{3}}[1+o(1)].$$

This means that for the corresponding equation (1.1) the solution $w_+(x, i)$ is a constant multiple of a function which for large x is of the form

$$[1+o(1)]\, e^{-i[\log(x+c)]^{\frac{1}{3}}}$$

so that the spiral $S_+(i)$ has an *asymptotic circle.* Naturally such a circle can arise only when $F(x)\in L(0, \infty)$. The corresponding solution $w_1(x, i)$ may be shown to be $O(x)$, the minimal order, with the aid of (4.5) and (4.6).

Let us finally consider the class $\mathbf{L}$ of all linear differential equations of type (1.1) with $xF(x)\notin L(0, \infty)$. For each equation in $\mathbf{L}$ there are two indices $\sigma_n(\lambda; F)$ and $\sigma_s(\lambda; F)$. For a given $\lambda\in\Lambda$ we consider inf $\sigma_n(\lambda; F)$ and sup $\sigma_s(\lambda; F)$ where $F(x)$ ranges over all admissible functions. These two quantities depend only on $\arg\lambda=\varphi$ and not on $|\lambda|=\varrho$ since

$$\sigma(\varrho e^{i\varphi}; F)=\sigma(e^{i\varphi}; \varrho F).$$

Therefore we define

$$\omega_s(\varphi)=\sup\sigma_s(e^{i\varphi}; F), \quad \omega_n(\varphi)=\inf\sigma_n(e^{i\varphi}; F). \tag{6.5}$$

Evidently

$$\omega_s(\varphi)\equiv 0, \quad \omega_n(\varphi)\equiv 1, \qquad -\tfrac{1}{2}\pi\leqq\varphi\leqq\tfrac{1}{2}\pi. \tag{6.6}$$

In the remainder of Λ the situation is different, but we are unable to determine the exact values of the extremal functions. We shall prove merely the following inequalities:

$$\omega_s(\varphi)\geqq\tfrac{1}{2}[1-\sin|\varphi|], \quad \omega_n(\varphi)\leqq\tfrac{1}{2}[1+\sin|\varphi|], \quad \tfrac{1}{2}\pi<|\varphi|<\pi. \tag{6.7}$$

For $F(x)=(1+ax)^{-2}$ the solutions are linear combinations of powers $(1+ax)^{\alpha}$ where α satisfies the indicial equation $a\,\alpha(\alpha-1)-\lambda=0$. Here we set $\lambda=\varrho e^{i\varphi}$ and keep φ fixed, $\frac{1}{2}\pi<|\varphi|<\pi$. There is a root α_1 with $\Re(\alpha_1)>\frac{1}{2}$ and $\Re(\alpha_1)$ will be a minimum if $\varrho=-\frac{1}{2}a^2\cos\varphi$ and simultaneously the real part of the other root will reach its maximum. Calculating the values of these extrema we obtain (6.7). We have no means of telling if these inequalities are actually equalities; at least the right hand sides have the growth and convexity properties as functions of φ as we expect the extremal functions $\omega_s(\varphi)$ and $\omega_n(\varphi)$ to have.

If it should really turn out that the functions of the form $(ax+b)^{-2}$ give the solution of the extremal problems defined by (6.5), it might be of some interest to observe that the same class of functions is connected with condition $M(a)$ of (5.5). One might ask if the related functional equation

$$\log[1+aF_1(x)]=aF_{\frac{1}{2}}(x)$$

has a solution. Here we have replaced $F_1(x)$ by $1+aF_1(x)$ to get the appropriate normalization at the origin. Twofold differentiation gives a first order differential equation for $F(x)$ and

$$F(x)=(1-ax)^{-2}$$

satisfies the equation for $0\leqq x<1/a$. The fact that the solution has only a finite range of existence, underlies the fact observed above that the inferior limit of the quotient in (5.5) is zero for increasing functions.

REFERENCES. [1] **M. Bôcher,** On regular singular points of linear differential equations of the second order whose coefficients are not necessarily analytic, Trans. Amer. Math. Soc. 1, 39—52 (1900). — [2] **E. Hille,** On the zeros of Sturm-Liouville functions, Arkiv för Mat., Astr. o. Fysik, 16, no. 17, 20 pp. (1922). — [3] ——, Non-oscillation theorems, Trans, Amer. Math. Soc., 64, 234—252 (1948). — [4] **E. L. Ince,** Ordinary Differential Equations. Dover, New York, 1944, viii + 558 pp. — [5] **H. Weyl,** Über gewöhnliche lineare Differen, tialgleichungen mit singulären Stellen und ihre Eigenfunktionen, Göttinger Nachrichten, Math.-phys. Klasse, 1909, pp. 37—63. — [6] **A. Wiman,** Über die reellen Lösungen der linearen Differentialgleichungen zweiter Ordnung, Arkiv för Mat., Astr. o. Fysik, 12. no. 14. 22 pp. (1917). — [7] **A. Wintner,** On almost free linear motions. Amer. J. of Math., 71, 595—602 (1949).

Tryckt den 10 december 1951

Uppsala 1951. Almqvist & Wiksells Boktryckeri AB

Remarques sur les systèmes des équations différentielles linéaires à une infinité d'inconnues;

PAR EINAR **HILLE**.

(New Haven, Connecticut).

---

1. La théorie des équations différentielles linéaires a une beauté et une simplicité frappantes quand on se restreint aux systèmes finis. Les solutions sont uniquement déterminées par leurs valeurs initiales, les singularités sont fixées en avance par les singularités des coefficients si ceux-ci sont des fonctions analytiques. En particulier, pour des coefficients constants, les solutions sont des fonctions exponentielles multipliées par des polynomes.

Tout cela change s'il s'agit de systèmes infinis. De tels systèmes se présentent d'une manière naturelle par exemple dans la théorie des variables aléatoires quand les valeurs possibles de la variable forment un ensemble dénombrable. Bien que l'existence des solutions nulles soit connue dans cette théorie par les travaux de M. Feller (*voir* [1], p. 393, par exemple), il me semble que la recherche systématique de la nature de l'ensemble des solutions reste à faire.

Le but de cette petite Note est de donner quelques indications sur les phénomènes qui peuvent se présenter même dans des cas assez simples. Pour éviter des longueurs inutiles, je considérerai seulement deux systèmes spéciaux qui présentent néanmoins un assortiment de singularités frappantes.

Journ. de Math., tome XXXVII. — Fasc. 4, 1958.

2. En premier lieu, soit $\mathbf{A}$ la matrice infinie, $\mathbf{A}=(a_{jk})$, où $a_{jk}=1$, $-j$ ou 0 selon que $k<j$, $k=j$ ou $k>j$ et considérons le système différentiel

$$\mathbf{z}'(t)=\mathbf{z}(t)\mathbf{A} \tag{1}$$

ou explicitement

$$z'_k(t)+k\,z_k(t)=\sum_{m=k+1}^{\infty} z_m(t) \qquad (k=1,\,2,\,3,\,\ldots). \tag{2}$$

Ici t est une variable complexe et l'on cherche des solutions $\mathbf{z}(t)=\{z_k(t)\}$ qui appartiennent à l'espace (l) des séries absolument convergentes.

Le système adjoint

$$y'_k(t)+k\,y_k(t)=\sum_{m=1}^{k-1} y_m(t), \qquad y_k(0)=y_{k0}$$

a une solution unique

$$y_k(t)=\sum_{m=1}^{k}(y_{m0}-y_{m-1,0})\,e^{-mt}, \qquad y_{00}=0,$$

qui appartient à l'espace (m) des suites bornées pour $\mathcal{R}(t)\geqq 0$ si les y_{k0} sont bornés.

Le système (2) est d'une nature tout à fait différente. Ici on peut trouver des solutions nulles (*voir* E. Hille [3]). La plus simple est peut-être

$$\mathbf{z}_0(t)=\{z_k(t)=e^{-kt}-e^{-(k+1)t}\}.$$

On voit que $\mathbf{z}_0(t)\in(l)$ pour chaque t tel que $\mathcal{R}(t)>0$ et

$$\|\mathbf{z}_0(t)\|=e^{-t} \quad \text{si} \quad t>0.$$

Le vecteur

$$\mathbf{z}_1(t)=\int_0^t \mathbf{z}_0(s)\,ds \tag{4}$$

est aussi une solution nulle avec

$$\|\mathbf{z}_1(t)\|=1-e^{-t} \qquad (t>0),$$

qui tend vers zéro avec t. En fait, on peut construire une infinité de solutions nulles dont les propriétés sont plus ou moins à notre disposition.

Soit $\mathbf{z}(t)$ une solution nulle de (1) qui existe et appartienne à (l) dès que $t \geqq 0$ et soit $\| \mathbf{z}(t) \| \leqq M e^{at}$. Alors

$$\mathbf{x}(\lambda) = \int_0^\infty e^{-\lambda t} \mathbf{z}(t)\, dt, \qquad \mathcal{R}(\lambda) > a, \tag{5}$$

est aussi un élément de (l), de plus c'est une fonction bornée et holomorphe de λ dans chaque demi-plan $\mathcal{R}(\lambda) \geqq a + \varepsilon$, $\varepsilon > 0$. On a

$$\mathbf{x}(\lambda)\mathbf{A} = \lambda \mathbf{x}(\lambda) \tag{6}$$

ou, explicitement,

$$(\lambda + k)\, x_k(\lambda) = \sum_{n=k+1}^{\infty} x_m(\lambda), \qquad \mathbf{x}(\lambda) = \{ x_k(\lambda) \}. \tag{7}$$

Réciproquement, si $\mathbf{x}(\lambda)$ est une solution de (6) qui existe comme fonction holomorphe de λ dans (l) pour chaque λ avec $\mathcal{R}(\lambda) \geqq a + \varepsilon$, et si, par exemple, $|\lambda|^3 \| \mathbf{x}(\lambda) \| \leqq M(\varepsilon)$ dans ce demi-plan, alors la formule

$$\mathbf{z}(t) = \frac{1}{2\pi i} \int_{\gamma - i\infty}^{\gamma + i\infty} e^{\lambda t} \mathbf{x}(\lambda)\, d\lambda \qquad (\gamma > a), \tag{8}$$

donne une solution de (1) dont la norme n'excède pas $M(\varepsilon) e^{(a+\varepsilon)t}$ et qui tend fortement vers zéro avec t.

La solution générale de (7) est fournie par

$$x_k(\lambda) = \frac{(\lambda + 1)(\lambda + 2)}{(\lambda + k)(\lambda + k + 1)} x_1(\lambda) \qquad (k > 1), \tag{9}$$

où $x_1(\lambda)$ est parfaitement arbitraire. Le choix

$$x_k(\lambda) = \frac{1}{(\lambda + k)(\lambda + k + 1)} \tag{10}$$

donne la solution (3).

Il faut ajouter quelques remarques sur les solutions nulles. Les coefficients de $\mathbf{z}_0(t)$ et de $\mathbf{z}_1(t)$, comme fonctions entières de t, sont définis pour toutes valeurs finies de t, mais elles satisfont au système (2) seulement pour $\mathcal{R}(t) > 0$, car les séries au deuxième membre divergent pour $\mathcal{R}(t) < 0$. Il en résulte que les fonctions $\mathbf{z}_0(t)$

et $\mathbf{z}_1(t)$, en tant que fonctions analytiques de t dans (l), existent seulement dans le demi-plan droit, l'axe imaginaire étant une ligne singulière pour ces fonctions, quoique aucun point sur l'axe ne soit singulier pour les coefficients. Cela montre que pour les systèmes infinis aux coefficients analytiques il ne suffit pas de faire le prolongement analytique de la solution. Même si l'on se débarrasse de la restriction que la solution doit appartenir à l'espace (l), on voit que le prolongement d'une solution ne reste pas toujours une solution. Nous verrons d'autres exemples de ce phénomène dans la suite.

3. Passons maintenant au problème de construire une solution de de l'équation (1) dont les coefficients ont des points singuliers finis. Soit α un nombre positif irrationnel, qui sera précisé plus tard, et essayons de construire une solution $\mathbf{z}(t)$ de (1) de la forme

$$z_k(t) = \sum_{n=0}^{\infty} c_{kn} e^{-n\alpha t}, \qquad \mathbf{z}(t) = \{ z_k(t) \}.$$

Un calcul facile montre que, pour n fixe, les c_{kn} doivent satisfaire à (7) avec $\lambda = -n\alpha$. Grâce à (10), on peut donc prendre

$$z_k(t, \alpha) = \sum_{n=0}^{\infty} \frac{e^{-n\alpha t}}{(k - n\alpha)(k+1-n\alpha)}. \tag{11}$$

Les quantités $|k - n\alpha|$ sont différentes de zéro, mais elles ne sont pas bornées inférieurement. En faisant une hypothèse convenable sur le caractère arithmétique de α [1] on peut établir la convergence absolue et uniforme de la série $\sum_{1}^{\infty} z_k(t, \alpha)$ dans $\mathcal{R}(t) \geqq \varepsilon > 0$, d'où l'on peut conclure que $\mathbf{z}(t, \alpha) = \{ z_k(t, \alpha) \} \in (l)$, et satisfait à (1) si $\mathcal{R}(t) > 0$. De plus, $z_k(t, \alpha)$ étant une série entière de $e^{-\alpha t}$ dont les coefficients sont ultérieurement positifs, on voit que $t = 0$ est nécessairement un point singulier de chaque $z_k(t, \alpha)$. Nous avons aussi

$$\mathbf{z}\left(t + \frac{2\pi i}{\alpha}, \alpha\right) = \mathbf{z}(t, \alpha), \tag{12}$$

---

[1] Il suffit que α soit un nombre algébrique.

de sorte que chaque nombre

$$(13) \qquad m\frac{2\pi i}{\alpha} \qquad (m = 0, \pm 1, \pm 2, \ldots)$$

sera un point singulier de $z_k(t, \alpha)$.

En effet, il n'y a pas d'autres singularités de $z_k(t, \alpha)$ à distance finie. Traçons des coupures rectilignes parallèles à l'axe réel négatif, de chaque point (13) jusqu'à l'infini. En suivant la méthode de E. Lindelöf ([4], n° 59), on montre que $z_k(t, \alpha)$ est holomorphe dans le plan coupé. En particulier, on trouve la représentation

$$(14) \quad z_k(t, \alpha) = -\sum_{n=0}^{\infty} \frac{e^{n\alpha t}}{(k+n\alpha)(k+1+n\alpha)} - \frac{2\pi i}{\alpha}\left[\frac{e^{-kt}}{\omega^k - 1} - \frac{e^{-(k+1)t}}{\omega^{k+1} - 1}\right]$$

dans la bande $\mathcal{R}(t) < 0$, $0 < \mathcal{I}(t) < \frac{2\pi}{\alpha}$, avec $\omega = e^{\frac{2\pi i}{\alpha}}$. Dans la bande homologue où $m\frac{2\pi}{\alpha} < \mathcal{I}(t) < (m+1)\frac{2\pi}{\alpha}$, la représentation s'obtient en remplaçant t par $t + m\frac{2\pi i}{\alpha}$ dans (14).

De ces représentations on conclut que $\{z_k(t, \alpha)\}$ ne peut pas appartenir à (l) ou satisfaire à (2) dans le demi-plan gauche; c'est-à-dire comme élément de (l) la fonction $\mathbf{z}(t, \alpha)$ existe seulement dans le demi-plan droit et admet l'axe imaginaire comme ligne singulière. Une sommation par parties dans (11) donne que

$$|z_k(i\tau, \alpha)| < C\left|\sin\frac{\alpha\tau}{2}\right|^{-1} k^{-2}$$

de sorte que $\mathbf{z}(i\tau, \alpha) \in (l)$, sauf pour $\tau = m\frac{2\pi}{\alpha}$.

La fonction $z_k(t, \alpha)$ à l'autre côté peut être prolongée dans tout le plan sans rencontrer d'autres singularités que les points (13) et $t = \infty$ dont les premiers sont des points critiques logarithmiques telles que deux déterminations de $z_k(t, \alpha)$ diffèrent par une fonction de la forme

$$\frac{2\pi i}{\alpha}\sum_m N_m[e^{-kt_m} - e^{-(k+1)t_m}], \qquad \left(t_m = t + m\frac{2\pi}{\alpha}i\right),$$

où les N_m sont des entiers qui sont différents de zéro seulement pour un nombre fini de valeurs de m.

4. On peut aussi construire des solutions pour lesquelles chaque fonction $z_k(t)$ admette l'axe imaginaire comme ligne singulière. Il suffit de prendre

$$z_k(t) = \sum_p A_p z_k(t, \sqrt{p}), \tag{15}$$

où p parcourt les nombres premiers, $z_k(t, \sqrt{p})$ est défini par (11) et

$$A_p > 0, \qquad \sum_p \sqrt{p}\, A_p < \infty.$$

Un calcul facile donne que

$$z''_k(t) = \sum_p \frac{A_p}{1 - e^{-t\sqrt{p}}} + O\left[\log \frac{1}{\sigma}\right] \qquad (t = \sigma + i\tau).$$

Pour $\tau = m \dfrac{2\pi}{\sqrt{q}}$ on a

$$\frac{A_q}{1 - e^{-t\sqrt{q}}} = \frac{A_q}{\sigma\sqrt{q}} + O(1),$$

tandis que

$$\sum_{p \neq q} \frac{A_p}{\left|1 - e^{-t\sqrt{p}}\right|} \leq \frac{1}{2} \sum_p (4m\sqrt{pq} + q) A_p < \infty,$$

car, en notant $\{u\}$ la distance de u à l'entier le plus proche de u, on a

$$\left|1 - \exp\left[-m\sqrt{\frac{p}{q}}\, 2\pi i - \sqrt{p}\,\sigma\right]\right| > \left|\sin\left(m\sqrt{\frac{p}{q}}\, 2\pi\right)\right| > 2\left\{2m\sqrt{\frac{p}{q}}\right\}$$
$$> 2[4m\sqrt{pq} + q]^{-1}.$$

Il en résulte que

$$\lim_{\sigma \to 0+} \sigma z''_k\left(\sigma + m\frac{2\pi}{\sqrt{q}} i\right) = \frac{A_q}{\sqrt{q}} \tag{16}$$

pour chaque entier m et nombre premier q. Par conséquence, l'axe imaginaire est une ligne singulière de chaque $z_k(t)$.

Bien entendu les solutions de (1) obtenues ci-dessus n'épuisent pas toutes les possibilités, mais nous n'insisterons pas sur une étude plus détaillée.

5. Pour le deuxième exemple nous prenons la matrice $\mathbf{B}=(b_{jk})$, avec $b_{jk}=1, \ -j^2$, o selon que $k<j, \ k=j, \ k>j$. L'équation caractéristique

$$\mathbf{x}(\lambda)\mathbf{B}=\lambda\mathbf{x}(\lambda) \tag{17}$$

donne le système

$$(\lambda+k^2)x_k(\lambda)=\sum_{m=k+1}^{\infty}x_m(\lambda), \tag{18}$$

dont une solution est fournie par

$$x_k(\lambda)=\frac{(\lambda+1)(\lambda+4)\ldots(\lambda+(k-1)^2)}{(\lambda+2)(\lambda+5)\ldots(\lambda+1+(k-1)^2)(\lambda+1+k^2)}\frac{\operatorname{sh}\pi\sqrt{\lambda+1}}{\pi\sqrt{\lambda+1}}. \tag{19}$$

Ici chaque $x_k(\lambda)$ est une fonction entière de λ d'ordre $\frac{1}{2}$. Soit

$$x_k(\lambda)=\sum_{n=0}^{\infty}\alpha_{nk}\lambda^n, \tag{20}$$

où

$$\alpha_{nk}\geqq 0, \qquad n^2\sqrt{\alpha_{nk}}<C_k.$$

Posons alors

$$z_k(t;\,t_0)=\int_0^{\infty}e^{\lambda(t-t_0)}x_k(\lambda)\,d\lambda, \qquad \mathcal{R}(t-t_0)<0, \tag{21}$$

où, pour simplifier, t_0 sera réel. On a

$$z_k(t;\,t_0)=\sum_{n=0}^{\infty}n!\,\alpha_{nk}(t_0-t)^{-n-1}, \tag{22}$$

une fonction entière de $(t_0-t)^{-1}$ d'ordre 1. Si t est réel et $t<t_0$, on voit que $z_k(t;\,t_0)$ est positif et

$$\sum_{k=1}^{\infty}z_k(t;\,t_0)=\int_0^{\infty}e^{\lambda(t-t_0)}\sum_{k=1}^{\infty}x_k(\lambda)\,d\lambda$$
$$=\int_0^{\infty}e^{-\lambda(t_0-t)}\frac{\operatorname{sh}\pi\sqrt{\lambda+1}}{\pi\sqrt{\lambda+1}}\,d\lambda<\frac{C}{\sqrt{t_0-t}}\exp\frac{\pi^2}{4(t_0-t)}.$$

Pour d'autres valeurs de t, la série (22) montre que

$$|z_k(t;\,t_0)|\leqq z_k(t_0-|t-t_0|;\,t_0)$$

de sorte que $\mathbf{z}(t\,;\,t_0) = \{z_k(t\,;\,t_0)\} \in (l)$ pour chaque valeur de $t \neq t_0$. Si t tend vers t_0 par valeurs inférieures, on voit de (22) que chaque $z_k(t;\,t_0)$ tend vers l'infini.

Un calcul simple montre que $z_k(t;\,t_0)$ vérifie le système

$$z'_k(t) + k^2 z_k(t) = \sum_{m=k+1}^{\infty} z_m(t) \qquad (k = 1, 2, 3, \ldots) \tag{23}$$

ou

$$\mathbf{z}'(t) = \mathbf{z}(t)\mathbf{B}. \tag{24}$$

Dans la construction de $\mathbf{z}(t;\,t_0)$ nous pouvons évidemment multiplier les $x_k(\lambda)$ de (19) par une fonction entière $\mu(\lambda)$, d'ordre < 1, mais autrement arbitraire, sans changement essentiel dans les propriétés de la solution. Mais cela entraîne que l'équation (24) possède une infinité non dénombrable de solutions, linéairement indépendantes deux par deux, qui sont des fonctions entières de $(t_0 - t)^{-1}$ et qui appartiennent à (l) pour chaque $t \neq t_0$, où t_0 est un nombre réel ou complexe arbitraire. En ajoutant des solutions de ce genre correspondantes à différentes valeurs du paramètre t_0, on peut construire des solutions possédant un nombre fini ou infini de singularités essentielles.

Soit E un ensemble dans le plan de t tel que

$$\int_0^{\infty} e^{-\lambda t} f(\lambda)\, d\lambda = 0 \quad \text{pour tout } t \text{ de E}$$

entraîne que $f(\lambda) \equiv 0$, si $f(\lambda)$ est une fonction bornée continue dans $[0, \infty)$. Soit $z_k(t;\,t_0)$ la fonction définie par (19) et (21). Je dis que l'ensemble $[\mathbf{z}(0,\,t_0),\ t_0 \in \mathrm{E}]$ est fondamental dans l'espace (l). Si non, on pourrait trouver une fonctionnelle linéaire bornée x^* sur (l) telle que $x^* \neq 0$ et $x^*[\mathbf{z}(0,\,t_0)] = 0$, $t_0 \in \mathrm{E}$. Il y aurait donc une suite $\{\alpha_k\}$ non nulle, $|\alpha_k| \leq \mathrm{M}$, telle que

$$\sum_{k=1}^{\infty} \alpha_k z_k(0,\,t_0) = 0 \qquad (t_0 \in \mathrm{E}),$$

de sorte que

$$\sum_{k=1}^{\infty} \alpha_k x_k(\lambda) = 0, \qquad (0 \leq \lambda < \infty).$$

Mais la série

$$\sum_{k=1}^{\infty} \alpha_k \frac{(\lambda+1)(\lambda+4)\ldots(\lambda+(k-1)^2)}{(\lambda+2)(\lambda+5)\ldots(\lambda+1+(k-1)^2)(\lambda+1+k^2)}$$

converge pour chaque valeur de λ différent de $-k^2-1$. Alors en posant successivement $\lambda=-1, -4, \ldots, -k^2, \ldots$, on voit que chaque $\alpha_k=0$ et $x^\star=0$, contrairement à l'hypothèse.

L'ensemble $[\mathbf{z}(0; t_0), t_0 \in E]$ étant fondamental dans (l) entraîne la conséquence suivante. Soient $\mathbf{c}$ un élément de (l), $\mathbf{z}(t)$ une solution de (24) ou $\mathbf{z}(t)\in(l)$, $\mathbf{z}(0)=\mathbf{c}$. Alors dans chaque voisinage de $\mathbf{c}$ il y a des éléments $\mathbf{c}_j$ et des solutions correspondantes $\mathbf{z}_j(t)$, avec $\mathbf{z}_j(0)=\mathbf{c}_j$ telles que $\mathbf{z}_j(t)$ admette un point singulier essentiel dans E, où $\mathbf{z}_j(t)$ n'appartient pas à (l). Il en résulte que dans le cas présent on ne peut pas affirmer l'existence d'une solution, donnée pour $t=0$, dans un intervalle $0<t<\delta$ arbitrairement petit.

Pour le principe employé dans la construction de solutions « explosives », *voir* E. Hille ([2], n° 2).

On peut étudier les systèmes plus généraux

$$z'_k(t)+m_k a_k z_k(t)=\sum_{n=k+1}^{\infty} a_n z_n(t) \qquad (a_n>0,\ m_k \geqq k-1), \tag{25}$$

par la même méthode. On trouve des solutions explosives si $\sum(m_k a_k)^{-1}$ converge.

BIBLIOGRAPHIE.

[1] W. Feller, *An introduction to probability theory and its applications*, I, John Wiley and Sons, New-York, 1950.

[2] E. Hille, *The abstract Cauchy problem and Cauchy's problem for parabolic differential equations* (*J. Anal. Math.*, t. 3, 1953-1954, p. 81-196).

[3] E. Hille, *Quelques remarques sur les équations de Kolmogoroff* (*Bull, Soc. Math. Phys.*, Serbie, t. 5, 1953-1954, p. 1-14).

[4] E. Lindelöf, *Le calcul de résidu et ses applications à la théorie des fonctions*, Collection Borel, Gauthier-Villars, Paris, 1905.

(Manuscrit reçu le 25 janvier 1957.)

HILLE, EINAR
1961
Annali di Matematica pura ed applicata
(IV), Vol. LV, pp. 133-148

Pathology of infinite systems of linear first order differential equations with constant coefficients (*).

Memoria di EINAR HILLE (a New Haven, Conn. U.S.A.)

To Enrico Bompiani on his Scientific Jubilee

Summary. - *The properties of solutions of finite systems are analyzed and it is shown by examples how these properties may be distorted and ultimately lost in passing from finite to infinite systems.*

1. **Introduction.** - There are few fields of mathematics which have such intrinsic beauty, simplicity, and symmetry as the theory of systems of linear differential equations as long as the system is a finite one. Suppose that the system is

$$y'_j(t) = \sum_{k=1}^{n} a_{jk} y_k(t), \qquad j = 1, 2, \ldots, n, \tag{1.1}$$

or in equivalent vector and matrix form

$$\boldsymbol{y}'(t) = \boldsymbol{A}(t)\boldsymbol{y}(t), \tag{1.2}$$

where $\boldsymbol{A}(t) = (a_{jk}(t))$ is an n by n matrix and $\boldsymbol{y}(t)$ is the column vector whose components are $y_1(t), y_2(t), \ldots, y_n(t)$.

We list below some of the properties of the solutions, but in order to obtain a better perspective we start with nonlinear vector equations and gradually specialize. The following table gives the desired information.

(*) This research was supported in part by United States Army through its Office of Ordnance Research under Grant DA-ORD-12. - Most of the results have been presented to various organisations such as Section A of the American Association for the Advancement of Science and mathematics colloquia at the universities of Leningrad, Moscow, and Toronto.

Equation	Property
(i) $y' = f(t, y)$ $f(t, y)$ continuous	(1) Existence of solutions (2) Holding of parity principle (3) Solutions involve arbitrary constants
(ii) Same assumptions + Lipschitz condition	(4) Uniqueness
(iii) Linear equation $y' = A(t)y$ $A(t)$ is continuous	(5) No null solutions exist (6) Solutions form a linear manifold
(iv) Linear equation $A(t)$ is analytic	(7) Solutions are analytic (8) Permanency of functional equations (9) Fixed singular points
(v) Linear equation $A(t)$ is entire	(10) Solutions are entire functions
(vi) Linear equation $A(t)$ is constant	(11) Solutions are exponential functions

Some comments are in order. It is well known that continuity of the prescribed slope implies the existence of solutions. More precisely, if t_0 and y_0 are in the domain of definition of $f(t, y)$ and if $f(t, y)$ is continuous in some neighborhood of $(t_0\ \boldsymbol{y}_0)$, then there exists at least one solution of the equation

$$\boldsymbol{y}' = \boldsymbol{f}(t, \boldsymbol{y})$$

which tends to $\boldsymbol{y}_0$ when $t \to t_0$. This explains properties (1) and (3).

In connection with (3) the important point is the dependence of the solution on arbitrary *constants* and not on arbitrary *functions* as in the vase of partial differential equations. Further we note that the equation does not distinguish between left and right: the solution exists on both sides of $t = t_0$ and is continuous for left-handed as well as for right-handed approach. This is what we mean by saying that a *principle of parity* holds.

Uniqueness appears when $\boldsymbol{f}(t, \boldsymbol{y})$ satisfies a LIPSCHITZ condition, or similar restraint, with respect to the second argument. If the equation is linear and the matrix $\boldsymbol{A}(t)$ is continuous, then the LIPSCHITZ condition is trivially satisfied and uniqueness expresses itself in the absence of non-trivial null solutions. Further the solutions form a linear manifold spanned by n base vectors whose Wronskian does not vanish.

If $\boldsymbol{A}(t)$ is analytic, the three properties (7) - (9) come into play: the solutions are also analytic, the analytic continuation of a solution remains a solution of the differential equation provided the coefficients are continued along the same path, and, finally, the singularities of the solutions are located at the points where the coefficients become singular. To this set we may have to add the point at infinity. If the coefficients have no finite singularities, that is, are entire functions of t then the solutions have the same property. In particular, if $\boldsymbol{A}(t) = \boldsymbol{A}$ is a constant matrix, then

$$\boldsymbol{y}(t;\ 0,\ y_0) = \exp\ (t\boldsymbol{A})\boldsymbol{y}_0 \tag{1.3}$$

is an exponential function. This implies that the components of the solution are ordinary exponential functions, possibly multiplied by polynomials if the matrix has repeated characteristic roots.

Our problem now is to discuss what happens to these properties when n becomes infinite. Thus we are concerned with the equation

$$\boldsymbol{y}' = \boldsymbol{A}\,\boldsymbol{y} \tag{1.4}$$

where $\boldsymbol{A}$ is an infinite matrix of constants. Here questions of convergence loom large. A vector

$$\boldsymbol{y}(t) = \{y_1(t),\ y_2(t),\ \dots,\ y_n(t),\ \dots\}$$

is a solution of (1.4) if and only if all the series

(1.5) $$\sum_{k=1}^{\infty} a_{jk}(t)y_k(t)$$

are convergent for the values of t under consideration. Normally there is a profusion of solutions in this sense and it is often desirable to restrict the field further by requiring that the solution belong to some particular metric linear space. Sometimes either the system or the problem which gives rise to the system severely limits what linear spaces may be considered, but there are also cases where the choice is almost arbitrary.

The richest source of differential equations of type (1.4) is the theory of stochastic processes. Here the $y_j(t)$ are probabilities that is, $0 \leq y_j(t) \leq 1$ and normally $\sum_{j=1}^{\infty} y_j(t) \leq 1$. In this case the underlying problem gives preference to the space (l), and more precisely, to the positive cone of this space. But if we do not pretend to do stochastic theory and if we do not care whether or not our results admit of probabilistic interpretation, then we should feel free to study the properties of the solutions of the system regardless of the provenience of the latter. The phenomena that we shall encounter are associated with particular equations in particular spaces and the character of admissible solutions of an equation may depend sharply upon the space. A trivial example will illustrate this.

We take $\boldsymbol{A}$ to be a diagonal matrix with

$$a_{jk} = j\delta_{jk}$$

so that the system becomes

(1.6) $$y'_j(t) = jy_j(t), \qquad j = 1, 2, 3, \ldots,$$

with the trivial solution

(1.7) $$y_j(t) = e^{jt}y_{j0}.$$

There are no convergence questions and the choice of a suitable space X as universe of discourse is arbitrary. Let us take $X = (m)$, the space of bounded sequences $\boldsymbol{x} = \{x_n\}$ with

$$\| \boldsymbol{x} \| = \sup |x_n| .$$

If the initial vector $\boldsymbol{y}_0 = \{y_{j0}\}$ is in (m), there is no reason why $\{e^{jt}y_{j0}\}$ should be in (m) for any positive value of t. On the other hand, we note that the vector certainly belongs to (m) for $t < 0$. Thus, this system violates the principle of parity in the space (m) and the existence problem splits into two: does there exist a solution, defined in an interval (a, b) and belonging to (m) for such values of t, which tends to a preassigned limit in (m) when (1) t decreases to a, (2) when t increases to b? The first problem usually does not have a solution while the second one does. Each component is analytic, an exponential function, the solution in (m) is analytic as an element of (m), and can always be continued to the left. To the right there is normally a barrier, the position of which is in no ways determined by the equation but only by the initial values. These statements hold for (m) and need no longer hold if we replace (m) by another space.

2. Existence in B-spaces.

Let us return to equation (1.4) where we look for solutions which belong to a given (B)-space X. There are three distinct possibilities.

Case I. $\boldsymbol{A}$ defines a linear bounded transformation $\mathfrak{A}$ on X to X. This means that to every $\boldsymbol{x} \in X$ there corresponds a vector $\boldsymbol{y} \in X$ such that

$$\boldsymbol{y} = \mathfrak{A}\boldsymbol{x}. \tag{2.1}$$

Further, there is a constant M such that

$$\| \boldsymbol{y} \| \leq M \| \boldsymbol{x} \|$$

for every $\boldsymbol{x} \in X$. We can then define $\mathfrak{A}^n \boldsymbol{x}$ by

$$\mathfrak{A}^n \boldsymbol{x} = \mathfrak{A}[\mathfrak{A}^{n-1}\boldsymbol{x}], \quad n = 2, 3, \ldots \tag{2.2}$$

These powers are also bounded linear transformations and

$$\| \mathfrak{A}^n \boldsymbol{x} \| \leq M^n \| \boldsymbol{x} \| .$$

We can also define

$$\exp(t\mathfrak{A})\boldsymbol{x}$$

by the exponential series, convergent for all t. In this case we have complete analogy with the finite case, formula (1.3) holds and gives the solution for every real or complex value of t and for every initial vector $\boldsymbol{y}_0$ of X. Moreover, it is the only solution in X having the given initial value $\boldsymbol{y}_0$ at $t = 0$. Since the system is autonomous, we have merely to replace t in (1.3) by $(t - t_0)$ to get the solution which takes on the value $\boldsymbol{y}_0$ at $t = t_0$.

It should be noted, however, that the transformation $x \to \mathfrak{A}x$ may change character from one space to another. Thus it may be unbounded in one space and bounded in another.

CASE II. - $D[\mathfrak{A}]$, the domain of the operator $\mathfrak{A}$, is dense in X but does not coincide with X. $\mathfrak{A}$ is supposed to be the infinitesimal generator of a semi-group of linear bounded operators $\mathfrak{S}(t)$ on X to itself (see E. HILLE and R. S. PHILLIPS [5]). This implies that for $0 < t_1,\ t_2 < \infty$ we have

$$\mathfrak{S}(t_1)\mathfrak{S}(t_2)x = \mathfrak{S}(t_1 + t_2)x \tag{2.3}$$

for all x, further

$$\lim_{t \to 0+} \mathfrak{S}(t)x = x \tag{2.4}$$

for all x, while for $x \in D[\mathfrak{A}]$ we also have

$$\frac{d}{dt}[\mathfrak{S}(t)x] = \mathfrak{A}[\mathfrak{S}(t)x], \tag{2.5}$$

where differentiation and limits are taken in the sense of the normed metric. In this case

$$y(t;\ 0,\ y_0) = \mathfrak{S}(t)[y_0], \qquad y_0 \in D[\mathfrak{A}] \tag{2.6}$$

is a solution of the equation which tends strongly to y_0 as t decreases to 0. Here normally we are limited to initial vectors in $D[\mathfrak{A}]$ so the arbitrariness of the «constants» is restricted. The solution is unique when it exists.

As an example to illustrate this case we may take A as the negative of the diagonal matrix in (1.6) above so that

$$\mathfrak{S}(t)[y_0] = \{e^{-jt}y_{j0}\}.$$

For X we may again take the space (m). In this case y_0 can be any element of (m) since

$$\{je^{-jt}y_{j0}\} \in (m),\ 0 < t.$$

We note that in this case the solution normally ceases to exist as an element of (m) for $t < 0$. We can always expect violation of the principle of parity in Case II.

CASE III. - In the given space X the operator $\mathfrak{A}$ corresponding to the matrix A is unbounded and does not generate a bounded semigroup. Here anything can happen as will be seen from the examples given below.

3. Null solutions.

Uniqueness is a more sensitive property than existence and is more easily lost. Let us first examine the implications of non-uniqueness in the case of the equation

$$y' = \mathfrak{A}y. \tag{3.1}$$

This implies that there exists a non-trivial vector $z(t)$ in X which satisfies (3.1) and tends to the zero element as t decreases to 0.

We shall impose some restrictions on $z(t)$. We suppose that $z(t)$ is defined for $t > 0$, that in the given normed topology $\| (zt) \|$ is a measurable function of t, and that there are numbers α (real) and $M(\varepsilon)$ (positive), such that for each $\varepsilon > 0$ we have

$$\| z(t) \| < M(\varepsilon)\; e^{(\alpha+\varepsilon)t}. \tag{3.2}$$

We say that such a null solution is of *normal type*. This property implies that $z(t)$ has a LAPLACE transform in X

$$x(\lambda) = \int_0^\infty e^{-\lambda t} z(t)\,dt \tag{3.3}$$

which exists for $R(\lambda) > \alpha$ and is a bounded holomorphic function of λ for each λ with $R(\lambda) \geq \alpha + \varepsilon$. Further we have

$$A[x(\lambda)] = \lambda\, x\,(\lambda). \tag{3.4}$$

Conversely, if this characteristic equation of the operator $\mathfrak{A}$ has a solution in X which is a bounded holomorphic function of λ in some right half-plane, then we can also find a solution of (3.4) which, in addition, has the property that $\| \lambda x(\lambda) \|$ is integrable along vertical lines in the half-plane. With the aid of such a solution $x(\lambda)$ we can construct a null solution of (3.1) by setting

$$z(t) = \frac{1}{2\pi i}\int_{c-i\infty}^{c+i\infty} e^{\lambda t}\; x(\lambda)\,d\lambda\,. \tag{3.5}$$

This tells us, in particular, that the point spectrum of $\mathfrak{A}$ in the space X must contain a right half-plane.

We observe further that we can always «improve» a null solution by multiplication by a suitable kernel $K(t-s)$ and forming

$$\int_0^t K(t-s)\mathbf{z}(s)ds. \tag{3.6}$$

This is again a null solution and if $K(u)$ is suitably chosen, this solution has derivatives of all orders which tend to zero with t. For further elaboration of these facts see E. HILLE [2].

4. Example 1. Birth and death process.

We shall illustrate the preceding considerations and bring out more of the pathology by means of an example taken from the theory of birth and death processes. For the probabilistic back ground see Chapter XVII of W. FELLER [1], especially p. 422. See also W. LEDERMANN and G. E. H. REUTER [6]. The so called *backward equations* in this theory take the form

$$P'_{i,n}(t) = -(\lambda_i + \mu_i)P_{i,n}(t) + \lambda_i P_{i+1,n}(t) + \mu_i P_{i-1,n}(t). \tag{4.1}$$

Our discussion applies to all equations of this class, but we can save computation and better bring out the essential features by treating a special case. We take

$$\lambda_i = 1, \ \mu_i = i - 1. \tag{4.2}$$

Thus, the equations to be considered are

$$\begin{aligned} y'_1(t) &= -y_1(t) + y_2(t), \\ y'_2(t) &= y_1(t) - 2y_2(t) + y_3(t), \\ &\cdots\cdots\cdots \\ y'_n(t) &= (n-1)y_{n-1}(t) - ny_n(t) + y_{n+1}(t), \\ &\cdots\cdots\cdots\cdots \end{aligned} \tag{4.3}$$

For the time being we disregard completely the origin of these equations and we aim to study them from the point of view of systems of differential equations.

The characteristic equation (3.4) now leads to the following system:

$$
\begin{aligned}
&x_2(\lambda) = (\lambda + 1)x_1(\lambda), \\
&x_3(\lambda) = (\lambda + 2)x_2(\lambda) - x_1(\lambda), \\
(4.4) \qquad &\cdots\cdots\cdots\cdots \\
&x_{n+1}(\lambda) = (\lambda + n)x_n(\lambda) - (n - 1)x_{n-1}(\lambda), \\
&\cdots\cdots\cdots\cdots\cdots\cdots
\end{aligned}
$$

It is clear that $x_1(\lambda)$ is arbitrary and that we can determine the other components uniquely in terms of $x_1(\lambda)$. Thus,

$$(4.5) \qquad x_{n+1}(\lambda) = P_n(\lambda)x_1(\lambda),$$

where $P_n(\lambda)$ is polynomial of degree n. The coefficients of $P_n(\lambda)$ are positive integers whose sum is $< (n+1)!$ if $n > 2$.

It is clear that no matter how we choose $x_1(\lambda)$ we cannot force $\boldsymbol{x}(\lambda)$ to belong to one of the conventional spaces (l) or (m) considered in stochastic theory. But we can easily find weighted (l) - or (m)-spaces to which $\boldsymbol{x}(\lambda)$ will belong. Let L be the space of vectors $\boldsymbol{x}$ such that

$$(4.6) \qquad \| \boldsymbol{x} \| \equiv \sum_{n=1}^{\infty} e^{-n^2} | x_n | < \infty .$$

For $| \lambda | > 1$ we have

$$| P_n(\lambda) | < (n+1)! \, | \lambda |^n$$

and if we choose

$$x_1(\lambda) = \exp(-\lambda^{1/2}),$$

then for $R(\lambda) \geq 0$

$$| x_n(\lambda) | < 2^n (n+1)! \, (2n)!,$$

whence it follows that $\boldsymbol{x}(\lambda) \in L$ and there are null solutions in L.

In the present case it is of greater interest to note that the general solution of (4.3) can be given explicitly. We choose arbitrarily

$$(4.7) \qquad y_1(t) = f(t),$$

where $f(t)$ is a function having derivatives of all orders for the values of t under consideration. Thus, we assume either that t is a real variable and that $f(t) \in C^\infty[a, b]$ on some interval $[a, b]$ or else that t is complex and that $f(t)$ is an analytic function holomorphic in some domain D. With these conventions we get

$$\begin{aligned} y_2(t) &= f'(t) + f(t), \\ y_3(t) &= f''(t) + 3f'(t) + f(t), \\ &\cdots\cdots\cdots \\ y_{n+1}(t) &= P_n\left(\frac{d}{dt}\right)[f(t)], \\ &\cdots\cdots\cdots \end{aligned} \tag{4.8}$$

where $P_n(.)$ is the polynomial introduced above. All these functions exist by virtue of the restrictions imposed upon $f(t)$. We note that the general solution of (4.3) involves a function $f(t)$ which is an arbitrary element of C^∞.

If $f(t)$ is analytic, this solution always belongs to the space L. Suppose that $f(t)$ is holomorphic and bounded in some domain D and let Δ be a compact subset of D. Then for $t \in \Delta$ we have

$$| f^{(n)}(t) | \leqq M d^{-n} n!$$

where d is the distance from Δ to the boundary of D. Combining this with the estimate of $P_n(\lambda)$ we obtain

$$| y_n(t) | < M_1 \, n! \, (n+1)! \max (1, d^{-n}) \tag{4.9}$$

and this shows that the solution is in L.

If, on the other hand, $f(t) \in C^\infty[a, b]$, then we have to impose some restrictions on the rate of growth of the derivatives to make sure that the solution belong to L. It suffices that we can find a fixed k such that

$$| f^{(n)}(t) | \leq M(n!)^k, \qquad a < t < b, \tag{4.10}$$

for all n. Such a condition is satisfied by

$$f(t) = \exp(-t^{-2}), \; 0 < t < 1,$$

where any $k > \frac{3}{2}$ will do. The corresponding solution is a null solution which vanishes together with all its derivatives when t decreases to 0.

Since (4.8) defines the general solution of (4.3), it will also give the solutions which are of probabilistic interest. In stochastic theory it is required to find a matrix of transition probabilities $\boldsymbol{P}(t) = (p_{jk}(t))$ such that

$$\boldsymbol{P}'(t) = \boldsymbol{BP}(t),\ 0 < t, \tag{4.11}$$

where $\boldsymbol{B}$ is the matrix of the system (4.3) which is of the Jacobi type. Here $0 \leq p_{jk}(t) \leq 1$, $\sum_{k=1}^{\infty} p_{jk}(t) \equiv 1$ for each j, $p_{jk}(0) = \delta_{jk}$, and

$$\boldsymbol{P}(s + t) = \boldsymbol{P}(s)\boldsymbol{P}(t).$$

This matrix is known to be unique. We would expect its elements to be expressible in terms of the differential operator $P_n\ (d/dt)$ and this is indeed the case. G.E.H. REUTER, to whom I had communicated my results, informs me that

$$p_{jk}(t) = P_{j-1}\left(\frac{d}{dt}\right)P_{k-1}\left(\frac{d}{dt}\right)g(t) \tag{4.12}$$

where $g(t)$ is a LAPLACE transform.

We have similar results for any birth and death process. In particular, the general solution of the system (4.1) depends upon an arbitrary function in C^∞.

This class of equations shows that many of the eleven properties of our list are no longer valid:

Re (3). Instead of arbitrary constants, the solution depends upon an arbitrary function in C^∞.

Re (4) and (5). In suitably chosen (B)-spaces there are null solutions and hence non-uniqueness.

Re (7). There are non-analytic solutions since $f(t)$ need not be analytic.

Re (9) - (11). An analytic solution may have arbitrary singularities, namely those of $f(t)$ and of its derivatives.

5. Example II. Summatory systems.

In a note [3] the author considered an infinite system of equations of the form

$$y'_n(t) + na_n y_n(t) = \sum_{k=n+1}^{\infty} a_k y_k(t),\ n = 1, 2, 3, \dots, \tag{5.1}$$

where the a_n's are positive. This system is of the type called « Summengleichungen » by O. PERRON. The matrix is not of the KOLMOGOROFF type but its transpose has this property, at least in the wider sense where the sum of the elements in a row may be negative instead of zero.

Convergence questions arise right away and the most natural space to consider is a weighted (l)-space, (l_a), where

$$\| x \| = \sum_{n=1}^{\infty} a_n \mid x_n \mid < \infty.$$

In [3] the existence of null solutions was proved. They will occur if

$$\sum_{n=1}^{\infty} \frac{1}{n^2 a_n} < \infty.$$

In a later paper [4] the special case $a_n \equiv 1$ was treated and the existence of analytic solutions with logarithmic branch points or with singular lines was proved. G.E.H. REUTER called my attention to the fact that the general solution of

$$y'_n(t) + ny_n(t) = \sum_{k=n+1}^{\infty} y_k(t),\ n = 1,\ 2,\ 3, \dots, \tag{5.2}$$

may be found and that it involves an arbitrary function. With his kind permission, the solution is reproduced below.

Before embarking on this, let me observe that similar results may be obtained by a slightly different method which applies to all equations of class (5.1). We give an initial vector $\{y_k(0)\}$ in (l_a), that is such that

$$\sum_{n=1}^{\infty} a_n \mid y_n(0) \mid < \infty, \tag{5.3}$$

and we assume the existence of a solution $\{ y_n(t) \} \in (l_a)$ for all $t > 0$. We can then introduce the function

$$\sum_{n=1}^{\infty} a_n y_n(t) \equiv F(t). \tag{5.4}$$

Since the series converges for all $t > 0$ by assumption and the terms of the series are continuous, $F(t)$ can be at most pointwise discontinuous. Now if $F(t)$ is supposed to be known, we can recover the individual terms of the series with the aid of the system (5.1) and the given initial conditions. As a first step we find

$$y_1(t) = e^{-2a_1 t} \left\{ y_1(t) + \int_0^t e^{2a_1 s} F(s) ds \right\} \tag{5.5}$$

and if $y_1(t)$, $y_2(t)$, ..., $y_{n-1}(t)$ have been found, then

$$y_n(t) = e^{-(n+1)a_n t}\left\{ y_n(0) + \int_0^t e^{(n+1)a_n s}[F(s) - \sum_{k=1}^{n-1} a_k y_k(s)]ds \right\}. \tag{5.6}$$

It is not difficult to prove that this procedure leads to a vector $y(t) = \{y_n(t)\}$ which belongs to (l_a). Further, the system (5.1) is satisfied as well as the initial conditions and condition (5.4) holds. It is clear that every solution of the system in the space (l_a) can be obtained by this method and that the solution is uniquely determined by the initial condition together with the arbitrary function $F(t)$ but not by either of these data alone.

We now return to equation (5.2) and the space (l). We give an initial vector $y(0) = \{y_n(0)\}$ in (l) and suppose that the equation has at least one solution $y(t) = \{y_n(t)\}$ in (l) defined for positive values of t which tends to $y(0)$ as t decreases to 0. We set

$$z_n(t) = \sum_{k=n+1}^{\infty} y_k(t) \tag{5.7}$$

where the series in convergent by assumption. We have

$$y_n(t) = z_{n-1}(t) - z_n(t)$$

which we substitute in the nth equation obtaining

$$z'_{n-1}(t) - z'_n(t) + n[z_{n-1}(t) - z_n(t)] = z_n(t),$$

whence

$$z'_{n-1}(t) + nz_{n-1}(t) = z'_n(t) + (n+1)z_n(t)$$

is independent of n. We denote the common value by $f(t)$.

We now have to find $z_n(t)$ from the differential equation

$$z'_n(t) + (n+1)\, z_n(t) = f(t), \tag{5.8}$$

the solution of which is

$$z_n(t) = z_n(0)e^{-(n+1)t} + \int_0^t e^{(n+1)(u-t)}f(u)du. \tag{5.9}$$

Here we replace n by n-1 and subtract, obtaining

$$(5.10)\qquad \begin{aligned} y_n(t) &= z_{n-1}(0)e^{-nt} - z_n(0)e^{-(n+1)t} \\ &+ \int_0^t e^{n(u-t)}[1 - e^{u-t}]f(u)du. \end{aligned}$$

The only condition which this discussion imposes on $f(t)$ is that it be integrable over each finite interval. If this is true, then $y_n(t)$ is well defined for $t > 0$. As t decreases to 0, $y_n(t)$ tends to the limit

$$z_{n-1}(0) - z_n(0) = y_n(0).$$

Further,

$$\sum_{n=1}^{\infty} | y_n(t) | \leq \sum_{n=1}^{\infty} | z_{n-1}(0) - e^{-t}z_n(0) | e^{-nt} + \int_0^t e^{u-t} | f(u) | du$$

so that we are dealing with an element of (l). Finally, if we compute the derivative as the limit of the difference quotient we find

$$(5.11)\qquad \begin{aligned} y'_n(t) &= -nz_{n-1}(0)e^{-nt} + (n+1)z_n(0)e^{-(n+1)t} \\ &- \int_0^t e^{n(u-t)}[n - (n+1)e^{u-t)}]f(u)du \end{aligned}$$

at every point $t > 0$. For $t = 0$ this is the right hand derivative. From this we conclude that (5.10) is indeed a solution of (5.2) for $t > 0$ which belongs to (l) and tends to the preassigned initial value when t decreases to 0. The existence of the first derivative in (5 11) depends upon the fact that the integrand in (5.10) vanishes for $u = t$. This is no longer the case in (5.11) and as a consequence $y''_n(t)$ exists only almost everywhere unless $f(t)$ is continuous.

Thus we see that the system (5.2) has a number of extreme properties. The general solution in (l) involves an arbitrary function $f(t)$. If this function is defined merely for real values of t and integrable over finite intervals, then the solution is non-analytic, its first derivative is absolutely continuous but the second derivative exists only almost everywhere. We are thinking here of derivatives in the componentwise sense, but extensions to strong derivatives may also be considered. If $f(t)$ is analytic then we can dispose of its singularities *ad lib.* This means that we can place singularities of the solution wherever we want and, within wide limits, we can also prescribe the nature of the singularities. In view of this situation, the results obtained in [4] cease to be astonishing.

Let us return to the real case. Since we have an arbitrary function $f(t)$ at our disposal, it is clear that we have null solutions and nonuniqueness. Moreover, the principle of parity is obviously violated. The equation so to speak prefers the future and is very recluctant to give information about the past. It does not help if we assume that $f(t)$ is integrable over every finite interval. Normally we cannot continue the vector $\mathbf{y}(t)$ to the left of the origin though we can continue each component. Since the series

$$\sum_{n=1}^{\infty} y_n(t)$$

whose terms are defined by (5.10), usually diverges for $t < 0$, the continuation of the components no longer gives a solution of the system, that is, we have a form of violation of property (8), the permanency of functional equations. This phenomenon arises even if $f(t) \equiv 0$ for then we have

$$y_n(t) = [z_{n-1}(0) - z_n(0)e^{-t}]e^{-nt}$$

and the infinite series whose general term is given by this expression normally diverges for $t < 0$ unless $z_n(0)$ tends at least exponentially to 0.

The proper concept of analyticity here is that of the space (l). We see that if $f(t)$ is a constant or, more generally an entire function of t, then $\mathbf{y}(t)$ is holomorphic in the right half-plane as a function in (l) and normally the imaginary axis is a singular line.

This discussion applies with minor modifications to the solution of the general equation (5.1) which is given by (5.6). There are several features of these examples which reminds one of a partial differential equation, such as the heat equation. The dependence of the solution on an arbitrary function is one such feature, the preference for right rather than left is another.

REFERENCES

[1] W. FELLER, *An Introduction to Probability Theory and Its Applications.* Volume 1, Second Edition, John Wiley & Sons, New York, 1957.

[2] E. HILLE, *The abstract Cauchy problem and Cauchy's problem for parabolic differential equations.* « Journal d'Analyse Mathématique », 3 (1953, 54) 81-196.

[3] E. HILLE, *Quelques remarques sur les équations de Kolmogoroff.* «Bull. de la Société des mathématiciens et physiciens de la R. P. de Serbie», 5 (1953) 3-14.

[4] — —, *Remarques sur les systèmes des équations différentielles linéaires à une infinité d'inconnues.* «Journal de Math. pures et appliquées», (9) 37 (1958) 375 - 383.

[5] E. HILLE and R.S. PHILLIPS. *Functional Analysis and Semi-Groups.* «Colloquium Publications», Vol. 31 Revised Edition. Amer. Math. Soc., Providence, R.I., 1958.

[6] W. LEDERMANN and G.E.H. REUTER, *Spectral Theory for the differential equations of simple birth and death processes.* «Phil. Trans. R. Society of London», (A) 246 (1954) 321-369.

Green's transforms and singular boundary value problems [1];

By Einar HILLE

(New Haven, Connecticut).

1. Introduction. — More than forty years ago I introduced a class of identities in the theory of linear second order differential equations for which the name of Green's transforms was proposed (*see* [3] in the Bibliography at the end of this paper). Such a transform serves to delimit in the complex plane the regions where a given solution can have zeros. But the transform can be used for other purposes than to obtain zero free regions. Thus in [4] it was used to study integrability and growth properties of solutions. In the present paper I propose to utilize the same technique for a study of the singular boundary problem of Hermann Weyl [12].

Given the differential equation

$$u'' - [\lambda + q(x)]u = 0, \tag{1}$$

(1) This research was supported in part by the United States Air Force through the Office of Scientific Research of the Air Research and Development Command under Contract No. SAR/AF 49-(638) 224. Most of the results of this paper were presented to the International Congress of Mathematicians in Edinburgh 1958. For a brief summary, *see* p. 80-81 of the Abstracts of Short Communications distributed at the Congress.

where $q(x)$ is real and belongs to $C[0, \infty)$, together with the side conditions

$$(2) \qquad \cos\alpha\, u(0, \lambda) + \sin\alpha\, u'(0, \lambda) = 0, \qquad u(x, \lambda) \in L_2(0, \infty),$$

find $u(x, \lambda)$. This is Weyl's problem.

Multiplying (1) by $\overline{u(x, \lambda)}$ and integrating, one obtains

$$(3) \quad \left[\overline{u(x, \lambda)}\, u'(x, \lambda)\right]_a^b - \int_a^b |u'(s, \lambda)|^2 ds - \int_a^b [\lambda + q(s)]\, |u(s, \lambda)|^2 ds = 0$$

which is Green's transform of (1). If $\lambda = \mu + i\nu$, the imaginary part of (3) equals

$$(4) \qquad \mathfrak{J}\left\{\left[\overline{u(x, \lambda)}\, u'(x, \lambda)\right]\right\}_a^b - \nu \int_a^b |u(s, \lambda)|^2 ds = 0.$$

This identity is basic in the treatment of singular boundary value problems and was used by Weyl for such purposes. I want to point our, however, that the real part of the identity can also be used to great advantage in this discussion. The purpose of this paper is to illustrate how this is done. Little or no claim can be made for novelty of the results, the only claim that is made is that here is a useful method which, apparently, has been overlooked. In some portions of the paper I supplement the method by various considerations of an elementary or function theoretical nature. Such devices may possibly have been used by other authors as well.

2. The case where $q(x)$ is bounded below. — As an illustration of the method, the following theorem will be proved :

Theorem 1. — *If $q(x)$ is bounded below, if $\nu \neq 0$, and if $u(x, \lambda) \in L_2(0, \infty)$ is a solution of (1), then*

$$(5) \qquad |q(x)|^{\frac{1}{2}} u(x, \lambda) \in L_2(0, \infty),$$

$$(6) \qquad u'(x, \lambda) \in L_2(0, \infty),$$

$$(7) \qquad \lim_{x \to +\infty} \overline{u(x, \lambda)}\, u'(x, \lambda) = 0.$$

This solution is unique save for a constant multiplier and the limit-point case holds at infinity.

Proof. — For the terminology used in this paper, *see* for instance E. A. Coddington and N. Levinson ([2], chapter 9, or E. C. Titchmarsh [10]. We start out by taking $a = 0$, $b = x$ in (3). Thus for the particular solution $u(x, \lambda)$ in $L_2(0, \infty)$ we have

$$(8) \quad \overline{u(x, \lambda)}\, u'(x, \lambda) = A + \int_0^x |u'(s, \lambda)|^2\, ds + \int_0^x [\lambda + q(s)]\, |u(s, \lambda)|^2\, ds,$$

where

$$A = A_1 + iA_2 = \overline{u(0, \lambda)}\, u'(0, \lambda).$$

Taking the imaginary part of (8) we see that

$$(9) \quad \lim_{x \to \infty} \Im [\overline{u(x, \lambda)}\, u'(x, \lambda)] = A_2 + \nu \int_0^\infty |u(s, \lambda)|^2\, ds$$

exists as a finite number. This, of course, is well known. The real part gives

$$(10) \quad \frac{1}{2} \frac{d}{dx} |u(x, \lambda)|^2 = A_1 + \int_0^x |u'(s, \lambda)|^2\, ds + \int_0^x [\mu + q(s)]\, |u(s, \lambda)|^2\, ds.$$

Since $q(x)$ is supposed to be bounded below, we may assume

$$(11) \quad \inf q(x) = q_0 > 0$$

without restricting the generality. If $\mu \geqq 0$, the right hand side of (10) is an increasing function of x and consequently has a finite limit or tends to $+\infty$ with x. If $\mu < 0$, we have

$$\int_a^b [\mu + q(s)]\, |u(s, \lambda)|^2\, ds > (\mu + q_0) \int_a^b |u(s, \lambda)|^2\, ds.$$

If a is large, μ fixed, the right hand side is as close to 0 as we wish. It follows again that the right hand side of (10) either has a finite limit or tends to $+\infty$ with x. Thus

$$\lim_{x \to +\infty} \frac{d}{dx} |u(x, \lambda)|^2 \equiv 2 l_1$$

exists and $-\infty < l_1 \leqq +\infty$. But $u(x, \lambda) \in L_2(0, \infty)$ and this requires that $l_1 = 0$. It follows that (5) and (6) hold. On the other hand, if $u(x, \lambda)$ and $u'(x, \lambda)$ are in $L_2(0, \lambda)$, then clearly

$\overline{u(x, \lambda)}\, u'(x, \lambda) \in L(0, \infty)$. If the limit of the product is $i\, l_2$, then we must have $l_2 = 0$. Thus (7) also holds.

In order to complete the proof, we have only to observe that there are solutions of (1) for which the right hand side of (9) does not tend to the limit zero. Such a solution is given by

$$u_1(x) = u_1(x; \alpha, \lambda)$$

determined by the initial conditions

$$u_1(0) = \cos \alpha, \qquad u'_1(0) = \sin \alpha. \tag{12}$$

Here

$$\mathfrak{I}[\overline{u_1(x)}\, u'_1(x)] = \nu \int_0^x |u_1(s)|^2\, ds$$

and this obviously does not tend to 0 as $x \to +\infty$.

3. Properties of $m(\lambda, \alpha)$. — We recall that for $\nu \neq 0$ the equation (1) always has a solution in $L_2(0, \infty)$. This solution may be written in the form

$$U_\alpha(x; \lambda) = u_1(x; \alpha, \lambda) + m(\lambda, \alpha)\, u_2(x; \alpha, \lambda), \tag{13}$$

$u_1(x; \alpha, \lambda)$ is determined by the initial conditions (12) and $u_2(x) = u_2(x; \alpha, \lambda)$ is determined by

$$u_2(0) = \sin \alpha, \qquad u'_2(0) = -\cos \alpha. \tag{14}$$

For fixed x and α the solutions $u_k(x; \alpha, \lambda)$ are entire functions of λ of order $\frac{1}{2}$. Weyl showed that $m(\lambda, \alpha)$ is a holomorphic function of λ in each of the half-planes $\nu > 0$ and $\nu < 0$,

$$m(\bar{\lambda}, \alpha) = \overline{m(\lambda, \alpha)}, \tag{15}$$

and

$$\operatorname{sgn} \mathfrak{I}[m(\lambda, \alpha)] = -\operatorname{sgn} \mathfrak{I}(\lambda). \tag{16}$$

We also have the interesting functional equation

$$\frac{m(\lambda, \alpha) - m(\lambda, \beta)}{1 + m(\lambda, \alpha)\, m(\lambda, \beta)} = \tan(\alpha - \beta) \tag{17}$$

which expresses that $U_\alpha(x, \lambda)$ and $U_\beta(x, \lambda)$ are linearly dependent.

We shall now use identity (3) to prove

THEOREM 2. — *If* $q(x)$ *is bounded below,* $\inf q(x) = q_0$, *then* $m(\lambda, \alpha)$ *is holomorphic in the* λ*-plane cut along the real axis from* $-\infty$ *to* $-q_0$, *save for at most one pole on the real axis to the right of* $-q_0$. *In particular, there is no such exceptional pole if* $\frac{1}{2}\pi < \alpha < \pi$ *or* $-\frac{1}{2}\pi < \alpha < 0$.

Proof. — We substitute $U_\alpha(x, \lambda)$ in (3), take $a = 0$, and let $b \to +\infty$. In view of theorem 1 and the initial conditions

$$(18)\qquad -\frac{1}{2}\sin 2\alpha[1 - |m(\lambda, \alpha)|^2] + \cos 2\alpha \mathcal{R}[m(\lambda, \alpha)] + i\mathcal{I}[m(\lambda, \alpha)]$$
$$= \int_0^\infty |U'_\alpha(s, \lambda)|^2\, ds + \int_0^\infty [\lambda + q(s)]\, |U_\alpha(s, \lambda)|^2\, ds.$$

Here we set $\alpha = \frac{3}{4}\pi$ and take the real part, obtaining

$$(19)\qquad 1 - \left|m\left(\lambda, \frac{3}{4}\pi\right)\right|^2 = 2\int_0^\infty \left|U'_{\frac{3}{4}\pi}(s, \lambda)\right|^2 ds$$
$$+ 2\int_0^\infty [\mu + q(s)] \left|U_{\frac{3}{4}\pi}(s, \lambda)\right|^2 ds.$$

This shows that

$$(20)\qquad \left|m\left(\lambda, \frac{3}{4}\pi\right)\right| < 1 \qquad \text{when } \mu \geqq -q_0, \qquad \nu \neq 0.$$

Further

$$(21)\qquad 2(\mu + q_0)\int_0^\infty \left|U_{\frac{3}{4}\pi}(s, \lambda)\right|^2 ds < 1.$$

Repeated use will be made of these inequalities.

First, from

$$\left|\mathcal{I}\left[m\left(\lambda, \frac{3}{4}\pi\right)\right]\right| = |\nu| \int_0^\infty \left|U_{\frac{3}{4}\pi}(s, \lambda)\right|^2 ds < \frac{|\nu|}{2(\mu + q_0)},$$

we conclude that

$$(22)\qquad \lim_{\nu \to 0} \mathcal{I}\left[m\left(\mu + i\nu, \frac{3}{4}\pi\right)\right] = 0, \qquad \mu > -q_0,$$

uniformly in μ for $\mu \geqq -q_0 + \varepsilon$.

Secondly, we can compute the value of the integral

$$\int_0^\infty U_\alpha(s, \lambda_1)\, U_\alpha(s, \lambda_2)\, ds = \frac{m(\lambda_1, \alpha) - m(\lambda_2, \alpha)}{\lambda_2 - \lambda_1} \tag{23}$$

by the classical method of Sturm. In particular,

$$\int_0^\infty [U_\alpha(s, \lambda)]^2\, ds = -\frac{\partial}{\partial \lambda} m(\lambda, \alpha). \tag{24}$$

Combining this formula with (21), we see that

$$2(\mu + q_0)\left|\frac{\partial}{\partial\lambda} m\left(\lambda, \frac{3}{4}\pi\right)\right| < 1. \tag{25}$$

It follows from (20) and (25) that the functions

$$\left\{ m\left(\mu + \frac{i}{n}, \frac{3}{4}\pi\right) \middle| n = 1, 2, 3, \ldots \right\}$$

are uniformly bounded and equicontinuous in the interval $(-q_0 + \varepsilon, \infty)$. Thus there is at least one subsequence which converges uniformly to a bounded continuous limit, $m\left(\mu, \frac{3}{4}\pi\right)$ say. By (22) this limit function must be real. Using (25) once more we see that for $\nu > 0$;

$$\left| m\left(\mu + i\nu, \frac{3}{4}\pi\right) - m\left(\mu, \frac{3}{4}\pi\right)\right| < \frac{\nu}{2(\mu + q_0)}.$$

Hence

$$\lim_{\nu \to 0} m\left(\mu + i\nu, \frac{3}{4}\pi\right) \equiv m\left(\mu, \frac{3}{4}\pi\right) \tag{26}$$

exists as a real continuous function for $\mu > -q_0$ and its absolute value does not exceed one. Here we have supposed $\nu > 0$, but formula (15) shows that we get the same limit for approach from the lower half-plane.

By Schwarz's principle of symmetry it follows that $m\left(\lambda, \frac{3}{4}\pi\right)$ is holomorphic also on the real axis for $\mu > -q_0$ and (20) holds everywhere in the half-plane $\mathcal{R}(\lambda) > -q_0$. Using (21) we conclude that

$$\lim_{\nu \to 0} U_{\frac{3}{4}\pi}(x; \mu + i\nu) \equiv U_{\frac{3}{4}\pi}(x; \mu), \qquad \mu > -q_0, \tag{27}$$

exists, uniformly in any finite interval, and $U_{\frac{3}{4}\pi}(x, \mu) \in L_2(0, \infty)$. Further, formulas (19), (23), (24), and (25) hold also for real values of the parameter λ greater than $-q_0$. In particular, $m\left(\mu, \frac{3}{4}\pi\right)$ is a strictly decreasing function of μ.

Using formula (17) we can now extend these results to arbitrary values of α, for we have

$$m\left(\lambda, \frac{1}{4}\pi\right) = -\frac{1}{m\left(\lambda, \frac{3}{4}\pi\right)}, \tag{28}$$

and if $\alpha \not\equiv \frac{1}{4}\pi \pmod{\pi}$, then

$$m(\lambda, \alpha) = \frac{m\left(\lambda, \frac{3}{4}\pi\right) + \tan\left(\alpha - \frac{3}{4}\pi\right)}{1 - m\left(\lambda, \frac{3}{4}\pi\right)\tan\left(\alpha - \frac{3}{4}\pi\right)}. \tag{29}$$

These relations show that $m(\lambda, \alpha)$ can be extended to the real axis for $\mu > -q_0$ and is holomorphic there save for poles. Now $m(\lambda, \alpha)$ has a pole at $\lambda = \lambda_0$ provided $m\left(\lambda, \frac{3}{4}\pi\right) = \operatorname{cotan}\left(\alpha - \frac{3}{4}\pi\right)$. Since $m\left(\mu, \frac{3}{4}\pi\right)$ is strictly decreasing in $(-q_0, \infty)$, there can be at most one such pole and, since $\left|m\left(\lambda, \frac{3}{4}\pi\right)\right| \leq 1$, the function $m(\lambda, \alpha)$ has no pole in $(-q_0, \infty$ as) long as $\frac{1}{2}\pi < \alpha < \pi$ or $-\frac{1}{2}\pi < \alpha < 0$. This completes the proof.

4. Asymptotic properties of $m(\lambda, \alpha)$. — It is possible to obtain fairly accurate information concerning the asymptotic properties of $m(\lambda, \alpha)$ from formula (18). We restrict ourselves to the imaginary axis, but similar results hold in any sector

$$\left|\arg\lambda \pm \frac{\pi}{2}\right| \leq \frac{\pi}{2} - \varepsilon.$$

The following theorem will be proved.

THEOREM 3. — *If* $q(x) \in C[0, \infty)$ *and if* $\alpha \not\equiv 0 \pmod{\pi}$, *then*

$$\text{(30)} \qquad R(\nu) \equiv \nu^{\frac{1}{2}} | m(i\nu, \alpha) + \operatorname{cotan} \alpha |$$

is bounded away from 0 *and* ∞ *as* $\nu \to +\infty$, *while for* $\alpha = 0$,

$$\text{(31)} \qquad S(\nu) \equiv \nu^{-\frac{1}{2}} | m(i\nu, 0) |$$

is bounded away from 0 *and* ∞.

The point of departure for the proof is the relation

$$\text{(32)} \qquad \mathfrak{J}[m(i\nu, \alpha)] = \nu \int_0^\infty | U_\alpha(s, i\nu) |^2 \, ds,$$

which is obtained from (18), and the following two lemmas :

LEMMA 1. — *If* $|q(x)| \leq q$ *for* $0 \leq x \leq 1$, *and if* $u(x, i\nu)$ *is a solution of* (1) *for* $\lambda = i\nu$ *such that*

$$\text{(33)} \qquad u(0, i\nu) = a, \qquad u'(0, i\nu) = b,$$

then we have

$$\text{(34)} \qquad u(x, i\nu) = a(1 + \theta_1) + bx(1 + \theta_2),$$

for $0 \leq x \leq \delta \equiv [2(\nu + q)]^{-\frac{1}{2}}$. *Here* $|\theta_k| \leq \frac{1}{3}$, $k = 1, 2$, *and* $\nu > 1$.

LEMMA 2. — *Under the same assumptions*

$$\text{(35)} \qquad \int_0^\infty | u(s, i\nu) |^2 \, ds > \frac{4}{9} \delta \left[| a |^2 - 4 | ab | \delta + \frac{1}{3} | b |^2 \delta^2 \right].$$

Proof of lemma 1. — From the obvious integral equation

$$\text{(36)} \qquad u(x, i\nu) = a + bx + \int_0^x (x - s) [i\nu + q(s)] u(s, i\nu) \, ds,$$

setting

$$M(x) = \max_{0 \leq s \leq x} | u(s, i\nu) |, \qquad 0 \leq x \leq 1,$$

we obtain

$$M(x) \leq | a | + | b | x + (\nu + q) \frac{1}{2} M(x) x^2.$$

If x is restricted to the interval $[0, \delta]$, δ defined above, we get

$$M(x) \leq \frac{4}{3} (a | + | b | x).$$

Using (36) once more, we find that

$$|u(x, i\nu) - a - bx| \leqq \frac{4}{3}(\nu + q) \int_0^x (x - s)(|a| + |b|s)\, ds$$
$$= \frac{4}{3}(\nu + q)\left[\frac{1}{2}|a|x^2 + \frac{1}{3}|b|x^3\right]$$
$$< \frac{1}{3}(|a| + |b|x)$$

which is the desired result.

Proof of lemma 2. — From (34) we get for $0 \leqq x \leqq \delta$, after some simplification,

$$|u(x, i\nu)|^2 > \frac{4}{9}|a|^2 - \frac{32}{9}|ab|x + \frac{4}{9}|b|^2 x^2$$

and the left side of (35) exceeds the result of integrating the last member from o to δ.

Proof of theorem 3. — We now combine formulas (32) and (35). We take $u(x, i\nu) = U_\alpha(x, i\nu)$ so that

$$a = \cos\alpha + m(i\nu, \alpha)\sin\alpha, \qquad b = \sin\alpha - m(i\nu, \alpha)\cos\alpha.$$

Hence

$$(37) \quad |m(i\nu, \alpha)| \geqq |\mathfrak{J}[m(i\nu, \alpha)]|$$
$$> \frac{4}{9}\delta\nu\Big\{|\cos\alpha + m(i\nu, \alpha)\sin\alpha|^2$$
$$- 4\delta|\cos\alpha + m(i\nu, \alpha)\sin\alpha|.|\sin\alpha - m(i\nu, \alpha)\cos\alpha|$$
$$+ \frac{1}{3}\delta^2|\sin\alpha - m(i\nu, \alpha)\cos\alpha|^2\Big\}.$$

To simplify we set

$$|m(i\nu, \alpha)| = Q(\nu), \qquad |\cos\alpha| = \gamma, \qquad |\sin\alpha| = \sigma.$$

and obtain the inequality

$$(38) \quad Q(\nu) > \frac{4}{9}\delta\nu\Big\{[\gamma - \sigma Q(\nu)]^2$$
$$- 4\delta[\gamma + \sigma Q(\nu)][\sigma + \gamma Q(\nu)] + \frac{1}{3}\delta^2[\sigma - \gamma Q(\nu)]^2\Big\}.$$

This shows that for $\nu > 1$ the function $Q(\nu)$ lies between the roots of a certain quadratic equation. If we let $\nu \to \infty$, then $\delta \to 0$, $\delta\nu \to \infty$, and the quadratic equation becomes in the limit

$$(\gamma - \sigma Q)^2 = 0.$$

Here we assume $\alpha \not\equiv 0 \pmod{\pi}$ so that $\sigma = |\sin\alpha| \neq 0$. The root of the quadratic is then $|\operatorname{cotan}\alpha|$. It follows that

$$\lim |m(i\nu, \alpha)| = |\operatorname{cotan}\alpha|, \qquad \alpha \not\equiv 0 \qquad (\operatorname{mod} \pi).$$

Thus $|m(i\nu, \alpha)|$ for fixed α is a bounded function of ν for $\nu > 1$.

We now go back to formula (37) and conclude that

$$\delta\nu\, |\cos\alpha + m(i\nu, \alpha)\sin\alpha|^2$$

is bounded. Here $\delta\nu$ becomes infinite with ν as $\nu^{\frac{1}{2}}$ so that

$$\lim m(i\nu, \alpha) = -\operatorname{cotan}\alpha.$$

To prove the stronger result (30), we observe that

$$|m(i\nu, \alpha) + \operatorname{cotan}\alpha| \geqq |\mathfrak{I}[m(i\nu, \alpha)]|.$$

We can then replace $|m(i\nu, \alpha)|$ in the first member of (37) by $|m(i\nu, \alpha) + \operatorname{cotan}\alpha|$, introduce the function $R(\nu)$ of (30), and obtain the inequality

$$(39) \qquad R(\nu) > \frac{4}{9} d(\nu) \sin^2\alpha [R(\nu)]^2 - \frac{16}{9} [d(\nu)]^2 |\sin\alpha| M(\nu) R(\nu) + \frac{4}{27} [d(\nu)]^3 [M(\nu)]^2,$$

where

$$d(\nu) = \delta\sqrt{\nu} \to \frac{1}{2}\sqrt{2},$$

$$M(\nu) = |\sin\alpha - m(i\nu, \alpha)\cos\alpha| \to |\operatorname{cosec}\alpha|$$

as $\nu \to \infty$. Again $R(\nu)$ lies between the roots of a quadratic equation with real positive roots. In the limit this equation becomes

$$(40) \qquad R^2 - \frac{17}{4}\sqrt{2}\operatorname{cosec}^2\alpha R + \frac{1}{6}\operatorname{cosec}^4\alpha = 0.$$

The roots of this equation are also real positive. This proves (30).

If $\alpha = 0$, the inequality (37) reduces to

$$|m(i\nu, 0)| > \frac{4}{9}\delta\nu\left\{1 - 4\delta|m(i\nu, 0)| + \frac{1}{3}\delta^2|m(i\nu, 0)|^2\right\}$$

and if we set

$$T(\nu) = \delta|m(i\nu, 0)||$$

and assume $\nu > 1$, then T (ν) always lies between the roots of the quadratic equation

$$2T^2 - (33 + 9q)T + 6 = 0. \tag{41}$$

This implies (31) and completes the proof of the theorem.

Corollary. — *$m(\lambda, \alpha)$ is not rational function of λ.*

For the asymptotic behavior of $m(\lambda, \alpha)$ on the imaginary axis is not consistent with such an assumption.

5. $q(x)$ becomes infinite with x. — We shall prove the following well known proposition.

Theorem 4. — *If $q(x) \to +\infty$ with x, then $m(\lambda, \alpha)$ is a meromorphic function of λ having infinitely many poles on the real axis of which at most one lies to the right of $-\inf q(x)$.*

Proof. — It is enough to give the argument for $\alpha = \frac{3}{4}\pi$. Suppose that we want to show that this function is meromorphic in the half-plane $\mathcal{R}(\lambda) > \sigma$ where $\sigma < -q_0$, $q_0 = \inf q(x)$. We choose an a so large that

$$\sigma + q(x) \geq 0 \quad \text{for} \quad x \geq a > 0.$$

Define two solutions of (1), $\varphi_1(x, \lambda)$ and $\varphi_2(x, \lambda)$, by the initial conditions

$$\begin{cases} \varphi_1(a, \lambda) = -\frac{1}{2}\sqrt{2}, \; \varphi_2(a, \lambda) = \frac{1}{2}\sqrt{2}, \\ \varphi'_1(a, \lambda) = \frac{1}{2}\sqrt{2}, \; \varphi'_2(a, \lambda) = \frac{1}{2}\sqrt{2}. \end{cases} \tag{42}$$

The conditions are chosen so that $\varphi_1(x, \lambda)$ and $\varphi_2(x, \lambda)$ satisfy the same conditions at $x = a$ as the solutions $u_1\left(x; \frac{3}{4}\pi, \lambda\right)$ and

$u_2\left(x; \frac{3}{4}\pi, \lambda\right)$ satisfy at $x = 0$. There is a uniquely defined solution of (1) of the form

$$U(x, \lambda) = v_1(x, \lambda) + M(\lambda)\, v_2(x, \lambda) \tag{43}$$

which is in $L_2(a, \infty)$ and hence also in $L_2(0, \infty)$. The proof given by Weyl applies to this solution and shows that $M(\lambda)$ is holomorphic for $\mathcal{I}(\lambda) \neq 0$ and satisfies (15). The argument presented in Section **2** above shows that $M(\lambda)$ is holomorphic in the half-plane $\mathcal{R}(\lambda) > \sigma$. But $U_{\frac{3}{4}\pi}(x, \lambda)$ and $U(x, \lambda)$ are linearly dependent so that their Wronskian is zero. For $x = a$ this gives

$$m\left(\lambda, \frac{3}{4}\pi\right) = \frac{A(\lambda)\, M(\lambda) + B(\lambda)}{C(\lambda)\, M(\lambda) + D(\lambda)}, \tag{44}$$

where

$$A = u_1 - u'_1, \qquad B = u_1 + u'_1,$$
$$C = -u_2 + u'_2, \qquad D = -u_2 - u'_2,$$

and

$$u_k = u_k\left(a; \frac{3}{4}\pi, \lambda\right).$$

It follows that the coefficients are entire functions of λ so that $m\left(\lambda, \frac{3}{4}\pi\right)$ is meromorphic in the half-plane $\mathcal{R}(\lambda) > \sigma$. Since σ is arbitrary, this implies that $m\left(\lambda, \frac{3}{4}\pi\right)$ is a meromorphic function of λ whose poles, if any, are located on the real axis to the left of $-q_0$. The same is true for $m(\lambda, \alpha)$ for arbitrary values of α by (29) except that here one pole may exceed $-q_0$ if α lies outside of the intervals $\left(-\frac{1}{2}\pi, 0\right)$ and $\left(\frac{1}{2}\pi, \pi\right)$.

Suppose that $\lambda = \mu_0$ is a pole of $m(\lambda, \alpha)$ of order p. Then we consider the following analogue of (32):

$$\mathcal{I}[m(\mu_0 + i\nu, \alpha)] = \nu \int_0^\infty |U_\alpha(s, \mu_0 + i\nu)|^2\, ds \tag{45}$$

and let $\nu \to 0$. The left member then becomes infinite at most as ν^{-p} while the right member grows at least as fast as ν^{1-2p}. This requires that $p = 1$ and, hence, that all poles are simple.

Actually the preceding argument does not show the existence of any poles at all. In the present case H. Weyl ([12], p. 252-256). proved the existence of infinitely many real poles by a discussion of the zeros of a solution $u(x, \lambda)$ determined by a fixed boundary condition at $x = 0$. Here we propose to use formula (19) for this purpose together with some simple function theoretical considerations. The argument goes as follows. It was shown, formula (20), that

$$\left| m\left(\mu + i\nu, \frac{3}{4}\pi\right) \right| < 1, \qquad -q_0 \leqq \mu, \qquad \nu \neq 0.$$

Without restricting the generality we may assume that $q_0 \geqq 0$. Going back to formula (19) and taking $\mu < -q_0$, we see that

$$1 - \left| m\left(\lambda, \frac{3}{4}\pi\right) \right|^2 > 2\mu \int_0^\infty \left| U_{\frac{3}{4}\pi}(s, \lambda) \right|^2 ds$$
$$= -2\left|\frac{\mu}{\nu}\right| . \left| \mathfrak{I}\left[m\left(\lambda, \frac{3}{4}\pi\right)\right]\right| > -2\left|\frac{\mu}{\nu}\right| . \left| m\left(\lambda, \frac{3}{4}\pi\right)\right|.$$

Thus $\left| m\left(\lambda, \frac{3}{4}\pi\right) \right|$ does not exceed the positive root of the quadratic equation

$$M^2 - 2\left|\frac{\mu}{\nu}\right| M - 1 = 0.$$

A fortiori

$$\left| m\left(\mu + i\nu, \frac{3}{4}\pi\right) \right| \leqq 2\left|\frac{\mu}{\nu}\right| + 1 \tag{46}$$

and this is clearly valid for all values of μ.

Suppose now that $m\left(\lambda, \frac{3}{4}\pi\right)$ has only a finite number of poles. We can then find a rational function $F(\lambda)$ with the same poles and principal parts, say

$$F(\lambda) = \sum_{k=1}^{n} \frac{a_k}{\lambda - \lambda_k},$$

such that

$$E(\lambda) \equiv m\left(\lambda, \frac{3}{4}\pi\right) - F(\lambda)$$

is an entire function. Since $F(\lambda) \to 0$ as $\lambda \to \infty$, the estimate (46) shows the existence of a constant C such that

$$|E(re^{i\theta})| < 2|\cot\theta| + C$$

for all θ and all large values of r. This gives

$$\sin^2\theta\,|E(re^{i\theta})| < 2 + C$$

for all θ and all large values of r. By the extension of Liouville's theorem due to M. H. Stone [9], it follows that $E(\lambda)$ is a constant. But this makes $m\left(\lambda, \frac{3}{4}\pi\right)$ into a rational function and this contradicts the corollary of theorem 3. It follows that $m\left(\lambda, \frac{3}{4}\pi\right)$ has infinitely many poles.

Since formula (24) is valid whenever the integral exists, we conclude that $m\left(\mu, \frac{3}{4}\pi\right)$ is strictly decreasing from $+\infty$ to $-\infty$ as μ increases from one pole to the next. For any given value of α the function $m\left(\mu, \frac{3}{4}\pi\right)$ will assume the value $\cot\left(\alpha - \frac{3}{4}\pi\right)$ once and only once between consecutive poles. At such a point $m(\lambda, \alpha)$ will have a pole. Thus, for any fixed value of α, the function $m(\lambda, \alpha)$ has infinitely many poles and these poles are separated by those of $m\left(\lambda, \frac{3}{4}\pi\right)$. The poles of $m(\lambda, \alpha)$ are the characteristic values of the corresponding boundary value problem (2). Thus the latter has an infinite discrete point spectrum. This completes the proof of theorem 4.

6. The resolvent. — If $q(x)$ is bounded below, an analogue of (3) can render good service in the discussion of the resolvent of the differential operator. The Green's function of the singular boundary values problem (1) + (2) is defined by

$$G_\alpha(x, s; \lambda) = \begin{cases} u_2(x; \alpha, \lambda)\, U_\alpha(s, \lambda), & 0 \leqq x < s < \infty, \\ u_2(s; \alpha, \lambda)\, U_\alpha(x, \lambda), & 0 \leqq s < x < \infty. \end{cases} \tag{47}$$

Here the factor $u_2(x, \alpha, \lambda)$ satisfies the boundary condition at $x = 0$ and $U_\alpha(x, \lambda)$ is in $L_2(0, \infty)$. Form

$$R_\alpha(\lambda; L)[f] \equiv Y_\alpha(x, \lambda; f) = \int_0^\infty G_\alpha(x, s; \lambda) f(s)\, ds, \tag{48}$$

where $f(s) \in L_2(0, \infty)$ and $\mathfrak{J}(\lambda) \neq 0$. This is the resolvent of the operator L which is the closure in $L_2(0, \infty)$ of the operator L^0 defined by

$$L^0[F] = F''(x) - q(x)F(x). \tag{49}$$

The domain of L^0 is the set of functions in $L_2(0, \infty)$, which are absolutely continuous together with the first derivative, such that $L^0[F] \in L_2(0, \infty)$ and

$$\cos\alpha F(0) + \sin\alpha F'(0) = 0. \tag{50}$$

The function $Y_\alpha(x, \lambda; f)$ has the following properties:

$$Y_\alpha(., \lambda; f) \in L_2(0, \infty) \quad \text{when} \quad f(.) \in L_2(0, \infty), \tag{51}$$

$$Y''_\alpha(x, \lambda; f) - [\lambda + q(x)]Y_\alpha(x, \lambda; f] = -f(x) \qquad \text{(almost all } x) \tag{52}$$

$$\cos\alpha Y_\alpha(0, \lambda; f) + \sin\alpha Y'_\alpha(0, \lambda; f) = 0. \tag{53}$$

If the limit-point case holds at infinity, in particular, if $q(x)$ is bounded below, then $Y_\alpha(x, \lambda; f)$ is the only solution of

$$Y'' - [\lambda + q(x)]Y = -f(x) \tag{54}$$

which satisfies (51) and (53) for $\mathfrak{J}(\lambda) \neq 0$.

The last equation gives the identity

$$\left[\overline{Y(x)}\, Y'(x)\right]_a^b - \int_a^b |Y'(s)|^2 ds - \int_a^b [\lambda + q(s)]\,|Y(s)|^2 ds = -\int_a^b f(s)\overline{Y(s)}\, ds. \tag{55}$$

We now take for $f(x)$ a continuous function of compact support, choose $Y(x) = Y_\alpha(x, \lambda; f)$, and note that

$$\overline{Y_\alpha(0, \lambda; f)}\, Y'_\alpha(0, \lambda; f) = -\frac{1}{2}\sin 2\alpha \left|\int_0^\infty f(s) U_\alpha(s, \lambda)\, ds\right|^2 \tag{56}$$

and that

$$Y_\alpha(x, \lambda; f) \to 0, \qquad Y'_\alpha(x, \lambda; f) \to 0 \qquad \text{as} \quad x \to +\infty. \tag{57}$$

Taking the imaginary part of (55), setting $a = 0$ and letting $b \to \infty$, we get

$$\nu \int_0^\infty |Y_\alpha(s, \lambda; f)|^2 ds = \int_0^\infty f(s)\overline{Y_\alpha(s, \lambda; f)}\, ds. \tag{58}$$

This implies that the L_2-norms satisfy the inequality

$$\| Y_\alpha(.,\lambda;f) \| \leq \frac{1}{|\nu|} \|f\|. \tag{59}$$

Since continuous functions of compact support are dense in $L_2(0, \infty)$, it follows that the resolvent of the linear transformation L is linear and bounded with

$$\| R_\alpha(\mu + i\nu; L) \| \leq \frac{1}{|\nu|}. \tag{60}$$

Incidentally, this shows that iL satisfies the conditions of the Hille-Yosida theorem ([5], p. 363-364). Hence iL is the infinitesimal generator of a (semi) group of contraction operators $\{ G_\alpha(t; iL) \mid -\infty < t < \infty \}$ say, and $G_\alpha(t, iL)[f]$ is the solution of the Cauchy problem for the partial differential equation

$$H_{xx}(x,t) - q(x) H(x,t) = -i H_t(x,t) \tag{61}$$

with the side conditions

$$\cos\alpha\, H(0,t) + \sin\alpha\, H_x(0,t) = 0, \qquad H(x,0) = f(x) \tag{62}$$

and $H(.,t) \in L_2(0, \infty)$ for all t.

We now return to the case in which $q(x)$ is bounded below. The following proposition is analogous to theorem 1.

THEOREM 5. — *If* $q(x) \geq q_0 > -\infty$, *then*

$$Y'_\alpha(x,\lambda;f) \in L_2(0,\infty), \qquad |q(x)|^{\frac{1}{2}} Y_\alpha(x,\lambda;f) \in L_2(0,\infty), \tag{63}$$

$$\lim_{x\to\infty} \overline{Y_\alpha(x,\lambda;f)}\, Y'_\alpha(x,\lambda;f) = 0. \tag{64}$$

If $\frac{1}{2}\pi < \alpha < \pi$ *or* $-\frac{1}{2}\pi < \alpha < 0$, *then* $R_\alpha(\lambda; L)$ *is holomorphic in the half-plane* $\mathcal{R}(\lambda) > -q_0$ *as well as in* $\mathcal{I}(\lambda) \neq 0$ *and*

$$\| R_\alpha(\lambda; L \| \leq \min\left\{ \frac{1}{\mu + q_0}, \frac{1}{|\nu|} \right\}. \tag{65}$$

Proof. — We return to (55). Taking $a = 0$, $b = x$, and the real part of the identity, we obtain for $Y(x) = Y_\alpha(x, \lambda; f)$

$$(66)\quad \frac{1}{2}\frac{d}{dx}|Y_\alpha(x, \lambda; f)|^2$$
$$= -\frac{1}{2}\sin 2\alpha \left| \int_0^\infty f(s)\, U_\alpha(s, \lambda)\, ds \right|^2 + \int_0^x |Y'_\alpha(s, \lambda; f)|^2\, ds$$
$$+ \int_0^x [\mu + q(s)]\, |Y_\alpha(s, \lambda; f)|^2\, ds - \mathcal{R}\left\{ \int_0^x f(s)\, \overline{Y_\alpha(s, \lambda; f)}\, ds \right\}$$

Here the first term on the right is fixed and finite, the last term tends to a finite limit when $x \to \infty$ since $f \in L_2(0, \infty)$, while the second and third terms are strictly increasing for $\mu > -q_0$. Thus the right member either tends to a finite limit or to $+\infty$. Now $Y_\alpha(x, \lambda; f) \in L_2(0, \infty)$ if $\mathcal{J}(\lambda) \neq 0$. It follows that the limit must be finite and can have no other value than 0. This proves (63) for $\mathcal{J}(\lambda) \neq 0$. Further we know that the limit in (64) exists for we can now use the full identity (55) with $a = 0$, $b = x \to \infty$, $Y(x) = Y_\alpha(x, \lambda; f)$, and $\mathcal{J}(\lambda) \neq 0$. We know moreover that the limit is purely imaginary or zero and only the second alternative is compatible with the integrability conditions.

If α lies in the indicated intervals, then theorem 2 asserts that $m(\lambda, \alpha)$ is holomorphic on $(-q_0, \infty)$ and so is $U_\alpha(x, \lambda)$ so that $G_\alpha(x, s; \lambda)$ has the same property for fixed x and s. From this fact together with the estimate given below it follows that $R_\alpha(\lambda; L)$ is a holomorphic operator in $L_2(0, \infty)$ in the half-plane $\mathcal{R}(\lambda) > -q_0$. We restrict ourselves here to giving the estimate.

From (66) we get for $\nu \neq 0$,

$$-\frac{1}{2}\sin 2\alpha \left| \int_0^\infty f(s)\, U_\alpha(s, \lambda)\, ds \right|^2 + \int_0^\infty |Y'_\alpha(s, \lambda; f)|^2\, ds$$
$$+ \int_0^\infty [\mu + q(s)]\, |Y_\alpha(s, \lambda; f)|^2\, ds = \mathcal{R}\left\{ \int_0^\infty f(s)\, \overline{Y_\alpha(s, \lambda; f)}\, ds \right\}.$$

The first term on the left is positive if $\frac{1}{2}\pi < \alpha < \pi$ or $-\frac{1}{2}\pi < \alpha < 0$ while the third term is positive and exceeds

$$(\mu + q_0)\, \| Y_\alpha(\cdot, \lambda; f) \|^2$$

if $-q_0 < \mu$. The absolute value of the right member is dominated by

$$\| Y_\alpha(.,\lambda; f)\| . \|f\|.$$

This gives the inequality

$$(67) \qquad \| Y_\alpha(., \mu + i\nu; f)\| \leq \frac{1}{\mu + q_0} \|f\|.$$

Since this holds independently of ν, it will also hold for any limit obtained by letting $\nu \to 0$. Thus we have

$$(68) \qquad \| R_\alpha(\mu + i\nu; L)\| \leq \frac{1}{\mu + q_0}, \qquad -q_0 < \mu.$$

This implies (65).

The estimate also implies that the operator $G_\alpha(t; iL)$, defined above for real values of t, can be extended to the upper half of the t-plane as an analytic semi-group operator.

7. Additional remarks. — The singular boundary value problem (1) + (2) determines uniquely a spectral function $\rho_\alpha(\xi)$. Since

$$(69) \qquad \mathcal{I}[m(i\nu, \alpha)] = -\nu \int_{-\infty}^{\infty} \frac{d\rho_\alpha(\xi)}{\xi^2 + \nu^2},$$

theorem 3 can be used to estimate $\rho_\alpha(\xi)$, but the results obtained in this fashion cannot compete with the estimates given by B. M. Levitan [7] and the asymptotic formulas of V. A. Marčenko[8] both ultimately based on a general method due to M. G. Krein [6]. I am indebted to M. G. Krein for these references.

On the other hand, theorem 3 itself does not seem to have been noticed although Göran Borg [1] has $m(\lambda, \alpha) = \mathcal{O}(\lambda^{\frac{1}{2}})$ which for $\alpha = 0$ is a best possible result. It is amusing to note that equation (17) which is satisfied by $m(\lambda, \alpha)$ as a function of the parameter α, is the functional equation of $\tan \alpha$. It is also satisfied by

$$-\operatorname{cotan} \alpha = \lim_{|\nu| \to \infty} m(\mu + i\nu, \alpha),$$

as we have seen.

A brief and elegant proof of theorem 4 was given by E. C. Titchmarsh [11].

BIBLIOGRAPHIE.

[1] G. BORG, *Uniqueness Theorems in the Spectral Theory of* $y'' + (\lambda - q(x))\, y = 0$ (*Comptes rendus du XIe Congrès de Mathématiciens scandinaves*, Trondheim, 1949, p. 276-287).

[2] E. A. CODDINGTON and N. LEVINSON, *Theory of Ordinary Differential Equations*, Mc Graw-Hill Book Company, New York, 1955.

[3] E. HILLE, *An Integral Equality and its Applications* (*Proc. Nat. Acad. Sc.*, vol. 7, 1921, p. 303-305).

[4] E. HILLE, *Behavior of Solutions of Linear Second Order Differential Equations* (*Ark. f. Mat.*, vol, 2, 1951, p. 25-41).

[5] E. HILLE and R. S. PHILLIPS, *Functional Analysis and Semi-Groups* (*Amer. Math. Soc. Coll. Publ.*, vol. 31, Rev. Ed. Providence, 1957).

[6] M. G. KREIN, *On a general Method of decomposing positive Definite Kernels into Elementary Products* (Russian). (*D.A.N., S.S.S.R.*, vol. 53, 1946, p. 3-6).

[7] B. M. LEVITAN, *On a Decomposition Theorem for Spectral Functions of Differential Equations of the Second Order* (Russian). *D. A. N., S.S.S.R.*, vol. 71, 1950, p. 605-608).

[8] V. A. MARČENKO, *Some Questions of the Theory of One-dimensional Linear Differential Equations of the Second Order* (Russian). I, II. (*Trudy Moskov. Mat. Obsč.*, vol. 1, 1952, p. 327-420; vol. 2, 1953, p. 3-83).

[9] M. H. STONE, *On a theorem of Pólya.* (*J. Indian Math. Soc.*, vol. 12, 1948, p. 1-7).

[10] E. C. TITCHMARSH, *Eigenfunction Expansions associated with Second-Order Differential Equations*, Oxford, 1946.

[11] E. C. TITCHMARSH, *On the Discreteness of the Spectrum associated with certain Differential Equations* [*Annali di Mat.*, (4), vol. 28, 1949, p. 141-147.

[12] H. WEYL, *Uber gewöhnliche lineare Differentialgleichungen mit Singularitäten und die zugehörigen Entwicklungen willkürlicher Funktionen* (*Math. Ann.*, vol. 68, 1910, p. 220-269).

(Manuscrit reçu le 25 novembre 1961.)

Reprinted from the PROCEEDINGS OF THE NATIONAL ACADEMY OF SCIENCES
Vol. 62, No. 1, pp. 7–10. January, 1969.

*ON THE THOMAS-FERMI EQUATION**

BY EINAR HILLE

UNIVERSITY OF NEW MEXICO, ALBUQUERQUE

Communicated October 28, 1968

Abstract and Summary.—A study has been made of some mathematical aspects of the Thomas-Fermi equation. This is a preliminary report on the results obtained, including (1) convergence of relevant series, (2) existence of unbounded solutions, (3) existence of solutions having an arbitrary branch point, (4) determination of a class of solutions bounded for large values of the variable, and (5) determination of a class of solutions unbounded for small values.

0. *Orientation.*—In 1927, L. H. Thomas[1] and E. Fermi[2] independently gave a method of studying the electron distribution in an atom, using the statistics for a degenerate gas. This led to a nonlinear second-order differential equation

$$x^{1/2}\cdot y''(x) = [y(x)]^{3/2}. \tag{0.1}$$

The physicists were interested in three boundary value problems for this equation,

$$y(0) = 1, \qquad r\, y'(r) = y(r), \tag{0.2}$$

$$y(0) = 1, \qquad \lim_{x\to\infty} y(x) = 0, \tag{0.3}$$

$$y(0) = 1, \qquad y(a) = 0. \tag{0.4}$$

In the first problem, r is the Bohr atom radius. The second problem corresponds to the neutral atom, whereas (0.4) is the ion case.

The first boundary condition leads to formal series of the type

$$y(x) = 1 + b_2 x + b_3\, x^{3/2} + \ldots + b_k x^{1/2k} + \ldots. \tag{0.5}$$

There is a critical value $b_2 = \omega = -1.588\ldots$ that leads to $y_0(x)$, the solution of (0.3). Solutions of problem (0.2) correspond to $b_2 > \omega$, those of problem (0.4) to $b_2 < \omega$. The first 11 coefficients of the series (0.5) have been computed. They are polynomials in b_2. The series appears to be regarded as semiconvergent.[3]

The asymptotic behavior of $y_0(x)$ for large values of x was determined by A. Sommerfeld.[4] We shall write his representation in the form

$$y_0(x) \sim 144\, x^{-3}\{1 + (12^{2/3}x^{-1})^{\sigma}\}^{-1/2\tau} \tag{0.6}$$

with

$$\sigma = {}^1/_2[\sqrt{73} - 7], \qquad \tau = {}^1/_2[\sqrt{73} + 7]. \tag{0.7}$$

The factor

$$144\, x^{-3} \tag{0.8}$$

is a solution in its own right, the "singular" solution noted by Thomas. The solution $y_0(x)$ was determined by numerical integration by Fermi and Thomas.

The existence of such a solution follows from a theorem by A. Mambriani,[5] published in 1929.

The physicists presumably have got out of the Thomas-Fermi theory all that is of interest to them. They have enriched the mathematical literature by the equation (0.1) above, and it seems to be time for mathematicians to react to the challenge by discovering the astounding properties of the solutions. The present note is a preliminary account of results obtained for which proofs will be published elsewhere.

We start by listing some questions suggested by the preceding summary of the classical results.

(1) Is the series (0.5) actually convergent?

(2) Does the solution of (0.2) stay bounded for $x > r$?

(3) What happens to the solution of (0.4) near $x = a$?

(4) What is the analytical nature of the solutions that are bounded for large values of x?

(5) What is the analytical nature of solutions that are unbounded for small values of x?

All these questions are posed directly by the physical problems. If they can be answered, a function theoretical investigation of the solutions in the complex plane is the next order of business. We shall give at least partial answers to questions (1)–(5).

1. *Convergence.*—The series (0.5) is actually convergent for small values of $|x|$. A somewhat tricky application of the Cauchy-Lindelöf majorant method[6] leads to absolute convergence of the series for

$$|x| \leq r = {}^2/_5 \, ({}^3/_5)^{3/2} \, (|b_2| + 2)^{-1}. \tag{1.1}$$

The constant is most unlikely to be the best possible.

The equation also has a solution of the form

$$x + {}^1/_6 \, x^3 + {}^1/_{80} \, x^5 + \ldots, \tag{1.2}$$

where the exponents appear to be positive odd integers. This series also has a positive radius of convergence.

2. *Unbounded Solutions.*—The solutions of problem (0.2) become unbounded as x grows beyond r. Moreover, this holds for any solution satisfying an initial condition of the form

$$y(x_0) = a, \qquad y'(x_0) = b, \qquad a \geq 0, \, b \geq 0, \, a + b > 0. \tag{2.1}$$

In each case there exists a finite $x_1 = x_1(x_0,a,b)$ such that $y(x)$ becomes infinite as x increases to x_1. The rate of growth is bounded above inasmuch as

$$y(x) < 400 \, x_1 \, (x - x_1)^{-4}. \tag{2.2}$$

In view of the continuous dependence of x_1 upon the initial conditions, an infinitude can be arranged at any preassigned positive number. Moreover, there exists a value r, possibly not unique, and a problem (0.2) for which the "explosion" occurs at the preassigned point.

The right member of (2.2) appears to be the correct dominant term for approach to x_1 from below. Attempts to prove that an infinitude at a point $x_1 \neq 0$ is a pole of order 4 have so far failed. It is possible that the singularity also involves a branch point.

3. *Movable Branch Points.*—The solution of problem (0.4), the ion case, has a branch point at $x = a$ with a local expansion

$$y(x) = b(a - x) + b_7(a - x)^{7/2} + \ldots, \tag{3.1}$$

where b and b_7 are positive numbers. Thus the solution necessarily becomes complex-valued if $x > a$. A branch point can be placed at any point $a > 0$. If the leading coefficient b is negative, the solution is real for $a < x$ and is ultimately infinite.

4. *Ultimately Bounded Solutions.*—The theorem of Mambriani shows that the boundary value problem

$$y(a) = b, \quad \lim_{x \to \infty} y(x) = 0, \qquad a \geq 0, b > 0, \tag{4.1}$$

has a unique solution $Y_0(x; a,b)$. In particular, if

$$C: \; a^3b = 144, \tag{4.2}$$

then $Y_0(x; a,b)$ coincides with the singular solution (0.8). The curve C separates the first quadrant into two regions, D_+ above C and D_- below. A solution $Y_0(x; a,b)$ stays in the region where (a,b) is located. Thus

$$Y_0(x; a,b) > 144\, x^{-3},\ x > a, \quad \text{if } (a,b) \in D_+, \tag{4.3}$$

$$Y_0(x; a,b) < 144\, x^{-3},\ x > 0, \quad \text{if } (a,b) \in D_-. \tag{4.4}$$

Moreover, any solution $Y_0(x; a,b)$ admits of a convergent expansion of the form

$$Y_0(x) = 144\, x^{-3}\{1 + \sum_{n=1}^{\infty} c_n x^{-n\sigma}\}, \quad \sigma = {}^1\!/_2(\sqrt{73} - 7). \tag{4.5}$$

Here c_1 is arbitrary and the higher coefficients are polynomials in c_1. The convergence of the series is proved by the Cauchy-Lindelöf majorant method. In particular, such an expansion is valid for $y_0(x)$. A plausible value of c_1 for this solution can be read from the Sommerfeld approximation (0.7).

5. *Unbounded Solutions at the Origin.*—The expansion (4.5) is obtained from a study of the differential equation

$$x^2v'' - 6xv' - 6v = v^{3/2}, \tag{5.1}$$

which is satisfied by x^3y. For large values of x we obtain (4.5), and for small values of x a similar expansion where $-\sigma$ is replaced by $\tau = {}^1\!/_2(\sqrt{73} + 7)$. This gives a class of solutions of (0.1)

$$y(x) = 144\, x^{-3}\{1 + \sum_{n=1}^{\infty} d_n x^{n\tau}\}. \tag{5.2}$$

Again the first coefficient is arbitrary and convergence can be proved by the Cauchy-Lindelöf method.[7]

* This research was supported by National Science Foundation grant GP 8856.

[1] Thomas, L. H., "The calculation of atomic fields," *Proc. Cambridge Phil. Soc.*, **23**, 542 (1927).

[2] Fermi, E., "Un metodo statistico per la determinazione di alcune priorietà dell'atome," *Rend. Accad. Naz. Lincei*, (6) **6**, 602 (1927).

[3] For such questions see, e.g., Feynman, R. P., N. Metropolis, and E. Teller, "Equations of state of elements based on the generalized Fermi-Thomas theory," *Phys. Rev.*, **75**, 1561 (1949).

[4] Sommerfeld, A., "Asymptotische Integration der Differentialgleichung des Thomas-Fermischen Atoms," *Z. Physik*, **78**, 283 (1932).

[5] Mambriani, A., "Su un teorema relativo alle equazioni differenziali ordinarie del 2º ordine," *Rend. Accad. Naz. Lincei*, (6) **9**, 620 (1929). See also *ibid.*, p. 142.

[6] See, e.g., Hille, E., *Lectures on Ordinary Differential Equations* (Reading, Mass.: Addison-Wesley, 1969), pp. 48–57.

[7] *Note added in proof:* The methods apply also to Emden's equation and give similar results.

Chapter 14
Partial Differential Equations

The papers included in this chapter are

[98] On the integration problem for Fokker-Planck's equation in the theory of stochastic processes

[100] "Explosive" solutions of Fokker-Planck's equation

[106] Une généralisation du problème de Cauchy

[115] Some aspects of Cauchy's problem

An introduction to Hille's work on the abstract Cauchy problem and semi-groups can be found starting on page 695.

On the Integration Problem for Fokker-Planck's Equation in the Theory of Stochastic Processes.

By

Einar Hille.

1. Introduction. The names of FOKKER and PLANCK are attached to a class of partial differential equations of the second order and parabolic type which were introduced in statistical dynamics by MAX PLANCK in 1917 (see [5] of the appended References where the earlier work of A. EINSTEIN and A. FOKKER [2] is quoted). In the case of a dynamical system with one degree of freedom, the equation reads

$$\frac{\partial^2}{\partial x^2}[b(x)\,U]-\frac{\partial}{\partial x}[a(x)\,U]=\frac{\partial U}{\partial t}, \tag{1.1}$$

where $b(x)$ is positive and, normally, large in comparison with $[a(x)]^2$. In 1931 A. KOLMOGOROFF [4] proved that the transition frequency of a stochastic process satisfies a pair of partial differential equations; in the case of a one parameter process which is temporally homogeneous, the second (adjoint) equation coincides with (1.1).

The existence and uniqueness problems for KOLMOGOROFF'S equations and various generalizations were solved by W. FELLER [1] in 1936 under the assumption that the solution converges pointwise to continuous initial values. In our case convergence will be in the mean of order one instead.

Let $L(-\infty, \infty)$ denote the usual Lebesgue space with the customary metric. By L_{02} we denote the subspace made up of twice differentiable functions which vanish outside of some finite interval. It is well known that L_{02} is dense in $L(-\infty, \infty)$. We can now formulate the integration problem as follows.

Problem A. *What conditions should $a(x)$ and $b(x)$ satisfy in order that for every given function $g(x)$ in L_{02} the equation (1.1) shall have a unique solution $U(x, t)$ for $t > 0$ with the following properties.*

(i) $\int_{-\infty}^{\infty} | U(x, t) - g(x) | \, dx \to 0 \text{ when } t \to 0;$

(ii) $\int_{-\infty}^{\infty} | U(x, t) | \, dx \leqq \int_{-\infty}^{\infty} | g(x) | \, dx;$

(iii) *if* $g(x) \geqq 0$ *for all* x, *then* $U(x, t)$ *has the same property;*

(iv) *if* $g(x) \geqq 0$ *for all* x, *then the sign of equality holds in* (ii).

Problem A with omission of condition (iv) is called Problem A_0 in the following. It is apparently of no interest in the theory of stochastic processes, but it would seem to be of some importance to the theory of partial differential equations of parabolic type. We shall prove below that Problem A_0 is always solvable if

$$b''(x) - a'(x) \leqq M, \quad -\infty < x < \infty. \tag{1.2}$$

This result could not be inferred from FELLER's general existence theorem which is based upon the construction of a fundamental solution under the basic assumption that

$$\int_{0}^{\infty} [b(t)]^{-\frac{1}{2}} dt = \infty, \quad \int_{-\infty}^{0} [b(t)]^{-\frac{1}{2}} dt = \infty. \tag{1.3}$$

In a recent note K. YOSIDA [7] has shown that Problem A is solvable whenever

$$| a(x) | + | b(x) | + | b'(x) | + | b''(x) - a'(x) | \leqq B \tag{1.4}$$

for all x. YOSIDA bases his discussion upon the theory of semigroups of linear operators developed by him [6] and the present writer [3]. It is the purpose of the present paper to call attention to the possibilities of this method, to replace YOSIDA's analysis by a simpler argument, and to extend his results. Assuming condition (1.2), we give various additional conditions which ensure the existence of solutions of Problem A. The simplest sufficient condition is

$$| b'(x) - a(x) | \leqq K(| x | + 1), \quad \int_{0}^{\infty} \frac{t\,dt}{b(t)} = \infty, \quad \int_{-\infty}^{0} \frac{t\,dt}{b(t)} = -\infty. \tag{1.5}$$

If the inequalities hold in (1.2) and (1.5), the divergency condition becomes necessary as well as sufficient for the solvability of Problem A.

2. *Background of the problem.* In order to appreciate the significance of the various conditions imposed on the solution in Problem A, one must know the origin of the problem, that is, the stochastic problem which leads to equations of type (1.1). A stochastic variable X which can take on all real values is subjected to a stochastic process which is homogeneous with

respect to the time. The probability that X takes on a value $\leqq x$ if its value t seconds earlier was ξ is supposed to be

$$P(t;\xi,x)=\int_{-\infty}^{x} f(t;\xi,s)\,ds,\qquad t>0, \tag{2.1}$$

so that

$$f(t;\xi,s)\geqq 0,\quad \int_{-\infty}^{\infty} f(t;\xi,s)\,ds=1. \tag{2.2}$$

Further it is supposed that

$$\lim_{t\to 0} P(t;\xi,x)=E(x-\xi),\quad E(s)=\begin{cases}1, & s>0,\\ 0, & s<0.\end{cases} \tag{2.3}$$

In addition the frequency function $f(t;\xi,x)$ satisfies the CHAPMAN-SMOLUCHOWSKI equation

$$f(t_1+t_2;\xi,x)=\int_{-\infty}^{\infty} f(t_1;\xi,s)\,f(t_2;s,x)\,ds \tag{2.4}$$

which is the origin of the semi-group properties used below.

If $g(x)$ is an arbitrary function in $L(-\infty,\infty)$ then

$$T(t)[g]=g(t;x)=\int_{-\infty}^{\infty} f(t;\xi,x)\,g(\xi)\,d\xi \tag{2.5}$$

defines a linear bounded transformation on $L(-\infty,\infty)$ to itself. The norm of this transformation is one because

$$\int_{-\infty}^{\infty} |g(t;x)|\,dx\leqq\int_{-\infty}^{\infty} |g(x)|\,dx \tag{2.6}$$

with equality if $g(x)$ is non-negative almost everywhere in which case $g(t;x)$ has same property. We say that $T(t)$ is a *transition operator*. If $g(x)$ is the frequency function of the variable X at the time $t=0$, then $g(t;x)$ is the frequency function at the time $t>0$. Formula (2.4) shows that

$$T(t_1+t_2)=T(t_1)\,T(t_2), \tag{2.7}$$

that is, the operators $T(t)$ form a *semi-group*.

KOLMOGOROFF showed in [4] that $f(t;\xi,x)$ as a function of t and x satisfies an equation of type (1.1) and it is clear that $g(t;x)$ satisfies the same equation. It follows from what was said above that $g(t;x)$ has properties (ii), (iii) and (iv) and with the aid of (2.3) one shows that (i) also holds.

From this brief survey we see that if equation (1.1) corresponds to a stochastic process of the type mentioned above, then Problem A must possess a solution. Moreover, this solution $U(x, t)$ is obtained from the initial value $g(x)$ by operating on $g(x)$ with a transition operator $T(t)$ at the time t and the operators $\{T(t)\}$ form a semi-group S. Such a transformation semi-group has an *infinitesimal generator* A which in the present case is (possibly an extension of) the differential operator

$$A[f] = \frac{d^2}{dx^2}[b(x)f(x)] - \frac{d}{dx}[a(x)f(x)]. \tag{2.8}$$

The domain of A contains L_{02}. For $g(x)$ in the domain of A we have

$$T(t)A[g] = AT(t)[g] = \frac{\partial}{\partial t}T(t)[g] \tag{2.9}$$

which expresses that $T(t)[g]$ satisfies (1.1).

The resolvent of A, that is,

$$[\lambda I - A]^{-1} = R(\lambda; A) \tag{2.10}$$

exists for $\Re(\lambda) > 0$ and is given by

$$R(\lambda; A)[g] = \int_0^\infty e^{-\lambda t}T(t)[g]\,dt. \tag{2.11}$$

This formula shows that $\lambda R(\lambda; A)$ is a transition operator in $L(-\infty, \infty)$ for positive values of λ. Conversely, the inversion formula

$$U(x, t) = T(t)[g] = \operatorname*{l.i.m.}_{n \to \infty} \left\{\frac{n}{t}R\left(\frac{n}{t}; A\right)\right\}^n [g] \tag{2.12}$$

shows that $T(t)$ is a transition operator if $\lambda R(\lambda; A)$ is one. Moreover, the two formulas show that if one of the operators *preserves positivity* so does the other and if one of them is a *contraction operator*, that is, does not increase the norm, then the other has the same property.

Conversely, it has been shown by the author ([3] p. 238) and, with a much simpler proof, by K. YOSIDA [6] that if a linear operator G has its domain dense in the space under consideration and if $\lambda R(\lambda; G)$ exists as a contraction operator for large positive values of λ, then G is the infinitesimal generator of a semi-group of linear contraction operators $T(t)$ and $T(t)$ is strongly continuous for $t \geq 0$. In particular, if $\lambda R(\lambda; G)$ preserves positivity or is a transition operator, then, as pointed by YOSIDA, $T(t)$ has the same properties. This situation was utilized by YOSIDA in his attack on Problem A mentioned above and it serves as the basis for our discussion of the same problem and of Problem A_0 below.

3. Reduction to ordinary differential equations. According to the preceding discussion the solvability of Problem A is equivalent to showing that the coefficients $a(x)$ and $b(x)$ can be so chosen that $\lambda R(\lambda; A)$ becomes a transition operator for large positive values of λ while for Problem A_0 we require merely that $\lambda R(\lambda; A)$ is a positive contraction operator. Translating these statements into classical terminology, we see that Problem A can be replaced by the following equivalent problem.

Problem B. *What conditions should $a(x)$ and $b(x)$ satisfy in order that for every given function $g(x)$ in $L(-\infty, \infty)$ the non homogeneous differential equation*

$$\frac{d^2}{dx^2}[b(x)\,Y] - \frac{d}{dx}[a(x)\,Y] - \lambda Y = -g(x) \tag{3.1}$$

shall have a unique solution $Y(x, \lambda)$ with the following properties.

(i) *There exists a fixed $\lambda_0 \geqq 0$ such that $Y(x, \lambda) \in L(-\infty, \infty)$ for $\lambda > \lambda_0$;*

(ii)
$$\lambda \int_{-\infty}^{\infty} |\,Y(x, \lambda)|\, dx \leqq \int_{-\infty}^{\infty} |g(x)\, dx;$$

(iii) *if $g(x) \geqq 0$ for almost all x, then $Y(x, \lambda)$ has the same property;*

(iv) *if $g(x) \geqq 0$ for almost all x, then the sign of equality holds in* (ii).

Problem B with condition (iv) omitted is called Problem B_0 in the following and the latter is equivalent with Problem A_0.

Theorem 1. *A necessary and sufficient condition in order that Problem B_0 be solvable is that the homogeneous differential equation*

$$\frac{d^2}{dx^2}[b(x)y] - \frac{d}{dx}[a(x)y] - \lambda y = 0 \tag{3.2}$$

have the following properties for $\lambda > \lambda_0$. (1) There is no solution $\neq 0$ in $L(-\infty, \infty)$. (2) There exists a fundamental system of solutions $y_+(x, \lambda)$ and $y_-(x, \lambda)$ which are positive for all x and have the positive Wronskian $y_+(x, \lambda)\,y'_-(x, \lambda) - y_-(x, \lambda)\,y'_+(x, \lambda)$. (3) $y_+(x, \lambda) \in L(0, \infty)$ and $y_-(x, \lambda) \in L(-\infty, 0)$. (4) The following limits exist

$$\begin{aligned} &\lim_{x \to \infty} \left\{ \frac{d}{dx}[b(x)\,y_+(x, \lambda)] - a(x)\,y_+(x, \lambda) \right\} \equiv \varphi_+(\lambda) \leqq 0, \\ &\lim_{x \to -\infty} \left\{ \frac{d}{dx}[b(x)\,y_-(x, \lambda)] - a(x)\,y_-(x, \lambda) \right\} \equiv \varphi_-(\lambda) \geqq 0. \end{aligned} \tag{3.3}$$

A necessary and sufficient condition that Problem B be solvable is that in addition $\varphi_+(\lambda) = \varphi_-(\lambda) = 0$.

Proof. Let us first show that the conditions are necessary. Condition (1) is clearly necessary for the uniqueness of the solution of (3.1) in $L(-\infty, \infty)$. According to a classical formula, the solutions of (3.1) are represented by

$$(3.4)\quad Y(x,\lambda)=y_1(x,\lambda)\int_{\alpha}^{x}\frac{y_2(s,\lambda)}{b(s)D(s,\lambda)}g(s)\,ds+y_2(x,\lambda)\int_{x}^{\beta}\frac{y_1(s,\lambda)}{b(s)D(s,\lambda)}g(s)\,ds,$$

where y_1 and y_2 form a fundamental system of solutions of (3.2) with Wronskian $D(x,\lambda)$ and $\alpha,\beta,\ \alpha<\beta$, are arbitrary real numbers. Suppose in particular that $g(x)=\chi(x;\alpha,\beta)$ is the characteristic function of the finite interval $[\alpha,\beta]$ and that $y_1(x,\lambda)$ and $y_2(x,\lambda)$ have been chosen in such a manner that formula (3.4) represents the corresponding unique solution of (3.1) in $L(-\infty,\infty)$. Fo $x<\alpha$ the solution is a constant multiple of $y_2(x,\lambda)$, for $x>\beta$ a constant multiple of $y_1(x,\lambda)$. It is consequently necessary that $y_2(x,\lambda)\,\varepsilon\,L(-\infty,0)$ and $y_1(x,\lambda)\,\varepsilon\,L(0,\infty)$, that is, (3.2) has a fundamental system $\{y_+(x,\lambda),y_-(x,\lambda)\}$ with $y_-(x,\lambda)\,\varepsilon\,L(-\infty,0)$ and $y_+(x,\lambda)\,\varepsilon\,L(0,\infty)$. Moreover, these solutions are unique save for constant multipliers and may be normalized by taking $y_-(0,\lambda)=y_+(0,\lambda)=1$. Condition (iii) of Problem B now shows that these solutions must be positive for all x and, in view of (3.4), that the corresponding Wronskian $D(x,\lambda)$ must be positive. Thus (2) is also necessary and we have merely to prove (3.3).

If $g(x)\,\varepsilon\,L(-\infty,\infty)$ and $Y(x,\lambda)$ is the corresponding solution of (3.1) in $L(-\infty,\infty)$, then

$$(3.5)\qquad \lambda\int_{-\infty}^{\infty}Y(x,\lambda)\,dx-\int_{-\infty}^{\infty}A[Y(x,\lambda)]\,dx=\int_{-\infty}^{\infty}g(x)\,dx.$$

Suppose now that condition (ii) of Problem B holds for non-negative functions $g(x)$ in which case $Y(x,\lambda)$ is also non-negative. This requires

$$(3.6)\qquad \int_{-\infty}^{\infty}A[Y(x,\lambda)]\,dx\leqq 0.$$

But

$$\int_{-\sigma}^{\omega}A[Y(x,\lambda)]\,dx=\left\{\frac{d}{dx}[b(x)\,Y(x,\lambda)]-a(x)\,Y(x,\lambda)\right\}_{-\sigma}^{\omega}$$

and this must tend to a finite limit when σ and ω tend to infinity independently of each other. Applying this to the case $g(x)=\chi(x;\alpha,\beta)$ we see that the limits $\varphi_+(\lambda)$ and $\varphi_-(\lambda)$ must exist and be finite. Moreover, if (3.6) is to hold for every choice of α and β we must have $\varphi_+(\lambda)\leqq 0$, $\varphi_-(\lambda)\geqq 0$. Thus (3) and (4) are necessary conditions for Problem B_0. If problem B is to be solvable, we must have the sign of equality in (3.6) for positive $g(x)$ and this requires that $\varphi_+(\lambda)=\varphi_-(\lambda)=0$. Thus the stated conditions are necessary.

Suppose conversely that the conditions of Theorem 1 are satisfied. Let $g_n(x) \varepsilon L(-\infty, \infty)$ be zero outside the interval $(-n, n)$. Then formula (3.4) with $y_1(x, \lambda) = y_+(x, \lambda)$, $y_2(x, \lambda) = y_-(x, \lambda)$, $\alpha = -n$, $\beta = n$ gives a solution of (3.1) having all the desired properties. Next suppose that $g(x) \varepsilon L(-\infty, \infty)$ and is never negative. Set $g_n(x) = g(x)\chi(x; -n, n)$ so that the sequence $\{g_n(x)\}$ is a Cauchy sequence converging to $g(x)$. Then the corresponding sequence $\{Y_n(x, \lambda)\}$ is monotone increasing and is made up of positive functions. Since

$$\lambda \int_{-\infty}^{\infty} |Y_m(x, \lambda) - Y_n(x, \lambda)| \, dx \leqq \int_{-\infty}^{\infty} |g_m(x) - g_n(x)| \, dx,$$

it is also a Cauchy sequence converging to a limit $Y(x, \lambda)$ in $L(-\infty, \infty)$ and the limit is clearly

$$Y(x, \lambda) = y_+(x, \lambda) \int_{-\infty}^{x} \frac{y_-(s, \lambda)}{b(s) D(s, \lambda)} g(s) \, ds + y_-(x, \lambda) \int_{x}^{\infty} \frac{y_+(s, \lambda)}{b(s) D(s, \lambda)} g(s) \, ds, \tag{3.7}$$

where the integrals exist for every x. $Y(x, \lambda)$ is positive and

$$\lambda \int_{-\infty}^{\infty} Y(x, \lambda) \, dx \leqq \int_{-\infty}^{\infty} g(x) \, dx$$

with equality if $\varphi_+(\lambda) = \varphi_-(\lambda) = 0$. That $Y(x, \lambda)$ is a solution of (3.1) may be verified by direct differentiation. The extension to arbitrary $g(x)$ in $L(-\infty, \infty)$ is immediate and the solution is always given by (3.7). This completes the proof.

Since $Y(x, \lambda) = R(\lambda; A)[g]$ we note that (3.7) gives an explicit representation of the resolvent of A.

4. *Discussion of the homogeneous equation.* We shall now investigate the possibilities of satisfying the conditions of Theorem 1. We take as our first assumption

I. *There exists a finite M such that $b''(x) - a'(x) \leqq M$ for all x.*

Theorem 2. *Assumption I implies that Problems A_0 and B_0 are solvable.*

Proof. Writing equation (2.3) in the form

$$b(x) y'' + [2b'(x) - a(x)] y' + [b''(x) - a'(x) - \lambda] y = 0, \tag{4.1}$$

we see that for $\lambda > M$ assumption I implies that no solution of (4.1) can have a positive maximum or a negative minimum. This observation is the basis of the following discussion. As a first consequence it implies condition (1) of Theorem I, that is, the non-existence of solutions in $L(-\infty, \infty)$. Secondly it implies that if $y(x) \not\equiv 0$, then the product $y(x) y'(x)$ can vanish at most once when $\lambda > M$. This observation can be used for a classification of the integral curves of (4.1) passing through $(0, 1)$. They fall into five classes: (1) curves intersecting the negative x-axis, (2) monotone increasing curves in the upper half-plane, (3) curves having a positive minimum, (4) monotone decreasing curves in the upper half-plane, and (5) curves intersecting the positive x-axis. Only curves of the second and the fourth classes are of interest for Problem B_0. Each class contains at least one member. We define $y_-(x, \lambda)$ as that member of the second class for which $\lim_{x \to -\infty} y(x)$ is a minimum and likewise $y_+(x, \lambda)$ as that member of the fourth class for which $\lim_{x \to \infty} y(x)$ is a minimum.

We shall first prove that these minima are zero. It is enough to consider $y_+(x, \lambda)$. Integrating (3.2) we obtain

$$[b'(x) - a(x)]y_+(x, \lambda) + b(x)y'_+(x, \lambda) - C_0 = \lambda \int_0^x y_+(t, \lambda)\,dt, \tag{4.2}$$

where $y'_+(x, \lambda) < 0$ and $\lambda > M$. On the other hand, assumption I gives

$$b'(x) - a(x) \leqq Mx + C, \qquad x \geqq 0. \tag{4.3}$$

Hence, dividing by x in (4.2) and letting $x \to \infty$, we obtain

$$M\,d(\lambda) \geqq \lambda\,d(\lambda), \quad d(\lambda) = \lim_{x \to \infty} y_+(x, \lambda).$$

Since $\lambda > M$, this implies that $d(\lambda) = 0$.

Next we prove that $y_+(x, \lambda) \in L(0, \infty)$. Since $y_+(x, \lambda)$ is positive and monotone decreasing, (4.2) plus (4.3) give

$$(Mx + C)y_+(x, \lambda) - C_0 > \lambda \int_0^x y_+(t, \lambda)\,dt > \lambda x y_+(x, \lambda)$$

or

$$[(M - \lambda)x + C]y_+(x, \lambda) - C_0 > 0$$

for all $x > 0$. Thus $xy_+(x, \lambda)$ must be bounded. This makes the left member of (4.2) bounded above whence it follows that $y_+(x, \lambda) \in L(0, \infty)$, the limit $\varphi_+(\lambda)$ exists and is finite. Since

$$\limsup_{x \to \infty} [b'(x) - a(x)]y_+(x, \lambda) \leqq 0, \quad b(x)y'_+(x, \lambda) < 0,$$

it follows that $\varphi_+(\lambda) \leqq 0$. The discussion of $y_-(x, \lambda)$ follows the same lines. Thus equation (3.2) satisfies the conditions of Theorem 1 for $\lambda > M$ so that Problem B_0 is solvable for $\lambda > M$ and hence also Problem A_0.

In general we cannot assert that Problem B is solvable under assumption I. For this to be the case, we must have

$$\begin{aligned} &\lim_{x \to \infty} [b'(x) - a(x)]y_+(x, \lambda) = 0, \quad \lim_{x \to \infty} b(x)y'_+(x, \lambda) = 0, \\ &\lim_{x \to -\infty} [b'(x) - a(x)]y_-(x, \lambda) = 0, \quad \lim_{x \to -\infty} b(x)y'_-(x, \lambda) = 0. \end{aligned} \tag{4.4}$$

In order to satisfy (4.4) we may impose the following simple assumption.

II. *There exists a finite positive K such that for all x*

$$|a(x)|+|b'(x)|\leqq K(|x|+1).$$

Theorem 3. *Problems B and A are solvable under assumptions* I *and* II.

Proof. Since $xy_+(x,\lambda)\to 0$ when $x\to\infty$, assumption II implies that the upper left relation of (4.4) holds and the same argument applies to the lower left one. Assumption II also implies that $b(x)=O(x^2)$ when $x\to\infty$. But $b(x)y'_+(x,\lambda)\to\varphi_+(\lambda)<0$, $y_+(x,\lambda)\,\varepsilon\,L(0,\infty)$ give $x/b(x)\,\varepsilon\,L(0,\infty)$, a contradiction. This implies the validity of the upper right relation and the lower right is handled in the same manner.

Assumption II is unnecessarily restrictive, but it has the advantage of being explicit and easy to verify. A less explicit but necessary and sufficient additional condition is obtained by the following considerations.

We say that *a differential operator $L[y]$ belongs to the first or to the second class in the interval (α,β) according as the differential equation $L[y]=0$ does or does not have a solution which is positive for $\alpha<x<\beta$ and belongs to $L(\alpha,\beta)$.* We set

$$\varphi[y]=\frac{d}{dx}[b(x)y(x)]-a(x)y(x), \tag{4.5}$$

$$L_+[y]=\varphi[y]+1,\quad L_-[y]=\varphi[y]-1. \tag{4.6}$$

III. *$L_+[y]$ and $L_-[y]$ belong to the second class in $(0,\infty)$ and $(-\infty,0)$ respectively.*

Theorem 4. *If condition* I *holds, then* III *is a necessary and sufficient condition in order that Problems B and A be solvable.*

Proof. We know that Problem B_0 is solvable. We shall show that $\varphi_+(\lambda)<0$ if and only $L_+[y]$ belongs to the first class in $(0,\infty)$. Suppose that $L_+[y]$ belongs to the first class in $(0,\infty)$ and let $y_0(x)$ be the positive solution of $L_+[y]=0$ in $L(0,\infty)$, the existence of which is postulated. A study of the explicit form of the solutions of $L_+[y]=0$ reveals that solutions positive for large positive values of x exist if and only if

$$exp\left\{-\int_0^x\frac{a(t)}{b(t)}dt\right\}\varepsilon\,L(0,\infty). \tag{4.7}$$

If this condition is satisfied we may assume that

$$y_0(x)=\frac{1}{b(x)}\,exp\left\{\int_0^x\frac{a(t)}{b(t)}dt\right\}\int_x^\infty exp\left\{-\int_0^s\frac{a(t)}{b(t)}dt\right\}ds \tag{4.8}$$

since this is a positive solution having the least rate of growth when $x \to \infty$. We can then solve the functional equation

$$L_+[y] = -\lambda \int_x^\infty y(t)\,dt, \qquad \lambda > 0, \tag{4.9}$$

by the method of successive approximations setting

$$y_n(x) = \frac{1}{b(x)} \exp\left\{\int_0^x \frac{a(t)}{b(t)}dt\right\} \int_x^\infty \left\{1 + \lambda \int_s^\infty y_{n-1}(t)\,dt\right\} \exp\left\{-\int_0^s \frac{a(t)}{b(t)}dt\right\} ds$$

for $n = 1, 2, 3, \ldots$ The approximations form an increasing sequence of positive functions belonging to $L(0, \infty)$ and converge rapidly to a positive limit $y(x)$, also in $L(0, \infty)$, which satisfies (4.9). Differentiation with respect to x shows that $y(x)$ also satisfies (3.2) whence it follows that $y(x)$ is a constant multiple of $y_+(x, \lambda)$. Since $L_+[y(x)] \to 0$ when $x \to \infty$, we see that $\varphi_+(\lambda) < 0$ and Problem B is not solvable.

Conversely, suppose that $\varphi_+(\lambda) < 0$. We shall then show that $L_+[y]$ belongs to the first class in $(0, \infty)$. Denoting the left member of (4.7) by $W(x)$, we see that for large x the assumption $\varphi_+(\lambda) < 0$ implies

$$C_1(\lambda)\, W(x) < \frac{d}{dx}[b(x)\, W(x) y_+(x, \lambda)] < C_2(\lambda)\, W(x) \tag{4.10}$$

with $C_1(\lambda) < C_2(\lambda) < 0$. In particular, the product in the square brackets is decreasing and tends to a finite limit c as $x \to \infty$, $c \geqq 0$. If $c > 0$, this implies $[b(x)\, W(x)]^{-1} \varepsilon L(0, \infty)$. In any case, integration of (4.10) shows that (4.7) holds. Formula (4.8) then makes sense and the integrated inequality may be written

$$c[b(x)\, W(x)]^{-1} - C_2(\lambda) y_0(x) < y_+(x, \lambda)$$
$$< c[b(x)\, W(x)]^{-1} - C_1(\lambda) y_0(x),$$

whence it follows that $y_0(x) \varepsilon L(0, \infty)$, so that $L_+[y] = 0$ has a positive solution in $L(0, \infty)$. Hence $L_+[y]$ is of the first class in $(0, \infty)$. The interval $(-\infty, 0)$ is handled in the same manner. This completes the proof of Theorem 4.

The criterion given in III is fairly easy to apply in practice. Some special cases are worthy of separate mention. We first introduce the following conditions.

IV. $$\int_0^\infty \frac{t}{b(t)}dt = -\int_{-\infty}^0 \frac{t}{b(t)}dt = +\infty.$$

$$\text{V.} \quad \int_0^\infty \frac{1}{b(t)}\,dt = \int_{-\infty}^0 \frac{1}{b(t)}\,dt = +\infty .$$

Theorem 5. *Problems B and A are solvable if either*

(i) $a(x)=0,\ b''(x)<M$, *or*

(ii) $a(x)=b'(x)$ *and IV holds, or*

(iii) $a(x)=b(x)$, *I and V hold.*

In (ii) and (iii) the divergency conditions are also necessary if the other conditions hold. The result under (ii) may be generalized as follows.

Theorem 6. *If I holds and if* $|b'(x)-a(x)|<K[|x|+1]$ *for all* x, *then IV is necessary as well as sufficient for the solvability of Problems B and A.*

For the proof we observe that the left relations of (4.4) are satisfied and the discussion of the right relations can be reduced to the study of simpler differential operators $by'+1$ and $by'-1$. Condition IV merely expresses that these operators are of the second class in the intervals $(0,\infty)$ and $(-\infty,0)$ respectively. The details of the proof parallel those of Theorem 4 and will be omitted.

We can interchange right and left in this argument. Assuming $b(x)=O(x^2)$, the right relations in (4.4) are satisfied and the behaviour of the left relations becomes a question of the properties of $[a(x)-b'(x)]^{-1}$ for large x. If this function has the sign of x for large values of x and is not integrable over any infinite range, then Problems B and A are solvable. We suppress further details.

5. Final remarks. The reader may have noticed in the formulation of Problem A that a solution is demanded only for $g(x)$ in L_{02} which is the cross section of the domains of definition of all operators A here considered. This is due to the fact that (2.9) holds only in the domain of A. Actually we are not interested in all of (2.9) but only in the relation

$$AT(t)[g]=\frac{\partial}{\partial t}T(t)[g] \tag{5.1}$$

which expresses that $T(t)[g]$ satisfies the differential equation (1.1). Now if Problem A is solvable, $T(t)[g]$ is defined for every $g\,\varepsilon\,L(-\infty,\infty)$ and not merely for $g(x)$ in the domain of A or in L_{02}. It is then natural to expect that $T(t)[g]$ will always satisfy the differential equation. One way of proving this is to show that $T'(t)$ is a bounded operator in $L(-\infty,\infty)$ and this will be the case, in particular, if $T(t)$ is analytic. In the trivial case, $a(x) = 0$, $b(x) = 1$, the equation of linear heat conduction, $T(t)$ is actually analytic in the right t-halfplane and there is no reason for assuming that this is an exceptional case. In the general case it would be sufficient to show that the spectrum of A (which is likely to be real and negative) is located in a sector of opening less than π and that $\lambda R(\lambda; A)$ is bounded outside such a sector. The actual verification of this surmise would require methods beyond the scope of the present paper so we restrict ourselves to pointing out the possibilities.

In concluding, I want to recall that K. Yosida [8, 9] has extended his investigations to stochastic processes with n degrees of freedom and, in particular, to the Brownian motion on a sphere.

References.

[1] W. Feller, Zur Theorie der stochastischen Prozesse. (Existenz- und Eindeutigkeitssätze.) Math. Annalen, 113 (1936) 113—160.

[2] A. Fokker, Die mittlere Energie rotierender elektrischer Dipole im Strahlungsfeld. Ann. d. Physik, (4) 43 (1914) 810—820.

[3] E. Hille, Functional analysis and semi-groups. Amer. Math. Soc. Coll. Publ., XXXI. New York, 1948, xii + 528 pp.

[4] A. Kolmogoroff, Über die analytischen Methoden in der Wahrscheinlichkeitsrechnung. Math. Annalen, 104 (1931) 415—458.

[5] M. Planck, Über einen Satz der statistischen Dynamik und seine Erweiterung in der Quantentheorie. Sitzungsberichte d. Preuss. Akad. d. Wiss 1917, 324—341.

[6] K. Yosida, On the differentiability and the representation of one-parameter semi-group of linear operators. Journal Math. Soc. Japan, 1 (1948) 15—21.

[7] — An operator-theoretical treatment of temporally homogeneous Markoff process. Journal Math. Soc. Japan, 1 (1949) 244—253.

[8] — Brownian motion on the surface of the 3-sphere. Annals of Math. Stat., 20 (1949) 292—296.

[9] — Integration of Fokker-Planck's equation in a compact Riemannian space. Arkiv f. Mat., 1:9 (1949) 71—75.

Note added in proof, Nov. 1951. Necessary and sufficient conditions for the solution of Problem A (and the adjoint problem) have appeared in my note Les probabilités continués en chaine, C. R. Acad. Sci. Paris, 230 (1950) 34—35.

Reprinted from Vol. I, PROCEEDINGS OF THE
INTERNATIONAL CONGRESS OF MATHEMATICIANS, 1950
Printed in U.S.A.

"EXPLOSIVE" SOLUTIONS OF FOKKER-PLANCK'S EQUATION

EINAR HILLE

This note deals with partial differential equations of parabolic type having certain singular solutions implying non-uniqueness of corresponding boundary value problems and the existence of solutions "exploding" after a preassigned time.

We consider the equations of Kolmogoroff connected with temporally homogeneous stochastic processes of one degree of freedom, involving the arbitrary functions $a(x)$ and $b(x)$ where the variance $b(x) > 0$ for all x. If $a(x) \equiv 0$ for the sake of simplicity, the progressive (Fokker-Planck) equation becomes $[b(x)\,T]_{xx} = T_t$ with $T(x, t)$ to be defined for all $t > 0$ and tending in the mean of order one to a preassigned function $g(x) \in L(-\infty, \infty)$ when $t \to +0$. We have shown elsewhere that the solution is not unique if $x/b(x) \in L(-\infty, 0)$ or $L(0, \infty)$, though the adjoint problem for the retrospective equation in $C[-\infty, \infty]$ has a unique solution.

If $x/b(x) \in L(-\infty, \infty)$, the function $P(x, \lambda)$, satisfying the equation

$$[b(x)y]_{xx} = \lambda y$$

with $P(0, \lambda) \equiv 0$, $P'(0, \lambda) \equiv 1$, is in $L(-\infty, \infty)$. Here $P(x, \lambda)$ is an entire function of λ of order 1/2 for fixed x; the L-norm of $P(\cdot\,; \lambda)$ has a similar property, at least if $b(x)$ is suitably restricted. Under the same assumptions $\| P(\cdot\,; \lambda) \| \to 0$ when $\lambda \to -\infty$. Let $\mu(\lambda)$ be any entire function of order <1, positive when $\lambda > 0$, and let $S(x, t)$ be the Laplace transform of $P(x, \lambda)\mu(\lambda)$. This is an entire function of $1/t$ and $\| S(\cdot\,; t) \| \leqq \exp(C/|t|)$. Further $S(x, t) \to \infty \operatorname{sgn} x$ and $\| S(\cdot\,; t) \| \to \infty$ when $t \to +0$.

Now $S(x, t_0 - t)$ is a solution of Fokker-Planck's equation for any t_0. Taking $t_0 > 0$ and $g(x) = S(x, t_0)$, we see that among the corresponding solutions of the boundary problem, there is also $S(x, t_0 - t)$ which "explodes" after t_0 seconds. The initial values producing such "explosive" solutions may be dense in $L(-\infty, \infty)$ in view of the arbitrariness of $\mu(\lambda)$. Related phenomena have been noted by J. L. Doob for the equations of Kolmogoroff connected with Markoff chains.

YALE UNIVERSITY,
NEW HAVEN, CONN., U. S. A.

UNE GÉNÉRALISATION DU PROBLÈME DE CAUCHY[1]

par Einar **HILLE** (Nancy et New Haven, Conn.)

1. — Introduction.

Il y a un grand nombre de problèmes que nous devons au génie de Cauchy : le problème dont il s'agit ici est le problème des valeurs initiales pour les équations aux dérivées partielles. Un cas spécial de ce problème nous sert comme point de départ pour les considérations suivantes.

Soit donc R_ν l'espace euclidien à ν dimensions, $y(P, t)$ un vecteur à m composantes, dépendant du point P de R_ν et du temps t. De plus, soit U un opérateur différentiel linéaire indépendant de t dont les coefficients sont des fonctions continues de P. Le problème de Cauchy demande de déterminer $y(P, t)$ pour chaque t positif comme solution du système différentiel

$$(1) \qquad \frac{\partial}{\partial t} y(P, t) = U[y(P, t)], \qquad t > 0,$$

$$(2) \qquad \lim_{t \to 0} y(P, t) = y_0(P),$$

où $y_0(P)$ est un vecteur donné à l'avance. Le problème est dit *bien posé* s'il admet une solution et une seule.

On peut généraliser ce problème de différentes manières. Pour plus de précision dans la généralisation suivante on se sert du langage des espaces normés. Soit donc Y un espace complexe de Banach, U un opérateur linéaire faisant l'application d'un sous-espace $D = D[U]$

[1] Conférence faite au IIIe Congrès Autrichien de Mathématiciens à Salzbourg le 10 septembre 1952. Cette conférence a été rédigée pendant que l'auteur avait une bourse Fulbright et une bourse de la Fondation Guggenheim.

de Y sur un sous-espace $R = R[U]$ et considérons le système

$$\frac{d}{dt} y(t) = U[y(t)], \qquad t > 0, \tag{3}$$

$$\lim_{t \to 0} \|y(t) - y_0\| = 0, \tag{4}$$

où y_0 est donné à l'avance. Ici il faut que, pour chaque $t > 0$, $y(t)$ soit un élément de $D[U]$ et que l'application $t \to y(t)$ définisse une fonction absolument continue dont la dérivée au sens fort existe et satisfasse à (3).

Ce problème est étroitement lié à la théorie des semi-groupes d'opérations linéaires et bornées([2]). Soit $\{T(t) | 0 \leqslant t\}$ un tel semi-groupe où $T(0) = I$ et $T(t) \to I$ au sens fort quand $t \to 0$ et soit A la génératrice infinitésimale de $T(t)$. Alors on a

$$\frac{d}{dt}[T(t) y_0] = A[T(t) y_0], \qquad t > 0, \tag{5}$$

$$\lim_{t \to 0} \|T(t) y_0 - y_0\| = 0 \tag{6}$$

pour tous les y_0 du domaine de A, c'est-à-dire un système de type (3)+(4). Il en résulte que dans le cas où U est génératrice infinitésimale d'un semi-groupe $\{T(t)\}$, une solution de notre système est fournie par $T(t) y_0$ quand $y_0 \in D[U]$. Y a-t-il d'autres solutions?

C'est une question assez difficile dont la solution générale nous échappe, mais si l'on impose une restriction convenable sur l'ordre de croissance de la norme de $y(t)$ les difficultés s'évanouissent dans des cas étendus et le problème admet au plus une solution. Nous disons que la solution $y(t)$ est du *type normal* ω si l'on a

$$\limsup_{t \to \infty} \frac{1}{t} \log \|y(t)\| = \omega < \infty. \tag{7}$$

Cela étant on peut démontrer([3]):

([2]) Cf. mon livre [1] pour tout ce qui concerne cette théorie.

([3]) J'ai énoncé un théorème moins général dans un travail qui va paraître dans un autre recueil [3] et dont la connaissance n'est pas nécessaire pour la lecture de la présente note. Je dois à mon ami M. R. S. Phillips l'observation qu'on peut supprimer une des conditions restrictives dans l'énoncé original en s'appuyant sur le lemme du paragraphe 2 ci-dessous. Il s'agit de l'ordre de croissance de la résolvante de U; en y réfléchissant je me suis convaincu qu'on peut se débarrasser de chaque hypothèse sur l'existence ou sur les propriétés de la résolvante sans modifications essentielles du raisonnement. Voir la démonstration du théorème 3 ci-dessous.

THÉORÈME 1. — *Soit* U *un opérateur linéaire et clos dont les valeurs propres ne sont denses dans aucun demi-plan* $\Re(\lambda) > \lambda_0$. *Alors, pour chaque* y_0 *de* Y, *le système* (3) + (4) *a au plus une solution du type normal.*

Les conditions du théorème sont satisfaites si, par exemple, U est un opérateur borné ou, plus généralement, si U est une génératrice infinitésimale d'un semi-groupe $T(t)$. Dans ce cas, $T(t)y_0$ donne la seule solution du type normal et la solution existe pour chaque y_0 dans D, mais peut cesser d'exister en dehors de D.

2. — Unicité et caractère de la solution.

L'existence d'une seule solution du type normal a des conséquences très importantes. En effet on a :

THÉORÈME 2. — *Soit* U *un opérateur linéaire et clos de domaine dense dans* Y *et supposons que, pour chaque* y_0 *de* D, *le système* (3) + (4) *ait une solution* $y(t) = y(t; y_0)$ *et une seule qui soit du type normal au sens plus restreint que voici. Il y a des constantes* M *et* ω, *indépendantes de* y_0, *telles que*

$$\|y(t;\, y_0)\| \leqslant M e^{\omega t} \|y_0\|, \qquad t > 0. \tag{8}$$

Alors la résolvante $R(\lambda;\, U)$ *existe pour* $\Re(\lambda) > \omega$, U *est une génératrice infinitésimale d'un semi-groupe* $T(t)$, *et* $y(t;\, y_0) = T(t) y_0$ [4].

Il nous faut le lemme suivant :

LEMME 1. — *Soit* U *un opérateur linéaire et clos de domaine* D. *Soit* S *un ensemble mesurable dans l'espace euclidien* R_ν, $x(s)$ *une fonction définie dans* S *à valeurs dans* D, *telle que* $x(s)$ *et* $U[x(s)]$ *sont intégrables au sens de Bochner dans* S. *Alors*

$$U\left\{\int_S x(s)\, ds\right\} = \int_S U[x(s)]\, ds. \tag{9}$$

Dans le cas où S est borné et les fonctions $x(s)$ et $U[x(s)]$ sont continues et bornées on peut évidemment trouver des sommes finies

$$\Sigma \mu_k x(s_k) \qquad \text{et} \qquad \Sigma \mu_k U[x(s_k)] = U\{\Sigma \mu_k x(s_k)\}$$

telles que la première soit voisine de l'intégrale de $x(s)$ tandis que la

[4] Le théorème 2 est du même caractère et découle du même principe que le théorème 5 de notre travail cité [3]. Nous en donnerons ici la démonstration pour faciliter la lecture.

seconde soit voisine de l'intégrale de $U[x(s)]$ dans S. U étant clos, il en résulte que (9) est vrai. Dans le cas général on recourt au théorème de Lusin d'après lequel on peut approximer les intégrales dans S par des intégrales étendues à un sous-ensemble borné où les fonctions $x(s)$ et $U[x(s)]$ sont bornées et continues. Alors la démonstration s'achève de la même manière.

Cela étant, nous démontrons le théorème 2 comme il suit : Avec la solution $y(t, y_0)$ formons la transformée de Laplace

$$(10) \qquad L(\lambda; y) = \int_0^\infty e^{-\lambda t} y(t; y_0)\, dt,$$

qui est une fonction analytique de λ dans $\Re(\lambda) > \omega$ où elle satisfait à l'inégalité

$$(11) \qquad \|L(\lambda; y_0)\| \leqslant M[\Re(\lambda) - \omega]^{-1} \|y_0\|.$$

L'application $y_0 \to L(\lambda; y_0)$ est définie dans D comme une transformation linéaire et bornée. D étant dense dans Y, on peut prolonger la transformation dans tout Y; soit $y_0 \to R(\lambda)\, y_0$ l'application prolongée.

$L(\lambda; y_0)$ est un élément de Y, nous verrons qu'il est aussi dans D. Pour montrer cela, remarquons que si $0 < \alpha < \beta < \infty$ on a

$$\int_\alpha^\beta e^{-\lambda t} U[y(t; y_0)]\, dt = \int_\alpha^\beta e^{-\lambda t} y'(t; y_0)\, dt$$
$$= e^{-\lambda\beta} y(\beta; y_0) - e^{-\lambda\alpha} y(\alpha; y_0) + \lambda \int_\alpha^\beta e^{-\lambda t} y(t; y_0)\, dt.$$

Les conditions du lemme 1 étant vérifiées, il en résulte que

$$U\left\{\int_\alpha^\beta e^{-\lambda t} y(t; y)\, dt\right\} = \int_\alpha^\beta e^{-\lambda t} U[y(t; y_0)]\, dt.$$

U étant clos, un passage aux limites nous donne enfin

$$(12) \qquad (\lambda I - U)\, L(\lambda; y_0) = y_0$$

pour $\Re(\lambda) > \omega$, $y_0 \epsilon D$. Vu que D est dense dans Y et que U est clos, la relation s'étend à Y, c'est-à-dire que

$$(13) \qquad (\lambda I - U)\, R(\lambda) y_0 = y_0$$

pour $\Re(\lambda) > \omega$, $y_0 \epsilon Y$.

Jusqu'ici nous n'avons fait aucun recours à l'unicité de la solution. Maintenant nous remarquons que si $y(t)$ est une solution de l'équation (3) pour $t > 0$ et si t_0 est fixe, $t_0 > 0$, alors $y(t + t_0)$ en est une autre.

Il en résulte que $y(t+t_0; y_0)$, $y_0 \in D$, est une solution, évidemment du type normal, qui tend vers $y(t_0; y_0)$ quand $t \to 0$. Mais alors il faut que

$$(14) \qquad y(t+t_0; y_0) = y(t; y(t_0; y_0))$$

parce que le membre de droite est une solution ayant les mêmes propriétés. Cela veut dire qu'il existe une famille d'opérations linéaires $\{T(t)\}$ avec les propriétés suivantes :

$$(15) \qquad T(t)y_0 = y(t; y_0), \qquad 0 < t, \qquad y_0 \in D,$$

$$(16) \qquad T(t_1+t_2) = T(t_1)T(t_2), \qquad 0 < t_1, t_2 < \infty,$$

$$(17) \qquad \|T(t)\| < Me^{\omega t},$$

$$(18) \qquad \|T(t)y_0 - y_0\| \to 0, \qquad t \to 0, \qquad y_0 \in D.$$

L'opération $T(t)$ étant définie, bornée et linéaire dans le domaine D dense dans Y, elle peut se prolonger dans tout Y en conservant ses propriétés.

Le semi-groupe $\{T(t)\}$ possède une génératrice infinitésimale, que nous désignons par A, dont la résolvante est donnée par

$$R(\lambda; A)y_0 = \int_0^\infty e^{-\lambda t} T(t)y_0 \, dt, \qquad \Re(\lambda) > \omega.$$

La comparaison avec (10) montre que

$$R(\lambda; A)y_0 = R(\lambda)y_0, \qquad y_0 \in D,$$

d'où il résulte que l'identité est valable pour tout y_0 dans Y. Mais alors il faut que $A = U$. En effet, soit $z_0 = R(\lambda; A)y_0 = R(\lambda)y_0$ un point arbitraire du domaine de A. Alors

$$(\lambda I - A)z_0 = y_0 = (\lambda I - U)z_0 \qquad \text{ou} \qquad Az_0 = Uz_0.$$

La proposition est ainsi démontrée.

3. — Équations d'ordre supérieur.

Si l'opérateur U engendre un semi-groupe $\{T(t)\}$, alors on a aussi

$$(19) \qquad \frac{d^n}{dt^n}[T(t)y_0] = U^n[T(t)y_0], \quad t > 0, \quad y_0 \in D[U^n], \quad n = 1, 2, 3, \ldots,$$

où $D[U^n]$ désigne le domaine de U^n.

Il est donc naturel de considérer aussi le système

$$(20) \qquad y^{(n)}(t) = \mathrm{U}^n[y(t)], \qquad t > 0,$$

$$(21) \qquad \lim_{t \to 0} \|y^{(k)}(t) - y_k\| = 0, \qquad k = 0, 1, \ldots, n-1,$$

où $y_0, \ldots, y_{n-1}$ sont des éléments donnés de Y. Ici il faut supposer que, pour chaque $t > 0$, $y(t)$ soit un élément de $\mathrm{D}[\mathrm{U}^n]$ et que l'application $t \to y(t)$ définisse une fonction de t qui a des dérivées au sens fort jusqu'à l'ordre n.

Pour ce système on a un théorème d'unicité tout à fait analogue de théorème 1.

Théorème 3. — *Soit* U *tel que* U^n *soit linéaire et clos et tel que les valeurs propres de* U^n *ne soient denses dans aucun demi-plan* $\Re(\lambda) > \lambda_0$. *Alors pour chaque choix de* $y_0, \ldots, y_{n-1}$ *dans* Y *il y a au plus une solution du système* (20)+(21) *dont la dérivée d'ordre* $n - 1$ *est du type normal.*

Remarquons que si la dérivée d'ordre $n - 1$ est du type normal, les dérivées d'ordre k, $0 \leqslant k < n - 1$, en sont aussi, mais on ne peut rien conclure sur les dérivées d'ordre plus grand que $n - 1$.

S'il y a deux solutions du type normal de notre système, le système (20)+(21_0) où tous les y_k sont nuls a une solution $y(t)$ non-nulle. Alors

$$(22) \qquad \mathrm{L}(\lambda\,;y) \equiv \int_0^\infty e^{-\lambda t} y(t)\,dt$$

existe pour $\Re(\lambda) > \omega$, ω étant le type de $y(t)$ défini par (7) ; et dans ce demi-plan $\mathrm{L}(\lambda\,;y)$ est une fonction analytique qui n'est pas identiquement nulle. En suivant la même méthode que ci-dessus on voit que

$$\begin{aligned}\int_\alpha^\beta e^{-\lambda t}\mathrm{U}^n[y(t)]\,dt &= \int_\alpha^\beta e^{-\lambda t} y^{(n)}(t)\,dt \\ &= \left\{ e^{-\lambda t} \sum_{k=0}^{n-1} \lambda^{n-1-k} y^{(k)}(t) \right\}_\alpha^\beta + \lambda^n \int_\alpha^\beta e^{-\lambda t} y(t)\,dt \\ &= \mathrm{U}^n \left\{ \int_\alpha^\beta e^{-\lambda t} y(t)\,dt \right\}.\end{aligned}$$

En passant aux limites on trouve que

$$(23) \qquad (\lambda^n \mathrm{I} - \mathrm{U}^n)\mathrm{L}(\lambda\,;y) = 0, \qquad \Re(\lambda) > \omega,$$

c'est-à-dire, pour presque chaque λ dans ce demi-plan, λ^n est une valeur propre de U^n et cela n'est pas en accord avec les hypothèses de notre théorème. Il en résulte que $\mathrm{L}(\lambda\,;y) \equiv 0$, ce qui entraîne $y(t) \equiv 0$ et la démonstration est achevée.

4. — Questions d'existence.

Dans le cas $n > 1$, on n'est pas sûr, en général, de l'existence d'une solution du problème de Cauchy. C'est seulement dans le cas où U est borné qu'on a toujours une solution, évidemment donnée par

$$(24) \qquad y(t) = \sum_{k=0}^{n-1} \sum_{m=0}^{\infty} \frac{t^{mn+k}}{(mn+k)!} U^{mn} y_k.$$

Cela peut s'écrire d'une façon plus suggestive

$$(25) \qquad y(t) = \sum_{k=0}^{n-1} \exp(\eta^k t U) z_k, \qquad \eta = e^{\frac{2\pi i}{n}},$$

où les z_k sont assujettis aux conditions

$$(26) \qquad \sum_{p=0}^{n-1} \eta^{kp} U^k z_p = y_k, \qquad k = 0, 1, \ldots, n-1.$$

Ces conditions ne suffisent pas, en général, pour la complète détermination des z_k, mais on voit immédiatement que les quantités $U^{n-1} z_k$ sont univoquement déterminées

$$(27) \quad U^{n-1} z_k = \sum_{p=0}^{n-1} \alpha_{p,k}^{(n)} U^{n-p-1} y_p, \qquad k = 0, 1, \ldots, n-1$$

où les $\alpha_{p,k}^{(n)}$ sont des constantes numériques, de sorte que la solution $y(t)$ est univoquement déterminée par les équations (25) et (26).

Le cas où U est non borné mais engendre un semi-groupe $T(t) = T(t|U)$ est d'un intérêt considérable. Si $y_0 \in D[U^n]$ la fonction $T(t|U) y_0$, comme nous venons d'observer, donne une solution de (20), la première condition de (21) est vérifiée, les autres seulement si on a $y_k = U^k y_0$. Remarquons que $T(t|U) = \exp(tU)$ si U est borné.

D'après la relation (25), on sent d'une manière assez vague qu'il faut avoir recours aux opérateurs $T(t|\eta^k U)$, $k \neq 0$, pour aller plus loin. L'opérateur $T(t|\eta^k U)$ est bien déterminé si $\eta^k U$ est la génératrice infinitésimale d'un semi-groupe fortement continu. Ici il y a deux cas distincts : (1) $\eta^k = -1$, (2) $\eta^k \neq \pm 1$.

Dans le premier cas n est pair, $n = 2p$, et nous avons :

Théorème 4. — *Si $n = 2p$ et si* U *et* $-$U *engendrent les semi-*

groupes $T(t|U)$ *et* $T(t|-U)$, *continus au sens fort à l'origine avec* $T(o|U)=T(o|-U)=I$, *alors on a*

$$T(t|-U)\,T(t|U)=T(t|U)\,T(t|-U)=I.$$

En posant $T(t|-U)=T(-t|U)$ *pour* $t>o$, *on obtient un groupe* $\{T(t|U)|-\infty<t<\infty\}$.

Il suffit de montrer que $T(t|U)\,T(t|-U)=I$. Pour faire voir cela, notons que

$$R(\lambda\,;\,-U)=-R(-\lambda\,;\,U),\qquad \Re(\lambda)>\omega_2,$$

où ω_2 est le type de $T(t|-U)$, c'est-à-dire que $R(\lambda;\,U)$ existe dans deux demi-plans $\Re(\lambda)>\omega_1$ et $\Re(\lambda)<-\omega_2$. Alors, pour chaque y de $D[U]$ on a

$$T(t|U)\,T(t|-U)y$$
$$=-\lim_{\rho\to\infty}\lim_{\sigma\to\infty}\frac{1}{(2\pi i)^2}\int_{\gamma-i\rho}^{\gamma+i\rho}\int_{\delta-i\sigma}^{\delta+i\sigma}e^{(\lambda+\mu)t}R(\lambda\,;\,U)\,R(-\mu\,;\,U)\,y\,d\mu\,d\lambda$$

où $\gamma>\omega_1$, $\delta>\omega_2$. Cf. [1] p. 232, formule (11. 7. 2). Ici on peut écrire l'intégrale double comme la différence de deux intégrales plus simples en s'appuyant sur la première équation fonctionnelle satisfaite par la résolvante qui prend la forme

$$-(\lambda+\mu)\,R(\lambda\,;\,U)\,R(-\mu\,;\,U)=R(\lambda\,;\,U)-R(-\mu\,;\,U)$$

dans le cas présent. Dans la première de ces intégrales on passe à la limite avec σ en obtenant

$$\lim_{\rho\to\infty}\frac{1}{2\pi i}\int_{\gamma-i\rho}^{\gamma+i\rho}R(\lambda;\,U)\,y\,d\lambda=\frac{1}{2}y.$$

La seconde donne de la même manière

$$\lim_{\sigma\to\infty}\frac{1}{2\pi i}\int_{\delta-i\sigma}^{\delta+i\sigma}R(-\mu;\,U)\,y\,d\mu=-\frac{1}{2}y.$$

Il en résulte que

$$T(t|U)\,T(t|-U)\,y=y$$

pour chaque $y\in D[U]$. Mais $D[U]$ étant dense dans Y et l'opérateur dans le membre de gauche étant borné pour chaque t fixe, la relation vaut pour tous les y.

Théorème 5. — *Si* $n=2$ *et si* U *engendre le groupe*

$\{T(t|U)|-\infty < t < \infty\}$, *continu à l'origine où* $T(0|U) = I$, *alors le problème de Cauchy pour le système* (20) + (21) *a une solution et une seule dont la dérivée première est du type normal pour chaque choix de* y_0 *dans* $D[U^2]$ *et de* y_1 *dans* $D[U] \cap R[U]$, *à savoir*

$$(28)\quad y(t) = \frac{1}{2}[T(t|U)(y_0+z_1)+T(-t|U)(y_0-z_1)], \quad Uz_1 = y_1.$$

On vérifie sans peine que toutes les conditions sont satisfaites. L'unicité de la solution serait démontrée si l'on pouvait appliquer le théorème 3. Nous savons que le spectre de U est renfermé dans la bande $-\omega_2 \leqq \Re(\lambda) \leqq \omega_1$ d'où il résulte que le spectre de U^2 est renfermé dans un domaine bordé par deux hyperboles. L'opérateur U étant clos, U^2 a la même propriété grâce au lemme, à peine nouveau, que voici :

LEMME 2. — *Soit* U *un opérateur linéaire et clos de domaine dense et admettons l'existence de la résolvante* $R(\lambda; U)$ *pour une valeur de* λ [5]. *Alors les opérateurs* U^n *sont clos et leurs domaines denses pour chaque* n.

Donnons la démonstration pour $n = 2$. Posons $Ux_n = y_n$, $Uy_n = z_n$ et supposons que $x_n \to x_0$, $z_n \to z_0$. Il est entendu que $x_n \in D[U^2]$, il faut démontrer que $x_0 \in D[U^2]$ et que $U^2 x_0 = z_0$. De l'équation liant U et $R(\lambda_0; U)$ il résulte que

$$\lambda_0^2 R(\lambda_0; U)x_n = \lambda_0 x_n + Ux_n + R(\lambda_0; U)U^2 x_n$$

ou

$$y_n = \lambda_0^2 R(\lambda_0; U)x_n - \lambda_0 x_n - R(\lambda_0; U)z_n.$$

Mais alors $\lim y_n \equiv y_0$ existe et, U étant clos, il s'en suit que $y_0 = Ux_0$. En s'appuyant sur la clôture de U une fois de plus, on voit que $y_0 \in D[U]$, $z_0 = Uy_0$, ce qu'il faut démontrer. La transformation $R(\lambda_0; U)$ donne l'application de Y sur $D[U]$ et l'application de $D[U]$ sur $D[U^2]$. $D[U]$ étant dense, il en résulte que $D[U^2]$ l'est aussi.

Retournons au théorème 5. Alors, U comme génératrice infinitésimale d'un groupe, continu au sens fort à l'origine, est clos et son domaine est dense dans Y, de plus $R(\lambda; U)$ existe en dehors de la bande $-\omega_2 \leqq \Re(\lambda) < \omega_1$. Il en résulte que U^2 est aussi clos et le théorème 3 montre que la solution fournie par (28) est la seule de

(5) L'existence de $R(\lambda_0; U)$ entraîne l'existence de $R(\lambda; U)$ au moins dans le cercle $|\lambda - \lambda_0| \, \|R(\lambda_0; U)\| < 1$.

type normal. Cela est un peu étonnant parce que dans le cas où $\lambda = 0$ est une valeur propre de U, la quantité z_1 n'est pas univoquement déterminée. Mais lorsque $Uz = 0$ on a $T(t|U)z = z$ pour chaque t, $-\infty < t < \infty$, c'est-à-dire que l'expression dans (28) ne dépend pas effectivement du choix particulier de z_1 pourvu que $Uz_1 = y_1$. C'est le même phénomène d'indétermination formelle, non-réelle que nous avons observé ci-dessus pour les opérateurs bornés. Le théorème est ainsi démontré.

Pour $n > 1$, le cas traité dans le théorème 5 est le seul où nous puissions donner la solution complète du problème de Cauchy. Remarquons que la solution (28) est tout à fait analogue à celle donnée par (25) dans le cas où U est borné et $n = 2$.

Soit maintenant $n > 2$ et soient U et $\eta^k U$, $\eta^k \neq \pm 1$, les génératrices infinitésimales des semi-groupes $T(t|U)$ et $T(t|\eta^k U)$ continus au sens fort à l'origine où $T(0|U) = T(0|\eta^k U) = I$. Alors $T(t|U)$ a nécessairement un secteur d'analyticité :

Théorème 6. — *Supposons que* U *et* $\eta^k U$, $\eta^k \neq \pm 1$, *engendrent les semi-groupes* $T(t|U)$ *et* $T(t|\eta^k U)$, *tous les deux continus au sens fort à l'origine où* $T(0|U) = T(0|\eta^k U) = I$. *Soit* S *le plus petit secteur du plan complexe de* t *déterminé par les rayons* $\arg t = 0$ *et* $\arg t = 2k\pi/n = \theta_k$. *Alors il existe un opérateur* $T(t)$, *analytique dans* S, *tel que*

$$(29) \qquad \lim_{\theta \to 0} T(re^{i\theta}) = T(r|U), \qquad \lim_{\theta \to \theta_k} T(re^{i\theta}) = T(r|\eta^k U).$$

Pour chaque t_1 *et* t_2 *dans* S *on a*

$$(30) \qquad T(t_1 + t_2) = T(t_1)\, T(t_2).$$

Pour montrer cela, observons que

$$R(\lambda;\ U)y = \int_0^\infty e^{-\lambda r} T(r|U)y\, dr, \qquad \Re(\lambda) > \omega_0,$$
$$R(\lambda;\ \eta^k U)y = \int_0^\infty e^{-\lambda r} T(r|\eta^k U)y\, dr, \qquad \Re(\lambda) > \omega_k,$$

où ω_0 et ω_k sont les types de $T(r|U)$ et de $T(r|\eta^k U)$. Mais on a

$$R(\lambda;\ \eta^k U) = \eta^{-k} R(\eta^{-k}\lambda;\ U),$$

d'où il résulte que $R(\lambda;\ U)$ existe et est analytique dans le domaine Δ, réunion des deux demi-plans $\Re(\lambda) > \omega_0$ et $\Re(\eta^k \lambda) > \omega_k$. Nous avons aussi $\delta(\lambda) \| R(\lambda;\ U) \| \leq M(\varepsilon)$ où $\delta(\lambda)$ désigne la distance de λ à la frontière de Δ et $\delta(\lambda) \geq \varepsilon$. Pour fixer les idées, supposons que $0 < \theta_k < \pi$. Nous posons

$$T(t)y = \frac{1}{2\pi i} \int_\Gamma e^{\lambda t} R(\lambda;\ U) y \, d\lambda, \tag{31}$$

où $t \in S$ et Γ est la ligne brisée dans Δ sur laquelle λ est constamment à la distance $\delta > 0$ de la frontière, l'argument de λ étant supposé croissant sur Γ, On voit sans peine que $T(t)$ est analytique dans S et indépendant de δ. De plus on a

$$\begin{aligned} T(t_1)T(t_2)y &= \frac{1}{(2\pi i)^2} \int_{\Gamma_1} \int_{\Gamma_2} e^{\lambda t_1 + \mu t_2} R(\lambda;\ U) R(\mu;\ U) y \, d\lambda \, d\mu \\ &= \frac{1}{(2\pi i)^2} \int_{\Gamma_1} \int_{\Gamma_2} e^{\lambda t_1 + \mu t_2} [R(\lambda;\ U) - R(\mu;\ U)] y \frac{d\mu \, d\lambda}{\mu - \lambda} \\ &= \frac{1}{2\pi i} \int_{\Gamma_1} e^{\lambda(t_1 + t_2)} R(\lambda;\ U) y \, d\lambda = T(t_1 + t_2) y, \end{aligned}$$

où nous supposons que λ trace la ligne Γ_1, $\delta = \delta_1$, et que μ trace Γ_2, $\delta = \delta_2$, et $\delta_1 > \delta_2$.

Les relations (29) s'obtiennent, en observant que pour $y \in D[U]$ et t sur l'un des bords de S, l'intégrale dans (31) se décompose en deux parties, une pour chaque chemin rectiligne ; l'une de ces intégrales est absolument convergente, l'autre existe seulement comme valeur principale. La valeur de l'intégrale est connue (voir [1], p. 232, formule (11. 7. 2)) et est égale à $T(r|U)y$ ou $T(r|\eta^k U)y$, selon le le bord où t se trouve. Supposons, pour fixer les idées, que la valeur est $T(r|U)y$. Alors on peut approximer $T(r|U)y$, uniformément par rapport à r dans un intervalle fini donné, par une intégrale de $(2\pi i)^{-1} e^{\lambda r} R(\lambda;\ U) y$ prise le long d'une partie finie fixe de Γ, cette intégrale étant la limite pour $t \to r$ d'une intégrale approximante de $T(t)y$; il en résulte que $\lim T(re^{i\theta})y = T(r|U)y$ pour $\theta \to 0$, ce qui prouve notre proposition.

On déduit de (30) que

$$\lim_{r \to \infty} \frac{1}{r} \log \| T(re^{i\theta}) \| \leq \frac{\omega_0 \sin(\theta_k - \theta) + \omega_k \sin\theta}{\sin\theta_k}, \tag{32}$$

c'est-à-dire que $T(t)$ est du type normal sur chaque rayon de S.

Corollaire. — *Si $\eta^k U$ engendre un semi-groupe fortement continu pour trois valeurs distinctes de k et si les trois angles entre les* *directions correspondantes,* $\arg t = 2\pi k_\nu/n$, *sont plus petits que* π, *alors* U *est une opération bornée.*

En effet, il résulte de la démonstration du théorème précédent que la fonction $\lambda R(\lambda\,;\,U)$ est holomorphe à l'infini, ce qui entraîne que U est borné.

En conséquence, si U est non-borné, il y a au plus $\frac{1}{2}n + 1$ valeurs de k, distinctes modulo n, qui peuvent donner des semi-groupes bornés et continus. Ce cas se présente, par exemple, si $n = 4p$ et si U engendre un semi-groupe $T(t|U)$, analytique pour $\Re(t) > 0$ et continu pour $\Re(t) \geqq 0$. Dans ce cas on peut imposer $2p + 1$ conditions initiales de caractère général à la solution de l'équation (20), c'est-à-dire une pour chaque direction admissible, mais non pas les $4p$ conditions demandées dans le problème de Cauchy.

5. — Problème réduit.

Il résulte de l'observation que nous venons de faire qu'il faut remplacer le problème de Cauchy par un problème réduit. La définition suivante semble convenable :

Définition. — *Le problème aux valeurs initiales de l'équation* (20) *est d'ordre* n *et de défaut* $d = n - m$ *s'il existe un entier* $m \leqslant n$ *tel qu'on peut toujours trouver une solution de* (20) *satisfaisant aux conditions*

$$\lim_{t \to 0} \| y^{(k)}(t) - y_k \| = 0, \qquad k = 0,\ 1,\ \dots,\ m - 1, \tag{33}$$

où y_k *est un élément arbitraire de* $D[U^{n-k}] \cap R[U^k]$, *tandis qu'on ne peut pas disposer de la limite de* $y^{(m)}(t)$.

Cela étant nous avons :

Théorème 7. — *Soit* U *la génératrice infinitésimale d'un semi-groupe* $T(t|U)$, *analytique dans un secteur* S *et continu au sens fort à l'origine. Soit* m *le nombre de rayons* $\arg t = 2k\pi/n$ *contenus dans* S *où un rayon sur le bord de* S *est inclus seulement si* $T(t|U)$ *est continu sur le bord en question. Alors le problème aux conditions initiales de l'équation* (20) *est d'ordre* n *est de défaut* $d \leqslant n - m$.

En effet, nous pouvons poser

$$y(t) = \Sigma_k T(\eta^k t\,|\,U) z_k, \tag{34}$$

où la sommation s'étend aux valeurs admissibles de k et où les z_k

sont des éléments de $D[U^n]$ à déterminer. Les conditions initiales (33) donnent le système d'équations

$$(35) \qquad y_p = \Sigma_k \eta^{kp} U^p z_k, \qquad p = 0, 1, \ldots, m-1,$$

tout à fait analogue au système (26). Par hypothèse,

$$y_k \in D[U^{n-k}] \cap R[U^k].$$

Alors le système (35) donne

$$U^{m-p-1} y_p = \Sigma_k \eta^{kp} U^{m-1} z_k, \qquad p = 0, 1, \ldots, m-1.$$

Le déterminant du système étant différent de zéro pour $m \leqq n$, il en résulte que

$$U^{m-1} z_k = \Sigma_p \alpha_{pk}^{(m)} U^{m-1-p} y_p$$

et

$$(36) \qquad z_k = \Sigma_p \alpha_{pk}^{(m)} v_p, \qquad U^p v_p = y_p.$$

Enfin

$$(37) \qquad y(t) = \Sigma_p [\Sigma_k \alpha_{pk}^{(m)} T(\eta^k t \mid U)] v_p.$$

C'est une solution de (20) satisfaisant aux conditions (33). Est-elle unique? Nous ignorons la réponse. Dans le cas où $\lambda = 0$ est une valeur propre de l'opérateur U^p, l'élément v_p n'est pas univoquement déterminé mais cela n'introduit aucune ambiguïté dans la valeur de $y(t)$. En effet, on peut écrire la solution sous la forme

$$(38) \quad y(t) = \sum_{k=0}^{m-1} \frac{t^k}{k!} y_k + \frac{1}{(m-1)!} \int_0^t (t-s)^{m-1} [\Sigma_k \eta^{mk} T(\eta^k s \mid U) U^m z_k] ds$$

où toutes les quantités sont parfaitement déterminées par les conditions initiales. Il en résulte que le défaut $d \leqq n - m$. Dans le cas où S est le domaine d'existence exact de $T(t \mid U)$, on peut espérer démontrer que $d = n - m$ mais nous ne savons pas le faire.

6. — Exemples.

Nous prenons $Y = L(-\infty, \infty)$ sauf pour l'exemple 5.

EXEMPLE 1. — *L'équation de Cauchy-Riemann sous la forme complexe*

$$(39) \qquad \frac{\partial w}{\partial y} = i \frac{\partial w}{\partial x}, \qquad z = x + iy.$$

Il s'agit de trouver une solution de (39) dans le demi-plan $y > 0$ qui tende en moyenne d'ordre un vers une fonction donnée $f(x)$ quand $y \to 0$. On a $U = i\, d/dx$, opérateur clos dont le spectre ponctuel est vide. Alors le théorème 3 dit qu'il y a au plus une solution du type normal. Mais les fonctions $f(x)$ de Y donnant des solutions du type normal dont les types ne surpassent pas un nombre fixe ω, ne sont pas denses dans Y car il faut que

$$(40) \qquad F(t) \equiv \frac{1}{\sqrt{2\pi}} \int_{-\infty}^{\infty} f(s)\, e^{-its}\, ds$$

soit identiquement nulle quand $t \leqq -\omega$. Pour le voir observons que la solution est une fonction analytique de z dans $y > 0$, soit $f(z)$, alors on peut intégrer $e^{-itz} f(z)$, $t < -\omega$, le long d'un rectangle $\pm a + i\varepsilon$, $\pm a + iB$, puis on passe à la limite, $a \to \infty$, $B \to \infty$, $\varepsilon \to 0$ dans cet ordre. Il en résulte que le théorème 2 ne s'applique pas, de plus sa conclusion aurait été fausse dans le cas présent : En effet, la résolvante peut s'évaluer et on trouve que $R(\lambda;\, U)$ existe dans tout le plan sauf sur l'axe réel. Alors on conclut que $i\, d/dx$ ne peut pas engendrer un semi-groupe continu dans $L(-\infty, \infty)$ et le problème de Cauchy correspondant est de défaut un, c'est-à-dire mal posé.

De plus, il y a autre chose dans certains sous-espaces de Y. Soit $\omega \geqq 0$ fixe et considérons le sous-espace Y_ω de Y dont les éléments sont des fonctions $f(x)$ telles que la transformée de Fourier $F(t)$, définie par (40), soit identiquement nulle pour $t \leqq -\omega$. Y_ω est complet dans la métrique de $L(-\infty, \infty)$. Alors la formule

$$(41) \qquad w(z) = \frac{1}{\sqrt{2\pi}} \int_{-\omega}^{\infty} e^{itz}\, F(t)\, dt$$

donne une solution du problème de Cauchy qui est du type $\leqq \omega$ pour chaque $f(x)$ dans Y_ω. On voit que l'équation (39) a des solutions de type normal de chaque type fini mais il y a aussi des solutions de type infini [6].

[6] Soit $E_\alpha(z)$ la fonction entière de Mittag-Leffler. On voit sans peine que $E_\alpha(-(x+iy)^2) \in L(-\infty, \infty)$ comme fonction de x pour chaque y fixe, si $0 < \alpha < \frac{1}{2}$, et que

$$C_1[1 + |y|^{2/\alpha}] \leqslant \log \|E_\alpha(-(x+iy)^2)\| \leqslant C_2[1 + |y|^{2/\alpha}].$$

Alors on peut choisir les coefficients a_n dans la série

$$\sum_{3}^{\infty} a_n E_{1/n}(-z^2)$$

de manière que la série converge en norme pour chaque y et que la norme de la somme croisse plus vite qu'une fonction donnée de y.

EXEMPLE 2. — *Équation des ondes*

$$\frac{\partial^2 y}{\partial t^2}=\frac{\partial^2 y}{\partial x^2}. \tag{42}$$

On prend $U=d/dx$ qui engendre le groupe unitaire

$$T\left(t\,\middle|\,\frac{d}{dx}\right)[f]=f(x+t), \qquad -\infty<t<\infty. \tag{43}$$

On trouve $n=2$, $m=2$, $d=0$ et le théorème 5 donne la solution classique

$$y(x,\,t)=\frac{1}{2}[y_0(x+t)+y_0(x-t)]+\frac{1}{2}\int_{x-t}^{x+t}y_1(s)\,ds. \tag{44}$$

EXEMPLE 3. — *Équation de Laplace*

$$\frac{\partial^2 y}{\partial t^2}+\frac{\partial^2 y}{\partial x^2}=0. \tag{45}$$

L'opérateur U peut se choisir de deux manières différentes, soit $U=i\dfrac{d}{dx}$, soit $U=\dfrac{\tilde{d}}{dx}$ où le tilde signifie l'opération de conjugaison au sens de la théorie du potentiel logarithmique. Tous deux sont clos et leurs spectres ponctuels sont vides, par conséquent le théorème d'unicité s'applique ; mais cela ne vaut pas grand'chose parce qu'il est bien connu que le problème de Cauchy est mal posé pour l'équation de Laplace. L'opérateur $\dfrac{\tilde{d}}{dx}$ engendre un semi-groupe mais pas un groupe, ce qui donne $d\leqq 1$, ici $d=1$ est la vraie valeur.

EXEMPLE 4. — *Équation du troisième ordre*

$$\frac{\partial^3 y}{\partial t^3}=\frac{\partial^3 y}{\partial x^3}. \tag{46}$$

L'opérateur $U=\dfrac{d}{dx}$ engendre le groupe (43) mais ni ηU ni $\bar{\eta}U$, $\eta=\exp(2\pi i/3)$, n'engendrent des semi-groupes. Alors on a $n=3$, $d\leqq 2$. Ici on peut démontrer que $d=2$ par l'observation suivante. Les combinaisons linéaires des fonctions $(x-k\eta)^{-2}$ et $(x-k\bar{\eta})^{-2}$, $k=1,\ 2,\ 3,\ \ldots$, plus $(x-\eta)^{-1}(x-\bar{\eta})^{-1}$ sont denses dans $L(-\infty,\,\infty)$. Nous proposons de trouver une solution de (46) telle que les trois données initiales y_0, y_1, y_2 soient

$$(x-k\eta)^{-2}, \qquad a\,(x-k\eta)^{-3}, \qquad b\,(x-k\eta)^{-4},$$

où a et b sont des constantes données. La solution formelle est

$$\alpha(x+t-k\eta)^{-2}+\beta(x+\eta t-k\eta)^{-2}+\gamma(x+\bar{\eta}t-k\eta)^{-2},$$

les α, β, γ étant des formes linéaires de a et b. Si

$$3(a+2)\eta+(b-6)\bar{\eta}-(3a+b)\neq 0$$

on a $\beta\neq 0$ et la solution formelle n'est pas dans $L(-\infty, \infty)$ quand $t=k$ parce qu'elle devient infinie pour $x=0$. Le même résultat s'obtient en échangeant η et $\bar{\eta}$ et la méthode s'applique aussi au troisième cas. Donc il y a un ensemble fondamental de fonctions $y_0(.)$ pour lesquelles on ne peut choisir ni $y_1(.)$ ni $y_2(.)$ d'une manière arbitraire. Alors il faut que le défaut soit égal à 2.

Exemple 5. — *Équation du quatrième ordre*

$$\frac{\partial^4 y}{\partial t^4}=\frac{\partial^4 y}{\partial x^4}. \tag{47}$$

Nous prenons $Y=L_2(-\infty, \infty)$, $U=\frac{\tilde{d}}{dx}\cdot$ Alors U engendre le semi-groupe de Poisson (7).

$$P(t)[f]=\frac{t}{\pi}\int_{-\infty}^{\infty}\frac{f(x+\xi)\,d\xi}{\xi^2+t^2}. \tag{48}$$

La transformation de Fourier nous donne le semi-groupe isométrique plus simple

$$P^*(t)[F]=e^{-t|s|}F(s). \tag{49}$$

Évidemment $P^*(t)$ est holomorphe dans $\Re(t)>0$ et continu pour $\Re(t)\geqq 0$, d'où il résulte que $P(t)$ a les mêmes propriétés. Alors on trouve

$$P(i\tau)[f]=\frac{1}{2}[f(x+\tau)+f(x-\tau)]+\frac{i}{2}[\tilde{f}(x+\tau)-\tilde{f}(x-\tau)] \tag{50}$$

en calculant la fonction inverse de $e^{-i\tau|s|}F(s)$. Ici τ est réel, et $\|P(i\tau)\|=\|P^*(i\tau)\|=1$. Les opérateurs $\{P(i\tau)|-\infty<\tau<\infty\}$ forment un groupe, avec $P(0)=I$, fortement continu. Alors le théorème 7 donne $n=4$, $d\leqq 1$ et la solution du problème réduit

(7) Pour ce semi-groupe voir [1] p. 385. Il y a plus de détails dans [2] où la discussion s'applique à l'espace L_p, $1<p<\infty$.

correspondant s'obtient des formules (34) et (36) :

$$y(t) = P(t)z_0 + P(it)z_1 + P(-it)z_{-1}, \tag{51}$$

où

$$z_0 = \frac{1}{2}[y_0 + v_2], \quad Uv_1 = y_1,$$

$$z_1 = \frac{1}{4}(1+i)y_0 - \frac{1}{2}iv_1 - \frac{1}{4}(1-i)v_2, \quad U^2v_2 = y_2, \tag{52}$$

$$z_{-1} = \frac{1}{4}(1-i)y_0 + \frac{1}{2}iv_1 - \frac{1}{4}(1+i)v_2.$$

On objectera, peut-être, que, en remplaçant l'équation (47) par

$$\left(\frac{\partial}{\partial t}\right)^4 y = \left(\frac{\tilde{\partial}}{\partial v}\right)^4 y, \tag{53}$$

comme nous l'avons fait, on perdra des solutions et qu'il faut aussi considérer l'opérateur $U = d/dx$ et le groupe (43) engendré par lui. Cela revient à dire que la solution générale de l'équation des ondes est une solution de (47) en supposant les données suffisamment dérivables. Mais on peut écrire (44) sous la forme

$$\frac{1}{2}P(it)[y_0 + iY_1] + \frac{1}{2}P(-it)[y_0 - iY_1], \qquad \frac{\tilde{d}}{dx}Y_1 = y_1, \tag{54}$$

ce qui est seulement un cas spécial de (51). Alors le groupe $T(t|d/dx)$ n'apportera rien de nouveau au problème de Cauchy pour (47). Ce résultat négatif rendra peut-être la conjecture $d = 1$ plus vraisemblable. Remarquons en passant que la solution générale de l'équation de Laplace satisfait aussi à (47), c'est le terme $P(t)z_0$ de (51). Alors, le premier terme de cette somme vient de l'équation de Laplace, tandis que les autres viennent de l'équation des ondes, et, si $d = 1$ pour l'équation (47) cela découlerait du fait que $d = 1$ pour l'équation de Laplace.

Il est assez facile de donner des exemples où interviennent des opérateurs non-différentiels mais ce qui précède suffit pour illustrer la méthode.

BIBLIOGRAPHIE

[1] HILLE. Functional Analysis and Semi-Groups. *American Mathematical Society, Colloquium Lectures XXXI*, New-York, 1948.

[2] HILLE. On the Generation of Semi-Groups and the Theory of Conjugate Functions. *Proc. R. Physiographical Society, Lund*, t. 21, nº 14, 1951, 130-142.

[3] HILLE. A Note on Cauchy's Problem. *Annales de la Société Polonaise de Mathématiques*, t. 25, 1952, 13 p.

(Parvenu aux Annales le 8 octobre 1952.)

SOME ASPECT OF CAUCHY'S PROBLEM [1])

EINAR HILLE

1. *Introduction.*

The subject matter of this paper is closely related to that of Professor K. Yosida's hour addres to which I refer for back ground, further problems, and literature. Professor Yosida having kindly placed a copy of his manuscript at my disposal, I shall try to avoid topics which are covered in his addres, but naturally semi-group theory will crop up steadily in my talk and I shall have to discuss the one dimensional case of the temporally homogeneous diffusion equations.

2. *The abstract Cauchy problem.*

In a number of publications (see [7] — [12] in the References at the end of this paper) the author has proposed and discussed an abstract Cauchy problem which may be formulated as follows:

ACP^1. *Given a complex (B)-space X and a linear operator U with domain D[U] and range R[U] in X and given an element* $y_0 \in X$, *find a function* $y(t) = y(t; y_0)$ *such that*

(i) $y(t)$ *is strongly absolutely continuous and continuously differentiable in each finite subinterval of* $[0, \infty)$ [*or* $(0, \infty)$];

(ii) *for each* $t > 0$, $y(t) \in D[U]$ *and*

$$U[y(t)] = y'(t); \tag{2.1}$$

(iii) $\lim_{t \to 0+} ||y(t; y_0) - y_0|| = 0.$

This is the formulation given by R. S. Phillips [14] and differs somewhat from earlier versions of mine in the phrasing of condition (i), but it has certain advantages, in particular, it leads to a much stronger form of Theorem 4 below than I was able to prove. The reader will note that there are two alternatives in condition (i); the corresponding problems will be denoted by ACP_1^1 and ACP_2^1 respectively. The symbol ACP^n will be used in section 8 in referring to corresponding problems for the equation

$$U^n[y(t)] = y^{(n)}(t). \tag{2.2}$$

[1]) This paper was written under the sponsorship of the Air Research and Development Command [Contract AF 18 (600)—1127]. Credit should also be given to the National Science Foundation whose generous support made it possible for the author to deliver the paper in person.

These problems are evidently suggested by the classical Cauchy problem especially for diffusion equations, where U is a second order differential operator, and so far the most important applications have been to the integration problem of these equations. But in principle the considerations apply to much wider classes of differential and functional equations (see for instance Joanne Elliott [1]). In these applications the space X is one of the common function spaces of analysis, C or L_p, with the usual metric. R. S. Phillips (personal communication) has pointed out that in such cases numerically-valued representations can be found for the abstract valued function $y(t)$ so that an abstract solution becomes a solution of the differential equation in the ordinary sense.

3. *Uniqueness and non-uniqueness.*

If for a given y_0 and a given operator U the corresponding ACP^1 has two distinct solutions their difference is a *nul solution* $y(t; 0)$. The existence of such solutions is regulated by the two following theorems (see E. Hille [11, pp. 95—97]) where a solution is said to be of *normal type* ω if

$$\limsup_{t\to+\infty} t^{-1} \log \|y(t)\| = \omega < \infty. \tag{3.1}$$

Theorem 1. *For each $y_0 \in X$ the ACP^1 has at most one solution of normal type if U is a closed operator whose point spectrum is not dense in any right half-plane.*

Theorem 2. *If U is closed a necessary and sufficient condition that the ACP^1 shall have a nul solution of normal type $\leqq \omega$ is that the characteristic equation*

$$U[x(\lambda)] = \lambda\, x(\lambda) \tag{3.2}$$

have a solution $x(\lambda) \not\equiv 0$, bounded and holomorphic in each halfplane $\Re(\lambda) \geqq \omega + \varepsilon$, $\varepsilon > 0$.

Since a solution of (3.2) may be multiplied by any numerically-valued bounded holomorphic function, we may assume that $(1 + |\nu|)^3 \|x(\omega + \varepsilon + i\nu)\| \in L(-\infty, \infty)$. The desired nul solution is then given by

$$y(t) = \frac{1}{2\pi i} \int_{\gamma-i\infty}^{\gamma+i\infty} e^{\lambda t} x(\lambda)\, d\lambda, \quad \gamma > \omega. \tag{3.3}$$

If the point spectrum of U covers the whole complex plane, "*explosive*" solutions may exist. Thus if X is an L-space, if $x(\lambda)$ is an entire function of order < 1, positive for $\lambda < 0$, while $\|x(\lambda)\| \notin L(0, \infty)$, then to every real t_0 there exist solutions of (2.1) which are entire functions of $(t - t_0)^{-1}$ but belong to X for every $t \neq t_0$. Here $\|y(t)\| \to +\infty$ as $t \to t_0 -$ but may very well tend to zero as $t \to t_0 +$. Cf. [6] and [11, pp. 97—99].

4. *Applications to the heat equation.* Theorem 2 may be used to construct nul solutions of the ordinary heat equation of a nature different from those found by A. Tychonoff. Here (2.1) and (3.2) take the form

$$\frac{\partial^2 y}{\partial s^2} = \frac{\partial y}{\partial t}, \quad \frac{d^2 x}{ds^2} = \lambda x \tag{4.1}$$

respectively. We consider the interval $(-\infty, \infty)$ and the L-space of measurable function $f(s)$ with finite norm defined by

$$(4.2) \qquad \| f \|_{\varrho} = \int_{-\infty}^{\infty} | f(s) | \exp(- | s |^{\varrho})\, ds$$

where $\varrho \geqq 0$. For $0 \leqq \varrho \leqq 2$ one finds that the characteristic equation has no solution which is holomorphic and bounded in a right half-plane but for $\varrho > 2$ suitable multipliers may be found and nul solutions are constructed with the aid of (3.3). For $\varrho > 2$ we have also explosive solutions belonging to the space, an example being given by the classical integral

$$\frac{1}{\sqrt{\pi}} \int_0^{\infty} e^{\lambda(t_0-t)} ch(\sqrt{\lambda} s)\, d\lambda = \frac{1}{t_0 - t} \sum_{n=0}^{\infty} \frac{1}{\Gamma(n + \frac{1}{2})} \left[\frac{s^2}{4(t_0-t)} \right]^n .$$

5. *Relations to semi-group theory.*

Let $\{S(t);\ 0 < t\}$ be a semi-group of linear bounded transformations on X to itself so that $S(t_1 + t_2) = S(t_1)\, S(t_2)$ for $t_1, t_2 > 0$. We assume $S(t)$ to be continuous in the strong operator topology for $t > 0$. In this case $S(t)$ is of normal type in the sense of (3.1). A semi-group is said to be of class (C_0) if

$$(5.1) \qquad \lim_{t \to 0+} \| S(t)x - x \| = 0$$

for each $x \in X$. $S(t)$ is of class $(0, A)$ in the sense of R. S. Phillips if $\int_0^1 \| S(t)x \| dt < \infty$ for each $x \in X$ and if the linear bounded operator

$$(5.2) \qquad R(\lambda)x \equiv \int_0^{\infty} e^{-\lambda t}\, S(t)x\, dt,$$

defined for $\Re(\lambda) > \omega$, has the property

$$(5.3) \qquad \lim_{t \to +\infty} \| \lambda\, R(\lambda)x - x \| = 0$$

for each x. For semi-groups of both classes

$$(5.4) \qquad \lim_{t \to 0+} t^{-1}[S(t)x - x] \equiv A_0 x$$

exists for x in a dense set $D[A_0]$. A_0 is called the *infinitesimal operator* of $S(t)$; A_0 is *closed* if $S(t)$ is of class (C_0), otherwise A_0 has a smallest closed extension A called the *infinitesimal generator of* $S(t)$. In either case the resolvent $R(\lambda; A)$ exists and $R(\lambda; A) = R(\lambda)$ for $\Re(\lambda) > \omega$. If $S(t)$ is of class $(0, A)$ relation (5.1) holds for $x \in D[A]$. Further

$$(5.5) \qquad \frac{d}{dt}[S(t)x] = A\, S(t)\, x = S(t)\, A\, x, \qquad x \in D[A].$$

From these relations one obtains, see R. S. Phillips [14]:

Theorem 3. *If U is the infinitesimal generator of a semi-group $\{S(t)\}$ of* *class* $(0, A)$, *then the corresponding* ACP_2^1 *has a unique solution* $y(t; y_0) \equiv S(t)[y_0]$ *for each* $y_0 \in D[U]$.

Phillips has also proved the converse proposition:

Theorem 4. *Let* U *be a closed operator with dense domain and non-vacuous resolvent set* [*and* $\lambda||R(\lambda; U)|| = O(1)$ *as* $\lambda \to +\infty$]. *Suppose that for each* $y_0 \in D[U]$ *there is a unique solution to* ACP_1^1 [*to* ACP_2^1]. *Then* U *generates a semi-group* $\{S(t)\}$ *of class* (C_0) [*of class* $(0, A)$] *and* $y(t; y_0) = S(t)[y_0]$ *for all* $y_0 \in D[U]$.

For the case in which the ACP^1 corresponds to a genuine physical problem. Theorem 4 gives the sharpest known formulation of the implications of what J. Hadamard once called the *major premise of Huygens' principle.*

The two theorems show that solving the ACP^1 for the operator U and finding if U generates a semi-group are practically equivalent problems. In view of this situation there is evidently an urgent call for criteria enabling one to recognize whether or not a given operator U is the infinitesimal generator of a semi-group $\{S(t)\}$. For the investigations of Hille, Yosida, and Phillips on this question we refer to Yosida's address [16]. It is evident that U must be closed, its domain dense, and its resolvent $R(\lambda; U)$ must exist in a half-plane. Formula (5.2) shows that a necessary condition is that $R(\lambda ; U)\, x$ be the Laplace transform of $S(t)\, x$, where $S(t)$ is strongly continuous for $t > 0$, is of normal type, and satisfies (5.1) for $x \in D[U]$. Conversely, if this condition is satisfied one verifies easily that $S(t) = S(t; U)$ is the semi-group generated by U and $S(t)$ is of class $(0, A)$. This fact serves to vindicate the time-honored device of applying the Laplace transformation to the characteristic equation.

It may very well happen, however, that a given operator U cannot generate a semi-group of class $(0, A)$. Before giving up this mode of attack as useless, one had better look into the reason for failure. A common reason for failure is that the spectrum of U covers the right half-plane. If this is the case, some restriction U_0 of U may still have all the desirable properties and the theory should be built around U_0 instead. This simple observation underlies the work of W. Feller in [2] of which more below.

6. *Diffusion equations with one degree of freedom.*

The first partial differential equations with variable coefficients to which the newly created theory of semi-groups was applied was the Fokker-Planck equation of one degree of freedom, studied by K. Yosida in [15]. Reading his work in manuscript led me to the investigations [4, 5, 11] on diffusion equations as well as to the work on the abstract Cauchy problem. Indirectly it also started Feller on his basic work in [2]. Yosida proceeded to develop the theory of diffusion equations with more degrees of freedom and for these investigations the reader is referred to his address. My results and those of Feller will be briefly summarized below.

Following the notation of [11] we write the formally adjoint diffusion equations

$$C[S] \equiv b(\xi)\frac{\partial^2 S}{\partial \xi^2} + a(\xi)\frac{\partial S}{\partial \xi} = \frac{\partial S}{\partial t}, \tag{6.1}$$

$$L[T] \equiv \frac{\partial}{\partial x}\left\{\frac{\partial}{\partial x}[b(x)T] - a(x)T\right\} = \frac{\partial T}{\partial t}, \tag{6.2}$$

where $a(x)$, $b(x)$ are continuous in (α, β) and $b(x) > 0$. We assume $-\infty \leqq \alpha < 0 < \beta \leqq +\infty$. Feller's notation is different, in particular, $a(x)$ and $b(x)$ are interchanged, and his equations involve terms in $c(\xi)S$ and $c(x)T$. We introduce the following auxiliary functions

$$W(x) = \exp\left\{-\int_0^x \frac{a(s)}{b(s)}\,ds\right\},\quad q(x) = \int_0^x [b(s)\,W(s)]^{-1}\,ds, \tag{6.3}$$

$$W_1(x) = \int_0^x W(\xi)\,d\xi,\ W_2(x) = \int_0^x W_1(s)\,dq(s),\ \Omega(x) = \int_0^x [W_1(x) - W_1(s)]dq(s).$$

We are concerned with the Cauchy problems for the operator C in $C(\alpha, \beta]$ and the operator L in $L(\alpha, \beta)$. The corresponding characteristic equations

$$C[u] - \lambda u = 0,\quad L[y] - \lambda y = 0 \tag{6.4}$$

are ordinary differential equations and for the decision problem it suffices to take λ positive. It turns out that C will generate a semi-group in $C[\alpha, \beta]$ if and only if the first equation under (6.4) has no solution in $C[\alpha, \beta]$ and similarly for L. The resulting semi-groups are made up of positive contraction operators. In terms of the functions (6.3) we can state the result as

Theorem 5. *The Cauchy problem for* (6.1) *in* $C[\alpha, \beta]$ *has a unique solution for every* $y_0 \in D[C]$ *if and only if* $\Omega(\alpha) = \Omega(\beta) = +\infty$. *The Cauchy problem for* (6.2) *in* $L[\alpha, \beta]$ *has a unique solution for every* $y \in D[L]$ *if and only if* $W_2(\alpha) = W_2(\beta) = +\infty$. *The solution of* (6.2) *is defined by a transition operator if and only if both conditions are satisfied.*

Equivalent results have been found by Feller, but the latter's main contribution to this problem is the study of the cases in which C or L or both fail to generate semi-groups. This situation arises when the point spectrum of the operator in question covers the whole complex plane. It then becomes necessary to restrict the operator so that its point spectrum releases the positive real axis. This is done by imposing *lateral conditions* and Feller set himself the task of determining all lateral conditions leading to semi-groups of positive contraction operators. This he based upon a classification of the boundaries into four types. In our notation the point β is

a *regular* boundary if $\Omega(\beta) < \infty$, $W_2(\beta) < \infty$,
an *exit* boundary if $\Omega(\beta) < \infty$, $W_2(\beta) = \infty$,
an *entrance* boundary if $\Omega(\beta) = \infty$, $W_2(\beta) < \infty$,
a *natural* boundary if $\Omega(\beta) = \infty$, $W_2(\beta) = \infty$.

Lateral conditions have to be prescribed if at least one boundary is not natural. If both boundaries are regular, the lateral conditions are immediate generalizations of the classical boundary conditions, but if exit or entrance boundaries occur great complications are possible. Thus for an exit boundary at β and a natural boundary at α, the solution $S(t; x)$ of (6.1) has to satisfy a condition of the form

$$p_2 S(t; \beta) = p_1 S(t; \alpha) + \tau \int_\alpha^\beta S(t; x)\, dp(x) - \sigma \lim_{x \to \beta} \left[b(x) \frac{\partial^2 S}{\partial x^2} + a(x) \frac{\partial S}{\partial x} \right]$$

where p_1, p_2, σ, τ are given constants and $p(x)$ a given function of bounded variation, all subject to certain conditions. In this case the true adjoint of (6.1) is no longer (6.2) but a more general type of functional equation.

In spite of the formidable nature of the lateral conditions, the solutions of (6.1) and (6.2) appear to have simpler properties when the conditions of Theorem 5 are not satisfied than when they are. For the case of two entrance boundaries Hille [11] has carried through a detailed analysis of the unique solution $S(t)[g]$ of (6.1). Here C has a pure point spectrum $\{\lambda_n\}$, $\lambda_0 = 0$, $\lambda_n < 0$, with corresponding characteristic functions $\omega_n(\xi)$ in terms of which the solution may be represented by the "theta series"

$$S(t)[g] = \sum_0^\infty g_n \omega_n(\xi)\, e^{\lambda_n t}, \quad g(\xi) \sim \sum_0^\infty g_n \omega_n(\xi). \tag{6.5}$$

Here $g(\xi)$ is any element of $C[\alpha, \beta]$ and $S(t)$ is analytic in $\Re(t) > 0$.

The reason for the relative simplicity of this case appears to be that the solutions of $C[u] = \lambda u$ tending to $+1$ as $\xi \to \alpha$ and β respectively, which are used in constructing the resolvent, are entire functions of λ and the characteristic values λ_n are simply the zeros of their Wronskian at $\xi = 0$, again an entire function of λ, usually of order $\frac{1}{2}$. If $a(x) \equiv 0$, $\alpha = -\infty$, $\beta = +\infty$, this case will arise if and only if

$$x[b(x)]^{-1} \in L(-\infty, \infty). \tag{6.6}$$

Cases of two regular boundaries with generalized classical boundary conditions have also been examined by Hille [11] with similar results. The case of exit boundaries has been considered by Joanne Elliot [see Trans. Amer. Math. Soc., 78 (1955) 406—425; added in proof]. Here again theta series arise.

7. *Equations of higher order with coincident characteristics.* The attack on partial differential equations of higher order than the second has not proceeded very far. Nevertheless the Cauchy problem for the equations

$$L_{2n}[T] \equiv \frac{\partial^n}{\partial x^n} \left\{ b(x) \frac{\partial^n T}{\partial x^n} \right\} = (-1)^{n-1} \frac{\partial T}{\partial t}, \quad b(x) > 0, \tag{7.1}$$

in $L(-\infty, \infty)$ seems to be within grasp, at least for the case

$$x^{2n-1}[b(x)]^{-1} \in L(-\infty, \infty) \tag{7.2}$$

which is the analogue of (6.6). Here again the n solutions of $L_{2n}[y] + (-1)^n\lambda y = 0$ which are in $L(-\infty, 0)$ and the n solutions in $L(0, \infty)$ used in constructing the resolvent, are entire functions of λ, apparently of order $1/(2n)$. The Wronskian at the origin is also an entire function of low order and all its zeros are real and negative. One would expect L_{2n} to have a pure point spectrum and the solutions of (7.1) to be given by theta series analogous to (6.5). In this discussion one has to use the known results for self-adjoint operators in $L_2(-\infty, \infty)$ (see K. Kodaira [13] and I. M. Glazman [3]) which apply since the resolvents involve the same solutions in both spaces.

For the case $n = 2$ some primary results have been found by J. L. Howell [unpublished Yale dissertation]. Among other things he proved, without imposing integrability conditions on $b(x)$, that the equation $L_4[y] + \lambda y = 0$ has no solutions in $L(-\infty, \infty)$ or $L_2(-\infty, \infty)$.

8. *Cauchy problems of higher order.* The problem formulated in section 2 may be generalized as follows:

ACP^n. *Given a complex (B)-space X and a linear operator U with domain D[U] and range R[U] in X and given n elements* $y_0, y_1, \ldots, y_{n-1}$ *in X, find a function* $y(t) = y(t; y_0, y_1, \ldots, y_{n-1})$ *such that*

(i) $y(t)$ *is strongly absolutely continuous and n times continuously differentiable in each finite subinterval of* $[0. \infty)$;

(ii) *for each* $t > 0$, $y(t) \in D[U^n]$ *and*

$$U^n[y(t)] = y^{(n)}(t); \tag{8.1}$$

(iii) $\lim_{t\to 0+} ||y^{(k)}(t; y_0, y_1, \ldots, y_{n-1}) - y_k|| = 0,\ k = 0, 1, \ldots, n-1.$

For the theory of this problem see [8, 9]. There are uniqueness theorems and theorems on nul solutions analogous to Theorems 1 and 2 above. For $n = 2$ there is an interesting analogue of Theorem 3:

Theorem 6. *If* $n = 2$ *and U generates a group* $\{T(t; U);\ -\infty < t < \infty\}$, *strongly continuous at the origin with* $T(0; U) = I$, *then the corresponding* ACP^2 *has a unique solution for each* $y_0 \in D[U^2]$ *and* $y_1 \in D[U] \cap R[U]$, *namely*

$$y(t) = \tfrac{1}{2}[T(t; U)(y_0 + z_1) + T(-t; U)(y_0 - z_1)],\ Uz_1 = y_1. \tag{8.1}$$

This formula reduces to the solution of the equation for the vibrating string when $U = d/ds$.

For $n > 2$ the ACP^n does not seem to be solvable for arbitrary initial values. If U generates a semi-group $T(t; U)$, holomorphic in a sector having its vertex at the origin and if $T(t; U)$ is strongly continuous on the boundary, then one can satisfy $m \leqq [\frac{1}{2}n] + 1$ of the conditions (iii) if the closed sector contains m n^{th} roots of unity, but the remaining $n - m$ conditions are no longer at our disposal.

References

[1] Joanne Elliott, The boundary value problems and semi-groups associated with certain integro-differential equations. Trans. Amer. Math. Soc., 76 (1954) 300—331.

[2] W. Feller, The parabolic differential equations and the associated semi-groups of transformations. Ann. of Math., (2) 55 (1952) 468—519.

[3] I. M. Glazman, On the theory of singular differential equations (Russian). Uspehi Mat. Nauk, (N.S.) 5, no. 6 (40) (1950) 102—135. Amer. Math. Soc. Translation No. 96 (1953).

[4] E. Hille, On the integration problem for Fokker-Planck's equation in the theory of stochastic processes. Onzième Congrès des Math. Scandinaves, Trondheim 1949, (1952) 183—194.

[5] E. Hille, Les probabilités continues en chaîne. Comptes Rendus Acad. Sci. Paris, 230 (1950) 34—35.

[6] E. Hille "Explosive" solutions of Fokker-Planck's equation. Proc. Internat. Congres Math. Cambridge, Mass. 1950, I. p. 435.

[7] E. Hille, A note on Cauchy's problem. Ann. Soc. Polonaise de Math., 25 (1952) 56—68.

[8] E. Hille, Une généralisation du problème de Cauchy. Ann. Inst. Fourier, 4 (1953) 31—48.

[9] E. Hille, Sur le problème abstrait de Cauchy. Comptes Rendus Acad. Sci. Paris, 236 (1953) 1466—1467.

[10] E. Hille, Le problème abstrait de Cauchy, Rendiconti Sem. Mat. Università di Torino, 12 (1953) 95—103.

[11] E. Hille, The abstract Cauchy problem and Cauchy's problem for parabolic differential equations. Journ. d'Analyse Math., 3 (1954) 81—196, Corrigenda.

[12] E. Hille, An abstract formulation of Cauchy's problem. Proc. Twelfth Congress Scand. Math. Lund 1953, (1954) 79—89.

[13] K. Kodaira, On ordinary differential equations of any even order and the corresponding eigenfunction expansions. Amer. Journal of Math. 72 (1950) 502—544.

[14] R. S. Phillips, A note on the abstract Cauchy problem. Proc. Nat. Acad. Sci. U.S.A., 40 (1954) 244—248.

[15] K. Yosida, An operator theoretical treatment of temporally homogeneous Markoff process. Journal Math. Soc. Japan, 1 (1949) 244—363.

[16] K. Yosida, Semi-group theory and the integration problem of diffusion equations. These Proceedings, I.

[Added in proof: For further development of this Cauchy problem see § 23.3 of E. Hille and R. S. Phillips, Functional Analysis and Semi-Groups, Amer. Math. Soc. Coll., Vol. XXXI, second edition, to appear. When $n > 2$ the existence of a unique solution of the ACP^n for certain restricted classes of prescribed data implies that U is a bounded operator.]

Yale University.
New Haven, Conn.,
U.S.A.

Chapter 15
Differential Equations in B-Algebras

The papers included in this chapter are

LINEAR DIFFERENTIAL EQUATIONS IN BANACH ALGEBRAS*

EINAR HILLE
Yale University, New Haven

1. Introduction.

The classical theory of linear differential equations in the complex domain is concerned with equations of the form

$$w'(\zeta) = f(\zeta)\, w(\zeta). \tag{1.1}$$

Here ζ is a complex variable, $f(\zeta)$ and $w(\zeta)$ are n by n matrices of analytic functions, $f(\zeta)$ is given and it is desired to find the solution $w(\zeta)$. The main local problems are the study of the solutions in a neighbourhood of a point $\zeta = \zeta_0$ where either $f(\zeta)$ is holomorphic or $(\zeta - \zeta_0)^k f(\zeta)$ has this property. In the latter case we have a singularity at $\zeta = \zeta_0$ which is regular singular if $k = 1$ otherwise irregular singular. The rate of growth and the analytic properties of the solutions are totally different in these two cases.

The set of n by n matrices over the complex field is a non-commutative algebra with unit element. A norm is easily found in terms of which this set becomes a Banach algebra. This suggests an immediate generalization of the classical problem. Suppose that B is a non-commutative (B)-algebra over the complex field with unit element e. Suppose that $f(\zeta)$ is a B-valued analytic function of the complex variable ζ. What are now the properties of the solutions of (1.1): (i) in a simply-connected domain of holomorphism of $f(\zeta)$, (ii) in the neighbourhood of a simple pole of $f(\zeta)$, and (iii) in a partial neighbourhood of a multiple pole?

Case (i) is elementary and the classical existence theorems carry over without any difficulties. This is well known. Less well known, perhaps, is the fact that a priori estimates for the rate of growth may be obtained in cases (ii) and (iii) which are exact analogues of the classical ones. In case (ii) the Fuchs–Frobenius technique applies up to a certain point, but since the spectral

* This research was supported by the United States Air Force through the Office of Scientific Research of the Air Research and Development Command under Contract No. SAR/AF 49–(638) 224.

properties of elements of B may be very much more complicated than those of finite matrices this mode of approach breaks down in the general case.

We concentrate on case (ii) and start the discussion with a review of some properties of commutators.

2. Commutators.

Suppose that $c \in B$ but is not in the centre of B. Then

$$T(c)[x] = cx - xc \tag{2.1}$$

defines a linear transformation on B to B which is bounded and whose norm does not exceed $2\|c\|$. We refer to $T(c)$ as the c-commutator or simply the commutator if the value of c is clear from the context.

We denote the spectrum of an element of B or of an operator on B to itself by the symbols $\sigma(a)$ and $\sigma(T)$ respectively. We denote resolvents by the letter R so that

$$R(\lambda; T) = (\lambda I - T)^{-1}$$

when the inverse operator exists.

We have

$$\sigma[T(c)] \subset \{\alpha - \beta \mid \alpha \in \sigma(c), \beta \in \sigma(c)\} \equiv \Delta. \tag{2.2}$$

In particular, 0 is always in $\sigma[T(c)]$. Suppose that B is a prime ring, that is,

$$xBy = 0 \quad \text{implies either } x = 0 \text{ or } y = 0. \tag{2.3}$$

In this case it is possible to make more precise statements about $\sigma[T(c)]$. Suppose there are elements x_α and x_β in B, not 0, such that

$$cx_\alpha = \alpha x_\alpha, \quad x_\beta c = \beta x_\beta,$$

and suppose that B is a prime ring, then there exists a z, $z \neq 0$, such that

$$cz - zc = (\alpha - \beta)z.$$

An element γ of $\sigma[T(c)]$ can be the difference of spectral values of c in infinitely many ways. Suppose there is only a finite number so that $\gamma = \alpha_j - \beta_j$, $j = 1, 2, \ldots, k$, and that α_j and β_j are poles of $R(\lambda; c)$ of order μ_j and ν_j respectively. Then γ is a pole of $R(\lambda; T(c))$ of order $\leqq \max(\mu_j + \nu_j - 1)$. If B is a prime ring, equality holds. For these spectral properties, see S. R. Foguel [3].

There are a number of representations available for the resolvent $R(\lambda;T(c))$. We mention two only. Suppose that α is an isolated point of $\sigma(c)$ and suppose that e_α and q_α are the corresponding idempotent and quasi-nilpotent elements respectively. Here e_α is the residue of $R(\lambda;c)$ while q_α is the residue of $(\lambda-\alpha)R(\lambda;c)$. Suppose that $x \in Be_\alpha$ and that $\alpha+\lambda$ belongs to the resolvent set of c. Then

$$R(\lambda;T(c))[x] = \sum_{m=0}^{\infty}(-1)^m[R(\alpha+\lambda;c)]^{m+1}xq_\alpha^m. \tag{2.4}$$

If α is a pole of $R(\lambda;c)$, q_α is nilpotent so the series breaks off after a finite number of terms. In any case it is rapidly convergent.

The second representation is by formulas of the Cauchy type. I am indebted to M. G. Krein for this information. Credit should be given to S. G. Krein and Yu. L. Daletsky [2]. Suppose that the point sets $\Delta-\lambda$, Δ, and $\Delta+\lambda$ have no points in common. Then there exists a rectifiable contour Γ which contains Δ in its interior while $\Delta-\lambda$ and $\Delta+\lambda$ are exterior to Γ. Γ may conceivably consist of several simple closed curves. Then

$$R(\lambda;T(c))[x] = \frac{1}{(2\pi i)^2}\int_\Gamma\int_\Gamma \frac{R(\alpha;c)\,x\,R(\beta;c)}{\lambda-\alpha+\beta}\,d\alpha\,d\beta, \tag{2.5}$$

where the integral exists in the Riemann–Graves sense.

3. Regular points.

Let $f(\zeta)$ be holomorphic in a simply-connected domain Δ and consider the differential equation

$$w'(\zeta) = f(\zeta)\,w(\zeta). \tag{3.1}$$

Let $\zeta_0 \in \Delta$ and let $w_0 \in B$. It is well known that the equation has a unique solution $w(\zeta;\zeta_0,w_0)$ which is holomorphic in Δ and tends to w_0 when $\zeta \to \zeta_0$. Such a solution can be constructed by the usual method of successive approximations.

If Δ is not simply-connected, the value of the solution at a point ζ will in general depend upon the path from ζ_0 to ζ. In any case, the solution can be continued indefinitely in Δ and the various resulting determinations of $w(\zeta)$ are locally holomorphic.

We note that

$$w(\zeta;\zeta_0,w_0) = w(\zeta;\zeta_0,e)\,w_0. \tag{3.2}$$

Here $w(\zeta;\zeta_0,e)$ is a *regular* element of the algebra B, that is, it has an inverse in B. This is certainly true if $|\zeta-\zeta_0|$ is so small that $\|w(\zeta;\zeta_0,e)-e\| < 1$.

But this implies that every determination of $w(\zeta;\ \zeta_0,\ e)$ is regular in Δ for we have

$$w(\zeta;\zeta_0,e) = w(\zeta;\zeta_1,e)\,w(\zeta_1;\zeta_0,e) \tag{3.3}$$

by an obvious extension of (3.2). Here ζ_1 is an arbitrary point on the path of integration leading from ζ_0 to ζ. By intercalating sufficiently many points on the path we can represent $w(\zeta;\ \zeta_0,\ e)$ as the product of regular elements whence it follows that the product is also regular in B. This shows that $w(\zeta;\ \zeta_0,\ w_0)$ is a regular element of B if and only if w_0 is regular. An algebraically regular solution is called *fundamental.*

4. A priori estimates at singular points.

We shall now be concerned with the rate of growth of the solution when ζ approaches a singular point of $f(\zeta)$ which we may place at the origin. Suppose then that $f(\zeta)$ is holomorphic in a sector Σ with $\zeta = 0$ as a vertex. Suppose that ζ_0 and ζ are on the same ray in Σ. Then

$$w(\zeta) = w_0 + \int_{\zeta_0}^{\zeta} f(\sigma)\, w(\sigma)\, d\sigma, \tag{4.1}$$

where the integral exists in the Riemann-Graves sense. This implies

$$\| w(\zeta) \| \leqq \|w_0\| + \int_{\zeta}^{\zeta_0} \|f(\sigma)\|\ \|w(\sigma)\|\ |d\sigma|, \tag{4.2}$$

and the estimate

$$\| w(\zeta) \| \leqq \|w_0\| \exp\left\{\int_{\zeta}^{\zeta_0} \|f(\sigma)\|\ |d\sigma|\right\}. \tag{4.3}$$

See N. Bourbaki [1] p. 10. Since

$$w_0 = w(\zeta_0;\zeta, w(\zeta)),$$

we can interchange initial and terminal values and obtain also

$$\|w_0\| \exp\left\{-\int_{\zeta}^{\zeta_0} \|f(\sigma)\|\ |d\sigma|\right\} \leqq \|w(\zeta)\|. \tag{4.4}$$

Here and above it is not necessary that we integrate radially.

Suppose now that $\|f(\rho e^{i\theta})\|$ can be integrated down to 0 and that

$$\sup \int_0^{\alpha} \|f(\rho e^{i\theta})\|\, d\rho = A(\Sigma_0) < \infty, \tag{4.5}$$

where the supremum refers to an arbitrary interior sector Σ_0 of Σ. In this case the estimates show that $\|w(\zeta)\|$ is bounded away from 0 and ∞ in Σ_0. Referring back to (4.1), we conclude that

$$\lim_{\zeta \to 0} w(\zeta; \zeta_0, w_0) \equiv w_1, \tag{4.6}$$

exists for approach in Σ. A Phragmén–Lindelöf argument shows that the limit is unique and exists uniformly in every fixed interior sector Σ_0. Moreover, we can solve the initial value problem for the sector Σ. If $w_1 \in B$ is given, there exists a unique solution of (3.1) defined in Σ which approaches w_1 as $\zeta \to 0$.

The situation changes radically when $f(\zeta)$ is no longer integrable down to 0. The most important case is that in which $f(\zeta)$ has a pole at $\zeta = 0$ and this case splits into two subcases according as the pole is simple or multiple. If the pole is simple we have

$$\|w_0\| A^{-1} |\zeta|^M \leqq \|w(\zeta)\| \leqq \|w_0\| A |\zeta|^{-M} \tag{4.7}$$

where

$$M = \sup |\zeta| \, \|f(\zeta)\|, \quad |\zeta| < \rho,$$

and A is bounded provided $\arg \zeta$ is bounded.

For the case of a k-fold pole, $k > 1$, we obtain instead

$$\|w_0\| A^{-1} \exp\left\{-\frac{|\zeta|^{1-k}}{k-1}\right\} \leqq \|w(\zeta)\| \leqq \|w_0\| A \exp\left\{\frac{|\zeta|^{1-k}}{k-1}\right\}. \tag{4.8}$$

In analogy with the classical case, we speak of a *regular singular point* in the first case and of an *irregular singularity of rank* $k-1$ in the second.

5. The regular singular point, Case A.

We pass to a study of the differential equation (3.1) in the neighbourhood of the origin where $f(\zeta)$ shall have a simple pole. Suppose that

$$\zeta w'(\zeta) = \left\{\sum_{n=0}^{\infty} c_n \zeta^n\right\} w(\zeta), \quad c_0 \neq 0, \tag{5.1}$$

where the series converges for $|\zeta| \leqq \rho_0$. We try to satisfy the equation by a series of the form

$$w(\zeta) = \sum_{m=0}^{\infty} a_m \zeta^{c_0 + me}, \tag{5.2}$$

where the abstract power is defined by the operational calculus. A motivation for this substitution is the fact that

$$\zeta w'(\zeta) = c_0 w(\zeta) \quad \text{implies} \quad w(\zeta) = \zeta^{c_0} w_0.$$

Substitution of the series in (6.1) and combination of terms leads to the following system of equations

$$\begin{aligned} &c_0 a_0 - a_0 c_0 = 0, \\ (5.3) \qquad &m a_m - (c_0 a_m - a_m c_0) = \sum_{k=1}^{m} c_k a_{m-k}, \quad m = 1, 2, 3, \ldots. \end{aligned}$$

We can satisfy the first equation by setting

$$(5.4) \qquad a_0 = e.$$

In the notation of section **2** the second set of equations becomes

$$(5.5) \qquad [mI - T(c_0)]\, a_m = \sum_{k=1}^{m} c_k a_{m-k} \equiv b_m.$$

According to the character of the spectrum of c_0 we have three distinct possibilities:

Case A. No positive integer belongs to the spectrum of the commutator $T(c_0)$.

Case B. Some positive integers belong to the spectrum of $T(c_0)$ but they are all poles of $R(\lambda;\ T(c_0))$.

Case C. The spectrum of $T(c_0)$ contains positive integers which are not poles of $R(\lambda;\ T(c_0))$.

We proceed with a discussion of case A. According to section **2** this case presents itself when no two spectral values of c_0 differ by integers. This corresponds to the classical case of an ordinary linear differential equation of order n whose indicial equation at $\zeta = 0$ has roots which are not congruent modulo 1.

For $m = 1$ we get

$$a_1 = R(1;\ T(c_0))[c_1],$$

so that a_1 is uniquely determined. We can then proceed step by step and obtain $a_1, a_2, \ldots, a_m, \ldots$, each of which is uniquely determined. In particular, we note that a_m depends upon $c_0, c_1, \ldots, c_m$ and upon $a_1, a_2, \ldots, a_{m-1}$. If $m > 2\,\|c_0\|$ we have

$$\|a_m\| < \frac{\|b_m\|}{m - 2\|c_0\|}$$

On the basis of this estimate and the recurrence relations, it is not difficult to prove that the terms of the series (5.2) are bounded on the circle $|\zeta| = \rho_0$ whence it follows that the series is absolutely convergent inside the circle. For such values of ζ the product series and the derived series are absolutely convergent and may be rearranged ad lib. It follows that the formal series (5.2) is an actual solution of the equation (5.1). Since

$$w(\zeta) = \left[e + \sum_{m=1}^{\infty} a_m \zeta^m\right] \zeta^{c_0},$$

where the first factor is a regular element of B for small values of $|\zeta|$ and the power is regular in $0 < |\zeta| < \infty$, it follows that the solution is regular in $0 < |\zeta| < \rho_0$, no matter how it be continued in this punctured disk. Thus $w(\zeta)$ is a fundamental solution and we get the general solution by multiplying $w(\zeta)$ on the right by an arbitrary element w_0 of B. The resulting solution is regular if and only if w_0 has this property.

6. Case B, The method of Frobenius.

We shall now suppose that the positive integers $n_1, n_2, \ldots, n_p$ are poles of $R(\lambda; T(c_0))$ of order $\mu_1, \mu_2, \ldots, \mu_p$ respectively, and that no other positive integers belong to the spectrum of $T(c_0)$. We write

$$\mu_1 + \mu_2 + \ldots + \mu_p = N.$$

The assumption implies that we can find a positive number δ such that each of the circular disks

$$|\lambda - n| < \delta, \quad n = 1, 2, 3, \ldots, [2 \|c_0\|],$$

contains at most one point of the spectrum of $T(c_0)$. Suppose now that $0 < |\eta| < \delta$ and form the series

$$w(\zeta, \eta) \equiv \sum_{m=0}^{\infty} a_m(\eta) \zeta^{c_0 + (m+\eta)e}, \tag{6.1}$$

where the coefficients are functions of η to be determined. We substitute this series in the differential equation (5.1) and impose the following conditions on the coefficients:

$$\begin{aligned} c_0 a_0(\eta) &= a_0(\eta) c_0, \\ [(m+\eta)I - T(c_0)]\, a_m(\eta) &= \sum_{k=1}^{m} c_k a_{m-k}(\eta), \quad m = 1, 2, 3, \ldots \end{aligned} \tag{6.2}$$

The first of these equations we can satisfy by taking

$$a_0(\eta) = \eta^N e. \tag{6.3}$$

In view of the hypotheses governing the spectrum of $T(c_0)$, we can then solve all the equations (6.2) uniquely. If we assume

$$1 \leqq n_1 < \ldots < n_p,$$

we find that

$$a_k(\eta) = \eta^N b_k(\eta), \quad k = 1, 2, \ldots, n_1 - 1,$$

where $b_k(\eta)$ is a holomorphic function of η in $|\eta| < \delta$ and, normally,

$$\lim_{\eta \to 0} b_k(\eta) \neq 0.$$

Here and below we utilize the fact that $R(\lambda; T(c_0))$ is a bounded linear transformation on B to itself which is locally analytic in λ.

$R(\lambda; T(c_0))$ has a pole of order μ_1 at $\lambda = n_1$ and

$$R(n_1 + \eta; T(c_0)) = E_1 \eta^{-1} + \sum_{j=2}^{\mu_1} Q_1^{j-1} \eta^{-j} + \sum_{k=0}^{\infty} (-\eta)^k S_1^{k+1},$$

where E_1, Q_1, S_1 are three bounded operators of which E_1 is idempotent and Q_1 is nilpotent. The resolvent operator now acts on

$$\sum_{k=1}^{n_1} c_k a_{n_1 - k}(\eta) \equiv \eta^N s_{n_1}(\eta),$$

where $s_{n_1}(\eta)$ is holomorphic in $|\eta| < \delta$ and, in general, does not vanish at $\eta = 0$. The result is, by definition, $a_{n_1}(\eta)$ and we see that

$$a_{n_1}(\eta) = \eta^{N - \mu_1} b_{n_1}(\eta),$$

where $b_{n_1}(\eta)$ is holomorphic and, normally, $b_{n_1}(0) \neq 0$. We can then proceed and find that

$$a_k(\eta) = \eta^{N - \mu_1} b_k(\eta), \quad n_1 \leqq k < n_2.$$

At $\lambda = n_2$ the function $R(\lambda; T(c_0))$ has a pole of order μ_2. As a consequence

$$a_k(\eta) = \eta^{N - \mu_1 - \mu_2} b_k(\eta), \quad n_2 \leqq k < n_3,$$

and so on. The last pole of $R(\lambda; T(c_0))$ is found at $\lambda = n_p$ and the holomorphic function $a_k(\eta)$ normally does not tend to 0 when $\eta \to 0$ for $k \geqq n_p$.

The corresponding series (6.1) may be shown to converge for $0 < |\zeta| < \rho_0$, uniformly with respect to η for $|\eta| \leqq \delta_1 < \delta$. The corresponding function $w(\zeta,\eta)$ satisfies the differential equation

$$\zeta\frac{\partial}{\partial\zeta}w(\zeta,\eta) = \left\{\sum_{k=0}^{\infty} c_k\zeta^k\right\} w(\zeta,\eta) - \eta^{N+1}\zeta^{c_0+\eta e} \tag{6.4}$$

uniformly with respect to η in $|\eta| \leqq \delta_1$. It follows that

$$\lim_{\eta\to 0} w(\zeta,\eta)$$

is a solution of (5.1). The limit may be identically zero, however, and normally it is not a fundamental solution. We therefore resort to the time-honoured device of differentiating with respect to the parameter η. The terms of the series (6.1) are holomorphic functions of η and the series converges uniformly with respect to ζ and η. Hence the series defines a holomorphic function of both variables and may be differentiated termwise with respect to either variable as often as we please. Further, the operators $\partial/\partial\zeta$ and $\partial/\partial\eta$ commute. It follows that

$$\lim_{\eta\to 0} \frac{\partial^k}{\partial\eta^k} w(\zeta,\eta)$$

is a solution of (5.1) for $k = 0, 1,\ldots, N$. In particular the Nth derivative has this property. Carrying out the differentiations and collecting terms, we find that

$$W(\zeta) \equiv \sum_{j=0}^{N} \binom{N}{j} (\log\zeta)^{N-j} \sum_{m=0}^{\infty} a_m^{(j)}(0)\zeta^{c_0+me} \tag{6.5}$$

is a solution. Here the term of lowest order corresponds to $m = 0$, $j = N$ and equals

$$n!\,\zeta^{c_0}.$$

This shows that the solution is fundamental. This completes the discussion of case B.

7. Miscellaneous remarks.

In case C there is very little of significance that I can say. If it should happen that the coefficients

$$c_1, c_2,\ldots, c_k$$

commute with c_0, where k is the last integer which is a non-polar spectral value of $T(c_0)$, then the first k equations under (5.3) have trivial solutions and for $m > k$ we can then continue as above.

Further, we can always find a solution of the differential equation by the following device. Let k be the last integer in the spectrum of $T(c_0)$ and let $a_k \neq 0$ satisfy

$$c_0 a_k - a_k c_0 = k a_k. \tag{7.1}$$

We can then determine a solution of the form

$$\sum_{m=0}^{\infty} a_{m+k} \zeta^{c_0 + (m+k)e} \tag{7.2}$$

using the method under case A. The corresponding equations of type (5.3) can always be solved and the resulting series is absolutely convergent and hence an actual solution. It is not a fundamental solution, however, since an a_k satisfying (7.1) must be nilpotent.

It is always possible to reduce equation (5.1) to the form

$$\zeta W'(\zeta) = P(\zeta)\, W(\zeta), \tag{7.3}$$

where $P(\zeta)$ is a polynomial, by means of a substitution of the form

$$w(\zeta) = E(\zeta)\, W(\zeta),$$

where $E(0) = e$ and $E(\zeta)$ is holomorphic in $|\zeta| < \rho_0$. If the spectrum of $T(c_0)$ does not contain any integers, that is, in case A, we can take $P(\zeta) = c_0$ as reduced form of the equation. In general we can get rid of all powers of ζ except those whose exponents belong to $\sigma[T(c_0)]$.

We speak of the Fuchsian class if equation (3.1) has only regular singular points in the extended plane. For this to be the case, $f(\zeta)$ must be a rational function with simple poles which vanishes at infinity. In the hypergeometric case the equation is

$$w'(\zeta) = \left\{\frac{a_0}{\zeta} + \frac{a_1}{\zeta - 1}\right\} w(\zeta). \tag{7.4}$$

The theory of Kummer carries over to the abstract case, but it is not to be expected that the theory will be as rich as in the classical case.

We end with a remark concerning irregular singular points. The equation

$$w'(\zeta) = (b_0 \zeta + b_1)\, w(\zeta) \tag{7.5}$$

has an irregular singularity of rank one at $\zeta = \infty$. We can satisfy (7.5) by a Laplace transform

$$w(\zeta) = \int_L \exp(\zeta\sigma)\, u(\sigma)\, d\sigma. \tag{7.6}$$

Here $u(\sigma)$ is a solution of the differential equation

$$u'(\sigma) = R(\sigma; b_0)(b_1 - e)u(\sigma) \tag{7.7}$$

and the contour L should be so chosen that

$$(\sigma e - b_1)u(\sigma)\exp(\zeta\sigma)$$

tends to 0 at the endpoints for the values of ζ under consideration. If $R(\sigma; b_0)$ has only a finite number of spectral singularities each of which is a simple pole, then (7.7) has only regular singular points. If these present cases A or B, then the preceding theory gives a complete discussion of equation (7.7). We can then choose paths L and obtain the asymptotic representation of $w(\zeta)$ by divergent series in different sectors at infinity. The details are quite complicated and additional difficulties arise if the rank is taken to be two or higher.

REFERENCES

[1] N. Bourbaki, *Eléments de Mathématique. Livre IV. Fonctions d'une Variable Réelle Chap. IV. Equations Différentielles*. Actualités Scient. et Industr., 1132, Hermann, Paris (1952).

[2] Yu. L. Daletski,, On the asymptotic solution of a vector differential equation. *Doklady Akad. Nauk SSSR*, **92,** (1953), 881–884. (Russian).

[3] S.R. Foguel, Sums and products of commuting spectral operators. *Ark. Mat.*, **3, (41),** (1957), 449–461.

Published by

JERUSALEM ACADEMIC PRESS THE ISRAEL ACADEMY OF SCIENCES AND HUMANITIES PERGAMON PRESS, LONDON

Jerusalem 1961

Some Aspects of Differential Equations in B-Algebras

EINAR HILLE

In the main I shall deal with differential equations in a complex non-commutative algebra B with unit element e. Coefficients and solutions will be B-valued functions of a complex variable and the problems studied are those which are suggested in a natural manner by the classical theory of differential equations in the complex plane and by the matrix case. Most of the time first order linear equations will be in the focus but some time will be devoted to the second order case and to nonlinear problems, in particular Riccati's equation.

1. THE HOMOGENEOUS LINEAR FIRST ORDER EQUATION

That the equation is homogeneous means that every term contains either a factor $w(z)$ or $w'(z)$, but the coefficients may multiply these factors on the left, on the right or one on either side. In the general case we have something completely unmanageable. We have to restrict ourselves and the most interesting and most promising case is that of

$$w'(z) = F(z)\,w(z) \tag{1.1}$$

where $F(z)$ is a B-valued function holomorphic in a neighborhood of a point $z = z_0$ of the complex plane. The theory of this equation does not differ essentially from that of

$$w'(z) = w(z)\,F(z). \tag{1.2}$$

On the other hand

$$F(z)\,w'(z) = w(z), \tag{1.3}$$

where $F(z)$ is not regular in the algebra, poses entirely new problems. For an example let me mention that if q is a nilpotent element of the algebra, then the equation

$$q\,w'(z) = w(z) \tag{1.4}$$

has $w(z) \equiv 0$ as its only solution and this extends also to certain classes of quasinilpotent elements q.

Let us return to equation (1.1) with an initial condition

$$w(z_0) = w_0, \tag{1.5}$$

a given element of B. The method of successive approximations applies and gives a unique solution $w(z; z_0, w_0)$ which is holomorphic not merely in a neighborhood of $z = z_0$ but actually in the Mittag-Leffler star $A(z_0, F)$ of $F(z)$ with respect to z_0. We have

$$w(z, z_0, w_0) = w(z; z_0, e) w_0 \tag{1.6}$$

and the algebraic nature of the solution is determined entirely by that of the initial element w_0.

It is obvious that $w(z; z_0, w_0)$ is a singular element of the algebra for every z in the domain of existence of the solution if w_0 is singular. But here it is "iff." It is clear that $w(z; z_0, e)$ is regular in some neighborhood of $z = z_0$. Actually it is regular everywhere in $A(z_0; F)$. This follows essentially from the relation

$$w(z; z_0, e) = w(z; z_1, e)\, w(z_1; z_0, e), \quad z_1 \in A(z_0; F), \tag{1.7}$$

which follows from (1.6) and serves as the basis for the analytic as well as the algebraic continuation of the solution. We say that *a solution $w(z; z_0, w_0)$ is fundamental if it is algebraically regular.* A nasc for this to be the case is that w_0 be regular.

Formula (1.7) gives the analytic continuation of $w(z; z_0, e)$ in $A(z_1; F)$. If $z_2 \in A(z_1; F)$ we can then get the analytic continuation in $A(z_2; F)$ etc. This shows that analytic continuation is possible along any path that does not encounter a singular point of F. If the path is a closed curve, beginning and ending at $z = z_0$, then if we start with the value e we may very well return with a different value. There is an element $m(C)$ of B, determined by the path C such that traversing C in one direction carries

$$w(z; z_0, e) \quad \text{into} \quad w(z_0, e)\, m(C) \tag{1.8}$$

and traversing $-C$ instead carries

$$w(z; z_0, e) \quad \text{into} \quad w(z; z_0, e)[m(C)]^{-1}. \tag{1.9}$$

There substitutions $m(C)$ form a group, the *group of monodromy* of the equation. The dependence of the group on the initial point z_0 is only apparent; changing initial point merely introduces an isomorphic group. The group is finitely generated if F has only a finite number of isolated singular points plus a finite number of singular lines.

Let us return to the notion of "fundamental solution." If $B = M_n$, the algebra of $n \times n$ matrices, a solution matrix is fundamental if and only if the *Wronskian* is $\neq 0$. If

$$W'(z) = F(z)\, W(z), \tag{1.10}$$

the Wronskian is the determinant of $W(z)$ and

$$\det [W(z)] = \det [W_0] \exp \left\{ \int_{z_0}^{z} Tr[F(s)]\, ds \right\}, \tag{1.11}$$

where Tr stands for the trace of the matrix F.

We can generalize this as follows. Suppose that μ is a functional defined on B and having the following properties:

(i) μ is multiplicative and neither identically zero nor identically one.
(ii) μ is continuous and bounded.
(iii) μ is Fréchet analytic at $x = e$.
(iv) μ vanishes on the singular elements of B.

Since $\mu(e) = 1$ it follows that $\mu(a) \neq 0$ if a is regular. We have then

$$\mu[w(z; z_0, w_0)] = \mu(w_0) \exp \left\{ \int_{z_0}^{z} \delta\mu[e, F(s)]\, ds \right\} \tag{1.12}$$

and for any $g \in B$

$$\delta\mu[e, g] = \lim_{\eta \to 0} \frac{1}{\eta} [\mu(e + \eta g) - \mu(e)].$$

In the case $B = M_n$ the Fréchet derivative at E in the direction F is simply the trace of F. Condition (iv) above ensures that the formula is trivially true for singular elements w_0. If this condition does not hold we must restrict ourselves to nonsingular initial values.

If μ is linear as well as multiplicative and if B is commutative then the formula becomes

$$\mu[w(z; z_0, w_0)] = \mu(w_0) \exp \left\{ \int_{z_0}^{z} \mu[F(s)]\, ds \right\}, \tag{1.13}$$

and this holds even if B has no unit element.

The definition of a fundamental solution as one which is algebraically regular lacks meaning in an algebra without unit element. But even in such an algebra there would seem to be solutions which are *more equal* than others, but I am at a loss how to single out such elements. Perhaps the circle product $p \quad q = p + q - pq$ could be used. Another guess, possibly a better one, is to say that *a solution is fundamental if it annihilates no nontrivial multiplicative functional.*

2. APPROACH TO A SINGULAR POINT

The singularities of the solutions are determined by those of F to which the point at infinity may have to be added. In general the singularities of thc solutions are much more complicated than those of F. For a study of the behavior of a solution as z tends to a singular point, we fall back on what is known as Gronwall's Lemma:

Let $f(t)$ and $g(t) \in C^{+}[0, \omega]$, let $K(t) \in C^{+}(0, \omega] \cap L(0, \omega)$ and let

$$f(t) \leqq g(t) + \int_0^t K(s) f(s)\, ds, \tag{2.1}$$

then

$$f(t) \leqq g(t) + \int_0^t K(s) \exp\left[\int_s^t K(u)\, du\right] g(s)\, ds. \tag{2.2}$$

Suppose now that the given equation is

$$w'(t) = F(t)\, w(t), \tag{2.3}$$

where $F(t)$ is B-valued and continuous in the interval $(0, \tau]$. It is desired to find bounds for $w(t)$ as $t \downarrow 0$. Here the Lemma gives the double inequality

$$\| w(\tau) \| \exp\left\{- \int_t^\tau \| F(s) \|\, ds\right\}$$

$$\leqq \| w(t) \| \leqq \| w(\tau) \| \exp\left\{\int_t^\tau \| F(s) \|\, ds\right\}. \tag{2.4}$$

This shows that the integrability properties of $\| F(t) \|$ are decisive.

Note that the inequalities can be extended to complex variables. Consider the equation

$$w'(z) = F(z)\, w(z) \tag{2.5}$$

where $F(z)$ is holomorphic in a sector

$$\alpha \leqq \arg z \leqq \beta, \qquad 0 < | z | \leq R.$$

Set

$$\max \| F(re^{i\theta}) \| \equiv K(r), \qquad \begin{matrix}\max \\ \min\end{matrix} \| w(Re^{i\theta}) \| = \begin{matrix}M \\ m\end{matrix}.$$

Then the inequality reads

$$m \exp\left\{- \int_r^R K(s)\,ds\right\} \leqq \| w(re^{i\theta}) \| \leqq M \exp\left\{\int_r^R K(s)\,ds\right\}. \tag{2.7}$$

Here there are two alternatives.

(i) $K(s) \in L(0, \omega)$. Then the equation (2.5) has a unique solution which tends to e in the sector as $z \to 0$ and all solutions tend to finite limits, the limit being 0 iff $w(z) \equiv 0$.

(ii) $K(s) \notin L(0, \omega)$. Here there are many possibilities. The most interesting one or at least the one best studied is that where $K(s)$ becomes infinite as an integral power of $1/s$. Again there are two distinct cases.

(iia) $K(s) = a/s$. Here

$$m\left(\frac{r}{R}\right)^a \leqq \| w(r\,e^{i\theta}) \| \leqq M\left(\frac{R}{r}\right)^a. \tag{2.8}$$

This case includes in particular that of a *regular singular point* where

$$F(z) = \frac{1}{z}\{a_0 + a_1 z + \cdots\}, \qquad a_n \in B, \tag{2.9}$$

where the power series converges for $|z| < R$.

(iib) $K(s) = a\,s^{-p-1}$. Here

$$m \exp\left\{\frac{a}{p}[R^{-p} - r^{-p}]\right\} \leqq \| w(r\,e^{i\theta}) \| \leqq M \exp\left\{\frac{a}{p}[r^{-p} - R^{-p}]\right\}. \tag{2.10}$$

This includes in particular an *irregular singular point of rank p*.

3. THE REGULAR SINGULAR CASE

We return to equation (2.9). Here the behavior of the solution is essentially determined by the leading coefficient a_0. In the case of an "Euler equation" with

$$F(z) = a/z$$

a solution would be given by

$$z^a = \exp(a \log z).$$

This suggests the classical approach

$$w(z) = \sum_{m=0}^{\infty} c_m\, z^{a_0 + me}. \tag{3.1}$$

Here it turns out that the spectral properties of a_0 as an element of B and of the *commutator operator*

$$C_{a_0} x = a_0 x - x a_0 \tag{3.2}$$

are decisive. If we substitute the series (3.1) into the differential equation, we obtain equations of the form

$$a_0 c_0 = c_0 a_0, \tag{3.3}$$

$$m c_m - a_0 c_m + c_m a_0 = \sum_{k=1}^{m} a_k c_{m-k}, \quad m = 1, 2, \ldots, \tag{3.4}$$

to be solved for the c_m's. We can take $c_0 = e$ and knowing c_0 we can solve successively for $c_1, c_2, \ldots$, if, and it is a big IF, no integer belongs to the spectrum of the operator C_{a_0}. Now the spectrum of the operator is contained in the difference set

$$\Delta \equiv \{\gamma \mid \gamma = \alpha - \beta, \quad \alpha \in \sigma(a), \beta \in \sigma(a)\}. \tag{3.5}$$

Thus an integer can belong to $\sigma(C_{a_0})$ only if two spectral values of a_0 differ by an integer. If this does not happen, then we can solve successively for the coefficients and the resulting series converges for $|z| < R$.

If integers should belong to $\sigma(C_a)$ there are two alternatives.

(1) All integers in question are poles of the resolvent of C_{a_0} and are finitely generated as spectral differences. This will happen if, e.g., the spectral values of a_0 which give rise to integral differences are themselves poles of the resolvents of a_0 as a left and right multiplier.

(2) Other singularities than poles are present and/or singularities are not finitely generated.

In case (1) a modification of a method of Frobenius combined with a theorem of Foguel and a representation theorem of the resolvent of the commutator due to Daletsky can be used and leads to solutions involving powers of $\log z$. In case (2) no effective procedure has been found.

At an irregular singular point representations by generalized Laplace integerals are available at least in important special cases. This is an extension of a classical method of G. D. Birkhoff given by J. B. Miller. The case $p = 1$ which is more elementary, had been discussed by the author (unpublished).

4. THE GENERALIZED BESSEL EQUATION

Here is a linear second order problem. Given the equation

$$w''(z) + [e - F(z)]\, w(z) = 0, \tag{4.1}$$

where $F(z)$ is holomorphic outside a circle $|z| = R \geqq 0$ and

$$\int_{z_1}^{z_2} \|F(t)\| \, |dt| \leqq M \tag{4.2}$$

for any choice of two points z_1 and z_2 in $|z| > R$ which can be joined by a straight line segment all points of which are outside the disk. Note that Bessel's equation corresponds to $B = C$ and

$$F(z) = \left(\alpha^2 - \frac{1}{4}\right) z^{-2}.$$

Moreover, an equation of this type arises whenever the transformation of Liouville is applied to an equation of the form

$$[P(t)\, y'(t)] + Q(t)\, y(t) = 0 \tag{4.3}$$

where $P(t)$ and $Q(t)$ are polynomials in t and deg $(P) >$ deg $(Q) - 2$. More general cases also lead to the same normal form so it is one of the most important equations in classical theory.

The problem is to discuss the asymptotic behavior of the solutions of (4.1) outside of the disk. This is a perturbation problem and the unperturbed equation

$$w_0'' + w_0 = 0 \tag{4.4}$$

has the independent solutions

$$e \exp(iz) \text{ and } e \exp(-iz). \tag{4.5}$$

Using Gronwall's Lemma one proves that any solution of (4.1) is bounded in any horizontal strip omitting a disk $|z| \leqq R + \epsilon$. This shows the existence of

$$w(z) + \int_z^\infty \sin(t - z)\, F(t)\, w(t)\, dt.$$

Differentiation shows that this function is a solution of (4.4). If this solution is $w_0(z)$ we see that $w(z)$ satisfies the singular Volterra equation

$$w(z) = w_0(z) - \int_z^\infty \sin(t - z)\, F(t)\, w(t)\, dt \tag{4.6}$$

where the integral is taken along the horizontal from z to $+\infty + iy$. To this equation we again apply Gronwall's Lemma and obtain

$$\| w(z) - w_0(z) \| \leqq M(b) \left\{ \exp\left[\int_z^\infty \|F(s)\|\, ds \right] - 1 \right\} \tag{4.7}$$

where $M(b) = \max \| w_0(z) \|$ in the strip $-b \leqq y \leqq b$. This is for approach to infinity in the right half of the strip. A similar formula holds in the left half but usually with a different solution $w_0(z)$.

The special solutions $E^+(z)$ and $E^-(z)$ which correspond to

$$w_0(z) = e \exp(iz) \text{ and } e \exp(-iz),$$

respectively, are the analogues of Hankel's functions in the theory of the Bessel equation in the strict sense. These solutions have asymptotic representations valid in angles of opening $3\pi - \epsilon$ centered on $\arg z = \frac{1}{2}\pi$ in the first case and on $\arg z = -\frac{1}{2}\pi$ in the second. The representation in the first case is of the form

$$\| E^+(z) \exp(-iz) - e \| \leqq \exp\left[\int_z^\infty \| F(s) \| \, |ds| \right] - 1 \tag{4.8}$$

where the path of integration is a straight line in the (2π)-angle symmetric to the central line. A similar formula holds for $E^-(z)$ in its sector.

The solution $w_0(z)$ in (4.7) is in general of the form

$$w_0(z) = a \exp(iz) + b \exp(-iz), \qquad a, b \in B. \tag{4.9}$$

In the classical case $B = C$ such a solution would be oscillatory in some horizontal strip and from (4.9) one could conclude that $w(z)$ is also oscillatory and its zeros approach those of $w_0(z)$. In the general case this does not make sense but multiplicative functionals may still be oscillatory and there may conceivably be asymptotic relations between zeros. Some examples show that this will indeed happen in the matrix case using the determinant as the functional.

5. EQUATIONS OF HIGHER ORDER. THE RICCATI EQUATION

In the classical theory Riccati's equation

$$w'(z) = A_0(z) + A_1(z) w(z) + A_2(z) [w(z)]^2 \tag{5.1}$$

plays an outstanding role for two distinct reasons:

(1) The equation is closely related to the linear second-order equation. The logarithmic derivative of a solution of such an equation satisfies a Riccati equation and introducing $y'/y = w$ in (5.1) gives a second order linear equation.

(2) Of all equations of the form

$$w' = P(z, w) \tag{5.2}$$

with P a polynomial in w, the Riccati equation is the only nonlinear case where the critical points are fixed, i.e. given by the coefficients of the equation. Riccati's equation does have movable singularities but they are simple poles. No point can be a branch-point of a solution unless it is a singularity of A_0 to A_2 or a zero of A_2.

Here there is a big question of how much of this remains true in the B-algebra case or even in the matrix case. The first property is sufficiently formal so it remains valid and thanks to this fact a considerable part of the theory of the abstract Riccati equation can be derived from corresponding second-order linear equations which are easier to handle. But property (2) goes by the board at least in the strict sense that it holds for (5.1).

Such evidence as is available to me shows that movable singularities in the Riccati case are probably isolated singularities in the neighborhood of which the solution is single valued and can be expanded in convergent Laurent series where the negative powers of order < -1 have quasinilpotent coefficients. In the higher order case there are movable singularities which are branch points combined with an essential singularity. In the matrix case the quasinilpotents are nilpotents and instead of essential singularities we have poles which may be of any order.

The following simple example is of some interest in itself and may possibly be typical. Take

$$(k - 1)\, w'(z) = -[w(z)]^k, \; k > 1. \tag{5.3}$$

We start with the case $k = 2$

$$W'(z) = -[W(z)]^2. \tag{5.4}$$

This is the differential equation satisfied by the resolvent of any element of B, more generally, by any locally bounded solution of the first resolvent equation. No solution of this equation can vanish for a finite value of z unless it vanishes identically. By the theorem of Nagumo any point of the plane can be a singular point of a solution and at such a point there is a convergent Laurent series where the holomorphic as well as the principal parts are solutions in their own rights. If the singular point is at $z = 0$ the principal part has the form

$$W(z) = j\,z^{-1} + q\,z^{-2} + \cdots + q^{m-1}\,z^{-m} + \cdots \tag{5.5}$$

which converges for all $z \neq 0$. Here j is an idempotent, q a quasinilpotent, $jq = qj = q$. In the matrix case, $B = M_n$, we have $q^n = 0$ and we have a pole of an order not exceeding n.

We can reduce the general case (5.3) to the special one (5.4) by setting

$$W = w^{k-1}. \tag{5.6}$$

It follows that (5.3) also has singularities at preassigned points. In particular, there is a solution given by

$$w(z) = (j/z)^{1/(k-1)}\left[\sum_{m=0}^{\infty} (q/z)^m\right]^{1/(k-1)} \tag{5.7}$$

Since the series inside the brackets represent a function holomorphic for $z \neq 0$ and different from zero, the formal expansion of the $(k - 1)$th root converges for all values of $z \neq 0$ and the resulting Laurent series has coefficients which are quasinilpotents. In the matrix case it reduces to a polynomial in $1/z$ of degree $n - 1$ or less. The outside factor contributes an algebraic branch point and is not single-valued. Thus in this case there are movable branch points which in general are combined with essential singular points.

Chapter 16
Banach Algebras

The papers included in this chapter are

[124] On roots and logarithms of elements of a complex Banach algebra

[125] On the inverse function theorem in Banach algebras

[141] Some properties of the Jordan operator

An introduction to Hille's work on Banach algebras can be found starting on page 702.

On Roots and Logarithms of Elements of a Complex Banach Algebra*

By

EINAR HILLE in New Haven (Conn., USA)

To HEINRICH BEHNKE on his sixtieth anniversary October 9, 1958

1. Introduction. Let $\mathfrak{B}$ be a complex non-commutative Banach algebra having unit element e. Let $\mathfrak{G}$ be the group of regular (= inversible) elements, $\mathfrak{G}_0$ the principal component of $\mathfrak{G}$ containing e. Further, let $\mathfrak{R}_k (k \geqq 2)$, $\mathfrak{R}_\infty$, and $\mathfrak{L}$ denote the set of elements of $\mathfrak{B}$ having kth roots, roots of all orders, and logarithms respectively.

We say that $a \in \mathfrak{B}$ has a kth root if there exists an $x \in \mathfrak{B}$ such that

$$x^k = a \,, \tag{1.1}$$

and a has a logarithm if there exists a $y \in \mathfrak{B}$ such that

$$\exp(y) = a \,. \tag{1.2}$$

It is well known that

$$[x \mid \|x - e\| < 1] \subset \mathfrak{G}_0 \cap \mathfrak{R}_\infty \cap \mathfrak{L} \,. \tag{1.3}$$

Further $\mathfrak{L}$ is connected via the unit element and

$$\mathfrak{L} \subset \mathfrak{G}_0 \,, \quad \mathfrak{L} \subset \mathfrak{R}_\infty \,. \tag{1.4}$$

The first relation says merely that every element of $\mathfrak{L}$ has an inverse and hence is in $\mathfrak{G}_0$ since $\mathfrak{L}$ is connected. The second relation says that every element having a logarithm must have roots of all orders. This inclusion is always proper for there are singular elements having roots of all orders, any idempotent ($\neq e$) being an example. In particular, the zero element is in $\mathfrak{R}_\infty$.

The first inclusion under (1.4) is more doubtful. We have $\mathfrak{L} = \mathfrak{G}_0$ for commutative B-algebras and also for finite complex matrix algebras. If $\mathfrak{L}$ is a group then $\mathfrak{L} = \mathfrak{G}_0$. IRVING KAPLANSKY, according to SHIZUO KAKUTANI, has observed that the inclusion is proper for real 2 by 2 matrices. For the case of the algebra $\mathfrak{E}[\mathfrak{H}]$ of linear bounded operators on an infinite dimensional Hilbert space, P. HALMOS, G. LUMER and J. J. SCHÄFFER [3] have exhibited elements of $\mathfrak{G} = \mathfrak{G}_0$ which are not in $\mathfrak{R}_2$ and AUREL WINTNER [9] has shown that $\mathfrak{L}$ is not a group and hence that $\mathfrak{L}$ is properly contained in $\mathfrak{G}_0$.

---

* This research was supported by the United States Air Force through the Office of Scientific Research of the Air Research and Development Command under Contract No. AF 18 (600) — 1127.

Moreover, G. LUMER [8] has recently sharpened this result by showing that in $\mathfrak{E}[\mathfrak{H}]$ the set $\mathfrak{G}_0 \ominus \mathfrak{R}_2$ has interior points and $\mathfrak{R}_2$ is not even dense in $\mathfrak{G}_0$.

Two problems are discussed in the present paper:

(1) Suppose that x_1 and x_2 are two kth roots of a. What relations hold between x_1 and x_2? We ask the same question for logarithms.

(2) If a has a kth root (or logarithm), do elements in some neighborhood of a admit kth roots (or logarithms)?

We restrict ourselves to regular elements a even in the case of roots.

In this investigation the author has had the benefit of long discussions with his colleague Professor SHIZUO KAKUTANI. In particular, Section 5 below represents joint work with KAKUTANI.

2. Notation and auxiliary lemmas. Lower case italics represent elements of $\mathfrak{B}$ except for j, k, m, n which stand for positive integers. If $\mathfrak{B}$ is an operator algebra, $\mathfrak{B} = \mathfrak{E}[\mathfrak{X}]$ the algebra of linear bounded operators mapping the complex B-space $\mathfrak{X}$ into $\mathfrak{X}$, then elements are denoted by capital italics and for the unit element we write I instead of e. Complex numbers are denoted by Greek letters.

With respect to a given element a of $\mathfrak{B}$ the complex numbers λ fall into two classes, $\varrho(a)$ the resolvent set of a and $\sigma(a)$ the spectrum of a, according as $\lambda e - a$ has an inverse in $\mathfrak{B}$ or not. For $\lambda \in \varrho(a)$ we write

$$(\lambda a - a)^{-1} = R(\lambda; a) . \tag{2.1}$$

The set $\varrho(a)$ is open but not necessarily connected and $R(\lambda; a)$ is holomorphic in each component of $\varrho(a)$. The set $[\lambda \mid |\lambda| > ||a||]$ belongs to $\varrho(a)$. The spectrum is closed. At an isolated point $\lambda = \lambda_0$ of $\sigma(a)$ the principal part of $R(\lambda; a)$ has the form

$$e_0(\lambda - \lambda_0)^{-1} + \sum_{n=2}^{\infty} q_0^{n-1}(\lambda - \lambda_0)^{-n} , \tag{2.2}$$

where e_0 is idempotent, q_0 quasi-nilpotent, and $e_0 q_0 = q_0 e_0 = q_0$.

If $f(\lambda)$ is an analytic function of λ, holomorphic in a domain D containing $\sigma(a)$, then we may define $f(a)$ by

$$f(a) = \frac{1}{2\pi i} \oint_{\Gamma} R(\lambda; a) f(\lambda)\, d\lambda , \tag{2.3}$$

where Γ is a contour in D surrounding $\sigma(a)$. The spectral mapping theorem asserts that

$$\sigma[f(a)] = f[\sigma(a)] . \tag{2.4}$$

If $\mathfrak{B} = \mathfrak{E}[\mathfrak{X}]$, we may subject $\sigma[A]$ to further analysis. In particular, we say that $\lambda_0 \in P\sigma[A]$, the point spectrum of A, if there exists a characteristic vector $x_0 \in \mathfrak{X}$, $x_0 \neq 0$, such that

$$(\lambda_0 I - A)\,[x_0] = 0 . \tag{2.5}$$

The fine structure theorem then asserts that

$$P\sigma[f(A)] = f[P\sigma[A]] , \quad [f(\lambda_0)\, I - f(A)]\,[x_0] = 0 . \tag{2.6}$$

For further details concerning these questions, see E. HILLE and R. S. PHILLIPS [6].

Let $F(x)$ be a function on $\mathfrak{B}$ to $\mathfrak{B}$ defined in a domain $\mathfrak{D}$ of $\mathfrak{B}$. We say that $F(x)$ is Fréchet analytic in $\mathfrak{D}$ if $F(x)$ is single-valued, locally bounded and (G)-differentiable, that is,

$$\lim_{\alpha\to 0} \frac{1}{\alpha} [F(x+\alpha h) - F(x)] \equiv \delta F(x,h) \tag{2.7}$$

exists for every $h \in \mathfrak{B}$. Under the stated assumptions this (G)-differential is actually an (F)-differential and for fixed x there exists a linear bounded operator $F'(x)$ on $\mathfrak{B}$ to $\mathfrak{B}$ such that

$$\delta F(x,h) = F'(x)[h]. \tag{2.8}$$

Finally we shall need the abstract implicit function theorem due to T. H. HILDEBRANDT and L. M. GRAVES [5]. We formulate what we need as

Lemma 1. *Let $G(x,y)$ be a function on $\mathfrak{B}\times\mathfrak{B}$ to $\mathfrak{B}$, defined and Fréchet analytic in both variables in $\|x-x_0\| < \varrho$, $\|y-y_0\| < \varrho$. Define the partial Fréchet differential with respect to y at (x_0, y_0) by*

$$\delta_y G(x_0, y_0; h) \equiv \lim_{\alpha\to 0} \frac{1}{\alpha} [G(x_0, y_0 + \alpha h) - G(x_0, y_0)] \equiv D_0[h]. \tag{2.9}$$

Suppose that the linear bounded transformation D_0 has a bounded inverse. Then there exists a $\varrho_0 \leqq \varrho$ and a function $y(x)$ defined in $\|x-x_0\| < \varrho_0$ having values in $\mathfrak{B}$ and being Fréchet analytic in x such that

$$G(x, y(x)) \equiv G(x_0, y_0), \quad y(x_0) = y_0, \tag{2.10}$$

and this solution is unique.

I. Roots

3. The operator $S(b)$. Suppose that $b \in \mathfrak{G} \subset \mathfrak{B}$ and consider the linear bounded operator $S(b) \in \mathfrak{E}[\mathfrak{B}]$ defined by

$$S(b)[x] = b^{-1} x b, \quad x \in \mathfrak{B}. \tag{3.1}$$

If a_1, a_2 are arbitrary elements of $\mathfrak{B}$, we define

$$L(a_1)[x] = a_1 x, \quad R(a_2)[x] = x a_2 \tag{3.2}$$

in terms of which

$$S(b) = L(b^{-1}) R(b). \tag{3.3}$$

We note that

$$L(a_1) R(a_2) = R(a_2) L(a_1). \tag{3.4}$$

We note the relations

$$\sigma[L(a_1)] = \sigma[a_1], \ \sigma[R(a_2)] = \sigma[a_2] \tag{3.5}$$

where on the left occurs the spectrum of the operator relative to the algebra $\mathfrak{E}[\mathfrak{B}]$ and on the right the spectrum of the element in question with respect to the algebra $\mathfrak{B}$.

Lemma 2[1]**.** *The spectrum of $S(b)$ with respect to $\mathfrak{E}[\mathfrak{B}]$ satisfies*

$$\sigma[S(b)] \subset [\lambda \mid \lambda = \lambda_1^{-1}\lambda_2;\ \lambda_1, \lambda_2 \in \sigma[b]]\,. \tag{3.6}$$

Proof. Let $\mathfrak{A}$ be the commutative subalgebra of $\mathfrak{E}[\mathfrak{B}]$ generated by I, $L(b^{-1})$, and $R(b)$ to which $S(b)$ belongs by (3.3). Let $\mathfrak{A}^c$ be the first commutant of $\mathfrak{A}$, that is, the set of all elements of $\mathfrak{E}[\mathfrak{B}]$ which commute with every element of $\mathfrak{A}$, and let $\mathfrak{A}^{cc} = [\mathfrak{A}^c]^c$. Then $\mathfrak{A} \subset \mathfrak{A}^{cc} \subset \mathfrak{A}^c$ and $\mathfrak{A}^{cc}$ is abelian. Suppose $A \in \mathfrak{A}$ and that A is regular in $\mathfrak{E}[\mathfrak{B}]$. If $T \in \mathfrak{A}^c$

$$A^{-1}T = A^{-1}TA\,A^{-1} = A^{-1}A\,T A^{-1} = T\,A^{-1},$$

so that $A^{-1} \in \mathfrak{A}^{cc}$, that is A is also regular as element of $\mathfrak{A}^{cc}$. It follows that the spectra of A with respect to $\mathfrak{E}[\mathfrak{B}]$ and $\mathfrak{A}^{cc}$ are identical. But $\mathfrak{A}^{cc}$ is abelian and the spectral theory in a commutative algebra may be studied with the aid of the Gelfand representation theory. Thus if $\mathfrak{m}$ is any maximal ideal of $\mathfrak{A}^{cc}$ and if $A \equiv A[\mathfrak{m}]\,I$ (mod. $\mathfrak{m}$), then by (3.3) and (3.5)

$$S(b)\,[\mathfrak{m}] = L(b^{-1})\,[\mathfrak{m}]\,R(b)\,[\mathfrak{m}] = \lambda_1^{-1}\lambda_2\,,\ \lambda_1, \lambda_2 \in \sigma(b)\,,$$

so that (3.6) holds. This is all that we can expect to prove, it may very well happen that the spectrum of $S(b)$ reduces to $\lambda = 1$.

The following observations deal with point spectra.

Suppose that $\alpha \in P\sigma[L(b^{-1})]$, $\beta \in P\sigma[R(b)]$ and that

$$b^{-1}x_1 = \alpha\,x_1\,,\quad x_2 b = \beta\,x_2\,.$$

Then

$$b^{-1}(x_1 x_2)\,b = \alpha\,\beta\,(x_1 x_2)\,,$$

so that either

$$\alpha\,\beta \in P\sigma[S(b)] \ \text{ or } \ x_1 x_2 = 0\,. \tag{3.7}$$

If $\mathfrak{B}$ is a prime ring (see N. Jacobson [7], p. 196), than there exists at least one $z \in \mathfrak{B}$ such that $x_1 z\,x_2 \neq 0$ and now

$$b^{-1}(x_1\,z\,x_2)\,b = \alpha\,\beta\,(x_1\,z\,x_2)$$

so that the first alternative under (3.7) holds.

Suppose that

$$\lambda_0 \in P\sigma[S(b)]\,,\ S(b)\,[x_0] = \lambda_0 x_0\,.$$

If n is a positive integer, then

$$S(b)\,[x_0^n] = \{S(b)\,[x_0]\}^n = \lambda_0^n\,x_0^n\,,$$

that is, either

$$\lambda_0^n \in P\sigma[S(b)] \ \text{ or } \ x_0^n = 0\,. \tag{3.8}$$

We see in particular that if $|\lambda_0| \neq 1$, then x_0 must be nilpotent since $S(b)$ is bounded and has the bounded inverse

$$[S(b)]^{-1} = S(b^{-1})\,. \tag{3.9}$$

[1]) This result is not new. For a detailed study of the spectra of sums and products of operators, see R. S. Foguel [2]. The proof of Lemma 2 is included to facilitate the reading of this paper. For the properties of the commutants and of functions on maximal ideals see E. Hille and R. S. Phillips [6], p. 21 and p. 134 respectively.

4. Distinct roots. Suppose that $a \in \mathfrak{G} \cap \mathfrak{R}_k \subset \mathfrak{B}$ and that x is a solution of

$$x^k = a\,. \tag{4.1}$$

Obviously $x \in \mathfrak{G}$. We say that $\sigma(x)$ is *irrotational* (mod. $2\pi/k$) if

$$\sigma(x) \cap \sigma(\omega^j x) = \theta\,,\quad j = 1, 2, \ldots, k-1\,,\quad \omega = e^{2\pi i/k}\,. \tag{4.2}$$

Theorem 1. *Suppose that x_1 and x_2 are two distinct roots of* (4.1) *and that x_1 satisfies condition* (4.2). *Then x_1 and x_2 commute and there exists a set of k idempotents $e_1, \ldots, e_k$, some of which may be zero, such that the e_α's commute with x_1 and x_2 and*

$$e_\alpha e_\beta = \delta_{\alpha\beta} e_\alpha\,,\quad \Sigma\, e_\alpha = e\,,\quad \Sigma\, \omega^\alpha e_\alpha = x_1 x_2^{-1}\,. \tag{4.3}$$

Proof. The assumption that $\sigma(x_1)$ is irrotational (mod. $2\pi/k$) implies that $\sigma[S(x_1)]$ does not contain any kth root of unity besides $\lambda = 1$. Now

$$[S(x_1)]^k [x_2] = x_1^{-k} x_2 x_1^k = x_2^{-k} x_2 x_2^k = x_2\,.$$

Thus $\lambda = 1$ is a characteristic value of $[S(x_1)]^k$ with x_2 as a characteristic vector. Thus

$$\prod_{\alpha=1}^{k} [\omega^{\alpha-1} I - S(x_1)]\,[x_2] = 0\,. \tag{4.4}$$

Since $R(\omega^{\alpha-1}, S(x_1))$ exists for $\alpha = 2, 3, \ldots, k$, we get

$$[I - S(x_1)]\,[x_2] = 0 \text{ or } x_1 x_2 = x_2 x_1$$

as asserted.

We shall now examine $x_1 x_2^{-1}$. Since

$$(x_1 x_2^{-1})^k = x_1^k x_2^{-k} = e\,,$$

the spectral mapping theorem shows that the spectrum of $x_1 x_2^{-1}$ is contained in the set of the kth roots of unity. Hence there exist idempotents e_α, some of which may be zero, such that

$$e_\alpha e_\beta = \delta_{\alpha\beta} e_\alpha\,,\quad \Sigma\, e_\alpha = e$$

and corresponding quasi-nilpotents q_α such that

$$e_\alpha q_\alpha = q_\alpha e_\alpha = q_\alpha\,,\quad q_\alpha q_\beta = 0\,,\quad \alpha \neq \beta\,,$$

in terms of which

$$R(\lambda\,; x_1 x_2^{-1}) = \sum_{\alpha=1}^{k} e_\alpha \sum_{n=1}^{\infty} q_\alpha^{n-1} (\lambda - \omega^\alpha)^{-n}\,, \tag{4.5}$$

where $q_\alpha^0 = e_\alpha$ by definition. For the series on the right is the sum of the principal parts of the resolvent; the difference between the two sides must be an entire function of λ, but both sides are holomorphic at infinity and vanish there, hence the two sides must be equal. By formula (2.3)

$$e = (x_1 x_2^{-1})^k = \frac{1}{2\pi i} \oint_\Gamma \lambda^k R(\lambda\,; x_1 x_2^{-1})\, d\lambda\,, \tag{4.6}$$

where Γ is $|\lambda| = \varrho > 1$. Substituting the series and integrating termwise we get

$$e = \sum_{\alpha=1}^{k} e_a \sum_{n=1}^{k+1} \binom{k}{n-1} \omega^{-\alpha(n-1)} q_\alpha^{n-1}.$$

Thus the q's are nilpotents. We shall show that they are actually zero. Multiplying both sides by e_α we get

$$\sum_{n=2}^{k+1} \binom{k}{n-1} \omega^{-(n-1)\alpha} q_\alpha^{n-1} = 0 , \quad \alpha = 1, 2, \ldots, k .$$

Here q_α is a factor of the left side and the remaining factor has an inverse, since $e + q$ has an inverse if q is nilpotent. Hence every $q_\alpha = 0$ and

$$R(\lambda ; x_1 x_2^{-1}) = \sum_{\alpha=1}^{k} e_\alpha (\lambda - \omega^\alpha)^{-1}. \tag{4.7}$$

Expanding both sides in powers of λ^{-1} and comparing the coefficients of λ^{-2} we get (4.3).

It remains to prove that the idempotents e_α commute with x_1 and x_2. To prove this we observe that any power of $x_1 x_2^{-1}$ commutes with x_1 as well as with x_2. Hence we have identities of the form

$$\sum_{\alpha=1}^{k} \omega^{n\alpha} (e_\alpha x_1 - x_1 e_\alpha) = 0 , \quad n = 1, 2, 3, \ldots , \tag{4.8}$$

and similar identities with x_1 replaced by x_2. It follows that each parenthesis must be zero. This completes the proof.

For the validity of this theorem it is essential that the spectrum of at least one of the two kth roots be irrotational. The two matrices

$$\begin{pmatrix} 0 & 1 \\ 1 & 0 \end{pmatrix} \quad \text{and} \quad \begin{pmatrix} 1 & 0 \\ 0 & -1 \end{pmatrix}$$

are non-commuting square roots of the unit matrix in the algebra of two by two matrices and their spectra are identical, namely $\lambda = 1$ and -1 which are carried into each other by a rotation of π.

5. Interior points of $\mathfrak{R}_k$. We turn now to the second problem and shall prove

Theorem 2. *If $a \in \mathfrak{G}$ is the kth power of b and if $\sigma(b)$ is irrotational* (mod. $2\pi/k$), *then a is an interior point of $\mathfrak{R}_k$. There exists a neighborhood of $x = a$ all the points of which are kth powers of elements in a neighborhood of b.*

Proof. We shall study the mapping

$$(b + y)^k = a + x \tag{5.1}$$

in the neighborhood of (0, 0). We have

$$x = \sum_{j=1}^{k} B_j (b, y) , \tag{5.2}$$

where $B_j(b, y)$ is a homogeneous polynomial in y of degree j and in b of degree $k - j$. In particular

$$B_1(b, y) = \sum_{n=0}^{k} b^n \, y \, b^{k-n-1} = b^{k-1} \sum_{j=0}^{k-1} [S(b)]^j \, [y] \tag{5.3}$$

so that

$$B_1(b, y) = b^{k-1} \prod_{j=1}^{k-1} [\omega^j I - S(b)] \, [y] , \quad \omega = e^{2\pi i/k}. \tag{5.4}$$

This is a linear bounded transformation on $\mathfrak{B}$ to $\mathfrak{B}$ and, since $\sigma(b)$ is irrotational, the operator has a linear bounded inverse given by

$$b^{1-k} \prod_{j=1}^{k-1} R(\omega^j;\ S(b)) . \tag{5.5}$$

Putting

$$G(x, y) \equiv (b+y)^k - a - x\ ,\ x_0 = y_0 = 0\ , \tag{5.6}$$

we see that

$$\delta_y G(0, 0; h) = B_1(b, h) \equiv D_0[h]$$

and the inverse of D_0 is given by (5.5). Thus the assumptions of Lemma 1 are satisfied. Hence there exists a sphere $||x|| < \varrho_0$ and a function $y(x)$, Fréchet analytic in this sphere, such that

$$[b + y(x)]^k = a + x\ ,\ ||x|| < \varrho_0\ ,\ y(0) = 0\ .$$

This proves the theorem.

6. Disjoint spectra. The condition that $\sigma(b)$ be irrotational (mod. $2\pi/k$) in order that $a = b^k$ be an interior point of $\mathfrak{R}_k$ is unnecessarily restrictive. It suffices that $\sigma(b)$ breaks up into disjoint spectral sets σ_α distributed between k congruent sectors. We can then find another kth root of a, whose spectrum is confined to one of these sectors, and Theorem 2 applies to this root.

More precisely, we shall say that $\sigma(b)$ may be *sectorized* (mod. $2\pi/k$) if there exists a rectifiable arc C leading from $\lambda = 0$ to a distant point of the plane such that the k arcs

$$C,\ \omega C,\ \omega^2 C,\ \ldots,\ \omega^{k-1} C\ , \qquad \omega = e^{2\pi i/k} \tag{6.1}$$

do not contain any points of $\sigma(b)$. These k arcs then subdivide the spectrum of b into k spectral set σ_α, one for each sector, and it is understood that one or more of these sets may be void. The spectral set σ_α lies between $\omega^\alpha C$ and $\omega^{\alpha+1} C$.

Theorem 3. *If* $a \in \mathfrak{S}$ *is the kth power of* b *and if* $\sigma(b)$ *may be sectorized* (mod. $2\pi/k$), *then* a *is an interior point of* $\mathfrak{R}_k$. *There exists an element* c *of* $\mathfrak{S}$ *such that* $a = c^k$ *and a neighborhood of* $x = a$, *all the elements of which are kth powers of elements in some neighborhood of* c.

Proof. Let C_1 and C_2 be two circles with center at $\lambda = 0$ such that $\sigma(b)$ lies in the open annulus bounded by C_1 and C_2. Let Γ_α be the closed positively oriented contour surrounding σ_α which is formed by the subarcs of $\omega^\alpha C$ and $\omega^{\alpha+1} C$ between C_1 and C_2 joined by the two arcs of these circles. Here $\alpha = 0, 1, \ldots, k-1$. A positive rotation through an angle of $\alpha\, 2\pi/k$ will take Γ_0 into Γ_α. We set

$$e_\alpha = \frac{1}{2\pi i} \oint_{\Gamma_\alpha} R(\lambda; b)\, d\lambda = \frac{1}{2\pi i} \oint_{\Gamma_0} R(\lambda; b\,\omega^{-\alpha})\, d\lambda\ , \tag{6.2}$$

$$b_\alpha = \frac{1}{2\pi i} \oint_{\Gamma_\alpha} \lambda R\,(\lambda; b)\, d\lambda = \frac{\omega^\alpha}{2\pi i} \oint_{\Gamma_0} \lambda R\,(\lambda; b\,\omega^{-\alpha})\, d\lambda\ . \tag{6.3}$$

Then

$$\sum_{\alpha=0}^{k-1} e_\alpha = e\ ,\ e_\alpha e_\beta = \delta_{\alpha\beta} e_\alpha\ ,\ \sum_{\alpha=0}^{k-1} b_\alpha = b\ ,\ b_\alpha b_\beta = 0\ ,\ \alpha \neq \beta,\ b_\alpha = e_\alpha b = b e_\alpha\ . \tag{6.4}$$

For these properties and for the facts regarding the spectra of b_α used below, see E. HILLE and R. S. PHILLIPS [6] Theorem 5.6.1. We set

$$c = b \sum_{\alpha=0}^{k-1} \omega^{-\alpha} e_\alpha = \sum_{\alpha=0}^{k-1} \omega^{-\alpha} b_\alpha . \tag{6.5}$$

By (6.4)

$$\lambda e - \sum_{\alpha=0}^{k-1} \omega^{-\alpha} b_\alpha = \lambda^{-k+1} \prod_{\alpha=0}^{k-1} (\lambda e - \omega^{-\alpha} b_\alpha) , \quad \lambda \neq 0 .$$

Thus the left side has an inverse if and only if each of the factors on the right has an inverse. But

$$\sigma[\omega^{-\alpha} b_\alpha] = \omega^{-\alpha} \sigma[b_\alpha] = \omega^{-\alpha} \sigma_\alpha \cup \{0\}$$

and $\omega^{-\alpha}\sigma_\alpha$ is confined to the interior of Γ_0. The point $\lambda = 0$ is not in $\sigma[c]$ for in the second member of (6.5) both factors have inverses. In the case of b this was a part of the assumption since $a \in \mathfrak{G}$ and a simple calculation shows that the resolvent of the second factor is

$$R(\lambda; \sum_{\alpha=0}^{k-1} \omega^{-\alpha} e_\alpha) = \sum_{\alpha=0}^{k-1} e_\alpha (\lambda - \omega^{-\alpha})^{-1}$$

which is holomorphic at $\lambda = 0$. Hence $\sigma[c]$ lies in the interior of Γ_0. Finally

$$c^k = b^k \left[\sum_{\alpha=0}^{k-1} \omega^{-\alpha} e_\alpha \right]^k = b^k \sum_{\alpha=0}^{k-1} \omega^{-k\alpha} e_\alpha = b^k = a .$$

Thus c is a kth root of a and Theorem 2 applies to c. This completes the proof.

II. Logarithms

7. The operator $T(a)$. In problems involving the exponential function the *commutator operator* $T(a)$ plays a similar role to that of the operator $S(b)$ in the study of powers. We set

$$T(a)\,[x] = a\,x - x\,a \tag{7.1}$$

so that

$$T(a) = L(a) - R(a) . \tag{7.2}$$

The spectral properties of $T(a)$ are listed in

Lemma 3. *We have*

$$\sigma[T(a)] \subset [\lambda \mid \lambda = \lambda_1 - \lambda_2 ;\ \lambda_1, \lambda_2 \in \sigma(a)] . \tag{7.3}$$

We omit the proof which follows the same lines as that of Lemma 2. The following remarks refer to point spectra. Suppose that

$$\lambda_1 \in P\sigma[L(a)] , \quad \lambda_2 \in P\sigma[R(a)]$$
$$a\,x = \lambda_1 x , \quad y\,a = \lambda_2 y .$$

Then

$$a\,x\,y - x\,y\,a = (\lambda_1 - \lambda_2)\,x\,y ,$$

that is, either

$$\lambda_1 - \lambda_2 \in P\sigma[T(a)] \text{ or } xy = 0 . \tag{7.4}$$

If $\mathfrak{B}$ is a prime ring, we can find a z such that $xzy \neq 0$ and then

$$a\,xzy - xzya = (\lambda_1 - \lambda_2)\,xzy$$

so that the first alternative under (7.4) must hold.

Suppose that $\lambda_0 \neq 0$ is in $P\sigma[T(a)]$ and

$$a x_0 - x_0 a = \lambda_0 x_0, \quad x_0 \neq 0,$$

then

$$a x_0^n - x_0^n a = n \lambda_0 x_0^n,$$

so that either

$$n \lambda_0 \in P\sigma[T(a)] \text{ or } x_0^n = 0. \tag{7.5}$$

It is clear that the second alternative must hold for all large n, that is, every characteristic vector corresponding to a $\lambda_0 \neq 0$ must be nilpotent.

We list finally the important formula

$$\exp(a)\, x \exp(-a) = \{\exp[T(a)]\}\,[x], \tag{7.6}$$

due to J. E. CAMPBELL ([1], p. 385—386) and F. HAUSDORFF ([4], p. 26). This formula is basic for the following discussion.

8. Distinct logarithms. We suppose now that $a \in \mathfrak{L}$ and that y is a solution of

$$\exp(y) = a. \tag{8.1}$$

We say that $\sigma(y)$ is *incongruent* (mod. $2\pi i$) if

$$\sigma(y) \cap [\sigma(y) + 2k\pi i] = \theta, \quad k = \pm 1, \pm 2, \pm 3, \ldots. \tag{8.2}$$

Theorem 4. *Suppose that y_1 and y_2 are two distinct solutions of* (8.1) *and that $\sigma(y_1)$ is incongruent* (mod. $2\pi i$). *Then y_1 and y_2 commute and there exist idempotents $e_1, e_2, \ldots, e_n$, commuting with y_1 and y_2, and distinct integers $k_1, k_2, \ldots, k_n$ such that*

$$y_1 - y_2 = 2\pi i \sum_{\alpha=1}^{n} k_\alpha e_\alpha, \quad \sum_{\alpha=1}^{n} e_\alpha = e, \quad e_\alpha e_\beta = \delta_{\alpha\beta} e_\alpha. \tag{8.3}$$

Proof. We start by assuming that neither y_1 nor y_2 is zero. Since $\sigma(y_1)$ is incongruent (mod. $2\pi i$), the spectrum of $T(y_1)$ does not contain any multiple of $2\pi i$ different from zero. Formula (7.6) asserts that

$$\exp(y_1)\, y_2 \exp(-y_1) = \{\exp[T(y_1)]\}\,[y_2]. \tag{8.4}$$

Since $\exp(y_1) = \exp(y_2)$ and y_2 commutes with $\exp(y_2)$, the left member reduces to y_2 so that

$$\{\exp[T(y_1)]\}\,[y_2] = y_2. \tag{8.5}$$

This states that $\lambda = 1$ is a characteristic value of the operator $\exp[T(y_1)]$ and that y_2, $y_2 \neq 0$, is a corresponding characteristic vector. By (2.6), the fine structure theorem, there exists a characteristic value α of $T(y_1)$ such that $\exp \alpha = 1$ and the assumption on $\sigma(y_1)$ implies that $\alpha = 0$. We have to show that y_1 is a corresponding characteristic vector. Let us set

$$E(\alpha) = [1 - \exp(-a)]/\alpha. \tag{8.6}$$

Then

$$\{\exp[T(y_1)] - I\}\,[y_2] = E[-T(y_1)]\, T(y_1)\,[y_2].$$

Since

$$\sigma\{E[-T(y_1)]\} = E\{-\sigma[T(y_1)]\}$$

does not contain the origin, $E[-T(y_1)]$ has a bounded inverse, $F(-y_1)$ say, so that

$$F(-y_1)\{\exp[T(y_1)]-I\}[y_2]=T(y_1)[y_2].$$

Here the left side is zero so that $T(y_1)[y_2]=0$ or y_1 commutes with y_2 as asserted.

We have then

$$\exp(y_1-y_2)=\exp(y_1)\exp(-y_2)=e,$$

so that

$$\sigma(y_1-y_2)\equiv 0 \ (\text{mod.}\, 2\pi i)$$

by the spectral mapping theorem. This implies

$$R(\lambda; y_1-y_2)=\sum_{\alpha=1}^{n} e_\alpha \sum_{m=1}^{\infty} q_\alpha^{m-1} \quad (\lambda-2k_\alpha\pi i)^{-m},$$

where $\Sigma e_\alpha = e$ and the e_α's and q_α's have the usual orthogonality properties. With the aid of (2.3) we then obtain

$$\exp(y_1-y_2)=\sum_{\alpha=1}^{n} e_\alpha \exp(q_\alpha)=\exp\left[\sum_{\alpha=1}^{n} q_\alpha\right]\equiv\exp(q).$$

From $\exp q = e$ we get

$$q\left\{e+\sum_{1}^{\infty}\frac{q^m}{(m+1)!}\right\}=0.$$

The quantity inside the braces differs from e by a quasi-nilpotent and consequently has an inverse. This implies $q=0$ and in the same manner we prove that

$$q_\alpha=0, \quad \alpha=1,2,\ldots,n.$$

Thus the spectral singularities of $R(\lambda; y_1-y_2)$ are all simple poles. Expanding the resolvent in powers of λ^{-1} and comparing coefficients of λ^{-2} we get (8.3). Here the k_α's are distinct integers one of which may be zero.

If $y_1=0$, it is clear that y_1 and y_2 commute and there is nothing to change in the second part of the proof.

Since any power of (y_1-y_2) commutes with y_1 as well as with y_2 we obtain identities of the form

$$\sum_{\alpha=1}^{n}(k_\alpha)^m[e_\alpha y_1-y_1 e_\alpha]=0, \quad m=1,2,3,\ldots,$$

whence it follows that every bracket must be zero, that is, the e_α's commute with y_1 and here we may replace y_1 by y_2. This completes the proof.

If neither y_1 nor y_2 satisfies condition (8.2), the conclusion of Theorem 4 need not hold. S. Kakutani has called my attention to the fact that in the algebra of two by two matrices the unit matrix has non-commuting logarithms. If (8.2) does not hold, there exists a polynomial $P(\alpha)$ such that $P[T(y_1)][y_2]=0$ but in general we cannot conclude from this that $T(y_1)[y_2]=0$.

9. Interior points of L. We turn now to the second problem for logarithms and shall prove

Theorem 5. *If $a \in L$, if b is a logarithm of a, and if $\sigma(b)$ is incongruent* (mod. $2\pi i$), *then a is an interior point of L. There exists a neighborhood of $x = a$, all the points of which have logarithms in some neighborhood of b.*

Proof. In this case we set

$$G(x, y) \equiv \exp(-b)\exp(b + y) - \exp(x)\,, \quad x_0 = y_0 = 0\,. \tag{9.1}$$

Here $G(0, 0) = 0$ and

$$\delta_y G(0, 0; h) = \exp(-b) \sum_{n=1}^{\infty} \frac{1}{n!} \sum_{j=0}^{n-1} b^j\, h\, b^{n-1-j} \equiv D(b)\,[h]\,. \tag{9.2}$$

A simple calculation gives

$$D(b)\,[h] = \{E\,[T(b)]\}\,[h] \tag{9.3}$$

with $E(\alpha)$ defined by (8.6). Just as in the preceding section we see that $\sigma\{E\,[T(b)]\} = E\,\{\sigma[T(b)]\}$ does not contain $\lambda = 0$ since none of the zeros of $E(\alpha)$ belongs to $\sigma[T(b)]$. It follows that $D(b)$ has a bounded inverse and Lemma 1 applies. Thus there exists a $\varrho_0 = \varrho_0(b)$ and a Fréchet analytic function $y(x) = y(x; 0, b)$ such that

$$\exp(b)\exp(x) = \exp[b + y(x; 0, b)]\,, \quad ||x|| < \varrho_0(b)\,. \tag{9.4}$$

The mapping $x \to a \exp x$ being open, we conclude the existence of a neighborhood of a, all the points of which have logarithms, that is, belong to $\mathfrak{L}$, and these logarithms may be found in some neighborhood of b as asserted.

10. Disjoint spectra. The condition that $\sigma(b)$ be incongruent (mod. $2\pi i$) in order that $a = \exp(b)$ be an interior point of L is of course unnecessarily restrictive. Just as in the case of roots, it suffices that $\sigma(b)$ breaks up into disjoint spectral set σ_α which are distributed between the period strips of the exponential function. We can then find another logarithm of a whose spectrum is confined to a single period strip, and Theorem 5 applies to this logarithm.

We shall say that $\sigma(b)$ may be *stratified* (mod. $2\pi i$) if rectifiable arcs $C_0, C_1, \ldots, C_n$ exist with the following properties:

(1) $C_\alpha = C_{\alpha-1} + 2\pi i$, $\alpha = 1, 2, \ldots, n$.

(2) Every vertical line in the λ-plane intersects C_0 in at most one point.

(3) $\sigma(b)$ is confined to the domain bounded by C_0 and C_n and two vertical lines. No point of $\sigma(b)$ lies on any of the arcs C_α.

Theorem 6. *If $a = \exp(b)$ and if $\sigma(b)$ may be stratified* (mod. $2\pi i$), *then a is an interior point of L.*

Proof. We proceed as in the proof of Theorem 3. Let σ_α be the spectral set between $C_{\alpha-1}$ and C_α and let Γ_α be the closed positively oriented contour surrounding σ_α which is formed by $C_{\alpha-1}$ and C_α and the vertical line segments joining their endpoints. A translation by $2\alpha\pi i$ takes Γ_1 into $\Gamma_{\alpha+1}$. We now define e_α and b_α by formulas (6.2) and (6.3) giving Γ_α its new meaning and in each case omitting the third member which is no longer relevant. These quantities again satisfy (6.4). We now set

$$c_\alpha = b_\alpha - (\alpha - 1)\, 2\pi i\, e_\alpha\,, \quad c = \sum_{\alpha=1}^{n} c_\alpha\,. \tag{10.1}$$

Here

$$c_\alpha = \frac{1}{2\pi i} \int_{\Gamma_\alpha} [\lambda - (\alpha - 1)\, 2\, \pi\, i]\, R(\lambda; b)\, d\lambda$$
$$= \frac{1}{2\pi i} \int_{\Gamma_1} \lambda\, R(\lambda + (\alpha - 1)\, 2\, \pi\, i\, ; b)\, d\lambda$$
$$= \frac{1}{2\pi i} \int_{\Gamma_1} \lambda\, R(\lambda; b - (\alpha - 1)\, 2\, \pi\, i\, e)\, d\lambda\, .$$

The spectrum of $b - (\alpha - 1)\, 2\, \pi\, i\, e$ is also stratified; one of its spectral sets lies interior to Γ_1 and is congruent to σ_α. It follows that $\sigma(c_\alpha) = \{\sigma_\alpha - (\alpha - 1)\, 2\, \pi\, i\} \cup \{0\}$. If σ_α is void, $c_\alpha = 0$ and gives no contribution to the value of c. We note that in any case $c_\alpha c_\beta = 0$ when $\alpha \neq \beta$. This implies that for every $\lambda \neq 0$

$$e\, \lambda - \sum_{\alpha=1}^{n} c_\alpha = \lambda^{1-n} \prod_{\alpha=1}^{n} (\lambda\, e - c_\alpha)\, .$$

It follows that the left side has an inverse if and only if each factor on the right has one, that is,

$$\sigma(c) = \cup_1^n\, \sigma(c_\alpha)\, .$$

Without restricting the generality we may suppose that $\lambda = 0$ is interior to Γ_1. This can always be achieved by subtracting a suitable multiple of $2\, \pi\, i\, e$ from b. It follows that $\sigma(c)$ is interior to Γ_1 and is a fortiori incongruent (mod. $2\, \pi\, i$). Since

$$c = b - 2\, \pi\, i \sum_{\alpha=2}^{n} (\alpha - 1)\, e_\alpha\, , \tag{10.2}$$

we have

$$\exp c = \exp b = a.$$

Here c satisfies the conditions of Theorem 5 and the desired conclusion follows.

References

[1] Campbell, J. E.: On a law of combination of operators bearing on the theory of continuous transformation groups. Proc. London math. Soc. **28**, 381—390 (1897); **29**, 14—32 (1898). — [2] Foguel, R. S.: Sums and products of commuting spectral operators. Ark. f. Mat. **3**, no. 41, 449—461 (1957). — [3] Halmos, P., G. Lumer and J. J. Schäffer: Square roots of operators. Proc. Amer. math. Soc. **4**, 142—149 (1953); Part II [Halmos and Lumer), Proc. Amer. math. Soc. **5**, 589—595 (1954). — [4] Hausdorff, F.: Die symbolische Exponentialformel in der Gruppentheorie. S.-B. kgl. sächs. Ges. Wiss. Leipzig, Math.-phys. Kl. **58**, 19—48 (1906). — [5] Hildebrandt, T. H., and L. M. Graves: Implicit functions and their differentials in general analysis. Trans. Amer. math. Soc. **29**, 127—153 (1927). — [6] Hille, E., and R. S. Phillips: Functional Analysis and Semi-Groups. Amer. math. Soc. Coll. Publ. **31**, 2nd ed. (1957), xii + 808 pp. — [7] Jacobson, N.: Structure of Rings. Amer. math. Soc. Coll. Publ. 37 (1956) vii + 263 pp. — [8] Lumer, G.: The range of the exponential function. Fac. Ingen. Agrimens. Montevideo. Publ. Inst. Mat. Estadist. **3**, 53—55 (1957). — [9] Wintner, A.: Bounded matrices and linear differential equations. Amer. J. Math. **79**, 139—151 (1957).

(Eingegangen am 21. März 1958)

ON THE INVERSE FUNCTION THEOREM IN BANACH ALGEBRAS[1]

By

Einar Hille, *Yale University*

(*Received—October, 11, 1958*)

1. Formulation. Let $\mathfrak{B}$ be a complex, non-commutative Banach algebra. For $n = 1, 2, 3, \ldots$ let $\mathfrak{F}_n(\mathfrak{z}_1, \mathfrak{z}_2, \ldots, \mathfrak{z}_n)$ be an n-linear bounded form on $\mathfrak{B}\times\mathfrak{B}\times\ldots\times\mathfrak{B}$ into $\mathfrak{B}$. Thus, $\mathfrak{F}_n$ $(\mathfrak{z}_1, \mathfrak{z}_2, \ldots, \mathfrak{z}_n)$ is linear in each variable separately and there exists a positive number M_n such that

$$\|\mathfrak{F}_n(\mathfrak{z}_1, \mathfrak{z}_2, \ldots, \mathfrak{z}_n)\| \leqslant M_n\|\mathfrak{z}_1\| \, \|\mathfrak{z}_2\| \cdots \|\mathfrak{z}_n\| \tag{1.1}$$

We do not assume that $\mathfrak{F}_n$ is a symmetric function of the variables.

Set

$$\mathfrak{F}_n(\mathfrak{z}) \equiv \mathfrak{F}_n(\mathfrak{z}, \mathfrak{z}, \ldots, \mathfrak{z})$$

and

$$\mathfrak{F}(\mathfrak{z}) = \sum_{n=1}^{\infty} \mathfrak{F}_n(\mathfrak{z}). \tag{1.3}$$

In order for $\mathfrak{F}(\mathfrak{z})$ to be well defined, we shall assume the power series

$$g(y) \equiv \sum_{n=2}^{\infty} M_n y^n \tag{1.4}$$

has a positive radius of convergence r.

Theorem. *Suppose that the linear transformation on $\mathfrak{B}$ into $\mathfrak{B}$ defined by*

$$\mathfrak{w} = \mathfrak{F}_1(\mathfrak{z}) \tag{1.5}$$

has a linear bounded inverse

$$\mathfrak{z} = L(\mathfrak{w}) \tag{1.6}$$

and set $A = max\ (1,\|L\|)$. Then there exists a number s, $0<s<r$, and a function $\mathfrak{G}(\mathfrak{w})$ on $\mathfrak{B}$ having the following properties :

$$(i) \qquad \mathfrak{G}(\mathfrak{w}) = \sum_{n=1}^{\infty} \mathfrak{G}_n(\mathfrak{w})$$

where each $\mathfrak{G}_n$ is a homogeneous bounded polynomial of degree n in $\mathfrak{w}$ and

$$\text{Sup} \left\{ \sum_{n=1}^{\infty} \|\mathfrak{G}_n(\mathfrak{w})\| \right\} < \infty \qquad \text{for} \qquad \|\mathfrak{w}\| \leqslant s.$$

(1) This research was supported by the United States Air Force through the Office of Scientific Research of the Air Research and Development Command under Contract No. 49 (638)—224.

(*ii*) $\mathfrak{w} \equiv \mathfrak{F}(\mathfrak{G}(\mathfrak{w}))$ *for* $\|\mathfrak{w}\| \leqslant s$.

(*iii*) $\mathfrak{G}(\mathfrak{w})$ *is the only solution of* $\mathfrak{w} = \mathfrak{F}(\mathfrak{z})$ *which tends to zero with* $\mathfrak{w}$.

We can take $s = A^{-2}R$, *where* $R = r - g(r)$ *if* $\lim_{y \to r} g'(y) \leqslant 1$, *otherwise* $R = u - g(u)$ *where* $g'(u) = 1, 0 < u < r$.

This theorem is obviously a special case of abstract implicit function theorems first discussed by L. M. Graves (1) alone and together with T. H. Hildebrandt (2) under very general assumptions. In view of the importance of analytic function theory in Banach algebras, it seems worth while to give a formulation of the theorem which is appropriate for this theory. The restrictive assumptions on $\mathfrak{F}(\mathfrak{z})$ permit a simpler formulation of the inverse function theorem and gives more properties of the inverse function $\mathfrak{G}(\mathfrak{w})$. It also lends itself to the use of the classical methods of function theory including the time-honoured "calcul des limites" of Cauchy. It is clear that the method applies to any Fréchet analytic function $\mathfrak{F}(\mathfrak{z})$ with $\mathfrak{F}(0) = 0$. We prefer to start from the series expansion (1.3) which is available for such a function.

For the terminology used above and for the properties of bounded n-linear forms, the reader is referred to E. Hille and R. S. Phillips (3. Chapter XXVI).

2. Proof. We introduce an auxiliary complex parameter in the equation $\mathfrak{w} = \mathfrak{F}(\mathfrak{z})$ by replacing $\mathfrak{z}$ and $\mathfrak{w}$ by $\alpha\, \mathfrak{z}$ and $\alpha\, \mathfrak{w}$ respectively. The result is

$$\mathfrak{w} = \sum_{n=1}^{\infty} \alpha^{n-1} \mathfrak{F}_n(\mathfrak{z}) \tag{2.1}$$

by the homogeneity of $\mathfrak{F}_n(\mathfrak{z})$. This equation we try to solve for $\mathfrak{z}$ by a power series in α having coefficients in $\mathfrak{B}$:

$$\mathfrak{z} = \sum_{k=0}^{\infty} \mathfrak{z}_k \alpha^k. \tag{2.2}$$

We substitute this series in (2.1) and use the multilinearity to obtain

$$\mathfrak{w} = \sum_{n=1}^{\infty} \alpha^{n-1} \left\{ \sum_{k_1=0}^{\infty} \cdots \sum_{k_n=0}^{\infty} \alpha^{k_1+k_2+\cdots+k_n} \mathfrak{F}_n(\mathfrak{z}_{k_1}, \mathfrak{z}_{k_2}, \ldots, \mathfrak{z}_{k_n}) \right\}. \tag{2.3}$$

Assuming the convergence of the series involved, we treat this as an identity in α and equate "coefficients" obtaining

$$\mathfrak{F}_1(\mathfrak{z}_0) = \mathfrak{w},$$

$$\mathfrak{F}_1(\mathfrak{z}_1) = -\mathfrak{F}_2(\mathfrak{z}_0, \mathfrak{z}_0),$$

$$\mathfrak{F}_1(\mathfrak{z}_2) = -\mathfrak{F}_2(\mathfrak{z}_0, \mathfrak{z}_1) - \mathfrak{F}_2(\mathfrak{z}_1, \mathfrak{z}_0) - \mathfrak{F}_3(\mathfrak{z}_0, \mathfrak{z}_0, \mathfrak{z}_0),$$

. . . .

the n-th equation being

$$\mathfrak{F}_1(\mathfrak{z}_n) = -\Sigma \mathfrak{F}_m(\mathfrak{z}_{k_1}, \mathfrak{z}_{k_2}, \ldots, \mathfrak{z}_{k_m}), \tag{2.4}$$

where the summation extends over all integers $m \geqslant 2$ and $k_1, k_2 \ldots, k_m \geqslant 0$ such that

$$m+k_1+k_2+ \ldots +k_m = n+1. \tag{2.5}$$

Since by assumption $\mathfrak{F}_1$ has a bounded linear inverse L, we obtain successively

$$\mathfrak{z}_0 = -L[\mathfrak{w}] \equiv \mathfrak{G}_1(\mathfrak{w}),$$

$$\mathfrak{z}_1 = -L[\mathfrak{F}_2(\mathfrak{z}_0)] \equiv \mathfrak{G}_2(\mathfrak{w}),$$

$$\mathfrak{z}_2 = -L[\mathfrak{F}_2(\mathfrak{z}_0, \mathfrak{z}_1)]-L[\mathfrak{F}_2(\mathfrak{z}_1, \mathfrak{z}_0)]-L[\mathfrak{F}_3(\mathfrak{z}_0)] \equiv \mathfrak{G}_3(\mathfrak{w}),$$

. . . .

and

$$\mathfrak{z}_n = -\Sigma L[\mathfrak{F}_m(\mathfrak{z}_{k_1}, \mathfrak{z}_{k_2}, \ldots, \mathfrak{z}_{k_m})] \equiv \mathfrak{G}_{n+1}(\mathfrak{w}). \tag{2.6}$$

Here each $\mathfrak{z}_n = \mathfrak{G}_{n+1}(\mathfrak{w})$ is a uniquely determined homogeneous polynomial in $\mathfrak{w}$ of degree $n+1$. This is evidently true for $n = 0$ and is proved by complete induction on n using (2.5).

To prove that the resulting series (2.1) and the series (2.3) are convergent when each term is replaced by its norm for sufficiently small values of $|\alpha|$, we argue as follows. From (2.6) together with (1.1) and the definition of A we get

$$||\mathfrak{z}_n|| \leqslant A \ \Sigma \ M_m||\mathfrak{z}_{k_1}|| \ ||\mathfrak{z}_{k_2}|| \ldots ||\mathfrak{z}_{k_m}||. \tag{2.7}$$

Since we have

$$||\mathfrak{z}_0|| \leqslant A \ ||\mathfrak{w}||, \qquad ||\mathfrak{z}_1|| \leqslant M_2 \ A^3 \ ||\mathfrak{w}||^2,$$

we shall assume that

$$||\mathfrak{z}_k|| \leqslant c_k \ A^{2k+1} \ ||\mathfrak{w}||^{k+1} \tag{2.8}$$

for all k, where the c_k's are numerical constants to be determined with $c_0 = 1$. Substituti this estimate in the right member of (2.7) we find that

$$||\mathfrak{z}_n|| \leqslant A \ \Sigma \ M_m \ c_{k_1} \ c_{k_2} \ldots c_{k_m} A^{m+2(k_1+k_2+-+k_m)}||\mathfrak{w}||^{m+k_1+k_2+ \ldots +k_m}$$

$$\leqslant A^{2n+1}||\mathfrak{w}||^{n+1}\Sigma \ M_m c_{k_1} c_{k_2} \ldots c_{k_m}$$

where we have used (2.5) and the fact that $A \geqslant 1$.

It follows that we can satisfy (2.8) for all values of k if we choose the c_k's so that they satisfy the recurrence relations

$$c_n = \Sigma\ M_m c_{k_1} c_{k_2} \dots c_{k_m},\quad n = 1, 2, 3, \dots \tag{2.9}$$

where $c_0 = 1$ and the summation is subject to (2.5) with $m \geqslant 2$ and the k_j's non-negative. Since on the right in (2.9) no subscript k_j can exceed $n-1$, the relations determine the coefficients c_n uniquely. But the coefficients c_n correspond to an inverse function problem of classical analysis. If we set

$$h(x) = \sum_{n=0}^{\infty} c_n x^{n+1}, \tag{2.10}$$

then $y = h(x)$ is the inverse function of

$$x = y - g(y), \tag{2.11}$$

where $g(y)$ is defined by (1.4) for $|y| < r$. This is easily verified by substituting (2.10) in (2.11) and equating coefficients. We obtain the recurrence relations (2.9).

This shows that the series (2.10) has a positive radius of convergence, R say. The value of R is found by the following argument. Since $g(y)$ has positive coefficients, the function $g'(y)$ is positive and monotone increasing in the interval $(0, r)$. Thus $g'(y)$ tends to a limit, j say, as $y \to r$. There are two cases to distinguish. If $j \leqslant 1$, then $y - g(y)$ is a positive monotone increasing function of y on the interval $(0, r)$ and has a unique inverse

$$y = h(x) \qquad \text{for} \qquad 0 \leqslant x \leqslant r - g(r).$$

Since the inverse is given by (2.10) for small values of x and this series has positive coefficients, we conclude that $R = r - g(r)$. If $j > 1$, then there exists a unique quantity u, $0 < u < r$, such that $g'(u) = 1$, and $y - g(y)$ is increasing in the interval $(0, u)$ but not beyond u. A unique inverse $y = h(x)$ then exists for $0 \leqslant x \leqslant u - g(u)$ and $h'(x)$ becomes infinite as $x \to u - g(u)$ while $h(x)$ tends to a finite limit. Now $R = u - g(u)$. We note that the series (2.10) converses for $x = R$, since it has positive coefficients and $h(x)$ tends to a finite limit as $x \to R$.

Using these estimates we see that

$$\sum_{n=0}^{\infty} |\alpha|^m \|\mathfrak{G}_{n+1}(\mathfrak{w})\| \leqslant (A|\alpha|)^{-1}\, h(A^2\|\mathfrak{w}\|\,|\alpha|), \tag{2.12}$$

provided

$$A^2\|\mathfrak{w}\|\ |\alpha| \leqslant R. \tag{2.13}$$

If this condition is satisfied, then the series (2.3) with $\mathfrak{z}_k$ replaced by $\mathfrak{G}_{k+1}(\mathfrak{w})$ will remain convergent when each term is replaced by its norm since $R < r$. This justifies the rearrangements necessary to obtain the recurrence relations (2.4) It follows that the series

$$\mathfrak{z} = \mathfrak{G}(\mathfrak{w}, \alpha) \equiv \sum_{n=0}^{\infty} \alpha^n \mathfrak{G}_{n+1}(\mathfrak{w}) \tag{2.14}$$

is indeed convergent in norm and satisfies equation (2.1). We can now set $\alpha = 1$ and see that $\mathfrak{z} = \mathfrak{G}(\mathfrak{w}, 1) \equiv \mathfrak{G}(\mathfrak{w})$ is a solution of the equation $\mathfrak{w} = \mathfrak{F}(\mathfrak{z})$ provided

$$\|\mathfrak{w}\| \leqslant A^{-2}R. \tag{2.15}$$

This proves assertions (*i*) and (*ii*).

In order to prove (*iii*) the uniqueness of the solution, it suffices to show the existence of a positive quantity a such that

$$\mathfrak{F}(\mathfrak{z}_1) = \mathfrak{F}(\mathfrak{z}_2), \quad \|\mathfrak{z}_1\| < a, \ \|\mathfrak{z}_2\| < a, \tag{2.16}$$

implies

$$\mathfrak{z}_1 = \mathfrak{z}_2. \tag{2.17}$$

To prove this, we choose a so that $A\, g'(a) < 1$. This is possible since $g'(y) \to 0$ with y. Using the multilinearity and (1.1) we get

$$\|\mathfrak{F}_n(\mathfrak{z}_1) - \mathfrak{F}_n(\mathfrak{z}_2)\| < n\, M_n a^{n-1} \|\mathfrak{z}_1 - \mathfrak{z}_2\| \tag{2.18}$$

if $\|\mathfrak{z}_1\| < a$, $\|\mathfrak{z}_2\| < a$. From (2.16) we get successively

$$\mathfrak{F}_1(\mathfrak{z}_1) - \mathfrak{F}_1(\mathfrak{z}_2) = \Sigma_2^\infty \, [\mathfrak{F}_n(\mathfrak{z}_2) - \mathfrak{F}_n(\mathfrak{z}_1)],$$

$$\mathfrak{z}_1 - \mathfrak{z}_2 = \Sigma_2^\infty \, L[\mathfrak{F}_n(z_2) - \mathfrak{F}_n(\mathfrak{z}_1)],$$

$$\|\mathfrak{z}_1 - \mathfrak{z}_2\| \leqslant \Sigma_2^\infty \, \|L[\mathfrak{F}_n(\mathfrak{z}_2) - \mathfrak{F}_n(\mathfrak{z}_1)]\| \leqslant \Sigma_2^\infty \, An\, M_n a^{n-1} \, \|\mathfrak{z}_2 - \mathfrak{z}_1\|,$$

and, finally,

$$\|\mathfrak{z}_1 - \mathfrak{z}_2\| \leqslant A\, g'(a)\, \|\mathfrak{z}_1 - \mathfrak{z}_2\|,$$

which implies (2.17). This completes the proof.

YALE UNIVERSITY

References

1. Graves, L. M., Implicit functions and differential equations in general analysis, *Trans. Amer. Math. Soc.*, **29**, (1927) 514-552.

2. Hildebrandt, T. H. and Graves, L. M., Implicit functions and their differentials in general analysis, *Ibid.*, 127-153.

3. Hille, E. and Phillips, R. S., *Functional Analysis and Semi-groups*, Coll. Publ. Amer. Math. Soc., **31**, Providence, R. I., 1957.

Reprint from
aequationes mathematicae
Vol. 2, fasc. 1, 1968 BIRKHÄUSER VERLAG BASEL pages 105–110

Some Properties of the Jordan Operator

EINAR HILLE (Eugene, Oregon)

To Professor Alexander Ostrowski on his 75th birthday

There is an extensive theory of Jordan algebras for which see, e.g., H. BRAUN and M. KOECHER [1]. This theory indicates the lively interest taken by the algebraist in the Jordan operator. In this brief note we aim to show that Jordan operators are also of some interest to the functional analyst.

1. Let $\mathfrak{B}$ be a non-commutative Banach algebra with unit element e and let $a \in \mathfrak{B}$ but not to the center of $\mathfrak{B}$. Define four linear bounded operators L_a, R_a, J_a and S_a on $\mathfrak{B}$ to $\mathfrak{B}$ by

$$L_a(x) = a\,x\,, \quad R_a(x) = x\,a\,, \quad J_a(x) = \tfrac{1}{2}(a\,x + x\,a)\,, \quad S_a(x) = a\,x\,a \tag{1}$$

so that

$$J_a = \tfrac{1}{2}(L_a + R_a)\,, \quad S_a = L_a R_a = R_a L_a\,. \tag{2}$$

These operators are elements of the algebra $\mathfrak{E}(\mathfrak{B})$ of linear bounded operators on $\mathfrak{B}$ to $\mathfrak{B}$. Here L_a is the image of the element a in the left regular representation of $\mathfrak{B}$ while R_a is the image in the right regular representation. J_a is the Jordan operator defined by a. On the other hand, S_a is the quadratic representation of $\mathfrak{B}$ as defined at a. For a general non-associative algebra the quadratic representation is given by

$$P_a = L_a(L_a + R_a) - L_{a^2} = R_a(L_a + R_a) - R_{a^2}$$

which in the associative case reduces to S_a.

We shall be concerned with the spectral properties of these operators. Spectra will be denoted by the letter 'σ', point spectra by the symbol $P\sigma$'. It is well known that the spectra of the operators L_a and R_a with respect to the algebra $\mathfrak{E}(\mathfrak{B})$ are identical and coincide with the spectrum of a with respect to $\mathfrak{B}$.

Since L_a and R_a commute the Gelfand representation theorem may be used to get a grip on the spectra of J_a and S_a. Their spectra with respect to $\mathfrak{E}(\mathfrak{B})$ are the same as the spectra with respect to the commutative algebra $\mathfrak{B}_0$ obtained by taking the second commutant of the algebra generated by L_a, R_a and the identity operator I. This gives

THEOREM 1. *The spectrum of J_a with respect to $\mathfrak{E}(\mathfrak{B})$ is contained in the mid-point set*

$$S_1 \equiv [\tfrac{1}{2}(\alpha + \beta) \mid \alpha \in \sigma(a),\, \beta \in \sigma(a)] \tag{3}$$

Received November 22, 1967 and, in revised form, February 3, 1968

while the spectrum of S_a is contained in the product set

$$S_2 \equiv [\alpha\beta \mid \alpha \in \sigma(a), \beta \in \sigma(a)]. \tag{4}$$

This result is well known to the algebraists in the finite dimensional case (See [1], Chapter VIII, Theorem 1.3.). In this case necessary and sufficient conditions are known in order that a particular combination $\frac{1}{2}(\alpha+\beta)$ or $\alpha\beta$ shall be a characteristic value. The following result is closely related to this criterion.

THEOREM 2. *Let $\alpha \in P\sigma[L_a]$, $\beta \in P\sigma[R_a]$ and suppose that $L_a(u)=\alpha u$, $R_a(v)=\beta v$, $u \neq 0$, $v \neq 0$. Then either the mapping $x \to uxv$ annihilates all x in $\mathfrak{B}$ or $\frac{1}{2}(\alpha+\beta) \in P\sigma[J_a]$* and $\alpha\beta \in P\sigma[S_a]$.

This follows from

$$J_a(uxv) = \tfrac{1}{2}(auxv + uxva) = \tfrac{1}{2}(\alpha+\beta)\,uxv, \tag{5}$$

$$S_a(uxv) = auxva = \alpha\beta\,uxv. \tag{6}$$

There is one case in which we can be sure that the second alternative holds. N. JACOBSON [4, p. 196] has introduced the notion of a prime ring. Such a ring has the important property that if $uxv=0$ for all x in the ring, then either $u=0$ or $v=0$. Finite matrix algebras and the operator algebras $\mathfrak{E}(\mathfrak{X})$ of linear bounded transformations on a Banach space into itself are examples of prime rings. This concept is eminently suitable for our problem and leads to

THEOREM 3. *If $\mathfrak{B}$ is a prime ring and the assumptions of Theorem 2 are valid, then $\frac{1}{2}(\alpha+\beta) \in P\sigma[J_a]$, $\alpha\beta \in P\sigma[S_a]$.*

COROLLARY. If $\mathfrak{B}=\mathfrak{M}_n$, the algebra of n by n matrices over the complex field, then $\sigma(J_a)=P\sigma(J_a)=S_1$ and $\sigma(S_a)=P\sigma(S_a)=S_2$.

For $\mathfrak{M}_n$ is a prime ring as well as a B-algebra under the usual definition of the algebraic operations and a suitable norm. Further the spectral values of a matrix are isolated and belong to the point spectrum of the corresponding L and R operators. All this is of course well known to the algebraists.

The following result belongs to the same range of ideas.

THEOREM 4. *If $\alpha \in P\sigma(L_a) \cap P\sigma(R_a)$ and if $au=ua=\alpha u$ where u is an idempotent $\neq 0$, then $\alpha \in P\sigma(J_a)$ and $\alpha^2 \in P\sigma(S_a)$.*

For

$$J_a(u) = \tfrac{1}{2}(au + ua) = \alpha u, \quad S_a(u) = aua = \alpha^2 u. \tag{7}$$

The following special case of Theorem 1 is of some independent interest.

THEOREM 5. *A necessary condition for the equation*

$$ax + xa = 0 \tag{8}$$

to have a non-trivial solution x is that a is either singular or has two spectral values of sum 0. *The condition is sufficient if $\mathfrak{B}$ is a prime ring.*

2. The resolvents of J_a and S_a may be found by a construction given by Yu. L. DALETSKY [2] in 1953. Let us cover the spectrum of a by a finite number of closed ε-disks. Let Σ_ε be the union of these disks and form the mid-point set Δ_ε^1 and product set Δ_ε^2.

$$\Delta_\varepsilon^1 = [\tfrac{1}{2}(\alpha+\beta)],\ \Delta_\varepsilon^2 = (\alpha\beta),\ \alpha\in\Sigma_\varepsilon,\ \beta\in\Sigma_\varepsilon\,. \tag{9}$$

Let Λ_ε^1 be the complement of Δ_ε^1 and Λ_ε^2 the complement of Δ_ε^2. We denote the resolvent of a by $R(\lambda, a)$ with similar notation for other resolvents.

THEOREM 6. *For any $\lambda\in\Lambda_\varepsilon^1$ the solution v of the equation*

$$\lambda y - \tfrac{1}{2}(a y + y a) = x \tag{10}$$

is given by

$$y = R(\lambda, J_a)[x] = \frac{1}{(2\pi i)^2}\int_{\Gamma_\varepsilon}\int_{\Gamma_\varepsilon}\frac{R(\alpha, a)\,\chi\,R(\beta, a)}{\lambda-\frac{1}{2}(\alpha+\beta)}\,d\alpha\,d\beta \tag{11}$$

where $\Gamma_\varepsilon=\partial\Delta_\varepsilon$. Similarly the solution z of the equation

$$\lambda z - a z a = x \tag{12}$$

for $\lambda\in\Lambda_\varepsilon^2$ is given by

$$z = R(\lambda, S_a)[x] = \frac{1}{(2\pi i)^2}\int_{\Gamma_\varepsilon}\int_{\Gamma_\varepsilon}\frac{R(\alpha, a)\,x\,R(\beta, a)}{\lambda-\alpha\beta}\,d\alpha\,d\beta\,. \tag{13}$$

We shall sketch the argument for the first case. The second is handled in a similar manner. Denoting the integrand in (11) by Q we can write $(\lambda I - J_a)[y]$ as

$$\iint[\lambda-\tfrac{1}{2}(\alpha+\beta)]\,Q\,d\alpha\,d\beta + \iint\tfrac{1}{2}(\alpha e - a)\,Q\,d\alpha\,d\beta + \iint\tfrac{1}{2}Q(\beta e - a)\,d\alpha\,d\beta\,.$$

The first integral reduces to x, the second and the third are both 0 as is seen with the aid of the identities

$$(\alpha e - a)\,R(\alpha, a) = e\,,\qquad R(\beta, a)\,(\beta e - a) = e$$

which are valid on the boundary. This shows that

$$(\lambda I - J_a)\,R(\lambda, J_a)[x] = x$$

for $\lambda\in\Lambda_\varepsilon^1$. In a similar manner one verifies that

$$R(\lambda, J_a)[\lambda x - J_a(x)] = x\,.$$

In the matrix case $\mathscr{R}(\alpha, \mathscr{A})$ and $\mathscr{R}(\beta, \mathscr{A})$ are rational functions of α and β, respectively. Thus if we substitute their well known expansions in partial fractions in (11) and (13), the evaluation of the double integrals is reduced to the evaluation of numerical integrals which act as coefficients for certain matrix products. In the case of $\mathscr{R}(\lambda, J_{\mathscr{A}})$ the numerical integrals are of the form

$$I_{jk\mu\nu}(\lambda) = \frac{1}{(2\pi i)^2} \int_{\gamma_j}\int_{\gamma_k} (\alpha - \lambda_j)^{-\mu} (\beta - \lambda_k)^{-\nu} [\lambda - \tfrac{1}{2}(\alpha + \beta)]^{-1} \, d\alpha \, d\beta . \tag{14}$$

Here λ_j is the jth characteristic value of the matrix $\mathscr{A}$, $1 \leqq j, k \leqq n$. Further μ is a positive integer $\leqq m_j$, the multiplicity of λ_j as a pole of $\mathscr{R}(\alpha, \mathscr{A})$. The integral is multiplied by the factor $(\mathscr{Q}_j)^{\mu-1} \mathscr{X} (\mathscr{Q}_k)^{\nu-1}$ where $(\mathscr{Q}_j)^0 = \mathscr{P}_j$ with $\mathscr{P}_j$ the idempotent and $\mathscr{Q}_j$ the nilpotent associated with the characteristic value. Further, γ_j is an ε-circle centered at λ_j. A simple calculation gives

$$I_{jk\mu\nu}(\lambda) = (\tfrac{1}{2})^{\mu+\nu-2} \frac{(\mu + \nu - 2)!}{(\mu - 1)!\,(\nu - 1)!} [\lambda - \tfrac{1}{2}(\lambda_j + \lambda_k)]^{-(\mu+\nu-1)} . \tag{15}$$

This shows that $\mathscr{R}(\lambda, J_{\mathscr{A}})$ is also a rational function of λ and its poles are located at all the points $\frac{1}{2}(\lambda_j + \lambda_k)$. A particular pole γ will usually come from several terms in the expansion and the multiplicity of the pole is $\leqq \max(m_j + m_k - 1)$ where j and k run through those integers for which

$$\gamma = \tfrac{1}{2}(\lambda_j + \lambda_k).$$

Actually the maximum is reached since we can choose a matrix $\mathscr{X}_0$ so that a particular

$$(\mathscr{Q}_j)^{m_j - 1} \mathscr{X}_0 (\mathscr{Q}_k)^{m_k - 1} \equiv \mathscr{X} \neq 0$$

while this $\mathscr{X}$ annihilates all other terms of the same degree in $(\lambda - \gamma)^{-1}$.

We get similar results for the operator $S_{\mathscr{A}}$. Here

$$\mathscr{R}(\lambda, S_{\mathscr{A}})[\mathscr{X}] = \sum \frac{(\mu + \nu - 2)!}{(\mu - 1)!\,(\nu - 1)!} \lambda_k^{\mu-1} \lambda_j^{\nu-1} (\lambda - \lambda_j \lambda_k)^{-(\mu+\nu-1)} (\mathscr{Q}_j)^{\mu-1} \mathscr{X} (\mathscr{Q}_k)^{\nu-1} \tag{16}$$

The summation extends over the spectral values λ_j of $\mathscr{A}$ and $\mu \leqq m_j$, $\nu \leqq m_k$. The $\mathscr{P}_j$ and $\mathscr{Q}_j$ are as previously defined. All the products $\lambda_j \lambda_k$ are exhibited as poles of the resolvent and the multiplicity of a particular pole $\gamma = \lambda_j \mu_k$ is again $\leqq \max(m_j + m_k - 1)$ where j and k run through the integers for which $\gamma = \lambda_j \lambda_k$. Again the maximum is reached since we can choose $\mathscr{X}$ as a 'buffer' so as to keep any particular term in the expansion while annihilating the rest.

These results for $\mathfrak{M}_n$ extend, at least locally, to arbitrary B-algebras by virtue of a theorem which is essentially due to S. R. FOGUEL [3].

THEOREM 7. *Let γ be an isolated point of the set S_1 and suppose that the equation*

$$\gamma = \tfrac{1}{2}(\alpha + \beta) \tag{17}$$

has only a finite number of solutions (α, β) where α and β are in $\sigma(a)$, say

$$(\alpha_1, \beta_1), (\alpha_2, \beta_2), \ldots, (\alpha_p, \beta_p). \tag{18}$$

Suppose that for each j the numbers α_j and β_j are poles of $R(\lambda, a)$ of order m_j and n_j, respectively. Then γ is a pole of $R(\lambda, J_a)$ of order $\leqq \max(m_j+n_j-1)$ and if $\mathfrak{B}$ is a prime ring the maximum is reached. If in this announcement S_1 is replaced by S_2 and $\frac{1}{2}(\alpha+\beta)$ by $\alpha\beta$, we obtain the corresponding result for $R(\lambda, S_a)$.

A proof can be based upon formulas (11) and (13) and the double decomposition of the spectrum suggested by (18). In the first case we get a number of integrals of type (14) which are holomorphic save for a single pole at $\lambda = \frac{1}{2}(\alpha_j + \beta_k)$ of order $\leqq m_j + n_k - 1$. In addition there are integrals arising from the complementary spectral sets all of which are holomorphic at the points under consideration. Only the integrals corresponding to $k = j$ give poles at γ. A similar argument applies to $R(\lambda, S_a)$.

3. In the preceding discussion we have obtained precise results only for the case in which the constituent singularities belong to the point spectrum of L_a and R_a. If they are isolated they are then poles of $R(\lambda, a)$. If $\mathfrak{B}$ is an operator algebra, the resolvents $R(\lambda, J_a)$ and $R(\lambda, S_a)$ may have other singularities than poles. The following simple example presents a case with a singular line.

We take $\mathfrak{B} = \mathfrak{E}[\mathfrak{C}]$ where $\mathfrak{C} = C[0, 1]$ so that $\mathfrak{B}$ is the algebra of linear bounded transformations on $C[0, 1]$ into itself. Take the two operators U and V defined by

$$U[f](t) = tf(t), \; V[g](t) = \int_0^t g(s)\, ds. \tag{19}$$

We consider J_U and aim to evaluate $R(\lambda, J_U)\, V$ operating on f in $C[0, 1]$. Here

$$R(\alpha, U)[f](t) = \frac{f(t)}{\alpha - t}, \tag{20}$$

so that $\sigma[U]$ is the interval $[0, 1]$ of the real axis. This interval is also the set S_1 for the operator J_U and the set S_2 for S_U. To prove this it is enough to exhibit an operator T in $\mathfrak{E}[\mathfrak{C}]$ such that $R(\lambda, J_U)\, T$ is not an element of $\mathfrak{E}[\mathfrak{C}]$ for any λ in S_1 and $R(\lambda, S_U)\, T$ is not an element of $\mathfrak{E}[\mathfrak{C}]$ for any λ in $S_2 = S_1$. Here we can take $T = V$ since

$$R(\lambda, J_U)\, V[f](t) = \frac{1}{(2\pi i)^2} \int_\Gamma \int_\Gamma \frac{1}{\alpha - t} \int_0^t \frac{f(s)\, ds}{\beta - s} \frac{d\alpha\, d\beta}{\lambda - \frac{1}{2}(\alpha + \beta)}. \tag{21}$$

Here Γ surrounds S_1 and a simple calculation gives

$$R(\lambda, J_U)\, V[f](t) = \int_0^t \frac{f(s)\, ds}{\lambda - \frac{1}{2}(s+t)}. \tag{22}$$

Similarly we get

$$R(\lambda, S_U)\, V[f](t) = \int_0^t \frac{f(s)\, ds}{\lambda - st}. \tag{23}$$

Both expressions define holomorphic functions in $C[0, 1]$ for λ not in $S_1 = S_2$. For λ on the interval $[0, 1]$ the integrals normally do not exist at least not as elements of $C[0, 1]$. Thus every point of this interval belongs to the spectrum of J_U as well as the spectrum of S_U.

The operator V is quasi-nilpotent and so are the corresponding operators J_V and S_V. In this case $S_1 = S_2$ reduces to a single point, $\lambda = 0$, which is an isolated essential singularity of the resolvents.

Finally it is a pleasure to acknowledge help from a referee who supplied the reference to [1] and thus led me to revise the paper and sharpen the results.

REFERENCES

[1] Braun, H. and Koecher, M., *Jordan-Algebren* (Springer-Verlag, 1966 [Grundlehren d. Math. Wiss., No. 128]).

[2] Daletsky, Yu. L., *On the Asymptotic Solution of a Vector Differential Equation* (Russ.), Doklady Akad. Nauk SSSR *92*, 881–884 (1953).

[3] Foguel, S. R., Sums and Products of Commuting Spectral Operators, *Ark. Mat. 3*, 449–461 (1957).

[4] Jacobson, N., *Structure of Rings* (Amer. Math. Soc., Providence R.I. 1956 [Coll. Publications No. *37*]).

University of Oregon

Chapter 17
Transfinite Diameters

The paper included in this chapter is

[139] Some geometric extremal problems

An introduction to Hille's papers on averaging processes and transfinite diameters can be found starting on page 705.

SOME GEOMETRIC EXTREMAL PROBLEMS *

EINAR HILLE

(Received 15 May 1965)

1. Background and notation

In a recent study of generalized transfinite diameters [4, 5] some geometric extremal problems were encountered. These form the subject matter of this note.

A generalized transfinite diameter is based on an averaging process defined by a strictly monotone and continuous function $F(u)$, $0 < u$. If X is a complete metric space, E a closed bounded set in X, let $P_1, P_2, \ldots, P_n$ be n points of E and set $\delta_{jk} = d(P_j, P_k)$. The F-average of the $N = \frac{1}{2}n(n-1)$ numbers δ_{jk} is defined by

$$(1) \qquad F(a) = \frac{1}{N} \sum_{1 \leq j < k \leq n} F(\delta_{jk}).$$

The first problem is to maximize this average. If for a given n the supremum of a is denoted by $F\text{-}\delta_n(E)$ then the sequence $\{F\text{-}\delta_n(E)\}$ is decreasing. Its limit $F\text{-}\delta_0(E)$ is by definition the transfinite F-diameter of E. The case $F(u) = u^p$ will figure prominently in the following. Here we write $\delta_0^{(p)}(E)$ instead.

In the following we restrict ourselves essentially to inner product spaces such as m-dimensional Euclidean space R^m and Hilbert space H. The set E will be the unit ball of such a space. We write U^m in R^m and U^H in H.

2. The basic identity

A straightforward computation shows that for any n vectors X_j in an inner product space

$$(2) \qquad \sum_{1 \leq j < k \leq n} ||X_j - X_k||^2 + ||\sum_{j=1}^{n} X_j||^2 = n \sum_{j=1}^{n} ||X_j||^2.$$

* This note represents work done at the 1965 S.R.I. of the Australian Mathematical Society whose invitation and support is gratefully acknowledged. The research was supported in part by the U.S. Air Force through the Office of Scientific Research of the Air Research and Development Command, under Grant No. AF-AFOSR 776-65.

The particular case $n = 2$ with

$$||X_1+X_2||^2+||X_1-X_2||^2 = 2(||X_1||^2+||X_2||^2) \tag{3}$$

is very well known and is often used as a characterization of an inner product space.

For the case of real or complex numbers, the identity has been known for some thirty odd years and has been used in the geometry of numbers by authors such as R. Remak, J. G. van der Corput, and G. Schaake [1]. I am indebted to Professor Kurt Mahler for this information. For the case of vectors in R^m priority appears to be due to L. Fejes Thót [2] who, however, gives credit to E. Makai. Closely related work was done by J. Schopp [6]. Some of the results given below (especially Theorems 1 and 2 for $p = 1$) are included in the work of these authors. I am indebted to Professor A. Rényi for this information.

3. The mean square case

From the identity (2) we obtain

THEOREM 1. *If the* X's *are unit vectors in an inner product space, then*

$$\sum_{1 \leqq j<k \leqq n} ||X_j-X_k||^2 \leqq n^2 \tag{4}$$

with equality if and only if $\sum X_j = 0$.

Let us now take $F(u) = u^2$ in formula (1). Using (4) we see that

$$\max \left(\frac{1}{N} \sum \delta_{jk}^2\right)^{\frac{1}{2}} = \sqrt{2} \sqrt{\frac{n}{n-1}}$$

and that the maximum value is attained for any choice of points P_j, $j = 1, 2, \ldots, n$, on the unit sphere under consideration such that the corresponding unit vectors $X_j = \overrightarrow{OP_j}$ satisfy $\sum_1^n X_j = 0$. Hence

$$\delta_n^{(2)}(U^m) = \delta_n^{(2)}(U^H) = \sqrt{2} \sqrt{\frac{n}{n-1}}, \tag{5}$$

and, letting $n \to \infty$,

$$\delta_0^{(2)}(U^m) = \delta_0^{(2)}(U^H) = \sqrt{2}. \tag{6}$$

Since for any admissible function F we have

$$\sqrt{2} \leqq F\text{-}\delta_0(U^H) \leqq 2, \tag{7}$$

we see that the minimum value $\sqrt{2}$ is attained for $F(u) = u^2$.

4. The pth power case, $0 < p < 2$.

Here we use Hölder's inequality to get

$$\sum_{j<k} \delta_{jk}^p \leqq N^{1-p/2} \Big(\sum_{j<k} \delta_{jk}^2 \Big)^{p/2} \leqq N^{1-p/2} n^p \tag{8}$$

with equality in the first place if and only if all δ_{jk} are equal and in the second place if and only if $\sum_1^n X_j = 0$. For each n the maximizing configuration is unique up to a rotation. This gives

THEOREM 2. *For $0 < p < 2$ the p-th power mean of the lengths of the edges of an $(n+1)$-simplex inscribed in U^n is an absolute maximum if the simplex is regular.*

Formula (8) gives upon letting $n \to \infty$

$$\delta_0^{(p)}(U^H) = \sqrt{2} \tag{9}$$

but gives no information about U^m except the trivial

$$\delta_0^{(p)}(U^m) \leqq \sqrt{2}.$$

5. The pth power case, $2 < p$

Here the discussion is based on the observation that $u^p \leqq u^2$ for $0 \leqq u \leqq 1$ with equality if and only if $u = 0$ or 1. This leads to the following sequence of inequalities

$$\begin{aligned} \sum_{j<k} \delta_{jk}^p &= 2^p \sum_{j<k} (\tfrac{1}{2}\delta_{jk})^p \leqq 2^p \sum_{j<k} (\tfrac{1}{2}\delta_{jk})^2 \\ &= 2^{p-2} \sum_{j<k} \delta_{jk}^2 \leqq 2^{p-2} n^2. \end{aligned} \tag{10}$$

Here equality holds in the first place if and only if all δ_{jk} are 0 or 2 and in the second place if and only if $\sum_1^n X_j = 0$. If n is even, $n = 2\nu$, both conditions can be satisfied by choosing an arbitrary unit vector X and setting $X_j = X$ for $1 \leqq j \leqq \nu$, and $X_j = -X$ for $\nu+1 \leqq j \leqq 2\nu$. It follows that

$$\delta_{2\nu}^{(p)}(U^m) = \delta_{2\nu}^{(p)}(U^H) = 2^{1-1/p} \left(\frac{2\nu}{2\nu - 1} \right)^{1/p}$$

Hence

$$\delta_0^{(p)}(U^m) = \delta_0^{(p)}(U^H) = 2^{1-1/p}, \quad 2 < p. \tag{11}$$

The last result has a bearing on the problem of finding $\delta_0^{(p)}(U_p)$ where U_p is the unit ball in the space (l_p) of sequences (ξ_j) such that

$$||x|| = (\sum |\xi_j|^p)^{1/p} < \infty.$$

If in the preceding argument we choose $X = (1, 0, 0, \ldots)$, then this is a unit vector in (l_p) and we see that

$$\delta_{2\nu}^{(p)}(U_p) \geqq 2^{1-1/p}\left(\frac{2\nu}{2\nu-1}\right)^{1/p},$$

so that

$$\delta_0^{(p)}(U_p) \geqq 2^{1-1/p}, \; p > 2. \tag{12}$$

Here it is not unlikely that the sign of equality holds, but the argument does not show it.

6. Convex F-functions

The results obtained above hold also for certain classes of convex functions $F(u)$.

THEOREM 3. *If $F(u)$ is decreasing and strictly convex then*

$$F\text{-}\delta_0(U^H) = \sqrt{2}. \tag{13}$$

Moreover, the analogue of Theorem 2 holds for such averages.

PROOF. We have

$$F(a) = \frac{1}{N}\sum_{j<k} F(\delta_{jk}) \geqq F\left(\frac{1}{N}\sum \delta_{jk}\right) \geqq F\left(\sqrt{2}\sqrt{\frac{n}{n-1}}\right),$$

so that

$$a \leqq \sqrt{2}\sqrt{\frac{n}{n-1}}$$

with equality if and only if $\sum_1^n X_j = 0$ and δ_{jk} are equal. This means that for each n the maximizing configuration is given by the regular n-simplex which is unique up to a rotation. Passing to the limit with n we get formula (13). This completes the proof.

The same type of proof gives

THEOREM 4. *If $F(u)$ is increasing and strictly concave, then formula* (13) *holds as well as the analogue of Theorem* 2.

This includes the case $F(u) = u^p$, $0 < p \leqq 1$, as well as $F(u) = \log u$ which leads to the geometric mean. Professor Basil Rennie of the RAAF Academy has called my attention to the fact that the theorem remains valid if we replace $F(u)$ by $F(\sqrt{u})$. It then includes $F(u) = u^p$ for $0 < p < 2$. The proof of the extension is immediate.

7. Other cases

We have a less favorable situation if $F(u)$ is increasing and strictly concave. The special case $F(u) = u^p$, $2 < p$, indicates that the transfinite F-diameter of U^H depends effectively on F. In the general case all we can get are inequalities.

THEOREM 5. *If F_1, F_2, F are increasing and strictly convex and if for* $0 \leqq u \leqq 2$

$$F_1(u) \leqq F(u) \leqq F_2(u),$$

then

$$F_1\text{-}\delta_0(U^H) \leqq F\text{-}\delta_0(U^H) \leqq F_2\text{-}\delta_0(U^H). \tag{14}$$

Since the graph of $v = F(u)$ lies below the chord joining $[0, F(0)]$ with $[2, F(2)]$ we get

$$\begin{aligned}\frac{1}{N}\sum_{j<k} F(\delta_{jk}) &\leqq F(0)+\tfrac{1}{2}[F(2)-F(0)]\frac{1}{N}\sum_{j<k}\delta_{jk} \\ &\leqq F(0)+\tfrac{1}{2}[F(2)-F(0)]\left(2\frac{n}{n-1}\right)^{\frac{1}{2}} \\ &\to F(0)+\tfrac{1}{2}\sqrt{2}[F(2)-F(0)]\end{aligned}$$

whence

$$F\text{-}\delta_0(U^H) \leqq F^{-1}\{(1-\tfrac{1}{2}\sqrt{2})F(0)+\tfrac{1}{2}\sqrt{2}F(2)\}. \tag{15}$$

Using the convexity of F we see that the right member exceeds $\sqrt{2}$ and since F is increasing we have

$$F\text{-}\delta_0(U^H) < 2 \tag{16}$$

for this class of means.

In the opposite direction we observe that

$$F\text{-}\delta_0(U^H) > \sqrt{2} \tag{17}$$

if $F(\sqrt{u})$ is increasing and strictly convex in $(0, 4)$. This implies that

$$F(\sqrt{2}) < \tfrac{1}{2}[F(0)+F(2)] \tag{18}$$

and this is the only implication of the convexity that we need. To establish (17) we proceed as in the proof of formula (11). We take $n = 2\nu$ and choose ν unit vectors X and ν unit vectors $-X$. In formula (1) we have $\nu^2-\nu$ terms $F(0)$ and ν^2 terms $F(2)$ while $N = \nu(2\nu-1)$. It follows that $F(a)$ is at least equal to

$$\frac{\nu-1}{2\nu-1}F(0)+\frac{\nu}{2\nu-1}F(2) \to \tfrac{1}{2}[F(0)+F(2)] > F(\sqrt{2})$$

and this implies (17).

8. Remarks on other polyhedra

It was found in Section 4 that the extremal configuration in U^H for a given n is produced by n unit vectors whose sum is 0 and whose end points are equidistant from each other. This is a regular n-simplex which can be inscribed in U^{n-1}. In particular, for $n=4$ we are dealing with extremal properties of a regular tetrahedron: Of all 4-simplexes inscribed in U^3 the regular tetrahedron shows the largest pth power average for the lengths of its sides. This holds for $p < 2$. For $p = 2$ the regular tetrahedron gives $\sum \delta_{jk}^2$ its maximum value 16 but now there are infinitely many 4-simplexes which give the same square sum.

This raises the question whether or not other regular polyhedra have extremal properties with respect to sums of squares of lengths of the sides. In the case of the octahedron the square sum is 24 and for the cube 16 if the circum radius is 1. Here it turns out that the regular polyhedra are far from maximizing the square sums. There exist degenerate "octahedra" with square sum 32 and degenerate "hexahedra" with square sum 48. The latter are of some interest and are obtained by the following construction. In an ordinary cube let the eight vertices be given by the unit vectors X_1 to X_8. Let the twelve edges be denoted by the symbols

$$\begin{matrix} (1,2), & (2,3), & (3,4), & (4,1), & (1,6), & (3,8) \\ (5,6), & (6,7), & (7,8), & (8,5), & (4,5), & (2,7). \end{matrix} \tag{19}$$

Here (j, k) is the line joining the endpoints of the vectors X_j and X_k. Here each side has the length $\frac{2}{3}\sqrt{3}$ and the square sum is $12 \cdot \frac{4}{3} = 16$. Let us now deform the cube by moving X_1, X_3, X_5, X_7 into coincidence, say with the vector $(0, 0, 1)$, and by moving X_2, X_4, X_6, X_8 into the antipodal position $(0, 0, -1)$. Let (j, k) still denote the "edges" of the new "hexahedron" where (j, k) runs over the list (19). Each of these edges will now have the length 2 and the square sum is 48. This shows that

$$\max \sum \delta_{jk}^2 \geqq 48. \tag{20}$$

Moreover

$$\max \sum \delta_{jk} \geqq 24 \tag{21}$$

while the regular cube would give only $8\sqrt{3}$. Thus the regular cube does not even maximize the sum of the lengths of the edges of hexahedra inscribed in a fixed sphere. I have not examined corresponding properties of dodecahedra and icosahedra.

References

[1] J. G. van der Corput and G. Schaake, "Anwendung einer Blichfeldschen Beweismethode in der Geometrie der Zahlen", *Acta Arithmetica*, 2 (1937) 152-160.
[2] L. Fejes Thót, "On the sum of distances determined by a point set", *Acta Mathematica Academiae Scientiarum Hungaricae*, 7 (1957) 397-401.
[3] L. Fejes Thót, "Über eine Punktverteilung auf der Kugel", *Ibid*, 10 (1959) 13-19.
[4] E. Hille, "A note on transfinite diameters", *Journal d'Analyse*, Jerusalem, 14 (1965) 209-224.
[5] E. Hille, "Topics in classical analysis", in T. L. Saaty, *Lectures on Modern Mathematics*, Volume III, Wiley, New York, 1965. See pp. 34-43.
[6] J. Schopp, "Simplexungleichungen", *Elemente der Mathematik*, 16 (1961) 13-16.

University of California, Irvine
Irvine, California, U.S.A.

Commentaries on Hille's Papers

(Bracketed numbers are from the Bibliography; items mentioned by author's name with a superscript can be found in the References at the end of each commentary.)

Chapter 1. Functional Equations

A commentary on [8] A Pythagorean functional equation

For simplicity of writing we denote the set of all entire functions of a complex variable z by E.

In this paper the following functional equation is considered.

$$|f(s+it)|^2 = |f(s)|^2 + |f(it)|^2, \tag{1}$$

where s,t are variables and f is in E.

Equation (1) is closely related to Cauchy's following functional equation (Aczél[1]) which is fundamental in the theory of functional equations:

$$f(x+y) = f(x) + f(y), \tag{2}$$

where x,y are complex variables and f belongs to E and is real on the real axis.

We shall prove without use of the explicit form of the solution of (2) that (2) implies (1). Since f belongs to E and is real on the real axis, f can be expanded in a power series with real coefficients in $|z| < \infty$. So we have $\overline{f(s+it)} = f(s-it)$, where s, t are real variables. Hence we have $2|f(s+it)|^2 = |f(s+it)|^2 + |f(s-it)|^2$. By (2) and by the parallelogram law we have $|f(s+it)|^2 + |f(s-it)|^2 = |f(s)+f(it)|^2 + |f(s)-f(it)|^2 = 2|f(s)|^2 + 2|f(it)|^2$, which implies (1).

Hille has proved the following theorem:

Theorem 1. *If* $f \in E$, *then the only solutions of* (1) *are* $f(z) = C \sin cz$, $f(z) = C \sinh cz$, and $f(z) = Cz$, *where* C *is an arbitrary complex constant and* c *is an arbitrary real constant.*

To prove Theorem 1 he applied the fact that $\log|f|$ is a harmonic function in the domain where $f(z) \neq 0$. Here we shall give an alternative proof of Theorem 1. Operating on the equation (1) with $\partial^2/\partial s \partial t$ and using the Cauchy-Riemann equations, we have $\operatorname{Im}(f''(z)/f(z)) = 0$ in the domain where $f(z) \neq 0$. Hence, by a property of regular functions and by the Identity Theorem we have the differential equation $f''(z) = Kf(z)$ in $|z| < \infty$, where K is a real constant. Solving this differential equation and taking into account the fact that $f(0) = 0$ completes the proof of the theorem.

A simple calculation shows that (1) implies the following functional equation:

$$|f(x+y)|^2 + |f(x-y)|^2 = |f(x+\bar{y})|^2 + |f(x-\bar{y})|^2, \tag{3}$$

where f belongs to E and x, y are complex variables. Therefore, (3) is a generalization of (1). Putting $x = y = (s + it)/2$ (s, t real) in (3) yields the functional equation

$$|f(s+it)|^2 + |f(0)|^2 = |f(s)|^2 + |f(it)|^2, \tag{4}$$

where f belongs to E and s, t are real variables. Therefore also, (4) is a generalization of (1). Furthermore, a simple calculation shows that (4) implies (3), and so (3) is equivalent to (4).

The following theorem has been proved (see Haruki[6,9]):

Theorem 2. *If f is in E, then the only solutions of* (3) *(and so* (4)*) are* $f(z) = A \sin cz + B \cos cz$, $f(z) = A \sinh cz + B \cosh cz$, *and* $f(z) = Az + B$, *where A, B are arbitrary complex constants and c is an arbitrary real constant.*

Generalizations of (3) to functional inequalities have been investigated in Haruki[9].

The following two functional equations are consequences and thus generalizations of (1):

$$|f(x+y) + f(x-y)| = |f(x+\bar{y}) + f(x-\bar{y})|, \tag{5}$$

$$|f(x+y) - f(x-y)| = |f(x+\bar{y}) - f(x-\bar{y})|, \tag{6}$$

where f belongs to E and x, y are complex variables. It has been proved that (3) and (5) are equivalent (see Haruki[6,9]) and that (5) implies (6) (see Haruki[9]). Furthermore, the following theorem has been proved in Haruki[2,6,7,8,10,11]:

Theorem 3. *If f is in E, then the only solutions of* (6) *are* $f(z) = A \sin cz + B \cos cz + C$, $f(z) = A \sinh cz + B \cosh cz + C$, *and* $f(z) = Az^2 + Bz + C$, *where A, B, C are arbitrary complex constants and c is an arbitrary real constant.*

Equation (6) is related to Ivory's following theorem in geometry (see Haruki[2,5,7,8,10,11]): For a family of confocal conics, let P, Q, R, S be the four vertices of a curvilinear rectangle formed by any four conics arbitrarily chosen. Then $\overline{PR} = \overline{QS}$ holds. Geometrically speaking, (6) is the above

Ivory property and the three solutions of (6) listed in Theorem 3 characterize the confocal conics. Thus, Ivory's Theorem characterizes these conics.

Generalizations of (6) to functional inequalities have been investigated in Haruki[11].

We shall prove that the following functional equation is a consequence of (1):

$$|f(s+it)| = |f(s) - f(it)|, \tag{7}$$

where f belongs to E and s,t are real variables. Putting $x = y = (s+it)/2$ in (6) and using the fact that $f(0) = 0$, which follows from (1), yields (7). The following theorem has been proved in Haruki[7]:

Theorem 4. *If f is in E, then the only solutions of* (7) *are $f(z) = A \sin cz + B \cos cz - B$, $f(z) = A \sinh cz + B \cosh cz - B$, and $f(z) = A z^2 + Bz$, where A, B are arbitrary complex constants and c is an arbitrary real constant.*

We shall prove that (1) implies Robinson's following functional equation (Robinson[13]):

$$|f(s+it)| = |f(s) + f(it)|, \tag{8}$$

where f belongs to E and s, t are real variables. By the parallelogram law and by (1) we have $|f(s) + f(it)|^2 + |f(s) - f(it)|^2 = 2|f(s)|^2 + 2|f(it)|^2 = 2|f(s+it)|^2$. On the other hand, (7) holds, as already proved. Hence, we have (8). The following theorem has been proved in Robinson[13]:

Theorem 5. *If f is in E, then the only solutions of* (8) *are $f(z) = A \sin cz$, $f(z) = A \sinh cz$, and $f(z) = Az$, where A is an arbitrary complex constant and c is an arbitrary real constant.*

Now we shall give a geometric meaning of (1) by proving the following theorem:

Theorem 6. *For a family of confocal conics, let P be a point of intersection of two arbitrary conics which intersect the principal axis of this family at Q, R and let F be a common focus. Then and only then $\overline{FP} = \overline{QR}$ holds.*

We discuss two cases:

Case 1, confocal ellipses and hyperbolas: Consider the mapping $w = \sin^2 z$ from the z-plane into the w-plane. We find that the mapping $w = \sin^2 z$ transforms horizontal and vertical lines in the z-plane, respectively, into confocal ellipses and hyperbolas in the w-plane. The common foci of all these conics are at $w = 0$ and $w = 1$. Since $f(z) = \sin z$ is a solution of (1), we have $|\sin^2 (s + it)| = |\sin^2 s| + |\sin^2 it|$, where s, t are real variables. Furthermore, we have $|\sin^2 (s + it)| = \overline{FP}$ and $|\sin^2 s| + |\sin^2 it| = \overline{FQ} + \overline{FR}$. Hence, we have $\overline{FP} = \overline{FQ} + \overline{FR} = \overline{QR}$.

Case 2, confocal parabolas: In this case we have, in a similar manner, $\overline{FP} = \overline{QR}$ by use of the mapping $w = z^2$. The converse follows from Theorem 1.

It is easy to prove the following theorem with the aid of Theorem 6:

Theorem 7. *For a family of confocal conics, let P, Q, R, S be the four vertices of a curvilinear rectangle formed by any four conics arbitrarily chosen from the family, and let F be a common focus of this family. Then* $\overline{FP} + \overline{FR} = \overline{FQ} + \overline{FS}$ *holds.*

Hille's proof of Theorem 1 includes a proof of the following theorem:

Theorem 8. *If f belongs to E and satisfies the functional equation*

$$|f(s + it)| = |f(s)| + |f(it)|, \tag{9}$$

where s, t are real variables, then the only solutions of (9) *are* $f(z) = A \sin^2 cz$, $f(z) = A \sinh^2 cz$, *and* $f(z) = Az^2$, *where A is an arbitrary complex constant and c is an arbitrary real constant.*

It has been proved in Haruki[4] that the equality in Theorem 7 characterizes the confocal conic sections. To prove this fact the following functional equation was considered:

$$|f(x + y)| + |f(x - y)| = |f(x + \bar{y})| + |f(x - \bar{y})|, \tag{10}$$

where f belongs to E and x, y are complex variables. A simple calculation shows that (10) is a consequence of (9).
J. Aczél and H. Haruki

A commentary on [14] A class of functional equations

In this paper Hille generalizes the method of [8]. The equation (1) is evidently a special case of the following general problem which is considered in this paper: Given a rational function R of u and v, is it possible to determine a meromorphic function f in $|z| < \infty$ such that

$$|f(s+it)| = R(|f(s)|, |f(it)|), \tag{11}$$

where s, t are real variables? As illustrations of his method, Hille studies the following three functional equations:

$$|f(s+it)| = u \tag{12}$$

$$|f(s+it) = uv, \tag{13}$$

$$|f(s+it)| = (u+v)/(1+uv), \tag{14}$$

where $u = |f(s)|$, $v = |f(it)|$.

Let M $(\supset E)$ be the set of all the nonconstant functions each of which is meromorphic in $|z| < \infty$. We put $\phi(s, t) = |f(z)| = |f(s+it)|$, where $z = s + it$ (s, t real) and $f \in M$. Now we shall consider the following two conditions:

(C1) There exists a neighborhood N where $\phi \neq 0$, $\phi_t \neq 0$, and the quotient $\phi_s(s, t)/\phi_t(s, t) = A(s)B(t)$.

(C2) There exists a neighborhood N where $\phi \neq 0$, $\phi_s \neq 0$, and the quotient $\phi_t(s, t)/\ \phi_s(s, t) = A(s)B(t)$.

We shall consider the set of all the functions each of which satisfies either the condition (C1) or the condition (C2) and denote this set by S $(\subset M)$. The following two theorems have been proved in Haruki[5] :

Theorem 9. *The only functions that belong to S are given by:*

$$f(z) = C\left[\frac{\mu + \lambda \mathrm{sn}^2\left[\frac{1}{2}\left(\frac{\mu}{n}z + A\right); \frac{\lambda}{\mu}\right]}{\mu - \lambda \mathrm{sn}^2\left[\frac{1}{2}\left(\frac{\mu}{n}z + A\right); \frac{\lambda}{\mu}\right]}\right]^n \quad (n = \pm 1, \pm 2, \pm 3, \ldots), \tag{i}$$

where sn *is the elliptic sine function, C $(\neq 0)$ and A are arbitrary complex constants, and λ^2 and μ^2 are the two roots of the quadratic equation $at^2 + bt + c = 0$ with real coefficients a $(\neq 0)$, b, c $(\neq 0)$;*

$$f(z) = C \exp(A \sin \alpha z + B \cos \alpha z), \tag{ii}$$

$$f(z) = C \exp(A \sinh \alpha z + B \cosh \alpha z),$$

$$f(z) = C \exp(A_1 z^2 + A_2 z + A_3),$$

where $C\ (\neq 0)$, A, B, $(|A| + |B| > 0)$, A_2, A_3 *are arbitrary complex constants,* $\alpha\ (\neq 0)$ *is an arbitrary real constant, and* A_1 $(|A_1| + |A_2| > 0)$ *is an arbitrary real or purely imaginary constant; and*

$$f(z) = C \tan^n (\alpha z + A) \qquad (n = \pm 1, \pm 2, \pm 3, \ldots), \tag{iii}$$

$$f(z) = C \tanh^n (\alpha z + A) \qquad (n = \pm 1, \pm 2, \pm 3, \ldots),$$

and

$$f(z) = (Cz + A)^n \qquad (n = \pm 1, \pm 2, \pm 3, \ldots),$$

where $C\ (\neq 0)$, A *are arbitrary complex constants and* $\alpha\ (\neq 0)$ *is an arbitrary real constant.*

Theorem 10. *The only entire functions that belong to S are given by:*

$$f(z) = C \cosh^n (\alpha z + A) \qquad (n = 1, 2, 3, \ldots), \tag{i}$$

$$f(z) = C \cos^n (\alpha z + A) \qquad (n = 1, 2, 3, \ldots),$$

where $C\ (\neq 0)$, A *are arbitrary complex constants and* $\alpha\ (\neq 0)$ *is an arbitrary real constant;*

$$f(z) = C \exp(A \sin \alpha z + B \cos \alpha z), \tag{ii}$$

$$f(z) = C \exp(A \sinh \alpha z + B \cosh \alpha z),$$

$$f(z) = C \exp(A_1 z^2 + A_2 z + A_3),$$

where $C\ (\neq 0)$, A, B $(|A| + |B| > 0)$, A_2, A_3 *are arbitrary complex constants,* $\alpha\ (\neq 0)$ *is an arbitrary real constant, and* A_1 $(|A_1| + |A_2| > 0)$ *is an arbitrary real or purely imaginary constant; and*

$$f(z) = (Cz + A)^n \qquad (n = 1, 2, 3, \ldots), \tag{iii}$$

where $C\ (\neq 0)$, A *are arbitrary complex constants.*

A simple calculation shows that every nonconstant entire solution of (1) or (9) or (12) belongs to S. Hence, applying Theorem 10 and selecting the appropriate solutions in each case from the functions listed in Theorem 10 yields Theorem 1, Theorem 8, and the following two results:

(I) *The only entire solution of* (12) *is* $f(z) = A \exp(cz)$, *where* A *is an arbitrary complex constant and* c *is an arbitrary real constant.*
(II) *The only entire solutions of* (13) *are* $f(z) = 0$ *or* $f(z) = \exp(cz^2 + Az + i\theta)$, *where* c, θ *are arbitrary real constants and* A *is an arbitrary complex constant.*

Suppose that $f \in M$ satisfies (14). Then a calculation of some length shows that $f \in S$. However, it is difficult to find the most general solution of (14) (obtained by Hille) by the use of Theorem 9.

If we write $g(z) = |f(s + it)|$, then (11) goes over into

$$g(s + it) = R[g(s), g(it)]. \tag{15}$$

A more restrictive equation, of which (15) is a special case, is

$$g(x + y) = R[g(x), g(y)], \tag{16}$$

where x, y are complex variables. For the general solution of this equation under weak regularity suppositions see A. Kuwagaki[12]. He has proved the following theorem:

Theorem 11. *The only nonconstant continuous functions satisfying a functional equation of the form* (15) *are given by*

$$f(z) = (ae^{kz} + b)/(ce^{kz} + d) \tag{17}$$

and

$$f(z) = (az + b)/(cz + d), \tag{18}$$

where a, b, c, d, k *are arbitrary complex constants, except that* $k(ad - bc) \neq 0$.

In (15) and (16), just as in (11), R is a rational function. Further generalizations have been examined (see Aczél[1]). We mention only the following result (Haruki[3,6]):

Theorem 12. *All functions of the class M (nonconstant and meromorphic) that satisfy functional equations of the form*

$$g(x + y) = F[g(x), g(y)],$$

where F is a one-valued complex function of two variables, are given by (17) *and* (18).

J. Aczél and H. Haruki

References

1. J. Aczél, *Lectures on Functional Equations and Their Applications,* Academic Press, New York and London, 1966, pp. 31-49.
2. H. Haruki, *On Ivory's theorem,* Math. Japon. vol. 1 (1949) p. 151.
3. H. Haruki, *On the functional equation* $f(x + y) = F\{f(x), f(y)\}$, Sci. Rep. Osaka Univ. no. 6 (1957) pp. 11-12.
4. H. Haruki, *On a characteristic property of confocal conic sections,* Proc. Japan Acad. vol. 39 (1963) pp. 564-565.
5. H. Haruki, *On a certain family of meromorphic functions,* Proc. Japan Acad. vol. 40 (1964) pp. 88-93.
6. H. Haruki, *Studies on certain functional equations from the standpoint of analytic function theory,* Sci. Rep. Coll. Gen. Ed. Osaka Univ. vol. 14, no. 1 (1965) pp. 1-40.
7. H. Haruki, *On the functional equations* $|f(x + iy)| = |f(x) + f(iy)|$ *and* $|f(x + iy)| = |f(x) - f(iy)|$ *and on Ivory's theorem,* Canad. Math. Bull. vol. 9 (1966) pp. 473-480.
8. H. Haruki, *On the "rectangle functional equation" and the functional equation* $|f(x + y) - f(x - y)| = |f(x + \bar{y}) - f(x - \bar{y})|$ *connected with Ivory's theorem,* Univ. Beograd. Publ. Elektrotehn. Fak. Ser. Mat. Fiz. no. 175-179 (1967) pp. 9-14.
9. H. Haruki, *On inequalities generalizing the Pythagorean functional equation and Jensen's functional equation,* Pacific J. Math. vol. 26 (1968) pp. 85-90.
10. H. Haruki, *On parallelogram functional equations,* Math. Z. vol. 104 (1968) pp. 358-363.
11. H. Haruki, *On inequalities generalizing a functional equation connected with Ivory's theorem,* Amer. Math. Monthly vol. 75 (1968) pp. 624-627.
12. A. Kuwagaki, *Sur l'équation fonctionnelle* $f(x + y) = R[f(x), f(y)]$, Mem. Coll. Sci. Univ. Kyoto Ser. A vol. 26 (1951) pp. 139-144.
13. R. M. Robinson, *A curious trigonometric identity,* Amer. Math. Monthly vol. 64 (1957) pp. 83-85.

Chapter 2. Zero Point Problems and Asymptotic Integration of Second Order Linear Differential Equations

An introduction to Hille's four papers on complex oscillation theory

Of the four papers in this section, three ([5], [15], and [23]) deal with the distribution of the complex zeros of the solutions of the complex linear differential equation of the second order:

$$\frac{d}{dz}\left[K(z)\,\frac{dw}{dz}\right] + G(z)w = 0, \tag{1}$$

where $K(z) \neq 0$ and $G(z)$ are assumed to be single-valued analytic functions in a simply connected region D in the complex plane. The remaining paper [12] presents an existence theorem for a certain singular integral equation of Volterra's type which Hille used to study the complex zeros of the solutions of (1) near irregular points.

In [5], methods of constructing various types of zero-free regions in the complex plane (in which a solution and its derivative cannot vanish) are given; it is also shown that the zeros of a solution form sequences of a regular type and that the argument of the difference between two zeros satisfies certain bounds determined by the coefficients of the equation. Distribution of zeros in the neighborhood of an irregular-singular point is treated in another paper [3], using the integral equation in [12].

The basic tool employed in constructing zero-free regions in paper [5] is the following integral expression, called the Green's transform of the equation (1):

$$[\overline{w}_1\, w_2]_{z_1}^{z_2} - \int_{z_1}^{z_2} \frac{|w_2|^2}{\overline{K(z)}}\,\overline{dz} + \int_{z_1}^{z_2} |w_1|^2\; G(z)dz = 0, \tag{2}$$

where $w_1 = w$, $w_2 = K(z)\,(dw/dz)$, and the integration is taken along a suitable path C in the simply connected region D. The Green's transform (2) is obtained by integration from equation (1) after multiplication by $\overline{w}$. The path C of integration from z_1 to z_2 is assumed to have a continuous tangent everywhere, except possibly at a finite number of points. Let there be an angle θ satisfying the following conditions along the path C for a given solution $w(z)$:

$$\theta \leqslant \arg[-\overline{w}_1(z_1)w_2(z_1)] \leqslant \theta + \pi,$$

$$\theta \leqslant \arg[G(z)dz] \leqslant \theta + \pi,$$

$$\theta \leqslant \arg[-dz/K(z)] \leqslant \theta + \pi,$$

except possibly at a finite number of points and with the signs of equality either not holding in the first condition or else not holding in the second or the third condition over points having a positive measure. The Green's transform (2) thus implies that $\theta < \arg[-\overline{w}_1(z_2)w_2(z_2)] < \theta + \pi$, and hence $\overline{w}_1 w_2 = K\overline{w}(dw/dz) \neq 0$ at z_2. Hille called such a path C a line of influence on which neither the solution $w(z)$ nor its derivative dw/dz has a zero. Varying and refining this idea, Hille constructed various types of zero-free regions in paper [5], and thus the zeros of a solution are located in the complement of these regions.

Paper [12] treats the following singular integral equation of Volterra's type:

$$w(z) = w_0(z) + \int_z^\infty \sin(t - z)f(t)w(t)dt, \tag{3}$$

where $w_0(z)$ is any solution of the differential equation

$$w_0''(z) + w_0(z) = 0. \tag{4}$$

The path of integration is the ray $\arg(t - z) = 0$. The function $f(z)$ is single-valued and analytic in the sector S defined by $0 < \gamma \leqslant |z|; -\theta \leqslant \arg z \leqslant \theta$ and satisfies the inequality $|f(z)| < M/|z|^{1+a}$ in S, with M and a being positive constants. Using the method of successive approximations, it is proved that (3) has a unique analytic solution $w(z)$ in S for each given solution $w_0(z)$ of (4). Furthermore, in any strip in S defined by lines y = constant, a bound for $w(z) - w_0(z)$ is established, and $w(z) - w_0(z)$ converges to 0 as $z \to \infty$ in the strip. Thus $w_0(z)$ reflects the asymptotic behavior of $w(z)$ in such a strip S. Hille had used this asymptotic representation of a solution to study the distribution of zeros of solutions of the equation (1) in the neighborhood of an irregular-singular point. His results were published in [3].

Papers [15] and [23] give short accounts of Hille's investigations on the distribution of complex zeros of the solutions of equation (1); these are based on the construction of zero-free regions from the associated Green's transforms and on the asymptotic representation of a solution in the neighborhood of an irregular-singular point. His investigation of this subject extended over the period 1918-1924 and were reported in a series of about ten papers. He treated the general problems by using Green's transforms, asymptotic representation, and variation of parameters, and he applied these methods to Mathieu's, Legendre's, and Hermite-Weber's equations.

Hille's work represents substantial progress on this subject whose literature is not extensive. Two general methods were introduced to the problems after Hille: the Schwarzian derivative and univalence, employed by Z. Nehari[1] and his students P. Beesack and B. Schwartz, and the comparison theorems, used by C. T. Taam[2]. An excellent digest of the methods of Hille, Nehari, Beesack, Schwartz, and Taam is now available in [B8], Chapter 11 (see also Bieberbach[3], Section 12).

No general method on the subject was developed before Hille. There were only isolated investigations of particular equations. For more references to these problems, consult [B8], Chapter 11.

For the correspondent oscillatory and nonoscillatory behavior on the real line of real solutions of real linear second order equations, there is an extensive literature with various interesting developments, starting with the fundamental work of Sturm and Liouville in the 1830s. Consult [B8], Swanson[4], Hartman[5], or Reid[6] for further references.
C. T. Taam

References

1. Z. Nehari, *The Schwarzian derivative and schlicht functions,* Bull. Amer. Math. Soc. vol. 55 (1949) pp. 545-551.
2. C. T. Taam, *Oscillation theorems,* Amer. J. Math. vol. 74 (1952) pp. 317-324.
3. L. Bieberbach, *Theorie der gwoehnlichen Differentialgleichungen,* Springer-Verlag, Berlin, 1965.
4. C. A. Swanson, *Comparison and Oscillation Theory of Linear Differential Equations,* Academic Press, New York, 1968.
5. P. Hartman, *Ordinary Differential Equations,* John Wiley, New York, 1964.
6. W. T. Reid, *Ordinary Differential Equations,* John Wiley, New York, 1971.

Chapter 3. Dirichlet Series

Hille's work on Dirichlet series

Forty years ago the theory of Dirichlet series was one of the major branches of analysis, and it is to be expected that Hille should have contributed to it. Two of his papers in this domain are reproduced in this volume. One is typical of the work being done at the time and now seems rather dated; the second, written with H. F. Bohnenblust, contains one of the deep and interesting results of the subject, and is still waiting to be elaborated and simplified.

Some remarks on Dirichlet's series [17] gives necessary and sufficient conditions for a Dirichlet series of general type

$$\sum_{n \geqslant 1} a_n \exp(-\lambda_n s) \quad (0 < \lambda_n < \lambda_{n+1}, \ \lambda_n \to \infty) \tag{1}$$

to represent an entire function. A representation for the sum $f(s)$ is found, based on a formula of Cahen and Perron. Under appropriate hypotheses, this representation exhibits the analytic continuation of $f(s)$ throughout the plane. Hardy had discussed the case of ordinary Dirichlet series. The general case is difficult, and Hille's paper is a contribution of considerable power to classical summability theory.

The paper with Bohnenblust, *On the absolute convergence of Dirichlet series* [41], is an answer to a problem of Bohr: What is the maximum width of a strip in which a Dirichlet series can converge uniformly but not absolutely? The answer depends on the type $\{\lambda_n\}$, and the paper is mainly concerned with ordinary Dirichlet series

$$\sum_{n \geqslant 1} a_n n^{-s}. \tag{2}$$

Let B denote the upper bound of widths of such strips for series (2). It is trivial that $B \leqslant 1/2$; the main result of this paper is that $B = 1/2$.

Bohr had already transformed the problem in two steps. (References are given in the paper under review.) For ordinary Dirichlet series, the half-plane of uniform convergence is the same as the half-plane in which the sum is almost-periodic. Thus the problem can be restated on the infinite-dimensional torus T^∞. Let P denote the family of continuous functions on T^∞ whose Fourier series have the form

$$F(z_1, z_2, \ldots) \approx \sum a(n_1, n_2, \ldots)\, z_1^{\,n_1} z_2^{\,n_2} \ldots, \tag{3}$$

where the n_j are restricted to nonnegative values. Here $|z_j| = 1$ for each j, and only finitely many n_j are different from zero in any term. A correspondence between series of types (2) and (3) is established by identifying $(n_1, n_2, \ldots)$ with $n = p_1^{\,n_1} p_2^{\,n_2} \ldots$, where $p_1, p_2, \ldots$ is the sequence of prime numbers. Now Bohr's constant B is the lower bound of numbers α such that

$$\sum_{n \geqslant 1} |a(n)| n^{-\alpha} < \infty \tag{4}$$

whenever (3) belongs to P.

The growth of the primes appears to be involved in (4), but actually it is not. Bohr showed that the lower bound α is the reciprocal of the upper bound of β such that for positive numbers $r_1, r_2, \ldots$

$$\sum_{n \geqslant 1} r_n^{\,\beta} < \infty \text{ implies } \sum_{n \geqslant 1} |a(n)|\, r_1^{\,n_1} r_2^{\,n_2} \ldots < \infty \tag{5}$$

for every member of P.

For any probability measure μ on T^{∞}, the convolution of μ with F in P is another element of P whose coefficients are $\hat{\mu}(n)a(n)$, where $\hat{\mu}(n)$ is the Fourier-Stieltjes coefficient of μ. Among such measures are two types of special importance, whose coefficients are

$$\begin{aligned} \hat{\mu}_m(n) &= 1 \quad \text{if } n_{m+1} = n_{m+2} = \ldots = 0; \\ &= 0 \quad \text{otherwise;} \end{aligned} \tag{6}$$

$$\begin{aligned} \hat{\nu}_m(n) &= 1 \quad \text{if } n = 1; \\ &= \tfrac{1}{2} \quad \text{if } \textstyle\sum n_j = m; \\ &= 0 \quad \text{otherwise.} \end{aligned} \tag{7}$$

Convolution with μ_m selects the terms of (3) in which only $z_1, \ldots, z_m$ appear. Convolution with ν_m has the very different effect of isolating terms of (3) having total degree m. The first operation provides a useful approximation technique, but nothing more; Bohr's problem itself is trivial for power series in finitely many variables. The reduction to homogeneous forms, on the contrary, gives an interesting problem for each positive integer m.

For $m = 2$, for example, we are led to the class of bounded quadratic forms in infinitely many variables

$$\sum b_{ij} z_i z_j, \tag{8}$$

where we may suppose $b_{ij} = b_{ji}$ for all i, j. The associated bilinear form

$$\sum b_{ij} z_i w_j \tag{9}$$

has the same bound. Thus one can attack Bohr's problem by constructing bounded bilinear or quadratic forms as far as possible from being absolutely convergent. Toeplitz had shown in this way that $B \geqslant 1/4$.

A good deal later Littlewood studied bilinear forms and obtained results applicable to the problem. The present paper accomplishes the difficult program of extending Littlewood's results to m-linear forms for each m, and using them to prove that Bohr's constant for homogeneous m-forms is $(m-1)/2m$. In particular, the constant found by Toeplitz is the best possible for quadratic forms, and at the other extreme $B = 1/2$.

Even this outline is not simple, and the paper itself is a remarkable piece of work.

H. Helson

Chapter 4. Integral Equations

A comment on Hille's work on
integral equations

The contents of this paper were worked out and full proofs were presented by the joint authors in a large paper having the same title [39]. Meanwhile, the authors had also published another paper [37] in which it was shown that, although the Fredholm theory can be extended to Hilbert-Schmidt kernels in L^2, this is no longer true for kernels $K(x,y)$ satisfying

$$\iint |K(x,y)|^p \, dx dy < \infty$$

for some $p \neq 2$ in the interval $(p\colon 1 \leqslant p < \infty)$. Several years later (1934), they discovered the "right" generalization from $p = 2$ to more general p. The generalization concerns the kernels of finite double-norm, also sometimes called Hille-Tamarkin kernels. For $1 < p < \infty$, let q be the conjugate exponent of p, that is, let $p^{-1} + q^{-1} = 1$. The kernel $K(x,y)$ is now said to be of finite double-norm (in L^p) whenever

$$|||K|||_p = \left[\int \left(\int |K(x,y)|^q \, dy \right)^{p/q} dx \right]^{1/p} < \infty,$$

and the number $|||K|||_p$ is then called the double-norm of $K(x,y)$. It was shown by Hille and Tamarkin [59] that for a kernel of finite double-norm the corresponding operator is compact (completely continuous) in L^p. In 1940 the Fredholm theory for Hilbert-Schmidt kernels was reinvestigated by F. Smithies[6]. Instead of using the approach by means of infinite determinants, Smithies approximates the given transformation K by transformations of the form PKP, where P is the orthogonal projection in L^2 on an appropriate subspace of finite dimension. A similar approach in L^2 is due to S. Mihlin[3], and this was extended to general p $(1 < p < \infty)$ by I. A. Ickovič[2]. Explicitly, the Fredholm formulas hold for kernels that are completely of finite double-norm in L^p, that is, for kernels $K(x,y)$ satisfying $|||K|||_p < \infty$ and also

$$|||K|||_p^{\text{inv}} = \left[\int \left(\int |K(x,y)|^p \, dx \right)^{q/p} dy \right]^{1/q} < \infty.$$

The notion of a kernel of finite double-norm can be extended to transformations in Orlicz spaces, and it was proved by A. C. Zaanen that the Fredholm theory remains valid for kernels that are completely of finite double-norm in Orlicz spaces. Recently, the Fredholm theory was extended to linear integral equations with kernels of finite double-norm in general Köthe spaces by J. J. Grobler[1]. It is an essential feature of all these successive generalizations that if the original kernel is completely of finite double-norm, then the same holds for the resolvent kernel and for all the kernels occurring in the power series expansion of the resolvent kernel.

There also exist abstract Fredholm formulas for certain bounded linear operators (Fredholm operators) in an arbitrary Banach space (A. F. Ruston, A. Grothendieck, T. Lezański, R. Sikorski; for extensive bibliographies see A. F. Ruston[4] and R. Sikorski[5]).

A. C. Zaanen

References

1. J. J. Grobler, *Non-singular linear integral equations in Banach function spaces,* dissertation, Leiden University, 1970.
2. I. A. Ickovič, *On the Fredholm series,* Dokl. Akad. Nauk. SSSR (N. S.) vol. 59 (1948) pp. 423-425.
3. S. Mihlin, *On the convergence of the Fredholm series,* Dokl. Akad. Nauk. SSSR (N. S.) vol. 42 (1944) pp. 373-376.
4. A. F. Ruston, *Fredholm formulae and the Riesz theory,* Compositio Math. vol. 18 (1967) pp. 25-48.
5. R. Sikorski, *The determinant theory in Banach spaces,* Colloq. Math. vol. 8 (1961) pp. 141-198.
6. F. Smithies, *The Fredholm theory of integral equations,* Duke Math. J. vol. 8 (1941) pp. 107-130.

Chapter 5. Summability

A commentary on [48] (with J. D. Tamarkin)
On the summability of Fourier series. II

This paper is the second of a series of three memoirs with the same title by Einar Hille and J. D. Tamarkin. (The first is [43] and the third is [49].)

While the first and third articles deal with the questions of special, Norlund-, and Hausdorff-summability of Fourier series, this second one investigates not only certain questions concerning Fourier series, but also arbitrary Toeplitz-summability of general orthogonal expansions.

Let $\{\phi_n(x)\}$ $(n = 1, 2, \ldots)$ be an orthonormal system of continuous functions defined on the interval $[a, b]$, closed in the space $L^\infty(a, b)$ of continuous functions on $[a, b]$. For $f(x)$ in $L^p(a, b)(1 \leqslant p \leqslant \infty)$, let $f(x) \approx \Sigma_{n\geqslant 1} f_n\phi_n(x)$ express the fact that the series $\Sigma_{n\geqslant 1} f_n\phi_n(x)$ is the generalized Fourier expansion of the function $f(x)$ associated with the set $\{\phi_n(x)\}$; further, let $\Sigma_{n\geqslant 1} f_n\phi_n(x) \in L^p(a, b)$ mean that there exists a function $f(x) \in L^p(a, b)$ such that $f(x) \approx \Sigma_{n\geqslant 1} f_n\phi_n(x)$. For a matrix $A = (a_{m,n})$ $(m, n = 1, 2, \ldots)$ and a given expansion $f(x) \approx \Sigma_{n\geqslant 1} f_n\phi_n(x)$, consider the transformed sums $\Sigma_{n\geqslant 1} a_{m,n} f_n\phi_n(x)$ $(m = 1, 2, \ldots)$. The transformation A so defined is called L^p-effective if, for every function $f(x)$ in $L^p(a, b)$, we have $\tau_m(x; f) \approx \Sigma_{n\geqslant 1} a_{m,n} f_n\phi_n(x) \in L^p(a, b)$ $(m = 1, 2, \ldots)$ and $\|\tau_m(x; f) - f(x)\|_{L^p} \to 0$ as $m \to \infty$. Let p' denote the conjugate exponent to p (i.e., $1/p + 1/p' = 1$).

It is proved in the paper that the classes of L^p- and $L^{p'}$-effective transformations are identical and that, if the transformation A is L^p-effective, then A is also L^q-effective for an arbitrary q between p and p'. Furthermore, the authors give the normal representation of those matrices and regular matrices A which are L^p-or L^∞-effective with respect to Fourier expansion in the real (i.e., narrow) sense. Finally, they prove a necessary and sufficient condition for the series $\Sigma_{n\geqslant 1} q_n \sin nx$ with positive and monotonically decreasing coefficients to be the Fourier expansion of an integrable function.

In the paper *Addition to the paper "On the summability of Fourier series. II"* Hille and Tamarkin prove that the classes of B-effective, of L^1-effective, and of L^∞-effective transformations are identical. (B denotes the space of essentially bounded functions on the interval $[a,b]$, and the transformation given by the matrix $A = (a_{m,n})$ is called B-effective if, for every function $f(x)$ in B, we have $\tau_m(x;f) \approx \Sigma_{n\geqslant 1} a_{m,n} f_n\phi_n(x) \in B$ $(m = 1, 2, \ldots)$, and the sequence $\{\tau_m(x;f)\}$ converges in measure to $f(x)$ on $[a,b]$ and is bounded in the space B.)

These results are obtained through the use of the theory of Banach spaces; the authors apply the then new results of this theory in a masterly way.

The papers, with both their results and their methods, decisively contributed to the theory of the so-called multiplicator problems of both Fourier and general orthogonal series. These problems are, in general, related to the question of the conditions imposed upon the matrix A under which we can assert that, by transforming an expansion belonging to a certain class, we obtain an expansion in another function class. Certain results of these memoirs have also been incorporated in the classical material of the theory of Fourier series; this is shown in particular by the fact that they are included in standard monographs such as A. Zygmund, *Trigonometrical Series* (Warsaw-Lwow, 1935) or *Trigonometrical Series I and II* (Cambridge, 1959) and S. Kaczmarcz and H. Steinhaus, *Theorie der Orthogonalreihen* (Warsaw-Lwow, 1935). Similarly, the above-mentioned result concerning pure sine series also belongs to the body of classical knowledge today.
K. Tandori

Chapter 6. Fourier Transforms and Analytic Function Theory

A commentary on [50] (with J. D. Tamarkin)
On a theorem of Paley and Wiener

One version of the Paley-Wiener theorem (Trans. Amer. Math. Soc. vol. 35 (1933) pp. 348-355) gives a necessary and sufficient condition that a non-negative L_2 function ω on R be the absolute value of the Fourier transform of an L_2 function that vanishes on the negative half-line: it is that

$$\int_{-\infty}^{\infty} \log(|\omega(t)|)\,(1+t^2)^{-1}\,dt < \infty. \tag{*}$$

Hille and Tamarkin generalize this theorem to all values of p such that $1 \leqslant p \leqslant 2$. Given a nonnegative function ω in $L_p(R)$, the condition (*) is necessary and sufficient for there to exist a function $g \in L_p(R)$ whose absolute value is equal to ω and for which the Fourier transform $\hat{g}$ of g vanishes on the negative half-line. The paper also contains a novel description of Fourier transforms, which the authors remark can be generalized "in the direction of the Hahn-Wiener-Bochner theory of generalized trigonometric integrals." So far as the commentator knows, no one has pursued this generalization. The theorem proved by Hille and Tamarkin in this paper has apparently not been studied for groups other than R. Analogues of the theorem almost surely exist for some other groups (at least for R^n and quite possibly for all local fields), and such analogues would be very interesting.
Edwin Hewitt

A commentary on [54] (with J. D. Tamarkin)
On the theory of Fourier transforms

This famous paper extends the $L_2(R)$ Fourier inversion formula, due to Plancherel (Rend. Palermo vol. 30 (1910) pp. 289-335), to all functions in $L_p(R)$ for $1 < p < 2$. The authors first quote Titchmarsh's extension of the direct part of Plancherel's theorem (Proc. London Math. Soc. (2) vol. 23 (1925) pp. 279-289), which states that if $f \in L_p(R)$, then the functions

$$\hat{f}_n(y) = (2\pi)^{-1/2} \int_{-n}^{n} f(x)\exp(-iyx)\,dx$$

converge in the $L_{p'}(R)$ norm ($1/p + 1/p' = 1$) to a function, nowadays almost universally written $\hat{f}$ and called the L_p Fourier transform of f. Curiously enough, the authors do not cite the fundamental paper of M. Riesz (Acta Math. vol. 49 (1926) pp. 465-497), in which the real method behind Titchmarsh's theorem is given. Prior to the appearance of the paper under discussion, no one had proved that the Fourier transform $\hat{f}$ can itself be inverted. The functions

$$\hat{\hat{f}}_n(x) = (2\pi)^{-\frac{1}{2}} \int_{-n}^{n} \hat{f}(y)\exp(iyx)dy$$

also converge, this time in the $L_p(R)$ norm, to the original function f. This theorem now appears in standard treatises on Fourier analysis, e.g., Titchmarsh's *Introduction to the Theory of Fourier Integrals* (Oxford: Oxford University Press, 1937, p. 148) and in Zygmund's *Trigonometric Series* (2nd ed., Cambridge, England: Cambridge University Press, 1959, vol. II, p. 254). No complete generalization to locally compact Abelian groups is known to the commentator. For what is known, see Hewitt and Ross's *Abstract Harmonic Analysis* (New York: Springer-Verlag, 1970, vol. II, pp. 240-241). A complete analogue exists for all compact groups, Abelian or not. See Hewitt and Ross, pp. 228-229.
Edwin Hewitt

A commentary on [55] (with J. D. Tamarkin)
A remark on Fourier transforms and functions analytic in a half-plane

This paper is based on and closely related to [54]. Let p and q be real numbers greater than or equal to 1. Let a function f on R be in $L_p(R)$ and also equal to the Fourier transform of some function ϕ in $L_q(R)$. This transform is defined, when it exists at all, as the L_p limit of the integrals

$$\hat{\phi}_n(y) = (2\pi)^{-\frac{1}{2}} \int_{-n}^{n} \phi(x)\exp(-ixy)dx.$$

Then f is the nontangential limit of a function analytic in the upper half-plane if and only if ϕ vanishes on the negative half-line. So far as the commentator is aware, this theorem has not been generalized, although it could certainly be considered at least in R^n: see for example the treatise *Introduction to*

Fourier Analysis on Euclidean Spaces by E. M. Stein and G. Weiss (Princeton: Princeton University Press, 1971).
Edwin Hewitt

A commentary on [64] (with J. D. Tamarkin)
On the absolute integrability of Fourier transforms

This paper was apparently inspired by the important paper of Hardy and Littlewood *Some new properties of Fourier constants* (Math. Ann. vol. 97 (1927) pp. 159-209). In this paper Hardy and Littlewood showed that the Fourier coefficients $\hat{f}(n)$ of a function f in $L_p(-\pi, \pi)$ $(1 < p < \infty)$ satisfy the remarkable inequality

$$\sum_{n=-\infty}^{\infty} |\hat{f}(n)|^p (|n| + 1)^{p-2} \leqslant C_1 \, ||f||_p^p .$$

For $p = 1$, the result fails. However, if $f \in H_1$, then one has

$$\sum_{n=0}^{\infty} |\hat{f}(n)| (n + 1)^{-1} \leqslant C_1 \, ||f||_1 .$$

The authors explore the analogue of this result for the line. The main result of the paper is the following: Let g and its conjugate function $\tilde{g}$ (defined as the Cauchy principal value of the convolution of g with $1/t$) be in $L_1(R)$. Let $f = g + i\tilde{g}$. Then we have

$$\int_0^{\infty} |\hat{f}(t)| t^{-1} \, dt \leqslant \left(\frac{\pi}{2}\right)^{1/2} ||f||_1 ,$$

and the constant $(\pi/2)^{1/2}$ is the best possible. This is also an analogue of a theorem of M. Riesz about functions in $H_p(R)$ (Math. Z. vol. 27 (1927) pp. 218-244). So far as the commentator can find, this theorem has not made its way into any standard treatises, nor has it been studied for groups other than R.
Edwin Hewitt

Chapter 7. Laplace Integrals

A commentary on [62] On Laplace integrals

There are two principal ideas set forth in this article. The one involves the analytic continuation of a function $f(s)$ defined by a Laplace integral

$$f(s) = \int_0^\infty \exp(-st)\, d\alpha(t) \tag{1}$$

beyond the region of convergence of the integral. The other treats the representation of meromorphic functions by Laplace integrals. They are not unrelated, for each involves the quotient of integrals, as we shall see.

The first idea may be thought of as an outgrowth of the following considerations. If the integral (1) has abscissa of convergence σ_0, then $f(s)$ is analytic for $\sigma > \sigma_0$, $s = \sigma + i\tau$. But it may be analytic in a larger region. For example, if

$$g(s) = \int_0^\infty \exp(-st) \sin(\exp t)\, dt = \int_1^\infty \sin x / x^{s+1}\, dx, \tag{2}$$

then $\sigma_0 = -1$, whereas $g(s)$ can be shown to be entire. Indeed, integration by parts in (1) gives

$$f(s) = s\int_0^\infty \exp(-st)\, \alpha(t)\, dt = s\int_0^\infty \exp(-st)\, d\alpha_1(t), \tag{3}$$

$$\text{where } \alpha_1(t) = \int_0^t \alpha(y)\, dy = \int_0^t (t-y)\, d\alpha(y).$$

The integral (3) may well converge in a larger region than the integral (1); as in the case of example (2) when it converges for $\sigma > -2$.

More generally, equation (3) may be replaced by

$$f(s) = s^p \int_0^\infty \exp(-st) d\alpha_p(t), \quad \alpha_p(t) = \int_0^t (t-y)^p/\Gamma(p+1)\, d\alpha(y), \tag{4}$$

where p is any positive number. In example (2), this new integral (4) serves to extend $g(s)$ analytically into the half-plane $\alpha > -p - 1$ $(p = 1, 2, \ldots)$, thus confirming that $g(s)$ is entire. Observe that the reciprocal of the multiplier s^p in equation (4) is itself a Laplace transform,

$$s^{-p} = (1/\Gamma(p)) \int_0^\infty \exp(-st)\, t^{p-1}\, dt. \tag{5}$$

Since $f(s)$ is thus replaced by the quotient of two transforms, the procedure is naturally described as the quotient method of continuation. Of course many other Laplace transforms may be used instead of (5). In particular, if (5) itself is used, the procedure is known as the method of typical means, originally devised by M. Riesz for the analytic continuation of functions defined by Dirichlet series.

The second idea exploits a criterion, previously devised by the Nevanlina brothers, for the representation of a meromorphic function as the quotient of bounded functions. By its use it is possible to state necessary and sufficient conditions for the representation of such a function as the quotient of Laplace integrals. For, if $B(s)$ is analytic and bounded in a half-plane $\sigma > 0$, for example, then $B(s)/s$ is known to have a Laplace representation there. Accordingly, if $f(s)$ satisfies the Nevanlina conditions and is consequently the quotient of two bounded analytic functions $B_1(s)$ and $B_2(s)$, then $f(s) = (B_1(s)/s)/(B_2(s)/s)$, so that $f(s)$ is also the quotient of Laplace integrals. The converse is equally evident.

The exposition of the paper is somewhat in the style of a lecture; thus the main ideas are stressed and the proofs are de-emphasized or omitted.
D. V. Widder

Chapter 8. Factorisatio Numerorum and Möbius Inversion

These papers introduce some of the ideas on inversion formulas which were later to receive a broad generalization in combinatorial theory. See, for example, Doubilet, Rota, Stanley, *On the foundations of combinatorial theory. VI, The idea of a generating function,* Proc. Sixth Berkeley Symposium on Probability and Statistics, University of California Press, 1972, vol. 2.
G.-C. Rota

Chapter 9. Hermitian Series and Differential Operators of Infinite Order Which Are Entire Functions of Finite Operators

Introduction to Hille's papers on Hermitian series

As a result of Hille's work in functional analysis and related fields, his is a name familiar to all analysts, both young and old. In mathematics, as in all aspects of science as well as other areas of human activity, fashions play a strong role. In recent years functional analysis has been a vigorous discipline in which active research has been flourishing. Hille's treatise on *Functional Analysis and Semi-groups,* first published in 1948, served as one of the first modern introductions to the subject, and it has been a basic element in the education of many analysts.

An analysis of Hille's research areas shows, however, that he was an active participant not only in the above-mentioned areas, but also in such basic subjects as integral equations, Fourier series, properties of second order differential equations and solutions of such equations, and special functions that satisfy such differential equations.

The articles under discussion here may indeed come as a surprise to many of the younger analysts. Here we are concerned with properties of Hermite functions. These are eigenfunctions of the self-adjoint differential operator $\delta_z = z^2 - d^2/dz^2$ defined on the space $L^2(-\infty, \infty)$. In [76], [77], and [78], a study is made of the characteristics of series expansions of analytic functions $f(z)$ in terms of the orthogonal Hermite functions $h_n(z)$. In particular, Hille is concerned with the region of convergence of such expansions in the complex plane. Necessary and sufficient conditions on $f(z)$ are established that guarantee the convergence of its expansion in the strip $|y| < \tau$ in the complex plane. By means of explicit examples, it is shown that many of the results are the best possible.

The above results are analogous to results concerning Jacobi polynomials. In the latter case, one deals with functions analytic on $[-1, 1]$ and investigates under which conditions the expansions converge in ellipses whose foci are the points $-1, 1$. In a sense, Hille's results can be viewed as limiting cases of these, but the analysis becomes far more delicate. In addition to the above results, he also investigates factor sequences. The latter are sequences $\{a_n\}$ such that the series $\Sigma\, a_n f_n h_n(z)$ (where $\Sigma\, f_n h_n(z)$ represents $f(z)$) can be analytically continued along any path through the origin along which $f(z)$ can also be continued. Various gap theorems are derived under which the strip of convergence $|y| = \tau$ represents a natural boundary.

A second theme that is developed in these papers as well as others (see [76], [77], and [79]) concerns operators of the type $G(\delta_z)$. Here δ_z is the operator $z^2 - d^2/dz^2$ and $G(t)$ is a suitable entire function. The operator $G(\delta_z)$ is said to be applicable to $f(z)$ if the series $G(\delta_z) = \Sigma_{k \geqslant 1}\, g_k \delta_z^k f(z)$ converges at every z where $f(z)$ is holomorphic. Necessary and sufficient conditions on $G(t)$ and $f(z)$ are derived which guarantee that $G(\delta_z)$ is applicable to $f(z)$. Both functions are entire functions belonging to certain well-defined classes.

The results concerning operators of the form $G(\delta_z)$ are analogous to those of operators like $G(d/dz)$. There is, however, a great difference between the two operators d/dz and δ_z. The operator d/dz has all complex λ as eigenvalues, with the eigenfunctions $\exp(\lambda z)$. The operator δ_z also has all complex λ as eigenvalues, but the odd integers play a distinguished role because the equation $(\delta_z - 2n - 1)f(z) = 0$ then has the solutions $h_n(z)$ in $L^2(-\infty, \infty)$. For other values of λ, this equation has solutions that must become vanishingly small as $z \to \infty$ in suitable sectors of the complex plane. In regard to these operators, one might well pose the question whether one could not develop such a theory for more general operators of degree higher than two, with analytic coefficients.

Finally, in [127] Hille addresses himself to a related but different problem. How can expansions of the type $\Sigma\, f_n h_n(z)$ be constructed so that the resultant function is analytic? If the A_n are suitable constants, depending only on $h_n(z)$, and if $G(t)$ is regular in the right half-plane (and subject to some other conditions), then series of the form $\Sigma\, G(n)\, h_n(z)/A_n$ necessarily represent entire functions, and converge in suitable strips.

In summary, Hille has given us a comprehensive theory for the operator δ_z. We now have a complete picture of the region of convergence of $\Sigma\, f_n h_n(z)$, which represents an analytic function, and the coefficients f_n are defined only by the values of $f(z)$ for real z. Furthermore, we have a theory that describes the domains of operators of the form $G(\delta_z)$, where $G(t)$ is a suitable entire function. The resultant theory is a beautiful amalgam of complex function theory and differential operator theory.

H. Hochstadt

Chapter 11. Ergodic Theory

A commentary on [89] Remarks on ergodic theorems

1. This paper studies the relations between Cesàro convergence and Abel convergence of iterates of power-bounded operators in Banach spaces. Einar Hille realizes that essentially the problem is one in summability theory of Banach-space-valued sequences, rather than in ergodic theory. Generalizing arguments known from the classical case, he simply shows that for an arbitrary sequence in a Banach space the Cesàro convergence implies the Abel convergence, and conversely—here a Banach-valued version of Wiener's Tauberian theorem is needed and established—Abel convergence of bounded sequences implies Cesàro convergence. These are the main results of the paper, but an unaveraged convergence theorem is also proved for self-adjoint positive-definite contractions in Hilbert space, and finally a detailed discussion is given of the relations between Abel summability and Cesàro summability for an operator defined by a fractional integration formula. This discussion yields examples of operators which are (i) Abel-ergodic, but not power-bounded and not Cesàro ergodic and (ii) not power-bounded, but Cesàro ergodic.

2. *An Abelian-Tauberian Theorem and Ergodic Theory.* Let X be a complex Banach space with elements $x, y, z, \ldots$. Let

$$C_n^k = \binom{n+k}{n} = \frac{\Gamma(n+k+1)}{\Gamma(n+1)\cdot\Gamma(k+1)},$$

$$z_n^k = \frac{1}{C_n^k} \sum_{i=0}^{n} C_{n-i}^{k-1} z_i$$

Thus $z_n^1 = [1/(n+1)]\, \Sigma_{i=0}^{n} z_i$, and z_n^0 is simply z_n.

Set, whenever the limit exists, $(C, k) - \lim z_n = \lim_n z_n^k$. Let $R(\gamma,(z_n)) = \Sigma_{n=0}^{\infty} z_n \gamma^{-n-1}$ and, whenever the limit exists,

$$(A) - \lim z_n = \lim_{\gamma \to 1+} (\gamma - 1) R(\gamma, (z_n)).$$

The main result of the paper is the following:

Theorem. *For every sequence* (z_n) *in a Banach space, if* $(C,k) - \lim z_n = z$

for some positive constant k, then $(A) - \lim z_n = z$, *and moreover* $\lim_n n^{-k} z_n = 0$. *Conversely, if* $(A) - \lim z_n = z$, *and also* $\sup_n \|z_n\| < \infty$, *then for every positive k one has* $(C,k) - \lim z_n = z$.

Let T be a power-bounded linear operator on X. From the theorem it follows at once that the ergodic theorem with Abel limit for T is equivalent to such a theorem with Cesàro limit of any order k, $k > 0$ (it suffices to let $z_n = T^n x$ for each fixed x and $n = 0, 1, \ldots$). Also, the equivalence of uniform ergodic theorems, in the power-bounded case, for the two modes of convergence, Abel and Cesàro, can be easily proved by applying the theorem to the Banach space of linear bounded operators on X and the sequences $z_n = T^n$, T fixed with $\sup_n \|T^n\| < \infty$.

These results and their continuous parameter analogues (see [B2]) were so conclusive that little work in ergodic theory has since been done on Abel convergence, or on Cesàro convergence of positive order $k \neq 1$. G.-C. Rota[13] and L. Baez-Duarte[1] have given Abelian versions of E. Hopf's maximal ergodic theorem and of the "recurrent means" variety of the Chacon-Ornstein theorem. A claim of simplicity is made for the Abelian approach, but the papers of Brunel[2], Garsia[9], and Chacon[6] seem to show that the Cesàro versions with $k = 1$ are no less simple. In a recent paper, P.L. Butzer and U. Westphal[5] have studied the speed of convergence in the Cesàro ($k = 1$) and Abel mean ergodic theorems. A continuation, as yet unpublished, considers the Cesàro case with positive $k \neq 1$.

3. *Unaveraged Convergence.* The following theorem in Hille's work is easily derived from the spectral theorem:

Theorem. *Let* (T_s), $s > 0$ *be a semi-group of self-adjoint contraction operators on a Hilbert space; then* $\lim_n T_{ns}$ *exists in the strong operator topology.*

An immediate proof of this result is obtained if the elegant multiplication form of the spectral theorem, popularized by Halmos[10], is used. The semi-group property implies that T_s is the square of the operator $T_{s/2}$; therefore, T_s is positive-definite in the Hilbert-space sense. Thus T_s is unitarily equivalent to the multiplication by a positive function bounded a.e. by 1; the theorem follows. Another simple proof is sketched in Starr[14], p. 109.

If the mean convergence is easy, the almost everywhere convergence is not, and here important work has been done recently. After the pioneering paper of Burkholder and Chow[4] solving a problem about alternating conditional expectations asked by Doob, the deepest positive results were the following theorems obtained by Rota[12] and Stein[15]:

Stein's Theorem. *Let T be a self-adjoint contraction operator on L_2 of a measure space, and assume that T is both positive-definite in the Hilbert-space sense and also positive in the sense that $f \in L_2$, $f \geqslant 0$ implies $Tf \geqslant 0$. Then* $\lim_n T^n f$ *exists almost everywhere.*

Rota's Theorem (as improved by Doob[7] and Starr[14]). *Let T_n be for each n a positive contraction operator on each space L_p, $1 < p < \infty$ of a measure space, and assume that $f \in L_1$ and $\int |f| \log^+ |f| < \infty$. Then*

$$\lim_n T_1^* T_2^* \cdots T_{n-1}^* T_n^* T_n T_{n-1} \cdots T_2 T_1 f$$

exists almost everywhere. (T^* is the adjoint of T.)

That the assumption $\int |f| \log^+ |f| < \infty$ cannot be omitted was shown by Burkholder[3]; see also D. Ornstein[11].

4. *Fractional Integration.* Let C be the space of functions $f(t)$ continuous for $0 \leqslant t \leqslant 1$ and such that $f(0) = 0$; set $||f|| = \max_t |f(t)|$. Let L_1 be the Lebesgue space of integrable functions on $[0,1]$. For each positive α set

$$J_\alpha f = \frac{1}{\Gamma(\alpha)} \int_0^t (t-u)^{\alpha-1} f(u)\, du\,.$$

It is easily checked that $J_\alpha J_\beta = J_{\alpha+\beta}$ (cf., for example, Zaanen[16], p. 59). T_α is defined by $T_\alpha(f) = f - J_\alpha f$. A very complete discussion of the behavior of T_α acting as a linear operator on either C or L_1 is given. The Abelian limit $(A) - \lim T_\alpha^n = 0$ exists for $0 < \alpha < 2$ in the strong operator topology of the two spaces, but, while for $0 < \alpha < 1$ $||T_\alpha^n||$ is bounded and hence the Cesàro limits also exist in the strong topology, for $\alpha > 1$ $||T_\alpha^n||$ is unbounded and the Cesàro ergodic theorems fail. The case $\alpha = 1$ is intermediate: $||T_\alpha^n|| = 0(n^{1/4})$ and $(C,k) - \lim T_\alpha^n$ exists and equals zero in strong operator topology of both spaces for every $k > 1/2$, but for no $k \leqslant 1/2$.

I do not know of any recent related work on fractional integration. The limits found here for the parameters α and k are sharp, and no improvement seems possible.

L. Sucheston

References

1. Luis Baez-Duarte, *An ergodic theorem of Abelian type,* J. Math. Mech. vol. 15 (1966) pp. 599-607.

2. A. Brunel, *Sur un lemme ergodique voisin du lemme de E. Hopf et sur une de ses applications,* C. R. Acad. Sci. Paris vol. 256 (1963) pp. 5481-5484.

3. D. L. Burkholder, *Successive conditional expectations of an integrable function,* Ann. Math. Statist. vol. 33 (1962) pp. 887-893.

4. D. L. Burkholder and Y. S. Chow, *Iterates of conditional expectation operators,* Proc. Amer. Math. Soc. vol. 12 (1961) pp. 490-495.

5. P. L. Butzer and U. Westphal, *The mean ergodic theorem and saturation,* Indiana Univ. Math. J. vol. 20 (1971) pp. 1163-1174.

6. R. V. Chacon, *Ordinary means imply recurrent means,* Bull. Amer. Math. Soc. vol. 70 (1964) pp. 796-797.

7. J. L. Doob, *A ratio operator limit theorem,* Z. Wahrscheinlichkeitstheorie und Verw. Gebeite vol. 1 (1963) pp. 288-294.

8. N. Dunford and J. T. Schwartz, *Linear Operators,* part II, Interscience, New York, 1963.

9. A. Garsia, *A simple proof of Eberhard Hopf's maximal ergodic theorem,* J. Math. Mech. vol. 14 (1965) pp. 381-382.

10. P. R. Halmos, *What does the spectral theorem say?* Amer. Math. Monthly vol. 70 (1963) pp. 241-247.

11. D. Ornstein, *On pointwise behavior of iterates of a self-adjoint operator,* J. Math. Mech. vol. 18 (1968) pp. 473-477.

12. G. -C. Rota, *An "alternierendes Verfahren" for general positive operators,* Bull. Amer. Math. Soc. vol. 68 (1962) pp. 95-102.

13. G. -C. Rota, *On the maximal ergodic theorem for Abel limits,* Proc. Amer. Math. Soc. vol. 14 (1963) pp. 722-723.

14. N. Starr, *Operator limit theorems,* Trans. Amer. Math. Soc. vol. 121 (1966) pp. 90-115.

15. E. M. Stein, *On the maximal ergodic theorem,* Proc. Nat. Acad. Sci. U.S.A. vol. 47 (1961) pp. 1894-1897.

16. A. C. Zaanen, *Linear Analysis,* North-Holland Publ. Co., Amsterdam, 1956.

Hille's work on abstract summability in connection with orthogonal expansions, multipliers, and approximation theory

One of the great perceptions of Hille's paper [89] on ergodic theory is that classical summability may readily be carried over to the vector-valued situation (see also the previous commentary). His many papers on pointwise Abel summability of Laguerre [20] and Hermite [22] series, and on pointwise convergence by Hausdorff means of trigonometric series ([43], [46], [47], [48], and [49]), surely led him to this conclusion. The purpose of this commentary is to view this connection in the light of Hille's work on multipliers for concrete orthogonal expansions and approximation.

Hille made use of the notion of factor (= multiplier) sequences for orthogonal complete (= total) systems from one Banach space X into another Y (see [B1], p. 343; [B2], p. 544); he carefully examined multiplier conditions for trigonometric and Hermite series and observed ([B1], p. 372) that similar considerations also apply to the other classical orthogonal series (actually with respect to their approximation behavior).

On the other hand, Hille was the first to study the rate with which an arbitrary (C_0)-semi-group approximates the identity operator in the strong topology (see, for example, [B1], p. 351). For the particular instances of trigonometric or Hermite series, he constructed semi-groups by using the Abel-Cartwright means of these series (see [B1], p. 371).

If one develops the basic ideas of these two fields of Hille in all consequences, one arrives at the following question: Can summability theory be used to yield a multiplier theory which in turn allows a unified approach to fundamental problems in approximation? This is indeed the case, as has been carried out recently; here the notion of a multiplier operator (already considered by Hille in [B1], pp. 343, 371) turns out to be basic.

Let us briefly sketch this multiplier approach for the representative example of the comparison problem of Favard of 1963: Given two approximation processes $\{T(\rho); \rho > 0\}$, $\{S(\rho); \rho > 0\}$ on X into itself for $\rho \to \infty$, find an estimate of type

$$\|T(\rho)f - f\| \leqslant A \|S(\rho)f - f\| \qquad (f \in X; \rho > 0). \tag{1}$$

To this end, let $\{P_k\}$ $(k \geqslant 0)$ be a fundamental, total system of orthogonal, bounded linear projections on X. Then, to each $f \in X$ associate its Fourier expansion $f \approx \Sigma P_k f$ (see [B2], p. 544). If s is the set of all sequences of scalars, then $\tau \in s$ is a multiplier, $\tau \in M$ (see [B2], p. 544), if for each $f \in X$ there exists an element $f^\tau \in X$ such that $\tau_k P_k f = P_k f^\tau$. Obviously, one can define a bounded linear multiplier operator T by $Tf = f^\tau$ (and vice versa; see [B2], p. 544). Now let the above processes $\{T(\rho)\}$, $\{S(\rho)\}$ be families of multiplier operators with corresponding multipliers $\{\tau(\rho)\}$, $\{\sigma(\rho)\}$. Then the above comparison problem may be reduced to a condition upon the coefficients $\tau_k(\rho)$, $\sigma_k(\rho)$, namely that the existence of a family $\{\eta(\rho)\} \subset M$ with

$$\tau_k(\rho) - 1 = \eta_k(\rho)[\sigma_k(\rho) - 1], \qquad \|\eta(\rho)\|_M \leqslant A, \tag{2}$$

is sufficient for (1) to hold. But can one test this multiplier condition (2)? A useful multiplier criterion is based upon the uniform boundedness of the Cesàro means

$$(C, \alpha)_n f = \sum_{k=0}^{n} (A_{n-k}^{\alpha} / A_n^{\alpha}) P_k f, \qquad A_n^{\alpha} = \binom{n+\alpha}{n},$$

for some $\alpha \geqslant 0$, that is, upon the existence of a constant C_α such that

$$||(C,\alpha)_n f|| \leqslant C_\alpha ||f|| \qquad (f \in X). \tag{3}$$

Under this basic hypothesis one has $bv_{\alpha+1} \subset M$, where

$$bv_{\alpha+1} = \left\{ \eta \in 1^\infty; \qquad \sum_{k=0}^{\infty} A_k^\alpha \,|\Delta^{\alpha+1}\, \eta_k\,| < \infty \right\}, \tag{4}$$

the (fractional) difference operator Δ^β being defined by

$$\Delta^\beta \eta_k = \sum_{m=0}^{\infty} A_m^{-\beta-1} \eta_{k+m} :$$

Now, looking back at (2), it is difficult to check whether $\{\eta(\rho)\} \subset bv_{\alpha+1}$ uniformly in ρ. For this purpose it is convenient to extend $\{\eta_k(\rho)\}$ to a suitable function $e(x,\rho)$ defined on the whole positive half-axis. This yields the estimate

$$\sum_{k=0}^{\infty} A_k^\alpha \,|\Delta^{\alpha+1} \eta_k(\rho)| \leqslant B_\alpha \int_0^\infty x^\alpha |de^{(\alpha)}(x,\rho)|,$$

$e^{(\alpha)}$ being a suitable fractional derivative of order $\alpha \geqslant 0$. In this form the multiplier criterion is very sharp and still very practical for applications. It may be applied to such particular instances as one-dimensional and multidimensional trigonometric series, Laguerre and Hermite series, expansions into spherical harmonics, into Jacobi polynomials, into Walsh or Haar functions, etc. For the verifications of condition (3) for these examples, see the papers by R. Askey, A. Benedek, T. Koornwinder, C. Fefferman, G. Gasper, I. I. Hirschman, B. Muckenhoupt, R. Panzone, E. Poiani, H. Pollard, S. Wainger, G. M. Wing, and others, the majority of which have appeared in the last ten years. For the above approach, see Butzer et al.[1], Trebels[2], and Butzer[3].

Of course, one could also assume the Abel-boundedness of the expansion $f \approx \Sigma P_k f$; this would in turn lead to moment sequences (see [B1], p. 367). Here practically all of the necessary ingredients were supplied by Hille.

The author wishes to thank Dr. W. Trebels for constructive criticism.
P. L. Butzer

References

1. P. L. Butzer, R. J. Nessel, and W. Trebels, *On summation processes of Fourier expansions in Banach spaces, I: Comparison theorems; II: Saturation theorems,* Tôhoku Math. J. (2) vol. 24 (1972) pp. 127-140; pp. 551-570.

2. W. Trebels, *Multipliers for (C, α)-bounded Fourier Expansions in Banach Spaces and Approximation Theory,* Springer Lecture Notes, vol. 329, Springer-Verlag, New York, 1973.

3. P. L. Butzer, *A Survey of Work on Approximation at Aachen, 1968-1972,* Proc. Conf. Approximation, Austin, Texas, Jan. 22-24, 1973, Academic Press, New York, 1973.

Chapter 12. Semi-groups

An introduction to Hille's earlier papers on semi-groups

The first five lines of Hille's Foreword to [B1], published in 1948, well express his happy association with semi-groups: "The analytical theory of semi-groups is a recent addition to the ever-growing list of mathematical disciplines. It was my good fortune to take an early interest in this discipline and to see it reach maturity. It has been a pleasant association: I hail a semi-group when I see one and I seem to see them everywhere! Friends observed, however, that there are mathematical objects which are not semi-groups."

Indeed, Hille observed the semi-group property as early as 1936 [63], together with the continuity property at the origin of the parameter

$$F_\alpha[F_\beta[f]] = F_{\alpha+\beta}[f] \qquad (\alpha, \beta > 0),$$

$$\lim_{\alpha \to 0} F_\alpha[f] = f(x) \qquad \text{(for almost all } x)$$

for the Poisson transform

$$P_\alpha[f] = \frac{\alpha}{\pi} \int_{-\infty}^{\infty} \frac{f(u+x)}{u^2+\alpha^2} du$$

as well as for the Gauss-Weierstrass transform

$$W_\alpha[f] = (\pi\alpha)^{-1/2} \int_{-\infty}^{\infty} e^{-u^2/\alpha} f(u+x)\, du.$$

However, it seems that the term "semi-group" was first introduced into analysis in 1938 through Hille's paper [72]. The main result in this paper reads as follows: Let T_α, $\alpha > 0$, be a positive-definite, bounded linear operator on a Hilbert space H into H satisfying both the semi-group property $T_\alpha T_\beta = T_{\alpha+\beta}$ and the contraction property $\|T_\alpha\| \leqslant 1$. Then there exists a resolution of the identity $E(\lambda)$ such that $(T_\alpha f, f) = \int_0^\infty e^{-\alpha\lambda} d(E(\lambda)f, f)$. It follows in particular that, for fixed f, $(T_\alpha f, f)$ can be continued to be holomorphic for $\mathrm{Re}(\alpha) > 0$ and continuous for $\mathrm{Re}(\alpha) \geqslant 0$. It is shown that, for the case $T_\alpha = P_\alpha$, the operator $A_P = \int_0^\infty \lambda dE(\lambda)$ is given by $A_P f = \tilde{f}'$, where $\tilde{g}(x)$ is the conjugate function of $g(x)$.

As for semi-groups of bounded linear operators T_α in a general Banach space, Hille first investigated in 1939 [73] only those which can be continued to be holomorphic for $\mathrm{Re}(\alpha) > 0$; the general case was not attacked until 1942. In paper [73], he discussed the semi-group of bounded linear operators T_α bounded and holomorphic in every half-plane $\mathrm{Re}(\alpha) \geqslant \delta > 0$, with the object of characterizing T_α by means of the properties of its resolvents for small values of $\mathrm{Re}(\alpha)$. The paper culminates in a justification of the representation

$$W_\alpha = \exp\left(\frac{\alpha}{4}\frac{d^2}{dx^2}\right)$$

which shows the following relationship between the Gauss-Weierstrass transform and the Poisson transform:

$$A_W = \frac{1}{4}\frac{d^2}{dx^2} = \frac{1}{4}A_P^2 .$$

In 1942[83], Hille discussed the semi-group of bounded linear operators T_s $(s > 0)$ defined on a Banach space X into X satisfying four conditions:
(i) $T_sT_t = T_{s+t}$;
(ii) T_s is weakly measurable for $s > 0$;
(iii) $\|T_s\| \leqslant 1$; and
(iv) $T_s \cdot X$ is dense in X.
He defined $R(\lambda)f = \int_0^\infty e^{-\lambda s}\, T_s f\, ds$ for $f \in X$ and $\mathrm{Re}(\lambda) > 0$, and proved:
(v) $R(\lambda)$ is the resolvent $(\lambda I - A)^{-1}$ of a closed linear operator A which is defined through $Af = \lim_{h\to 0} h^{-1}(T_h - I)f$.
In this way, the notion of the infinitesimal generator A of the semi-group T_s was introduced, and Hille then proved the following three properties:
(vi) the domain $D(A)$ of A is dense in X;
(vii) the resolvent $R(\lambda) = (\lambda I - A)^{-1}$ of A satisfies the estimate $\|R(\lambda)\| \leqslant |\mathrm{Re}(\lambda)|^{-1}$; and
(viii) $T_s f = -\lim_{\omega\to\infty} (2\pi i)^{-1} \int_{c-i\omega}^{c+i\omega} e^{s\lambda} R(\lambda) f\, d\lambda \quad (f \in D(A))$.
However, the fact that (vi) and (vii) combined together give a characterization of the infinitesimal generator A of a semi-group T_s satisfying (i) — (iv) was proved only six years later in [B1]. Compare K. Yosida, J. Math. Soc. Japan vol. 1 (1948) pp. 15-21.

Paper [85] deals with the behavior of T_s in a semi-group for small positive s, and paper [88] with extensions to semi-groups of I. Gelfand's theory of characters of Abelian groups embedded in a normed commutative vector ring.

Paper [91], written jointly with N. Dunford, deals with the differentiability and the uniqueness of a complex-Banach-algebra-valued solution f of the equation $f(\xi+\eta)=G[f(\xi),f(\eta)]$ $(0\leqslant\xi,\eta,\xi+\eta\leqslant\omega)$ which generalizes the equation $T_\xi T_\eta=T_{\xi+\eta}$. Finally, paper [96] gives a natural extension to the n-parameter Lie group case of the case of a one-parameter semi-group of bounded linear operators on a Banach space X into X.

For the further development of the analytical theory of semi-groups, we refer the reader to the following four books: [B2]; N. Dunford and J. Schwartz, *Linear Operators,* part 1 (New York: Interscience, 1958); K. Yosida, *Functional Analysis* (2nd ed., New York: Springer-Verlag, 1968); and T. Kato, *Perturbation Theory for Linear Operators* (New York: Springer-Verlag, 1966).
K. Yosida

Hille's work on semi-group theory in connection with approximation theory

1. *Saturation.* It does not seem to be very well known that Hille's first paper on semi-groups [63] is especially concerned, in his terminology, with the "degree of approximation" of functions by linear transformations, in particular by the singular integrals of Weierstrass, Cauchy-Poisson, and Picard. Indeed, for the Weierstrass integral

$$(W(t)f)(x)=(1/\pi t)^{1/2}\int_{-\infty}^{\infty}f(x-u)\exp(-u^2/t)\,du,$$

for example, he showed that, if f is in $L^p(R)$, $1\leqslant p<\infty$, and

$$\lim_{t\to 0+}t^{-1}\,\|W(t)f-f\|_p=0,$$

then $f=0$. Moreover, for all $f=W(\tau)g$ with g in $L^p\,(R)$ and some $\tau>0$, it follows that $\|W(t)f-f\|_p\leqslant Ct$. In Hille's words, "the degree of approximation of a function by its Weierstrass transform is at best of the first order with respect to t, except for the fixed elements, and this order is actually reached for an infinite subclass of the space, namely by all the transforms." In modern terminology this is the small "o" as well as a "direct" approximation theorem for Weierstrass's integral.

Hille's next great advance in this direction is the generalization to arbitrary semi-groups. In fact, he showed in his book ([B1], p. 323; see also p. 351 ff.):

Theorem 1. *Let* $\{T(t); t \geqslant 0\}$ *be a semi-group of operators of class* (C_0) *acting on the Banach space X, A being its infinitesimal generator; then:*
(a) *if* $\liminf_{t\to 0+} t^{-1}\|T(t)f - f\| = 0$, *then* $Af = 0$ *and* $T(t)f = f$ *for each* $t \geqslant 0$;
(b) *for every f in* $D(A)$ *one has* $\|T(t)f - f\| \leqslant t\|Af\| \max_{0\leqslant u \leqslant t} \|T(u)\|$.
Thus the optimal degree of approximation of f by $T(t)f$, *namely* $O(t)$, *is reached for all elements of* $D(A)$ *(which are dense in X).*

The next step forward came in 1956-57, when Hille spent his sabbatical at the University of Mainz, where I was most fortunate to be at the same time. There we discussed whether the converse to (b) would hold. This led to

(c) *If X is reflexive, then* $\liminf_{t\to 0+} t^{-1}\|T(t)f - f\| < \infty$ *implies that* $f \in D(A)$.

This result is found in Butzer[1] as well as in the second edition of Hille's book coauthored with R. S. Phillips ([B2], p. 326). It opened the way to the development of saturation theory, originating in a problem posed by J. Favard in 1947/1949. In the terminology of this theory, the Favard or saturation class of the system $\{T(t); t \geqslant 0\}$ is the set of all f in X which belong to $D(A)$.

Hille's form of Taylor's theorem for semi-groups ([B2], p. 233) was then used to show that the family $\{T(t) - \Sigma_{k=0}^{r-1} t^k A^k/k!; t \geqslant 0\}$ is saturated with order $O(t^r)$, the Favard class given by $D(A^r)$, provided X is reflexive. This led to various generalizations of powers of the infinitesimal generator, to Riemann and Peano operators (see Butzer-Tillmann[3] and Butzer-Berens[4]).

Now what happens if the Banach space X is not reflexive? This situation is covered (at least for some examples of self-adjoint semi-groups) by a result on dual semi-groups due to K. de Leeuw[2] (see also Butzer-Berens[4], p. 90 ff. or, more directly, H. Berens[5], p. 42), namely:

Theorem 2. *Under the hypotheses of Theorem* 1, *one has* $\|T(t)f - f\| = O(t)$ *for an f in X if and only if f belongs to the completion of* $D(A)$ *relative to X.*

In this setting, the saturation class of Weierstrass's integral in $L^1(R)$ is the set of all f in $L^1(R)$ such that f is in $AC(R)$ with f' in $NBV(R)$ or for which f' is in Lip $(1, L^1(R))$.

2. *Nonoptimal Approximation.* The next goal was nonoptimal approximation and thus the precise characterization of those elements f in X for which $\|T(t)f - f\| = O(t^\alpha)$ $(t \to 0+)$ for fixed α, $0 < \alpha < 1$. In generalization of the integral $\int_0^b T(u)g\,du$ (g in X, $b > 0$) which belongs to $D(A)$ and is thus approximated with order $O(t)$, Hille ([B2], p. 324) chose the integral

$$\int_0^b (b-u)^{\alpha-1}\, T(u)g\, du \equiv f$$

of fractional order α, $0 < \alpha < 1$, and stated that $\|T(t)f - f\| = O(t^\alpha)(t \to 0+)$. This direct theorem of Hille may be said to have given impetus to the study of nonoptimal approximation by semi-group operators, also in connection with fractional powers of their infinitesimal generator.

If $\{T(t); t \geqslant 0\}$ is a holomorphic semi-group of class (C_0), Butzer-Berens[4], (p. 111 ff.) showed in 1964 that, for $0 < \alpha < 1$ and $t \to 0+$, $\|T(t)f - f\| = O(t^\alpha)$ if and only if $\|AT(t)f\| = O(t^{\alpha-1})$ if and only if $\|A^2T(t)f\| = O(t^{\alpha-2})$.

However, one desires a more direct characterization in terms of f. Westphal[6] showed that, for $0 < \alpha < 1$, f in $D((-A)^\alpha)$–$(-A)^\alpha$ being a suitable definition of a fractional power of $-A$–implies $\|T(t)f - f\| = O(t^\alpha)$ $(t \to 0+)$. This result being too strong, can one find a space Y with $D((-A)^\alpha) \subset Y \subset X$ such that all f in Y yield an approximation $T(t)f$ with order $O(t^\alpha)$, and conversely? Such a Y is the following intermediate space (see Butzer-Berens[4], p. 167):

$$(X, D((-A)^\gamma))_{\alpha/\gamma,\infty;K} = \left\{ f \text{ in } X:\ \sup_{0<t<\infty} t^{-\alpha/\gamma} K(t,f;X,D((-A)^\gamma)) < \infty \right\},$$

$$0 < \alpha < \gamma \leqslant 1,$$

$K(t,f)$ being Peetre's K-functional. Indeed (see Westphal[6] and Butzer-Scherer[7] for $\gamma < 1$; Butzer-Berens[4] for $\gamma = 1$), the following holds:

Theorem 3. *Let* $\{T(t); t \geqslant 0\}$ *be a uniformly bounded semi-group of class* (C_0). *One has, for* $0 < \alpha < \gamma \leqslant 1$, $f \in (X, D((-A)^\gamma))_{\alpha/\gamma,\infty;K}$ *if and only if* $\|T(t)f - f\| = O(t^\alpha)$ $(t \to 0+)$.

3. *Extensions.* The next question is whether the crucial semi-group property can be weakened in the foregoing results. In [63], Hille had already observed that Theorem 1 can be established for integral transforms satisfying certain "functional equations." In this respect, Theorem 2 may be generalized (Berens[5]) to families $\{T_t; t \geqslant 0\}$ of uniformly bounded, linear, commutative (replacing the semi-group property) operators tending to the identity for $t \to 0+$ and satisfying a Voronovskaja-type relation (replacing the infinitesimal generator). Concerning nonoptimal approximation, Theorem 3 can be formulated in the more general framework of function norms Φ, also giving a broader description of the order of approximation. As above, $\{T_t; t \geqslant 0\}$ is a commutative family, but here it satisfies Jackson and Bernstein-

type inequalities (of highest possible, or saturation order). This result not only covers Theorem 3 for holomorphic semi-groups (Butzer-Scherer[7]) and for resolvent operators of the infinitesimal generator of a semi-group, but also for practically all known summation methods of Fourier series or integrals (compare Butzer[8] and the literature cited there). In this connection, compare also E. Görlich[9] and J. Löfström[10].

Hille's Theorem 1 and Theorems 2 and 3 may be generalized in another direction concerned with semi-groups under perturbation. Indeed, Butzer-Köhnen[11] have shown that the approximation behavior, optimal as well as nonoptimal, of two arbitrary semi-groups of class (C_0) is left invariant provided that the domains of their infinitesimal generators coincide.

Finally, semi-group theory (and in part Theorem 1) has been generalized to locally convex spaces by L. Schwartz, H. G. Tillmann, H. Komatsu, K. Yosida, H. G. Garnir (see Butzer-Berens[4], pp. 76, 149, for literature), T. Komura[12], and M. Becker-U. Westphal[13], and to distribution semi-groups of operators by J. L. Lions, C. Foias, K. Yoshinaga and J. Löfström (see Butzer-Berens[4], pp. 77, 149). A very active direction of research is nonlinear semi-group theory, which is associated with the names of H. Brezis, F. Browder, M. G. Crandall, J. R. Dorroh, T. Kato, Y. Komura, T. Liggett, I. Miyadera, J. W. Neuberger, S. Oharu, A. Pazy, G. F. Webb, and others (see the survey article of Dorroh[14] for literature up to 1971). For Theorem 2 see U. Westphal[15].

The author wishes to thank Dr. U. Westphal for constructive criticism.

P. L. Butzer

References

1. P. L. Butzer, *Sur la théorie des semi-groupes et classes de saturation de certaines intégrales singulières,* C. R. Acad. Sci. Paris vol. 243 (1956) pp. 1473-1475 (see also Math. Ann. vol. 133 (1957) pp. 410-425).

2. K. de Leeuw, *On the adjoint semigroup and some problems in the theory of approximation,* Math. Z. vol. 73 (1960) pp. 219-234.

3. P. L. Butzer and H. G. Tillmann, *Approximation theorems for semi-groups of bounded linear transformations,* Math. Ann. vol. 140 (1960) pp. 256-262.

4. P. L. Butzer and H. Berens, *Semi-groups of Operators and Approximation,* Springer-Verlag, New York, 1967.

5. H. Berens, *Interpolationsmethoden zur Behandlung von Approximationsprozessen auf Banachraumen,* Springer Lecture Notes, vol. 64, Springer-Verlag, New York, 1968.

6. U. Westphal, *Ein Kalkül für gebrochene Potenzen von Operatoren.* Teil I: *Halbgruppenerzeuger.* Teil II: *Gruppenzeuger,* Compositio Math. vol. 22 (1970) pp. 67-103, 104-136.

7. P. L. Butzer and K. Scherer, *Jackson and Bernstein-type inequalities for families of commutative operators in Banach spaces,* J. Approximation Theory vol. 5 (1972) pp. 308-342.

8. P. L. Butzer, *A Survey of Work on Approximation at Aachen, 1968-1972,* Proc. Conf. Approximation, Austin, Texas, Jan. 22-24, 1973, Academic Press, New York, 1973.

9. E. Görlich, *Logarithmische und exponentielle Ungleichungen vom Bernstein-Typ und verallgemeinerte Ableitungen,* Habilitationsschrift, RWTH Aachen, 1971.

10. J. Löfström, *Some theorems on interpolation spaces with applications to approximation in* L^P, Math. Ann. vol. 172 (1967) pp. 176-196.

11. P. L. Butzer and W. Köhnen, *Approximation invariance of semigroup operators under perturbations,* J. Approximation Theory vol. 2 (1969) pp. 389-393.

12. T. Komura, *Semigroups of operators in locally convex spaces,* J. Functional Analysis vol. 2 (1968) pp. 258-296.

13. M. Becker and U. Westphal, *Approximation by families of linear operators in locally convex spaces* (in preparation).

14. J. R. Dorroh, *Semi-groups of nonlinear transformations,* in *Linear Operators and Approximation,* edited by P. L. Butzer, J. P. Kahane, B. Sz.-Nagy, ISNM, vol. 20, Birkhäuser, Basel, 1972.

15. U. Westphal, *Sur la saturation pur des semi-groupes non linéaires,* C. R. Acad. Sci. Paris Ser. A, vol. 274 (1972) pp. 1351-1353.

Chapter 13. Ordinary Differential Equations

Hille's work on ordinary differential equations

The papers included in this group are miscellaneous papers on ordinary differential equations. It should be noted though that Chapter 2 (Zero Point Problems and Asymptotic Integration of Second Order Linear Differential Equations) also contains papers on ordinary differential equations, that several important papers in this field are not included in this Selecta, and that two of the papers in the present section, namely [122] and [133], are devoted to infinite systems of differential equations.

Hille maintained an interest in differential equations throughout his career, and he eventually (in 1969) published a book [B8] on the subject. In the course of comments on the papers included in the present group, it will be appropriate to refer to cognate portions of that book.

At various periods of his career, Hille investigated linear differential equations of the second order, in the cases of both real and complex independent variables. His interest centered around zeros of solutions (see also the papers in Chapter 2), asymptotic integration, and spectral theory. The three papers reprinted in this volume ([92], [102], and [136]) are good examples of his later work in this field.

In [92], Hille discusses two related problems for

$$y'' + f(x)y = 0, \qquad x > 0,$$

with locally integrable f: (i) When does this equation have a solution approaching a finite nonzero limit as $x \to \infty$? (ii) If $f(x) \geqslant 0$, when is the equation nonoscillatory on a suitable interval (a, ∞)? Wintner had already given $f \in L(1, \infty)$ as a sufficient condition. This condition is also necessary if $f(x)$ has constant sign for large x. Problem (ii) is much more difficult. Its study was commenced by A. Kneser in 1892, and was continued by A. Wintner, whose work stimulated Hille's researches. Hille's principal tool is the integral equation

$$u(x) = x \int_x^\infty (u(t)/t)^2 \, dt + g(x),$$

which is satisfied in the nonoscillatory case by $u(x) = xy'(x)/y(x)$, with

$$g(x) = x \int_x^\infty f(t)\,dt;$$

and he also uses this tool to extend his investigations in part to complex x. Hille's work has, in turn, been followed up by others, including Wintner and Hartman (see P. Hartman, *Ordinary Differential Equations,* Wiley, New York, 1964, section 7.1).

In [102], the differential equation is

$$y'' - \lambda f(x) y = 0,$$

where f is positive and continuous for $0 \leqslant x < \infty$ and λ is a complex parameter. The equation is interpreted as describing the motion of a particle whose position at time x is $y = u + iv$, and which moves under the influence of a force of magnitude $|\lambda|\, f(x)|y|$ making an angle $\arg \lambda$ with the radius vector.

First it is proved that a necessary and sufficient condition for almost uniform motion (i.e., for the existence of solutions satisfying $y_1(x)/x \to 1$ and $y_2(x) \to 0$ as $x \to \infty$) is that $xf(x) \in L(0,\infty)$. Next it is assumed that $xf(x) \notin L(0,\infty)$ and that λ is neither 0 nor negative real. Spiral motion and growth properties are discussed. This work, which is connected with the investigations of Bôcher, Weyl, and Wintner, is taken up again in Chapter 9 of [B8], and is said to be of relevance in the study of Fokker-Planck-Kolmogoroff equations for temporally homogeneous stochastic processes.

The subject of [136] is the singular Sturm-Liouville problem for the differential equation

$$y'' + [\lambda - q(x)]\, y = 0.$$

This problem was first solved by H. Weyl in 1909, and has been subsequently studied by numerous research workers. Titchmarsh devoted a book to it in 1946 (with a second volume, on $\Delta V + (\lambda - q) V = 0$, published in 1958), and various ramifications remain of current interest. Hille's study of the problem is based on Green's identity

$$[\overline{y(x,\lambda)}\, y'(x,\lambda)]_a^b - \int_a^b |y'(x,\lambda)|^2\,dx - \int_a^b [\lambda + q(x)]\ |y(x,\lambda)|^2\,dx = 0,$$

and indeed he states that the purpose of his paper is to illustrate the use of this identity. The identity is also applied in [B8] in many other contexts.

In [113] and [114], Hille uses semi-group methods to investigate infinite systems of linear differential equations with constant coefficients, and he verifies the existence of null solutions, that is, of nontrivial solutions of $y'(t) = Ay(t)$ satisfying $y(0+) = 0$. Such solutions had been discovered earlier by W. Feller but had not been systematically investigated. In [122] and [133], reprinted in this volume, Hille studies in greater detail examples of infinite systems of linear differential equations with constant coefficients in order to exhibit some phenomena not encountered in finite systems. The results are reproduced in part in Section 5.3 of [B8].

The examples considered are

$$y_k'(t) + ka_k y_k(t) = \sum_{m \geqslant k+1} a_m y_m(t) \qquad (k = 1, 2, \ldots)$$

and, with $a_m = 1$ in [122] and $a_m > 0$ in [133],

$$y_k'(t) = (k - 1)\, y_{k-1}(t) - k\, y_k(t) + y_{k+1}(t) \qquad (k = 1, 2, \ldots).$$

It is shown that null solutions may exist, that general solutions may depend on arbitrary functions rather than on constants, that there are entire functions of t satisfying a system for $t > 0$ but not for $t < 0$, and that there are analytic solutions with singularities at arbitrary points. For the construction of such "explosive" solutions, reference is made to [109].

The nonlinear ordinary differential equation

$$x^{1/2} y''(x) = (y(x))^{3/2}$$

occurs in atomic physics and has been investigated by physicists, who have discovered curious properties of its solutions. The solutions of principal interest are of the form

$$y(x) = 1 + b_2 x + b_3 x^{3/2} + \ldots + b_k x^{k/2} + \ldots,$$

where b_2 is arbitrary and the other coefficients depend on b_2. For a special value ω of b_2, the solution vanishes at infinity, and the behavior of $y(x)$ for $b_2 < \omega$ differs sharply from that for $b_2 > \omega$.

Hille has undertaken a mathematical study of the solutions, and a more detailed publication is promised. The note reprinted in this volume is a brief

announcement of some results concerning convergence of the series given above (hitherto regarded as "semiconvergent") and the existence of "explosive" solutions, of solutions with a preassigned branch point, of solutions vanishing at infinity, and of solutions unbounded at $x = 0$.
A. Erdélyi

Chapter 14. Partial Differential Equations

An introduction to Hille's work on the abstract Cauchy problem and semi-groups

The theory of semi-groups of linear operators, developed simultaneously and independently by E. Hille [B1] and K. Yosida[15], has become a standard analytic tool in probability theory, partial differential equations, and functional analysis. This is clearly demonstrated by the large number of basic books on these subjects which devote substantial space to semi-group theory, and particularly to that part of the theory emanating from the famous Hille-Yosida theorem. For example, in probability one finds semi-group theory developed and used in the books of W. Feller[9] (vol. II), E. B. Dynkin[8] (vol. I), and P. A. Meyer[13]. In A. Friedman's book[10], semi-groups are presented as part of the necessary apparatus for solving partial differential equations, as also in Dunford-Schwartz[7] (Part I) and in R. Carroll's book[4] on abstract methods in partial differential equations. Applications in other areas of functional analysis will be found in the books of Hille-Phillips [B2] and Yosida[17].

In 1949 two papers appeared, one by Hille [98] and the other by Yosida[16], which initiated the application of their previously developed semi-group theory to problems arising in the theory of stochastic processes. The importance of their methods was recognized by W. Feller, who embarked in 1952 on a far-reaching program of research into the use of this new analytic tool in probability theory, especially in diffusion theory. The evolution of these ideas in the work of E. B. Dynkin, G. A. Hunt, K. Ito, H. P. McKean, D. Ray, A. Yushkhevich, G. Maruyama, and many others, has been of fundamental importance in modern probability theory. See the historical-bibliographical note at the end of vol. II of Dynkin's book[8], or Yosida's book[17], Ch. XIII, sec. 5, for the precise references.

Hille's paper [98] concerns the integration of the Fokker-Planck equation

$$\frac{\partial v}{\partial t} = \frac{\partial^2}{\partial x^2}[b(x)v] - \frac{\partial}{\partial x}[a(x)v] = U[v] \tag{1}$$

(where U is used to denote the differential operator) under the initial condition $v(x,t) \to g(x)$ as $t \to 0$ for a given g in some appropriate function space. By putting this problem in the general framework of a complex Banach space X with U a more general linear (but usually unbounded) operator, Hille was led to his "abstract Cauchy problem" (ACP for short), an approach inspired in part by the earlier work of Hadamard in 1903 on hyperbolic equations. (A sketch of the historical background of the abstract Cauchy problem will be found in [B2], pp. 617-618.) Hille's original formulation was presented in a series of papers between 1952 and 1954 ([103], [106], [107], [108], [110], and [115]). In 1954 R. S. Phillips gave a formulation which was in some ways more restrictive than Hille's version, but which had the advantage of leading to stronger results. This Hille-Phillips formulation can be found in [B2]; it is as follows: If we have a linear operator U in a complex Banach space X with domain $\mathcal{D}(U)$ and range $\mathcal{R}(U)$ in X and are given $y_0 \in X$, find a function $y(t) = y(t;y_0) \in \mathcal{D}(U)$ which is strongly absolutely continuous and continuously differentiable in each finite subinterval of $[0, \infty)$, or of $(0, \infty)$, and which satisfies

$$y'(t) = U[y(t)]$$

under the initial condition

$$\lim_{t \to 0^+} \|y(t;y_0) - y_0\| = 0.$$

The two alternatives indicated by brackets were called ACP_1 and ACP_2, respectively, by Hille.

In [115], Hille shows that, if U is a closed operator whose point spectrum is not dense in any right half-plane, then for each $y_0 \in X$ the ACP_1 has at most one solution of normal type, that is, of exponential growth at $t = \infty$. More precisely, a solution $y(t)$ is said to be of type ω_0 if

$$\omega_0 = \lim_{t \to \infty} t^{-1} \log \|y(t)\|. \tag{2}$$

He further shows that when U is closed the nonuniqueness of solutions of normal type $\leqslant \omega_0$ is equivalent to the existence of a nontrivial solution of $U[x(\lambda)] = \lambda x(\lambda)$ that is bounded and holomorphic in each half-plane $\mathrm{Re}(\lambda) \geqslant \omega_0 + \epsilon$, $\epsilon > 0$. On the other hand, if the spectrum of the operator U covers the whole complex plane, Hille had previously discovered the existence of "explosive" solutions $y(t)$ for which $\|y(t)\| \to \infty$ as $t \to t_0$ for some $t_0 < \infty$; cf. [100].

The connection between the ACP and semi-groups is summed up in the set of four theorems due to Phillips which are given in [B2], pp. 621-622. For simplicity, let us consider only the ACP_1. If U is the infinitesimal generator of a semi-group $\{T_t\}$ of class C_0 (i.e., strongly continuous at $t = 0$), then the

ACP_1 has a unique solution $y(t;y_0) = T_t y_0$, the infinitesimal generator meaning the operator

$$\lim_{t\to 0^+} t^{-1} [T_t y - y] = U[y],$$

defined wherever the limit exists in the strong sense in X. Conversely, if U is a closed, densely defined linear operator with nonempty resolvent set and if for each $y_0 \in \mathcal{D}(U)$ there is a unique solution to the ACP_1, then U generates a semi-group $\{T_t\}$ of class C_0 such that $y(t;y_0) = T_t y_0$. Thus, solving the ACP is in practice equivalent to ascertaining whether U generates a semi-group or not. The original Hille-Yosida theorem provided a very convenient criterion characterizing infinitesimal generators of contraction semi-groups, that is, of those C_0 semi-groups such that $\|T_t\| \leqslant 1$. Their result was generalized by W. Feller, R. S. Phillips, and I. N. Miyadera. The necessary and sufficient condition for a closed, densely defined linear operator U to generate a semi-group $\{T_t\}$ of class C_0 with $\|T_t\| \leqslant M$ is that there exist $\lambda_0 > 0$ such that

$$\|[R(\lambda)]^n\| \leqslant M\lambda^{-n} \qquad (\lambda > \lambda_0), \tag{3}$$

where $R(\lambda) = (\lambda I - U)^{-1}$ and $n = 1, 2, \ldots$. See [B2], Ch. 12, for more detail and generality. (It is to be noted that the condition $\|T_t\| \leqslant M$ is a matter of normalization, since a C_0 semi-group must be of exponential growth at $t = \infty$, that is, $\|T_t\|$ must satisfy a condition of the form (2), as proved in [B2], p. 306. Thus, we have normalized T_t to be of negative type.) Since

$$R(\lambda) = \int_0^\infty e^{-\lambda t} T_t \, dt,$$

one can view this theorem as giving an abstract Laplace-transform method of solution. The Hille-Yosida approach to solving the Fokker-Planck equation on the interval $[a,b]$ was to find conditions on the coefficients of the differential operator U which would assure that U is the infinitesimal generator of a semi-group of class C_0 on the space $L[a,b]$.

Feller's 1952 work was based on the observation that even if the operator U is not the infinitesimal generator of a semi-group of class C_0, certain restrictions of it may be. These restrictions are defined by "lateral conditions," special cases of which are the classical boundary conditions. He then set about finding all the restrictions of the one-dimensional diffusion operator U that generate contraction semi-groups of class C_0 in $C[r_1,r_2]$. His work, generalized to n dimensions by A. Wentzell, paved the way to a general study of the relation between Markoff processes and semi-groups. (See vol. I of Dynkin's book[8] for a detailed treatment of this subject.) The extreme usefulness of this abstract approach lay, of course, in the fact that U was not restricted to the class of differential operators, thus making it possible to study a much wider class of stochastic processes than just the diffusion processes.

In 1961 Phillips and Lumer gave a different characterization of the infinitesimal generators of contraction semi-groups in Banach spaces. Their characterization was: A necessary and sufficient condition for a densely linear operator U in a real or complex Banach space X to generate a C_0 contraction semi-group is that U be a dissipative operator and $\mathcal{R}(I - U) = X$, the term dissipative meaning that Re $\langle Ux, x'\rangle \leqslant 0$ whenever (1) $x \in \mathcal{D}(U)$ and (2) $x' \in X'$ satisfies $\langle x, x'\rangle = ||x||^2$, $||x|| = ||x'||$. Phillips devoted several papers (one jointly with P. D. Lax) to the application of dissipative operators in the theory of hyperbolic systems and of parabolic equations (see Yosida[17], Ch. IX, sec. 8, for specific references).

In [106] and [115], Hille considers a higher order Cauchy problem, which he abbreviates by ACP^n. The precise formulation, due to Hille and Phillips, will be found in [B2], p. 623. Roughly speaking, the ACP^n is the problem of finding functions $y(t;y_0, \ldots, y_{n-1})$ in a complex Banach space $X, t \geqslant 0$, such that y is a solution of

$$y^{(n)}(t) = U^n [y(t)]$$

under the initial conditions

$$\lim_{t\to 0^+} ||y^{(k)}(t;y_0, \ldots, y_{n-1}) - y_k|| = 0$$

($k = 0, 1, \ldots, n - 1$) for prescribed $y_k \in X$. For example, the wave equation is of this type when $n = 2$ and $U = d/dx$. The ACP we discussed above corresponds to $n = 1$. Another approach, still using semi-group theory, to the integration of the wave equation is found in Yosida[17], Ch. XIV, sec. 3. In 1959, using the theory of fractional powers of operators, A. V. Balakrishnan gave an interesting application of a higher order ACP^n for $n = 2$ (see the paper of Balakrishnan listed in the bibliography of Yosida's book).

The original formulation of the abstract Cauchy problem and of the Hille-Yosida theorem was in Banach spaces. In 1958 L. Schwartz gave the extension of the Hille-Yosida theorem to locally convex spaces, and the theory of

C_0 semi-groups is presented in this general setting in Yosida's book[17].

Another interesting extension is to the theory of distribution semi-groups and the abstract Cauchy problem in the sense of distributions, first studied by J. Lions in 1960 (see Lions[11], p. 253, for exact references). A distribution semi-group is a vector-valued distribution T on the interval $t > 0$ with values in $L(X,X)$, where X is a complex Banach space, such that

$$T(\Phi * \Psi) = T(\Phi) \cdot T(\Psi)$$

for each pair of test functions Φ and Ψ. Lions's definition of distribution semi-group (SGD) also includes regularity conditions governing behavior in the neighborhood of $t = 0$. The ordinary semi-groups studied by Hille-Yosida-Phillips are of exponential growth at $t = \infty$. However, examples were given by Foias (see the above reference in Lions's book) which show that an SGD may have arbitrarily rapid growth at ∞. Lions in his 1960 paper extended the Hille-Yosida theorem to the class of distribution semi-groups having exponential growth—SGDE's he called them. In this general theorem, condition (3) is replaced by

$$\|[R(\lambda)]^n\| \leqslant \mathrm{pol}(|\lambda|) \qquad (\mathrm{Re}(\lambda) \geqslant \lambda_0), \tag{4}$$

where pol(x) denotes a positive polynomial in x. In 1968 J. Chazarain[5,6] extended the Hille-Yosida theorem to arbitrary SGD's, and this theorem replaces (3) with the condition that $R(\lambda)$ exist in a logarithmic region Λ (i.e., a region of the type

$$\mathrm{Re}(\lambda) \geqslant \max[A \log|\mathrm{Im}(\lambda)| + B, \lambda_0],$$

where $0 \leqslant A, B, \lambda_0 < \infty$) and satisfy there the estimate (4). The theorem of Lions then reduces to the special case where Λ is a half-plane $\mathrm{Re}(\lambda) \geqslant \lambda_0$.

So far we have discussed the abstract Cauchy problem only for linear operators that are independent of t. The next natural step to take in the spirit of Hille's approach to initial value problems is to look at the temporally inhomogeneous "equation of evolution"

$$\frac{\partial y}{\partial t} = U(t)[y] \qquad (t > 0) \tag{5}$$

in an appropriate Banach space with given initial data. Here the linear operator $U(t)$ depends on t, in contrast to the situation we have heretofore considered. This more general type of equation would occur, for example, in a Cauchy problem involving a differential operator $U(t)$ whose coefficients depend on time as well as on spatial variables. The first successful attack on this more general problem was carried out by T. Kato in 1953. His method was an abstract version of the Cauchy polygon method for ordinary differential equations of the type $dy/dt = a(t)\, y(t)$. Since that time, much work has been done on the linear evolution equation (5). The book of Lions[11] gives an extensive list of references up to 1961. See also Yosida's book[17], particularly Ch. XIV on the integration of the equation of evolution. Chapters 3-5 of Carroll's book[4] and part 2 of Friedman's book[10] are also devoted to this subject.

Abstract Cauchy problems for nonlinear operators were formulated by F. E. Browder[1,2] (1962, 1964), and I. E. Segal[14] (1963). Research in the field of nonlinear semi-groups and nonlinear abstract Cauchy problems continues actively up to the present time. The complete analogue to the Hille-Yosida theorem for contraction semi-groups is known only in Hilbert spaces and was developed in 1969 by J. Dorroh, M. Crandall, and A. Pazy (for references see the paper by Brézis and Pazy cited in the bibliography of Lions's book[12]). The theorem is a generalization of the Phillips-Lumer formulation of the Hille-Yosida theorem in terms of dissipative operators. Sufficient conditions for a nonlinear operator U to generate a semi-group are known if X is a Banach space with uniformly convex dual (cf. Browder[3], sec. 9, or Lions[12], p. 300). The nonlinear operators U considered in current work are in general multivalued, and, in their paper cited above, Brézis and Pazy have introduced and discussed multivalued nonlinear semi-groups. For the many contributions in recent years to the theory of nonlinear equations of evolution and nonlinear semi-groups by numerous authors, we refer to Browder[3], or the references given in the paper by Brézis and Pazy cited above.
J. Elliott

References

1. F. E. Browder, *On nonlinear wave equations,* Math. Z. vol. 80 (1962) pp. 249-264.
2. F. E. Browder, *Non-linear equations of evolution,* Ann. of Math. (2) vol. 80 (1964) pp. 485-523.
3. F. E. Browder, *Nonlinear operators and nonlinear equations of evolution in Banach spaces,* Proc. Symp. Pure Math., vol. XVIII, part II (in preparation).
4. R. W. Carroll, *Abstract Methods in Partial Differential Equations,* Harper & Row, New York, 1969.
5. J. Chazarain, *Problèmes de Cauchy au sens de distributions vectorielles et applications,* C. R. Acad. Sci. Paris Ser. A vol. 226 (1968) pp. 10-13.

6. J. Chazarain, *Problèmes de Cauchy abstraits et applications,* Sem. Lions-Schwartz, exposé du Jan. 26, 1968.

7. N. Dunford and J. Schwartz, *Linear Operators,* part I, Interscience, New York, 1958.

8. E. B. Dynkin, *Markov Processes,* vols. I and II, Springer-Verlag, New York, 1965.

9. W. Feller, *Introduction to Probability Theory and Its Applications,* vol. II, 2nd ed., Wiley, New York, 1971.

10. A. Friedman, *Partial Differential Equations,* Holt, Rinehart and Winston, New York, 1969.

11. J. L. Lions, *Equations differentielles operationnelles et problèmes aux limits,* Springer-Verlag, Berlin, 1961.

12. J. L. Lions, *Quelques méthodes de resolution des problèmes aux limites non-linéaries,* Dunod, Paris, 1969.

13. P. A. Meyer, *Probability and Potentials,* Blaisdell, Waltham, Mass., 1966.

14. I. E. Segal, *Nonlinear semigroups,* Ann. of Math. (2) vol. 78 (1963) pp. 339-364.

15. K. Yosida, *On the differentiability and the representation of one-parameter semi-groups of linear operators,* J. Math. Soc. Japan vol. 1 (1948) pp. 15-21.

16. K. Yosida, *An operator-theoretic treatment of temporally homogeneous Markoff processes,* J. Math. Soc. Japan vol. 1 (1949) pp. 244-253.

17. K. Yosida, *Functional Analysis,* Springer-Verlag, New York, 1965 (2nd printing, corrected, 1966).

Chapter 16. Banach Algebras

Hille's work on Banach algebras

Einar Hille was one of a relatively small number of classical analysts who recognized early the importance of functional analysis, not only as a subject in its own right but as a tool for analysis in general. He also recognized immediately the importance of the Gelfand theory of normed rings and included in his *Functional Analysis and Semi-groups* [B1] a detailed elaboration of the theory which stood for some years as the standard reference to the subject. Although the theory of normed rings, or Banach algebras as they are now called, is not one of Hille's main areas of research, several of his papers ([88], [124], [125], [132], [135], [140], and [141]), in addition to his Colloquium Publication [B1], do involve the subject in a more or less substantial way. In paper [88], he generalizes a result of Gelfand on characters of groups in normed rings, and the papers [132], [135], and [140] are concerned with differential equations in Banach algebras. The remaining three papers, [124], [125], and [141], on which I will comment, are more directly involved with Banach algebras. The most important of these is paper [124], a portion of which represents joint work with S. Kakutani.

In paper [124], Hille studies certain properties of roots and logarithms of elements of a noncommutative complex Banach algebra B with unit element e. Two problems are considered. The first is concerned with relationships between two k^{th} roots (or logarithms) of a regular element of the Banach algebra, and the second concerns the question of when a regular element is interior to the set of elements having k^{th} roots (or logarithms).

If x_1 and x_2 are k^{th} roots of a regular element $a \in B$ (i.e., solutions in B of the equation $x^k = a$), then simple examples show that x_1 and x_2 need not commute. However, if they do commute, then Hille shows that there exist mutually orthogonal idempotents $e_1, \ldots, e_n$ in B such that $\Sigma e_\alpha = e$ and

$$x_1 x_2^{-1} = \sum_{\alpha=1}^{k} \omega^\alpha e_\alpha ,$$

where $\omega = \exp(2\pi i/k)$. Also, the idempotents commute with x_1 and x_2. If the spectrum of a kth root is "irrotational (mod $2\pi/k$)," in the sense that the spectra of the elements x, $\omega^j x$ in B are disjoint for $j = 1, \ldots, k-1$, then x commutes with every k^{th} root of a. Furthermore, in this case x covers a neighborhood of a; thus, in particular, a is interior to the set of elements with k^{th} roots in B. More generally, let C be a rectifiable curve in the complex plane joining the origin to a sufficiently distant point. If the spectrum of the kth root does not intersect any of the curves $C, \omega C, \ldots, \omega^{k-1}C$, where $\omega = \exp(2\pi i/k)$, then there exists a kth root of a whose spectrum is irrotational (mod $2\pi/k$); so again a is an interior point of the set of elements with k^{th} roots.

Results analogous to the above hold for logarithms. Thus, if y_1 and y_2 are logarithms of a (i.e., solutions in B of the equation $\exp(y) = a$) that commute, then there exist distinct integers $k_1, \ldots, k_n$ and mutually orthogonal idempotents $e_1, \ldots, e_n$ in B such that $\Sigma e_\alpha = e$ and

$$y_1 - y_2 = 2\pi i \sum_{\alpha=1}^{n} k_\alpha e_\alpha.$$

Also, the idempotents commute with y_1, y_2. If y is a logarithm whose spectrum is "incongruent (mod $2\pi i$)," in the sense that the elements y and $y + 2k\pi i$ have disjoint spectra for $k = \pm 1, \pm 2, \ldots$, then y commutes with every logarithm of a and a is interior to the set of elements with logarithms in B. A condition weaker than incongruence (mod $2\pi i$), analogous to the case of roots, will also imply that a is an interior point.

Proofs of the above results involve, in the case of roots, an analysis of the operator $S(b)$ acting on B as a Banach space, where b is a regular element of B and $S(b)(x) = b^{-1}xb$, $x \in B$. In the case of logarithms, it is the "Lie operator" $T(a)$, where $T(a)(x) = ax - xa$, $x \in B$, that must be analyzed. The proofs depend on a systematic use of spectral theory for linear operators along with various properties of the resolvent function and an abstract implicit function theorem due to Graves and Hildebrandt.

In paper [125], Hille states and proves a special case of the Graves-Hildebrandt implicit function theorem which is especially appropriate for applications to Banach algebras. It involves the inversion of functions of the form

$$F(z) = \sum_{n \geqslant 1} F_n(z), \qquad z \in B,$$

where F_n is an abstract polynomial (i.e., $F_n(z) = F_n(z, \ldots, a)$, where $F_n(z_1, \ldots, z_n)$ is an n-linear bounded form on $B \times \cdots \times B$ into B). Thus

$\|F_n(z_1, \ldots, z_n)\| \leqslant M_n\|z_1\| \cdots \|z_n\|$, and it is assumed that the series $\Sigma M_n y^n$ has a positive radius of convergence so that $F(z)$ is defined. If the linear operator $F_1(z)$ has a bounded inverse, then the inverse function exists and is of the same form as F. The proof involves a straightforward application of classical methods. Although the results are stated for Banach algebras, everything obviously goes through for an arbitrary Banach space. The present paper was no doubt motivated by the application of the abstract implicit function theorem in the preceding paper.

In paper [141], Hille studies the "Jordan operator" in a noncommutative Banach algebra. More precisely, he studies the spectral properties of the operator J_a on B, where $a \in B$ and $J_a(x) = \frac{1}{2}(ax + xa)$, $x \in B$. He also studies the "squaring operator" S_a, where $S_a(x) = axa$, $x \in B$. The spectrum of J_a is contained in the set $\{\frac{1}{2}(\alpha + \beta) : \alpha, \beta \in \sigma(a)\}$ and the spectrum of S_a is contained in $\{\alpha\beta : \alpha, \beta \in \sigma(a)\}$, where $\sigma(a)$ is the spectrum of the element a in B. A few more precise results can be obtained for the point spectra. Hille also calculates the resolvents of J_a and S_a. Methods used in this paper are similar to those used in [124].

C. E. Rickart

Chapter 17. Transfinite Diameters

An introduction to Hille's papers on averaging processes and transfinite diameters

The notions here under discussion start from old basic ideas on various types of means, pick up substance and depth from the theories of potential, complex variables, among others, and finally evolve into metric topology, Banach spaces, and their geometrical properties. It isn't surprising that Einar Hille, with his classic European training and his enthusiasm and zest for recent developments, should have become heavily involved with them over a considerable period of time. His first paper on the subject, *Remarks on transfinite diameters* [134], was delivered to the Prague Symposium on Topology in 1961. It would seem that this was his favorite field of research for the next four or five years.

The transfinite diameter of a bounded set in a metric space is defined by means of an averaging process, an extremal problem, and a limiting process. (Rapidly speaking, consider a fixed integer n. Choose n points and form their average (defined below). Maximize this by varying the points. Take the limit as $n \to \infty$.) Among averaging processes one can list all the obvious and classic ones. Hille's principal predecessors in this field were Fekete, Kolmogoroff, and Nagumo. In 1923 Fekete studied the transfinite diameter of compact sets in the complex plane. Using geometric means, he established the existence of this diameter and showed it to be equal to the Čebyšev constant and to the logarithmic capacity of the set. Kolmogoroff in 1930 defined abstractly an averaging process A. His conditions follow:

(i) Given any finite set of positive numbers $\{x_1, \ldots, x_m\}$, A assigns to them a positive average $A(x_1, \ldots, x_m)$.

(ii) $A(x_1, \ldots, x_m)$ is continuous, symmetric, and strictly increasing in each argument.

(iii) $A(x, \ldots, x) = x$.

(iv) If $y = A(x_1, \ldots, x_k)$, then
$A(x_1, \ldots, x_k, x_{k+1}, \ldots, x_m) = A(y, \ldots, y, x_{k+1}, \ldots, x_m)$.

Kolmogoroff and Nagumo proved that these conditions imply the existence of a continuous, strictly monotone function $F(u)$ such that

$$mF[A(x_1, \ldots, x_m)] = \sum_{j=1}^{m} F(x_j).$$

Important special cases for $F(u)$ are

$$u, \quad u^2, \quad \log \frac{1}{u}, \quad \frac{1}{u^{n-2}} \ (n > 2), \quad u^p (p > 0).$$

It seems to be just as easy to derive results from the axioms directly and to bypass the existence theorem for $F(u)$. This is the procedure adopted by Hille for the most part.

An immediate consequence of (ii) and (iii) is that

$$\min x_j \leqslant A(x_1, \ldots, x_m) \leqslant \max x_j.$$

In a compact metric space E one chooses n points and obtains the $m = \frac{1}{2} n(n-1)$ distances between them. These are the x_j. If the diameter of the set is $\delta(E)$, then the $A(x_1, \ldots, x_m)$ are bounded and there is a maximum called $\delta_n(E)$. Next comes the point of substance: $\delta_{n+1}(E) \leqslant \delta_n(E)$. Finally the transfinite diameter is defined by

$$\delta_0(E) = \lim_{n \to \infty} \delta_n(E).$$

For a compact set E in the complex plane, Fekete proves that the Čebyšev constant $\chi(E)$ is equal to the transfinite diameter $\delta_0(E)$. It is possible to extend the notion of the Čebyšev constant to compact sets E in metric spaces. Hille proves that in general

$$\chi(E) \leqslant \delta_0(E).$$

If $P, Q_1, \ldots, Q_n$ are points in E, set

$$f(P) = A[d(P, Q_1), \ldots, d(P, Q_n)],$$

where d stands for distance. Then write

$$M_n(E) = \inf \sup_{P \in E} f(P).$$

The sequence $\{M_n(E)\}$ can be proved (nontrivially) to converge to a limit $\chi(E)$ for which

$$\chi(E) \leqslant \delta_0(E),$$

with strict inequality possible.

In his next paper, *A note on transfinite diameters* [137], Hille calculates transfinite diameters of the unit spheres of some complex Banach spaces. Here U denotes the unit sphere. It is obvious that $\delta_0(U)$ $(= A - \delta_0(E)) \leqslant 2$, since 2 is the topological diameter. The proof that, for the spaces $C[0, 1]$, $L_1(0, 1)$, $L_\infty(0, 1)$, l, m, one has $\delta_0(U) = 2$ is very simple.

If U_n is the unit sphere in R_n and U_H is the unit sphere in Hilbert space H, then clearly

$$A - \delta_0(U_m) \leqslant A - \delta_0(U_{m+1}) \leqslant A - \delta_0(U_H)$$

because $U_m \subset U_{m+1} \subset U_H$ in the natural embedding of the spaces concerned.

Let S_{n+1} denote a regular simplex with $n + 1$ vertices inscribed in U_n. Let σ_n denote the length of an edge in S_{n+1}. One has $\sigma_n = \sqrt{2}\,\sqrt{1 + 1/n}$ and hence $\lim_{n\to\infty} \sigma_n = \sqrt{2}$.

A fundamental geometric question is: For what averaging processes A does one have

$$A - \delta_{n+1}(U_n) = \sigma_n\ ?$$

This is a highly interesting problem. For $n = 2$ the answer is yes for arithmetic, geometric, and harmonic means. Hille shows that with the addition of the condition on A:
(v) A is homogeneous of order 1,
then $\sigma_n = A - \delta_{n+1}(U_n)$ for all n,and the maximal configuration is a regular simplex providing this is true for $n = 2$. This is the simplicial theorem. He then shows that this theorem implies

$$A - \delta_0(U_H) = \sqrt{2}.$$

Finally, in his paper *Some geometric extremal problems* [139], published in Australia in 1966, Hille returns to the geometric questions raised above. He starts by showing that

$$\sqrt{2} \leqslant A - \delta_0(U_H) \leqslant 2$$

for any A,with the minimum value attained for the function $F(u) = u^2$.

For short, instead of $A - \delta_0(U_H)$, $F(u) = u^p$, $0 < p$, write $\delta_0^p(U_H)$. Then

$$\delta_0^p(U_H) = \sqrt{2}, \qquad 0 < p < 2.$$

For $p > 2$,

$$\delta_0^p(U_p) \geqslant 2^{1-(1/p)}.$$

Generalizations follow: If $F(u)$ is decreasing and strictly convex, $A - \delta_0(U_H) = \sqrt{2}$. Statements can also be made for increasing strictly concave functions.
E. R. Lorch

Commentators

J. Aczél, University of Waterloo
P. L. Butzer, Rheinisch-Westfälische Technische Hochschule, Aachen
J. Elliott, Rutgers University
A. Erdélyi, University of Edinburgh
H. Haruki, University of Waterloo
H. Helson, University of California, Berkeley
E. Hewitt, University of Washington
H. Hochstadt, Polytechnic Institute of Brooklyn
E. R. Lorch, Columbia University
C. E. Rickart, Yale University
G. -C. Rota, Massachusetts Institute of Technology
L. Sucheston, Ohio State University
C. T. Taam, George Washington University
K. Tandori, Bolyai Institute, Szeged
D. V. Widder, Harvard University
K. Yosida, Kyoto University
A. C. Zaanen, University of Leiden